Practical Skills in Food Science, Nutrition and Dietetics

PEARSON

We work with leading authors to develop the strongest educational materials in Food Science, bringing cutting-edge thinking and best learning practice to a global market.

Under a range of well-known imprints, including Printice Hall, we craft high quality print and electronic publications which help readers to understand and apply their content, whether studying or at work.

To find out more about the complete range of our publishing, please visit us on the World Wide Web at: www.pearsoned.co.uk

Practical Skills in Food Science, Nutrition and Dietetics

William Aspden
Fiona Caple
Rob Reed
Jonathan Weyers
Allan Jones

Prentice Hall
is an imprint of

Harlow, England • London • New York • Boston • San Francisco • Toronto • Sydney • Singapore • Hong Kong
Tokyo • Seoul • Taipei • New Delhi • Cape Town • Madrid • Mexico City • Amsterdam • Munich • Paris • Milan

Pearson Education Limited
Edinburgh Gate
Harlow
Essex CM20 2JE
England

and Associated Companies throughout the world

Visit us on the World Wide Web at:
www.pearsoned.co.uk

First published 2011

ISBN: 978-1-4082-2309-3

British Library Cataloguing-in-Publication Data
A catalogue record for this book is available from the British Library

Library of Congress Cataloging-in-Publication Data
A catalog record for this book is available from the Library of Congress

10 9 8 7 6 5 4 3 2 1
15 14 13 12 11

Typeset in 10/12 pt Times Roman by 73

Printed and bound in Malaysia (CTP-VVP)

Contents

List of Boxes

Guided tour

Definitions of key terms and concepts are highlighted in the text margin.

Tips and Hints provide useful hints and practical advice, and are highlighted in the text margin.

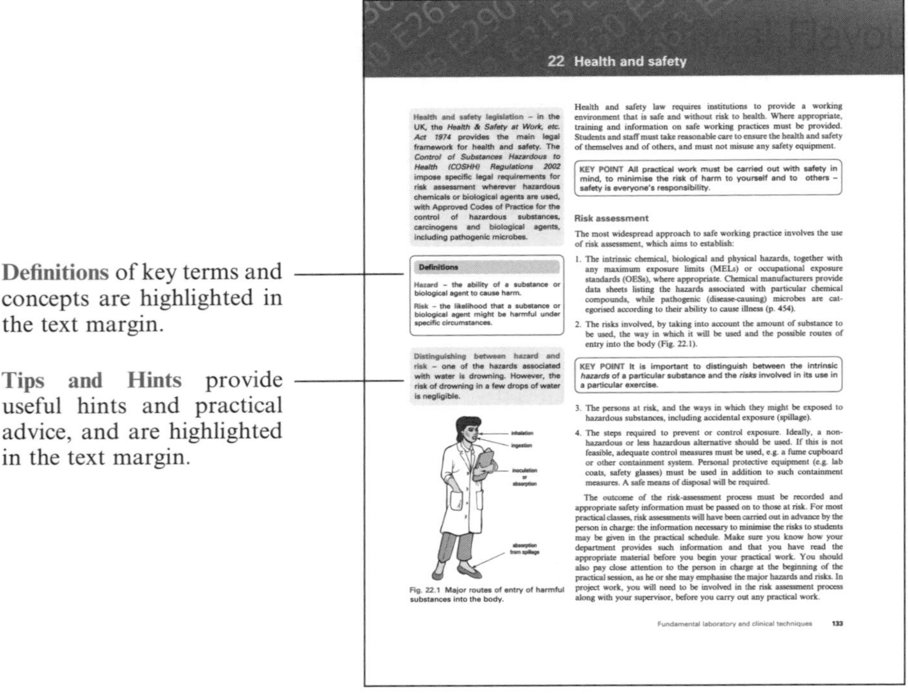

22 Health and safety

Health and safety legislation – in the UK, the *Health & Safety at Work, etc. Act 1974* provides the main legal framework for health and safety. The *Control of Substances Hazardous to Health (COSHH) Regulations 2002* impose specific legal requirements for risk assessment wherever hazardous chemicals or biological agents are used, with Approved Codes of Practice for the control of hazardous substances, carcinogens and biological agents, including pathogenic microbes.

Definitions

Hazard – the ability of a substance or biological agent to cause harm.

Risk – the likelihood that a substance or biological agent might be harmful under specific circumstances.

Distinguishing between hazard and risk – one of the hazards associated with water is drowning. However, the risk of drowning in a few drops of water is negligible.

Fig. 22.1 Major routes of entry of harmful substances into the body.

Health and safety law requires institutions to provide a working environment that is safe and without risk to health. Where appropriate, training and information on safe working practices must be provided. Students and staff must take reasonable care to ensure the health and safety of themselves and of others, and must not misuse any safety equipment.

KEY POINT All practical work must be carried out with safety in mind, to minimise the risk of harm to yourself and to others – safety is everyone's responsibility.

Risk assessment

The most widespread approach to safe working practice involves the use of risk assessment, which aims to establish:

1. The intrinsic chemical, biological and physical hazards, together with any maximum exposure limits (MELs) or occupational exposure standards (OESs), where appropriate. Chemical manufacturers provide data sheets listing the hazards associated with particular chemical compounds, while pathogenic (disease-causing) microbes are categorised according to their ability to cause illness (p. 454).
2. The risks involved, by taking into account the amount of substance to be used, the way in which it will be used and the possible routes of entry into the body (Fig. 22.1).

KEY POINT It is important to distinguish between the intrinsic *hazards* of a particular substance and the *risks* involved in its use in a particular exercise.

3. The persons at risk, and the ways in which they might be exposed to hazardous substances, including accidental exposure (spillage).
4. The steps required to prevent or control exposure. Ideally, a non-hazardous or less hazardous alternative should be used. If this is not feasible, adequate control measures must be used, e.g. a fume cupboard or other containment system. Personal protective equipment (e.g. lab coats, safety glasses) must be used in addition to such containment measures. A safe means of disposal will be required.

The outcome of the risk-assessment process must be recorded and appropriate safety information must be passed on to those at risk. For most practical classes, risk assessments will have been carried out in advance by the person in charge: the information necessary to minimise the risks to students may be given in the practical schedule. Make sure you know how your department provides such information and that you have read the appropriate material before you begin your practical work. You should also pay close attention to the person in charge at the beginning of the practical session, as he or she may emphasise the major hazards and risks. In project work, you will need to be involved in the risk assessment process along with your supervisor, before you carry out any practical work.

Fundamental laboratory and clinical techniques 133

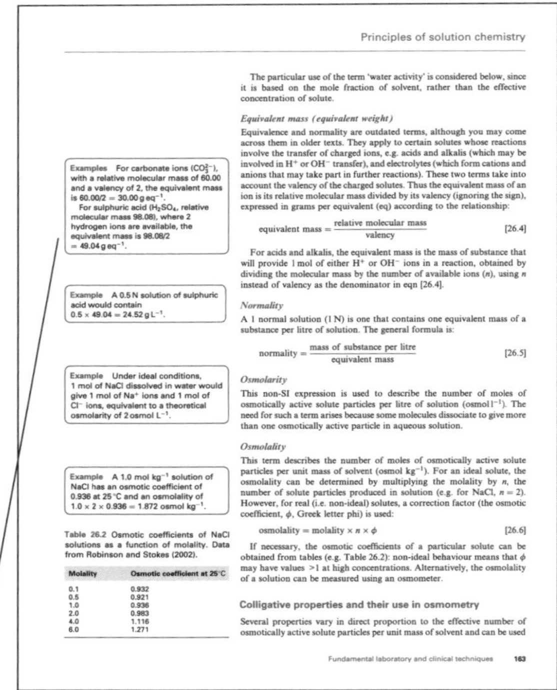

Principles of solution chemistry

Examples For carbonate ions (CO_3^{2-}), with a relative molecular mass of 60.00 and a valency of 2, the equivalent mass is 60.00/2 = 30.00 g eq^{-1}.
For sulphuric acid (H_2SO_4, relative molecular mass 98.08), where 2 hydrogen ions are available, the equivalent mass is 98.08/2 = 49.04 g eq^{-1}.

Example A 0.5 N solution of sulphuric acid would contain 0.5 × 49.04 = 24.52 g L^{-1}.

Example Under ideal conditions, 1 mol of NaCl dissolved in water would give 1 mol of Na^+ ions and 1 mol of Cl^- ions, equivalent to a theoretical osmolarity of 2 osmol L^{-1}.

Example A 1.0 mol kg^{-1} solution of NaCl has an osmotic coefficient of 0.936 at 25 °C and an osmolality of 1.0 × 2 × 0.936 = 1.872 osmol kg^{-1}.

Table 26.2 Osmotic coefficients of NaCl solutions as a function of molality. Data from Robinson and Stokes (2002).

Molality	Osmotic coefficient at 25 °C
0.1	0.932
0.5	0.921
1.0	0.936
2.0	0.983
4.0	1.116
6.0	1.271

The particular use of the term 'water activity' is considered below, since it is based on the mole fraction of solvent, rather than the effective concentration of solute.

Equivalent mass (equivalent weight)

Equivalence and normality are outdated terms, although you may come across them in older texts. They apply to certain solutes whose reactions involve the transfer of charged ions, e.g. acids and alkalis (which may be involved in H^+ or OH^- transfer), and electrolytes (which form cations and anions that may take part in further reactions). These two terms take into account the valency of the charged solutes. Thus the equivalent mass of an ion is its relative molecular mass divided by its valency (ignoring the sign), expressed in grams per equivalent (eq) according to the relationship:

$$\text{equivalent mass} = \frac{\text{relative molecular mass}}{\text{valency}} \qquad [26.4]$$

For acids and alkalis, the equivalent mass is the mass of substance that will provide 1 mol of either H^+ or OH^- ions in a reaction, obtained by dividing the molecular mass by the number of available ions (n), using n instead of valency as the denominator in eqn [26.4].

Normality

A 1 normal solution (1 N) is one that contains one equivalent mass of a substance per litre of solution. The general formula is:

$$\text{normality} = \frac{\text{mass of substance per litre}}{\text{equivalent mass}} \qquad [26.5]$$

Osmolarity

This non-SI expression is used to describe the number of moles of osmotically active solute particles per litre of solution (osmol l^{-1}). The need for such a term arises because some molecules dissociate to give more than one osmotically active particle in aqueous solution.

Osmolality

This term describes the number of moles of osmotically active solute particles per unit mass of solvent (osmol kg^{-1}). For an ideal solute, the osmolality can be determined by multiplying the molality by n, the number of solute particles produced in solution (e.g. for NaCl, $n = 2$). However, for real (i.e. non-ideal) solutes, a correction factor (the osmotic coefficient, ϕ, Greek letter phi) is used:

$$\text{osmolality} = \text{molality} \times n \times \phi \qquad [26.6]$$

If necessary, the osmotic coefficients of a particular solute can be obtained from tables (e.g. Table 26.2): non-ideal behaviour means that ϕ may have values >1 at high concentrations. Alternatively, the osmolality of a solution can be measured using an osmometer.

Colligative properties and their use in osmometry

Several properties vary in direct proportion to the effective number of osmotically active solute particles per unit mass of solvent and can be used

Fundamental laboratory and clinical techniques 163

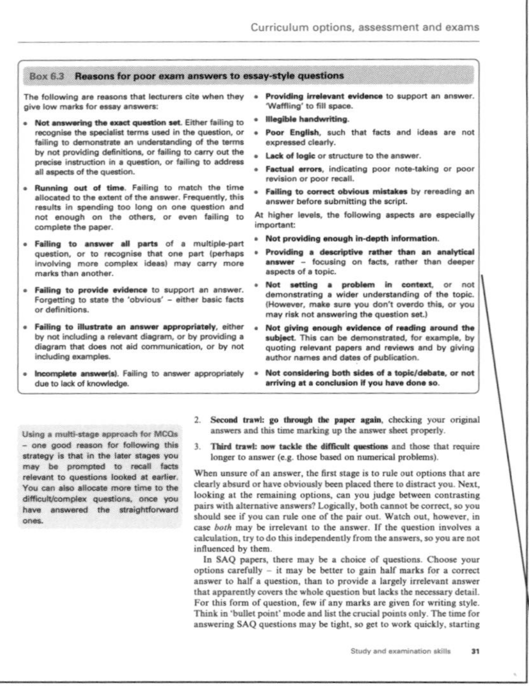

Curriculum options, assessment and exams

Box 6.3 Reasons for poor exam answers to essay-style questions

The following are reasons that lecturers cite when they give low marks for essay answers:

- **Not answering the exact question set**. Either failing to recognise the specialist terms used in the question, or failing to demonstrate an understanding of the terms by not providing definitions, or failing to carry out the precise instruction in a question, or failing to address all aspects of the question.
- **Running out of time**. Failing to match the time allocated to the extent of the answer. Frequently, this results in spending too long on one question and not enough on the others, or even failing to complete the paper.
- **Failing to answer all parts** of a multiple-part question, or to recognise that one part (perhaps involving more complex ideas) may carry more marks than another.
- **Failing to provide evidence** to support an answer. Forgetting to state the 'obvious' – either basic facts or definitions.
- **Failing to illustrate an answer appropriately**, either by not including a relevant diagram, or by providing a diagram that does not aid communication, or by not including examples.
- **Incomplete answer(s)**. Failing to answer appropriately due to lack of knowledge.
- **Providing irrelevant evidence** to support an answer. 'Waffling' to fill space.
- **Illegible handwriting**.
- **Poor English**, such that facts and ideas are not expressed clearly.
- **Lack of logic** or structure to the answer.
- **Factual errors**, indicating poor note-taking or poor revision or poor recall.
- **Failing to correct obvious mistakes** by rereading an answer before submitting the script.

At higher levels, the following aspects are especially important:

- **Not providing enough in-depth information.**
- **Providing a descriptive rather than an analytical answer** – focusing on facts, rather than deeper aspects of a topic.
- **Not setting a problem in context**, or not demonstrating a wider understanding of the topic. (However, make sure you don't overdo this, or you may risk not answering the question set.)
- **Not giving enough evidence of reading around the subject.** This can be demonstrated, for example, by quoting relevant papers and reviews and by giving author names and dates of publication.
- **Not considering both sides of a topic/debate, or not arriving at a conclusion if you have done so.**

Using a multi-stage approach for MCQs – one good reason for following this strategy is that in the later stages you may be prompted to recall facts relevant to questions looked at earlier. You can also allocate more time to the difficult/complex questions, once you have answered the straightforward ones.

2. **Second trawl: go through the paper again**, checking your original answers and this time marking up the answer sheet properly.
3. **Third trawl: now tackle the difficult questions** and those that require longer to answer (e.g. those based on numerical problems).

When unsure of an answer, the first stage is to rule out options that are clearly absurd or have obviously been placed there to distract you. Next, looking at the remaining options, can you judge between contrasting pairs with alternative answers? Logically, both cannot be correct, so you should see if you can rule one of the pair out. Watch out, however, in case *both* may be irrelevant to the answer. If the question involves a calculation, try to do this independently from the answers, so you are not influenced by them.

In SAQ papers, there may be a choice of questions. Choose your options carefully – it may be better to gain half marks for a correct answer to half a question, than to provide a largely irrelevant answer that apparently covers the whole question but lacks the necessary detail. For this form of question, few if any marks are given for writing style. Think in 'bullet point' mode and list the crucial points only. The time for answering SAQ questions may be tight, so get to work quickly, starting

Study and examination skills 31

Examples are included in the margin to illustrate important points without interrupting the flow of the main text.

Worked Examples and **'How to' boxes** set out the essential procedures in a step-by-step manner.

Box 6.1 Problem-based learning (PBL)

In this relatively new teaching method, you are presented with a 'real world' problem or issue, often working within a team. As you tackle the problem, you will gain factual knowledge, develop skills and exercise critical thinking (Chapter 9). Because there is a direct and relevant context for your work, and because you have to employ active learning techniques, the knowledge and skills you gain are likely to be more readily assimilated and remembered. This approach also more closely mimics workplace practices. PBL usually proceeds as follows:

1. **You are presented with a problem** (e.g. a case study, a hypothetical patient, a topical issue).
2. **You consider what issues and topics you need to research**, by discussion with others if necessary. You may need to identify where relevant resources can be found (Chapters 8 and 10).
3. **You then need to rank the issues and topics in importance**, allocating tasks to group members, if appropriate.
4. **Having carried out the necessary research, you should review what information has been obtained.** As a result, new issues may need to be explored and, where appropriate, allocated to group members.
5. **You will be asked to produce an outcome, such as a report, diagnosis, seminar presentation or poster.** An outline structure will be required and, for groups, further allocation of tasks to accomplish this goal.

If asked to carry out PBL as part of your course, it is important to get off to a good start. At first, the problem may seem unfamiliar. However, once you become involved in the work, you will quickly gain confidence. If working as part of a group, make sure that your group meets as early as possible, that you attend all sessions and that you do the necessary background reading. When working in a team, a degree of self-awareness is necessary regarding your 'natural' role in group situations (Table 3.1). Various methods are used for assessing PBL, and the assessment may involve peer marking.

Note relevant dates in your diary, and use this information to plan your work. Take special note if deadlines for different modules clash, or if they coincide with social or sporting commitments.

KEY POINT If, for some valid reason (e.g. illness), you will be late with an assessment, speak to your tutors as soon as possible. They may be able to take extenuating circumstances into account by not applying a marking penalty. They will let you know what paperwork you may require to submit to support your claim.

Key Points highlight critical features of methodology.

Summative exams – general points

Dealing with problems during an examination – is it important to raise any issue with the exam invigilator, so that the problem is fully documented. By waiting until after the exam, it may be too late to deal with the issue.

Summative exams (p. 22) normally involve you answering questions without being able to consult other students or your notes. Invigilators are present to ensure appropriate conduct, but departmental representatives may be present for some of the exam. Their role is to sort out any subject-related problems, so if you think something is wrong, ask at the earliest opportunity. It is not unknown for parts of questions to be omitted in error, or for double meanings to arise, for example.

Planning

When preparing for an exam, make a checklist of the items you'll need (see p. 33). On the day of the exam, give yourself sufficient time to arrive at the correct room, without the risk of being late. Double-check the times and places of your exams, both well before the exam and also on arrival. If you arrive at the exam venue early, you can always rectify a mistake if you find you've gone to the wrong place.

- **Form a revision group** to share ideas and discuss topics with other students.
- **Prepare answers to past papers**, e.g. write essays or, if time is limited, write essay plans (see Box 5.3).
- **Work through representative problems** if your subject involves numerical calculations.
- **Make up your own questions** – the act of putting yourself in the examiner's mindset by inventing questions can help revision. However, you should not rely on 'question-spotting': this is a risky practice!

Use revision checks – it is important to test yourself frequently during revision, to ensure that you have retained the information you are revising.

Final preparations – try to get a good night's sleep before an exam. Last-minute cramming will be counterproductive if you are too tired during the exam.

The evening before your exam should be spent in consolidating your material, and checking through summary lists and plans. Avoid introducing new material at this late stage: your aim should be to boost your confidence, putting yourself in the right frame of mind for the exam itself.

Text references

Bloom, B., Englehart, M., Furst, E., Hill, W. and Krathwohl, D. (1956) *Taxonomy of Educational Objectives: The Classification of Educational Goals. Handbook I: Cognitive Domain.* Longmans, Green, New York and Toronto.

Fleming, N.D. (2001) *Teaching and Learning Styles: VARK Strategies.* Neil Fleming, Christchurch.

Fleming, N.D. *VARK: A Guide to Learning Styles.* Available: http://www.vark-learn.com/ Last accessed 21/12/10.

Sources for further study

Burns, R. (1997) *The Student's Guide to Passing Exams.* Kogan Page, London.

Hamilton, D. (2003) *Passing Exams: A Guide for Maximum Success and Minimum Stress*, 2nd edn. Continuum Press, London.

Many universities host study skills websites; these can be found using 'study skills', 'revision' or 'exams' as keywords in a search engine.

Measuring pH – the pH of a neutral solution changes with temperature (Table 28.1), due to the enhanced dissociation of water with increasing temperature. This must be taken into account when measuring the pH of any solution and when interpreting your results.

KEY POINT While pH is strictly the negative logarithm (to the base 10) of H^+ activity, in practice H^+ concentration in mol L^{-1} (equivalent to kmol m^{-3} in SI terminology) is most often used in place of activity, since the two are virtually the same, given the limited dissociation of H_2O. The pH scale is not SI: nevertheless, it continues to be used widely in biological science.

The value where an equal amount of H^+ and OH^- ions are present is termed neutrality: at 25°C the pH of pure water at neutrality is 7.0. At this temperature, pH values below 7.0 are acidic while values above 7.0 are alkaline.

Always remember that the pH scale is a logarithmic one, not a linear one: a solution with a pH of 3.0 is not twice as acidic as a solution of pH 6.0, but one thousand times as acidic (i.e. contains 1000 times the amount of H^+ ions). Therefore, you may need to convert pH values into proton concentrations before you carry out mathematical manipulations (see Box 38.2, p. 254). For similar reasons, it is important that pH change is expressed in terms of the original and final pH values, rather than simply quoting the difference between the values: a pH change of 0.1 has little meaning unless the initial or final pH is known.

SAFETY NOTE Preparing a dilute acid solution using concentrated acid – always slowly add the concentrated acid to water, not the reverse, since the strongly exothermic process can trigger a violent reaction with water.

SAFETY NOTE Preparing an alkali solution – typically, the alkali will be in solid form (e.g. NaOH) and addition to water will rapidly raise the temperature of the solution: use only heat-resistant glassware, cooled with water if necessary.

Measuring pH

pH electrodes

Accurate pH measurements can be made using a pH electrode, coupled to a pH meter. The pH electrode is usually a combination electrode, comprising two separate systems: an H^+-sensitive glass electrode and a reference electrode which is unaffected by H^+ ion concentration (Fig. 28.2). When this is immersed in a solution, a pH-dependent voltage between the two electrodes can be measured using a potentiometer. In most cases, the pH electrode assembly (containing the glass and reference electrodes) is connected to a separate pH meter by a cable, although some hand-held instruments (pH probes) have the electrodes and meter within the same assembly, often using an H^+-sensitive field effect transistor in place of a glass electrode, to improve durability and portability.

Box 28.1 gives details of the steps involved in making a pH measurement with a glass pH electrode and meter.

Table 28.2 Properties of some pH indicator dyes

Dye	Acid-base colour change	Useful pH range
Thymol blue (acid)	red–yellow	1.2–6.8
Bromophenol blue	yellow–blue	1.2–6.8
Congo red	blue–red	3.0–5.2
Bromocresol green	yellow–blue	3.8–5.4
Resazurin	orange–violet	3.8–6.5
Methyl red	red–yellow	4.3–6.1
Litmus	red–blue	4.5–8.3
Bromocresol purple	yellow–purple	5.8–6.8
Bromothymol blue	yellow–blue	6.0–7.6
Neutral red	red–yellow	6.8–8.0
Phenol red	yellow–red	6.8–8.2
Thymol blue (alkaline)	yellow–blue	8.0–9.6
Phenol-phthalein	none–red	8.3–10.0

pH indicator dyes

These compounds (usually weak acids) change colour in a pH-dependent manner. They may be added in small amounts to a solution, or they can be used in paper strip form. Each indicator dye usually changes colour over a restricted pH range, typically 1–2 pH units (Table 28.2): universal indicator dyes/papers make use of a combination of individual dyes to measure a wider pH range. Dyes are not suitable for accurate pH measurement as they are affected by other components of the solution including oxidising and reducing agents and salts. However, they are useful for:

- estimating the approximate pH of a solution;
- determining a change in pH, for example at the end-point of a titration or the production of acids during bacterial metabolism (pp. 162–164);
- establishing the pH of a food (Chapter 49).

Sources for further study – every chapter is supported by a section giving printed and electronic sources for further study.

Safety Notes highlight specific hazards and appropriate practical steps to minimise risk.

Preface

While practical work forms the cornerstone of all scientific knowledge, the training required in food science, nutrition and dietetics is especially wide, covering relevant aspects of biology, chemistry, biochemistry, nutrition science and clinical practice. To be successful in these areas, students must develop a number of specific skills and abilities, ranging from those required to observe, measure, interview, record and calculate accurately, to those associated with operating up-to-date analytical laboratory equipment, as well as broader generic skills including team work, effective study and an ability to interact and consult with clients and allied health professionals. Students must develop an ability to communicate information effectively in an appropriate style, both in written and verbal form. This book aims to provide the support and guidance that will help students of food science, nutrition and dietetics to maximise their skills and abilities in all these aspects.

The book has been written for undergraduate students taking degree courses in food science, nutrition and dietetics; it will also be relevant to those taking related courses such as clinical dietetics, biomedical science and naturopathy. The content of the book is aimed mainly at the early years of the undergraduate programme, where there is the greatest exposure to new material and where the learning curve is steepest. However, some information is directly relevant to students taking postgraduate courses or undertaking project work in food science, nutrition or dietetics. As with the other books in the 'Practical Skills' series, we have tried to write in a concise but user-friendly style, giving key points and definitions, illustrations and worked examples, tips and hints, 'how to' boxes and checklists, where appropriate.

The material included in *Practical Skills in Food Science, Nutrition and Dietetics* has been selected on the basis of our teaching experience and extensive background in food science and analysis, nutrition and dietetics, highlighting those areas where students have asked for further guidance and support. We hope that students will find this book useful in the laboratory, during practical classes, in clinical and industry placements, and during project work. The book is not intended to replace conventional laboratory schedules and handouts, but to provide information that will help students to get the most out of their studies – e.g. by providing an explanation of the underlying principles on which a specific technique is based, e.g, spectrophotometry, or by giving advice on how to approach a particular practical procedure, such as taking skinfold measurements with calipers, or determining the lipid profile of foodstuffs. The book also covers more general procedures such as preparing a calibration graph, presenting research projects and tackling numerical problems. We hope that lecturers will find that the text is an effective way to supplement the information given in practical classes, where constraints on time and resources may lead to the under-performance of students.

While laboratory skills and nutritional assessment methods form a critical component of undergraduate work in food science, nutrition and dietetics, this book also aims to support the development of a wider range of skills. As a result, there are chapters dealing with the evaluation of information, the use of the Internet and email, revision and examination skills. Given the breadth of material covered, we have tried to focus on the broad principles and key points, rather than providing recipe-like solutions for every potential aspect. However, each chapter is supported by a section giving key sources for further study, including websites and conventional printed texts or papers.

We would like to take this opportunity to thank the following colleagues and friends who have provided assistance, comments, ideas and constructive feedback at various points during the writing of this book: Marc Barnbaum, Ian Brown, Karena Burke, Susan Ferguson, Keith Harrower, Carole Marshall, Peter Reaburn and Heather Smyth, together with the colleagues who supported the development of related texts within this series, including *Practical Skills in Biology* and *Practical Skills in Biomolecular Sciences*.

List of abbreviations

A	absorbance
ACDP	Advisory Committee on Dangerous Pathogens
ADP	adenosine diphosphate
AI	adequate intake
AMA	arm muscle area
AMDR	acceptable macronutrient
ANOVA	analysis of variance
ATP	adenosine triphosphate
BCAST	Basic Logic Alignment Search Tool
BEE	basal energy expenditure
BMI	body mass index
BMR	basal metabolic rate
BSA	bovine serum albumin
CCP	critical control point
CFU	colony-forming unit
CGS	centimetre gram second
COSHH	Control of Substances Hazardous to Health
CoV	coefficient of variation
CRP	C-reactive protein
DEFT	direct epifluorescence filter technique
DHEA	dehydroepiandrosterone
DIT	dietary induced thermogenesis
DNA	deoxyribonucleic acid
DPPH	diphenyl-1-picrylhydrazyl
DRI	dietary reference intakes (USA)
DRV	dietary reference value
EAR	estimated average requirement
EDTA	ethylenediaminetetraacetic acid
EER	estimated energy requirement (USA)
EIA	enzyme immunoassay
ELISA	enzyme-linked immunosorbent assay
EM	electron microscopy
EMR	electromagnetic radiation
EU	European Union
F	Faraday constant
FFQ	food frequency questionnaire
FOP	front of pack
FRAP	ferric reducing antioxidant power
FTP	file transfer protocol
g	acceleration due to gravity
GC	gas chromatography
GDA	guideline daily amounts
GI	gastrointestinal
GMP	good manufacturing practice
GRAS	generally regarded as safe
GWAS	genome-wide association studies
HACCP	hazard analysis and critical control points
HEPES	N-[2-hydroxyethyl]piperazine-N′-[2-ethanesulphonic acid]
HMB	hydroxymethylbutyrate
HPLC	high-performance liquid chromatography
HTST	high temperature short time
IADL	Lawton instrumental activities of daily living scale
IBM	ideal body mass
Ig	immunoglobulin
IR	infra-red (radiation)
IRMA	immunoradiometric assay
ISFET	ion-selective field effect transistor
ISP	Internet service provider
Kcal	kilocalorie
kJ	kilojoule
K_m	Michaelis constant
K_w	ionisation constant of water
LAN	local area network
LM	light microscopy
LRNI	lower reference nutrient intake
LSD	least significant difference
M	molar (mol L^{-1})
MAC	mid-arm circumference
MAMC	mid-arm muscle circumference
MC	mass change
MEL	maximum exposure limit
MFP	meat, fish, poultry factor
MNA	mini nutritional assessment
MPN	most probable number
M_r	relative molecular mass
MRD	maximum recovery diluent
MS	mass spectrometry
MSG	monosodium glutamate
MUST	malnutrition universal screening tool

List of abbreviations

NAD^+	nicotinamide adenine dinucleotide (oxidised form)
NADH	nicotinamide adenine dinucleotide (reduced form)
$NADP^+$	nicotinamide adenine dinucleotide phosphate (oxidised form)
NADPH	nicotinamide adenine dinucleotide phosphate (reduced form)
NH	null hypothesis
NRS	nutritional risk screening
NRV	nutrient reference values
OES	occupational exposure standards
ORAC	oxygen radical absorbance capacity
PABA	para-aminobenzoic acid
PAL	physical activity level
PAS	publically available specification
PCR	polymerase chain reaction
PDA	personal digital assistant
PDP	personal development planning
PEF	pulsed electric fields
PEM	protein-energy malnutrition
PFU	plaque-forming unit
pH	$-\log_{10}$ proton concentration (activity), in mol m^{-1}
pK_a	$\log_{10}$ acid dissociation constant
PKY	phenylketonuria
PLT	pulsed light technology
PPE	personal protection equipment
PSR	personal sweating rate
PTS	personal transferable skills
Q10	temperature coefficient
QMS	quality management system
QUID	quantitative ingredient declaration
R	universal gas constant
RAST	radioallergosorbent test
RBP	retinol-binding protein
RCF	relative centrifugal field
RDA	recommended daily amount (EU)
RDA	recommended dietary allowance (USA)
RDI	recommended dietary intake (Australia)
R_F	*relative frontal mobility*
RLU	relative light units
RMR	resting metabolic rate
rDNA	recombinant deoxyribonucleic acid
RIA	radioimmunoassay
RID	radioimmunodiffusion
RNA	ribonucleic acid
RNI	reference nutrient intake
ROS	reactive oxygen species
RP-HPLC	reverse phase high-performance liquid chromatography
r.p.m.	revolutions per minute
RQ	respiratory quotient
RV	residual lung volume
SCI	Science Citation Index
SD	standard deviation
SE	standard error (of the sample mean)
SEM	scanning electron microscopy
SGA	subjective global assessment
SGD	*Saccharomyces* Genome Database
SI	Système Internationale d'Unités
SNP	single nucleotide polymorphism
STP	standard temperature and pressure
T	Absolute temperature (in Kelvin)
TDT	thermal death time
TEAC	trolox equivalent antioxidant capacity
TEE	total energy expenditure
TEM	transmission electron microscopy
TLC	thin-layer chromatography
TRIS	tris(hydroxymethyl)aminomethane *or* 2-amino-2-hydroxymethyl-1,3-propanediol
TTE	total energy expenditure
TVC	total viable count
UBM	usual body mass
UHT	ultra-high temperature
UL	upper level of intake
UNICEF	United Nations Children's Fund
URL	uniform resource locator
USDA	United States Department of Agriculture
UV	ultraviolet (radiation)
VLDL	very low density lipoprotein
V_{max}	maximum velocity
VP	vacuum packaging
WHO	World Health Organisation
WHR	waist to hip ratio
WWW	World Wide Web
WYSIWYG	what you see is what you get

Acknowledgements

We are grateful to the following for permission to reproduce copyright material:

Figures

Figure 23.1 adapted from http://www.fao.org/docrep/W8088E/w8088e05.htm#section 3, the hazard analysis and critical control point (haccp) system; Figure 23.2 from http://www.fao.org/docrep/W8088E/w8088e01.jpg; Figure 23.3 from http://www.gdalabel.org.uk/gda/gdalabel.aspx, by the Food and Drink Federation; Figure 24.1 from *The Merck Index: An Encyclopedia of Chemicals, Drugs and Biologicals*, 14 ed. Merck & Co., Inc. (O'Neil, M.J., Heckelman, P.E., Koch, C.B., Roman, K.J., 2006), The Merck Index: An Encyclopedia of Chemicals, Drugs, and Biologicals, Fourteenth Edition, Maryadele J. O'Neil, Patricia E. Heckelman, Cherie B. Koch, Kristin J. Roman, Eds. (Merck & Co., Inc., Whitehouse Station, NJ, USA, 2006).; Figures 30.3 and 30.5 supplied by Microscopy an operating division of KeyMed (Medical & industrial Equipment) Ltd.; Figure 40.5 from Page 9 of: http://www.nhmrc.gov.au/_files_nhmrc/file/publications/synopses/n31.pdf, copyright Commonwealth of Australia, reproduced by permission; Figure 49.2 courtesy of Mettler-Toledo Ltd; Figure 57.4 from Marketing Department, Fisher Scientific UK Ltd., Bishop Meadow Road, Loughborough, Leicestershire, LE11 5RG, http://www.fisher.co.uk/, Thermo Fisher Scientific; Figure 60.3 adapted from http://www.rsc.org/chemistryworld/Issues/2007/September/ClassicKitSoxhletExtractor.asp, Andrea Sella, Chem. World, 4(9), Sep 2007, 77 – reproduced by permission of The Royal Society of Chemistry; Figure 69.2 reproduced by permission of Seward Ltd. Figure 69.3 from http://www.oxoid.com/UK/blue/prod_detail/prod_detail.asp?pr=DS0147&c=UK&lang=EN

Text

Box 63.1 adapted from 'Ferric reducing/antioxidant power assay: Direct measure of total antioxidant activity of biological fluids and modified version for simultaneous measurement of total antioxidant power and ascorbic acid concentration', *Methods in Enzymology*, 299 ed., pp. 15–23 (Benzie, I.F.F. and Strain, J.J. 1999)

In some instances we have been unable to trace the owners of copyright material, and we would appreciate any information that would enable us to do so.

Study and examination skills

1 The importance of transferable skills

Skills terminology – different phrases may be used to describe transferable skills, depending on place or context. These include: 'personal transferable skills' (PTS), 'key skills', 'core skills' and 'competences', or 'graduate attributes'.

This chapter outlines the range of transferable skills and their significance to students of food science, nutrition and dietetics. It also indicates where practical skills fit into this scheme. Having a good understanding of this topic will help you to place your work at university in a wider context. You will also gain an insight into the qualities that employers expect you to have developed by the time you graduate. Awareness of these matters will be useful when carrying out personal development planning (PDP) as part of your studies.

The range of transferable skills

Table 1.1 provides a comprehensive listing of university-level transferable skills under six skill categories. There are many possible classifications – and a different one may be used in your institution or field of study. Note particularly that 'study skills', while important, and rightly emphasised at the start of many courses, constitute only a subset of the skills acquired by most university students.

The phrase '*Practical Skills*' in the title of this book indicates that there is a special subset of transferable skills related to work in the laboratory or with clients. However, although this text deals primarily with skills and techniques required for laboratory practicals and associated studies, a broader range of material is included. This is because the skills concerned are important, not only within your degree but also in the wider world. Examples include time management, evaluating information and communicating effectively.

Using course materials – study your course handbook and the schedules for each practical session to find out what skills you are expected to develop at each point in the curriculum. Usually the learning objectives/outcomes (p. 22) will describe the skills involved.

KEY POINT Food science, nutrition and dietetics are essentially a practical subjects, and therefore involve highly developed laboratory and clinical skills. The importance that your lecturers place on practical skills will probably be evident from the large proportion of curriculum time you will spend on practical work in your course.

Example The skills involved in team-work cannot be developed without a deeper understanding of the inter-relationships involved in successful groups. The context will be dierent for every group and a flexible approach will always be required, according to the individuals involved and the nature of the task.

The word 'skill' implies much more than the robotic learning of, for example, a laboratory routine. Of course, some of the tasks you will be asked to carry out in practical classes *will* be repetitive. Certain techniques require manual dexterity and attention to detail if accuracy and precision are to be attained, and the necessary competence often requires practice to make perfect. However, a deeper understanding of the context of a technique is important if the skill is to be appreciated fully and then transferred to a new situation. That is why this text is not simply a 'recipe book' of methods and why it includes background information, tips and worked examples, as well as study exercises to aid your learning and test your understanding.

Transferability of skills

Transferable skills are those that allow someone with knowledge, understanding or ability gained in one situation to adapt or extend this for application in a different context. In some cases, the transfer of a skill is immediately obvious. Take, for example, the ability to use a

Table 1.1 Transferable skills identified as important in the biosciences. The list has been compiled from several sources, including the UK Quality Assurance Agency for Higher Education *Subject Benchmark Statement* (QAAHE, 2002). Particularly relevant chapters are shown for the skills covered by this book.

Skill category	Examples of skills and competences	Relevant chapters in this textbook
Generic skills	Having systematic knowledge of key disciplines underpinning safe and effective practice	40 49 52 53 54 55 56 57 66 67 68 70
	Ability to gather and evaluate evidence from a wide range of sources	8 9 10
	Ability to communicate a clear and accurate account of a topic, both verbally and in writing	14 15 16 17 18 19 20
	Applying critical and analytical skills to evaluate evidence	9 33
	Using a variety of investigative methods to study a topic	31 32 33 34 41 42 43 44 45 46 47 49 51 52 53 54 55 56 66 69 70
	Having the ability to think independently and solve problems	33 34 37 39
Intellectual skills	Recognising and applying theories, concepts and principles	9 33
	Analysing, synthesising and summarising information critically	35 36 37 38 39
	Obtaining evidence to formulate and test hypotheses; applying knowledge to address familiar and unfamiliar problems	31 32 33 34
	Recognising and explaining moral, ethical and legal issues	20 21 22 23
Experimental and observational skills	Carrying out basic laboratory procedures and understanding the principles that underlie them	21 22 23 24 25 26 27 28 29 30 49 50–71
	Working safely, responsibly and legally, with due attention to ethical aspects	21 22 23
	Designing, planning, conducting and reporting on investigations and data arising from them	14 17 33 34
	Obtaining, recording, collating and analysing data	31 32 33 34 35 36 37 38 39
Numeracy, communication and IT skills	Understanding and using data in several forms (e.g. numerical, textual, verbal and graphical)	31 35 36
	Communicating in written, verbal, graphical and visual forms	13 14 15 16 17 18 19 20 35 36
	Citing and referencing the work of others in an appropriate manner	8
	Obtaining data, including the concepts behind calibration and types of error	31 34 38 39 50
	Processing, interpreting and presenting data, and applying appropriate statistical methods for summarising and analysing data	35 36 37 38 39
	Solving problems with calculators and computers, including the use of tools such as spreadsheets	11 21 38 39
	Using computer technology to communicate and as a source of information	10 11 12
Interpersonal and teamwork skills	Working individually or in teams as appropriate; identifying individual and group goals and acting responsibly and appropriately to achieve them	3
	Recognising and respecting the views and opinions of others	3
	Evaluating your own performance and that of others	3 7 21
	Appreciating the interdisciplinary nature of food science, nutrition and dietetics	1 18
Self-management and professional development skills	Working independently, managing time and organising activities	2 32 34
	Identifying and working towards targets for personal, academic and career development	1 7
	Developing an adaptable and effective approach to study and work (including revision and exam technique)	4 5 6

spreadsheet to summarise experimental data and create a graph to illustrate results. Once the key concepts and commands are learned (Chapter 11), they can be applied to many instances outside the discipline where this type of output is used. This is not only true for similar data sets, but also in unrelated situations, such as making up a financial balance sheet and creating a pie chart to show sources of expenditure. Similarly, knowing the requirements for good graph drawing and tabulation (Chapters 35 and 36), perhaps practised by hand in earlier work, might help you use spreadsheet commands to make the output suit your needs.

Other cases may be less clear but equally valid. For example, towards the end of your undergraduate studies you may be involved in designing experiments as part of your project work. This task will draw on several skills gained at earlier stages in your course, such as preparing solutions (Chapters 21–26), deciding about numbers of replicates and experimental layout (Chapter 33) and perhaps carrying out some particular method of observation, measurement or analysis (Chapters 40–71). How and when might you transfer this complex set of skills? In the workplace, it is unlikely that you would be asked to repeat the same process, but in critically evaluating a problem or in planning a complex project for a new employer, you will need to use many of the time-management, organisational and analytical skills developed when designing and carrying out experiments and making measurements. The same applies to information retrieval and evaluation and writing essays and dissertations, when transferred to the task of analysing or writing a business report.

Opportunities to develop and practise skills in your private or social life – you could, for example, practise spreadsheet skills by organising personal or club finances using Microsoft Excel, or teamwork skills within any university clubs or societies you may join (see Chapter 7).

Types of PDP portfolio and their benefits – some PDP schemes are centred on academic and learning skills, while others are more focused on career planning. Some are carried out independently and others in tandem with a personal tutor or advisory system. Some PDP schemes involve creating an online portfolio, while others are primarily paper-based. Each method has specific goals and advantages, but whichever way your scheme operates, maximum benefit will be gained from being fully involved with the process.

Personal development planning

Many universities have schemes for personal development planning (PDP), which may have different names such as 'progress files' or 'professional development plans'. You will usually be expected to create a portfolio of evidence on your progress, then reflect on this, and subsequently set yourself plans for the future, including targets and action points. Analysis of your transferable skills profile will probably form part of your PDP. Other aspects commonly included are:

- your aspirations, goals, interests and motivations;
- your learning style or preference (see p. 20);
- your assessment transcript or academic profile information (e.g. record of grades in your modules);
- your developing CV (see p. 36).

Taking part in PDP can help focus your thoughts about your university studies and future career. This is important in biology, because most biological sciences degrees do not lead only to a specific occupation. The PDP process will introduce you to some new terms and will help you to describe your personality and abilities. This will be useful when constructing your CV and when applying for jobs.

Definition

Employability – the 'combination of in-depth subject knowledge, work awareness, subject-specific, generic and career management skills, and personal attributes and attitudes that enable a student to secure suitable employment and perform excellently throughout a career spanning a range of employers and occupations' (Anon., 2007).

What your future employer will be looking for

At the end of your course, which may seem some time away, you will aim to get a job and start on your chosen career path. You will need to sell yourself to your future employer, firstly in your application form and curriculum vitae (Chapter 7), and perhaps later at interview. Companies rarely employ graduates simply because they know how to carry out a particular lab routine or because they can remember specific facts about their chosen degree subject. Instead, employers tend to look for a range of qualities and transferable skills that together define an attribute known as 'graduateness'. This encompasses, for example, the ability to work in a team, to speak effectively and write clearly about your work, to understand complex data and to manage a project to completion. All of these skills can be developed at different stages during your university studies.

Finding out more about graduate attributes – many universities have defined a set of attributes that reflects their expectations of students who successfully complete their degree programmes. These can often be located through the university's website.

KEY POINT While factual knowledge is important in degrees with a strong vocational element, understanding how to find and evaluate information is usually rated more highly by employers than the ability to memorise facts.

Most likely, your future employer(s) will seek someone with an organised yet flexible mind, capable of demonstrating a logical approach to problems – someone who has a range of relevant skills and who can transfer these skills to new situations. Many competing applicants will probably have similar qualifications. If you want the job, you will have to show that your additional skills place you above the other candidates (Chapter 7).

Text references

Anon. *Define Employability in the Context of Teaching Bioscience*. Available: http://www.bioscience.heacademy.ac.uk/ftp/events/empforum/definition/pdf
Last accessed: 21/12/10.
[Part of HE Academy Centre for Bioscience website.]

QAAHE (2002) *Subject Benchmark Statement for the Biosciences*. Quality Assurance Agency for Higher Education, Gloucester.

Sources for further study

Drew, S. and Bingham, R. (2004) *The Student Skills Guide*, 2nd edn. Gower Publishing Ltd, Aldershot.

McMillan, K. and Weyers, J.D.B. (2006) *The Smarter Student: Study Skills and Strategies for Success at University*. Pearson Education, Harlow.

Quality Assurance Agency for Higher Education, UK. *Subject Benchmark Statements*. Available: http://www.qaa.ac.uk/academicinfrastructure/benchmark/default.asp
Last accessed: 21/12/10.
[Includes statements for aspects of dietetics, bioscience, food and consumer sciences.]

Race, P. (2007) *How to Get a Good Degree: Making the Most of Your Time at University*, 2nd edn. Open University Press, Buckingham.

2 Managing your time

One of the most important activities that you can do is to organise your personal and working time effectively. There is a lot to do at university and a common complaint is that there just isn't enough time to accomplish everything. In fact, research shows that most people use up a lot of their time without realising it through ineffective study or activities such as extended coffee breaks. Developing your time-management skills will help you achieve more in work, rest and play, but it is important to remember that putting time-management techniques into practice is an individual matter, requiring a level of self-discipline not unlike that required for dieting. A new system won't always work perfectly straight away, but through time you can evolve a system that is effective for you. An inability to organise your time effectively, of course, results in feelings of failure, frustration, guilt and being out of control in your life.

Definition

Time management – a system for controlling and using time as efficiently and as effectively as possible.

Setting your goals

The first step is to identify clearly what you want to achieve, both in work and in your personal life. We all have a general idea of what we are aiming for but, to be effective, your goals must be clearly identified and priorities allocated. Clear, concise objectives can provide you with a framework in which to make these choices. Try using the 'SMART' approach, in which objectives should be:

Definition

Goal (or aim) – a long-term end-point (for example 'to become a dietician').

Objective – a specific, defined activity that will bring you closer to your overall goal (for example, 'to take an active part in all lectures and lab classes in first year food science').

- **Specific** – clear and unambiguous, including what, when, where, how and why.
- **Measurable** – having quantified targets and benefits to provide an understanding of progress.
- **Achievable** – being attainable within your resources.
- **Realistic** – being within your abilities and expectations.
- **Timed** – stating the time period for completion.

Example The objective 'to spend an extra hour each week on directed study in food microbiology next term' fulfils the SMART criteria, in contrast with a general intention 'to study more'.

Having identified your goals, you can now move on to answer four very important questions:

1. Where does your time go?
2. Where should your time go?
3. What are your time-wasting activities?
4. What strategies can help you?

Advantages of time management – these include:

- a much greater feeling of control over your activities;
- avoidance of stress;
- improved productivity – achieve more in a shorter period;
- improved performance levels – work to higher standards because you are in charge;
- an increase in time available for non-work matters – work hard, but play hard too.

Analysing your current activities

The key to successful development of time management is a realistic knowledge of how you currently spend your time. Start by keeping a detailed time log for a typical week (Fig. 2.1), but you will need to be truthful in this process. Once you have completed the log, consider the following questions:

- How many hours do I work in total and how many hours do I use for 'relaxation'?
- What range of activities do I do?

Time slots	Activity										Notes
7.00–7.15											
7.15–7.30											
7.30–7.45											
7.45–8.00											
8.00–8.15											
8.15–8.30											
8.30–8.45											
8.45–9.00											
9.00–9.15											

Fig. 2.1 Example of how to lay out a time log. Write activities along the top of the page, and divide the day into 15-minute segments as shown. Think beforehand how you will categorise the different things you do, from the mundane (laundry, having a shower, drinking coffee, etc.) to the well timetabled (tutorial meeting, sports club meeting), and add supplementary notes if required. At the end of each day, place a dot in the relevant column for each activity and sum the dots to give a total at the bottom of the page. You will need to keep a diary like this for at least a week before you see patterns emerging.

Quality and time management – avoid spending a lot of time doing unproductive studying, e.g. reading a textbook without specific objectives for that reading.

- How long do I spend on each activity?
- What do I spend most of my time doing?
- What do I spend the least amount of my time doing?
- Are my allocations of time in proportion to the importance of my activities?
- How much of my time is ineffectively used, e.g. for uncontrolled socialising or interruptions?

If you wish, you could use a spreadsheet (Chapter 11) to produce graphical summaries of time allocations in different categories as an aid to analysis and management. Divide your time into:

- **Committed time** – timetabled activities involving your main objectives/goals.
- **Maintenance time** – time spent supporting your general life activities (shopping, cleaning, laundry, etc.).
- **Discretionary time** – time for you to use as you wish, e.g. recreation, sport, hobbies, socialising.

Recognise and avoid online time-wasting activities – these could include general 'surfing', visiting social networking sites and emailing friends.

Avoiding time-wasting activities

Look carefully at those tasks that could be identified as time-wasting activities. They include gossiping, over-long breaks, uninvited interruptions and even ineffective study periods. Try to reduce these to a minimum, but do not count on eliminating them entirely. Remember also that some relaxation *should* be programmed into your daily schedule.

Being assertive – if friends and colleagues continually interrupt you, find a way of controlling them, before they control you. Indicate clearly on your door that you do not wish to be disturbed and explain why. Otherwise, try to work away from disturbance.

Organising your tasks

Having analysed your time usage, you can now use this information, together with your objectives and prioritised goals, to organise your activities, on both a short-term and a long-term basis. Consider using a print or electronic diary/calendar system that will help you to plan ahead and analyse your progress.

WEEKLY DIARY

Week beginning:

	Sunday	Monday	Tuesday	Wednesday	Thursday	Friday	Saturday
DATE							
7–8 am		Breakfast	Breakfast	Breakfast	Breakfast	Breakfast	
8–9		Preparation	Preparation	Preparation	Preparation	Preparation	Breakfast
9–10	Breakfast	PE112(L)	PE112(L)	PE112(L)	PE112(L)	NU102(P)	Travel
10–11	FREE	FS101(L)	FS101(L)	FS101(L)	FS101(L)	NU102(P)	WORK
11–12	STUDY	STUDY	STUDY	STUDY	STUDY	NU102(P)	WORK
12–1 pm	STUDY	NU102(L)	NU102(L)	NU102(L)	NU102(L)	TUTORIAL	WORK
1–2	Lunch	Lunch	Lunch	Lunch	Lunch	Lunch	Lunch
2–3	(VOLLEY-	FS101(P)	STUDY	SPORT	PE112(P)	STUDY	WORK
3–4	BALL	FS101(P)	STUDY	(VOLLEY-	PE112(P)	STUDY	WORK
4–5	MATCH)	FS101(P)	STUDY	BALL	PE112(P)	SHOPPING	WORK
5–6	FREE	STUDY	STUDY	CLUB)	STUDY	TEA ROTA	WORK
6–7	Tea	Tea	Tea	Tea	Tea	Tea	Tea
7–8	FREE*	STUDY	STUDY	FREE*	STUDY	FREE*	FREE
8–9	FREE*	STUDY	STUDY	FREE*	STUDY	FREE*	FREE
9–10	FREE*	FREE*	STUDY	FREE*	STUDY	FREE*	FREE
Study (h)	2	10	11	4	11	6	0
Other (h)	13	5	4	11	4	9	15

Total study time = 44 h

Fig. 2.2 A weekly diary with an example of entries for a first-year science student with a Saturday job and active membership of a volleyball club. Note that 'free time' changes to 'study time', e.g. for periods when assessed work is to be produced or during revision for exams. Study time (including attendance at lectures, practicals and tutorials) thus represents between 42 and 50 per cent of the total time.

Divide your tasks into several categories, such as:

- **Urgent** – must be done as a top priority and at short notice (e.g. doctor's appointment).
- **Routine** – predictable and regular and therefore easily scheduled (e.g. preparation, lectures or playing sport).
- **One-off activities** – usually with rather shorter deadlines and which may be of high priority (e.g. a tutorial assignment or seeking advice).
- **Long-term tasks** – sometimes referred to as 'elephant tasks' that are too large to 'eat' in one go (e.g. learning a language). These are best managed by scheduling frequent small 'bites' to achieve the task over a longer timescale.

Matching your work to your body's rhythm – everyone has times of day when they feel more alert and able to work. Decide when these times are for you and programme your work accordingly. Plan relaxation events for periods when you tend to be less alert.

You should make a weekly plan (Fig. 2.2) for the routine activities, with gaps for less predictable tasks. This should be supplemented by individual daily checklists, preferably written at the end of the previous working day. Such plans and checklists should be flexible, forming the basis for most of your activities except when exceptional circumstances intervene. The planning must be kept brief, however, and should be scheduled into your activities. Box 2.1 provides tips for effective planning and working.

Use checklists as often as possible – post your lists in places where they are easily and frequently visible, such as in front of your desk. Ticking things off as they are completed gives you a feeling of accomplishment and progress, increasing motivation.

KEY POINT Review each day's plan at the end of the previous day, making such modifications as are required by circumstances, e.g. adding an uncompleted task from the previous day or a new and urgent task.

Box 2.1 Tips for effective planning and working

- **Set guidelines and review expectations regularly.**
- **Don't procrastinate:** don't keep putting off doing things you know are important – they will not go away but they will increase to crisis point.
- **Don't be a perfectionist** – perfection is paralysing.
- **Learn from past experience** – review your management system regularly.
- **Don't set yourself unrealistic goals and objectives** – this will lead to procrastination and feelings of failure.
- **Avoid recurring crises** – they are telling you something is not working properly and needs to be changed.
- **Learn to concentrate effectively** – do not let yourself be distracted by casual interruptions.
- **Learn to say 'no'** firmly but graciously when appropriate.
- **Know your own body rhythms:** e.g. are you a morning person or an evening person?
- **Learn to recognise the benefits of rest** and relaxation at appropriate times.
- **Take short but complete breaks from your tasks** – come back feeling refreshed in mind and body.
- **Work in a suitable study area** and keep your own workspace organised.
- **Avoid clutter** (physical and mental).
- **Learn to access and use information effectively** (Chapter 9).
- **Learn to read and write accurately and quickly** (Chapters 4 and 15).

Sources for further study

Anon. *Filofax*. Available: http://www.filofax.co.uk
Last accessed: 09/04/07.
[Website for products of Filofax UK, Unit 3, Victoria Gardens, Burgess Hill, West Sussex, RH15 9NB.]

Levin, P. (2007) *Skilful Time Management. Student-Friendly Guides*. Open University Press, McGraw-Hill, Maidenhead.

Prentice Hall. *Time Management*. Available: http://www.prenhall.com/success/StudySkl/timemanage.html
Last accessed: 21/12/10.

Zeller, D. (2008) *Successful Time Management for Dummies*. Wiley, Hoboken.

3 Working with others

> **Definitions**
>
> **Team** – a team is not a bunch of people with job titles, but a congregation of individuals, each of whom has a role that is understood by other members. Members of a team seek out certain roles and they perform most eectively in the ones that are most natural to them.
>
> **Team role** – a tendency to behave, contribute and interrelate with others in a particular way.
>
> (both after Belbin, 1993)

It is highly likely that you will be expected to work with fellow students during practicals and study exercises. This might take the form of sharing tasks or casual collaboration through discussion, or it might be formally directed teamwork such as problem-based learning (Box 6.1) or preparing a poster (Chapter 13). Interacting with others can be extremely rewarding and realistically represents the professional world, where teamworking is common. The advantages of working with others include:

- **Teamworking is usually synergistic** – it often results in better ideas, produced by the interchange of views, and better output, due to the complementary skills of team members.
- **Working in teams can provide support for individuals** within the team.
- **Levels of personal commitment can be enhanced** through concern about letting the team down.
- **Responsibilities for tasks can be shared.**

However, you can also feel both threatened and exposed if teamwork is not managed properly. Some of the main reasons for negative feelings towards working in groups include:

- **Reservations about working with strangers** – not knowing whether you will be able to form a friendly and productive relationship.
- **Worries over rejection** – a perception of being unpopular or being chosen last by the group.
- **Concerns over levels of personal commitment** – these can be enhanced through a desire to perform well, so the team as a whole achieves its target.
- **Fear of being held back by others** – especially for those who have been successful in individual work already.
- **Lack of previous experience** – worries about the kinds of personal interactions likely to occur and the team role likely to suit you best.
- **Concerns about the outcomes of peer assessment** – in particular, whether others will give you a fair mark for your efforts.

> **Peer assessment** – this term applies to marking schemes in which all or a proportion of the marks for a teamwork exercise are allocated by the team members themselves. Read the instructions carefully before embarking on the exercise, so you know which aspects of your work your fellow team members will be assessing. When deciding what marks to allocate yourself, try to be as fair as possible with your marking.

Teamwork skills

Some of the key skills you will need to develop to maximise the success of your teamworking activities include:

- **Interpersonal skills.** How do you react to new people? Are you able to both listen and communicate easily with them? How do you deal with conflicts and disagreements?
- **Delegation/sharing of tasks.** The primary advantage of teamwork is the sharing of effort and responsibility. Are you willing/able to do this? It involves trusting your team members. How will you deal with those group members who don't contribute fully?
- **Effective listening.** Successful listening is a skill that usually needs developing, e.g. during the exchange of ideas within a group.

> **Gaining confidence through experience** – the more you take part in teamwork, the more you know how teams operate and how to make teamwork effective for you.

- **Speaking clearly and concisely.** Effective communication is a vital part of teamwork, both between team members and when presenting team outcomes to others. Try to develop your communication skills through learning and practice (see Chapter 14).
- **Providing constructive criticism.** It is all too easy to be negative, but only constructive criticism will have a positive effect on interactions with others.

Studying with others – teaming up with someone else on your course for revision ('study buddying') is a potentially valuable activity and may especially suit some types of learners (p. 20). It can help keep your morale high when things get tough. You might consider:

- sharing notes, textbooks and other information;
- going through past papers together, dissecting the questions and planning answers;
- talking to each other about a topic (good for aural learners: see Box 5.1);
- giving tutorials to each other about parts of the course that have not been fully grasped.

Collaboration for learning

Much collaboration is informal and consists of pairs or groups of individuals getting together to exchange materials and ideas while studying. It may consist of a 'brainstorming' session for a topic or piece of work, or sharing efforts to research a topic. This has much to commend it and is generally encouraged. However, it is vital that this collaborative learning is distinguished from the collaborative writing of assessed documents: the latter is not usually acceptable and, in its most extreme form, is plagiarism, usually with a heavy potential punishment in university assessment systems. Make sure you know what plagiarism is, what unacceptable collaboration is, and how they are treated within your institution (see p. 49).

KEY POINT Collaboration is inappropriate during the final phase of an assessed piece of work unless you have been directed to produce a group report. Collaboration is often encouraged during learning activities and research but the final write-up must normally be your own work. The extreme of producing copycat write-ups is regarded as plagiarism (p. 49) and will be punished accordingly.

Web-based resources and support for brainstorming – websites such as http://www.brainstorming.co.uk give further information and practical advice for teamworking.

The dynamics of teamworking

It is important that team activities are properly structured so that all members knows what is expected of them. Allocation of responsibilities usually requires the clear identification of a leader. Several studies of groups have identified different team roles that derive from differences in personality. You should be aware of such categorisations, both in terms of your own predispositions and those of your fellow team members, as it will help the group to interact more productively. Belbin (1993) identified eight such roles, recently extended to nine, as shown in Table 3.1.

In formal team situations, your course organiser should deal with these issues; even if he or she does not, it is important that you are aware of these roles and their potential impact on the success or failure of teamwork. You should try to identify your own 'natural' role: if asked to form a team, bear the different roles in mind during your selection of colleagues and your interactions with them. The ideal team should contain members capable of adopting most of these roles. However, you should also note the following points:

Recording group discussions – make sure you structure meetings (including writing agendas) and note their outcomes (taking minutes and noting action points).

- **People will probably fit one of these roles naturally** as a function of their personality and skills.
- **Group members may be suited to more than one of these roles.**

Table 3.1 A summary of the team roles described by Belbin (1993). No one role should be considered 'better' than any other, and a good team requires members who are able to undertake appropriate roles at different times. Each role provides important strengths to a team, and its compensatory weaknesses should be accepted within the group framework.

Team role	Personality characteristics	Typical function in a team	Strengths	Allowable weaknesses
Coordinator	Self-confident, calm and controlled	Leading: causing others to work towards goals	Good at spotting others' talents and delegating activities	Often less creative or intellectual than others in the group
Shaper	Strong need for achievement; outgoing; dynamic; highly strung	Leading: generating action within a team; imposing shape and pattern to work	Providing drive and realism to group activities	Can be headstrong, emotional and less patient than others
Innovator[1]	Individualistic, serious-minded; often unorthodox	Generating new proposals and solving problems	Creative, innovative and knowledgeable	Tendency to work in isolation; ideas may not always be practical
Monitor–evaluator	Sober, unemotional and prudent	Analysing problems and evaluating ideas	Shrewd judgement	May work slowly; not usually a good motivator
Implementer	Well organised and self-disciplined, with practical common sense	Doing what needs to be done	Organising abilities and common sense	Lack of flexibility and tendency to resist new ideas
Teamworker	Sociable, mild and sensitive	Being supportive, perceptive and diplomatic; keeping the team going	Good listeners; reliable and flexible; promote team spirit	Not comfortable when leading; may be indecisive
Resource investigator	Extrovert, enthusiastic, curious and communicative	Exploiting opportunities; finding resources; external relations	Quick thinking; good at developing others' ideas	May lose interest rapidly
Completer–finisher	Introvert and anxious; painstaking, orderly and conscientious	Ensuring completion of activity to high standard	Good focus on fulfilling objectives and goals	Obsessive about details; may wish to do all the work to control quality
Specialist	Professional, self-motivated and dedicated	Providing essential skills	Commitment and technical knowledge	Contribute on a narrow aspect of project; tend to be single-minded

[1] May also be called 'plant' in some texts.

- **Team members may be required to adapt** and take a different role from the one that they feel suits them.
- **No one role is 'better' than any other.** For good teamwork, the group should have a balance of personality types present.
- **People may have to adopt multiple roles**, especially if the team size is small.

> **KEY POINT** In formal teamwork situations, be clear as to how individual contributions are to be identified and recognised. This might require discussion with the course organiser. Make sure that recognition, including assessment marks, is truly reflective of effort. Failure to ensure that this is the case can lead to disputes and feelings of unfairness within the team.

Your lab partner

Many laboratory sessions in the experimental sciences involve working in pairs. In some cases, you may work with the same partner for a series of practicals or for a complete module. The relationship you develop as a team is important to your progress, and can enhance your understanding of the material and the grades you obtain. Tips for building a constructive partnership include:

- **Introduce yourselves at the first session** and take a continuing interest in each other's interests and progress at university.
- **Discuss the practical** (both theory and tasks) and your understanding of what is expected of you.
- **Work jointly to complete the practical effectively,** avoiding the situation where one partner dominates the activities and gains most from the practical experience.
- **Share tasks according to your strengths,** but do this in such a way that one partner can learn new skills and knowledge from the other.
- **Make sure you ask questions of each other** and communicate any doubts about what you have to do.
- **Discuss other aspects of your course,** e.g. by comparing notes from lectures or ideas about in-course assessments.
- **Consider meeting up outside the practical sessions** to study, revise and discuss exams.

Text reference

Belbin, R.M. (1993) *Team Roles at Work.* Butterworth-Heinemann, Oxford.

Sources for further study

Belbin, R.M. *Homepage: Belbin Team Roles.* Available: http://www.belbin.com/ Last accessed: 21/12/10.

Lewin, P. (2004) *Successful Teamwork.* Open University Press, McGraw-Hill, Maidenhead.

4 Taking notes from lectures and texts

Choose note-taking methods appropriately – the method you choose to take notes might depend on the subject; the lecturer and his/her style of delivery; and your own preference.

Compare lecture notes with a colleague – looking at your notes for the same lecture may reveal interesting differences in approach, depth and detail.

Adjusting to the different styles of your lecturers – recognise that different approaches to lecture delivery demand different approaches to note-taking. For example, if a lecturer seems to tell lots of anecdotes or spend much of the time on examples during a lecture, do not switch off – you still need to be listening carefully to recognise the key messages. Similarly, if a lecture includes a section consisting mainly of images, you should still try to take notes – names of organisms, locations, key features, even quick sketches. These will help prompt your memory when revising. Do not be deterred by lecturers' idiosyncrasies; in every case you still need to focus and take useful notes.

Note-taking is an essential skill that you will require in many different situations, such as:

- listening to lectures and seminars;
- attending meetings and tutorials;
- reading texts and research papers;
- finding information on the World Wide Web.

KEY POINT Good performance in assessments and exams is built on effective learning and revision (Chapters 5 and 6). However, both ultimately depend on the quality of your notes.

Taking notes from lectures

Taking legible and meaningful lecture notes is essential if you are to make sense of them later, but many students find it difficult when starting their university studies. Begin by noting the date, course, topic and lecturer on the first page of each day's notes. Number every page in case they get mixed up later. The most popular way of taking notes is to write in a linear sequence down the page, emphasising the underlying structure via headings, as in Fig. 15.3. However, the 'pattern' and 'mind map' methods (Figs. 4.1 and 4.2) have their advocates: experiment, to see which method you prefer.

Whatever technique you use, do not try to take down all the lecturer's words, except when an important definition or example is being given, or when the lecturer has made it clear that he/she is dictating. Listen first, then write. Your goal should be to take down the structure and reasoning behind the lecturer's approach in as few words and phrases as possible. At this stage, follow the lecturer's sequence of delivery. Use headings and leave plenty of space, but don't worry too much about being tidy – it is more important that you get down the appropriate information in a readable form. Use abbreviations to save

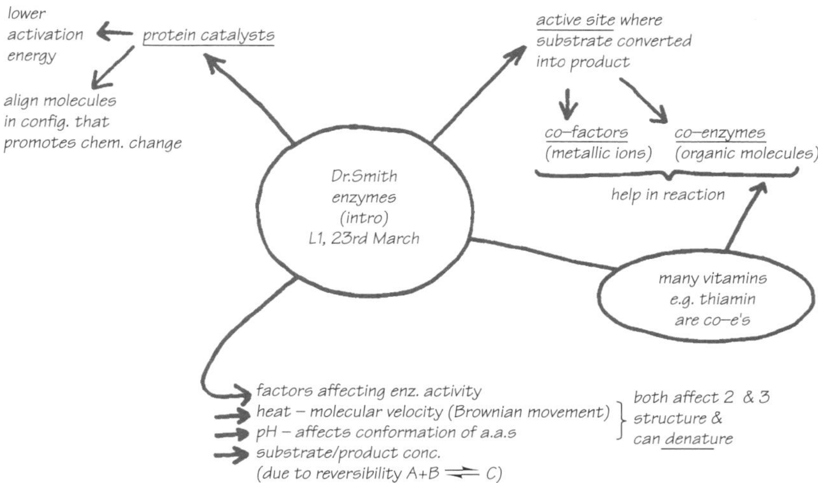

Fig. 4.1 An example of 'pattern' notes, an alternative to the more commonly used 'linear' format. Note the similarity to the 'spider diagram' method of brainstorming ideas (Fig. 15.2).

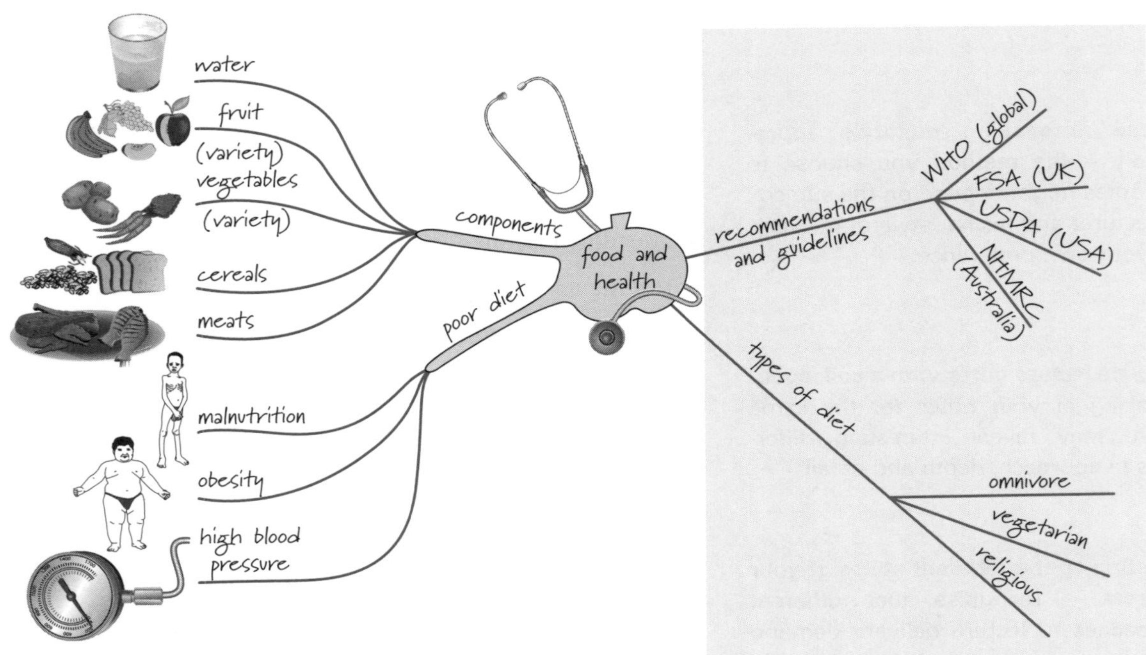

Fig. 4.2 Example of the 'mind map' approach to note-taking and 'brainstorming'. Start at the centre with the overall topic title, adding branches and sub-branches for themes and subsidiary topics. 'Basic' maps consist of a branched hierarchy overwritten with keywords (e.g. shaded portion). Connections should be indicated with arrows; numbering and abbreviations are encouraged. To aid recall and creativity, Buzan (2006) recommends use of colour, different fonts, 3-dimensional doodles and other forms of emphasis (e.g. non-shaded portion).

Example Commonly used abbreviations include:

$\exists$	there are, there exist(s)
$\therefore$	therefore
$\because$	because
$\propto$	is proportional to
$\rightarrow$	leads to, into
$\leftarrow$	comes from, from
$\rightarrow\rightarrow$	involves several processes in a sequence
1°, 2°	primary, secondary (etc.)
$\approx$, $\cong$	approximately, roughly equal to
$=$, $\neq$	equals, not equal to
$\equiv$, $\not\equiv$	equivalent, not equivalent to
$<$, $>$	smaller than, bigger than
$\gg$	much bigger than
[X]	concentration of X
$\sum$	sum
Δ	change
f	function
#	number
∞	infinity, infinite

You should also make up your own abbreviations relevant to the context, e.g. if a lecturer is talking about photsynthesis, you could write 'PS' instead.

time. Recognise that you may need to alter your note-taking technique to suit different lecturers' styles.

Make sure that you note down references to texts and take special care to ensure accuracy of definitions and numerical examples. If the lecturer repeats or otherwise emphasises a point, highlight (e.g. by underlining) or make a margin note of this – it could come in useful when revising. If there is something you do not understand, ask at the end of the lecture, and make an appointment to discuss the matter if there isn't time to deal with it then. Tutorials may provide an additional forum for discussing course topics.

Lectures delivered by PowerPoint or similar presentation programs

Some students make the mistake of thinking that lectures delivered as computer-based presentations with an accompanying handout or Web resource require little or no effort by way of note-taking. While it is true that you may be freed from the need to copy out large diagrams and the basic text may provide structure, you will still need to adapt and add to the lecturer's points. Much of the important detail and crucial emphasis will still be delivered verbally. Furthermore, if you simply listen passively to the lecture, or, worse, try to work from the handout alone, it will be far more difficult to understand and remember the content. Note-taking is an important aspect of active learning during lectures.

If you are not supplied with handouts, you may be able to print out the presentation beforehand, perhaps in the 'Handouts (3 slides per page)' format that allows space for notes beside each slide (Fig. 4.3).

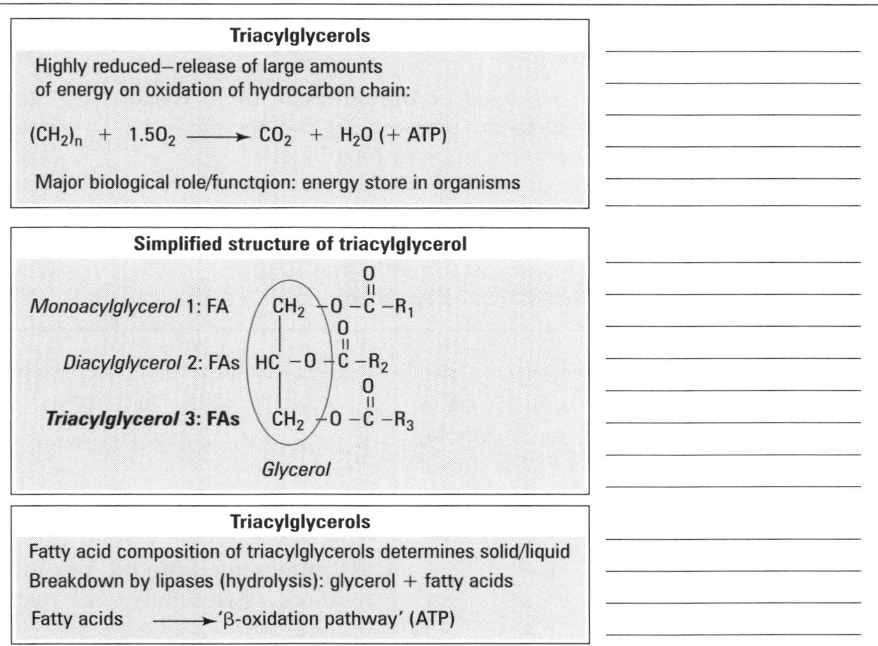

Fig. 4.3 An example of a printout from PowerPoint in 'Handouts (3 slides per page)' format.

Printing PowerPoint slides – use the 'Pure Black and White' option on the Print menu to avoid wasting ink on printing coloured backgrounds. If you wish to use colour, remember that slides can be difficult to read if printed in small format. Always print a sample page before printing the whole lecture.

Scan through this before the lecture if you can; then, during the presentation, focus on listening to what the lecturer has to say. Note down extra details, points of emphasis and examples. After lectures, you could also add notes from supplementary reading. The text in presentations can be converted to word processor format if you have access to the electronic file. In PowerPoint, this can be achieved from the *Outline* tab on the *Home* menu. You can copy and paste text between programs in the normal fashion, then modify font size and colour as appropriate.

'Making up' your notes

As soon as possible after each lecture, work through your notes, tidying them up and adding detail where necessary. Add emphasis to any headings you have made, so that the structure is clearer. If you feel it would be more logical for your purposes, change the order. Compare your notes with material in a textbook and correct any inconsistencies. Make notes from, or photocopy, any useful material you see in textbooks, ready for revision.

Taking notes from books and journal articles

Scanning and skimming are useful techniques to find sections to read in detail and to make notes from.

Scanning

Scanning effectively – you need to stay focused on your keywords, otherwise you may be distracted by apparently interesting but irrelevant material.

This involves searching for relevant content. Useful techniques are to:

- decide on keywords relevant to your search;
- check that the book or journal title indicates relevance;

Box 4.1 How to use the SQ3R technique to skim texts

Survey Get a quick overview of the contents of the book or chapter, perhaps by rapidly reading the contents page or headings.

Question Ask yourself what the material covers and how precisely it relates to your study objectives.

Read Now read the text, paying attention to the ways it addresses your key questions.

Recall Recite to yourself what has been stated every few paragraphs. Write notes of this if appropriate, paraphrasing the text rather than copying it.

Review Think about what you have read and/or review your notes as a whole. Consider where it all fits in.

- look through the contents page (either paper titles in a journal volume, or chapter titles in a text);
- look at the index if present.

Spotting sequences – writers often number their points (firstly, secondly, thirdly, etc.) and looking for these words in the text can help you skim it quickly.

Skimming

This is a valuable way to gain the maximum amount of information in the minimum amount of time, by reading as little of a text as is required. Essentially, the technique (sometimes termed 'surveying') requires you to look at the *structure* of the text, rather than the detail. In a sense, you are trying to see the writer's original plan and the purpose behind each part of the text. Look through the whole of the piece first, to gain an overview of its scope and structure. Headings provide an obvious clue to structure, if present. Next, look for the 'topic sentence' in each paragraph (p. 100), which is often the first. You might then decide that the paragraph contains a definition that is important to note, or it may contain examples, so may not be worth reading for your purpose.

Making sure you have all the details – when taking notes from a text or journal paper: (a) always take full details of the source (Chapter 8); (b) if copying word-for-word make sure you indicate this using quotes and take special care to ensure you do not alter the original.

Once you have found relevant material, note-taking fulfils the vital purpose of helping you understand and remember the information. If you simply read it, either directly or from a photocopy, you are at risk of doing neither. The act of paraphrasing (using different words to give the same meaning) makes you think about the meaning of the material and express this for yourself. It is an important active learning technique. A popular method of describing skimming and note-taking is called the SQ3R technique (Box. 4.1).

KEY POINT Obtaining information and understanding it are two distinct parts of the process of learning. As discussed in Chapter 5 (Table 5.1), you must be able to do more than recall facts to succeed.

Methods for finding and evaluating texts and articles are discussed further in Chapters 8 and 9.

Sources for further study

Belbin, R.M. *Homepage: Belbin Team Roles.* Available: http://www.belbin.com/ Last accessed: 21/12/10.

Buzan, T. (2006) *The Ultimate Book of Mind Maps.* Harper Thorsons, New York.

Buzan, T., Morris, S. and Smith, J. (1998) *Understanding Mind Maps . . . in a Week.* Hodder & Stoughton, London.

Lewin, P. (2004) *Successful Teamwork.* Open University Press, McGraw-Hill, Maidenhead.

5 Learning and revising

Taking an active approach to learning – adopting active methods of studying and revision (p. 20) that are suited to your personality can significantly improve your performance in assessments.

There are many different ways of learning; at university you have the freedom to choose which approach to study suits you best. You should tackle this responsibility with an open mind, and be prepared to consider new options. Understanding how you learn best and how you are expected to think about your discipline will help you to improve your approach to study and to understand your course materials at a deeper level. Your department will publish material that can help too. Taking account of learning outcomes and marking/assessment criteria, for example, will help you focus your revision.

KEY POINT At university, you are expected to set your own agenda for learning. There will be timetabled activities, assessments and exam deadlines, but it is your responsibility to decide how you will study and learn, how you will manage your time, and, ultimately, what you will gain from the experience. You should be willing to challenge yourself academically to achieve your full potential.

Learning styles

Significance of learning styles – no one learning style is 'better' than the others; each has its own strengths and weaknesses. However, since many university exams are conducted using 'reading and writing' modes of communication, you may need to find ways of expressing yourself appropriately using the written word (see Box 5.1).

Learning styles and teaching styles – there may be a mismatch between your preferred learning style and the corresponding 'teaching style' used by your lecturers, in which case you will need to adapt appropriately (see Box 5.1).

We do not all learn in the same way. Your preferred learning style is simply the one that suits you best for receiving, communicating and understanding information. It therefore involves approaches that will help you to learn and perform most effectively. There are many different ways of describing learning styles, and you may be introduced to specific schemes during your studies. Although methods and terminology may differ among these approaches, it is important to realise that the important thing is the *process* of analysing your learning style, together with the way you use the information to modify your approach to studying, rather than focussing on the type of learner that you may identify yourself to be within a particular scheme.

A useful approach for describing learning styles is the VARK system devised by Fleming (2001). By answering a short online questionnaire, you can 'diagnose' yourself as one of the types shown in Box 5.1, which also summarises important outcomes relating to how information and concepts can be assimilated, learned and expressed. People show different degrees of alignment with these categories, and research indicates that the majority of students are multi-modal learners – that is, falling into more than one category – rather than being only in one grouping. By carrying out an analysis like this, you can become more aware of your personal characteristics and think about whether the methods of studying you currently use are those that are best suited to your needs.

KEY POINT Having a particular learning preference or style does not mean that you are automatically skilled in using methods generally suited to that type of learner. You must work at developing your ability to take in information, study and cope with assessment.

Box 5.1 How to diagnose your learning preferences using the VARK learning styles scheme

Visit www.vark-learn.com to carry out the online diagnostic test, reflect on whether it is a fair description of your preferences, and think about whether you might change the way you study to improve your performance. None of the outcomes should be regarded as prescriptive – you should mix techniques as you see fit and only use methods that you feel comfortable adopting. Adapted with permission from material by Fleming (2001).

	Outcomes for your learning, studying and exam technique		
Learning style: ***Description of learning preferences***	**Advice for taking in information and understanding it**	**Best methods of studying for effective learning**	**Ways to cope with exams so you perform better**
Visual: *You are interested in colour, layout and design. You probably prefer to learn from sources with diagrams and charts. You tend to add doodles and use highlighters on lecture notes and express ideas and concepts as images.*	Use media incorporating images, flow-charts, etc. When constructing notes, use underlining, different colours and highlighters. Use symbols rather than words. Leave plenty of white space in notes. Experiment with 'mind maps' (p. 16).	Use similar methods to those described in column two. Reduce lecture notes to pictures. Try to construct your own images to aid understanding, then test your learning by redrawing these from memory.	Plan answers diagrammatically. Recall the images and doodles you used in your notes. Use diagrams in your answers (making sure they are numbered and fully labelled). As part of your revision, turn images into words.
Aural: *You prefer discussing subjects and probably like to attend tutorials and listen to lecturers, rather than read textbooks. Your lecture notes may be poor because you would rather listen than take notes.*	Make sure you attend classes and tutorials. Remember the interesting examples, stories, jokes. Leave spaces in notes for later recall and 'filling'. Discuss topics with others. Record lectures (with lecturer's permission).	Expand your notes by talking with others and making additional notes from the textbook. Ask others to 'hear' you talk about topics. Read your summarised notes aloud to yourself. Record your vocalised notes and listen to them later.	When writing answers, imagine you are talking to an unseen examiner. Speak your answers inside your head. Listen to your voice and write them down. Practise writing answers to old exam questions.
Read–Write: *You prefer using text in all formats. Your lecture notes are probably good. You tend to like lecturers who use words well and have lots of information in sentences and notes. In note-taking, you may convert diagrams to text and text to bullet points.*	Focus on note-taking. You may prefer the 'linear' style of note-taking (p. 15). Use lists; headings; glossaries and definitions. Expand notes by adding further information from handouts, textbooks and library readings.	Reduce notes to lists or headings. Write out and read the lists (silently). Turn actions, diagrams and flowcharts into words. Rewrite ideas and principles into other words. Organise diagrams/graphs into statements, e.g. 'The trend is . . .'.	Plan and write out exam answers using remembered lists. Arrange your words into hierarchies and points.
Kinesthetic: *You tend to recall by remembering real events and lecturers' 'stories'. You probably prefer lab work to theory and like lecturers who give real-life examples. Your lecture notes may be weak because the topics did not seem relevant.*	Focus on examples that illustrate principles. Concentrate on applied aspects and hands-on approaches, but try to understand the underlying principles. When taking in information, use all senses – sight, touch, taste, smell, hearing.	Put plenty of examples and pictures into your notes. Use case studies and applications to help with principles and concepts. Talk through your notes with others. Recall your experience of lectures, tutorials, experiments.	Write practice answers and paragraphs. Recall examples and things you did in the lab or field trip. Role-play the exam situation in your own room.
Multi-modal: *Your preferences fall into two or more of the above categories.*	If you have two dominant preferences or several equally dominant preferences, read the study strategies above that apply to each of these. You may find it necessary to use more than one strategy for learning and communicating, feeling less secure with only one.		

Thinking about thinking

The thinking processes that students are expected to carry out can be presented in a sequence, starting with shallower thought processes and ending with deeper processes, each of which builds on the previous level (see Table 5.1). The first two categories in this ladder apply to gaining basic knowledge and understanding – this is important when you first encounter a topic. Processes three to six are those carried out by high-performing university students, with the latter two being especially relevant to final-year students, researchers and professionals. Naturally, the tutors assessing you will want to reward the deepest thinking appropriate for your level of study. This is often signified by the words they use in assessment tasks and marking criteria (column four, Table 5.1,

Table 5.1 A ladder of thinking processes, moving from shallower thought processes (top of table) to deeper levels of thinking (bottom of table). This table is derived from research by Benjamin Bloom *et al.* (1956). When considering the cue words in typical question instructions, bear in mind that the precise meaning will always depend on the context. For example, while 'describe' is often associated with relatively simple processes of recall, an instruction such as 'describe how the human brain works' demands higher-level understanding. Note also that while a 'cue word' is often given at the start of a question/instruction, this is not universally so.

Thinking processes and description (in approximate order of increasing 'depth')	Example in food sciences	Example of typical question structure, with *cue word* highlighted	Other cue words used in question instructions
1. **Knowledge (knowing facts).** If you know information, you can *remember* or *recognise* it. This does not always mean that you understand it at a higher level.	You might have learned (memorised) the formula for a monosaccharide as $C_6H_{12}O_6$.	*Describe* the main features of the food pyramid dietary system used in the USA.	• Define • List • State • Identify • Label
2. **Comprehension.** If you comprehend a fact, you *understand* what it means.	You might understand the enzymic process by which polysaccharides are broken down to form monosaccharides.	*Explain* how antioxidants act to counteract the damage caused by aerobic respiration.	• Contrast • Compare • Distinguish • Interpret • Summarise
3. **Application.** To apply a fact means that you can *put it to use* in a particular context.	You might be able to apply your knowledge of enzyme biochemistry to predict the breakdown products of a novel polysaccharide.	*Calculate* the amount of energy available from a foodstuff that contains 30% protein, 30% saturated fat and 40% carbohydrate (0% fibre).	• Show • Illustrate • Solve • Demonstrate • Use
4. **Analysis.** To analyse information means that you are able to *break it down into parts* and *show how these components fit together.*	You might be able to identify literature-based evidence for and against the effect of a high-sugar diet on behaviour.	Drawing on primary literature sources, *defend* the theory of calorie restriction.	• Compare • Explain • Consider • Infer • Differentiate
5. **Synthesis.** To synthesise something you need to be able to *extract relevant facts* from a body of knowledge and use these to *address an issue in a novel way, or create something new.*	You might be able to develop a nutrition plan for an individual, based on recommended guidelines, together with information on his/her dietary preferences.	*Devise* an experiment to measure the degree of unsaturation of a fat, based on its reaction with iodine.	• Design • Integrate • Test • Create • Plan
6. **Evaluation.** If you evaluate information, you *arrive at a judgement* based on its importance relative to the topic being addressed.	You might be able to comment on the advantages and disadvantages of two different food labelling schemes.	*Evaluate* the relative merits of organic and non-organic foods	• Discuss • Review • Assess • Consider • Justify

Example A set of learning objectives taken from an introductory lecture on '*Carbohydrates*' might be:

'After this lecture, you should be able to:

- Define the following:
  - ✓ monosaccharide;
  - ✓ disaccharide;
  - ✓ alternative sweetner;
  - ✓ polysaccharide;
  - ✓ fibre (US fiber).
- Draw diagrams to show the general structure of glucose and glycogen.
- Demonstrate knowledge of the role of dietary sugar in dental caries and obesity.
- Explain the ways in which food labels present information on carbohydrates.
- Discuss the beneficial health effects of starch and fibre.'

and p. 105), and being more aware of this agenda can help you to gain more from your studies and appreciate what is being demanded of you.

KEY POINT When considering assessment questions, look carefully at words used in the instructions. These cues can help you identify what depth is expected in your answer. Take special care in multi-part questions, because the first part may require lower-level thinking, while in later parts marks may be awarded for evidence of deeper thinking.

The role of assessment and feedback in your learning

Your starting point for assessment should be the learning outcomes or objectives for each module, topic or learning activity. You will usually find them in your module handbook. They state in clear terms what your tutors expect you to be able to accomplish after participating in each part and reading around the topic. Also of value will be marking/assessment criteria or grade descriptors, which state in general terms what level of attainment is required for your work to reach specific grades. These are more likely to be defined at faculty/college/school/department level and consequently published in appropriate handbooks or websites. Reading learning outcomes and grade descriptors will give you a good idea of what to expect and the level of performance required to reach your personal goals. Relate them to both the material covered (e.g. in lectures and practicals, or online) and past exam papers. Doing this as you study and revise will indicate whether further reading and independent studying are required, and of what type. You will also have a much clearer picture of how you are likely to be assessed.

Definitions

Learning objectives/outcomes – statements of the knowledge, understanding or skills that a learner will be able to demonstrate on successful completion of a module, topic or learning activity.

Formative assessments – these may be mid-term or mid-semester tests and are often in the same format as later exams. They are intended to give you *feedback* on your performance. You should use the results to measure your performance against the work you put in, and to find out, either from grades or tutor's comments, how you could do better in future. If you don't understand the reason for your grade, contact your tutor.

Summative assessments – these include end-of-year or end-of-module exams. They inform others about the standard of your work. In continuous or 'in-course' assessment, the summative elements are spread out over the course. Sometimes these assessments may also include a formative aspect, if feedback is given.

KEY POINT Use the learning objectives for your course (normally published in the course guide or handbook) as a fundamental part of your study and revision planning. These indicate what you will be expected to be able to do after taking part in the course, so exam questions are often based on them. Check this by reference to past papers.

There are essentially two types of assessment – formative and summative, although the distinction may not always be clear-cut (see margin). The first way you can learn from formative assessment is to consider the grade you obtained in relation to the work you put in. If this is a disappointment to you, then there must be a mismatch between your understanding of the topic and the marking scheme and that of the marker, or a problem in the writing or presentation of your assignment. This element of feedback is also present in summative assessment.

The second way to learn from formative assessment is through the written feedback and notes on your work. These comments may be cryptic, or scribbled hastily, so if you don't understand or can't read them, ask the tutor who marked the work. Most tutors will be pleased to explain how you could have improved your mark. If you find that the same comments appear frequently, it may be a good idea to seek help from your university's academic support unit. Take along examples of your work and feedback comments so they can give you the best possible advice. Another suggestion is to ask to see the work of another student who obtained a good mark, and compare it with your own. This will help you judge the standard you should be aiming for.

Preparing for revision and examinations

Before you start revising, find out as much as you can about each exam, including:

- its format and duration;
- the date and location;
- the types of questions;
- whether any questions/sections are compulsory;
- whether the questions are internally or externally set or assessed;
- whether the exam is 'open book' and, if so, which texts or notes are allowed.

Time management when revising – this is a vital to success and is best achieved by creating a revision timetable (Box 5.2).

Your course tutor is likely to give you details of exam structure and timing well beforehand, so that you can plan your revision; the course handbook and past papers (if available) can provide further useful details. Always check that the nature of the exam has not changed before you consult past papers.

Organising and using lecture notes, assignments and practical reports

Filing lecture notes – make sure your notes are kept neatly and in sequence by using a ring binder system. File the notes in lecture or practical sequence, adding any supplementary notes or photocopies alongside.

Using tutors' feedback – it is always worth reading any comments on your work as soon as it is returned. If you don't understand the comments, or are unsure about why you might have lost marks in an assignment, ask for an explanation.

Given their importance as a source of material for revision, you should have sorted out any deficiencies or omissions in your lecture notes and practical reports at an early stage. For example, you may have missed a lecture or practical due to illness, etc., but the exam is likely to assume attendance throughout the year. Make sure you attend classes whenever possible and keep your notes up to date. Your practical reports and any assignment work will contain specific comments from the teaching staff, indicating where marks were lost, corrections, mistakes, inadequacies, etc. Most lecturers will readily discuss such details with students and this information may provide you with 'clues' to the expectations of individual lecturers that may be useful in exams set by the same members of staff. However, you should never 'fish' for specific information on possible exam questions, as this is likely to be counterproductive.

Revision

Recognise when your concentration powers are dwindling – take a short break when this happens and return to work refreshed and ready to learn. Remember that 20 minutes is often quoted as a typical limit to full concentration effort.

Begin early, to avoid last-minute panic. Start in earnest several weeks beforehand, and plan your work carefully:

- **Prepare a revision timetable** – an 'action plan' that gives details of specific topics to be covered (Box 5.2). Find out at an early stage when (and where) your examinations are to be held, and plan your revision around this. Try to keep to your timetable. Time management during this period is as important as keeping to time during the exam itself.

Box 5.2 How to prepare and use a revision timetable

1. **Make up a grid showing the number of days until your exams are finished.** Divide each day into several sections. If you like revising in large blocks of time, use a.m., p.m. and evening slots, but if you prefer shorter periods, divide each of these in two, or use hourly divisions.
2. **Write in your non-revision commitments,** including any time off you plan to allocate and physical activity at frequent intervals. Try to have about one-third or a quarter of the time off in any one day. Plan this in relation to your best times for useful work – for example, some people work best in the mornings, while others prefer evenings. If you wish, use a system where your relaxation time is a bonus to be worked for; this may help you motivate yourself.
3. **Decide on how you wish to subdivide your subjects** for revision purposes. This might be among subjects, according to difficulty (with the hardest getting the most time), or within subjects, according to topics. Make sure there is an adequate balance of time among topics and especially that you do not avoid working on the subject(s) you find least interesting or most difficult.
4. **Allocate the work to the different slots available on your timetable.** You should work backwards from the exams, making sure that you cover every exam topic adequately in the period just before each exam. You may wish to colour-code the subjects.
5. **As you revise, mark off the slots completed** – this has a positive psychological effect and will boost your self-confidence.
6. **After the exams, revisit your timetable** and decide whether you would do anything differently next time.

- **Study the learning objectives/outcomes for each topic** (usually published in the course handbook) to get an idea of what lecturers expect from you.
- **Use past papers as a guide to the form of exam** and the type of question likely to be asked (Box 5.3).
- **Remember to have several short (five-minute) breaks during each hour of revision** and a longer break every few hours. In any day, try to work for a maximum of three-quarters of the time.
- **Include recreation within your schedule:** there is little point in tiring yourself with too much revision, as this is unlikely to be profitable.
- Make your revision as *active* and *engaging* as possible (see below): the least productive approach is simply to read and reread your notes.
- **Ease back on revision near the exam:** plan your revision to avoid last-minute cramming and overload fatigue.

Active revision

The following techniques may prove useful in devising an active revision strategy:

- **'Distil' your lecture notes** to show the main headings and examples. Prepare revision sheets with details for a particular topic on a single sheet of paper, arranged as a numbered checklist. Wall posters are another useful revision aid.
- **Confirm that you know about the material by testing yourself** – take a blank sheet of paper and write down all you know. Check your full notes to see if you missed anything out. If you did, go back immediately to a fresh blank sheet and redo the example. Repeat, as required.

Aiding recall through effective note-taking – the mind map technique (p. 16), when used to organise ideas, is claimed to enhance recall by connecting the material to visual images or linking it to the physical senses.

Box 5.3 How to use past exam papers in your revision

Past exam papers are a valuable resource for targeting your revision.

1. **Find out where the past exam papers are kept**. Copies may be lodged in your department or the library; or they may be accessible online.
2. **Locate and copy relevant papers for your module(s)**. Check with your tutor or course handbook that the style of paper will not change for the next set of examinations.
3. **Analyse the design of the exam paper**. Taking into account the length in weeks of your module, and the different lecturers and/or topics for those weeks, note any patterns that emerge. For example, can you translate weeks of lectures/practicals into numbers of questions or sections of the paper? Consider how this might affect your revision plans and exam tactics, taking into account (a) any choices or restrictions offered in the paper, and (b) the different types of questions asked (i.e. multiple choice, short answer or essay).
4. **Examine carefully the style of questions**. Can you identify the expectations of your lecturers? Can you relate the questions to the learning objectives? How much extra reading do they seem to expect? Are the questions fact-based? Do they require a synthesis based on other knowledge? Can you identify different styles for different lecturers? Consider how the answers to these questions might affect your revision effort and exam strategy.
5. **Practise answering questions**. Perhaps with friends, set up your own mock exam when you have done a fair amount of revision, but not too close to the exams. Use a relevant past exam paper (don't study it beforehand). You need not attempt all of the paper at one sitting. You'll need a quiet room in a place where you will not be interrupted (e.g. a library). Keep close track of time during the mock exam and try to do each question in the length of time you would normally assign to it (see p. 104) – this gives you a feel for the speed of thought and writing required and the scope of answer possible. Mark each other's papers and discuss how each of you interpreted the question and laid out your answers and your individual marking schemes.
6. **Practise writing answer plans and starting answers**. This can save time compared with the 'mock exam' approach. Practise in starting answers can help you get over stalling at the start and wasting valuable time. Writing essay plans gets you used to organising your thoughts quickly and putting your thoughts into a logical sequence.

Question-spotting – avoid adopting this risky strategy to reduce the amount of time you spend revising. Lecturers are aware that this approach may be taken and try to ask questions in an unpredictable manner. You may find that you are unable to answer on unexpected topics that you failed to revise. Moreover, if you have a preconceived idea about what will be asked, you may also fail to grasp the nuances of the exact question set, and provide a response lacking in relevance.

- **Memorise definitions and key phrases** – definitions can be a useful starting point for many exam answers. Make up lists of relevant facts or definitions associated with particular topics. Test yourself repeatedly on these, or get a friend to do this. Try to remember *how many* facts or definitions you need to know in each case – this will help you recall them all during the exam.
- **Use mnemonics and acronyms** to commit specific factual information to memory. Sometimes, the dafter they are, the better they seem to work.
- **Use pattern diagrams or Mind Maps** as a means of testing your powers of recall on a particular topic (pp. 15–16).
- **Draw diagrams from memory** – make sure you can label them fully.
- **Try recitation as an alternative to written recall.** Talk about your topic to another person, preferably someone in your class. Talk to yourself if necessary. Explaining something out loud is an excellent test of your understanding.
- **Associate facts with images** or journeys if you find this method works.
- **Use a wide variety of approaches to avoid boredom during revision** (e.g. record information on audio tape, use cartoons, or any other method, as long as it's not just reading).

Use revision checks – it is important to test yourself frequently during revision, to ensure that you have retained the information you are revising.

Final preparations – try to get a good night's sleep before an exam. Last-minute cramming will be counterproductive if you are too tired during the exam.

- **Form a revision group** to share ideas and discuss topics with other students.
- **Prepare answers to past papers**, e.g. write essays or, if time is limited, write essay plans (see Box 5.3).
- **Work through representative problems** if your subject involves numerical calculations.
- **Make up your own questions** – the act of putting yourself in the examiner's mindset by inventing questions can help revision. However, you should not rely on 'question-spotting': this is a risky practice!

The evening before your exam should be spent in consolidating your material, and checking through summary lists and plans. Avoid introducing new material at this late stage: your aim should be to boost your confidence, putting yourself in the right frame of mind for the exam itself.

Text references

Bloom, B., Englehart, M., Furst, E., Hill, W. and Krathwohl, D. (1956) *Taxonomy of Educational Objectives: The Classification of Educational Goals. Handbook I: Cognitive Domain*. Longmans, Green, New York and Toronto.

Fleming, N.D. (2001) *Teaching and Learning Styles: VARK Strategies*. Neil Fleming, Christchurch.

Fleming, N.D. *VARK: A Guide to Learning Styles*. Available: http://www.vark-learn.com/ Last accessed 21/12/10.

Sources for further study

Burns, R. (1997) *The Student's Guide to Passing Exams*. Kogan Page, London.

Hamilton, D. (2003) *Passing Exams: A Guide for Maximum Success and Minimum Stress*, 2nd edn. Continuum Press, London.

Many universities host study skills websites; these can be found using 'study skills', 'revision' or 'exams' as keywords in a search engine.

6 Curriculum options, assessment and exams

Definition

Transcript – this is your record of achievement at university. Normally it will consist of details of each module or course you have taken, and an indication of the grade or mark achieved. It will also show your final (honours) classification, that is: first class, upper second class (2.1), lower second class (2.2), third class or unclassified (some UK universities do not differentiate second class degrees).

Many universities have adopted a modular system for their degree courses. This allows greater flexibility in subject choice and accommodates students studying on different degree paths. Modules also break a subject into discrete, easily assimilated elements. They have the advantage of spreading assessment over the academic year, but they can also tempt you to avoid certain difficult subjects or to feel that you can forget about a topic once the module is finished.

KEY POINT You should select your modules with care, mindful of potential degree options and how your transcript and CV will appear to a prospective employer. If you feel you need advice, consult your personal tutor or study adviser.

Aiming high – your goal should be to perform at your highest possible level and not simply to fulfil the minimum criteria for progression. This will lay sound foundations for your later studies. Remember too that a future employer might ask to see your academic transcript, which will detail *all* of your module grades including any fails/resits, and will not just state your final degree classification.

As you move between levels of the university system, you will be expected to have passed a certain number of modules, as detailed in the progression criteria. These may be expressed using a credit-point system. Students are normally allowed two attempts at each module exam and the resits often take place at the end of the summer vacation. If a student does not pass at the second attempt, they may be asked to 'carry' the subject in a subsequent year, and in severe cases of multiple failure, they may be asked to retake the whole year or even leave the course. Consequently, it is worth finding out about these aspects of your degree. They are usually published in relevant handbooks.

You are unlikely to have reached this stage in your education without being exposed to the examination process. You may not enjoy being assessed, but you probably want to do well in your course. It is therefore important to understand why and how you are being tested. Identifying and improving the skills required for exam success will allow you to perform to the best of your ability.

Studying an accredited degree programme – note that the accrediting authority (e.g. for UK dietetics programmes, the Health Professions Council) may specify *higher* minimum marks in core modules within the programme than those used in other degrees at the same university.

Assessed coursework

There is a component of assessed coursework in many modules. This often tests specific skills, and may require you to demonstrate thinking at deeper levels (see Table 5.1). The common types of coursework assessment are covered at various points in this book:

- **Practical exercises** (throughout and especially Chapters 49–71).
- **Essays** (Chapters 15 and 16).
- **Numerical problems** (Chapter 37).
- **Data analysis** (Chapters 35–39).
- **Poster and spoken presentations** (Chapters 13 and 14).
- **Literature surveys and reviews** (Chapter 18).
- **Project work** (Chapters 17 and 34).
- **Problem-based learning** (Box 6.1).

Avoiding plagiarism – this is a key issue for assessed coursework – see p. 48 for a definition and Chapter 8 for appropriate methods of referring to the ideas and results of others using citation.

At the start of each year or module, read the course handbook or module guide carefully to find out when any assessed work needs to be submitted.

Box 6.1 Problem-based learning (PBL)

In this relatively new teaching method, you are presented with a 'real world' problem or issue, often working within a team. As you tackle the problem, you will gain factual knowledge, develop skills and exercise critical thinking (Chapter 9). Because there is a direct and relevant context for your work, and because you have to employ active learning techniques, the knowledge and skills you gain are likely to be more readily assimilated and remembered. This approach also more closely mimics workplace practices. PBL usually proceeds as follows:

1. **You are presented with a problem** (e.g. a case study, a hypothetical patient, a topical issue).
2. **You consider what issues and topics you need to research,** by discussion with others if necessary. You may need to identify where relevant resources can be found (Chapters 8 and 10).
3. **You then need to rank the issues and topics in importance,** allocating tasks to group members, if appropriate.
4. **Having carried out the necessary research, you should review what information has been obtained.** As a result, new issues may need to be explored and, where appropriate, allocated to group members.
5. **You will be asked to produce an outcome, such as a report, diagnosis, seminar presentation or poster.** An outline structure will be required and, for groups, further allocation of tasks to accomplish this goal.

If asked to carry out PBL as part of your course, it is important to get off to a good start. At first, the problem may seem unfamiliar. However, once you become involved in the work, you will quickly gain confidence. If working as part of a group, make sure that your group meets as early as possible, that you attend all sessions and that you do the necessary background reading. When working in a team, a degree of self-awareness is necessary regarding your 'natural' role in group situations (Table 3.1). Various methods are used for assessing PBL, and the assessment may involve peer marking.

Note relevant dates in your diary, and use this information to plan your work. Take special note if deadlines for different modules clash, or if they coincide with social or sporting commitments.

KEY POINT If, for some valid reason (e.g. illness), you will be late with an assessment, speak to your tutors as soon as possible. They may be able to take extenuating circumstances into account by not applying a marking penalty. They will let you know what paperwork you may require to submit to support your claim.

Summative exams – general points

Dealing with problems during an examination – is it important to raise any issue with the exam invigilator, so that the problem is fully documented. By waiting until after the exam, it may be too late to deal with the issue.

Summative exams (p. 22) normally involve you answering questions without being able to consult other students or your notes. Invigilators are present to ensure appropriate conduct, but departmental representatives may be present for some of the exam. Their role is to sort out any subject-related problems, so if you think something is wrong, ask at the earliest opportunity. It is not unknown for parts of questions to be omitted in error, or for double meanings to arise, for example.

Planning

When preparing for an exam, make a checklist of the items you'll need (see p. 33). On the day of the exam, give yourself sufficient time to arrive at the correct room, without the risk of being late. Double-check the times and places of your exams, both well before the exam and also on arrival. If you arrive at the exam venue early, you can always rectify a mistake if you find you've gone to the wrong place.

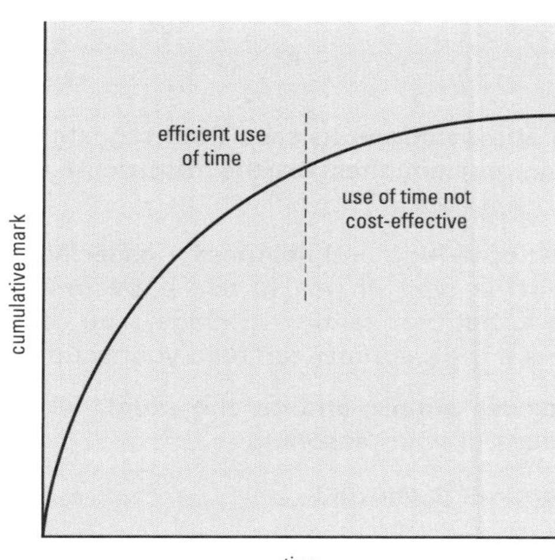

Fig. 6.1 Exam marks as a function of time. The marks awarded in a single answer will follow the law of diminishing returns – it will be far more difficult to achieve the final 25 per cent of the available marks than the initial 25 per cent. Do not spend too long on any one question.

Using the question paper – unless this is specifically forbidden, you should write on the question paper to plan your strategy, keep to time and organise answers.

Checklist for exam answers – *during the final phase, check for:*

- factual errors;
- missing information;
- grammatical and spelling errors;
- errors of scale and units;
- mathematical and calculation errors.

Tackling the paper

Begin by reading the instructions at the top of the exam paper very carefully, so that you do not make any errors based on lack of understanding of the exam structure. Make sure that you know:

- **How many questions are set.**
- **How many must be answered.**
- **Whether the paper is divided into sections.**
- **Whether any parts are compulsory.**
- **What each question/section is worth**, as a proportion of the total mark.
- **Whether different questions should be answered in different books.**

Do not be tempted to spend too long on any one question or section: the return in terms of marks will not justify the loss of time from other questions (see Fig. 6.1). Take the first 10 minutes or so to read the paper and plan your strategy, before you begin writing. Do not be put off by those who begin immediately; it is almost certain they are producing unplanned work of a poor standard.

Underline the key phrases in the instructions, to reinforce their message. Next, read through the set of questions. If there is a choice, decide on those questions to be answered and decide on the order in which you will tackle them. Prepare a timetable which takes into account the amount of time required to complete each section and which reflects the allocation of marks – there is little point in spending one-quarter of the exam period on a question worth only 5 per cent of the total marks. Use the exam paper to mark the sequence in which the questions will be answered and write the finishing times alongside; refer to this timetable during the exam to keep yourself on course.

Reviewing your answers

At the end of the exam, you should allow at least 10 minutes to read through your script, to check for errors. Make sure your name and/or ID number is on each exam book as required and on all other sheets of paper, including graph paper, even if securely attached to your script, as it is in your interest to ensure that your work is not lost.

> KEY POINT *Never* leave any exam early. Most exams assess work carried out over several months in a time period of 2–3 hours and there is always something constructive you can do with the remaining time to improve your script.

Special considerations for different types of exam question

Essay questions

Essay questions let examiners test the depth of your comprehension and understanding as well as your recall of facts. Essay questions give you plenty of scope to show what you know. They suit those with a good grasp of principles but who perhaps have less ability to recall specific details.

Before you tackle a particular question, you must be sure of what is required in your answer. Ask yourself 'What is the examiner looking for in

Box 6.2 Writing under exam conditions

Always go into an exam with a strategy for managing the available time.

- **Allocate some time (say 5 per cent of the total) to consider which questions to answer and in which order.**
- **Share the rest of the time among the questions, according to the marks available.** Aim to optimise the marks obtained. A potentially good answer should be allocated slightly more time than one you don't feel so happy about. However, don't concentrate on any one answer (see Fig. 6.1).
- **For each question divide the time into planning, writing and revision phases** (see p. 104).

Employ time-saving techniques as much as possible.

- **Use spider diagrams** (Fig. 15.2) **or mind maps** (Fig. 4.2) to organise and plan your answer.
- **Use diagrams and tables** to save time in making difficult and lengthy explanations, but make sure you refer to each one in the text.
- **Use standard abbreviations** to save time repeating text but always explain them at the first point of use.
- **Consider speed of writing and neatness** – especially when selecting the type of pen to use – ballpoint pens are fastest, but they tend to smudge. You can only gain marks if the examiner can read your script.
- **Keep your answer simple and to the point**, with clear explanations of your reasoning.

Make sure your answer is relevant.

- **Don't include irrelevant facts** just because you memorised them during revision, as this may do you more harm than good. You must answer the specific question that has been set.
- **Time taken to write irrelevant material is time lost from another question.**

Adopting different tactics according to the exam – you should adjust your exam strategy (and revision methods) to allow for the differences in question types used in each exam paper.

Penalties for guessing – if there is a penalty for incorrect answers in a multiple-choice test, the best strategy is not to answer questions when you know your answer is a complete guess. Depending on the penalty, it may be beneficial to guess if you can narrow the choice down to two options (but beware false or irrelevant alternatives). However, if there are no such penalties, then you should provide an answer to all questions.

this particular question?' and then set about providing a *relevant* answer. Consider each individual word in the question and highlight, underline or circle the keywords. Make sure you know the meaning of the terms given in Table 16.1 (p. 105) so that you can provide the appropriate information, where necessary. Spend some time planning your writing (see Chapter 15). Refer back to the question frequently as you write, to confirm that you are keeping to the subject matter. Box 6.2 gives advice on writing essays under exam conditions.

It is usually a good idea to begin with the question that you are most confident about. This will reassure you before tackling more difficult parts of the paper. If you run out of time, write in note form. Examiners are usually understanding, as long as the main components of the question have been addressed and the intended structure of the answer is clear. Common reasons for poor exam answers in essay-style questions are listed in Box 6.3.

Multiple-choice and short-answer questions

Multiple-choice questions (MCQs) and short-answer questions (SAQs) are generally used to test the breadth and detail of your knowledge. The various styles that can be encompassed within the SAQ format allow for more demanding questions than MCQs, which may emphasise specific factual knowledge.

A good approach for MCQ papers is to follow a three-stage strategy, as follows:

1. **First trawl: read through the questions fairly rapidly**, noting the 'correct' answer in those you can attempt immediately, perhaps in pencil.

Box 6.3 Reasons for poor exam answers to essay-style questions

The following are reasons that lecturers cite when they give low marks for essay answers:

- **Not answering the exact question set**. Either failing to recognise the specialist terms used in the question, or failing to demonstrate an understanding of the terms by not providing definitions, or failing to carry out the precise instruction in a question, or failing to address all aspects of the question.
- **Running out of time**. Failing to match the time allocated to the extent of the answer. Frequently, this results in spending too long on one question and not enough on the others, or even failing to complete the paper.
- **Failing to answer all parts** of a multiple-part question, or to recognise that one part (perhaps involving more complex ideas) may carry more marks than another.
- **Failing to provide evidence** to support an answer. Forgetting to state the 'obvious' – either basic facts or definitions.
- **Failing to illustrate an answer appropriately**, either by not including a relevant diagram, or by providing a diagram that does not aid communication, or by not including examples.
- **Incomplete answer(s)**. Failing to answer appropriately due to lack of knowledge.
- **Providing irrelevant evidence** to support an answer. 'Waffling' to fill space.
- **Illegible handwriting**.
- **Poor English**, such that facts and ideas are not expressed clearly.
- **Lack of logic** or structure to the answer.
- **Factual errors**, indicating poor note-taking or poor revision or poor recall.
- **Failing to correct obvious mistakes** by rereading an answer before submitting the script.

At higher levels, the following aspects are especially important:

- **Not providing enough in-depth information**.
- **Providing a descriptive rather than an analytical answer** – focusing on facts, rather than deeper aspects of a topic.
- **Not setting a problem in context**, or not demonstrating a wider understanding of the topic. (However, make sure you don't overdo this, or you may risk not answering the question set.)
- **Not giving enough evidence of reading around the subject**. This can be demonstrated, for example, by quoting relevant papers and reviews and by giving author names and dates of publication.
- **Not considering both sides of a topic/debate, or not arriving at a conclusion if you have done so.**

Using a multi-stage approach for MCQs – one good reason for following this strategy is that in the later stages you may be prompted to recall facts relevant to questions looked at earlier. You can also allocate more time to the difficult/complex questions, once you have answered the straightforward ones.

2. **Second trawl: go through the paper again**, checking your original answers and this time marking up the answer sheet properly.
3. **Third trawl: now tackle the difficult questions** and those that require longer to answer (e.g. those based on numerical problems).

When unsure of an answer, the first stage is to rule out options that are clearly absurd or have obviously been placed there to distract you. Next, looking at the remaining options, can you judge between contrasting pairs with alternative answers? Logically, both cannot be correct, so you should see if you can rule one of the pair out. Watch out, however, in case *both* may be irrelevant to the answer. If the question involves a calculation, try to do this independently from the answers, so you are not influenced by them.

In SAQ papers, there may be a choice of questions. Choose your options carefully – it may be better to gain half marks for a correct answer to half a question, than to provide a largely irrelevant answer that apparently covers the whole question but lacks the necessary detail. For this form of question, few if any marks are given for writing style. Think in 'bullet point' mode and list the crucial points only. The time for answering SAQ questions may be tight, so get to work quickly, starting

Answer the question as instructed – This is true for all questions, but especially important for SAQs. If the question asks for a diagram, make sure you provide one; if it asks for n aspects of a topic, try to list this number of points; if there are two or more parts, provide appropriate answers to all aspects. This may seem obvious, but many marks are lost for not following instructions.

Examples These are the principal types of question you are likely to encounter in a practical or information-processing exam:

Manipulative exercises Often based on work carried out during your practical course. Tests dexterity, specific techniques (e.g. sterile technique).

'Spot' tests Short questions requiring an identification, or brief descriptive notes on a specific item. Tests knowledge of seen material or ability to transfer this to a new example.

Calculations May include the preparation of aqueous solutions at particular concentrations (Chapter 26) and statistical exercises (Chapter 39). Tests numeracy.

Data analyses May include the preparation and interpretation of graphs (Chapter 35) and numerical information, from data either obtained during the exam or provided by the examiner. Tests problem-solving skills.

Preparing specimens for examination with a microscope Tests staining technique and light microscopy technique.

Terminology – the oral exams are sometimes known simply as 'orals' or, borrowing Latin, as '*viva voces*' (by or with the living voice) or '*vivas*'.

with answers that demand remembered facts. Stick to your timetable by moving on to the next question as soon as possible. Strategically, it is probably better to get part-marks for the full number of questions than good marks for only a few.

Practical and information-processing exams

The prospect of a practical or information-processing exam may cause you more concern than a theory exam. This may be due to a limited experience of practical examinations, or to the fact that practical and observational skills are tested, as well as recall, description and analysis of factual information. Your first thoughts may be that it is not possible to prepare for such exams but, in fact, you can improve your performance by mastering the various practical techniques described in this book.

You may be allowed to take your laboratory reports and other texts into the practical exam. Don't assume that this is a 'soft option', or that revision is unnecessary: you will not have time to read large sections of your reports or to familiarise yourself with basic principles, etc. The main advantage of 'open book' exams is that you can check specific details of methodology, reducing your reliance on memory, provided you know your way around your practical manual. In all other respects, your revision and preparation for such exams should be similar to theory exams. Make sure you are familiar with all of the practical exercises, including any work carried out in class by your partner (since exams are assessed on individual performance). If necessary, check with the teaching staff to see whether you can be given access to the laboratory, to complete any exercises that you have missed.

At the outset of the practical exam, determine or decide on the order in which you will tackle the questions. A question in the latter half of the paper may need to be started early on in the exam period (e.g. an assay requiring 2-hour incubation in a 3-hour exam). Such questions are included to test your forward-planning and time-management skills. You may need to make additional decisions on the allocation of material, e.g. if you are given 30 sterile test tubes, there is little value in designing an experiment that uses 25 of these to answer question 1, only to find that you need at least 15 tubes for subsequent questions.

Make sure you explain your choice of apparatus and experimental design. Calculations should be set out in a stepwise manner, so that credit can be given, even if the final answer is incorrect (see p. 238). If there are any questions that rely on recall of factual information and you are unable to remember specific details, e.g. you cannot identify a particular item, make sure that you describe it fully, so that you gain credit for observational skills. Alternatively, leave a gap and return to the question at a later stage, if there is time.

Oral exams and interviews

An oral interview is sometimes a part of final degree exams, representing a chance for the external examiner(s) to get to know the students personally and to test their abilities directly and interactively. In some departments, orals are used to validate the exam standard, or to test students on the borderline between exam grades. Sometimes an interview may form part of an assessment, as with project work or posters. This type of exam is often intimidating – many students say they don't know how to revise for an oral – and many candidates worry that they will be so nervous they won't be able to do themselves justice.

Preparation is just as important for orals as it is for written exams:

- **Think about your earlier performances** – if the oral follows your written papers, it may be that you will be asked about questions you did not do so well on. These topics should be revised thoroughly. Be prepared to say how you would approach the questions if given a second chance.
- **Read up a little about the examiner** – he/she may focus his/her questions in his/her area of expertise.
- **Get used to giving spoken answers** – it is often difficult to transfer between written and spoken modes. Write down a few questions and get a friend to ask you them, possibly with unscripted follow-up queries.
- **Research and think about topical issues in your subject area** – some examiners will feel this reflects how interested you are in your subject.

Your conduct during the oral exam is important, too:

- **Arrive promptly and wear reasonably smart clothing.** Not to do either might be considered disrespectful by the examiner.
- **Take your time before answering questions.** Even if you think you know the answer immediately, take a while to check mentally whether you have considered all angles. A considered, logical approach will be more impressive than a quick but ill-considered response.
- **Start answers with the basics**, then develop into deeper aspects. There may be both surface and deeper aspects to a topic and more credit will be given to students who mention the latter.
- **When your answer is finished, stop speaking**. A short, crisp answer is better than a rambling one.
- **If you don't know the answer, say so.** To waffle and talk about irrelevant material is more damaging than admitting that you don't know.
- **Make sure your answer is balanced.** Talk about the evidence and opinions on both sides of a contentious issue.
- **Don't disagree violently with the examiner.** Politely put your point of view, detailing the evidence behind it. Examiners will be impressed by students who know their own mind and subject area. However, they will expect you to support a position at odds with the conventional viewpoint.
- **Finally, be positive and enthusiastic about your topic.**

Counteracting anxiety before and during exams

Adverse effects of anxiety need to be overcome by anticipation and preparation well in advance (Box 6.4). Exams, with their tight time limits, are especially stressful for perfectionists. To counteract this tendency,

Allow yourself to relax in oral exams – external examiners are experienced at putting students at ease. They will start by asking 'simple-to-answer' questions, such as what modules you did, how your project research went, and what your career aspirations are. Imagine the external examiner as a friend rather than a foe.

Creating an exam action list – knowing that you have prepared well, checked everything on your list and gathered together all you need for an exam will improve your confidence and reduce anxiety. Your list might include:

- Verify time, date and place of the exam
- Confirm travel arrangements to exam hall
- Double-check module handbooks and past papers for exam structure
- Think through use of time and exam strategy
- Identify a quiet place near the exam hall to carry out a last-minute check on key knowledge (e.g. formulae, definitions, diagram labels)
- Ensure you have all the items you wish to take to the exam, e.g.
  - pens, pencils (with sharpener and eraser);
  - ruler;
  - correction fluid;
  - calculator (allowable type);
  - sweets and drink, if allowed;
  - tissues
  - watch or clock;
  - ID card;
  - texts and/or notes, if an open-book exam;
  - mascot.
- Lay out clothes (if exam is early in the morning)
- Set alarm and/or ask a friend or family member to check you are awake on time

Box 6.4 Symptoms of exam anxiety and strategies to combat and overcome them

Sleeplessness – this is commonplace and does little harm in the short term. Get up, have a snack, do some light reading or other activity, then return to bed. Avoid caffeine (e.g. tea, coffee and cola) for several hours before going to bed.

Lack of appetite – again commonplace. Eat what you can, but take sugary sweets into the exam to keep energy levels up in case you become tired.

Fear of the unknown – it can be a good idea to visit the exam room beforehand, so you can become familiar with the location. By working through the points given in the exam action list on p. 33 you will be confident that nothing has been left out.

Worries about timekeeping – get a reliable alarm clock or a new battery for an old one. Arrange for an alarm phone call. Ask a friend or relative to make sure you are awake on time. Make reliable travel arrangements, to arrive on time. If your exam is early in the morning, it may be a good idea to get up early for a few days beforehand.

Blind panic during an exam – explain how you feel to an invigilator. Ask to go for a supervised walk outside. Do some relaxation exercises (see below), then return to your work. If you are having problems with a specific question, it may be appropriate to speak to the departmental representative at the exam.

Feeling tense – shut your eyes, take several slow, deep breaths, do some stretching and relaxing muscle movements. During exams, it may be a good idea to do this between questions, and possibly to have a complete rest for a minute or so. Prior to exams, try some exercise activity or escape temporarily from your worries by watching TV or a movie.

Running out of time – don't panic when the invigilator says 'five minutes left'. It is amazing how much you can write in this time. Write note-style answers or state the areas you would have covered: you may get some credit.

focus on the following points during the exam:

- **Don't expect to produce a perfect essay** – this won't be possible in the time available.
- **Don't spend too long planning your answer** – once you have an outline plan, get started.
- **Don't spend too much time on the initial parts of an answer**, at the expense of the main message.
- **Concentrate on getting all of the basic points across** – markers are looking for the main points first, before allocating extra marks for the detail.
- **Don't be obsessed with neatness**, either in handwriting, or in the diagrams you draw, but make sure your answers are legible.
- **Don't worry if you forget something.** You can't be expected to know everything. Most marking schemes give a first class grade to work that misses out on up to 30 per cent of the marks available.

After the exam – try to avoid becoming involved in prolonged analyses with other students over the 'ideal' answers to the questions; after all, it is too late to change anything at this stage. Go for a walk, watch TV for a while, or do something else that helps you relax, so that you are ready to face the next exam with confidence.

KEY POINT Everyone worries about exams. Anxiety is a perfectly natural feeling. It works to your advantage, as it helps provide motivation and the adrenaline that can help you 'raise your game' on the day.

There is a lot to be said for treating exams as a game. After all, they are artificial situations contrived to ensure that large numbers of candidates can be assessed together, with little risk of cheating. They have conventions and rules, just like games. If you understand the rationale behind them and follow the rules, this will aid your performance.

Sources for further study

Burns, R. (1997) *The Student's Guide to Passing Exams*. Kogan Page, London.

Evans, M. (2004) *How to Pass Exams Every Time*, 2nd edn. How To Books, Oxford.

Hamilton, D. (1999) *Passing Exams: A Guide for Maximum Success and Minimum Stress*. Cassell, London.

Palmer, R. (2005) *Getting Straight As: A Student's Guide to Success*. Routledge, Oxford.

Many universities host study skills websites; these can be found using 'study skills', 'revision' or 'exams' as keywords in a search engine.

7 Preparing your curriculum vitae

> **Definition**
>
> **Curriculum vitae** (or CV for short) – a Latin phrase that means 'the course your life has taken'.

Many students only think about their curriculum vitae (CV) immediately before applying for a job. Reflecting this, chapters on preparing a CV are usually placed near the end of texts of this type. Putting the chapter near the beginning of this book emphasises the importance of focussing your thoughts on your CV at an early stage in your studies. There are four main reasons why this can be valuable:

1. Considering your CV and how it will look to a future employer will help you think more deeply about the direction and value of your academic studies.
2. Creating a draft CV will prompt you to assess your skills and personal qualities and how these fit into your career aspirations.
3. Your CV can be used as a record of all the relevant things you have done at university and then, later, will help you communicate these to a potential employer.
4. Your developing CV can be used when you apply for vacation or part-time employment.

Personal development planning (PDP) and your CV – many PDP schemes (p. 5) also include an element of career planning that may involve creating a draft or generic CV. The PDP process can help you improve the structure and content of your CV, and the language you use within it.

> **KEY POINT** Developing your skills and qualities needs to be treated as a long-term project. It makes sense to think early about your career aspirations so that you can make the most of opportunities to build up relevant experience. A good focus for such thoughts is your developing curriculum vitae, so it is useful to work on this from a very early stage.

Skills and personal qualities

Understanding skills and qualities – it may be helpful to think about how the skills and qualities in Tables 1.1 and 7.1 apply to particular activities during your studies, since this will give them a greater relevance.

Skills (sometimes called competences) are generally what you have learned to do and have improved with practice. Table 1.1 (p. 4) summarises some important and relevant skills. This list might seem quite daunting, but your tutors will have designed your courses to give you plenty of opportunities for developing your expertise. Personal qualities, on the other hand, are predominately innate. Examples include honesty, determination and thoroughness (Table 7.1). These qualities need not remain static, however, and can be developed or changed according to your experiences. By consciously deciding to take on new challenges and responsibilities, not only can you develop your personal qualities, but you can also provide supporting evidence for your CV.

Personal qualities and skills and are interrelated because your personal qualities can influence the skills you gain. For example, you may become highly proficient at a skill requiring manual dexterity if you are particularly adept with your hands. Being able to transfer your skills is highly important (Chapter 1) – many employers take a long-term view and look for evidence of the adaptability that will allow you to be a flexible employee and one who will continue to develop skills.

Focussing on evidence – it is important to be able to provide specific concrete information that will back up the claims you make under the 'skills and personal qualities' and other sections of your CV. A potential employer will be interested in your level of competence (what you can actually do) and in situations where you have used a skill or demonstrated a particular quality. These aspects can also be mentioned in your covering letter or at interview.

Developing your curriculum vitae

The initial stage involves making an audit of the skills and qualities you already have, and thinking about those you might need to develop.

Table 7.1 Some positive personal qualities

Adaptability
Conscientiousness
Curiosity
Determination
Drive
Energy
Enthusiasm
Fitness and health
Flexible approach
Honesty
Innovation
Integrity
Leadership
Logical approach
Motivation
Patience
Performance under stress
Perseverance
Prudence
Quickness of thought
Seeing others' viewpoints
Self-confidence
Self-discipline
Sense of purpose
Shrewd judgement
Social skills (sociability)
Taking initiative
Tenacity
Tidiness
Thoroughness
Tolerance
Unemotional approach
Willingness to take on challenges

Seeing yourself as others see you – you may not recognise all of your personal qualities and you may need someone else to give you a frank appraisal. This could be anyone whose opinion you value: a friend, a member of your family, a tutor or a careers adviser.

Setting your own agenda – you have the capability to widen your experience and to demonstrate relevant personal qualities through both curricular and extracurricular activities.

Tables 7.1 and 1.1 could form a basis of this self-appraisal. Assessing your skills may be easier than critically analysing your personal characteristics. In judging your qualities, try to take a positive view and avoid being overly modest. It is important to think of your qualities in a specific context, e.g. 'I have shown that I am trustworthy, by acting as treasurer for a university society', as this evidence will form a vital part of your CV and job applications.

If you can identify gaps in your skills, or qualities that you would like to develop, especially in relation to the needs of your intended career, the next step is to think about ways of improving them. This will be reasonably easy in some cases, but may require some creative thinking in others. A relatively simple example would be if you decided to learn a new language or to keep up with one you learned at school. There are likely to be many local college and university courses dealing with foreign languages at many different levels, so it would be a straightforward matter to join one of these. A rather more difficult case might be if you wished to demonstrate 'responsibility', because there are no courses available on this. One route to demonstrate this quality might be to put yourself up for election as an officer in a student society or club; another could be to take a leading role in a relevant activity within your community (e.g. voluntary work such as hospital radio). If you already take part in activities such as these, your CV should relate them to this context.

Basic CV structures and their presentation

Box 7.1 illustrates the typical parts of a CV and explains the purpose of each part. Employers are more likely to take notice of a well-organised and -presented CV, in contrast to one that is difficult to read and assimilate. They will expect it to be concise, complete and accurate. There are many ways of presenting information in a CV, and you will be assessed partly on your choices.

- **Order.** There is some flexibility as to the order in which you can present the different parts (see Box 7.1). A chronological approach within sections helps employers gain a picture of your experience.
- **Personality and 'colour'.** Make your CV different by avoiding standard or dull phrasing. Try not to focus solely on academic aspects: you will probably have to work in a team, and the social aspects of teamwork will be enhanced by your outside interests. However, make sure that the reader does not get the impression that these interests dominate your life.
- **Style.** Your CV should reflect *your* personality, but not in such a way that it indicates too idiosyncratic an approach. It is probably better to be formal in both language and presentation, as flippant or chatty expressions will not be well received.
- **Neatness.** Producing a well-presented, word-processed CV is very important. Use a laser-quality printer and good-quality paper; avoid poor-quality photocopying at all costs.
- **Layout.** Use headings for different aspects, such as personal details, education, etc. Emphasise words (e.g. with capitals, bold, italics or underlining) sparingly and with the primary aim of making the structure clearer. Remember that careful use of white space is important in design.

Box 7.1 The structure and components of a typical CV and covering letter

There is no right or wrong way to write a CV, and no single format applies. It is probably best to avoid software templates and CV 'wizards' as they can create a bland, standardised result, rather than something that demonstrates your individuality.

You should include the following with appropriate sub-headings, generally in the order given below:

1. **Personal details.** This section *must* include your full name and date of birth, your address (both home and term-time, with dates, if appropriate) and a contact telephone number at each address. If you have an email account, you might also include this. You need only mention gender if your name could be either male or female.
2. **Education.** Choose either chronological order, or reverse chronological order and make sure you take the same approach in all other sections. Give educational institutions and dates (month, year) and provide more detail for your degree course than for your previous education. Remember to mention any prizes, scholarships or other academic achievements. Include your overall mark for the most recent year of your course, if it seems appropriate. Make sure you explain any gap years.
3. **Work experience.** Include all temporary, part-time, full-time or voluntary jobs. Details include dates, employer, job title and major duties involved.
4. **Skills and personal qualities.** Tables 1.1 and 7.1 give examples of the aspects you might include under this heading. Emphasise your strengths, and tailor this section to the specific requirements of the post (the 'job description'): for example, you might emphasise the practical skills you have gained during your degree studies if relevant, but concentrate on generic transferable skills and personal qualities for other jobs. Provide supporting evidence for your statements in all cases.
5. **Interests and activities.** This is an opportunity to bring out the positive aspects of your personality, and explain their relevance to the post you are applying for. Aim to keep this section short, or it may seem that your social life is more important than your education and work experience. Include up to four separate items, and provide sufficient detail to highlight the *positive* aspects of your interests (e.g. positions of responsibility, working with others, communication, etc.). Use sections 4 and 5 to show that you have the necessary attributes for the post.
6. **Referees.** Include the names (and titles), job descriptions, full postal addresses, contact telephone numbers and email addresses of two referees (rarely, some employers may ask for three). It is usual to include your personal tutor or course leader at university (who among other things will verify your marks), plus another person – perhaps a current employer, or someone who runs a club or society and who knows your personal interests and activities. Unless you have kept in touch with a particular teacher since starting university, it is probably best to choose current contacts, rather than those from your previous education.

Some other points to consider:

- Try to avoid jargon and over-complicated phrases in your CV: aim for direct, active words and phrases (see Box 15.1, p. 100).
- Most employers will expect your CV to be word-processed (and spellchecked). Errors in style, grammar and presentation will count against you, so be sure to check through your final version (and ask a reliable person to second-check it for you).
- Aim for a *maximum* length of two pages, printed single-sided on A4 paper, using a 'formal' font (e.g. Times Roman or Arial) of no less than 12 point for the main text. Always print on good-quality white paper. Avoid fussy use of colour, borders or fonts.
- Don't try to cram in too much detail. Use a clear and succinct approach with short sentences and lists to improve 'readability' and create structure. Remember that your aim is to catch the eye of your potential employer, who may have many applications to work through.
- It is polite to check that people are willing to act as a referee for you and to provide them with an up-to-date copy of your CV.

Your covering letter should have four major components:

1. **Letterhead.** Include your contact details, the recipient's name and title (if known) and address, plus any job reference number.
2. **Introductory paragraph.** Explain who you are and state the post you are applying for.
3. **Main message.** This is your opportunity to sell yourself to a potential employer, highlighting particular attributes and experience. Keep it to three or four sentences at the most and relate it to the particular skills and qualities demanded in the job or person specification.
4. **Concluding paragraph.** A brief statement that you look forward to hearing the outcome of your application is sufficient.

Finally, add either 'Yours sincerely' (where the recipient's name is known) or 'Yours faithfully' (in a letter beginning 'Dear Sir or Madam') and then end with your signature.

Paying attention to the quality of your CV – your potential employer will regard your CV as an example of your very best work and will not be impressed if it is full of mistakes or badly presented, especially if you claim 'good written communication' as a skill!

- **Grammar and proofreading.** Look at your CV carefully before you submit it, as sloppy errors give a very poor impression. Even if you use a spellchecker, some errors may creep in. Ask someone whom you regard as a reliable proofreader to comment on it (many tutors will do this, if asked in advance).
- **Relevance.** If you can, slant your CV towards the job description and the qualifications required (see below). Make sure you provide evidence to back up your assertions about skills, qualities and experience.
- **Accuracy and completeness.** Check that all your dates tally; otherwise, you will seem careless. It is better to be honest about your grades and (say) a period of unemployment, than to cover this up or omit details that an employer will want to know. They may be suspicious if you leave things out.

Adjusting your CV

Creating a generic CV – as you may apply for several jobs, it is useful to construct a CV in electronic format (e.g. as a Word file) which includes all information of potential relevance. This can then be modified to fit each post. Having a prepared CV on file will reduce the work each time you apply, while modifying this will help you focus on relevant skills and attributes for the particular job.

You should fine-tune your CV for each post. Employers frequently use a 'person specification' to define the skills and qualities demanded in a job, often under headings such as 'essential' and 'desirable'. This will help you decide whether to apply for a position and it assists the selection panel to filter the applicants. Highlight relevant qualifications as early in your CV as possible. Be selective – don't include every detail about yourself. Emphasise relevant parts and leave out irrelevant details, according to the job. Similarly, your letter of application is not merely a formal document but is also an opportunity for persuasion (Box 7.1). You can use it to state your ambitions and highlight particular qualifications and experience. However, don't go over the top – always keep the letter to a single page.

KEY POINT A well-constructed and relevant CV won't necessarily guarantee you a job, but it may well get you on to the short list for interview. However, a poor-quality CV is a sure route to failure.

Sources for further study

Anon. (2007) *How to Write a Curriculum Vitae*, 5th edn. The Careers Group, University of London Careers Service, London.

Anon. *Graduate Prospects Website*. Available: http://www.prospects.ac.uk/cms/ShowPage/Home_page/p!eLaXi
Last accessed 21/12/10.
[Created and maintained by the Higher Education Careers Services Unit, contains good examples of typical CVs.]

Jackson, A. and Geckeis, K. (2003) *How to Prepare your Curriculum Vitae*. McGraw-Hill, New York.

McGee, P. (2007) *How to Write a CV That Works*. How To Books, Oxford.

Information technology and library resources

8 Finding and citing published information

Browsing in a library – this may turn up interesting material, but remember the books on the shelves are those not currently out on loan. Almost by definition, the latter may be more up to date and useful. To find out a library's full holding of books in any subject area, you need to search its catalogue (normally available as an online database).

Example The book *Clinical Nutrition,* a Nutrition Society text edited by M.J. Gibney in 2005 (Blackwell Scientific) is likely to be classified as follows:

Dewey Decimal system – 615.854

Where	**600**	Applied Sciences (technology)
	610	Medicine and Health
	615	Pharmacology and Therapeutics
	615.8	Specific therapies and kinds of therapies
	615.85	Miscellaneous therapies
	615.854	Diet therapy

Library of Congress – RM216

Where	R	Medicine
	RM	Therapeutics. Pharmacology
	RM216	Diet therapy. Dietary cookbooks

The ability to find scientific information is a skill required for many exercises in your degree programme. You will need to research facts and published findings as part of writing essays, literature reviews and project introductions, and when amplifying your lecture notes and revising for exams. You must also learn how to follow scientific convention in citing source material as the authority for the statements you have made.

Sources of information

For essays and revision

You are unlikely to delve into the primary literature (p. 50) for these purposes – books and reviews are much more readable. If a lecturer or tutor specifies a particular book, then it should not be difficult to find out where it is shelved in your library, using the computerised index system. Library staff will generally be happy to assist with any queries. If you want to find out which books your library holds on a specified topic, use the system's subject index. You will also be able to search by author or by keywords.

There are two main systems used by libraries to classify books: the Dewey Decimal system and the Library of Congress system. Libraries differ in the way they employ these systems, especially by adding further numbers and letters after the standard classification marks to signify, e.g., shelving position or edition number. Enquire at your library for a full explanation of local usage.

The World Wide Web is an ever expanding resource for gathering both general and specific information (see Chapter 10). Sites fall into analogous categories to those in the printed literature: there are sites with original information, sites that review information and bibliographic sites. One considerable problem is that websites may be frequently updated, so information present when you first looked may be altered or even absent when the site is next consulted. Further, very little of the information on the WWW has been monitored or refereed. Another disadvantage is that the site information may not state the origin of the material, who wrote it or when it was written.

For literature surveys and project work

You will probably need to consult the primary literature. If you are starting a new research project or writing a report from scratch, you can build up a core of relevant papers by using the following methods:

- Asking around: supervisors or their postgraduate students will almost certainly be able to supply you with a reference or two that will start you off.
- Searching an online database: these cover very wide areas and are a convenient way to start a reference collection, although a charge is often made for access and sending out a listing of the papers selected (your library may or may not pass this on to you).
- Consulting the bibliography of other papers in your collection: an important way of finding the key papers in your field. In effect, you are taking advantage of the fact that another researcher has already done all the hard work.

Web resources – your university library will provide you with access to a range of Web-based databases and information systems. The library Web pages will list these and provide links, which may be worth bookmarking on your Web browser. Resources especially useful to bioscientists include:

- ISI Web of Knowledge, *including the* Science Citation Index
- Ingentaconnect (*previously known as BIDS*), *including* Ingenta Medline
- CSA Illumina
- Dialog
- PubMed
- ScienceDirect
- Scopus
- Ovid, *including* Cinahl *and* Medline

Most of these electronic resources operate on a subscription basis and may require an 'Athens' username and password – for details of how to obtain these consult library staff or your library's website. Many of these resources can save search results directly to bibliographic databases such as Microsoft EndNote®.

Definitions

Journal/periodical/serial – any publication issued at regular intervals. In biosciences, usually containing papers (articles) describing original research findings and reviews of literature.

e-journal – a journal published online, consisting of articles structured in the same way as a paper-based journal. A valid username and password may be required for access (arranged via your library, if it subscribes to the e-journal).

The primary literature – this comprises original research papers, published in specialist scientific periodicals. Certain prestigious general journals (e.g. *Nature*) contain important new advances from a wide subject area.

Monograph – a specialised book covering a single topic.

e-book – a book published online in downloadable form.

ebrary – a commercial service offering e-books and other online resources.

HERON (Higher Education Resources ON demand) – a national service for UK higher education offering copyright clearance, digitisation and delivery of book extracts and journal articles.

- Referring to 'current awareness' online databases: these are useful for keeping you up to date with current research; they usually provide a monthly listing of article details (title, authors, source, author address) arranged by subject and cross-referenced by subject and author. Current awareness databases cover a wider range of primary literature than could ever be available in any one library.

  Examples relevant to the biosciences include: the *Current Advances* series (Elsevier) and *Biological Abstracts* (Thomson Scientific). Some online databases also offer a service whereby they will email registered users with updates based on saved search criteria. Consult library staff or your library website to see which of these databases and services are available to you.
- Using the *Science Citation Index* (SCI): this is a very valuable source for exploring the published literature in a given field, because it lets you see who has cited a given paper; in effect, SCI allows you to move forward through the literature from an existing reference. The Index is available online via ISI Web of Science.

Obtaining and organising research papers

Obtaining a copy

It is usually more convenient to have personal copies of key research articles for direct consultation when working in a laboratory or writing. The simplest way of obtaining these is to photocopy the originals or download and/or print off copies online (e.g. as pdf or HTML files). For academic purposes, this is normally acceptable within copyright law. If your library does not subscribe to the journal, it may be possible for them to borrow it from a nearby institute or obtain a copy via a national borrowing centre (an 'inter-library loan'). If the latter, you will have to fill in a form giving full bibliographic details of the paper and where it was cited, as well as signing a copyright clearance statement concerning your use of the copy.

Your department might be able to supply 'reprint request' postcards to be sent to the designated author of a paper. This is an unreliable method of obtaining a copy because it may take some time (allow at least 1–3 months) and some requests will not receive a reply. Taking into account the waste involved in postage and printing, it is probably best simply to photocopy or send for a copy via inter-library loan.

Organising papers

Although the number of papers you accumulate may be small to start with, it is worth putting some thought into their storage and indexing before your collection becomes disorganised and unmanageable. Few things are more frustrating than not being able to lay your hands on a vital piece of information, and this can seriously disrupt your flow when writing or revising.

Indexing your references

Whether you have obtained a printed copy, have stored downloaded files electronically or have simply noted the bibliographic details of a reference, you

Definitions

Review – an article in which recent advances in a specific area are outlined and discussed.

Proceedings – volume compiling written versions of papers read at a scientific meeting on a specific topic.

Abstracts – shortened versions of papers, often those read at scientific meetings. These may later appear in the literature as full papers.

Bibliography – a summary of the published work in a defined subject area.

Alternative methods of receiving information – RSS (really simple syndication) feeds and email updates from publishers are increasingly used to provide automated information services to academic clients, for example by supplying links to relevant contents of new editions of online journals.

Copyright law – In Europe, copyright regulations were harmonised in 1993 (Directive 93/98/EEC) to allow literary copyright for 70 years after the death of an author and typographical copyright for 25 years after publication. This was implemented in the UK in 1996, where, in addition, the Copyright, Designs and Patents Act (1988) allows the Copyright Licensing Agency to license institutions so that lecturers, students and researchers may take copies for teaching and personal research purposes – no more than a single article per journal issue, one chapter of a book, or extracts to a total of 5 per cent of a book.

Storing research papers – these can easily be kept in alphabetical order within filing boxes or drawers but, if your collection is likely to grow large, it will need to be refiled as it outgrows the storage space. You may therefore wish to add an 'accession number' to the record you keep in your database, and file the papers in sequence according to this as they accumulate. New filing space is only required at the 'end' and you can use the accession numbers to form the basis of a simple cross-referencing system.

will need to index each resource. This is valuable for the following reasons:

- You will probably need the bibliographic information for creating a reference list for an assignment or report.
- If the index also has database features, this can be useful, allowing you to search for keywords or authors.
- If you include an 'accession number' and if you then file printed material sequentially according to this number, then it will help you to find the hard copy.
- Depending on the indexing system used, you can add comments about the reference that may be useful at a later time, e.g. when writing an introduction or conclusion.

While the simplest way to create an index system is to put the details on reference cards, database software can be more convenient and faster to sort, once the bibliographic information has been entered. If you do not feel that commercial software is appropriate for your needs, consider using a word processor or spreadsheet; their rudimentary database sorting functions (see Chapters 11 and 12) may be all that you require.

If you are likely to store lots of references and other electronic resources digitally, then you should consider carefully how this information is kept, for example by choosing file names that indicate what the file contains and that will facilitate sorting.

Making citations in text

There are two main ways of citing articles and creating a bibliography (also referred to as 'references' or 'literature cited').

The Harvard system

For each citation, the author name(s) and the date of publication are given at the relevant point in the text. The bibliography is organised alphabetically and by date of publication for papers with the same authors. Formats normally adopted are, for example, 'Smith and Jones (2009) stated that ...' or 'it has been shown that ... (Smith and Jones, 2009)'. Lists of references within parentheses are separated by semi-colons, e.g. '(Smith and Jones, 2009; Jones and Smith, 2010)', normally in order of date of publication. To avoid repetition within the same paragraph, an approach such as 'the investigations of Smith and Jones indicated that' could be used following an initial citation of the paper. Where there are more than two authors it is usual to write '*et al.*'; this stands for the Latin *et alii*, meaning 'and others'. If citing more than one paper with the same authors, put, for example, 'Smith and Jones (1987, 1990)' and if papers by a given set of authors appeared in the same year, letter them (e.g. Smith and Jones, 1989a, 1989b). When citing chapters or sections from an edited book you should cite the author of the chapter in the text rather than the editor of the whole book, and include the page numbers of the chapter in the reference list. The page number should also be included within your in-text citation if you are including a direct quotation, e.g. '"Quotations should be relevant to your argument and used judiciously in your text" (Pears and Shields, 2008, p. 16)'.

Using commercial bibliographic database software to organise your references – for those with large numbers of references in their collection, and who may wish to produce lists of selected references in particular format, e.g. for inclusion in a project report or journal paper, systems such as EndNote, Reference Manager *or* ProCite can reward the investment of time and money required to create a personal reference catalogue. Appropriate bibliographic data must first be entered into fields within a database (some versions assist you to search online databases and upload data from these). The database can then be searched and used to create customised lists of selected references in appropriate citation styles.

Citation and referencing using Microsoft Word 2007 – the *References* tab provides a means of adding citations and links to EndNote.

Example Incorporating references in text – this sample shows how you might embed citations in text using the Harvard approach:
'... Brookes *et al.* (2001) proposed that protein A216 was involved in the degradation process. However, others have disputed this notion (Scott and Davis, 1997; Harley, 1998, 2000). Patel (1999a, 1999b) found that A216 is inactivated at pH values less than 5; while several authors (e.g. Hamilton, 1995; Drummond and Stewart, 2002) also report that its activity is strongly dependent on Ca^{2+} concentration ...'

Example Incorporating references in text – this sample shows how you might embed citations in text using the Vancouver approach:
'... Brookes *et al.* proposed that protein A216 was involved in the degradation process[1]. However, others have disputed this notion[2–4]. Patel[5,6] found that A216 is inactivated at pH values less than 5; while several authors[7,8] also report that its activity is strongly dependent on Ca^{2+} concentration ...'

The numerical or Vancouver system

Papers are cited via a superscript or bracketed reference number inserted at the appropriate point. Normal format would be, for example: 'DNA sequences[4,5] have shown that ...' or 'Jones [55,82] has claimed that ...'. Repeated citations use the number from the first citation. In the true numerical method (e.g. as in *Nature*), numbers are allocated by order of citation in the text, but in the alpha-numerical method (e.g. the *Annual Review* series), the references are first ordered alphabetically in the Bibliography, then numbered, and it is this number that is used in the text. Note that with this latter method, adding or removing references is tedious, so the numbering should be done only when the text has been finalised.

KEY POINT The main advantages of the Harvard system are that the reader might recognise the paper being referred to and that it is easily expanded if extra references are added. The main advantages of the Vancouver system are that it aids text flow and reduces length.

How to list your citations in a bibliography

Whichever citation method is used in the text, comprehensive details are required for the bibliography so that the reader has enough information to find the reference easily. Citations should be listed in alphabetical order with the priority: first author, subsequent author(s), date. Unfortunately, in terms of punctuation and layout, there are almost as many ways of citing papers as there are journals! Your department may specify an exact format for project work; if not, decide on a style and be consistent – if you do not pay attention to the details of citation you may lose marks. Take special care with the following aspects:

- Authors and editors: give details of *all* authors and editors in your bibliography, even if given as '*et al.*' in the text.
- Abbreviations for journals: while there are standard abbreviations for the titles of journals (consult library staff), it is a good idea to give the whole title, if possible.
- Books: the edition should always be specified as contents may change between editions. Add, for example, '(5th edition)' after the title of the book. You may be asked to give the International Standard Book Number (ISBN), a unique reference number for each book published.
- Unsigned articles, e.g. unattributed newspaper articles and instruction manuals – refer to the author(s) in text and bibliography as 'Anon.'.
- Websites: there is no widely accepted format at present. You should follow departmental guidelines if these are provided, but if these are not available, we suggest providing author name(s) and date in the text when using the Harvard system (e.g. Hacker, 2006), while in the bibliography giving the URL details in the following format: Hacker, A. (2006) *University of Anytown Homepage on Aardvarks*. Available: http://www.myserver.ac.uk/homepage. Last accessed: 23/02/07. In this example, the Web page was constructed in 2006, but was accessed in

Examples

Paper in journal:
Smith, A.B., Jones, C.D. and Professor, A. (1998) Innovative results concerning our research interest. *Journal of New Results*, 11, 234–235.

Book:
Smith, A.B. (1998). *Summary of My Life's Work.* Megadosh Publishing Corp., Bigcity. ISBN 0-123-45678-9.

Chapter in edited book:
Jones, C.D. and Smith, A.B. (1998). Earth-shattering research from our laboratory. In: *Research Compendium 1998* (ed. A. Professor), pp. 123–456. Bigbucks Press, Booktown.

Thesis:
Smith, A.B. (1995). *Investigations on my Favourite Topic.* PhD thesis, University of Life, Fulchester.

Note: if your references are handwritten, you should indicate italics by underlining text or numerals.

Finding dates on websites quickly – when visiting a particular page, you can find occurrences of year dates beginning '200' by pressing Control and F keys together, entering 200 in the Find window that appears, then carrying out a search using the Find next command.

Adding Web references in other systems – if you are using a different referencing system than Harvard, consult Pears and Shields (2008) or McMillan and Weyers (2006) for further information on how to cite websites in these systems.

February, 2007. If no author is identifiable, cite the sponsoring body (e.g. University of Anytown, 2006), and if there is no author or sponsoring body, write 'Anon.' for 'anonymous', e.g. Anon. (2006), and use Anon. as the 'author' in the bibliography. If the Web pages are undated, *either* use the 'Last accessed' date for citation and put no date after the author name(s) in the reference list, *or* cite as 'no date' (e.g. Hacker, no date) and leave out a date after the author name(s) in the reference list – you should be consistent whichever option you choose.

- Unread articles: you may be forced to refer to a paper via another without having seen it. If possible, refer to another authority who has cited the paper, e.g. '... Jones (1980), cited in Smith (1990), claimed that ...'. Alternatively, you could denote such references in the bibliography by an asterisk and add a short note to explain at the start of the reference list.
- Personal communications: information received in a letter, seminar or conversation can be referred to in the text as, for example, '... (Smith, pers. comm.)'. These citations are not generally listed in the bibliography of papers, though in a thesis you could give a list of personal communicants and their addresses.
- Online material: some articles are published solely online and others online ahead of publication in printed form. The article may be given a DOI (digital object identifier), allowing it to be cited and potentially tracked before and after it is allocated to a printed issue (see http://www.doi.org/). DOIs allow for Web page redirection by a central agency, and CrossRef (see http://www.crossref.org) is the official DOI registration organisation for scholarly and professional publications. DOIs can be used as 'live' hyperlinks in online articles, or cited in place of (and, when they become available, following) the volume and page numbers for the article, with the remainder of the details cited in the usual fashion: for example,

  'Smith, A. and Jones, B. (2006) Our latest important research in the form of a Web-published article. *Online Biosciences* 8/2006 (p. 781). Published Online: 26 March 2006. DOI: 10.1083/mabi.200680019'.

Text references and sources for further study

McMillan, K.M. and Weyers, J.D.B. (2006) *The Smarter Student: Study Skills and Strategies for Success at University*. Pearson Education, Harlow.

Pears, R. and Shields, G. (2008) *Cite Them Right! The Essential Guide to Referencing and Plagiarism.* Pear Tree Books, Durham.

9 Evaluating information

Checking the reliability of information, assessing the relative value of different ideas and thinking critically are skills essential to the scientific approach. You will need to develop your abilities to evaluate information in this way because:

- You will be faced with many sources of information, from which you will need to select the most appropriate material.
- You may come across conflicting sources of evidence and have to decide which is the more reliable.
- The accuracy and validity of a specific fact may be vital to your work.
- You may doubt the quality of the information from a particular source.
- You may wish to check the original source because you are not sure whether someone else is quoting it correctly.

Example A Web search for the letters 'BMI' (e.g. using the search engine Google, p. 64) will reveal that this acronym appears in several million websites. Not all of these are related to body mass index with the list including links to flight operators and finance firms. These examples are easy to identify as irrelevant when looking for information on body mass, but considering the many websites that do actually cover this topic, which might contain the exact information you seek? How valid is the information? Is it biased towards one viewpoint or hypothesis? Does it represent current mainstream thinking on the topic? These are some of the issues that an evaluation of the information sources might deal with.

KEY POINT Evaluating information and thinking critically are regarded as higher-order academic skills. The ability to think deeply in this way is greatly valued in the applied sciences and will consequently be assessed in coursework and exam questions (see Chapter 5).

The process of evaluating and using information can be broken down into four stages:

1. Selecting and obtaining material. How to find sources is covered in Chapter 8. Printed books and journals are important, but if you identify a source of this kind there may be delays in borrowing it or obtaining a photocopy. If the book or journal is available online, then downloading or printing sections or papers will be more convenient and faster. The Internet is often a first port of call if you wish to find something out quickly. For many websites, however, it can be difficult to verify the authenticity of the information given (see Box 10.5).
2. Assessing the content. You will need to understand fully what has been written, including any technical terms and jargon used. Establish the relevance of the information to your needs and assure yourself that the data or conclusions have been presented in an unbiased way.
3. Modifying the information. In order to use the information, you may need to alter it to suit your needs. This may require you to make comparisons, interpret or summarise. Some sources may require translation. Some data may require mathematical transformation before they are useful. There is a chance of error in any of these processes and also a risk of plagiarism.
4. Analysis. This may be your own interpretation of the information presented, or an examination of the way the original author has used the information.

Definition

Plagiarism – the unacknowledged use of another's work as if it were one's own. In this definition, the concept of 'work' includes ideas, writing, data or inventions and not simply words; and the notion of 'use' does not only mean copy 'word for word', but also 'in substance' (i.e. a copy of the ideas involved). Use of another's work is acceptable if you acknowledge the source (see Box 9.1).

Box 9.1 How to avoid plagiarism and copyright infringement

Plagiarism is defined on page 48. Examples of plagiarism include:

- Copying the work of a fellow student (past or present) and passing it off as your own.
- Using 'essay-writing services', such as those on offer on certain websites.
- Copying text or images from a source (book, journal article or website, for instance) and using this within your own work without acknowledgement.
- Quoting others' words without indicating who wrote or said them.
- Copying ideas and concepts from a source without acknowledgement, even if you paraphrase them.

Most students would accept that some of the above can only be described as cheating. However, many students, especially at the start of their studies, are unaware of the academic rule that they must *always* acknowledge the originators of information, ideas and concepts, and that not doing so is regarded a form of academic dishonesty. If you adopt the appropriate conventions that avoid such accusations, you will achieve higher marks for your work as it will fulfil the markers' expectations for academic writing.

Universities have a range of mechanisms for identifying plagiarism, from employing experienced and vigilant coursework markers and external examiners to analysing students' work using sophisticated software programs. Plagiarism is always punished severely when detected. Penalties may include awarding a mark of zero to all involved – both the copier(s) and the person whose work has been copied (who is regarded as complicit in the crime). Further disciplinary measures may be taken in some instances. In severe cases, such as copying substantive parts of another's work within a thesis, a student may be dismissed from the university.

If you wish to avoid being accused of plagiarism, the remedies are relatively simple:

1. **Make sure the work you present is always your own.** If you have been studying alongside a colleague, or have been discussing how to tackle a particular problem with your peers, make sure you write on your own when working on your assignments.
2. **Never be tempted to 'cut and paste'** from websites or online sources such as word-processed handouts. Read these carefully, decide what the important points are, express these *in your own words* and *provide literature citations to the original sources* (see Chapter 8). In some cases, further investigations may be required to find out details of the original sources. The lecturer's reading list or a book's references may help you here.
3. **Take care when note-taking**. If you decide to quote word for word, make sure you show this clearly in your notes with quotation marks. If you decide to make your own notes based on a source, make sure these are original and do not copy phrases from the text. In both cases, write down full details of the source at the appropriate point in your notes.
4. **Place appropriate citations throughout your text where required**. If you are unsure about when to do this, study reviews and articles in your subject area (see also Chapter 8).
5. **Show clearly where you are quoting directly from a source**. For short quotes, this may involve using quotation marks and identifying the source afterwards, as in the example '... as Samuel Butler (1877) wrote: "a hen is only an egg's way of making another egg". ' For longer quotes (say 40 words or more), you should create a separate paragraph of quoted text, usually identified by inverted commas, indentation, italicisation or a combination of these. A citation must *always* be included, normally at the end. Your course handbook may specify a layout. Try not to rely too much on quotes in your work. If a large proportion of your work is made up from quotes, this will almost certainly be regarded as lacking in originality, scoring a poor mark.

Copyright issues are often associated with plagiarism, and refer to the right to publish (and hence copy) original material, such as text, images and music. Copyright material is indicated by a symbol © and a date (see, for example, p. iv of this book). Literary copyright is the aspect most relevant to students in their academic studies. UK copyright law protects authors' rights for life and gives their estates rights for a further 70 years. Publishers have 'typographical copyright' that lasts for 25 years. This means that it is illegal to photocopy, scan or print out copyright material unless you have permission, or unless your copying is limited to an extent that could be considered 'fair dealing'. For educational purposes – private study or research – in a scientific context, this generally means:

- no more than 5 per cent in total of a work;
- one chapter of a book;
- one article per volume of an academic journal;
- 20 per cent of a short book;
- one separate illustration or map.

You may only take one copy within the above limits, may not copy for others, and may not exceed these amounts *even if you own a copy of the original*. These rules also apply to Web-based materials, but sometimes you will find sites where the copyright is waived. Some copying may be licensed; you should consult your library's website or helpdesk to see whether it has access to licensed material. Up-to-date copyright information is generally posted close to library and departmental photocopiers.

> **KEY POINT** Advances in communications and information technology mean that we can now access almost limitless knowledge. Consequently, the ability to evaluate information has become an extremely important skill.

Evaluating sources of information

One way of assessing the reliability of a piece of scientific information is to think about how it was obtained in the first place. Essentially, facts and ideas originate from someone's research or scholarship, whether they are numerical data, descriptions, concepts or interpretations. Sources are divided into two main types:

1. **Primary sources** – those in which ideas and data are first communicated. The primary literature is generally published in the form of 'papers' (articles) in journals whether printed or online. These are usually refereed by experts in the academic peer group of the author, who will check the accuracy and originality of the work and report their opinions back to the editors. This peer review system helps to maintain reliability, but it is not perfect. Books and, more rarely, websites and articles in magazines and newspapers, can also be primary sources but this depends on the nature of the information published rather than the medium. These sources are not formally refereed, although they may be read by editors and lawyers to check for errors and unsubstantiated or libellous allegations.
2. **Secondary sources** – those which quote, adapt, interpret, translate, develop or otherwise use information drawn from primary sources. It is the act of quoting or paraphrasing that makes the source secondary, rather than the medium. Reviews are examples of secondary scientific sources, and books and magazine articles are often of this type.

Distinguishing between primary and secondary sources – try the 'IMRaD test'. Many primary sources contain information in the order: ***I***ntroduction, ***M***aterials and ***M***ethods, ***R***esults ***a***nd ***D***iscussion. If you see this format, and particularly if *data* from an experiment, study or observation are presented, then you are probably reading a primary source.

Example If a journalist wrote an article about a new 'flesh-eating bug' for the *New York Times* that was based on an article in the *British Medical Journal*, the *New York Times* article would be the secondary source, while the *British Medical Journal* article would be the primary source.

When information is modified for use in a secondary source, alterations are likely to occur, whether intentional or unintentional. Most authors do not deliberately set out to change the meaning of the primary source, but they may unwittingly do so, e.g. in changing text to avoid plagiarism or by oversimplification. Others may consciously or unconsciously exert bias in their reporting, for example by quoting evidence that supports only one side of a debate. Therefore, the closer you can get to the primary source, the more reliable the information is likely to be. On the other hand, modification while creating a secondary source could involve correcting errors, or synthesising ideas and content from multiple sources.

Taking account of the changing nature of websites and wikis – by their very nature, these sources may change. This means that it is important to quote accurately from them and to give a 'Last accessed' date when citing (see p. 46).

Authorship and provenance

Clearly, much depends on who is writing the source and on what basis (e.g. who paid them?). Consequently, an important way of assessing sources is to investigate the ownership and provenance of the work (who and where it originated from, and why).

Can you identify who wrote the information? If it is signed or there is a 'byline' showing who wrote it, you might be able to make a judgement on the quality of what you are reading. This may be a simple decision, if you know or can assume that the writer is an authority in the area; otherwise a little research might help (e.g., by putting the name into a search engine).

Finding out about authors and provenance – these pieces of information are easy to find in most printed sources and may even be presented just below the title, for convenience. In the case of the Web, it may not be so easy to find what you want. Relevant clues can be obtained from 'homepage' links and the header, body and footer information. For example, the domain (p. 62) may be useful, while the use of the tilde symbol (~) in an address usually indicates a personal, rather than an institutional, website.

Assessing substance over presentation – just because the information is presented well (e.g. in a glossy magazine or particularly well-constructed website), this does not necessarily tell you much about its quality. Try to look below the surface, using the methods mentioned in this chapter.

Of course, just because Professor X thinks something, does not make it true. However, if you know that this opinion is backed up by years of research and experience, then you might take it a little more seriously than the thoughts of a school pupil. If an author is not cited, effectively nobody is taking responsibility for the content. Could there be a reason for this?

Is the author's place of work cited? This might tell you whether the facts or opinions given are based on an academic study. Is there a company with a vested interest behind the content? If the author works for a public body, there may be publication rules to follow and they may even have to submit their work to a publications committee before it is disseminated. They are certainly more likely to get into trouble if they include controversial material.

Evaluating facts and ideas

However reliable the source of a piece of information seems to be, it is probably a good idea to retain a slight degree of scepticism about the facts or ideas involved. Even information from impeccable primary sources may not be perfect – different approaches can give different outcomes, and interpretations can change with time and with further advances in knowledge.

Table 9.1 provides a checklist to use when evaluating sources.

Critically examining facts and ideas is a complex task depending on the particular issues involved, and a number of different general approaches can be applied. You will need to decide which of the following general tips are useful in your specific case:

- **Make cross-referencing checks** – look at more than one source and compare what is said in each. The cross-referenced sources should be as independent as possible (for example, do not compare a primary source together with a secondary review based on it). If you find that all the sources give a similar picture, then you can be more confident about the reliability of the information.
- **Look at the extent and quality of citations** – if references are quoted, these indicate that a certain amount of research has been carried out beforehand, and that the ideas or results are based on genuine scholarship. If you are doubtful about the quality of the work, these references might be worth looking at. How up to date are they? Do they cite independent work, or is the author exclusively quoting his/her own work, or solely the work of one person?
- **Consider the age of the source** – the fact that a source is old is not necessarily a barrier to truth, but ideas and facts may have altered since the date of publication, and methods may have improved. Can you trace changes through time in the sources available to you? What key events or publications have forced any changes in the conclusions?
- **Try to distinguish fact from opinion** – to what extent has the author supported a given viewpoint? Have relevant facts been quoted, via literature citations or the author's own researches? Are numerical data used to substantiate the points used? Are these reliable and can you verify the information, for example, by looking at the sources cited? Might the author have a reason for putting forward biased evidence to support a personal opinion?

Table 9.1 Checklist for assessing information in science. How reliable is the information you have been reading? The more 'yes' answers you can give below, the more trustworthy you can assume it to be.

Assessing sources

- Can you identify the author's name?
- Can you determine what relevant qualifications he/she holds?
- Can you say who employs the author?
- Do you know who paid for the work to be done?
- Is this a primary or secondary source?
- Is the content original or derived from another source?

Evaluating information

- Have you checked a range of sources?
- Is the information supported by relevant literature citation?
- Is the age of the source likely to be important regarding the accuracy of the information?
- Have you focused on the substance of the information presented rather than its packaging?
- Is the information fact or opinion?
- Have you checked for any logical fallacies in the arguments?
- Does the language used indicate anything about the status of the information?
- Have the errors associated with any numbers been taken into account?
- Have the data been analysed using appropriate statistics?
- Are any graphs constructed fairly?

Learning from examples – as your lecturers introduce you to case studies, you will see how scientists have applied critical thinking to understand the nature of cells, organisms and ecosystems. Some of your laboratory sessions may mimic the processes involved – observation, hypothesis, experimental design, data gathering and analysis, and formulating a conclusion (see Chapter 31). These skills and approaches can be applied in your course, e.g. when writing about a biological issue or carrying out a research project.

- **Analyse the language used** – words and their use can be very revealing. Subjective wording might indicate a personal opinion rather than an objective conclusion. Propaganda and personal bias might be indicated by absolute terms, such as 'everyone knows . . .'; 'It can be guaranteed that ...', or a seemingly one-sided consideration of the evidence. How carefully has the author considered the topic? A less studious approach might be indicated by exaggeration, ambiguity, or the use of 'journalese' and slang. Always remember, however, that content should be judged above presentation.
- **Look closely at any numbers** – if the information you are looking at is numerical in form, have statistical errors been taken into consideration and, where appropriate, quantified? If so, does this help you arrive at a conclusion about how genuine the differences are between important values?
- **Think carefully about any hypothesis-testing statistics used** – are the methods appropriate? Are the underlying hypotheses the right ones? Have the results of any tests been interpreted correctly in arriving at the conclusion? To deal with these matters, you will need at least a basic understanding of the 'statistical approach' and of commonly used techniques (see Chapters 38 and 39).

Critical thinking

Critical thinking involves the application of logic to a problem, issue or case study. It requires a wide range of skills. Key processes involved include: acquiring and processing information; creating appropriate hypotheses and formulating conclusions; and acting on the conclusions towards a specific objective.

KEY POINT Critical thinking needs reliable knowledge, but it requires you to use this appropriately to analyse a problem. It can be contrasted with rote learning – where you might memorise facts without an explicit purpose other than building your knowledge base.

Critical thinking is particularly important in biology, because the subject deals with complex and dynamic systems. These can be difficult to understand for several reasons:

- they are often multi-faceted, involving many interactions;
- it can be difficult to alter one variable in an experiment without producing confounding variables (see p. 205);
- many variables may be unmeasured or unmeasurable;
- heterogeneity (variability) is encountered at all scales from the molecular scale to the ecosystem;
- perturbation of the system can lead to unexpected ('counter-intuitive') results.

As a result, conclusions in biological research are seldom clear-cut. Critical thinking allows you to arrive at the most probable conclusion

'You can prove anything with statistics' – leaving aside the issue that statistical methods deal with probability, not certainty (Chapter 39), it *is* possible to analyse and present data in such a way that they support one chosen argument or hypothesis rather than another. Detecting a bias of this kind can be difficult, but the critical thinking skills involved are essential for all scientists (see, e.g., Box 35.3).

Analysing a graph – this process can be split into six phases:

1. Considering the context and purpose of the graph.
2. Recognising the type of presentation and examining the axes.
3. Looking closely at the scale on each axis.
4. Examining the data presented (e.g. data points, symbols, curves).
5. Considering errors and statistics associated with the graph.
6. Reaching conclusions based on the above.

These processes are amplified in Chapter 35.

Analysing a table – as with analysing a graph, this process can be split into six phases:

1. Considering the context and purpose.
2. Examining the sub-headings to see what information is contained in the rows and columns.
3. Considering the units used and checking any footnotes.
4. Comparing the data values across rows and/or down columns, looking for patterns, trends and unusual values.
5. Taking into account any statistics presented.
6. Reaching conclusions based on the above.

from the results at hand; however, it also involves acknowledging that other conclusions might be possible. It allows you to weigh up these possibilities and find a working hypothesis or explanation, but also to understand that your conclusions are essentially dynamic and might alter when new facts are known. Hypothesis-testing with statistics (Chapter 39) is an important adjunct to critical thinking because it demands the formulation of simple hypotheses and provides rational reasons for making conclusions.

Recognising fallacies in arguments is an important aspect of critical thinking. Philosophers and logicians recognise different forms of argument and many different fallacies in each form. Damer (2009) provides an overview of this wide-ranging and complex topic.

Interpreting data

Numerical data

Information presented in public, whether as a written publication or spoken presentation, is rarely in the same form as it was when first obtained. Chapter 50 deals with processes in which data are recorded and used, while Chapter 38 describes the standard descriptive statistics used to 'encapsulate' large data sets. Chapter 37 covers some relevant mathematical techniques. Sampling (essentially, obtaining representative measurements) is at the heart of many observational and experimental approaches (see Chapters 31 and 33), and analysis of samples is a key component of hypothesis-testing statistics (Chapter 39). Understanding these topics and carrying out the associated study exercises will help you improve your ability to interpret numerical data.

Graphs

Frequently, understanding and analysis in science depends on your ability to interpret data presented in graphical form. Sometimes, graphs may mislead. This may be unwitting, as in an unconscious effort to favour a 'pet' hypothesis of the author. Graphs may be used to 'sell' a product, e.g. in advertising, or to favour a viewpoint as, perhaps, in politics. Experience in drawing and interpreting graphs will help you spot these flawed presentations, and understand how graphs can be erroneously presented (Box 35.3) will help you avoid the same pitfalls.

Tables

Tables, especially large ones, can appear as a mass of numbers and thus be more daunting at first sight than graphs. In essence, however, most tables are simpler than most graphs. The construction of tables is dealt with in Chapter 36.

Evaluating nutritional information on food labels

Legislative requirements

Most countries now have strict legislative regulations such as the Food Labelling Regulations 1996 in the UK to control and standardise the information that food manufacturers include on food labels in terms of use-by date, list of ingredients, country of origin, etc. Where there is a particular nutritional claim made about the product (e.g. low fat, high fibre) then nutritional information must also be included, otherwise the inclusion of this information is voluntary. The information given in the nutritional panel is also standardised and in the UK includes either a Group 1 declaration of energy, protein, carbohydrate and fat (the 'Big Four') or a Group 2 declaration, which includes the Big Four plus sugars, saturates, fibre and sodium. The information is expressed per 100 g (or 100 mL) of the product and the data are derived from manufacturer's analysis, calculation from known values for the ingredients, or from calculations from generally established and accepted data such as food composition tables (see Chapter 41, p. 279).

Evaluating front-of-pack nutritional information

To help inform and educate the consumer, many food manufacturers include more than the minimum nutrition labelling, and this is increasingly appearing on so-called front-of-pack panels. The UK Food Standards Agency (FSA) recommends the use of traffic light signposting as the method of choice to provide consumers with clear, standardised guidance. Figure 9.1 shows examples of different formats for traffic light labels.

The information included in the traffic lights is based on values for each of the nutrients per 100 g of the product with clearly defined cut-off points for each colour banding as shown in Table 9.2

This type of nutritional signposting allows consumers to compare between products based on nutrient content.

An alternative method of nutritional labels, which has increasingly moved from back-of-pack to front-of-pack, is guideline daily amounts (GDAs). This is a method of expressing the nutrient content per portion of the product and is often expressed as a percentage of average requirements of nutrients based on dietary reference values (DRV, see Chapter 41).

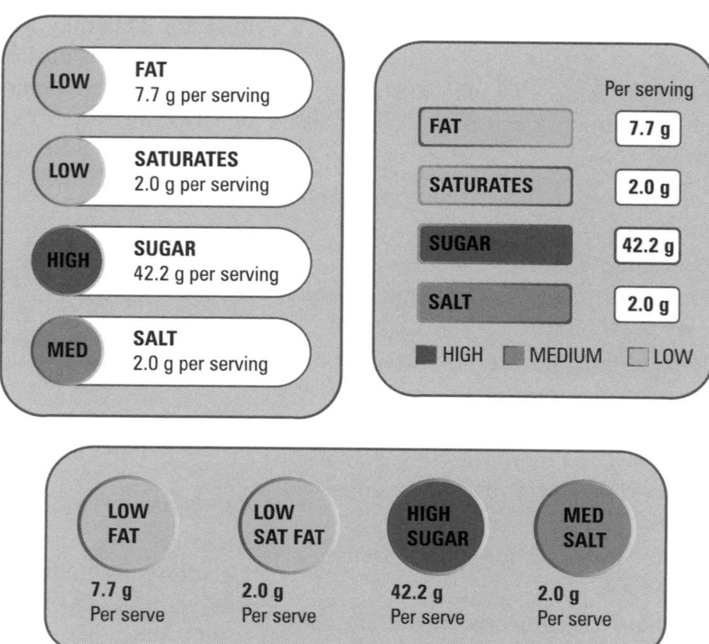

Fig. 9.1 Examples of different UK FSA traffic light labels.

Table 9.2 Colour banding criteria for traffic light labels

	Green (low)	Amber (medium)	Red (high)	
Fat	≤ 3.0 g/100 g	>3.0 to ≤ 20 g/100 g	>20.0 g/100 g	>21.0 g/portion
Saturated fat	≤ 1.5 g/100 g	>1.5 to ≤ 5.0 g/100 g	>5.0 g/100 g	>6.0 g/portion
Sugar	≤ 5.0 g/100 g	>5.0 to ≤ 12.5 g/100 g	>12.5 g/100 g	>15.0 g/portion
Salt	≤ 0.3 g/100 g	>0.30 to ≤ 1.5 g/100 g	>1.5 g/100 g	>2.4 g/portion

FSA (2007)

Table 9.3 Guideline daily amount (GDA) values

Typical values	Women	Men	Children (5–10 years)
Calories (kcal)	2000	2500	1800
Protein (g)	45	55	24
Carbohydrate (g)	230	300	220
Sugars (g)	90	120	85
Fat (g)	70	95	70
Saturates (g)	20	30	20
Fibre (g)	24	24	15
Salt (g)	6	6	4

FDA (2009)

Values for GDA are detailed in Table 9.3. On packs the values for women tend to be used, although this is usually made clear in the panel. Figure 9.2 illustrates what GDA labels tend to look like with, unlike the traffic light system, no colour coding being used. Some manufacturers have combined the two methods, however, using traffic light colours and percentage GDAs.

GDAs provide guidelines rather than target for intakes and require consumers to think about the size of the portion that they are likely to consume. For example, a sharing bag of crisps weighs 150 g, whereas the nutritional information given by the GDA label is based on a 40 g serving, therefore when using the GDA information the consumer will need to consider whether they are likely to eat more or less than this amount.

The 'Nutrition Facts' label in the USA is similar to GDA but expresses the nutrients per serving of product as a percentage of the Daily Value based on a 2000 kcal intake (http://www.fda.gov/Food/LabelingNutrition/ConsumerInformation/ucm078889.htm#see1).

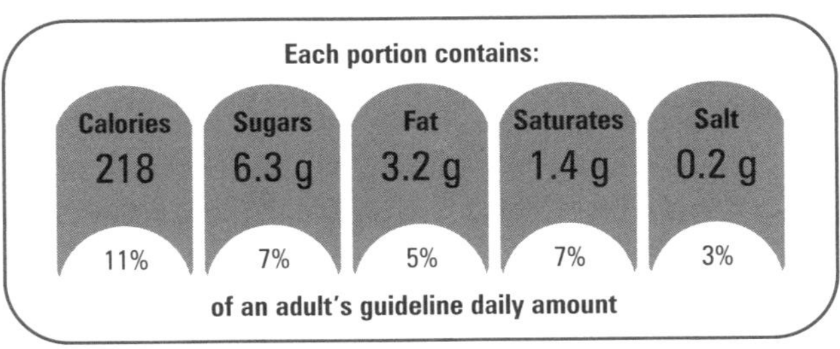

Fig. 9.2 Example of GDA label (FDF, 2009)

Although both traffic light and GDA-type labels are designed to inform the consumer, people may still need guidance on how to make best use of the information. Traffic lights are easily recognised by the public and require minimal reading to be able to make a decision between products of similar type but with different portion sizes. They are also very convenient for people who are controlling their intake of a particular nutrient, such as saturated fat or sodium. The drawback with this method is that it may put people off buying products that have red labels but which may be rich sources of other nutrients. GDA may need clearer guidance for use, as portion size is a critical factor and also the GDA values are not the same for everyone. As the nutrient requirements of an individual depend on sex, age and physical activity level (see Chapter 41), the GDA may lead to over or under consumption; however, regular use of GDA labels may also have the effect of educating people about appropriate portion sizes.

Evaluating weight-loss diets

There are literally thousands of weight-loss diets available to the general public; indeed a Google search for 'weight loss diets' results in over 32 000 000 hits. So how can you evaluate which will be effective? As discussed in Chapter 41, it is possible to analyse the nutrient content of dietary intakes using food composition tables or software. The diet that you advise a client to follow should be balanced in terms of meeting their micro- and macronutrient requirements, as well as reducing the amount of energy they consume. The fundamental aim of any weight-loss diet is to reduce energy intake, and the plethora of diets claiming to be effective will be so if they achieve this aim – and it will have very little to do with the macronutrient profile, e.g. low fat or high protein, and more to do with raising an awareness of personal intake and restricting the intake of particular foods. Chapter 44 explains how to calculate individual energy requirements, and this information will be important when advising a client on an appropriate intake to ensure that they have a negative energy balance (where their energy intake is less than the energy requirements of BMR and physical activity). Achieving and maintaining a negative energy balance is of course a challenge for the individual and regular monitoring of progress is important. If you have estimated their requirement and given detailed information on appropriate food choices and portion size, you should maintain accurate records of body weight, waist circumference and/or body composition (see Chapter 43). The same person should take all measurements using the same equipment to avoid any differences in method or readings, and it may be useful for both client and health professional to record and present these measurements in a table or as graphs.

Text reference

Damer, T.E. (2009) *Attacking Faulty Reasoning: A Practical Guide to Fallacy-Free Arguments*, 6th edn. Wadsworth, Belmont, California.

Food and Drink Federation (2009) *Guideline Daily Amounts*. Available: http://www.gdalabel.org.uk/gda/gda_values.aspx#item1. Last accessed 21/12/10.

Food Labelling Regulations 1996 (FLR) (SI 1996 No. 1499) (as amended). HMSO, London.

Food Standards Agency (2007) Front-of-pack traffic light signpost labelling technical guidance. Available: http://www.food.gov.uk/multimedia/pdfs/frontofpackguidance2.pdf Last accessed 21/12/10.

Sources for further study

Barnard, C.J., Gilbert, F.S. and MacGregor, P.K. (2001) *Asking Questions in Biology: A Guide to Hypothesis-testing Analysis and Presentation in Practical Work and Research,* 3nd edn. Pearson, Harlow.

Smith, A. *Evaluation of Information Sources.* Available: http://www.vuw.ac.nz/staff/alastair_smith/evaln/evaln.htm
Last accessed: 21/12/10.
[Part of the Information Quality WWW Virtual Library.]

Van Gelder, T. *Critical Thinking on the Web.* Available: http://www.austhink.org/critical/
Last accessed: 21/12/10.
[A useful directory of Web resources on the topic of critical thinking, associated with *Austhink,* a group based in Melbourne, Australia, that specialises in complex reasoning and argument, including critical thinking research, training and consulting.]

10 Using online resources

Definitions

Browser – a program to display Web pages and other Internet resources.

FAQ – frequently asked question; a file or Web page giving information on common queries, sometimes used as a file extension (.faq).

FTP – file transfer protocol; a mecha-nism of downloading files.

URL – uniform resource locator; the 'address' for WWW resources.

Information and communication technology (ICT) is vital in the modern academic world and 'IT literacy' is a core skill for all bioscientists. This involves a wide range of computer-based skills, including:

- **Accessing Web pages using a Web browser** such as Microsoft Internet Explorer, Mozilla Firefox or Opera.
- **Searching the Web for useful information and resources** using a search engine such as Google, or a meta-search engine such as Dogpile.
- **Finding what you need within online databases,** such as library cata-logues, or complex websites, such as your university's homepage.
- **Downloading, storing and manipulating files.**
- **Communicating via the Internet.**
- **Using e-learning facilities effectively.**
- **Working with 'office'-type programs and other software** (dealt with in detail in Chapters 11 and 12).

You will probably receive an introduction to your university's networked IT systems and will be required to follow rules and regulations that are important for the operation of these systems. Whatever your level of experience with PCs and the Internet, you should also follow the basic guidelines shown in Box 10.1. Reminding yourself of these from time to time will reduce your chances of losing data.

The Internet as a global resource

Understanding the technology – you do not need to understand the workings of the Internet to use it – most of it is invisible to the user. To ensure you obtain the right facilities, you may need to know some jargon, such as terms for the speed of data transfer (megabits) and the nature of Internet addresses. Setting up a modem and/or local wireless network can be complex, but instructions are usually provided with the hardware. White and Downs (2005) and Gralla (2003) are useful texts if you wish to learn more about computing and the Internet.

The Internet is a complex network of computer networks; it is loosely organised and no one group organises it or owns it. Instead, many private organisations, universities and government organisations fund and operate discrete parts of it.

The most popular application of the Internet is the WWW ('the Web'). It allows easy links to information and files which may be located on networked computers across the world. The WWW enables you to access millions of 'home pages' or 'websites' – the initial point of reference with many individuals, institutions and companies. Besides text and images, these sites may contain 'hypertext links', highlighted words or phrases that take you to another location via a single mouse click.

You can gain access to the Internet either through a local area network (LAN) at your university, at most public libraries, at a commercial 'Internet cafe', or from home via a modem connected to a broadband or dial-up Internet service provider (ISP).

KEY POINT Most material on the Internet has not been subject to peer review or vetting. Information obtained from the WWW or posted on newsgroups may be inaccurate, biased or spoof; do not assume that everything you read is true or even legal.

Box 10.1 Important guidelines for using PCs and networks

Hardware

- Don't drink or smoke around the computer.
- Try not to turn the computer off more than is necessary.
- Never turn off the electricity supply to the machine while in use.
- Switch off the computer and monitor when not in use (saves energy and avoids dangers of 'hijacking').
- Rest your eyes at frequent intervals if working for extended periods at a computer monitor. Consult Health and Safety Executive publications for up-to-date advice on working with display screens (http://www.hse.gov.uk/pubns/).
- Never try to reformat the hard disk without the help of an expert.

CDs and USB drives

- Protect CDs when not in use by keeping them in holders or boxes.
- Label USB (Universal Serial Bus) drives with your name and return details and consider adding these to a file stored on the drive.
- Try not to touch the surface of CDs, and if they need cleaning, do so carefully with a clean cloth, avoiding scratching. If floppy disks are used, keep these away from sources of magnetism (e.g. speakers).
- Keep disks and USB drives away from moisture, excess heat or cold.
- Don't use disks from others, unless you first check them for viruses.
- Don't insert or remove a disk or USB drive when it is operating (drive light on). Close all files before removing a USB drive and use the *Safely Remove Hardware* feature.
- Try not to leave a disk or USB drive in the drive when you switch the computer off.

File management

- Organise your files in an appropriate set of folders.
- Always use virus-checking programs on copied or imported files before running them.
- Make back-ups of all important files at frequent intervals (say, every ten minutes or half-hour), e.g. when using a word processor or spreadsheet.
- Periodically clear out redundant files.

Network rules

- Never attempt to 'hack' into other people's files.
- Do not give out any of your passwords to others. Change your password from time to time. Make sure it is not a common word, is longer than eight characters, and includes numerical characters and punctuation symbols, as well as upper- and lower-case letters.
- Never use network computers to access or provide financial or other personal information: spyware and Trojan programs may intercept your information.
- Never open email attachments without knowing where they came from; always virus-check attachments before opening.
- Remember to log out of the network when finished; others can access your files if you forget to log out.
- Be polite when sending email messages.
- Periodically reorganise your email folder(s). These rapidly become filled with acknowledgements and redundant messages that reduce server efficiency and take up your allocated filespace.
- Do not play games without approval – they can affect the operation of the system.
- If you are setting up your own network, e.g. in your flat, always install up-to-date firewall software, anti-spyware and anti-virus programs.

The Golden Rule – always make back-up copies of important files and store them well away from your working copies. Ensure that the same accident cannot happen to both copies.

Online communication

You will be allocated an email account by your university and should use this routinely for communicating with staff and fellow students, rather than using a personal account. You may be asked to use email to submit work as an attachment, or you may be asked to use a 'digital drop-box' within the university's e-learning system (Box 10.2). When using email at university, follow these conventions, including etiquette, carefully:

Box 10.2 Getting to grips with e-learning

Some key aspects of tackling e-learning are outlined below.

1. **Develop your basic IT skills, if required.** e-learning requires only basic IT skills, such as: use of keyboard and mouse; word processing; file management; browsing and searching. If you feel weak on any of these, seek out additional courses offered by the IT administration or your department.
2. **Visit your e-learning modules regularly.** You should try to get into a routine of doing this on a daily basis at a time that suits you. Staff will present up-to-date information (e.g. lecture-room changes) via the 'announcements' section, may post information about assessments, or links to the assessments themselves, and you may wish to provide feedback or look at discussion boards and their threads.
3. **Participate.** e-learning requires an active approach.
   - At the start of each new course, spend some time getting to know what's been provided online to support your learning. As well as valuable resources, this may include crucial information such as learning objectives (p. 22), dates of submission for coursework and weighting of marks for different elements of the course.
   - If you are allowed to download lecture notes (e.g. in the form of PowerPoint presentations), do not think that simply reading through these will be an adequate substitute for attending lectures and making further notes (see page 16).
   - Do not be tempted to 'lurk' on discussion boards: take part. Ask questions; start new threads; answer points raised by others if you can.
   - Try to gain as much as you can from formative online assessments (p. 22). If these include feedback on your answers, make sure you learn from this and if you do not understand it, consult your tutors.
   - Learn from the critical descriptions that your lecturers provide of linked websites. These pointers may help you to evaluate such resources for yourself in future (pp. 50–53).
   - Don't think that you will automatically assimilate information and concepts, just because you are viewing them online. The same principles apply as with printed media: you must apply active learning methods (p. 25).
   - Help your lecturers by providing constructive feedback when they ask for it. You may find this easier to do when using the computer interface, and it may be more convenient than hurriedly filling out a feedback sheet at the end of a session.
4. **Organise files and Web links.** Take the time to create a meaningful folder- and file-naming system for downloaded material in tandem with your own coursework files and set up folders on your browser for bookmarked websites (*Favorites* in Internet Explorer).
5. **Take care when submitting coursework.** Make sure you keep a back-up of any file you email or submit online and check the version you are sending carefully. Follow instructions carefully, for example regarding file type, or how to use your system's 'digital drop-box' and other assignment submission/handling functions.

Spam, junk mail and phishing – these should be relatively easy to identify, and should never be responded to or forwarded. Some may look 'official' and request personal or financial details (for example, they may pretend to come from, e.g., your bank, and ask for account details). Never send these details by email or your identity may be used illegally.

- **Check your email account regularly** (daily). Your tutors may wish to send urgent messages to you in this way.
- **Respond promptly to emails.** Even if you are just acknowledging receipt, it is polite to indicate that you have received and understood a message.
- **Be polite.** Email messages can seem to be abrupt and impersonal. Take care to read your messages through before sending and if you are at all in doubt, do not send your message right away: reread at a later time and consider how others might view what you say.
- **Consider content carefully.** Only send what you would be happy to hear being read out loud to classmates or family.
- **Take care with language and names** when communicating with tutors. Slang phrases and text message shorthand are unlikely to be understood. Overfamiliarity does not go down well.

- **Use email for academic purposes** – this includes discussing coursework with classmates, but not forwarding off-colour jokes, potentially offensive images, links to offensive websites, etc. In fact, doing so may break regulations and result in disciplinary action.
- **Beware of spam, junk and 'phishing'** via email.

Similar rules apply to discussion boards.

Newsgroups – these can be useful for getting answers to a specific problem: just post a query to the appropriate group and wait for someone to reply. Bear in mind that this may be the view of an individual person.

The Usenet Newsgroup service is an electronic discussion facility, and there are thousands of newsgroups representing different interests and topics. Any user can contribute to the discussion within a topic by posting his or her own message; it is like email but without privacy, since your message becomes available to all other subscribers. To access a newsgroup, your system must be running, or have access to, a newsgroup server that has subscribed to the newsgroup of interest. Obtain a list of newsgroups available on your system from the IT administration service and search it for those of interest, then join them. Contact your network administrator if you wish to propose the addition of a specific newsgroup, but expect a large amount of information to be produced.

Definition

Bookmark – a feature of browsers that allows you to save details of websites you have visited. This is termed 'add to favorites' in Internet Explorer. Bookmarks save you the trouble of remembering complex URL names and of typing them into the browser's address window.

Internet tools

The specific programs you will use for accessing the Internet will depend on what has been installed locally, on the network you are using, and on your ISP. The best way to learn the features of the programs is to try them out, making full use of whatever help services are available.

e-learning systems

Most university departments present their courses through a mixture of face-to-face sessions (e.g. lectures, tutorials, practicals) and online resources (e.g. lecture notes, websites, discussion boards, computerised tests and assessments). This constitutes 'blended learning' on your part, with the online component also being known as e-learning.

The e-learning element is usually delivered through an online module within a virtual learning environment (e.g. Blackboard, WebCT, Moodle). It is important not to neglect the e-learning aspects of your course just because it may not be as rigidly timetabled as your face-to-face sessions. This flexibility is to your advantage, as you can work when it suits you, but it requires discipline on your part. Box 10.2 provides tips for making the most of the e-learning components of your courses.

Social networking

Social networking sites such as Facebook can be useful for academic purposes when used as discussion forums. However, you should be cautious about the personal information that you post on these sites, as well as the images and content you post, as it is still unclear as to who owns the content once you make it publicly available. Although the Web may be an impermanent tool, you do not want images and content available to others even if you have deleted it.

Internet browsers

These are software programs that interact with remote server computers around the world to carry out the tasks of requesting, retrieving and displaying the information you require. Many different browsers exist, but the most popular are Internet Explorer, Mozilla Firefox and Opera. These three browsers dominate the market and have plug-ins and add-on programs available that allow, for example, video sequences to be seen online. Many browsers incorporate email and newsgroup functions. The standard functions of browsers include:

- accessing Web documents;
- following links to other documents;
- printing the current document;
- maintaining a history of visited URLs (including 'bookmarks' for key sites);
- searching for a term in a document;
- viewing images and image maps.

Examples Useful Web portals:
Biology Image Galleries – directory of online biology images, photos, microscopy and illustrations – http://www.academicinfo.net/bioimage.html
Intute: Health and Life Sciences – http://www.intute.ac.uk/healthandlifesciences

Browsers provide access to millions of websites. Certain sites specialise in providing catalogued links to other sites; these are known as portals and can be of enormous help when searching within a particular area of interest. Your university's library website will almost certainly provide a useful portal to catalogues and search services, often arranged by subject area, and this is often the first port of call for electronic resources; get to know your way around this part of the website as early as possible during your course.

When using a Web browser program to get to a particular page of information on the Web, all you require is the location of that page, i.e. the URL (uniform resource locator). Most Web-page URLs take the form http:// or https://, followed by the various terms (domains and sub-domains) that direct the system to the appropriate site. If you don't have a specific URL in mind but wish to explore appropriate sites, you will need to use a search tool with the browser.

Examples Common domains and sub-domains include:

.ac	academic
.com	commercial
.co	commercial
.edu	education (USA mainly)
.gov	government (USA and UK)
.mil	military (USA only)
.net	Internet-based companies
.org	organisation
.uk	United Kingdom

WWW search tools

With the proliferation of information on the Web, one of the main problems is finding the exact information you require. There is a variety of information services that you can use to filter the material on the network. These include:

- search engines (Boxes 10.3 and 10.4);
- meta-search engines;
- subject directories;
- subject gateways (portals).

Search engines such as Google (http://www.google.com/), Altavista (http://uk.altavista.com/) and Lycos (http://www.lycos.com/) are tools designed to search, gather, index and classify Web-based information. Searching is usually by keyword(s), although specific phrases can be defined. Many search engines offer advanced searching tools such as the use of Boolean operators to specify combinations of keywords to more precisely filter the sites. Box 10.3 provides tips for refining keyword searches while Box 10.4 provides tips for enhancing searches with Google.

Box 10.3 Useful tips for using search engines

- **Keywords should be chosen with care.** Try to make them as specific as possible, e.g. search for 'archaebacteria', rather than 'bacteria' or 'microorganisms'.
- **Most search engines are case-insensitive.** Thus 'Nobel Prize' will return the same number of hits as 'nobel prize'. If in doubt, use lower case throughout.
- **Check that your search terms have the correct spelling,** otherwise you may only find sites with the same mis-spelled word. In some cases, the search engine may prompt you with an alternative (correct) spelling. If a word has an alternative US spelling (e.g. color, hemoglobin), then a search may only find hits from sites that use the spelling you specify.
- **Putting keyword phrases in double quotes (e.g. "body mass index") will result in a search for sites with the phrase as a whole** rather than sites with both (all) parts of the phrase as separate words (i.e. 'body' and 'mass' and 'index' at different places within a site). This feature allows you to include common words normally excluded in the search, such as 'the'.
- **Use multiple words/phrases plus similar words to improve your search,** for example 'plant extract' 'cholesterol-lowering'. If you can, use scientific terms, as you are likely to find more relevant sites, e.g. search for the name of a particular compound such as 'resveratrol'.
- **Adding words preceded with + or – will add or exclude sites with that word present** (e.g. 'blood diseases – leukaemia' will search for all blood diseases excluding leukaemia). This feature can also be used to include common words normally excluded by the search engine.
- **Boolean operators (AND, OR, NOT) can be used with some search engines to specify combinations of keywords** to more precisely filter the sites identified (e.g. 'plant NOT engineering' will avoid sites about engineering plant and focus on botanical topics).
- **Some search engines allow 'wildcards' to be introduced with the symbol** *. For example, this will allow you to specify the root of a word and include all possible endings, as with anthropomorph*, which would find anthropomorphic, anthropomorphism, etc. If the search engine does not allow wildcards, then you will need to be especially careful with the key-words used, including all possible words of relevance.
- **Numbers can be surprisingly useful in search engines.** For example, typing in EC 1.1.1.1 will find sites concerned with alcohol dehydrogenase, as this is its code number. If you know the phone number for a person, institute or company or the ISBN of the book, this can often help you find relevant pages quickly.
- **If you arrive at a large site and cannot find the point at which your searched word or phrase appears, press Control + F together** and a 'local' search window will appear, allowing you to find the point(s) where it is mentioned.

'Dissecting' a Web address – if a URL is specified, you can often find out more about a site by progressively deleting sections of the address from the right-hand side. This will often take you to 'higher levels' of the site, or to the home page of the organisation or company involved.

It is important to realise that each search engine will cover at most about 40 per cent of the available sites; if you want to carry out an exhaustive search it is necessary to use several to cover as much of the Web as possible. Meta-search engines make this easier. These operate by combining collections of search engines. Examples include Mamma (http://www.mamma.com/), Dogpile (http://www.dogpile.com/index.gsp/) and Metacrawler (http://www.metacrawler.com/index.html/).

Some useful approaches to searching include the following:

- For a comprehensive search, use a variety of tools including search engines, meta-search engines and portals or directories.
- For a complex, finely specified search, employ Boolean operators and other tools to refine your keywords as fully as possible (Box 10.3). Some search engines allow you to include and exclude terms or restrict by date.
- Use 'cascading' searching when available – this is searching within the results of a previous search.

Box 10.4 Getting the most from Google searches

Google (http://www.google.com) has become the search engine of choice for millions of people, due to its simplicity and effectiveness. However, you may be able to improve your searches by understanding its default settings and how they can be changed.

- **Download the Google toolbar to your browser.** This is available from the Google homepage and will give you quick access to the Google search facility.
- **Understand how standard operators are used.** For combinations of keywords Google uses the minus operator '−' instead of NOT (exclude) and '+' instead of AND (include). Since Google usually ignores small words ('stop words' such as *in* or *the*), use '+' to include them in a search. Where no operator is specified, Google assumes that you are looking for both terms (i.e. '+' is default). If you want to search for alternative words, you can use 'OR' (e.g. *sulphur OR sulfur*). Google does not allow brackets and also ignores most punctuation marks.
- **While wildcard truncation of words using '*' is not allowed, you can use '*' to replace a whole word (or number).** For example, if you type the phrase *"a virus is approximately * nanometres"* your results will give you results for Web pages where the wildcard is replaced by a number.
- **Search for exact wording.** By placing text in double inverted commas ("), you can ensure that only websites with this exact phrasing will appear at the head of your search results.
- **Search within your results to improve the outcome.** If your first search has produced a large number of results, use the *Search within results* option near the bottom of each page to type in a further word or phrase.
- **Search for words within the title of a Web page.** Use the command *intitle:* to find a webpage, for example *intitle: "tissue culture"* returns Web pages with this phrase in the title (note that phrases must always be in double speech marks, not single quotes).
- **Search within a website.** Use the *site:* command to locate words/phrases on a specific website, for example *site:unicef.org dysentery* returns only those results for this disease on the UNICEF website (unicef.org). Pressing *Control+F* when visiting a Web page will give you a pop-up search window.
- **Locate definitions, synonyms and spellings.** The operator *define:* enables you to find the meaning of a word. If you are unsure as to the spelling of a word, try each possibility: Google will usually return more results for the correct spelling and will often also prompt you with the correct spelling (*Did you mean . . .?*).
- **Find similar Web pages.** Simply click the *Similar pages* option at the end of a Google search result to list other sites (note that these sites will not necessarily include the term(s) searched for).
- **If a Web link is unavailable, try the cached (stored) page.** Clicking on *Cached* at the end of a particular result should take you to the stored page, with the additional useful feature that the search term(s) will be highlighted.
- **Use the calculator functions.** Simply enter a calculation and press *Enter* to display the result, for example *'10+(2*4)'* returns 18. The calculator function can also carry out simple interconversion of units, e.g. *'2 feet 6 inches in metres'* returns 0.762 (see Box 32.1 for interconversion factors between SI and non-SI units).
- **Try out the advanced search features.** In addition to the standard operators these include the ability to specify the number of results per page (e.g. 50, to reduce the use of the *next* button), language (e.g. English), file format (e.g. for PDF files), recently updated Web pages (e.g. past three months), usage (e.g. free to use/share).
- **Find non-text material.** These include images, video and maps – always check that any material you use is not subject to copyright limitations (p. 52).
- **Use Google alerts to keep up to date.** This function (http://www.google.co.uk/alerts) enables you to receive regular updated searches by email.
- **Use Google Scholar to find articles and papers.** Go to http://scholar.google.com/ and type in either the general topic or specific details for a particular article, e.g. author names or words from the title. Results show titles/authors of articles, with links to either the full article, abstract or citation. A useful feature is the *Cited by . . .* link, taking you to those papers that have cited the article in their bibliography and enabling you to carry out forward citation searching to locate more recent papers. Also try out the advanced scholar search features to limit your search to a particular author, journal, date or subject area. However, you should note that Google Scholar provides only a basic search facility to easily accessible articles and should not be viewed as a replacement for your library's electronic journal holdings and searching software. For example, if you find the title of a paper via Google Scholar you may be able to locate the electronic version through your own library's databases, or request it via inter-library loan (p. 44). Another significant limitation is that older (more cited) references are typically listed first.
- **Use Google Earth to explore locations.** This allows you to zoom in on satellite images to find locations.

Box 10.5 How to evaluate information on the World Wide Web

It is often said that 'you can find anything on the Web'. The two main disadvantages of this are, firstly, that you may need to sift through many sources before you find what you are looking for and, secondly, that the sources you find will vary in their quality and validity. *It is important to realise that evaluating sources is a key aspect of using the Internet for academic purposes, and one that you will need to develop during the course of your studies.* The ease with which you can 'point and click' to reach various sources should not make you complacent about evaluating their information content. The following questions can help you to assess the quality of a website – the more times you can answer 'yes', the more credible the source is likely to be, and vice versa.

Authority

- Is the author identified?
- Are the author's qualifications or credentials given?
- Is the owner, publisher or sponsoring organisation identified?
- Is an address given (postal and/or email)?

It is sometimes possible to get information on authority from the site's metadata (try the 'View' 'Source' option in Internet Explorer, or look at the URL to see if it gives any clues as to the organisation, e.g. does the domain name end in .ac, .edu, .gov or .org, rather than .co or .com).

Content

- Is there any evidence that the information has been peer-reviewed (p. 50), edited or otherwise validated, or is it based on such sources?
- Is the information factual or based on personal opinions?
- Is the factual data original (primary) or derived from other sources (secondary)?
- Are the sources of specific factual information detailed in full (p. 52)
- Is there any indication that the information is up to date, or that the site has been recently updated?
- What is the purpose of the site and who is it aimed at?
- Is the content relevant to the question you are trying to answer?
- Is there any evidence of a potential conflict of interest, or bias? (Is the information comprehensive and balanced, or narrowly focused?)
- Did you find the information via a subject-specific website (e.g. a bioscience gateway such as Intute Health and Life Sciences), or through a more general source, such as a search engine (e.g. Google)?

The above questions are similar to those that you would use in assessing the value of a printed resource (pp. 49–52), and similar criteria should be applied to Web-based information. You should be especially wary of sites containing unattributed factual information or data whose primary source is not given.

Presentation

- What is your overall impression of how well the site has been put together?
- Are there many grammatical or spelling mistakes?
- Are there links to other websites, to support statements and factual information?

The care with which a site has been constructed can give you an indication of the credibility of the author/organisation. However, while a poorly presented site may cause you to question the credibility of the information, the reverse is not always necessarily true: don't be taken in by a slick, well-presented website – authority and content are *always* more important than presentation.

- Use advanced search facilities to limit your search, where possible, to the type of medium you are looking for (e.g. graphics, video), language, sites in a specific country (e.g. UK) or to a subject area (e.g. news only).

However well defined your search is, you will still need to evaluate the information obtained. Chapter 9 covers general aspects of this topic, while Box 10.5 provides specific advice on assessing the quality of information provided on websites.

Downloading files from the Internet and emails – read-only files are often available as 'pdf' files that can be viewed by Adobe reader software (available free from http://www.adobe.com), while other files may be presented as attachments to emails or as links from Web pages that can be opened by suitable software (e.g. Microsoft Word or 'paint' programs such as Paint Shop Pro). Take great care in the latter cases as the transfer of files can result in the transfer of associated viruses. Always check new files for viruses (especially .exe files) before running them, and make sure your virus-detecting software is kept up to date.

Understanding the impermanence of the Web – the temporary nature of much of the material on the Web is a disadvantage for academic purposes because it may change or even disappear after you have cited it. You may also find it difficult or impossible to find out who authored the material (p. 51). A case in point are wikis, such as Wikipedia (www.wikipedia.org). This online encyclopedia has many potential authors and the content may change rapidly as a result of new submissions or edits; nevertheless, it can be a useful resource for up-to-date general information about a wide range of topics, though it is not necessarily regarded as the best approach for researching assignments.

Using traditional sources – remember that using the Internet to find information is not a substitute for visiting your university library. Internet resources complement rather than replace CD-ROM and more traditional printed sources.

Directories

A directory is a list of Web resources organised by subject. It can usually be browsed and may or may not have a search facility. Directories often contain better-quality information than the lists produced by search engines, as they have been evaluated, often by subject specialists or librarians. The BUBL information service life sciences directory of links, at: http://bubl.ac.uk/link/lif.html/ is a good example.

Using the Internet as a resource

A common way of finding information on the Web is by browsing or 'surfing'. However, this can be time-consuming; try to restrict yourself to sites known to be relevant to the topic of interest. Some of the most useful sites are those that provide hypertext links to other locations. Some other resources you can use on the WWW are:

- Libraries, publishers and commercial organisations. Your university library is likely to subscribe to one or more databases providing access to scientific articles; these include BIDS (Bath Information and Database Services (http://www.bids.ac.uk), ISI Web of Science (http://wos.mimas.ac.uk/), and Science Direct (http://www.sciencedirect.com/). A password is usually required, especially for off-campus use; consult your library staff for further details. Some scientific database sites give free access, without subscription or password; these include National Center for Biotechnology Information (USA) (http://www.ncbi.nlm.nih.gov/) and the Highwire Press (http://highwire.stanford.edu/). Others allow free searching, but require payment for certain articles, e.g. The Scientific World (http://www.thescientificworld.com/) and Infotrieve (http://www4.infotrieve.com/). The Natural History Book Service (http://www.nhbs.co.uk) provides useful information on relatively inaccessible literature such as government reports. Publishers such as Pearson and booksellers such as Amazon.com provide online catalogues and e-commerce sites that can be useful sources of information (see http://vig.pearsoned.co.uk/ and http://www.amazon.com).
- Online journals and e-books. Most traditional journals have websites. You can keep up to date by visiting the websites of *Nature* (http://www.nature.com/), *New Scientist* (http://www.newscientist.com/) and *Scientific American* (http://www.sciam.com/), or the Elsevier Science Direct website (http://www.sciencedirect.com/science/journals). Some scientific societies make their journals and other publications available via their websites, e.g. the American Society for Microbiology, at: http://www.asm.org/. Journals solely published in electronic format are also available (e.g. *Molecular Vision*, http://www.molvis.org/molvis/) but some require a subscription password for access; check whether your institute is a subscriber. It is often possible to subscribe to updates from journals either through email alerts, where you can be sent a table of contents, or through RSS feeds. RSS stands for 'really simple syndication' and is a way of keeping up to date with changing content on websites.
- Data and images. Archives of text material, video clips and photographs can be accessed, and much of the material is readily available. The HEA Centre for Biosciences Image Bank (http://www.bioscience.heacademy.ac.uk/imagebank) is a good example. When downloading

Remembering useful websites – create a 'bookmark' (= add a 'favourite') for the ones you find of value, to make revisiting easy. This can be done from the menu of your browser program. Make a copy of your bookmark file occasionally, to avoid loss of this information. You can also make use of social bookmarking systems such as Delicious, Digg and reddit, which allow you to save your favourites list on the web so that you can access them from anywhere. Another advantage is that you can also share your favourites with others, which may help you to find other appropriate sources, much like looking in the reference lists of research papers to find new sources.

Examples Selected websites of food and nutritional interest:

http://www.who.int/nutrition/en/

http://www.eufic.org/

http://www.eufic.org/

http://www.food.gov.uk/

http://www.hero.ac.uk/uk/reference_and_subject_resources/resources/worldwide_library_resources3796.cfm

Note that URLs may change – make a keyword search using a search engine to find a particular site if the URL information you have does not lead you to an active page.

such material, you should (i) check that you are not breaching copyright and (ii) avoid potential plagiarism by giving a full citation of the source, if you use such images in an assignment (see p. 49).

- Databases. In addition to those covering the scientific literature, others focus on specific topics (e.g. World Health Organisation nutrition databases, http://www.who.int/nutrition/databases/en/index.html, or academic employment, http://www.jobs.ac.uk/).

Internet resources for bioinformatics

Bioinformatics is a term used to describe the application of computers in biology and, in particular, to the analysis of sequence data for biopolymers such as proteins and nucleic acids. These complex molecules contain a large amount of information within their structures, and the only practical approach to understanding this is to use a computer – generally, by comparing sequence data for all or part of a specific biomolecule with that in a database. For ease of access, many bioinformatics databases and programs have been made freely accessible via the Internet.

The use of post-genomic technologies is not new in nutrition and dietetics. For example, in the monogenetic condition phenylketonuria (PKU), variations in a single gene are known to have an effect on the metabolism of an amino acid, and being able to identify this variation allows the condition to be managed by diet. Most effects, however, are not monogenetic, and therefore require nutritionists and dietitians to be able to harness these new technologies in order to find links between genetic variation and response to diet.

The major databases holding primary sequence information for nucleic acids and proteins are operated by the European Bioinformatics Institute (http://www.ebi.ac.uk/Databases/) and the National Center for Biotechnology Information (http://www.ncbi.nlm.nih.gov). These allow you to:

- **find and retrieve a nucleotide or amino acid sequence** from the database;
- **translate a nucleotide sequence** into an amino acid sequence and vice versa;
- **search for similarity** between a particular sequence or sequences within the database, e.g. by comparing and aligning the sequences for several proteins (or nucleic acids), to identify regions of sequence similarity (see Fig. 10.1);
- **identify known single nucleotide polymorphisms (SNPs)** which may be relevant to metabolism; and
- **carry out phylogenetic analysis,** constructing 'ancestry trees' to show the most likely evolutionary relationships between sequences from various organisms.

In an attempt to standardise information in this fast-moving area, many journals now insist on the inclusion of accession numbers to allow researchers to identify molecular structures. For example, the GenBank accession number for the mouse mRNA sequence for leptin is NM_008493.

```
H. sapiens     atgtatggcattgagaatgaagtcttcctgagccttccatgtatcctcaatgcccgggg
               ||||| ||||| ||||| ||||||||||| || || || || ||||| ||||| |||||
R. norvegicus  atgtacggcatcgagaacgaagtcttcctcagtctcccgtgcatccttaatgctcgggg
```

Fig. 10.1 Representative output from a BLAST (Basic Local Alignment Search Tool) alignment search for DNA sequences from part of the lactate dehydrogenase genes of a human (*Homo sapiens*) and a laboratory rat (*Rattus norvegicus*). Here, identical bases are shown in coloured text, and non-identical bases are shown in black. This region shows 49 identical bases out of 59, i.e. $49 \div 59 \times 100 = 83\%$ identity (to the nearest integer).

Definitions

Note that terms ending in 'ome' can also be used to describe disciplines simply by changing the ending to 'omics'.

Genome – the entire complement of genetic information (coding and non-coding DNA) of an organism.

Transcriptome – the complement of mRNAs transcribed from the genome, weighted by the expression level of each ribonucleic acid (RNA).

Proteome – the expressed protein complement of the genome.

Sectretome – all of the secreted proteins from an organism.

Metabolome – all of the metabolites (low molecular mass biomolecules) of an organism.

Nutrigenomics – examines an individual's response to food compounds using the above technologies.

Nutrigenetics – examines how genetic variation influences the interaction between diet and health.

Secondary databases have been created using primary sequence data, to provide information on the patterns identified within particular types of biomolecules, e.g. PROSITE (http://www.expasy.org/prosite/) and PRINTS (http://www.bioinf.man.ac.uk/dbbrowser/index/index.html) databases, which identify protein families by their diagnostic 'signature' motifs, or the ExPaSy proteomics resource (http://www.expasy.org/), for analysis of protein structure. Other websites have been constructed to bring together information for a particular organism, e.g. The *Saccharomyces* Genome Database (SGD, http://www.yeastgenome.org), The *Arabidopsis* Information Resource (http://www.arabidopsis.org/) or The Caenorhabditis elegans Genome Project (http://www.sanger.ac.uk/Projects/C_elegans/) and the Human Genome Project (http://www.sanger.ac.uk/). Many of these sites are rather complex and newcomers can find them a little difficult to navigate, especially if you are just browsing. You are most likely to make use of bioinformatics databases in computer-based exercises: for example, where you are given a particular sequence and you then have to 'interrogate' a particular database, e.g. to identify similar sequences. Alternatively, you could try out the tutorials available at many of these sites.

Text references

Gralla, P. (2003) *How the Internet Works*, 7th edn. Pearson, Harlow.

White, R. and Downs, T. (2005) *How Computers Work*, 8th edn. Pearson, Harlow.

Sources for further study

Anon. *Web Search Help*. Available: http://www.google.com/support/websearch/?ctx=web
Last accessed 21/12/10.
[Advice for using the Google search engine.]

Brandt, D.S. *Why We Need to Evaluate What We Find on the Internet*. Available: http://www.lib.purdue.edu/research/techman/eval.html
Last accessed 21/12/10.

Dussart, G. (2002) *Biosciences on the Internet*. Wiley, Chichester.

Grassian, E. *Thinking Critically about World Wide Web Resources*. Available:
http://www.library.ucla.edu/libraries/college/11605–12337.cfm
Last accessed 21/12/10.

Isaacs, M. (2000) *Internet Users Guide to Network Resource Tools*, 2000 edn. Addison-Wesley, Harlow.

11 Using spreadsheets

Definitions

Spreadsheet – a display of a grid of cells into which numbers, text or formulae can be typed to form a worksheet. Each cell is uniquely identifiable by its column and row number combination (i.e. its two-dimensional coordinates) and can contain a formula that makes it possible for an entry to one cell to alter the contents of one or more other cells.

Template – a predesigned spreadsheet without data but including all formulae necessary for (repeated) data analysis.

Macro – a sequence of user-defined instructions carried out by the spreadsheet, allowing complex repeated tasks to be 'automated'.

Office suite – a package of complementary and integrated programs such as a word processor, a spreadsheet and a database.

The spreadsheet is one of the most powerful and flexible computer applications. It can be described as the electronic equivalent of a paper-based longhand calculation, where the sums are carried out automatically. Spreadsheets provide a dynamic method of storing, manipulating and analysing data sets. Advantages of spreadsheets include:

- **Ease and convenience** – especially when complex calculations are repeated on different sets of data.
- **Accuracy** – providing the entry data and cell formulae are correct, the result will be free of calculation errors.
- **Improved presentation** – data can be produced in graphical or tabular form to a very high quality.
- **Integration with other programs** – graphs and tables can be exported to other compatible programs, such as a word processor in the same office suite.
- **Useful tools** – advanced features include hypothesis-testing statistics, database features and macros.

Spreadsheets can be used to:

- manipulate raw data by removing the drudgery of repeated calculations, allowing easy transformation of data and calculation of statistics;
- graph out your data rapidly to get an instant evaluation of results. Printouts can be used in practical and project reports;
- carry out statistical analysis by built-in procedures or by allowing construction of formulae for specific tasks;
- model 'what if' situations where the consequences of changes in data can be seen and evaluated;

The spreadsheet (Fig. 11.1) is divided into rows (identified by numbers) and columns (identified by alphabetic characters). Each individual combination of column and row forms a cell that can contain either a data item, a formula or a piece of text. Formulae can include scientific and/or statistical functions and/or a reference to other cells or groups of cells (often called a range). Complex systems of data input and analysis can be constructed. The analysis, in part or complete, can be printed out. New data can be added at any time and the sheet will recalculate automatically. The power a spreadsheet offers is directly related to your ability to create arrays of formulae (models) that are accurate and templates that are easy to use.

Data output from analytical instruments – many devices provide output in spreadsheet-compatible form (e.g. a 'comma delimited' file). Once you have uploaded the information into a spreadsheet, you can manipulate, analyse and present it according to your needs. Consult instrument manuals and the spreadsheet help function for details.

Data entry

Spreadsheets have built-in commands that allow you to control the layout of data in the cells (see Fig. 11.2). These include number format, the number of decimal places to be shown (the spreadsheet always calculates using eight or more places), the cell width and the location of the entry within the cell (left, right or centre). An auto-entry facility assists greatly in entering large amounts of data by moving the entry cursor either vertically or horizontally as data is entered. Recalculation default is usually

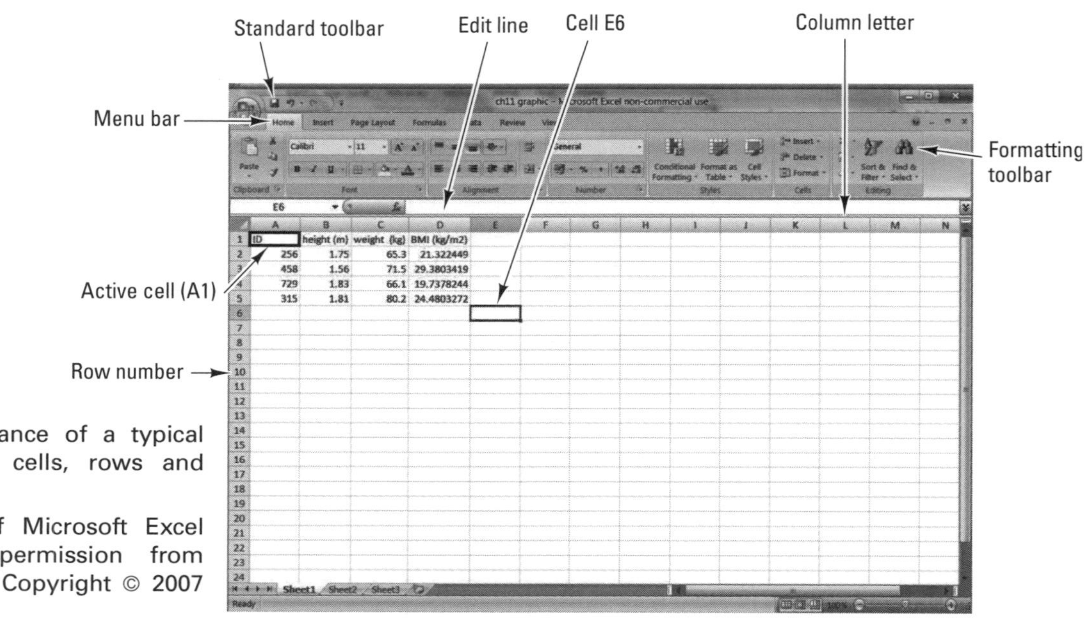

Fig. 11.1 The appearance of a typical spreadsheet, showing cells, rows and columns, toolbars, etc.

Source: Screenshot of Microsoft Excel 2007 reprinted by permission from Microsoft Corporation. Copyright © 2007 Microsoft Corporation.

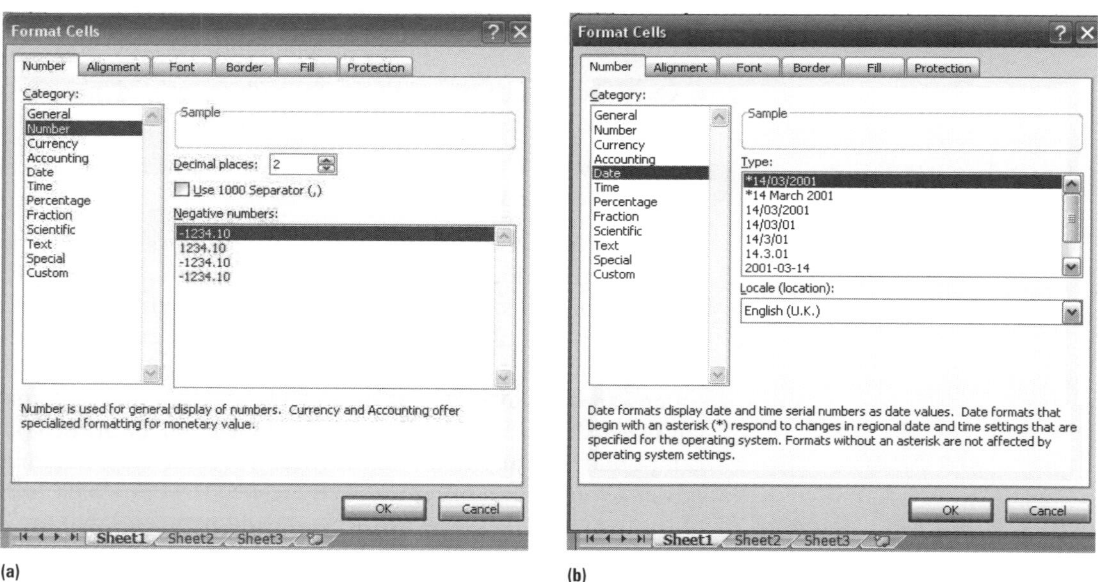

Fig 11.2 Example of cell-formatting options within a Microsoft Excel 2007 spreadsheet. These menus are accessed *via* the *Format > Format Cells* option and would apply to all of a range of selected cells. (a) Use of the number-formatting option to specify that data will be presented to three decimal places (the underlying data will be held to greater accuracy). (b) Use of the date-formatting option to specify that dates will be presented in day/month/year format (spreadsheet dates are stored numerically and converted to appropriate formats. This allows a period between two dates to be calculated more easily.)

Source: Screenshot of Microsoft Excel 2007 reprinted by permission from Microsoft Corporation, Copyright © 2007 Microsoft Corporation.

automatic so that when a new data value is entered the entire sheet is recalculated immediately.

Using hidden (or zero-width) columns – these are useful for storing intermediate calculations that you do not wish to be displayed on the screen or printout.

The parts of a spreadsheet

Labels

These should be used to identify parts of the spreadsheet – for example, stating what data is contained in a particular column or indicating that a cell's contents represent the end point of a calculation. It may be useful to use the *Format* > *Format Cells* > *Border* and *Format* > *Cells* > *Fill* functions to delimit numerical sections of your spreadsheet. Note that spreadsheet programs have been designed to make assumptions about the nature of data entry being made. If the first character is a number, then the entry is treated as numerical data; if it is a letter, then it is treated as a text entry; and if it is a specific symbol ('=' in Microsoft Excel), then what follows is a formula. If you wish to enter text that starts with a number, then you must type a designated character to show this (a single quote mark in Microsoft Excel).

Numbers

You can also enter numbers (values) in cells for use in calculations. Many programs let you enter numbers in more than one way and you must decide which method you prefer. The way you enter the number does not affect the way it is displayed on the screen as this is controlled by the cell format at the point of entry. There are usually special ways to enter data for percentages, currency and scientific notation for very large and small numbers.

Operators and brackets in spreadsheets – the standard mathematical operators ÷ and × are usually replaced by / and * respectively, while ^ signifies 'to the power'. In complex formulae, brackets should be used to separate the elements, otherwise the results may not be what you expect. For example, Excel will calculate = A1*B1/C1 – D1 differently from (A1*B1)/(C1 – D1).

Formulae

These are the 'power tools' of the spreadsheet because they do the calculations. A cell can be referred to by its alphanumeric code, e.g. A5 (column A, row 5) and the value contained in that cell manipulated within a formula, e.g. = (A5+10) or (A5+B22) in another cell. Formulae can include pre-programmed functions that can refer to a cell, so that if the value of that cell is changed, so is the result of the formula calculation. They may also include branching options through the use of logical operators (e.g. IF, TRUE, FALSE, OR, etc.).

Definition

Function – a pre-programmed code for the transformation of values (mathematical or statistical functions) or selection of text characters (string functions).

Functions

A variety of functions is usually offered, but only mathematical and statistical functions will be considered here.

Example – sin(A5) is an example of a function in Excel. If you write this in a cell, the spreadsheet will calculate the sine of the number in cell A5 (assuming it to be an angle in radians) and write it in the cell. Different programs may use a slightly different syntax.

Mathematical functions

Spreadsheets offer a wide range of functions, including trigonometrical functions, angle functions, logarithms and random-number functions. Functions are invaluable for transforming sets of data rapidly and can be used in formulae required for more complex analyses. Spreadsheets work with an order of preference of the operators in much the same way as a standard calculator and this must always be taken into account when operators are used in formulae. They also require a very precise syntax – the program should warn you if you break this.

Empty cells – note that these may be given the value 0 by the spreadsheet for certain functions. This may cause errors e.g. by rendering a minimum value inappropriate. Also, an 'error return' may result for certain functions if the cell content is zero.

Statistical calculations – make sure you understand whether any functions you employ are for populations or samples (see p. 259).

Using text functions – these allow you to manipulate text within your spreadsheet and include functions such as 'search and replace' and alphabetical or numerical 'sort'.

	Cell	Formula	
Original → cell	A1	=B1+C1	← Original formula
Copied cells ↓	A2	=B2+C2	Copied formulae (relative)
	A3	=B3+C3	
	A4	=B4+C4	

(a)

	Cell	Formula	
Original → cell	A1	=B1/$C$1	← Original formula
Copied cells ↓	A2	=B2/$C$1	Copied formulae (mixed relative and absolute)
	A3	=B3/$C$1	
	A4	=B4/$C$1	

(b)

Fig. 11.3 Illustration of relative (a) and absolute (b) copying. In Excel, the $ sign before and after the column letter makes the cell reference absolute, as shown in (b).

Statistical functions

Modern spreadsheets incorporate many sophisticated statistical functions and, if these are not appropriate, the spreadsheet can be used to facilitate the calculations required for most of the statistical tests found in textbooks. The descriptive statistics normally available include:

- the sum of all data present in a column, row or block;
- the minimum and maximum of a defined range of cells;
- counts of cells – a useful operation if you have an unknown or variable number of data values;
- averages and other statistics describing location;
- standard deviations and other statistics describing dispersion.

A useful function where you have large numbers of data allows you to create frequency distributions using predefined class intervals.

The hypothesis-testing statistical functions are usually reasonably powerful (e.g. *t*-test, ANOVA, regressions) and they often return the *probability* (*P*) of obtaining the test statistic when the null hypothesis (p. 259) is true (where $0 < P < 1$), so there may be no need to refer to statistical tables. Again, check on the effects of including empty cells.

Database functions

Many spreadsheets can be used as simple databases and offer a range of functions to support this, including filtering and sorting options. The rows and columns of the spreadsheet are used as the fields and records of the database (see Chapter 12). For many purposes, this form of database is perfectly adequate and should be seriously considered before using a full-feature database product.

Copying

All programs provide a means of copying (replicating) formulae or cell contents when required, and this is a very useful feature. This is usually accomplished by 'dragging' a cell's contents to a new range using the mouse. When copying, references to cells may be either relative, changing with the row/column as they are copied, or absolute, remaining a fixed cell reference and not changing as the formulae are copied (Fig. 11.3).

KEY POINT The distinction between relative and absolute cell references is very important and must be understood; it provides one of the most common forms of error when copying formulae.

In Excel, copying is normally *relative* and if you wish a cell reference to be *absolute* when copied, this is done by putting a dollar ($) sign before and after the column reference letter, e.g. $C$56.

Naming blocks

When a group of cells (a block) is carrying out a particular function, it is often easier to give the block a name that can then be used in all formulae referring to that block. This powerful feature also allows the spreadsheet to be more readable.

Templates – these should contain:

- a data input section
- data transformation and/or calculation sections
- a results section, which can include graphics
- text in the form of headings and annotations
- a summary section.

Constructing a spreadsheet – start with a simple model and extend it gradually, checking for correct operation as you go.

Spreadsheet templates

A template is a preconstructed spreadsheet containing the formulae required for repeated data analysis. Data is added when it become available, and results are available as soon as the last item is entered. To create a template, the sequence of operations is:

1. **Determine what information/statistics you want to produce.**
2. **Identify the variables you will need to use**, both for original data that will be entered and for any intermediate calculations that might be required.
3. **Set up separate areas of the spreadsheet** for data entry, calculation of intermediate values (statistical values such as sums of squares, etc.), calculation of final parameters/statistics and, if necessary, a summary area.
4. **Establish the format of the numeric data** if this is different from the default values. This can be done globally (affecting the entire spreadsheet) or locally (affecting only a specified part of the spreadsheet).
5. **Establish the column widths required** for the various activities.
6. **Add text (labels) to identify input**, intermediate formulae and output cells. This is valuable in error-tracking and when carrying out further development work. Text can be entered in designated cells, or cells can be annotated using the 'comments' feature (*Insert* > *Comment*).
7. **Enter a test set of values** to use during formula entry: use a fully worked example to check that formulae are working correctly.
8. **Enter the formulae required** to make all the calculations, both intermediate and final. Check that results are correct using the test data.

The spreadsheet is then ready for use. Delete all of the test-data values and you have created your template. Save the template to a disk and it is then available for repeated operations.

Graphics display

Most spreadsheets now offer a wide range of graphics facilities that are easy to use, and this represents an ideal way to examine your data sets rapidly and comprehensively (Chapter 35). The quality of the final graphics output (to a printer) is variable but is usually perfectly sufficient for data exploration and analysis. Many of the options are business graphics styles but there are usually histogram, bar charts, X – Y plotting, line and area graphics options available. Note that some spreadsheet graphics may not come up to the standards expected for the formal presentation of scientific data, unless you manipulate the initial output appropriately (see Box 35.2, p. 223).

Printing spreadsheets

This is usually a straightforward menu-controlled procedure, made difficult only by the fact that your spreadsheet may be too big to fit on one piece of paper. Try to develop an area of the sheet that contains only the data that you will be printing, e.g. perhaps a summary area. Remember that columns can usually be hidden for printing purposes and you can control whether the printout is in portrait or landscape mode, and for continuous paper or single sheets (depending on printer capabilities). Use a screen-preview option, if available, to check your layout before printing. A 'print to fit' option is also available in some programs, making the output fit the page dimensions.

Sources for further study

Hart-Davies, G. (2007) *How to Do Everything with Microsoft Office Excel 2007*. McGraw-Hill Education, Berkeley, California.

Harvey, G. (2007) *Excel 2007 for Dummies*. Wiley, New York.

(And similar texts for other release versions.)

12 Word processors, databases and other packages

Word processors

Word processing is a transferable skill valuable beyond the immediate requirements of your biology course. The word processor has facilitated writing because of the ease of revising text. Using a word processor should improve your writing skills and speed because you can create, check and change your text on the screen before printing it as 'hard copy' on paper. Once entered and saved, multiple uses can be made of a piece of text with little effort.

When using a word processor you can:

- refine material many times before submission;
- insert material easily, allowing writing to take place in any sequence;
- use a spellchecker to check your text;
- use a thesaurus when composing your text;
- carry out ongoing checks of the word count;
- produce high-quality final copies;
- reuse part or all of the text in other documents.

The potential disadvantages of using a word processor include:

- lack of ready access to a computer, software and/or printer;
- time taken to learn the operational details of the program;
- the temptation to make 'trivial' revisions;
- loss of files due to computer breakdown, or disk loss or failure.

The computerised office – many word processors are sold as part of an integrated suite, e.g. Corel WordPerfect Office and Microsoft Office, with the advantage that they share a common interface in the different components (word processor, spreadsheet, database, etc.) and allow ready exchange of information (e.g. text, graphics) between component programs.

Word processors come as 'packages' comprising the program and a manual, often with a tutorial program. Examples are WordPerfect and Microsoft Word. Most word processors have similar general features but differ in operational detail; it is best to pick one and stick to it as far as possible so that you become familiar with it. Learning to use the package is like learning to drive a car – you need only to know how to drive the computer and its program, not to understand how the engine (program) and transmission (data transfer) work, although a little background knowledge is often helpful and will allow you to get the most from the program.

Using textbooks, manuals and tutorials – most programs no longer come with paper-based manuals, and support information is usually provided in one or more of the following ways: as a help facility within the program; as a help file on the program CD; or as an online help support site. It is still often worthwhile investing in one of the commercial textbooks that support specific programs.

In most word processors, the appearance of the screen realistically represents what the printout on paper will look like (WYSIWYG – what you see is what you get). Because of variation in operational details, only general and strategic information is provided in this chapter: you must learn the details of your word processor through use of the appropriate manual and 'help' facilities.

Before starting you will need:

- the program (usually installed on a hard disk or available via a network);
- a medium for storage, retrieval and back-up of your own files when created;

- a draft page-layout design: in particular you should have decided on page size, page margins, typeface (font) and size, type of text justification and format of page numbering;
- an outline of the text content;
- access to a suitable printer: this need not be attached to the computer you are using since your file can be taken or sent to an office where a printer is available, providing that it has the same word-processing program.

Using a word processor – take full advantage of the differences between word processing and 'normal' writing (which necessarily follows a linear sequence and requires more planning):

- Simply jot down your initial ideas for a plan, preferably at paragraph topic level. The order can be altered easily and if a paragraph grows too much it can easily be split.
- Start writing wherever you wish and fill in the rest later.
- Just put down your ideas as you think, confident in the knowledge that it is the concepts that are important to note; their order and the way you express them can be adjusted later.
- Don't worry about spelling and use of synonyms – these can (and should) be checked during a separate revision run through your text, using the spell-checker first to correct obvious mistakes, then the thesaurus to change words for style or to find the *mot juste.*
- Don't forget that a draft printout may be required to check (a) for pace and spacing – difficult to correct on-screen; and (b) to ensure that words checked for spelling fit the required sense.

Laying out (formatting) your document

Although you can format your text at any time, it is good practice to enter the basic commands at the start of your document: entering them later can lead to considerable problems due to reorganisation of the text layout. If you use a particular set of layout criteria regularly, e.g. an A4 page with space for a letterhead, make a template containing the appropriate codes that can be called up whenever you start a new document. Note that various printers may respond differently to particular codes, resulting in a different spacing and layout.

Typing the text

If new to word processing, think of the screen as a piece of typing paper. The cursor marks the position where your text/data will be entered and can be moved around the screen by use of the cursor-control keys. When you type, don't worry about running out of space on the line because the text will wrap around to the next line automatically. Do not use a carriage return (usually the [ENTER] or [↵] key) unless you wish to force a new line, e.g. when a new paragraph is wanted. If you make a mistake when typing, correction is easy. You can usually delete characters or words or lines and the space is closed automatically. You can also insert new text in the middle of a line or word. You can insert special codes to carry out a variety of tasks, including changing text appearance such as underlining, **emboldening** and *italics*. Paragraph indentations can be automated using [TAB] or [⇆] as on a typewriter, but you can also indent or bullet whole blocks of text using special menu options. The function keys are usually pre-programmed to assist in many of these operations.

Deleting and restoring text – because deletion can sometimes be made in error, there is usually an 'undelete' or 'restore' feature that allows the last deletion to be recovered.

Editing features

Word processors usually have an array of features designed to make editing documents easy. In addition to the simple editing procedures described above, the program usually offers facilities to allow blocks of text to be moved ('cut and paste'), copied or deleted.

An extremely valuable editing facility is the find or search procedure: this can rapidly scan through a document looking for a specified word, phrase or punctuation. This is particularly valuable when combined with a replace facility so that, for example, you could replace the word 'test' with 'trial' throughout your document simply and rapidly.

Most WYSIWYG word processors have a command that reveals the normally hidden codes controlling the layout and appearance of the printed text. When editing, this can be a very important feature, since some changes to your text will cause difficulties if these hidden codes are not taken into account; in particular, make sure that the cursor is at the correct point before making changes to text containing hidden code, otherwise your text will sometimes change in apparently mystifying ways.

Fonts and line spacing

Most word processors offer a variety of fonts depending upon the printer being used. Fonts come in a wide variety of types and sizes, but they are defined in particular ways as follows:

Presenting your documents – it is good practice not to mix typefaces too much in a formal document; also the font size should not differ greatly for different headings, sub-headings and the text.

- **Typeface:** the term for a family of characters of a particular design, each of which is given a particular name. The most commonly used for normal text is Times Roman (as used here for the main text) but many others are widely available, particularly for the better-quality printers. They fall into three broad groups: serif fonts with curves and flourishes at the ends of the characters (e.g. Times Roman); sans serif fonts without such flourishes, providing a clean, modern appearance (e.g. Arial); and decorative fonts used for special purposes only, such as the production of newsletters and notices.
- **Size:** measured in points. A point is the smallest typographical unit of measurement, there being 72 points to the inch (about 28 points per cm). The standard sizes for text are 10, 11 and 12 point, but typefaces are often available up to 72 point or more.
- **Appearance:** many typefaces are available in a variety of styles and weights. Many of these are not designed for use in scientific literature but for desktop publishing.
- **Spacing:** can be either fixed, where every character is the same width, or proportional, where the width of every character, including spaces, is varied. Typewriter fonts such as Elite and Prestige use fixed spacing and are useful for filling in forms or tables, but proportional fonts make the overall appearance of text more pleasing and readable.
- **Pitch:** specifies the number of characters per horizontal inch of text. Typewriter fonts are usually 10 or 12 pitch, but proportional fonts are never given a pitch value since it is inherently variable.
- **Justification:** this is the term describing the way in which text is aligned vertically. Left justification is normal, but for formal documents, both left and right justification may be used (as here).

Preparing draft documents – use double spacing to allow room for your editing comments on the printed page.

You should also consider the vertical spacing of lines in your document. Drafts and manuscripts are frequently double-spaced. If your document has unusual font sizes, this may well affect line spacing, although most word processors will cope with this automatically.

Preparing final documents – for most work, use a 12-point proportional serif typeface, with spacing dependent upon the specifications for the work.

Table construction

Tables can be produced by a variety of methods:

- **Using the tab key** [⇆] as on a typewriter: this moves the cursor to predetermined positions on the page, equivalent to the start of each tabular column. You can define the positions of these tabs as required at the start of each table.
- **Using special table-constructing procedures** (see Box 36.2). Here the table construction is largely done for you and it is much easier than

using tabs, providing you enter the correct information when you set up the table.

- **Using a spreadsheet to construct the table** and then copying it to the word processor (see Box 36.2). This procedure requires considerably more manipulation than using the word processor directly and is best reserved for special circumstances, such as the presentation of a very large or complex table of data, especially if the data is already stored as a spreadsheet.

Graphics and special characters

Many word processors can incorporate graphics from other programs into the text of a document. Files must be compatible (see your manual) but if this is so, it is a relatively straightforward procedure. For professional documents this is a valuable facility, but for most undergraduate work it is probably better to produce and use graphics as a separate operation, e.g. using a spreadsheet (see Box 35.2).

You can draw lines and other graphical features directly within most word processors, and special characters (e.g. Greek characters and maths symbols) may be available dependent upon your printer's capabilities.

Inserting special characters – Greek letters and other characters are available using the 'Insert' and 'Symbol' features in Word.

Tools

Many word processors also offer you special tools, the most important of which are:

- **Macros:** special sets of files you can create when you have a frequently repeated set of keystrokes to make. You can record these keystrokes as a 'macro' so that it can provide a short cut for repeated operations.
- **Thesaurus** (*synonyms* function): used to look up alternative words of similar or opposite meaning while composing text at the keyboard.
- **Spellcheck:** a very useful facility that will check your spellings against a dictionary provided by the program. This dictionary is often expandable to include specialist words that you use in your work. The danger lies in becoming too dependent upon this facility, as they all have limitations: in particular, they will not pick up incorrect words that happen to be correct in a different context (i.e. 'was' typed as 'saw' or 'see' rather than 'sea'). Be aware of American spellings in programs from the USA, e.g. 'color' instead of 'colour'. The rule, therefore, is to use the spellcheck first and then carefully read the text for errors that have slipped through.
- **Word count:** useful when you are writing to a prescribed limit.

Using a spellcheck facility – do not rely on this to spot all errors. Remember that spellcheck programs do not correct grammatical errors.

Printing from your program

If more than one printer is attached to your PC or network, you will need to specify which one to use from the word processor's print menu. Most printers offer choices as to text and graphics quality, so choose draft (low) quality for all but your final copy since this will save both time and materials.

Use a print preview option to show the page layout if it is available. Assuming that you have entered appropriate layout and font commands, printing is a straightforward operation carried out by the word processor at your command. Problems usually arise because of some in-

Using the print preview mode – this can reveal errors of several kinds, e.g. spacing between pages, that can prevent you wasting paper and printer ink unnecessarily.

compatibility between the criteria you have entered and the printer's own capabilities. Make sure that you know what your printer offers before starting to type: although settings are modifiable at any time, changing the page size, margin size, font size, etc., all cause your text to be rearranged, and this can be frustrating if you have spent hours carefully laying out the pages.

KEY POINT It is vital to save your work frequently to a memory stick, hard drive or network drive. This should be done every 10 minutes or so. If you do not save regularly, you may lose hours or days of work. Many programs can be set to 'autosave' every few minutes.

Databases

A database is an electronic filing system whose structure is similar to a manual record-card collection. Its collection of records is termed a file. The individual items of information on each record are termed fields. Once the database is constructed, search criteria can be used to view files through various filters according to your requirements. The computerised catalogues in your library are just such a system; you enter the filter requirements in the form of author or subject keywords.

You can use a database to catalogue, search, sort and relate collections of information. The benefits of a computerised database over a manual card-file system are:

- The information content is easily amended/updated.
- Printout of relevant items can be obtained.
- It is quick and easy to organise through sorting and searching/selection criteria, to produce subgroups of relevant records.
- Record displays can easily be redesigned, allowing flexible methods of presenting records according to interest.
- Relational databases can be combined, giving the whole system immense flexibility. The older 'flat-file' databases store information in files that can be searched and sorted, but cannot be linked to other databases.

Choosing between a database and a spreadsheet – use a database only after careful consideration. Can the task be done better within a spreadsheet? A database program can be complex to set up and usually needs to be updated regularly.

Relatively simple database files can be constructed within spreadsheets using the columns and rows as fields and records respectively. These are capable of reasonably advanced sorting and searching operations and are probably sufficient for the types of databases you are likely to require as an undergraduate. You may also make use of a bibliographic database specially constructed for that purpose.

Statistical analysis packages

Using spreadsheet statistics functions – before using a specific statistics package, check whether your spreadsheet is capable of carrying out the form of analysis you require (see Boxes 38.3 and 39.3), as this can often be the simpler option.

Statistical packages vary from small programs designed to carry out very specific statistical tasks to large sophisticated packages (SYSTAT, SigmaStat, Minitab, etc.) intended to provide statistical assistance, from experimental design to the analysis of results. Consider the following features when selecting a package:

- The data entry and editing section should be user-friendly, with options for transforming data.

- Data exploration options should include descriptive statistics and exploratory data analysis techniques.
- Hypothesis-testing techniques should include ANOVA, regression analysis, multivariate techniques and parametric and non-parametric statistics.
- The program should provide assistance with experimental design and sampling methods.
- Output facilities should be suitable for graphical and tabular formats.

Some programs have very complex data entry systems, limiting the ease of using data in different tests. The data entry and storage system should be based on a spreadsheet system, so that subsequent editing and transformation operations are straightforward.

Presentation using computer packages – although many computer programs enhance presentational aspects of your work, there are occasions when they can make your presentation worse. Take care to avoid the following common pitfalls:

- Default or 'chart wizard' settings for graphs may result in output that is non-standard for the sciences (see Box 35.2).
- Fonts in labels and legends may not be consistent with other parts of your presentation.
- Some programs cannot produce Greek symbols (e.g. μ); do not use 'u' as a substitute. The same applies to scientific notation and superscripts: do not use 14C for ^{14}C, and replace, e.g., 1.4E+09 with 1.4×10^9. First try cutting and pasting symbols from Word or, if this fails, draw correct symbols by hand.

KEY POINT Make sure that you understand the statistical basis for your test and the computational techniques involved before using a particular program.

Graphics/presentation packages

Microsoft Office programs can be used to achieve most coursework tasks, e.g. PowerPoint is useful for creating posters (Box 13.1) and for oral presentations (Box 14.1). Should you need more advanced features, additional software may be available on your network: for example:

- SigmaPlot can produce graphs with floating axes;
- Macromedia Freehand is useful for designing complex graphics;
- DreamWeaver enables you to produce high-quality Web pages;
- MindGenius can be used to produce Mind Maps.

Important points regarding the use of such packages are:

- the learning time required for some of the more complex operations can be considerable;
- the quality of your printer will limit the quality of your output;
- not all files will readily import into a word processor such as Microsoft Word – you may need to save your work in a particular format. The different types of file are distinguished by the three-character filename extension, e.g. .jpg. and .bmp.

KEY POINT Computer graphics are not always satisfactory for scientific presentation. You should not accept the default versions produced – make appropriate changes to suit scientific standards and style. Box 35.1 gives a checklist for graph drawing and Box 35.2 provides guidelines for adapting Microsoft Excel output.

Image storage and manipulation

With the widespread use of digital images, programs that facilitate the storage and manipulation of electronic image files have become increasingly important. These programs create a library of your stored images and provide a variety of methods for organising and selecting images. The industry-standard program, Adobe Photoshop, is one of many programs for image manipulation that vary widely in capability, cost and associated learning time. Many are highly sophisticated programs intended for graphic artists. For most scientific purposes, however, relatively limited functions are required.

Sources for further study

Gookin, D. (2007) *Word 2007 for Dummies*. Wiley, New York.

Ulrich-Fuller, L., Cook, K. and Kaufield, J. (2007) *Access 2007 for Dummies*. Wiley, New York.

Wang. W. (2007) *Office 2007 for Dummies*. Wiley, New York.

(And similar texts for other packages and release versions.)

Communicating information

13 Organising a poster display

Learn from others – look at the various types of posters around your university and elsewhere; the best examples will be visual, not textual, with a clear structure that helps get the key messages across.

A scientific poster is a visual display of the results of an investigation, usually mounted on a rectangular board. Posters are used in undergraduate courses, to display project results or assignment work, and at scientific meetings to communicate research findings.

In a written report you can include a reasonable amount of specific detail and the reader can go back and reread difficult passages. However, if a poster is long-winded or contains too much detail, your reader is likely to lose interest.

> KEY POINT A poster session is like a competition – you are competing for the attention of people in a room. Because you need to attract and hold the attention of your audience, make your poster as interesting as possible. Think of it as an advertisement for your work and you will not go far wrong.

(a)

Title

1 2 3 4

5 6 7 8

(b)

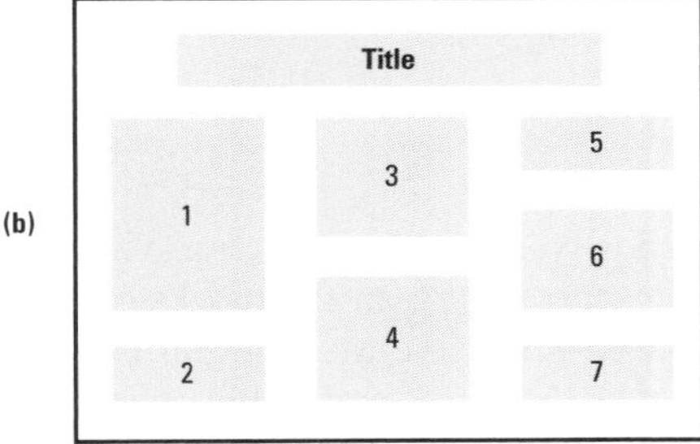

(c)

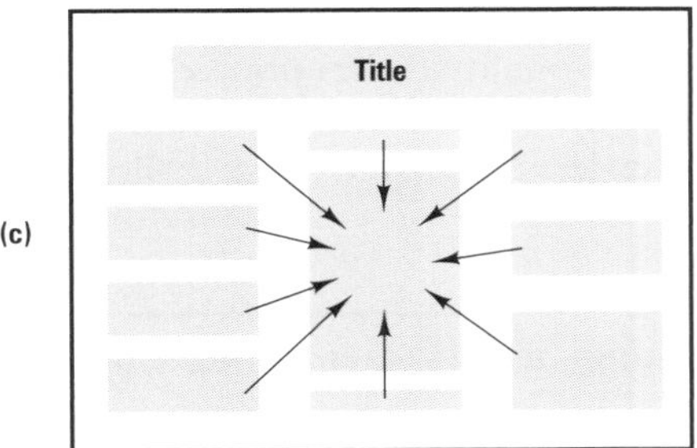

Fig. 13.1 Poster design. (a) An uninspiring design: sub-units of equal area, reading left to right, are not recommended. (b) This design is more interesting and the text will be easier to read (column format). (c) An alternative approach, with a central focus and arrows to guide the reader.

Preliminaries

Before considering the content of your poster, you should find out:

- the linear dimensions of your poster area, typically up to 1.5 m wide by 1 m high;
- the composition of the poster board and the method of attachment, whether drawing pins, Velcro tape or some other form of adhesive; and whether these will be provided – in any case, it is safer to bring your own;
- the time(s) when the poster should be set up and when you should attend;
- the room where the poster session will be held.

Design

Plan your poster with your audience in mind, as this will dictate the appropriate level for your presentation. Aim to make your poster accessible to a broad audience. Since a poster is a *visual* display, you must pay particular attention to the presentation of information: work that may have taken hours to prepare can be ruined in a few minutes by the ill-considered arrangement of items (Fig. 13.1). Begin by making a draft sketch of the major elements of your poster. It is worth discussing your intended design with someone else, as constructive advice at the draft stage will save a lot of time and effort when you prepare the final version (or consult Simmonds and Reynolds, 1994).

Layout

One approach is to divide the poster into several smaller areas, perhaps six or eight in all, and prepare each as a separate item on a piece of card. Alternatively, you can produce a single large poster on one sheet of paper or card and store it inside a protective cardboard tube. However, a single large poster may bend and crease, making it difficult to flatten out.

In addition, photographs and text attached to the backing sheet may work loose; a large poster with embedded images is an alternative approach.

Subdividing your poster means that each smaller area can be prepared on a separate piece of paper or card, of A4 size or slightly larger, making transport and storage easier. It also breaks the reading matter up into smaller pieces, looking less formidable to a potential reader. By using pieces of card of different colours you can provide emphasis for key aspects, or link text with figures or photographs.

You will need to guide your reader through the poster and headings/sub-headings will help with this aspect. It may be appropriate to use either a numbering system, with large, clear numbers at the top of each piece of card, or a system of arrows (or thin tapes), to link sections within the poster (see Fig. 13.1). Make sure that the relationship is clear and that the arrows or tapes do not cross.

Presenting a poster at a formal conference – it can be useful to include your photograph for identification purposes, e.g. in the top right-hand corner of the poster.

Title

Your chosen title should be concise (no more than eight words), specific and interesting, to encourage people to read the poster. Make the title large and bold – it should run across the top of your poster, in letters at least 4 cm high, so that it can be read from the other side of the room. Coloured spirit-based marker and block capitals drawn with a ruler work well, as long as your writing is readable and neat (the colour can be used to add emphasis). Alternatively, you can print out each word in large type, using a word processor. Details of authors, together with their addresses (if appropriate), should be given, usually across the top of the poster in somewhat smaller lettering than the title.

Making up your poster – text and graphics printed on good-quality paper can be glued directly onto a contrasting mounting card: use photographic spray mountant or Pritt rather than liquid glue. Trim carefully using a guillotine to give equal margins, parallel with the paper. Photographs should be placed in a window mount to avoid the tendency for their corners to curl. Another approach is to trim pages or photographs to their correct size, then encapsulate in plastic film: this gives a highly professional finish and is easy to transport.

Text

Write in short sentences and avoid verbosity. Keep your poster as visual as possible and make effective use of the spaces between the blocks of text. Your final text should be double-spaced and have a minimum capital-letter height of 8 mm (minimum type size 36 point), preferably greater, so that the poster can be read at a distance of 1 m. One method of obtaining text of the required size is to photo-enlarge standard typescript (using a good-quality photocopier), or use a high-quality (laser) printer. It is best to avoid continuous use of text in capitals, since it slows reading and makes the text less interesting to the reader. Also avoid italic, 'balloon' or decorative styles of lettering.

KEY POINT Keep text to a minimum – aim to have a *maximum* of 500 words in your poster.

Subtitles and headings

These should have a capital-letter height of 12–20 mm, and should be restricted to two or three words. They can be produced by word processor, photo-enlargement, stencilling or by hand, using pencilled guidelines (but make sure that no pencil marks are visible on your finished poster).

Producing composite material for posters – PowerPoint is generally more useful than Word when you wish to include text, graphics and/or images on the same page. It is possible to use PowerPoint to produce a complete poster (Box 13.1), although it can be expensive to have this printed out commercially to A1 or A0 size.

Colour

Consider the overall visual effect of your chosen display, including the relationship between your text, diagrams and the backing board. Colour can be used to highlight key aspects of your poster. However, it is very easy to ruin a poster by the inappropriate choice and application of colour. Careful use of two, or at most three, complementary colours and shades will be easier on the eye and should aid comprehension. Colour can be used to link the text with the visual images (e.g. by picking out a colour in a photograph and using the same colour on the mounting board for the accompanying text). For PowerPoint posters, careful choice of colours for the various elements will enhance the final product (Box 13.1). Use coloured inks or water-based paints to provide colour in diagrams and figures, as felt pens rarely give satisfactory results.

Presenting at a scientific meeting – never be tempted to spend the minimum amount of time converting a piece of scientific writing into poster format – the least interesting posters are those where the author simply displays pages from a written communication (e.g. a journal article) on a poster board.

Content

The typical format is that of a scientific report (see Box 17.1), i.e. with the same headings, but with a considerably reduced content. Keep references within the text to a minimum – interested parties can always ask you for further information. Also note that most posters have a summary/conclusions section at the end, rather than an abstract.

Introduction

This should give the reader background information on the broad field of study and the aims of your own work. It is vital that this section is as interesting as possible, to capture the interest of your audience. It is often worth listing your objectives as a series of numbered points.

Designing the materials and methods section – photographs or diagrams of apparatus can help to break-up the text of this section and provide visual interest. It is sometimes worth preparing this section in a smaller typeface.

Materials and methods

Keep this short, and describe only the principal techniques used. You might mention any special techniques or problems of general interest.

Results

Don't present your raw data: use data reduction wherever possible, i.e. figures and simple statistical comparisons. Graphs, diagrams, histograms and pie charts give clear visual images of trends and relationships and should be used in place of data tables (see pp. 221–222). Final copies of all figures should be produced so that the numbers can be read from a distance of 1 m. Each should have a concise title and legend, so that it is self-contained: if appropriate, a series of numbered points can be used to link a diagram with the accompanying text. Where symbols are used, provide a key on each graph (symbol size should be at least 5 mm). Avoid using graphs straight from a written version, e.g. a project report, textbook or a paper, without considering whether they need modification to meet your requirements.

Keeping graphs and diagrams simple – avoid composite graphs with different scales for the same axis, or with several trend lines (use a maximum of three trend lines per graph).

Conclusions

This is where many readers will begin, and they may go no further unless you make this section sufficiently interesting. This part needs to be the strongest part of your poster, summarising the main points. Refer to your figures here to draw the reader into the main part of your poster. A slightly

Listing your conclusions – a series of numbered points is a useful approach, if your findings fit this pattern.

Box 13.1 How to create a poster using PowerPoint 2007

Software such as PowerPoint can be used to produce a high-quality poster, providing you have access to a good colour printer. However, you should avoid the standard templates available on the Web as they encourage unnecessary uniformity and stifle creativity, leading to a less satisfying end result. The following steps give practical advice on creating a poster as a single PowerPoint slide:

1. **Sketch out your plans.** Decide on the main poster elements (images, graphs, tables and text sections) and their relationship with each other and draw out a one-page 'storyboard' (see Fig. 13.1). Think about colours for background, text and graphics (use two or three complementary colours): dark text on a light background is clearer (high contrast) and uses less ink when printing. Also consider how you will link the elements in sequence, to guide readers through your 'story'.
2. **Get your material ready.** Collect together individual files for pictures, figures and tables. Make any required adjustments to images, graphs or tables before you import them into your poster.
3. **Create a new/blank slide.** Open PowerPoint to open a new/blank slide. Select the *Design* tab then select either *Landscape* or *Portrait* from the *Slide Orientation* tab. Left-click the *Page Setup* button and set your chosen page size (input your *Width* and *Height* or choose a set size from the drop-down menu). Right-click on the slide and set *Ruler* and *Grid and Guides* (to help position elements within the slide – the horizontal and vertical guidelines can be dragged to different positions later, as required). Under *Grid and Guides* ensure that *Display grid on screen box* is ticked. Left-click on the *Background Styles* button, then left-click on *Format Background,* left-click on the *Color* button and choose a background colour. If necessary choose *More Colors*. In general, avoid setting a picture as your background as this tends to detract from the content of the poster. Before going further, save your work. Repeat this frequently and in more than one location (e.g. hard drive and USB memory stick).
4. **Add graphics.** For images, use the *Insert* tab, select *Picture* and browse to *Insert* the correct file. The *Insert, Object* command performs a similar function for Excel charts (graphs). Alternatively, use the copy-and-paste functions of complementary software. Once inserted, resize using the *sizing handles* in one of the corners (for photographs, take care not to alter one dimension relative to the other, or the image will be distorted). To reposition, put the mouse pointer over the image, left-click and hold, then drag to new location. While the *Drawing* group on the *Home* tab offers standard shapes and other useful features, you should avoid clipart (jaded and over-used) and poor-quality images from the Web (always use the highest resolution possible) – if you do not have your final images, use blank text boxes to show their position within the draft poster.
5. **Add text.** Under the *Insert* tab in the *Text* group select *Text Box.* Then hold down the left mouse button while dragging the pointer over the area of the slide where you want the text box, then either type in your text (use the *Enter* key to provide line spacing within the box) or copy-and-paste text from a word-processed file. You will need to consider the type size for the printed poster (e.g. for an A0 poster (size 1189 × 841 mm), a printed type size of 24 point is appropriate for the main text, with larger sizes for headings and titles. If you find things difficult to read on-screen, use the *Zoom* function (either select a larger percentage in the *Zoom box*, found under the *View* tab, or hold down the [Ctrl] key and use the mouse wheel to scroll up (*zoom*) or down (*reduce*). Use a separate text box for each element of your poster and don't be tempted to type too much text into each box – write in succinct phrases, using bullet points and numbered lists to keep text concise (aim for no more than 50 words per text box). Select appropriate font styles and colours using the *Font* group under the *Home* tab. For a background colour or surrounding line, under the *Format* tab, use *Shape Fill/Shape Outline*. Line thickness and colour can then be altered using the *Weight* settings under *Shape Outline* in the *Home* group or *Drawing* group. Present supplementary text elements in a smaller type: for example, details of methodology, references cited.
6. **Add boxes, lines and/or arrows** to link elements of the poster and guide the reader (see Fig. 13.1(c)). These features are available from the *Drawing* group. Note that new inserts are overlaid over older inserts – if this proves to be a problem, select the relevant item and use the *Bring to front* or *Send to back* functions to change its relative position.
7. **Review your poster.** Get feedback from another student or your tutor, e.g. on a small printed version, or use a projector to view your poster without printing (adjust the distance between projector and screen to give the correct size).
8. **Revise and edit your poster.** Revisit your work and remove as much unnecessary text as possible. Delete any component that is not essential to the message of the poster. Keep graphs simple and clear (p. 219 gives further advice). 'White space' is important in providing structure.
9. **Print the final version.** Use a high-resolution colour printer (this may be costly, so you should wait until you are sure that no further changes are needed).

larger or bolder typeface may add emphasis, although too many different typefaces can look messy. For references, smaller type can be used.

The poster session

Consider providing a handout – this is a useful way to summarise the main points of your poster, so that your readers have a permanent record of the information you have presented.

Coping with questions in assessed poster sessions – you should expect to be asked questions about your poster, and explain details of figures, methods, etc.

A poster session may be organised as part of the assessment of your coursework, and this usually mirrors those held at most scientific conferences and meetings. Staff and fellow students (delegates at conferences) will mill around, looking at the posters and chatting to their authors, who are usually expected to be in attendance. If you stand at the side of your poster throughout, you are likely to discourage some readers, who may not wish to become involved in a detailed conversation about the poster. Stand nearby. Find something to do – talk to someone else or browse among the other posters, but remain aware of people reading your poster and be ready to answer any queries they may raise. Do not be too discouraged if you aren't asked lots of questions: remember, the poster is meant to be a self-contained, visual story, without need for further explanation.

A poster display will never feel like an oral presentation, where the nervousness beforehand is replaced by a combination of satisfaction and relief as you unwind after the event. However, it can be a very satisfying means of communication, particularly if you follow these guidelines.

Text reference

Simmonds, D. and Reynolds, L. (1994) *Data Presentation and Visual Literacy in Medicine and Science*. Butterworth-Heineman, London.

Sources for further study

Alley, M. (2003) *The Craft of Scientific Presentations: Critical Steps to Succeed and Critical Errors to Avoid.* Springer-Verlag, New York.

Briscoe, M.H. (2000) *Preparing Scientific Illustrations: A Guide to Better Posters, Presentations and Publications,* 2nd edn. Springer-Verlag, New York.

Davis, M.F. (2005) *Scientific Papers and Presentations,* 2nd edn. Academic Press, New York.

Gosling, P.J. (1999) *Scientist's Guide to Poster Presentations*. Kluwer, New York.

Hess, G. and Liegel, L. *Creating Effective Poster Presentations.* Available: http://www.ncsu.edu/project/posters/
Last accessed: 21/12/10.

14 Giving a spoken presentation

Opportunities for practising speaking skills – these include:

- answering lecturers' questions;
- contributing in tutorials;
- talking to informal groups;
- giving your views at formal (committee) meetings;
- demonstrating or explaining to other students, e.g. during a practical class;
- asking questions in lectures/seminars;
- answering an examiner's questions in an oral exam.

Most students feel very nervous about giving talks. This is natural, since very few people are sufficiently confident and outgoing that they look forward to speaking in public. Additionally, the technical nature of the subject matter may give you cause for concern, especially if you feel that some members of the audience have a greater knowledge than you have. However, this is a fundamental method of scientific communication and an important transferable skill, therefore it forms an important component of many courses.

The comments in this chapter apply equally to informal talks, e.g. those based on assignments and project work, and to more formal conference presentations. It is hoped that the advice and guidance given below will encourage you to make the most of your opportunities for public speaking, but there is no substitute for practice. Do not expect to find all of the answers from this, or any other, book. Rehearse, and learn from your own experience.

KEY POINT The three 'Rs' of successful public speaking are: reflect – give sufficient thought to all aspects of your presentation, particularly at the planning stage; rehearse – to improve your delivery; revise – modify the content and style of your material in response to your own ideas and to the comments of others.

Learning from experience – use your own experience of good and bad lecturers to shape your performance. Some of the more common errors include:

- speaking too quickly;
- reading from notes or from slides and ignoring the audience;
- unexpressive, impersonal or indistinct speech;
- distracting mannerisms;
- poorly structured material with little emphasis on key information;
- factual information too complex and detailed;
- too few or too many visual aids.

Testing the room – if possible, try to rehearse your talk in the room in which it will be presented. This will help you to make allowance for layout of equipment, lighting, acoustics and sight lines that might affect the way you deliver your talk. It will also put you more at ease on the day, because of the familiarity of the surroundings.

Preparation

Preliminary information

Begin by marshalling the details needed to plan your presentation, including:

- the duration of the talk;
- whether time for questions is included;
- the size and location of the room;
- the projection/lighting facilities provided, and whether pointers or similar aids are available.

It is especially important to find out whether the room has the necessary equipment for digital projection (e.g. PC, projector and screen, black-out curtains or blinds, appropriate lighting) or overhead projection before you prepare your audio-visual aids. If you concentrate only on the spoken part of your presentation at this stage, you are inviting trouble later on. Have a look around the room and try out the equipment at the earliest opportunity, so that you are able to use the lights, projector, etc., with confidence. For digital projection systems, check that you can load/present your material. Box 14.1 gives advice on using PowerPoint.

Audio-visual aids

If you plan to use overhead transparencies, find out whether your department has facilities for their preparation, whether these facilities are

Box 14.1 Tips on preparing and using Microsoft PowerPoint 2007 slides in a spoken presentation

Microsoft PowerPoint can be used to produce high-quality visual aids, assuming that a computer and digital projector are available in the room where you intend to speak. The presentation is produced as a series of electronic 'slides' on to which you can insert images, diagrams and text. When creating your slides, bear the following points in mind:

- **Plan the structure of your presentation**. Decide on the main topic areas and sketch out your ideas on paper. Think about what material you will need (e.g. pictures, graphs) and what colours to use for background and text.
- **Choose slide layouts according to purpose**. Once PowerPoint is running, from the *Home* tab menu select *New Slide* from the *Slides* group. You can choose a format using the *Layout* function in the *Slides* group and then add material to each new slide to suit your requirements.
- **Select your background with care**. Many of the preset background templates available within the *Design* menu are best avoided, since they are overused and fussy, diverting attention from the content of the slides. Conversely, flat, dull backgrounds may seem uninteresting, while brightly coloured backgrounds can be garish and distracting. Choose whether to present your text as a light-coloured type on a dark background (more restful but less engaging if the room is dark) or a dark-coloured type on a light background (more lively).
- **Use visual images throughout**. Remember the old advertising slogan 'a picture is worth ten thousand words'. A presentation composed entirely of text-based slides will be uninteresting: adding images and diagrams will brighten up your talk considerably (use the *Insert* menu, *Picture* option). Images can be taken with a digital camera, scanned in from a printed version or copied and pasted from the Web, but you should take care not to break copyright regulations. 'Clipart' is copyright-free, but should be used sparingly, as most people will have seen the images before and they are rarely wholly relevant. Diagrams can be made from components created using the *Shapes* component of the *Insert* tab, and graphs and tables can be imported from other programs, e.g. Excel (Box 13.1 gives further specific practical advice on adding graphics, saving files, etc.).
- **Keep text to a minimum**. Aim for no more than 20 words on a single slide (e.g. four/five lines containing four/five words per line). Use headings and sub-headings to structure your talk: write only keywords or phrases as 'prompts' to remind you to cover a particular point during your talk – never be tempted to type whole sentences as you will then be reduced to reading these from the screen during your presentation, which is boring.
- **Use a large, clear font**. Use the *Slide Master* option within the *View* tab to set the default font to a non-serif style such as Arial, or Comic Sans MS. Default fonts for headings and bullet points are intentionally large, for clarity. Do not reduce these to anything less than 28-point type size (preferably larger), to cram in more words: if you have too much material, create a new slide and divide up the information.
- **Animate your material**. The *Animation* tab provides a *Custom Animation* function that enables you to introduce the various elements within a slide, e.g. text can be made to *Appear* one line at a time, to prevent the audience from reading ahead and to help maintain their attention.
- **Do not overdo the special effects**. PowerPoint has a wide range of features that allows complex slide transitions and animations, additional sounds, etc., but these quickly become irritating to members of an audience unless they have a specific purpose within your presentation.
- **Always edit your slides before use**. Check through your slides and cut out any unnecessary words, adjust the layout and animation. Remember the maxim 'less is more' – avoid too much text; too many bullet points; or too many distracting visual effects or sounds.

When presenting your talk:

- **Work out the basic procedures beforehand**. Practise, to make sure that you know how to move forwards and backwards, turn the screen on and off, hide the mouse pointer, etc.
- **Don't forget to engage your audience**. Despite the technical gadgetry, *you* need to play an active role in their presentation, as explained elsewhere in this chapter.
- **Don't go too fast**. Sometimes, new users tend to deliver their material too quickly: try to speak at a normal pace and practise beforehand.
- **Consider whether to provide a handout**. PowerPoint has several options, including some that provide space, for notes (e.g. Fig. 4.3). However, a handout should not be your default option, as there is a cost involved.

available for your use, and the cost of materials. Adopt the following guidelines:

- **Keep text to a minimum:** present only the key points, with up to 20 words per slide/transparency.
- **Make sure the text is readable:** try out your material beforehand.
- **Use several simpler figures rather than a single complex graph.**
- **Avoid too much colour on overhead transparencies:** blue and black are easier to read than red or green.
- **Don't mix slides and transparencies** as this is often distracting.
- **Use spirit-based pens for transparencies:** use alcohol for corrections.
- **Transparencies can be produced from typewritten or printed text using a photocopier**, often giving a better product than pens. Note that you must use special heat-resistant acetate sheets for photocopying.

Using audio-visual aids – don't let equipment and computer gadgetry distract you from the essential rules of good speaking (pp. 88, 95). Remember that you are the presenter.

Audience

You should consider your audience at the earliest stage, since they will determine the appropriate level for your presentation. If you are talking to fellow students you may be able to assume a common level of background knowledge. In contrast, a research lecture given to your department, or a paper at a meeting of a scientific society, will be presented to an audience from a broader range of backgrounds. An oral presentation is not the place for a complex discussion of specialised information: build up your talk from a low level. The speed at which this can be done will vary according to your audience. As long as you are not boring or patronising, you can cover basic information without losing the attention of the more knowledgeable members in your audience. Box 14.2 gives more details on audience analysis.

Pitching your talk at the right level – the general rule should be: 'do not overestimate the background knowledge of your audience'. This sometimes happens in student presentations, where fears about the presence of 'experts' can encourage the speaker to include too much detail, overloading the audience with facts.

Content

Although the specific details in your talk will be for you to decide, most spoken presentations share some common features of structure, as described below.

Introductory remarks

It is vital to capture the attention of your audience at the outset. Consequently, you must make sure your opening comments are strong, otherwise your audience will lose interest before you reach the main message. Remember it takes a sentence or two for an audience to establish a relationship with a new speaker. Your opening sentence should be some form of preamble and should not contain any key information. For a formal lecture, you might begin with 'Thank you for that introduction. My talk today is about . . .' then restate the title and acknowledge other contributors, etc. You might show a transparency or slide with the title printed on it, or an introductory photograph, if appropriate. This should provide the necessary settling-in period.

After these preliminaries, you should introduce your topic. Begin your story on a strong note – avoid timid or apologetic phrases.

Opening remarks are unlikely to occupy more than 10 per cent of the talk. However, because of their significance, you might reasonably spend up to 25 per cent of your preparation time on them.

Getting the introduction right – a good idea is to have an initial slide giving your details and the title of your talk, and a second slide telling the audience how your presentation will be structured. Make eye contact with all sections of the audience during the introduction.

What to cover in your introductory remarks – you should:

- explain the structure of your talk;
- set out your aims and objectives;
- explain your approach to the topic.

Box 14.2 How to analyse your audience

1. **Prepare your talk with your specific audience in mind.** Speaking to a group of fellow students requires a different approach than does, say, talking to a Weightwatchers™ group. Preparing a seminar on good nutrition for primary school students again would require a different strategy, such as a more active, colourful presentation to keep their interest. A group of fellow science students may understand your scientific terms and acronyms, whereas participants at a community seminar on Nutrition and Health will likely not. Ask yourself 'How much will the audience already know about my topic?' so that you can judge the level required. When required to give a talk before the class or staff for assessment, discuss with your instructor beforehand the audience context preparations required to meet the assessment criteria.
2. **Analyse your prospective audience**. Step into their shoes and imagine that you are a member of that audience. Ask yourself 'What aspects of the topic would I, as part of this audience, like to see covered?' What could you include in your talk that has relevance to this particular audience? Start your talk with material that directly relates to members of your audience's interests and experiences, to capture their attention and develop rapport. For example, in a talk about daily diet ask your audience to think about what they had for breakfast. After you have captured their interest you can lead them to consider points of view that may be different from those they currently hold, such as modifying their diet. Throughout your preparation, look at how you can make new points relate to something that your audience will already be familiar with.
3. **Monitor your audience during your presentation.** Develop the habit of being aware of the mood of your audience and its individuals by watching them. Key body language signals to watch out for include:
   - leaning forward in an open manner usually indicates interest;
   - leaning back with the hand covering part of the face may denote doubt or distrust;
   - crossed arms may mean defensiveness (or a cold room);
   - fidgeting and restlessness may signify boredom;
   - the eyes of a large percentage of the audiences focused on you strongly suggests interest.
4. **Seek direct feedback from your audience during your talk.** For instance, you might ask an audience member whose face is wrinkled up in puzzlement if she or he has a question or comment.
5. **Do not overreact to minor signals from your audience.** A person getting up and leaving the room probably just has another engagement. A person who appears asleep may just have their eyes closed to concentrate better.
6. **Do something if you appear to be losing the attention of your audience.** Raise your voice to emphasise a point, or vary the pace, volume and tone of your voice to attract attention. A good technique to regain attention is to include an example, anecdote or story connected with your topic, as almost everyone can relate to a story.
7. **Ask for feedback.** After your talk ask for feedback from selected members of the audience, either verbally or by providing an evaluation sheet. Then use this evaluation to help you improve your preparation and delivery next time.

KEY POINT Make sure you have practised your opening remarks so that you can deliver the material in a flowing style, with fewer chances of mistakes.

Allowing time for slides – as a rough guide you should allow at least two minutes per illustration, although some diagrams may need longer, depending on content. Make a note of the half-way point to help you check the timing/pace.

The main message

This section should include the bulk of your experimental results or literature findings, depending on the type of presentation. Keep details of methods to the minimum needed to explain your data. This is *not* the place for a detailed description of equipment and experimental protocol (unless it is a talk about methodology). Results should be presented in an easily digested format.

KEY POINT Do not expect your audience to cope with large amounts of data; use a maximum of six numbers per slide. Remember that graphs and diagrams are usually better than tables of raw data, since the audience will be able to see the visual trends and relationships in your data (pp. 221–222).

Present summary statistics (Chapter 38) rather than individual results. Show the final results of any analyses in terms of the statistics calculated, and their significance (p. 260), rather than dwelling on details of the procedures used. Figures should not be crowded with unnecessary detail. Every diagram should have a concise title and the symbols and trend lines should be clearly labelled, with an explanatory key where necessary. When presenting graphical data (Chapter 35) always 'introduce' each graph by stating the units for each axis and describing the relationship for each trend line or data set.

KEY POINT Use summary slides at regular intervals, to maintain the flow of the presentation and to emphasise the main points.

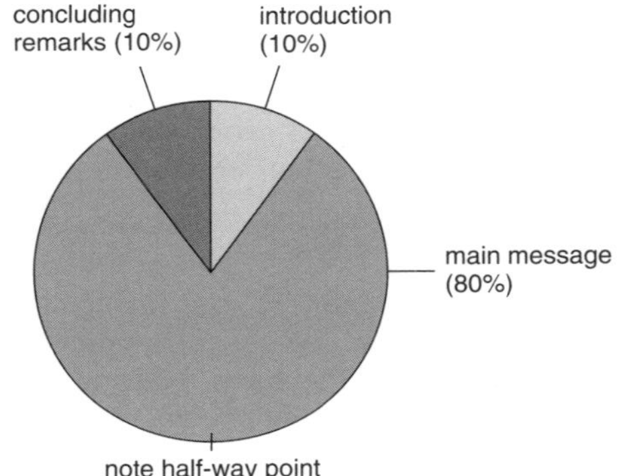

Fig. 14.1 Pie chart showing time allocation for a typical presentation.

Take the audience through your story step by step at a reasonable pace. Try not to rush the delivery of your main message due to nervousness. Avoid complex, convoluted storylines – one of the most distracting things you can do is to fumble backwards through PowerPoint slides or overhead transparencies. If you need to use the same diagram or graph more than once then you should make two (or more) copies. In a presentation of experimental results, you should discuss each point as it is raised, in contrast to written text, where the results and discussion may be in separate sections. The main message typically occupies approximately 80 per cent of the time allocated to an oral presentation (Fig. 14.1).

Concluding remarks

Having captured the interest of your audience in the introduction and given them the details of your story in the middle section, you must now bring your talk to a conclusion. Do not end weakly, e.g. by running out of steam on the last slide. Provide your audience with a clear 'take-home message' by returning to the key points in your presentation. It is often appropriate to prepare a slide or overhead transparency listing your main conclusions as a numbered series.

Signal the end of your talk by saying 'finally . . .', 'in conclusion . . .', or a similar comment and then finish speaking after that sentence. Your audience will lose interest if you extend your closing remarks beyond this point. You may add a simple end phrase (for example, 'thank you') as you put your notes into your folder, but do not say 'that's all folks!', or make any similar offhand remark. Finish as strongly and as clearly as you started. Box 14.3 gives further advice.

Final remarks – make sure you give the audience sufficient time to assimilate your final slide: some of them may wish to write down the key points. Alternatively, you might provide a handout, with a brief outline of the aims of your study and the major conclusions.

Box 14.3 Hints on spoken presentations

In planning the delivery of your talk, bear the following aspects in mind:

- **Using notes**. Many accomplished speakers use abbreviated notes for guidance, rather than reading word for word from a prepared script. When writing your talk:
  - (i) **Consider preparing your first draft as a full script**: write in spoken English and keep the text simple, to avoid a formal, impersonal style. Your aim should be to *talk* to your audience, not to *read* to them.
  - (ii) **If necessary, use notecards with keywords and phrases**: it is best to avoid using a full script in the final presentation. As you rehearse and your confidence improves, a set of note-cards may be an appropriate format. Mark the position of slides/key points, etc.: each note-card should contain details of structure as well as content. Your notes should be written/printed in text large enough to be read easily during the presentation (also check that the lecture room has a lectern light or you may have problems reading your notes if the lights are dimmed). Each note-card or sheet should be clearly numbered, so that you do not lose your place.
  - (iii) **Decide on the layout of your talk**: give each subdivision a heading in your notes, so that your audience is made aware of the structure.
  - (iv) **Memorise your introductory/closing remarks**: you may prefer to rely on a full written version for these sections, in case your memory fails, or if you suffer 'stage fright'.
  - (v) **Using PowerPoint** (Box 14.1): here, you can either use the 'notes' option (*View Tab > Notes Page*), or you may even prefer to dispense with notes entirely, since the slides will help structure your talk, with keywords acting as an *aide-memoire* for your material.
- **Work on your timing**. It is essential that your talk is the right length and the correct pace:
  - (i) **Rehearse your presentation**: ask a friend to listen and comment constructively on those parts that were difficult to follow, to improve your performance.
  - (ii) **Use 'split times' to pace yourself**: following an initial run-through, add the times at which you should arrive at the key points of your talk to your notes. These timing marks will help you keep to time during the final presentation.
  - (iii) **Avoid looking at your wristwatch when speaking**; this sends a negative signal to the audience. Use a wall clock (where available), or take off your watch and put it beside your notes so that you can glance at it without distracting the audience.
- **Consider your image**. Make sure that the image you project is appropriate for the occasion:
  - (i) **Think about what to wear**: aim to be respectable without 'dressing up', otherwise your message may be diminished.
  - (ii) **Develop a good posture**: it will help your voice projection if you stand upright, rather than slouching or leaning over a lectern.
  - (iii) **Deliver your material with expression**: project your voice towards the audience at the back of the room and make sure you look around to make eye contact with all sections of the audience. Arm movements and subdued body language will help maintain the interest of your audience. However, you should avoid extreme gestures (it may work for some TV personalities but it is not recommended for the beginner).
  - (iv) **Try to identify and control any repetitive mannerisms**: repeated 'empty' words/phrases, fidgeting with pens, keys, etc., will distract your audience. Note-cards held in your hand give you something to focus on, whereas laser pointers will show up any nervous hand tremors. Practising in front of a mirror may help.
- **Think about questions**. Once again, the best approach is to prepare beforehand:
  - (i) **Consider what questions are likely to come up, and prepare brief answers**. However, do not be afraid to say 'I don't know': your audience will appreciate honesty, rather than vacillation, if you don't have an answer for a particular question.
  - (ii) **If no questions are asked, you might pose a question yourself** and then ask for opinions from the audience: if you use this approach, you should be prepared to comment briefly if your audience has no suggestions, to avoid the presentation ending in an embarrassing silence.

Sources for further study

Alley, M. (2003) *The Craft of Scientific Presentations: Critical Steps to Succeed and Critical Errors to Avoid.* Springer-Verlag, New York.

Capp, C.C. and Capp, G.R. (1989) *Basic Oral Communication*, 5th edn. Prentice Hall, Harlow.

Learnhigher. *Oral Communication.* Brunel University, London. Available: http://www.brunel.ac.uk/learnhigher/oral-com-index.shtml
Last accessed: 21/12/10.

Matthews, C. and Marino, J. (1999) *Professional Interactions: Oral Communication Skills of Science, Technology and Medicine*. Pearson, Harlow.

Radel, J. *Preparing Effective Presentations.* Available: http://www.kumc.edu/SAH/OTEd/jrdel/effective.html
Last accessed: 21/12/10.

15 General aspects of scientific writing

Monday:	morning afternoon evening	Lectures (University) Practical (University) Initial analysis and brainstorming (Home)
Tuesday:	morning afternoon evening	Lectures (University) Locate sources (Library) Background reading (Library)
Wednesday:	morning afternoon evening	Background reading (Library) Squash (Sports hall) Planning (Home)
Thursday:	morning afternoon evening	Lectures (University) Additional reading (Library) Prepare outline (Library)
Friday:	morning afternoon evening	Lab class (University) Write first draft (Home) Write first draft (Home)
Saturday:	morning afternoon evening	Shopping (Town) Review first draft (Home) Revise first draft (Home)
Sunday:	morning afternoon evening	Free Produce final copy (Home) Proofread and print essay (Home)
Monday:	morning	Final read-through and check Submit essay (deadline mid-day)

Fig. 15.1 Example timetable for writing a short essay.

Written communication is an essential component of all sciences. Most courses include writing exercises in which you will learn to describe ideas and results accurately, succinctly, and in an appropriate style and format. The following features are common to all forms of scientific writing.

Organising your time

Making a timetable at the outset helps ensure that you give each stage adequate attention and complete the work on time (e.g. Fig. 15.1). To create and use a timetable:

1. Break down the task into stages.
2. Decide on the proportion of the total time each stage should take.
3. Set realistic deadlines for completing each stage, allowing some time for slippage.
4. Refer to your timetable frequently as you work: if you fail to meet one of your deadlines, make a serious effort to catch up as soon as possible.

> **KEY POINT** The appropriate allocation of your time to reading, planning, writing and revising will differ according to the task in hand (see Chapters 16–18).

Organising your information and ideas

Before you write, you need to gather and/or think about relevant material (Chapters 8 and 9). You must then decide:

- what needs to be included and what doesn't;
- in what order it should appear.

Start by jotting down headings for everything of potential relevance to the topic (this is sometimes called 'brainstorming'). A spider diagram (Fig. 15.2) or a mind map (Fig. 4.2) will help you to organise these ideas. The next stage is to create an outline of your text (Fig. 15.3). Outlines are valuable because they:

Creating an outline – an informal outline can be made simply by indicating the order of sections on a spider diagram (as in Fig. 15.2).

- force you to think about and plan the structure;
- provide a checklist so nothing is missed out;
- ensure the material is balanced in content and length;
- help you organise figures and tables by showing where they will be used.

> **KEY POINT** A suitable structure is essential to the narrative of your writing and should be carefully considered at the outset.

Talking about your work – Discussing your topic with a friend or colleague might bring out ideas or reveal deficiencies in your knowledge.

In an essay or review, the structure of your writing should help the reader to assimilate and understand your main points. Sub-divisions of the topic could simply be related to the physical nature of the subject

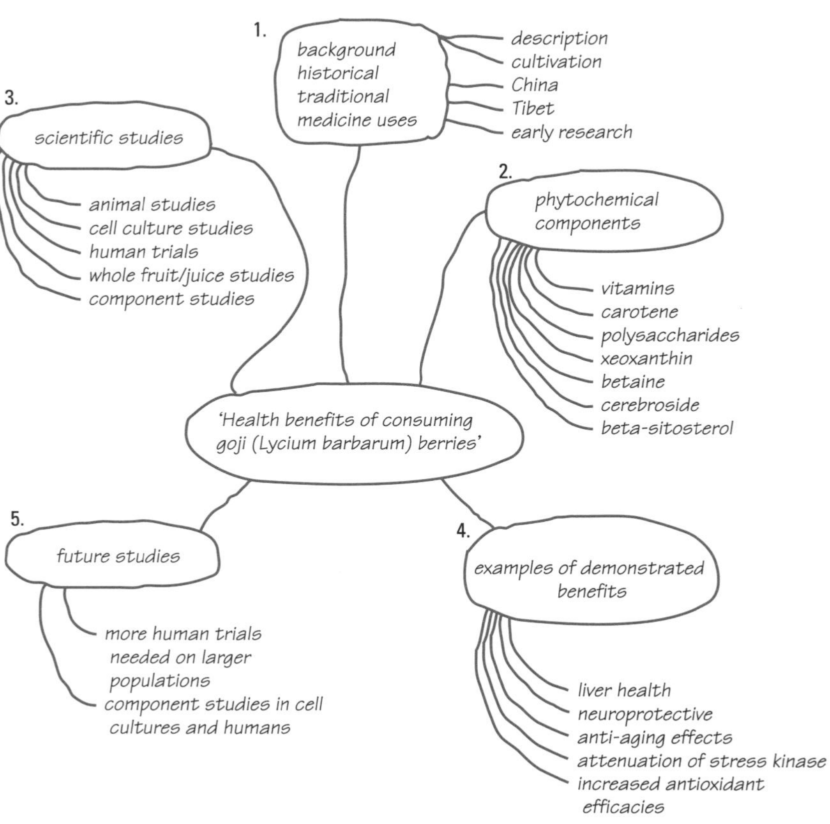

Fig. 15.2 Spider diagram showing how you might 'brainstorm' an essay with the title 'Health benefits of consuming goji (*Lycium barbarum*) berries'. Write out the essay title in full to form the spider's body, and as you think of possible content, place headings around this to form its legs. Decide which headings are relevant and which are not and use arrows to note connections between subjects. This may influence your choice of order and may help to make your writing flow because the links between paragraphs will be natural. You can make an informal outline directly on a spider diagram by adding numbers indicating a sequence of paragraphs (as shown). This method is best when you must work quickly, as with an essay written under exam conditions.

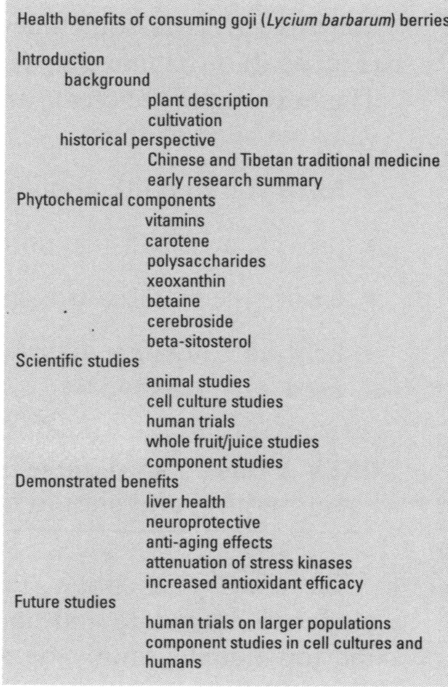

Health benefits of consuming goji (*Lycium barbarum*) berries

Introduction
- background
  - plant description
  - cultivation
- historical perspective
  - Chinese and Tibetan traditional medicine
  - early research summary

Phytochemical components
- vitamins
- carotene
- polysaccharides
- xeoxanthin
- betaine
- cerebroside
- beta-sitosterol

Scientific studies
- animal studies
- cell culture studies
- human trials
- whole fruit/juice studies
- component studies

Demonstrated benefits
- liver health
- neuroprotective
- anti-aging effects
- attenuation of stress kinases
- increased antioxidant efficacy

Future studies
- human trials on larger populations
- component studies in cell cultures and humans

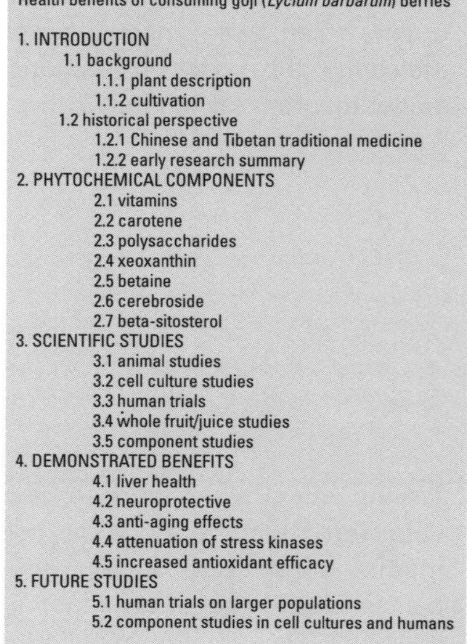

Health benefits of consuming goji (*Lycium barbarum*) berries

1. INTRODUCTION
   - 1.1 background
     - 1.1.1 plant description
     - 1.1.2 cultivation
   - 1.2 historical perspective
     - 1.2.1 Chinese and Tibetan traditional medicine
     - 1.2.2 early research summary
2. PHYTOCHEMICAL COMPONENTS
   - 2.1 vitamins
   - 2.2 carotene
   - 2.3 polysaccharides
   - 2.4 xeoxanthin
   - 2.5 betaine
   - 2.6 cerebroside
   - 2.7 beta-sitosterol
3. SCIENTIFIC STUDIES
   - 3.1 animal studies
   - 3.2 cell culture studies
   - 3.3 human trials
   - 3.4 whole fruit/juice studies
   - 3.5 component studies
4. DEMONSTRATED BENEFITS
   - 4.1 liver health
   - 4.2 neuroprotective
   - 4.3 anti-aging effects
   - 4.4 attenuation of stress kinases
   - 4.5 increased antioxidant efficacy
5. FUTURE STUDIES
   - 5.1 human trials on larger populations
   - 5.2 component studies in cell cultures and humans

Fig. 15.3 Formal outlines. These are useful for a long piece of work where you or the reader might otherwise lose track of the structure. The headings for sections and paragraphs are simply written in sequence with the type of lettering and level of indentation indicating their hierarchy. Two different forms of formal outline are shown: a minimal form (a) and a numbered form (b). Note that the headings used in an outline are often repeated within the essay to emphasise its structure. The content of an outline will depend on the time you have available and the nature of the work, but the most detailed hierarchy you should reasonably include is the subject of each paragraph.

matter (e.g. energy components of a food item) and should proceed logically (e.g. carbohydrates, from mono- through oligo- to polysaccharides). A chronological approach is good for evaluation of past work (e.g. the development of genetic engineering of foods), whereas a step-by-step comparison might be best for certain exam questions (e.g. 'Discuss the differences between so-called 'good' fats and 'bad' fats). There is little choice about structure for practical and project reports (see p. 108).

Writing

Adopting a scientific style

Your main aim in developing a scientific style should be to get your message across directly and unambiguously. Although you can try to achieve this through a set of 'rules' (see Box 15.1), you may find other requirements driving your writing in a contradictory direction. For instance, the need to be accurate and complete may result in text littered with technical terms, and the flow may be continually interrupted by references to the literature. The need to be succinct also affects style and readability through the use of, for example, stacked noun-adjectives (e.g. 'very low-density lipoprotein') and acronyms (e.g. 'VLDL'). Finally, style is very much a matter of taste and each tutor, examiner, supervisor or editor will have pet loves and hates that you may have to accommodate. Different assignments will need different styles; Box 15.2 gives further details.

Improving your writing skills – you need to take a long-term view if you wish to improve this aspect of your work. An essential preliminary is to invest in and make full use of a personal reference library (see Box 15.3).

Developing technique

Writing is a skill that can be improved, although not instantly. You should analyse your deficiencies with the help of feedback from your tutors, be prepared to change work habits (e.g. start planning your work more carefully), and willing to learn from some of the excellent texts that are available on scientific writing (p. 103).

Getting started

A common problem is 'writer's block' – inactivity or stalling brought on by a variety of causes. If blocked, ask yourself these questions:

Writing with a word processor – use the dynamic/interactive features of the word processor (Chapter 12) to help you get started: first make notes on structure and content, then expand these to form a first draft and finally revise/improve the text.

- **Are you comfortable with your surroundings?** Make sure you are seated comfortably at a reasonably clear desk and have minimised the possibility of interruptions and distractions.
- **Are you trying to write too soon?** Have you clarified your thoughts on the subject? Have you done enough preliminary reading?
- **Are you happy with the underlying structure of your work?** If you haven't made an outline, try making one. If you are unhappy because you can't think of a particular detail at the planning stage, just start writing – it is more likely to come to you while you are thinking of something else.
- **Are you trying to be too clever?** Your first sentence doesn't have to be earth-shattering in content or particularly smart in style. A short statement of fact or a definition is fine. If there will be time for revision, first get your ideas down on paper and then revise grammar, content and order later.

Box 15.1 How to achieve a clear, readable style

Words and phrases

- Choose short, clear words and phrases rather than long ones: e.g. use 'build' rather than 'fabricate'; 'now' rather than 'at the present time'. At certain times, technical terms must be used for precision, but don't use jargon if you don't have to.
- Don't worry too much about repeating words, especially when to introduce an alternative might subtly alter your meaning.
- Where appropriate, use the first person to describe your actions ('We decided to'; 'I conclude that'), but not if this is specifically discouraged by your supervisor.
- Favour active forms of writing ('the observer completed the survey in ten minutes') rather than a passive style ('the survey was completed by the observer in ten minutes').
- Use tenses consistently. Past tense is always used for Materials and Methods ('samples were taken from...') and for reviewing past work ('Smith (1990) concluded that...'). The present tense is used when describing data ('Fig. 1 shows...'), for generalisations ('Most authorities agree that...') and conclusions ('To conclude,...').
- Use statements in parentheses sparingly – they disrupt the reader's attention to your central theme.
- Avoid clichés and colloquialisms – they are usually inappropriate in a scientific context.

Punctuation

- Try to use a variety of types of punctuation, to make the text more interesting to read.
- Decide whether you wish to use 'closed' punctuation (frequent commas at the end of clauses) or 'open' punctuation (less frequent punctuation) – be consistent.
- Don't link two sentences with a comma. Use a full stop, this is an example of what *not* to do.
- Pay special attention to apostrophes, using the following rules:
  - To indicate possession, use an apostrophe before an 's' for a singular word (e.g. the rat's mass (weight) was...') and after the s for a plural word ending in s (e.g. the rats' masses were = the masses of the rats were). If the word has a special plural (e.g. woman ⟶ women) then use the apostrophe before the s (the women's masses (weights) were...).
  - When contracting words, use an apostrophe (e.g. do not = don't; it is = it's), but remember that contractions are generally *not* used in formal scientific writing.
  - Do *not* use an apostrophe for 'its' as the possessive form of 'it' (e.g. 'the university and its surroundings'). Note that 'it's' is reserved for 'it is'. This is an exception to the general rule and a very common mistake.
  - *Never* use an apostrophe to indicate plurals of any kind, including abbreviations.

Sentences

- Do not make them too long or too complicated.
- Introduce variety in structure and length.
- If unhappy with the structure of a sentence, try chopping it into a series of shorter sentences.

Paragraphs

- Get the paragraph length right – five sentences or so. Do *not* submit an essay that consists of a single paragraph, nor one that contains many single-sentence paragraphs.
- Make sure each paragraph is logical, dealing with a single topic or theme.
- Take care with the first sentence in a paragraph (the 'topic' sentence); this introduces the theme of the paragraph. Further sentences should then develop this theme, e.g. by providing supporting information, examples or contrasting cases.
- Use 'linking' words or phrases to maintain the flow of the text within a paragraph (e.g. 'for example'; 'in contrast'; 'however'; 'on the other hand').
- Make your text more readable by adopting modern layout style. The first paragraph in any section of text is usually *not* indented, but following paragraphs may be (by the equivalent of three character spaces). In addition, the space between paragraphs should be slightly larger than the space between lines. Follow departmental guidelines if they specify a format.
- Group paragraphs in sections under appropriate headings and sub-headings to reinforce the structure underlying your writing.
- Think carefully about the first and last paragraphs in any piece of writing: these are often the most important as they respectively set the aims and report the conclusions.

Note: If you're not sure what is meant by any of the terms used here, consult a guide on writing (see p. 103).

Box 15.2 Using appropriate writing styles for different purposes (with examples)

Note that courses tend to move from assignments that are predominantly descriptive in the early years to a more analytical approach towards the final year (see Chapter 5). Also, different styles may be required in different sections of a write-up, e.g. descriptive for introductory historical aspects, becoming more analytical in later sections.

Descriptive writing

This is the most straightforward style, providing factual information on a particular subject, and is most appropriate:

- in essays where you are asked to 'describe' or 'explain' (p. 105);
- when describing the results of a practical exercise, e.g.: 'The experiment shown in Figure 1 confirmed that enzyme activity was strongly influenced by temperature, as the rate observed at 37°C was more than double that seen at 20°C.'

However, in literature reviews and essays where you are asked to 'discuss' (p. 105) a particular topic, the descriptive approach is mostly inappropriate, as in the following example, where a large amount of specific information from a single scientific paper has been used without any attempt to highlight the most important points:

> A study on foodborne listeriosis was conducted in England and Wales from 1990 to 2004, involving almost 2,000 cases. In the 1990s, the mean yearly total cases was 110. However, the number of cases of listeriosis increased to 146, 136, 237 and 213 in 2001, 2002, 2003 and 2004 respectively. Multiple strains of *Listeria monocytogenes* were shown to be involved, with 14, 18 and 16 different subtypes reported in 2002, 2003 and 2004 respectively (Grey and Grey, 2006).

In the most extreme examples, whole paragraphs or pages of essays may be based on descriptive factual detail from a single source, often with a single citation at the end of the material, as above. Such essays often score low marks in essays where evidence of deeper thinking is required (Chapter 5).

Comparative writing

This technique is an important component of academic writing, and it will be important to develop your comparative writing skills as you progress through your course. Its applications include:

- answering essay questions and assignments of the 'compare and contrast' type (p. 105);
- comparing your results with previously published work in the Discussion section of a practical report.

To use this style, first decide on those aspects you wish to compare and then consider the material (e.g. different literature sources) from these aspects – in what ways do they agree or disagree with each other? One approach is to compare/contrast a different aspect in each paragraph. At a practical level, you can use 'linking' words and phrases to help orientate your reader as you move between aspects where there is agreement and disagreement. These include, for agreement: 'in both cases'; 'in agreement with'; 'is also shown by the study of'; 'similarly'; 'in the same way'; and for disagreement: 'however'; 'although'; 'in contrast to'; 'on the other hand'; 'which differs from'. The comparative style is fairly straightforward, once you have decided on the aspects to be compared. The following brief example compares two different studies using this style:

> Grey and Grey (2006) showed a steady increase in cases of foodborne listeriosis in England and Wales, from 110 cases in the 1990s to 213 cases in 2004. Similarly, East and West (2007), reviewing the literature, reported that this trend was also found in the data collected in other European countries.

Comparative text typically makes use of two or more references per paragraph.

Analytical writing

Typically, this is the most appropriate form of writing for:

- a review of scientific literature on a particular topic;
- an essay where you are asked to 'discuss' (p. 105) different aspects of a particular topic;
- evaluating a number of different published sources within the Discussion section of a final-year project dissertation.

By considering the significance of the information provided in the various sources you have read, you will be able to take a more critical approach. Your writing should evaluate the importance of the material in the context of your topic (see also Chapter 9). In analytical writing, you need to demonstrate critical thinking (p. 52) and personal input about the topic in a well-structured text that provides clear messages, presented in a logical order and demonstrating synthesis from a number of sources by appropriate use of citations (p. 46). Detailed information and relevant examples are used only to explain or develop a particular aspect, and not simply as 'padding' to bulk up the essay, as in the following example:

> Grey and Grey (2006) report an increase in foodborne listeriosis cases in England and Wales, from 110 cases in the 1990s to 213 cases in 2004. While this appeared to indicate an increased infection rate, Black and White (2008) suggested that improved referral and reporting of listeriosis, as well as improved laboratory methods for isolation and identification of the bacteria, were more likely to be responsible for the increase in recorded cases. A more detailed meta-analysis of the available data and literature is required to resolve the reason for the increase in reported cases.

Analytical writing is based on a broad range of sources, typically with several citations per paragraph.

Box 15.3 Improve your writing ability by consulting a personal reference library

Using dictionaries

We all know that a dictionary helps with spelling and definitions, but how many of us use one effectively? You should:

- Keep a dictionary beside you when writing and always use it if in any doubt about spelling or definitions.
- Use it to prepare a list of words that you have difficulty in spelling: apart from speeding up the checking process, the act of writing out the words helps commit them to memory.
- Use it to write out a personal glossary of terms. This can help you memorise definitions. From time to time, test yourself.

Not all dictionaries are the same! Ask your tutor or supervisor whether he/she has a preference and why. Try out the *Oxford Advanced Learner's Dictionary*, which is particularly useful because it gives examples of use of all words and helps with grammar, e.g. by indicating which prepositions to use with verbs. Dictionaries of biology tend to be variable in quality, possibly because the subject is so wide and new terms are continually being coined. *Henderson's Dictionary of Biological Terms* (Addison Wesley Longman) is a useful example.

Using a thesaurus

A thesaurus contains lists of words of similar meaning grouped thematically; words of opposite meaning always appear nearby.

- Use a thesaurus to find a more precise and appropriate word to fit your meaning, but check definitions of unfamiliar words with a dictionary.
- Use it to find a word or phrase 'on the tip of your tongue' by looking up a word of similar meaning.
- Use it to increase your vocabulary.

Roget's Thesaurus is the standard. Collins publish a combined dictionary and thesaurus and there are also online versions.

Using guides for written English

These provide help with the use of words.

- Use guides to solve grammatical problems such as when to use 'shall' or 'will', 'which' or 'that', 'effect' or 'affect', 'can' or 'may', etc.
- Use them for help with the paragraph concept and the correct use of punctuation.
- Use them to learn how to structure writing for different tasks.

- **Do you really need to start writing at the beginning?** Try writing the opening remarks after a more straightforward part. For example, with reports of practical work, the Materials and Methods section may be the easiest place to start.
- **Are you too tired to work?** Don't try to 'sweat it out' by writing for long periods at a stretch: stop frequently for a rest.

Revising your text

Learning from others – ask another student to read through your draft and comment on its content and overall structure.

Revising your text – to improve clarity and shorten your text, 'distil' each sentence by taking away unnecessary words and 'condense' words or phrases by choosing a shorter alternative.

Wholesale revision of your first draft is strongly advised for all writing, apart from in exams. Using a word processor, this can be a simple process. Where possible, schedule your writing so you can leave the first draft to 'settle' for at least a couple of days. When you return to it fresh, you will see more easily where improvements can be made. Try the following structured revision process, each stage being covered in a separate scan of your text:

1. **Examine content.** Have you included everything you need to? Is all the material relevant?
2. **Check the grammar and spelling.** Can you spot any 'howlers'?
3. **Focus on clarity.** Is the text clear and unambiguous? Does each sentence really say what you want it to say?

4. **Be succinct.** What could be missed out without spoiling the essence of your work? It might help to imagine an editor has set you the target of reducing the text by 15 per cent.
5. **Improve style.** Could the text read better? Consider the sentence and paragraph structure and the way your text develops to its conclusion.

Common errors

These include (with examples):

- **Problems over singular and plural words** ('a bacteria is'; 'the results shows').
- **Verbose text** ('One definition that can be employed in this situation is given in the following sentence').
- **Misconstructed sentences** ('Health and safety regulations should be made aware of').
- **Misuse of punctuation,** especially commas and apostrophes (for examples see Box 15.1).
- **Poorly constructed paragraphs** (for advice/examples see Box 15.1).

Sources for further study

Burchfield, R.W. (ed.) (2004) *Fowler's Modern English Usage*, revised 3rd edn. Oxford University Press, Oxford.

Clark, R. *The English Style Book. A Guide to the Writing of Scholarly English*. Available: http://www.litencyc.com/stylebook/stylebook.php
Last accessed: 21/12/10.

Kane, T.S. (1994) *The New Oxford Guide to Writing*. Oxford University Press, New York.
[This is excellent for the basics of English – it covers grammar, usage and the construction of sentences and paragraphs.]

Lindsay, D. (1995) *A Guide to Scientific Writing*, 2nd edn. Longman, Harlow.

McMillan, K.M. and Weyers, J.D.B. (2006) *The Smarter Student: Study Skills and Strategies for Success at University*. Pearson, Harlow.

Partridge, E. (1978) *You Have a Point There*. Routledge, London.
[This covers punctuation in a very readable manner.]

Tichy, H.J. (1988) *Effective Writing for Engineers, Managers and Scientists,* 2nd edn. Wiley, New York.
[This is strong on scientific style and clarity in writing.]

16 Writing essays

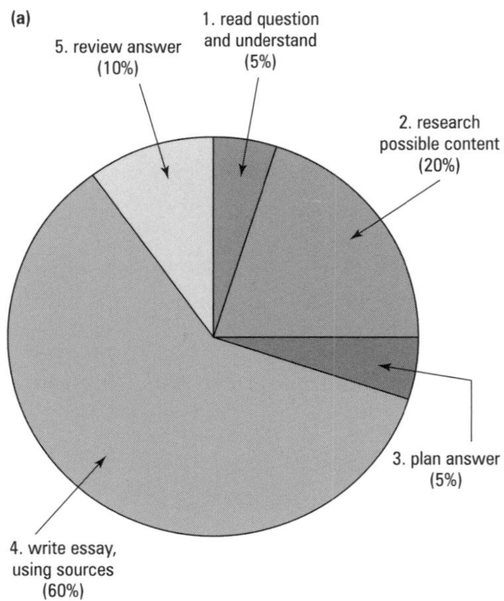

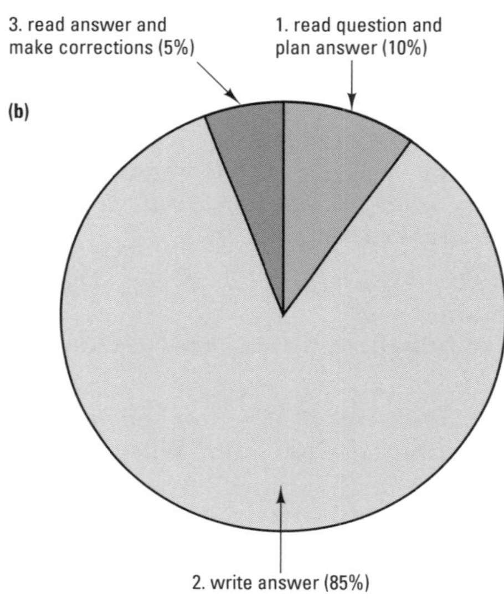

Fig. 16.1 Typical division of time for an essay written as part of in-course assessment (a) or under exam conditions (b).

Considering essay content – it is rarely enough simply to lay down facts for the reader – you must analyse them and comment on their significance (see p. 102).

The function of an essay is to show how much you understand about a topic and how well you can organise and express your knowledge.

Organising your time

The way you should divide your time when producing an essay depends on whether you are writing it for in-course assessment or under exam conditions (Fig. 16.1). Essays written over a long period with access to books and other resources will probably involve a research element, not only before the planning phase but also when writing (Fig. 16.1a). For exams, it is assumed that you have revised appropriately (Chapter 5) and essentially have all the information at your fingertips. To keep things uncomplicated, the time allocated for each essay should be divided into three components – planning, writing and reviewing (Fig. 16.1b), and you should adopt time-saving techniques whenever possible (Box 6.2).

Making a plan for your essay

Dissect the meaning of the essay question or title

Read the title very carefully and think about the topic before starting to write. Consider the definitions of each of the important nouns (this can help in approaching the introductory section). Also think about the meaning of the verb(s) used and try to follow each instruction precisely (see Table 16.1). Don't get sidetracked because you know something about one word or phrase in the title: consider the whole title and all its ramifications. If there are two or more parts to the question, make sure you give adequate attention to each part.

Consider possible content and examples

Research content using the methods described in Chapters 8 and 9. If you have time to read several sources, consider their content in relation to the essay title. Can you spot different approaches to the same subject? Which do you prefer as a means of treating the topic in relation to your title? Which examples are most relevant to your case, and why?

> **KEY POINT** Most marks for essays are lost because the written material is badly organised or is irrelevant. An essay plan, by definition, creates order and, if thought about carefully, should ensure relevance.

Construct an outline

Every essay should have a structure related to its title. Your plan should be written down (but scored through later if written in an exam book). Think about an essay's content in three parts:

1. **The introductory section**, in which you should include definitions and some background information on the context of the topic being considered. You should also tell your reader how you plan to approach the subject.

Ten Golden Rules for essay writing – these are framed for in-course assessments (p. 27), though many are also relevant to exams (see also Box 6.2).

1. Read the question carefully and decide exactly what the assessor wants you to achieve in your answer.
2. Make sure you understand the question by considering all aspects – discuss your approach with colleagues or a tutor.
3. Carry out the necessary research (using books, journals, WWW), taking appropriate notes. Gain an overview of the topic before getting involved with the details.
4. Always plan your work in outline before you start writing. Check that your plan covers the main points and that it flows logically.
5. Introduce your essay by showing that you understand the topic and stating how you intend to approach it.
6. As you write the main content, ensure it is relevant by continually looking back at the question.
7. Use headings and sub-headings to organise and structure your essay.
8. Support your statements with relevant examples, diagrams and references where appropriate.
9. Conclude by summarising the key points of the topic, indicating the present state of knowledge, what we still need to find out and how this might be achieved.
10. Always re-read the essay before submitting it. Check grammar and spelling and confirm that you have answered all aspects of the question.

Using diagrams – give a title and legend to each diagram so that it makes sense in isolation and point out in the text when the reader should consult it (e.g. 'as shown in Fig. 1 ...' or 'as can be seen in the accompanying diagram, ...').

Table 16.1 Instructions often used in essay questions and their meanings. When more than one instruction is given (e.g. compare and contrast; describe and explain), make sure you carry out *both* or you may lose a large proportion of the available marks (see also Table 5.1)

Account for:	give the reasons for
Analyse:	examine in depth and describe the main characteristics of
Assess:	weigh up the elements of and arrive at a conclusion about
Comment:	give an opinion on and provide evidence for your views
Compare:	bring out the similarities between
Contrast:	bring out dissimilarities between
Criticise:	judge the worth of (give both positive and negative aspects)
Define:	explain the exact meaning of
Describe:	use words and diagrams to illustrate
Discuss:	provide evidence or opinions about, arriving at a balanced conclusion
Enumerate:	list in outline form
Evaluate:	weigh up or appraise; find a numerical value for
Explain:	make the meaning of something clear
Illustrate:	use diagrams or examples to make clear
Interpret:	express in simple terms, providing a judgement
Justify:	show that an idea or statement is correct
List:	provide an itemised series of statements about
Outline:	describe the essential parts only, stressing the classification
Prove:	establish the truth of
Relate:	show the connection between
Review:	examine critically, perhaps concentrating on the stages in the development of an idea or method
State:	express clearly
Summarise:	without illustrations, provide a brief account of
Trace:	describe a sequence of events from a defined point of origin

2. **The middle of the essay**, where you develop your answer and provide relevant examples. Decide whether a broad analytical approach is appropriate or whether the essay should contain more factual detail.
3. **The conclusion**, which you can make quite short. You should use this part to summarise and draw together the components of the essay, without merely repeating previous phrases. You might mention such things as: the broader significance of the topic; its future; its relevance to other important areas of biology. Always try to mention both sides of any debate you have touched on, but beware of 'sitting on the fence'.

KEY POINT Use paragraphs to make the essay's structure obvious. Emphasise them with headings and sub-headings unless the material beneath the headings would be too short or trivial.

Start writing

- **Never lose track of the importance of content and its relevance.** Repeatedly ask yourself: 'Am I really answering this question?' Never waffle just to increase the length of an essay. Quality, rather than quantity, is important.

- **Illustrate your answer appropriately.** Use examples to make your points clear, but remember that too many similar examples can stifle the flow of an essay. Use diagrams where a written description would be difficult or take too long. Use tables to condense information.
- **Take care with your handwriting.** You can't get marks if your writing is illegible. Try to cultivate an open form of handwriting, making the individual letters large and distinct. If there is time, make out a rough draft from which a tidy version can be copied.

Reviewing your answer

Make sure that you leave enough time to:

- re-read the question to check that you have answered all points;
- re-read your essay to check for errors in punctuation, spelling and content. Make any corrections obvious. In an exam don't panic if you suddenly realise you've missed a large chunk out as the reader can be redirected to a supplementary paragraph if necessary.

Learning from lecturers' and tutors' comments – ask for further explanations if you don't understand a comment or why an essay was less successful than you thought it should have been.

Sources for further study

Anon. (2004) *Essay and Report Writing Skills*. Open University, Milton Keynes.

Anon. *Yahoo! Directory: Writing > Essays and Research Papers*. Available: http://dir.yahoo.com/Social_Science/Communications/Writing/Essays_and_Research_ Papers/
Last accessed: 21/12/10.
[An extensive directory of Web resources.]

Good, S. and Jensen, B. (1995) *The Student's Only Survival Guide to Essay Writing*. Orca Book Publishers, Victoria, USA.

17 Reporting practical and project work

Basic structure of scientific reports – this usually follows the 'IMRAD' acronym: **I**ntroduction, **M**aterials and Methods, **R**esults **a**nd **D**iscussion.

Practical reports, project reports, theses and scientific papers differ greatly in depth, scope and size, but they all have the same basic structure. Some variation is permitted, however (see Box 17.1), and you should always follow the advice or rules provided by your department.

Additional parts may be specified: for project reports, dissertations and theses, a Title Page is often required and a List of Figures and Tables as part of the Contents section. When work is submitted for certain degrees, you may need to include declarations and statements made by the student and supervisor. In scientific papers, a list of keywords is often added following the abstract: this information may be combined with words in the title for computer cross-referencing systems.

KEY POINT Department, school or faculty regulations may specify a precise format for producing your report or thesis. Obtain a copy of these rules at an early stage and follow them closely, to avoid losing marks.

Options for discussing data – the main optional variants of the general structure include combining *Results* and *Discussion* into a single section and adding a separate *Conclusions* section.

- The main advantage of a joint *results* and *discussion* section is that you can link together different experiments, perhaps explaining why a particular result led to a new hypothesis and the next experiment. However, a combined *Results* and *Discussion* section may contravene your department's regulations, so you should check before using this approach.
- The main advantage of having a separate *Conclusions* section is to draw together and emphasise the chief points arising from your work, when these may have been 'buried' in an extensive *Discussion* section.

Practical and project reports

These are exercises designed to make you think more deeply about your experiments, and to practise and test the skills necessary for writing up research work. Special features are:

- Introductory material is generally short and unless otherwise specified should outline the aims of the experiment(s) with a minimum of background material.
- Materials and methods may be provided by your supervisor for practical reports. If you make changes to this, you should state clearly what you did. With project work, your lab notebook (see p. 195) should provide the basis for writing this section.
- Great attention in assessment will be paid to presentation and analysis of data. Take special care over graphs (see p. 194 for further advice). Make sure your conclusions are justified by the evidence you present.

Theses and dissertations

These are submitted as part of the examination for a degree following an extended period of research. They act to place on record full details about your experimental work and will normally only be read by those with a direct interest in it – your examiners or colleagues. Note the following:

- You are allowed scope to expand on your findings and to include detail that might otherwise be omitted in a scientific paper.
- You may have problems with the volume of information that has to be organised. One method of coping with this is to divide your thesis into chapters, each having the standard format (as in Box 17.1). A General Introduction can be given at the start and a General Discussion at the end. Discuss this with your supervisor.

Oral assessments – there may be an oral exam (*'viva voce'*) associated with the submission of a thesis or dissertation. The primary aim of the examiners will be to ensure that you understand what you did and why you did it.

Box 17.1 The structure of reports of experimental work

Undergraduate practical and project reports are generally modelled on this arrangement or a close variant of it, because this is the structure used for nearly all research papers and theses. The more common variations include Results and Discussion combined into a single section and Conclusions appearing separately as a series of points arising from the work. In scientific papers, a list of keywords (for computer cross-referencing systems) may be included following the abstract. Acknowledgements may appear after the contents section, rather than near the end. Department or faculty regulations for producing theses and reports may specify a precise format; they often require a title page to be inserted at the start and a list of figures and tables as part of the contents section, and may specify declarations and statements to be made by the student and supervisor.

Part (in order)	Contents/purpose	Checklist for reviewing content
Title	Explains what the project was about	Does it explain what the text is about succinctly?
Authors plus their institutions	Explains who did the work and where; also where they can be contacted now	Are all the details correct?
Abstract/summary	Synopsis of methods, results and conclusion of work described. Allows the reader to grasp quickly the essence of the work	Does it explain why the work was done? Does it outline the whole of your work and your findings?
Contents	Shows the organisation of the text (not required for short papers)	Are all the sections covered? Are the page numbers correct?
Abbreviations	Lists all the abbreviations used (but not those of SI, chemical elements, or standard biochemical terms)	Have they all been explained? Are they all in the accepted form? Are they in alphabetical order?
Introduction	Orientates the reader, explains why the work has been done and its context in the literature, why the methods used were chosen, why the experimental organisms were chosen. Indicates the central hypothesis behind the experiments	Does it provide enough background information and cite all the relevant references? Is it of the correct depth for the readership? Have all the technical terms been defined? Have you explained why you investigated the problem? Have you outlined your aims and objectives? Have you explained your methodological approach? Have you stated your hypothesis?
Materials and methods	Explains how the work was done. Should contain sufficient detail to allow another competent worker to repeat the work	Is each experiment covered and have you avoided unnecessary duplication? Is there sufficient detail to allow repetition of the work? Are proper scientific names and authorities given for all organisms? Have you explained where you got them from? Are the correct names, sources and grades given for all chemicals?
Results	Displays and describes the data obtained. Should be presented in a form that is easily assimilated (graphs rather than tables, small tables rather than large ones)	Is the sequence of experiments logical? Are the parts adequately linked? Are the data presented in the clearest possible way? Have SI units been used properly throughout? Has adequate statistical analysis been carried out? Is all the material relevant? Are the figures and tables all numbered in the order of their appearance? Are their titles appropriate? Do the figure and table legends provide all the information necessary to interpret the data without reference to the text? Have you presented the same data more than once?
Discussion/ conclusions	Discusses the results: their meaning, their importance; compares the results with those of others; suggests what to do next	Have you explained the significance of the results? Have you compared your data with other published work? Are your conclusions justified by the data presented?
Acknowledgements	Gives credit to those who helped carry out the work	Have you listed everyone that helped, including any grant-awarding bodies?
Literature cited (bibliography)	Lists all references cited in appropriate format: provides enough information to allow the reader to find the reference in a library	Do all the references in the text appear on the list? Do all the listed references appear in the text? Do the years of publications and authors match? Are the journal details complete and in the correct format? Is the list in alphabetical order, or correct numerical order?

Steps in the production of a practical report or thesis

Choose the experiments you wish to describe and decide how best to present them

Choosing between graphs and tables – graphs are generally easier for the reader to assimilate, whereas tables can be used to condense a lot of data into a small space.

Try to start this process before your lab work ends because, at the stage of reviewing your experiments, a gap may become apparent (e.g. a missing control) and you might still have time to rectify the deficiency. Irrelevant material should be ruthlessly eliminated, at the same time bearing in mind that negative results can be extremely important (see p. 204). Use as many different forms of data presentation as are appropriate, but avoid presenting the same data in more than one form. Relegate large tables of primary data to an appendix and summarise the important points within the main text (with a cross-reference to the appendix). Make sure that the experiments you describe are representative: always state the number of times they were repeated and how consistent your findings were.

Repeating your experiments – remember, if you do an experiment twice, you have repeated it only once.

Make up plans or outlines for the component parts

The overall structure of practical and project reports is well defined (see Box 17.1), but individual parts will need to be organised as with any other form of writing (see Chapter 15).

Write

Presenting your results – remember that the order of results presented in a report need not correspond with the order in which you carried out the experiments: you are expected to rearrange them to provide a logical sequence of findings.

The materials and methods section is often the easiest to write once you have decided what to report. Remember to use the past tense and do not allow results or discussion to creep in. The results section is the next easiest as it should only involve description. At this stage, you may benefit from jotting down ideas for the discussion – this may be the hardest part to compose as you need an overview both of your own work and of the relevant literature. It is also liable to become wordy, so try hard to make it succinct. The introduction shouldn't be too difficult if you have fully understood the aims of the experiments. Write the abstract and complete the list of references at the end. To assist with the latter, it is a good idea as you write to jot down the references you use or to pull out their cards from your index system.

Revise the text

Using the correct tense – always use the past tense to describe the methodology used in your work, since it is now complete. Use the present tense only for generalisations and conclusions (p. 100).

Once your first draft is complete, try to answer all the questions given in Box 17.1. Show your work to your supervisors and learn from their comments. Let a friend or colleague who is unfamiliar with your subject read your text; he/she may be able to pinpoint obscure wording and show where information or explanation is missing. If writing a thesis, double-check that you are adhering to your institution's thesis regulations.

Prepare the final version

Markers appreciate neatly produced work but a well-presented document will not disguise poor science! If using a word processor, print the final version with the best printer available. Make sure figures are clear and in the correct size and format.

Submit your work

Your department will specify when to submit a thesis or project report, so plan your work carefully to meet this deadline or you may lose marks. Tell your supervisor early of any circumstances that may cause delay and check to see whether any forms must be completed for late submission, or evidence of extenuating circumstances.

Peer review – the process of evaluation and review of a colleague's work. In scientific communication, a paper is reviewed by two or more expert reviewers for comments on quality and significance as a key component of the validation procedure.

Producing a scientific paper

Scientific papers are the means by which research findings are communicated to others. Peer-reviewed papers are published in journals; each covers a well-defined subject area and publishes details of the format they expect.

KEY POINT Peer review is an important component of the process of scientific publication; only those papers whose worth is confirmed by the peer-review process will be published.

It would be very unusual for an undergraduate to submit a paper on his or her own – this would normally be done in collaboration with your project supervisor and only then if your research has satisfied appropriate criteria. However, it is important to understand the process whereby a paper comes into being (Box 17.2), as this can help you understand and interpret the primary literature.

Box 17.2 Steps in producing a scientific paper

Scientific papers are the lifeblood of any science and it is a major landmark in your scientific career to publish your first paper. The main steps in doing this should include the following:

Assessing potential content
The work must be of an appropriate standard to be published and should be 'new, true and meaningful'. Therefore, before starting, the authors need to review their work critically under these headings. The material included in a scientific paper will generally be a subset of the total work done during a project, so it must be carefully selected for relevance to a clear central hypothesis – if the authors won't prune, the referees and editors of the journal certainly will!

Choosing a journal
There are many journals covering food science, nutrition and dietetics and each covers a specific area (which may change through time). The main factors in deciding on an appropriate journal are the range of subjects it covers, the quality of its content, and the number and geographical distribution of its readers. The choice of journal always dictates the format of a paper since authors must follow to the letter the journal's 'Instructions to Authors'.

Deciding on authorship
In multi-author papers, a contentious issue is often who should appear as an author and in what order they should be cited. Where authors make an equal contribution, an alphabetical order of names may be used. Otherwise, each author should have made a substantial contribution to the paper and should be prepared to defend it in public. Ideally, the order of appearance will reflect the amount of work done rather than seniority. This may not happen in practice!

Writing
The paper's format will be similar to that shown in Box 17.1, and the process of writing will include outlining, reviewing, etc., as discussed elsewhere in this chapter. Figures must be finished to an appropriate standard and this may involve preparing photographs or digital images of them.

Submitting
When completed, copies of the paper are submitted to the editor of the chosen journal with a simple covering letter. A delay of one to two months usually follows while the manuscript is sent to two or more anonymous referees who will be asked by the editor to check that the paper is novel, scientifically correct and that its length is justified.

Responding to referees' comments
The editor will send on the referees' comments to the authors who will then have a chance to respond. The editor will decide on the basis of the comments and replies to them whether the paper should be published. Sometimes quite heated correspondence can result if the authors and referees disagree!

Checking proofs and waiting for publication
If a paper is accepted, it will be sent off to the typesetters. The next the authors see of it is the proofs (first printed version in style of journal), which have to be corrected carefully for errors and returned. Eventually, the paper will appear in print, but a delay of six months following acceptance is not unusual. Nowadays, papers are often available electronically, via the Web, in pdf format – see p. 46 for advice on how to cite 'online early' papers.

Sources for further study

Berry, R. (2004) *The Research Project: How to Write it*, 5th edn. Routledge, London.

Davis, M. (2005) *Scientific Papers and Presentations*. Academic Press, London.

Day, R.A. and Gastel, B. (2006) *How to Write and Publish a Scientific Paper*, 6th edn. Cambridge University Press, Cambridge.

Lobban, C.S. and Schefter, M. (1992) *Successful Lab Reports: A Manual for Science Students*. Cambridge University Press, Cambridge.

Luck, M. (1999) *Your Student Research Project*. Gower, London.

Luey, B. (2002) *Handbook for Academic Authors*, 4th edn. Cambridge University Press, Cambridge.

Matthews, J.R., Bowen, J.M. and Matthews, R. (2000) *Successful Science Writing: A Step-by-step Guide for the Biological and Medical Sciences*, 2nd edn. Cambridge University Press, Cambridge.

Valiela, I. (2001) *Doing Science: Design, Analysis and Communication of Scientific Research*. Oxford University Press, Oxford.
[Covers scientific communication, graphical presentations and aspects of statistics.]

18 Writing literature surveys and reviews

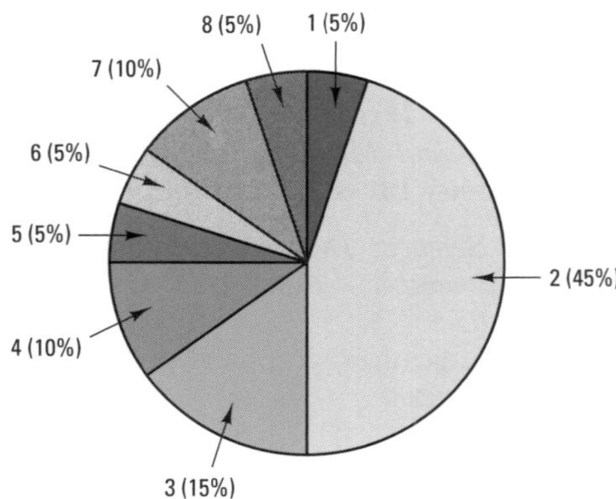

Fig. 18.1 Pie chart showing how you might allocate time for a literature survey:
1. select a topic;
2. scan the literature;
3. plan the review;
4. write first draft;
5. leave to settle;
6. prepare a structured review of text;
7. write final draft;
8. produce top copy.

The literature survey or review is a specialised form of essay that summarises and reviews the evidence and concepts concerning a particular area of research.

KEY POINT A literature review should *not* be a simple recitation of facts. The best reviews are those that analyse information rather than simply describe it.

Making up a timetable

Figure 18.1 illustrates how you might divide up your time for writing a literature survey. There are many subdivisions in this chart because of the size of the task: in general, for lengthy tasks, it is best to divide up the work into manageable chunks. Note also that proportionately less time is allocated to writing itself than with an essay. In a literature survey, make sure that you spend adequate time on research and revision.

Selecting a topic

Creating a glossary – one barrier to developing an understanding of a new topic is the jargon used. To overcome this, create your own glossary. You may wish to cross-reference a range of sources to ensure the definitions are reliable and context-specific. Remember to note your sources in case you wish to use the definition within your review.

You may have no choice in the topic to be covered but, if you do, carry out your selection as a three-stage process:

1. Identify a broad subject area that interests you.
2. Find and read relevant literature in that area. Try to gain a broad impression of the field from books and general review articles. Discuss your ideas with your supervisor.
3. Select a relevant and concise title. The wording should be considered very carefully as it will define the content expected by the reader. A narrow subject area will cut down on the amount of literature you will be expected to review, but will also restrict the scope of the conclusions you can make (and vice versa for a wide subject area).

Scanning the literature and organising your references

Using index cards – these can help you organise large numbers of references. Write key points and author information on each card – this helps when considering where the reference fits into the literature. Arrange the cards in subject piles, eliminating irrelevant ones. Order the cards in the sequence in which you wish to write.

You will need to carry out a thorough investigation of the literature before you start to write. The key problems are as follows:

- Getting an initial toe-hold in the literature. Seek help from your supervisor, who may be willing to supply a few key papers to get you started. Hints on expanding your collection of references are given on p. 43.
- Assessing the relevance and value of each article. This is the essence of writing a review, but it is difficult unless you already have a good understanding of the field. Try reading earlier reviews in your area and discussing the topic with your supervisor or other academic staff.
- Clarifying your thoughts. Subdividing the main topic and assigning your references to these smaller subject areas may help you gain a better overview of the literature.

Deciding on structure and content

The general structure and content of a literature survey are described below. The *Annual Review* series (available in most university libraries) provides good examples of the style expected in reviews of food science, nutrition and dietetics.

Defining terms – the introduction is a good place to explain the meaning of the key terms used in your survey or review.

Introduction

The introduction should give the general background to the research area, concentrating on its development and importance. You should also make a statement about the scope of your survey; as well as defining the subject matter to be discussed, you may wish to restrict the period being considered.

Main body of text

The review itself should discuss the published work in the selected field and may be subdivided into appropriate sections. Within each portion of a review, the approach is usually chronological, with appropriate linking phrases (e.g. 'Following on from this, ...'; 'Meanwhile, Bloggs (1980) tackled the problem from a different angle ...'). However, a good review is much more than a chronological list of work done. It should:

- allow the reader to obtain an overall view of the current state of the research area, identifying the key areas where knowledge is advancing;
- show how techniques are developing, and discuss the benefits and disadvantages of using particular analytical instruments or experimental systems;
- assess the relative worth of different types of evidence – this is the most important aspect (see Chapter 9). Do not be intimidated from taking a critical approach as the conclusions you may read in the primary literature aren't always correct;
- indicate where there is conflict in findings or theories, suggesting if possible which side has the stronger case;
- indicate gaps in current knowledge.

Balancing opposing views – even if you favour one side of a disagreement in the literature, your review should provide a balanced and fair description of all the published views of the topic. Having done this, if you do wish to state a preference, give reasons for your opinion.

You do not need to wait until you have read all of the sources available to you before starting to write the main body. Word processors allow you to modify and move pieces of text at any point and it will be useful to write paragraphs about key sources, or groups of related papers, as you read them. Try to create a general plan for your review as soon as possible. Place your draft sections of text under an appropriate set of sub-headings that reflects your plan, but be prepared to rearrange these and retitle or reorder sections as you proceed. Not only will working in this way help to clarify your thoughts, but it may help you avoid a last-minute rush of writing near to the submission date.

Conclusions

The conclusions should draw together the threads of the preceding parts and point the way forward, perhaps listing areas of ignorance or where the application of new techniques may lead to advances.

Making citations – a review of literature poses stylistic problems because of the need to cite large numbers of papers; in the *Annual Review* series this is overcome by using numbered references (see p. 46).

References, etc.

The references or literature cited section should provide full details of all papers referred to in the text (see p. 46). The regulations for your department may also specify a format and position for the title page, contents page, acknowledgements, etc.

Source for further study

Rudner, L.M. and Schafer, W.D. (1999) How to write a scholarly research report. *Practical Assessment, Research & Evaluation*, **6**(13). Available: http://pareonline.net/getvn.asp?v=6&n=13. Last accessed: 21/12/10.

19 Communicating with the public

While writing for academic purposes is covered in Chapters 13 to 18, students and scientists may also need to communicate their ideas and findings with the public. This involves important differences from formal scientific writing and this chapter provides guidance on how to tackle the main forms of communication with non-academic audiences.

Preparing a media release

Press releases are a good way to draw broad attention to your work or to a forthcoming event. This could include promoting an open day for your department or degree programme, making the public aware of the wider significance of your project outcome, attracting the attention of grant-awarding bodies or providing new nutrition or food science information of general interest. A press release must be newsworthy, noteworthy and attention-grabbing. It must be about a recent event (e.g. the latest discovery from a research project) or a forthcoming event (e.g. a new food product launch, or a community activity carried out by your class) to gain media attention.

> **Example**
>
> 1 April 2010
>
> MEDIA RELEASE – for immediate release
>
> **Title: Food for thought**
> **Who:** World expert on brain nutrition, Dr Paul Smith, is visiting the university.
> **What:** Public lecture on the latest findings on the influence of diet on mental function and neuroplasticity.
> **When:** 27 June 2010, 7.00 p.m.
> **Where:** Building 6, Room G.03, CQUniversity, Yaamba Road, Rockhampton.
> **Why:** The research suggests ways to improve the clarity of thought and prevent age-related mental deterioration. Dr Smith says 'By eating certain functional foods and leading an active and aware life it is possible to improve brain function. The latest research shows that the brain can improve in function and new cells and neural connections can be developed at any age.'
>
> (Note that the 'why' sub-section provides the attention-grabbing 'angle' to catch readers' interests in the first paragraph. Most people would be happy to increase their brain power, so that is also a potential 'hook' for the article, and journalists will often pick up on such cues.)
>
> ---
>
> **Contact:** John Gray on 0409 123 456 (mobile); email: j.gray@abc.edu.au for more information.
> Attached: digital image of Dr Smith.

> **KEY POINT** Presenting information in writing, rather than verbally, reduces the chance of communication errors.

Overview

- Use your organisational letterhead, if appropriate.
- Place the heading 'Press Release' or 'Media Release' at the top of the page.
- Date the release.
- Try to keep the release to a single page, neatly laid out. If you need to provide background information this is best done as a separate attachment.
- Keep your text brief and to the point – use sub-headings to structure your text, and make sure that your key messages are clear.
- Include appropriate contact details, e.g. your mobile phone number and your email, and then make sure that you monitor these in the days immediately after release. Busy press people may not call back if you are not available.
- If possible and appropriate, supply your own photograph(s) in a high resolution digital format. If appropriate specify acknowledgement for the photographer, and gain permission if the image is copyright.
- Timing your release in conjunction with either forthcoming or current related major events is more likely to result in success.

Detail and layout

- **Use a short catchy headline of six words or less** – your headline should be succinct and interesting, in contrast to the dry formal style of scientific writing (Chapter 15).

- **Base your press release around the five Ws of journalism:**
  1. Who?
  2. What?
  3. When?
  4. Where?
  5. Why?

Distribution of a media release – seek out help with distribution if it is available. Most universities and large organisations have an internal public relations division with up-to-date networks and media relationships. For news of national potential there are also commercially run release distribution options.

- **Make sure that your story is attractive to the public.** This is your 'angle' or USP (unique selling point). This should be clear and up-front in the first paragraph.
- **Prepare the release in a ready-to-print format,** ensuring it can be trimmed from the bottom without altering the meaning and flow. Overworked editors will often do a direct cut and paste if the material is newsworthy. Bullet points can sometimes be useful for a brief release format.
- **Use the third person to keep the release objective.** Personal opinions should be placed in quotes and may add to the interest of the reader.
- **Place a horizontal line beneath the text for publication.** Beneath this line goes your contact details and information referring to any attachments.
- **Double-check your spelling, grammar, numbers and proper nouns.** Have someone else read through the release and give you feedback.

Preparing for a media interview

Preparing for simple questions as well as difficult ones – sometimes, media questions are disarmingly untechnical, so you should be ready to answer very basic points and explain complex concepts in lay language.

In some cases, a press release may be picked up by radio or TV, resulting in a request for a media interview (this is one reason why mock interviews often form part of the assessment of university courses). An appropriate way to prepare for this is as follows:

1. **Make a list of questions you would like to be asked on your topic.** Then prepare yourself with interesting and honest answers. Be to the point. For a television or radio news broadcast 20-second answers would be all the time available. In other circumstances find out how long you will have. Provide the list of questions to the interviewer well ahead of the interview and see if he/she will accept them as the basis for the interview. Do not include humour unless you are very confident that it will work. Humour is often based on exaggeration or twist, and can easily be taken out of context or misinterpreted, as well as detracting from the seriousness of your message.
2. **Make a second list of questions you would not want the interviewer to ask you about your topic.** Prepare careful, honest and meaningful answers. It is suggested that you do not provide the interviewer with these questions, but also do not rule out the possibility that they may be asked during the session. Investigative journalists are used to probing for weaknesses, so you should be prepared.
3. **Have another student or colleague put all of the questions to you.** Practise answering, and then request feedback from your questioner. Try to keep your answers short and to the point, without too much hesitation or stumbling for the right words. Another useful approach is to record your answers and play them back to yourself.

Learning by publishing articles – carrying out this activity may help you to:

- achieve new insights by explaining your subject to others;
- gain knowledge and understanding from additional research carried out to back up your viewpoints;
- share important information with other students, staff and the public;
- encourage readers to take action that may be of benefit to them;
- promote your own professional image and add to your CV.

Example

(Explanation of technical term)

Neuroplasticity – this means flexibility in brain development; in other words, the capacity of the brain to modify its organisation, e.g. in response to damage or experience.

Writing articles for the Web – the basic principles also apply here, but keep the following in mind:

- People surfing the Web tend to skim pages and are easily distracted, so if you do not capture their immediate interest they are likely to click on the next link on their search page.
- The above sentence is an example of how not to write a crisp sentence for the Web – it is too long-winded.
- Use short, well-spaced, 'catchy' paragraphs.
- Use colourful images where this clearly illustrates your points.
- Make sure your page layout allows for easy printing.
- For how to compose an effective email, see Box 19.1.

Writing magazine and newsletter articles

Writing an article for the general public requires different skills from those used in formal scientific writing, and sometimes these are developed and assessed in university classes. The process is similar in some respects to writing a speech. In the first instance, consider the audience you are writing for (see also Box 14.2 for general guidance). To gain experience outside your coursework you could write an article on a project or topic of interest for your university or college magazine or newsletter.

Select your publication

- **Choose a magazine or newsletter to target.** It may be appropriate to first contact the editor with your article idea to see if he/she is interested.
- **Find and follow the publication's style requirements.**
- **Prepare your article with the publication's audience in mind** – in terms of its style and general content.
- **Avoid technical terms and abbreviations** that some of your readers will not understand, unless they are explained within the article. Use clear and simple language.
- **Keep the material timely.** This does not mean you cannot include historical information if you are putting a new or different focus on it, e.g. 'Goji berries have been used in traditional Chinese medicine for thousands of years. Now science is demonstrating their potential benefits.'

KEY POINT Read sample articles from current editions of your target publication to note the style and depth of content that the editors prefer and the word length of a typical article. Aim to produce an article that follows the same broad format.

Establish your purpose

Ask yourself what is to be the overall aim of your article. Write your purpose down, so that you stay focused.

Research your topic

You should choose a topic of interest to you. It may even be a class project, or something more personal/individual. Find out more about the topic even if you think you already know a lot about the area. Seek out any information that may oppose your view and include arguments against this, if appropriate.

Organise your thoughts

You could then use a spider diagram (Fig. 15.2) or a mind map (Fig. 4.2) to brainstorm and start organising your ideas; follow this by producing an informal or formal outline/plan.

Write your article

- Include a headline that is informative but also attention-grabbing (but note also that sub-editors will often modify, or choose their own).

Box 19.1 How to compose an effective email

1. **Design a concise, accurate and meaningful subject line.** Apart from the email address of the sender, this is the first thing the recipient notices. Recipients will usually make a decision based on the apparent subject whether to view an email or hit the delete key. People receive lots of emails these days and often delete anything seen as irrelevant.
2. **Greet the recipient appropriately.** Whether or not you use a formal approach, e.g. 'Dear Professor Greenwood', or something less formal instead, e.g. 'Hi Bob', depends on your familiarity with the person. For your first contact the more formal approach is probably best, and if the recipient then replies to you by first name it would probably be safe to use the informal style in reply.
3. **Keep messages short and focused on the point of the communication.** Begin your message with the main item, to capture attention. Rambling sentences may soon lose the interest of your reader and increase their interest in the delete key. Add lines between paragraphs to provide structure and make for less cluttered reading. Do not write long emails unless there is a good reason, e.g. replying to a detailed email with a number of points to cover.
4. **Write using correct grammar, especially in formal communications.** Emails with grammatical errors do not look professional. Proofread the email and correct any typos. If your email program does not have a spell checker, consider composing important messages using a word processor that does and copying the message once it has been spell-checked. Use standard upper and lower case lettering and do not drift into texting abbreviations. Capitalising and italicising whole words in an email can come across as 'SHOUTING', while using all lower case may seem like you could not be bothered to do it properly.
5. **Make your purpose clear, as well as the timing and nature of any response expected.** Always be polite when asking for something in return. Emails can sometimes appear offhand because there is no voice tone or body language to 'soften' the message. That means you need to be very careful in your phrasing.
6. **Be nice and do not 'flame'.** The saying 'A word spoken in haste is oft repented at leisure' also applies to emails. If you find you are angry when composing an email, or when replying to one that has annoyed you, save a draft and come back to it when you have cooled off. Ask yourself 'What will I gain from sending angry comments?' Probably nothing advantageous.
7. **Do not assume confidentiality.** It is best to not put any comments in an email that you would not be comfortable having posted on the student notice board. Remember also that it is so very easy for someone to forward an email, so do not defame or speak negatively about a third party in an email. It might seem unlikely that a hacker or cyber criminal will access your email, but it does happen. In many universities and companies an email administrator also checks a portion of email messages to monitor workplace compliance and appropriateness.
8. **Be sure to identify yourself clearly at the end of your email.** This is especially important if the recipient does not know you very well. Let them know your title, name, affiliations and contact details in a neat email signature. In Microsoft Outlook a signature can be set up to be automatically added at the end of each email.
9. **Take care with attachments.** Explain what you are attaching, and why – everyone is wary of computer viruses, and rightly so. Large attachments can slow recipient email downloads, especially for someone using a portable device. If you do plan to attach a file, make sure it *is* attached and then open it to check that it is the correct file and version.
10. **Always re-read your message before sending.** Take account of all the tips above and only hit 'send' if you are certain that the email will help you achieve your purpose.

- Write the introductory sentence or paragraph so that it motivates readers to keep reading, e.g. 'Eating a handful of walnuts a day has recently been shown to reduce the risk of heart attacks.'
- Use the first sentence of each paragraph to introduce the sub-topic that you will cover, then give examples (p. 105). Examples and stories keep attention. 'Show, not tell' is a good maxim to follow. Also, shorter paragraphs with shorter sentences make for easier reading.

- During the writing stage allow yourself to think creatively. Write first, then critically edit and 'polish' later.

Ask someone to read your article aloud to you – you will notice then where they stumble over the words or have difficulty following it.

- When writing to inspire or persuade your readers, you should finish on a thought-provoking point or a call to action. In contrast, an educational article will usually finish with a clear summary of the main points.
- After you have finished your draft article, put it aside for a day or two then re-read it. Ask others to read it (preferably people with little or no knowledge of the topic) and seek their feedback on how it might be improved.
- Make the final improvements to the article and submit it well before the deadline, as the editor will need to have the space to include it in their magazine or newsletter.

Understanding writing legislation – before submitting writing for publication make sure you are familiar with any legislation in your region, e.g. defamation or libel laws, and stay within the required bounds.

KEY POINT Ensure that any nutritional claims you make in an article are backed up by sufficient scientific research; look into any conflicting studies as well as those that support your case.

Different types of articles

The informative article

This is the most common article that you are likely to write for a general audience. The informative article should:

- contain new, interesting and useful information for the article's target audience;
- be unbiased and truthful;
- be organised so that it is easily understood and its key points are easily remembered;
- have a clear and logical structure: opening paragraph, introduction, main body, summary, closing paragraph;
- be written in a crisp, interesting style that makes the reader want to learn more;
- be pitched at the reader's knowledge level – associate the new and unfamiliar with the familiar.

The persuasive article

Nutritionists involved in public health promotion need to know how to write persuasively for the public, while making use of evidence-based research for their material. All the good research information in the world's journals will not necessarily get the message out there and acted upon, unless these key aspects are appreciated:

Disturb then solve –

1. Emphasise the problems associated with the issue you are addressing.
2. Make your readers feel even moreso the difficulty of not having a solution to their problem.
3. Offer the solution that will remove their problem.

- The reader has to believe what you say, so your good credentials and honesty are essential.
- You need to take a clear stand, e.g. 'Exercise is imperative for a long and healthy life.'

- Begin your article where the reader's beliefs are most likely to be and then work to gently modify those beliefs through the logic of your argument. Build a rational basis of belief and understanding for the reader.
- Strengthen your argument by acknowledging any contrasting views and explaining the flaws in such views.
- Repetition is important as you need to ensure the reader understands before they will agree with you. You can make a point in several ways, e.g. quote, example, use different contexts.
- To persuade, you must appeal to the self-interest of the readers. Ask yourself, 'How can I put this in a way in which the reader will see real and possibly immediate benefits from applying this information?'
- Emphasise the advantages and benefits of commitment for long-term solutions.
- Although anecdotal, specific case studies can add weight to statistics by providing human interest and motivation to readers. It may also help them empathise with the people involved, helping you to win them over to your viewpoint.
- The conclusion of a persuasive article is a positive call to action. e.g. 'Make one healthy change to your eating habits, starting today, and stick to it to improve your future health.'

Persevere – if at first you do not succeed in getting published, rewrite to improve and reissue.

Sources for further study

Owl Materials (2009) *Journalism and Journalistic Writing*. Purdue University, West Lafayette, Indiana. Available: http://owl.english.purdue.edu/owl/resource/735/01/
Last accessed: 21/12/10.

Ross, K. (2004) *How to Write a Press Release in One Easy Lesson*. Australian Government Research hub. Available: http://www.australiacouncil.gov.au/research/arts_marketing/promotion/how_to_write_a_press_release_in_one_easy_lesson
Last accessed: 21/12/10.

Summer, D.E. and Miller, H.G. (2009) *Feature and Magazine Writing: Action, Angle and Anecdotes*. Wiley, Hoboken, New Jersey.

20 Consultation skills

> **Definition**
>
> **Consultation** (from Latin '*consultare*' to discuss or take counsel) – typically, a meeting between two or more people to consider a specific question or topic.

The word 'consultation' can be used to describe a number of situations, from an exchange of opinions, to a meeting with an expert, to referring to a source of information (p. 48). In nutritional or dietetic practice it is used mostly to describe a meeting between a client and a nutritionist or dietitian, where the client is seeking expert advice or information. It could similarly mean a discussion with an expert in the food industry or it could refer to a meeting with others about a product being developed or a process being evaluated. As a student, you are more likely to begin your studies as the client doing the consulting (e.g. of tutors or information sources) rather than a consultant, but the principles can work from both perspectives. In later stages of your degree programme, you may practise acting as a consultant, e.g. in mock exercises, where questioning and answering skills can be refined. It is also useful to consider consultation as a part of your work with others (Chapter 3).

> **KEY POINT** Consultation skills can be developed in a teamworking assignment, in meetings with tutors and in job interviews, as well as in a professional context.

The consultation process

Preliminaries

Comfortable and appropriate surroundings can have a major impact on the effectiveness of the meeting. In an interview, if you are asked to sit in an uncomfortable chair in a room that is hot and stuffy or cold and draughty you are unlikely to be at your best.

Stating your qualifications – confirming your consultant status through your credentials means that clients can make a value judgement about whether or not you are indeed a credible expert or a valid source of the information they require. They may also use this information to decide whether or not to trust you – an important component of a client–consultant relationship.

When acting as a consultant, you should consider the orientation of the furniture, for example sitting across a desk or table may be too formal and may also contain too many distractions; consider how tidy your desk is and what message an untidy desk may convey to the client. In any consultation, it is important to introduce yourself and it may also be appropriate to state your position or qualifications. Stating the purpose of the meeting will set the scene and also provide some structure to the start of the meeting. By being clear about what the aims are at the outset, you help focus expectations and you may save time if it transpires that not everyone in the meeting is there with the same expectations. When you are asking someone for information it is good practice to explain *why* you are asking for it, otherwise he or she may be reluctant to give the answer.

Getting started

Ideally, most meetings will have a scheduled start and a defined end time, since this allows the meeting to be focused and more effective. Most meetings will readily fill the amount of time allocated for them so be clear and realistic about what can be achieved in the allocated time. This will mean that there is less chance of someone feeling that you haven't spent enough time with him or her, which may affect any rapport that may have developed.

Using an agenda within a meeting – do not be too constrained by this list of topics, as this may have a negative effect on the flow of the consultation. If necessary, adapt the agenda to reflect the development of the discussion.

It is useful to have an agenda, or list of topics to be covered in the consultation, prepared in advance and agreed among the participants, where applicable. This provides structure to the process and helps keep the meeting 'on track'. Try to be realistic about what is achievable in the time available – avoid over-long or over-complex agendas.

Verbal communication

Effective communication is a two-way process involving both the speaker and the listener. Listening should not be a passive process, as this increases the risk of not being fully attentive and therefore missing vital information. Active listening means that the speaker is more likely to feel that what he or she is saying is important, and this will encourage him or her to continue. Using Gerard Egans' SOLER acronym for active listening will help to show that you are interested and paying attention to what the other person is saying.

SOLER (five actions to support active listening):
S – *squarely* face the client
O – *open* your posture
L – *lean* towards the client
E – *eye-contact* is important
R – *relax*.
Source: from Egan (1986).

Use of appropriate language for your audience will reduce the risk of misunderstandings and confusion. It is easy to slip into 'jargon' and scientific terms with colleagues, even as a student, but this type of language is not appropriate for everyone, especially in a client consultation. Conversely, you should not patronise your audience by assuming that they have no knowledge of the subject – adapt your language in response to the feedback you receive from the others involved (see also Box 14.2).

The types of questions you ask will affect the responses you get and choosing which are appropriate will depend on the purpose of the meeting:

- 'Closed' questions limit the response of the other participant(s), e.g. yes/no answers.
- Open questions encourage others to provide more information.

Example (leading question)

'You don't add salt to your food, do you?'

This would be better rephrased in more neutral terms, e.g. 'How much salt do you add to your food?'

Try to avoid leading questions which reveal what your expectations are and are likely to result in a biased answer. Think about more neutral ways to ask the same questions. Also try to avoid asking several questions in a row as this may confuse the other person and lead to unclear answers. Instead, allow him or her to answer your primary question and then ask the secondary or supplementary question if required, depending on his or her first answer.

It is important that everyone is clear about what is being said in a consultation and that there should be no misunderstandings. It is common practice to reflect on what has been said. This can either be by repeating back to the speaker what he or she has said or by paraphrasing what he or she has said into your own words. This not only shows that you have been listening but also that you have understood what has been said. Taking the time to do this allows details to be clarified or misunderstandings to be rectified.

Non-verbal communication

We communicate information by our actions as well as what we say. This body language or non-verbal communication is potentially more informative than what someone is saying – learning to pay close attention to this in others is a good communication skill. Non-verbal communication includes tone of voice, eye-contact, appearance, posture and gestures, and these are likely to create conflict in your mind if they do not match up with what the person is saying. If this is the case then you

Using non-verbal communication – the more you become aware of your own non-verbal communication (body language) the more you will be able to interpret that of others.

may need to explore these issues with further questioning. There is evidence to suggest that non-verbal cues which we may associate with deception, such as lack of eye-contact, are not actually associated with lying, so all cues should be interpreted with caution (DePaulo *et al.*, 2003). You will also need to consider your own non-verbal messages as they may give away your personal feelings to the client. You should remain emotionally objective so that the client can express his or her opinions and feelings without being concerned about the response from the consultant. Reflect on your own prejudices and intolerances, as identifying these will help you avoid revealing them through your non-verbal actions. For further information, consult a specialist text, such as Riggio and Feldman (2005).

Developing assertiveness and recognising boundaries

Definition

Assertiveness – the ability to express oneself in a positive and self-assured manner, without being aggressive or disrespectful to other people.

Being assertive is delivering your message confidently. Assertive behaviour is not only about being aware of what you want but also about being aware of the views and sensitivities of others. It has been described as 'speaking to be heard' and is *not* simply about forcing your opinions onto others (Gable, 2007). When your self-esteem is high you are more likely to behave assertively, but it is not about being 'right' or about being superior to others. In consultation situations, you should always consider the values of the other person and while you may well disagree with his or her opinions or behaviour you should respect these and work to make acceptable progress for everyone.

Recognising your limits is a critical skill to develop prior to professional practice, and as a student of food science and/or nutrition you will no doubt be asked for dietary advice by friends or family. In such circumstances, as in all nutrition or dietetic consultations, you should be clear about what your limitations are and you should also make these clear to others. Similarly, in professional practice in nutrition you might be asked for information or advice on treating an illness while your scope of practice might be in prevention. In such cases the individual should be advised to see his or her physician or a specialist dietitian. This is not an indication of any lack of ability on your part, but it is about doing no harm, an important aspect of any code of ethics (see also p. 129).

Consultation and change

Although you may recognise the value of making healthy dietary choices, not everyone will share your enthusiasm. Changing health-related behaviour is challenging and the Transtheoretical Model is often used to help understand the process. The model includes Prochaska and DiClemente's 'Stages of Change' and variables which describe the processes of change between theses stages (CPRC, 2000). In this model it is proposed that behaviour change is made up of a series of steps along a continuum:

- **pre-contemplative stage** – there is no intention to change the behaviour in question;
- **contemplative stage** – the individual is considering changing his or her behaviour at some point in the future;
- **preparation stage** – this is where the individual intends to take action in the next few weeks;

- **action stage** – behaviour is changed;
- **maintenance stage** – new behaviour becomes the new norm.

Behavioural change is not necessarily a linear process where the individual will move from one stage to the next; more likely, he or she will move forward and backwards before fully maintaining the new habits. New Year's resolutions are a good example to apply to this model, where good intentions in early January (contemplative stage) may see some change in behaviour (action) only to see us back in contemplative (or worse, pre-contemplative) by early February. Recognising these stages of change can save a lot of time in consultations as you may simply choose to give the client some written information to take away rather than trying, at this stage, to help him/her to implement action.

Ending the consultation

Finalising a consultation meeting – take the time to thank the participant(s) for their involvement and comments, so that they feel appreciated.

This will be easier if you have a set timescale and/or have a fixed agenda or required outcome. In the final stage of the meeting, you should review the aims of the meeting and reflect on whether these have been achieved. It may be that it has not been possible to achieve all that was intended, in which case you should summarise the progress so far. Discuss any agreed actions so that all parties are aware of what is expected of them, giving timescales if appropriate. Plan when the next meeting or contact will be so that they leave knowing what will happen next.

Record keeping

Definition

Aide memoire – (French) literally, a memory aid.

Keeping accurate records is an important skill for you to develop, since it will be crucial in a professional setting. As a student it is necessary to keep records of project work or laboratory practical work so that you can prepare accurate reports (p. 107). During a consultation or interview it may be necessary to take notes as an *aide memoire*, but try to keep these to a minimum, particularly when the other person is speaking, as it may inhibit the flow of the conversation. You should write up reports immediately after the end of the meeting as you will quickly forget what has been said, particularly if you go to another meeting or work on a different topic.

Legislation on confidentiality – in the UK, this is covered by the *Data Protection Act 1998*, which gives clear guidance on the nature of data that you can collect and store, and the purpose for which you can use this information. The *Freedom of Information Act 2000* may also apply if the information is held by a public body.

Professional bodies such as the British Dietetic Association have strict requirements for confidentiality in record keeping, but you should be aware of relevant legislation, which varies from country to country. In the UK, a health record is defined in Section 68 (2) of the *Data Protection Act 1998* as 'any record which consists of information relating to the physical or mental health or condition of an individual, and has been made by or on behalf of a health professional in connection with the care of that individual'. If you are on placement in a hospital you should be aware of the organisation's policy on record keeping and also the Health Professions Council's standards for record keeping. In summary, records should be written legibly and clearly, avoiding the use of jargon or abbreviations which are not nationally recognised, the entries should be accurate and appropriate and should always be signed and dated. You should not delete previous entries and records should be protected from loss, damage and tampering (Health Professions Council, 2008).

Text references

Cancer Prevention Research Center (2000) *Summary Overview of the Transtheoretical Model.* Available: http://www.uri.edu/research/cprc/transtheoretical.htm Last Accessed: 21/12/10.

DePaulo, B.M., Lindsay, J.J., Malone, B.E., Muhlenbruck, L., Charlton, K. and Cooper, H. (2003) Cues to deception. *Psychological Bulletin,* **129**(1), 74–118.

Egan, G. (1986) *The Skilled Helper: a systematic approach to effective helping.* Brookes Cole Pub Co., California.

Gable, J. (2007) *Counselling Skills for Dietitians,* 2nd edn. Blackwell, Oxford.

Health Professions Council (2008) *Standards of Conduct, Performance and Ethics.* Available: http://www.hpc-uk.org/assets/documents/10002367FINALcopyof SCPEJuly2008.pdf Last accessed: 21/12/10.

Riggio, R.R. and Feldman, R.S. (eds) (2005) *Applications of Nonverbal Communication.* Lawrence Erlbaum Associates, New Jersey.

Sources for further study

Holli, B.B., Calabrese, R.J. and O'Sullivan Maillet, J. (2003) *Communication and Education Skills for Dietetics Professionals,* 4th edn. Lippencott, Williams & Wilkins, Baltimore, New Jersey.

Learnhigher (2009) *Listening and Interpersonal Skills.* Available: http://www.learnhigher.ac.uk/learningareas/listeningandinterpersonalskills/home.htm Last accessed: 21/12/10.

Fundamental laboratory and clinical techniques

21 Your approach to practical work

Developing practical skills – these will include:

- designing experiments;
- observing and measuring;
- recording data;
- analysing and interpreting data;
- reporting/presenting.

All knowledge and theory in science has originated from practical observation and experimentation: this is equally true for disciplines as diverse as dietetics research and nutritional genomics. Practical work is an important part of most courses and often accounts for a significant proportion of the assessment marks. The abilities developed in practical classes will continue to be useful throughout your course and beyond, some within nutrition/food science and others in any career you choose (see Chapter 1).

Being prepared

KEY POINT You will get the most out of practicals if you prepare well in advance. Do not go into a practical session assuming that everything will be provided, without any input on your part.

Using textbooks in the lab – take this book along to the relevant classes, so that you can make full use of the information during the practical sessions.

The main points to remember are:

- **Read any handouts in advance:** make sure you understand the purpose of the practical and the particular skills involved. Does the practical relate to, or expand upon, a current topic in your lectures? Is there any additional preparatory reading that will help?
- **Take along appropriate textbooks**, to explain aspects in the practical.
- **Consider what safety hazards might be involved**, and any precautions you might need to take, before you begin (p. 145).
- **Listen carefully to any introductory guidance** and note any important points: adjust your schedule/handout as necessary.
- **During the practical session, organise your bench space** – make sure your lab book is adjacent to, but not within, your working area. You will often find it easiest to keep clean items of glassware, etc. on one side of your working space, with used equipment on the other side.
- **Write up your work as soon as possible**, and submit it on time, or you may lose marks.
- **Catch up on any work you have missed as soon as possible** – preferably, before the next practical session.

SAFETY NOTE Mobile phones should never be used in a lab class, as there is a risk of contamination from hazardous substances. Always switch off your mobile phone before entering a laboratory. Conversely, they are an extremely useful accessory for fieldwork.

Getting to grips with bioethics – in addition to any moral implications of your lab practicals and project work, you may have the opportunity to address broader issues within your course (see Box 21.1). Professional scientists should always consider the consequences of their work, and it is therefore important that you develop your appreciation of these issues alongside your academic studies.

Ethical and legal aspects

You will need to consider the ethical and legal implications of work in food science and nutrition at several points during your studies, as detailed in Chapter 23:

- Safe working means following a code of safe practice, supported by legislation, alongside a moral obligation to avoid harm to yourself and others, as discussed in Chapter 22.
- Any laboratory work that involves working with animals must be carefully considered; in regard to legal and ethical requirements.

Box 21.1 Bioethics

Contemporary food science, nutrition and dietetics degrees place increasing emphasis on the ethical and social impacts of scientific advances, and on the need for scientists engaged in potentially controversial work to communicate their ideas to the general public. Bioscience research, including nutritional science, food science and food production, raises many moral and legal dilemmas, requiring difficult choices to be made (e.g. in relation to animal testing), and students are likely to be asked to reflect on bioethical topics, e.g. in group discussions and debates on current issues such as:

- environmental ethics (e.g. use of genetically modified food crop plants);
- animal ethics (e.g. testing in animals for possible toxicity of natural products);
- human ethics (e.g. where an experiment conduced to see if β-carotene could protect smokers from lung cancer showed a significant increase in lung cancer cases in the group taking the β-carotene).

In discussing such topics (sometimes referred to as ethical, legal and social issues, ELSI), you will find that there is rarely a 'right' or 'wrong' answer. However, it is important to be able to consider these issues in a logical manner, and provide a reasoned argument in support of a particular viewpoint. Gaining experience in such debates should also help you understand some of the issues linked to the public understanding of science, and how these can be addressed (see: http://www.copus.org.uk/pubs_guides.html). Although a full exposition is beyond the scope of this book, the following provides a framework of principles for considering particular topics:

- Beneficence – the obligation to do good (for example, if it is possible to prevent suffering by a particular course of action, then it should be carried out).
- Non-maleficence – the duty to cause no harm (contained within the Hippocratic Oath of medical practitioners).
- Justice – the obligation to treat all people fairly and impartially (for example, lack of discrimination between people on the grounds of race or sex).
- Autonomy – the duty to allow an individual to make his or her own choices, without constraints (this principle underlies the notion of informed consent in clinical dietetics and other human research).
- Respect – the need to show due regard for others (for example, by taking into account the rights and beliefs of all people equally).
- Rationality – the notion that a particular action or choice should be based on reason and logic (many scientists would argue that the scientific method is an example of rationality).
- Precautionary principle – the notion that it is better not to carry out an action if there is any risk of harm (for example, in deciding that the risks associated with genetically modified livestock outweigh the potential benefits).

Understanding the various theories of ethics may also help you formulate your ideas. These include:

- Utilitarianism – the notion that it is ethical to choose the action that produces the greatest good for the greatest number.
- Deontology – a theory that states than a particular action is either intrinsically good (right) or bad (wrong). According to deontological theory, decisions should be based on the actions themselves, rather than on their consequences.
- Virtue theory – the notion that making decisions according to established virtues (e.g. honesty, wisdom, justice) will lead to ethically valid choices.
- Objectivism – the theory that what is right and wrong is intrinsic and applies equally to all people, places and times (the alternative is that morality is subjective, being dependent on the views of each individual)

The principles and issues of bioethics are considered on a number of websites, including:

- introductory bioethics (e.g. http://www.accessexcellence.org/RC/AB/IE/);
- bioethics resources (e.g. http://bioethics.od.nih.gov/, http://www.ethicsweb.ca/resources/ and http://www.beep.ac.uk/content/130.0.html);
- ethics discussion groups (e.g. http://www.beep.ac.uk/discuss/index.php?c=6/ and http://www-hsc.usc.edu/~mbernste/).

The following introductory texts give further information and guidance: Bryant *et al.* (2005), Gert *et al.* (2006) and Mepham (2005).

- Client consultation, giving dietary advice, projects involving human subjects, clinical placements and food science practicals all have legal and ethical aspects since legislation is in place to protect students, workers, clients and the public. Ethical and human aspects might include mandatory qualifications for giving dietary advice; the safety of others within your project or laboratory group; and implications of your advice and treatments for others. These must all be considered before any work is carried out. For example, the use of bioelectrical impedance to measure body fat content should not be done in people with heart pacemakers; giving someone the general advice to drink plenty of water may be contraindicated if the person, unknown to you, has fluid retention; and some foods, such as grapefruit, interfere with the working of a number of medications. In classes, projects and clinical placements, carefully follow the directives of your instructor and also bear in mind the advice based on the classical Hippocratic Oath: 'I will apply dietetic measures for the benefit of the sick according to my ability and judgement; I will keep them from harm and injustice.'

Basic requirements

Presenting results – although you don't need to be a graphic designer to produce work of a satisfactory standard, presentation and layout are important and you will lose marks for poorly presented work. Chapter 35 gives further practical advice.

Recording practical results

Bound books should be used for recording experimental results. You can paste sheets of graph paper into a bound book, as required.

A good-quality ink pen should be used for recording your data in a lab book or clinical records book, for legal purposes. Rule a single line through any incorrect material. Buy a black, spirit-based (permanent) marker for labelling experimental glassware, Petri plates, etc. Fibre-tipped fine-line drawing/lettering pens are useful for preparing final versions of graphs and diagrams for assessment purposes. Use a see-through ruler (with an undamaged edge) for graph drawing, so that you can see data points and information below the ruler as you draw.

Using inexpensive calculators – many unsophisticated calculators have a restricted display for exponential numbers and do not show the 'power of 10', e.g. displaying 2.4×10^{-5} as 2.4^{-05}, or even 2.4E–05, or 2.4 – 05.

Using calculators for numerical problems – Chapter 37 gives further advice.

Calculators

These range from basic machines with no pre-programmed functions and only one memory, to sophisticated programmable portable computers with many memories. The following may be helpful when using a calculator:

- Power sources. Choose a battery-powered machine, rather than a mains-operated or solar-powered type. You will need one with basic mathematical/scientific operations, including powers, logarithms (p. 244), roots and parentheses (brackets), together with statistical functions such as sample means and standard deviations (Chapter 38).
- Mode of operation. The older operating system used by, e.g., Hewlett-Packard calculators is known as the reverse Polish notation: to calculate the sum of two numbers, the sequence is 2 [enter] 4 + and the answer 6 is displayed. The more usual method of calculating this equation is as 2 + 4 =, which is the system used by the majority of modern calculators. Most newcomers find the latter approach to be more straightforward. Spend some time finding out how a calculator operates, e.g. does it have true algebraic logic ($\surd$ then number, rather than number then $\surd$)? How does it deal with (and display) scientific notation and logarithms (pp. 243–244)?
- Display. Some calculators will display an entire mathematical operation (e.g. '2 + 4 = 6'), while others simply display the last number/operation. The former type may offer advantages in tracing errors.

- Complexity. In the early stages, it is usually better to avoid the more complex machines, full of impressive-looking, but often unused pre-programmed functions – go for more memory, parentheses, or statistical functions rather than engineering or mathematical constants. Programmable calculators may be worth considering for more advanced studies. However, it is important to note that such calculators are often unacceptable for exams.

Presenting graphs and diagrams – ensure these are large enough to be easily read: a common error is to present graphs or diagrams that are too small, with poorly chosen scales (see p. 220).

Printing on acetates – standard overhead transparencies are not suitable for use in laser printers or photocopiers: you need to make sure that you use the correct type.

Presenting more advanced practical work

In some practical reports and in project work, you may need to use more sophisticated presentation equipment. Computer-based graphics packages can be useful – choose easily read fonts such as Arial or Helvetica for posters and consider the layout and content carefully (p. 77). Alternatively, you could use fine-line drawing pens and dry-transfer lettering/symbols, although this can be more time-consuming than computer-based systems, e.g. using Microsoft Excel (pp. 73 and 223–227).

To prepare overhead transparencies for spoken presentations, you can use spirit-based markers and acetate sheets. An alternative approach is to print directly from a computer-based package, using a laser printer and special acetates, or directly on to 35 mm slides. You can also photocopy on to special acetates. Further advice on content and presentation is given in Chapter 14.

Text references

Bryant, J., Baggott le Velle, L. and Searle, J. (2005) *Introduction to Bioethics*. Wiley, Chichester.

Gert, B., Culver, C.M. and Clouser, K.D. (2006) *Bioethics: A Systematic Approach*, 2nd edn. Oxford University Press, Oxford.

Mepham, B. (2005) *Bioethics: An Introduction for the Biosciences*. Oxford University Press, Oxford.

Sources for further study

Barnard, C.J., Gilbert, F.S. and MacGregor, P.K. (2001) *Asking Questions in Biology: Key Skills for Practical Assessments and Project Work*, 2nd edn. Prentice Hall, Harlow.

Howell, J.H., Sale, W.F. and Callahan, D. (2000) *Life Choices: A Hastings Center Introduction to Bioethics*, 2nd edn. Georgetown University Press, Washington, DC.

Mier-Jedrzejowicz, W.A.C. (1999) *A Guide to HP Handheld Calculators and Computers*. Wilson-Barnett, Tustin, California.
[Provides further guidance on the use of Hewlett-Packard calculators (reverse Polish notation).]

Singer, P. and Kuhse, H. (2006) *Bioethics: An Anthology*. Blackwell, Oxford.

22 Health and safety

Health and safety legislation – in the UK, the *Health & Safety at Work, etc. Act 1974* provides the main legal framework for health and safety. The *Control of Substances Hazardous to Health (COSHH) Regulations 2002* impose specific legal requirements for risk assessment wherever hazardous chemicals or biological agents are used, with Approved Codes of Practice for the control of hazardous substances, carcinogens and biological agents, including pathogenic microbes.

Definitions

Hazard – the ability of a substance or biological agent to cause harm.

Risk – the likelihood that a substance or biological agent might be harmful under specific circumstances.

Distinguishing between hazard and risk – one of the hazards associated with water is drowning. However, the risk of drowning in a few drops of water is negligible.

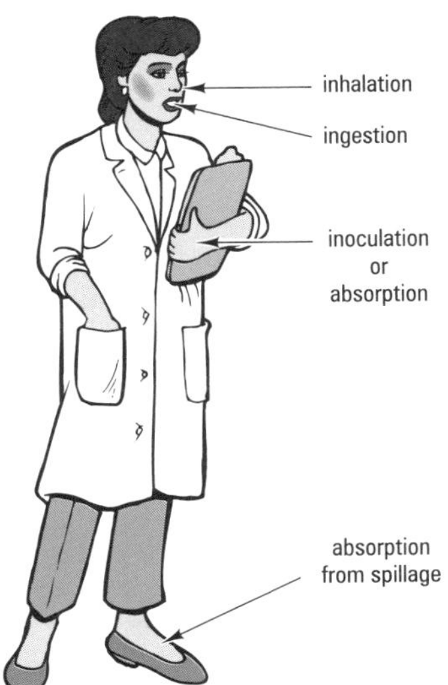

Fig. 22.1 Major routes of entry of harmful substances into the body.

Health and safety law requires institutions to provide a working environment that is safe and without risk to health. Where appropriate, training and information on safe working practices must be provided. Students and staff must take reasonable care to ensure the health and safety of themselves and of others, and must not misuse any safety equipment.

KEY POINT All practical work must be carried out with safety in mind, to minimise the risk of harm to yourself and to others – safety is everyone's responsibility.

Risk assessment

The most widespread approach to safe working practice involves the use of risk assessment, which aims to establish:

1. The intrinsic chemical, biological and physical hazards, together with any maximum exposure limits (MELs) or occupational exposure standards (OESs), where appropriate. Chemical manufacturers provide data sheets listing the hazards associated with particular chemical compounds, while pathogenic (disease-causing) microbes are categorised according to their ability to cause illness (p. 454).
2. The risks involved, by taking into account the amount of substance to be used, the way in which it will be used and the possible routes of entry into the body (Fig. 22.1).

KEY POINT It is important to distinguish between the intrinsic *hazards* of a particular substance and the *risks* involved in its use in a particular exercise.

3. The persons at risk, and the ways in which they might be exposed to hazardous substances, including accidental exposure (spillage).
4. The steps required to prevent or control exposure. Ideally, a non-hazardous or less hazardous alternative should be used. If this is not feasible, adequate control measures must be used, e.g. a fume cupboard or other containment system. Personal protective equipment (e.g. lab coats, safety glasses) must be used in addition to such containment measures. A safe means of disposal will be required.

The outcome of the risk-assessment process must be recorded and appropriate safety information must be passed on to those at risk. For most practical classes, risk assessments will have been carried out in advance by the person in charge: the information necessary to minimise the risks to students may be given in the practical schedule. Make sure you know how your department provides such information and that you have read the appropriate material before you begin your practical work. You should also pay close attention to the person in charge at the beginning of the practical session, as he or she may emphasise the major hazards and risks. In project work, you will need to be involved in the risk assessment process along with your supervisor, before you carry out any practical work.

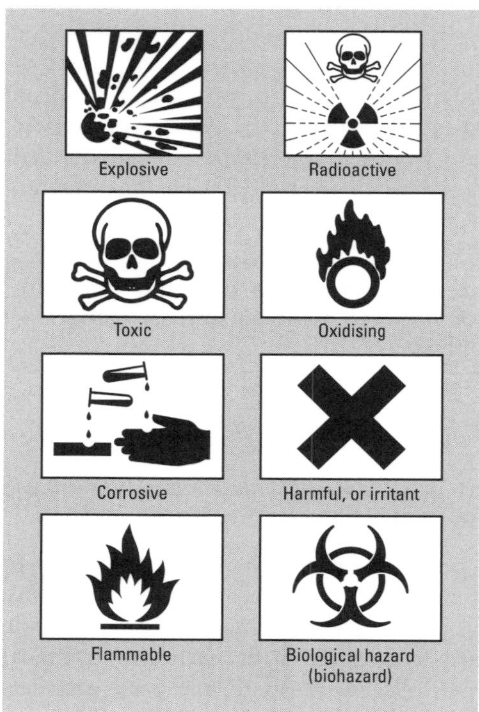

Fig. 22.2 Warning labels for specific chemical hazards. These appear on suppliers' containers and on tape used to label working vessels. New international symbols are being introduced and will be mandatory in 2015.

In addition to specific risk assessments, most institutions will have a safety handbook, giving general details of safe working practices, together with the names and telephone numbers of safety personnel, first aiders, hospitals, etc. Make sure you read this and follow any instructions given.

Basic rules for laboratory work

- Make sure you know what to do in case of fire, including exit routes, how to raise the alarm and where to gather on leaving the building. Remember that the most important consideration at all times is human safety: do not attempt to fight a fire unless it is safe to do so.
- All laboratories display notices telling you where to find the first aid kit and who to contact in case of accident/emergency. Report all accidents, even those appearing insignificant – your department will have a formal recording procedure to comply with safety legislation.
- Wear appropriate protective clothing at all times – a clean lab coat (buttoned up), plus safety glasses if there is any risk to the eyes.
- Never smoke, eat or drink in any laboratory, because of the risks of contamination by inhalation or ingestion (Fig. 22.1).
- Never mouth-pipette any liquid. Use a pipette filler (see p. 154) or, if appropriate, a pipettor (p. 155).
- Take care when handling glassware – see p. 158 for details.
- Know the chemical hazards warning symbols (Fig. 22.2).
- Use a fume cupboard for hazardous chemicals. Make sure that it is working and then open the front only as far as is necessary: many fume cupboards are marked with a maximum opening.
- Always use the minimum quantity of any hazardous materials.
- Work in a logical, tidy manner and minimise risks by thinking ahead.
- Always clear up at the end of each session. This is an important aspect of safety, encouraging responsible laboratory work.
- Dispose of waste in appropriate containers. Most labs will have bins for sharps, glassware, hazardous solutions and radioactive waste.

Basic food hygiene rules for food laboratory work

Many of the same principles for laboratory work also apply to food preparation areas; however, you will also need to follow basic food hygiene principles and take care with knives and hot objects, both for your own safety and the safety of others.

- Wash your hands thoroughly before entering the food lab.
- Wear appropriate protective clothing – clean lab coat (this should be a different coat to the one you wear in general laboratory classes) or plastic apron and hat or hairnet.
- If you have any cuts on your hands make sure these are covered. In food preparation areas dressings are usually blue or green so that they are easily detected in food should they drop off.
- You should remove all jewellery (a wedding band/ring may be permitted).

- Do not touch your face or hair without then washing your hands.
- If you have recently been suffering from vomiting or diarrhoea you should inform your tutor as it will not be appropriate for you to be preparing food.
- Take steps to avoid cross-contamination between raw foods and ready-to-eat foods by ensuring that you clean up as you go along; this includes washing utensils and surfaces as well as your hands.
- If you are tasting foods never return the utensil you used to taste the product back to the product.

Basic rules for fieldwork safety

Fieldwork can broadly be described as any activity undertaken as part of a student's programme of study which takes place away from the usual place of study. This would therefore cover visits to factories or other premises, placements as well as periods of work placement, or fieldwork associated with final year projects.

- Make sure you understand your objectives, the potential hazards and appropriate responses to them, before you set out.
- Your work must be designed carefully, to allow for the experience of the participants and the locations visited. Don't overestimate what can be achieved – fieldwork is sometimes physically demanding.
- Any physical disabilities must be brought to the attention of the organiser, so that appropriate precautions can be taken.
- A comprehensive first aid kit must be carried: at least two participants should have training in first aid.
- Never work alone without permission from your organiser.
- Your clothing and equipment must be suitable for all weather conditions likely to be encountered during the work.
- Leave full details of your intended working locations, routes and times. Never change these arrangements without informing someone.
- If you are on a visit to other premises such as a factory you must pay close attention to any health and safety information and specific instructions provided on the visit.

Sources for further study

American Chemical Society (2003) *Safety in Academic Chemistry Laboratories,* 7th edn. Available: http://membership.acs.org/c/ccs/pubs/SACL_Students.pdf
Last accessed: 21/12/10.

Anon. *BUBL Link – Biochemical Safety*. Available: http://bubl.ac.uk/link/b/biochemicalsafety.htm
Last accessed: 21/12/10.
[Provides a single-step link to a number of databases giving information on chemical and microbiological hazards.]

Food Standards Agency. *Safety and Hygiene Publications*. Available: http://www.food.gov.uk/aboutus/publications/safetyandhygiene/
Last accessed: 21/12/10.

Furr, A.K. (2000) *CRC Handbook of Laboratory Safety*. CRC Press, Boca Raton, Florida.

Health and Safety Executive (2005) *Control of Substances Hazardous to Health Regulations*. HSE, London.

23 Legal and ethical requirements in food safety and nutrition

Definitions

Standards and regulations – mandatory legal instruments (failure to comply is likely to result in prosecution).

Guideline – advisory document created by industry or regulatory agencies. Not legally binding, although compliance may form part of 'due diligence' procedures.

Commercial food manufacture, distribution and consumption is covered by a range of legal standards, regulations, guidelines and codes designed to protect the consumer from the dangers of unsafe food.

KEY POINT Food safety legislation is different in each country, although the fundamental aim of ensuring safe food for consumers lies at the heart of all legislation.

In line with the principle of natural justice, namely that humans should treat each other as they would like to be treated, food producers also have an ethical duty to consumers to not cause them any harm. This is often manifest in terms of industry codes of conduct and voluntary standards.

All students of food science, nutrition and dietetics need to have an understanding of the basic principles of legal and ethical conduct, and this is considered within this chapter from the perspective of issues related to food safety.

Food Standards Agency – this is an independent UK government department, established by the *Food Standards Act 1999.* It works independently to protect consumer interests and health in relation to food (see http://www.food.gov.uk/).

UK food safety legislation

The main legislation that applies to food businesses in the UK is the *Food Safety Act 1990*. This is an Act of Parliament, therefore the conditions set out in the document are legally binding. Any changes to this legislation need to be considered and approved at the parliamentary level. The details of the law are provided by Statutory Instruments, also known as regulations, where powers can be delegated to a government minister or other authority, e.g. local authorities.

The Food Safety Act 1990

Responsibilities of food businesses under the *Food Safety Act 1990*

- To ensure that food does not damage the health of consumers.
- To ensure that food is of expected quality, nature and substance.
- To ensure that product labelling and advertising are not misleading.

While this remains the main legislation associated with food safety in the UK, amendments have been made in response to European Union (EU) food safety legislation, specifically the *General Food Law Regulations* (Regulation (EC) 178/2002). The Act covers all food businesses throughout the food distribution chain, so is applicable from primary producers through to retail and catering industries.

The Act defines food as 'any substance or product, whether processed, partially processed or unprocessed, intended to be, or reasonably expected to be ingested by humans'. It is wide-ranging in that it covers the selling of (or intention to sell) food, the free supply of food in the course of a business, preparation, labelling and presentation, storing, transporting, import and export of food.

If a food business fails to comply with the requirements of the Act they will be deemed to have committed an offence. The Act is enforced by environmental health practitioners and trading standards officers. These officials carry out routine inspections and sampling as well as responding to complaints. They are also important sources of education and information for food businesses. There is a number of options available to enforcement officers if a business fails to comply with the conditions of the Act, including the following:

Chartered Institute of Environmental Health – this is the professional organisation for environmental health in the UK, setting standards and accrediting courses leading to qualification as an environmental health practitioner (see http://www.cieh.org/).

- **Providing informal advice** on the operation of a particular business.
- **Sending an informal letter**, e.g with recommendations for improvement of practice.
- **Serving an improvement notice**, setting out what steps the business must take to comply with the law within a specified time.
- **Serving a prohibition notice**, requiring a business immediately to stop an activity that contravenes the law. This might include stopping a specific process, or using a piece of equipment or prohibiting a person from working in the food business.
- **Serving an emergency prohibition notice** or order which immediately closes down a food business that is believed to be an imminent risk to health.

The choice of action will reflect the severity of non-compliance, in terms of the risk to the public. In the case of improvement or prohibition notices, these are usually issued under the *Food Hygiene Regulations 2006*.

Food Hygiene Regulations 2006

Taking *all* reasonable steps to ensure safe food – this is often termed *'due diligence'*, and can be used by a food business as a defence against prosecution under food safety legislation.

One of the most important aspects of these regulations is that food businesses must be able to show that they are doing everything possible to ensure that the foods they are producing are safe to eat. This involves keeping detailed records of food safety management procedures, such as hazard analysis critical control point analysis. Under these regulations, food businesses must register their premises with the relevant authority before starting production. The general requirements aim to ensure that the business premises are well-designed, well-maintained and clean. Personal hygiene is important (Chapter 22) and personnel working with food should be appropriately trained.

Hazard Analysis Critical Control Point (HACCP)

Training in food hygiene – many university degree programmes also include training to the level of basic certification in food hygiene as part of their curricula; take up this opportunity if available, as many jobs require applicants to have basic food hygiene training.

The HACCP system is an internationally recognised management approach that takes a systematic and preventative approach to address issues of food safety, based on the following seven principles:

1. **Conduct a hazard analysis** to identify potential biological, chemical and/or physical hazards which are likely to cause illness or injury if not controlled.
2. **Identify the critical control points** (CCPs). A critical control point is one at which a hazard can be reduced or eliminated. Note that not all hazards which are identified will be critical, for example if a hazard is eliminated at a later stage in production.
3. **Establish critical limits for each control point.** These are set in order to identify the difference between a safe and an unsafe operating condition at each CCP.
4. **Establish monitoring procedures for each control point**. This will ensure that the process is under control, and critical limits are not exceeded. The frequency and method of monitoring should be written into HACCP plans.
5. **Establish corrective actions** to identify the action required as a result of any monitored deviation from a critical limit at a CCP. This

Find out more about HACCP – principles and example applications are detailed in the Codex Alimentarius Recommended International *Code of Practice Codex General Principles of Food Hygiene* (http://www.codexalimentarius.net/web/index_en.jsp).

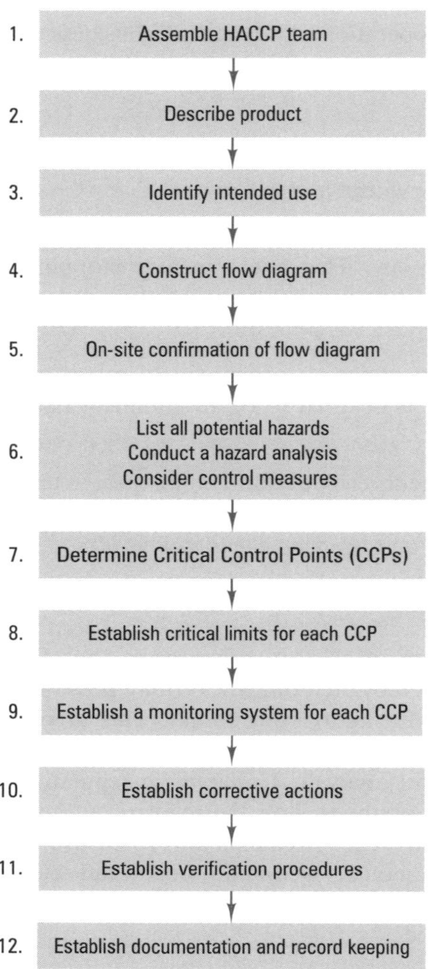

1. Assemble HACCP team
2. Describe product
3. Identify intended use
4. Construct flow diagram
5. On-site confirmation of flow diagram
6. List all potential hazards
Conduct a hazard analysis
Consider control measures
7. Determine Critical Control Points (CCPs)
8. Establish critical limits for each CCP
9. Establish a monitoring system for each CCP
10. Establish corrective actions
11. Establish verification procedures
12. Establish documentation and record keeping

Fig. 23.1 Flow chart showing the logical sequence for application of Hazards Analysis Critical Control Points (HACCP) (from FAO, 2003).

should prevent the release of unsafe food, and therefore safeguard the health of consumers.

6. **Establish appropriate verification procedures** which determine the validity of the plan. These should go beyond just monitoring in ensuring that the plan is effective.
7. **Establish documentation procedures** to provide a complete record of monitoring and other procedures.

A HACCP analysis is usually conducted by a team following the sequence of steps identified in Fig. 23.1 (FAO, 1998). While it takes training and experience in order to correctly identify critical control points, a decision tree can be used to standardise the process (Fig 23.2). Students who are on placements in any food related-industry will need to have a working knowledge of HACCP. It may also be an important part of assignments or case studies associated with food quality and safety.

ISO 9000

This is another internationally recognised quality management system, applicable to any industry, not only the food industry. It is maintained by the International Organisation for Standardisation (http://www.iso.org/). In a broadly similar way to HACCP, the ISO 9000 family of standards requires companies to have a set of procedures for key processes, with adequate records and appropriate corrective procedures and review processes. A company can be certified as having met these standards, allowing their customers to confirm that appropriate quality management systems are in place.

Food Labelling Regulations

There is a number of country-specific mandatory requirements for the labelling of pre-packed foods. For example, in the UK, these are detailed in the *Food Labelling Regulations (2004)*, which requires the following to appear on a label on the food product:

- **the name of the food**;
- **a list of ingredients,** including known allergens (see below);
- **quantitative ingredient declaration** (QUID) for certain specified ingredients and components, e.g. a product labelled strawberry yoghurt would be required to state the percentage of strawberries present in the product;
- **an appropriate durability indication**, typically either a 'use by' or 'best before' date;
- **any special storage conditions or conditions of use**;
- **the name or business name and an address or registered office** of either or both of (i) the manufacturer or packer, or (ii) a seller established within Europe;
- **particulars of the place of origin or provenance of the food** if failure to give such particulars might mislead a purchaser to a material degree as to the true origin or provenance of the food;
- **instructions for use** if it would be difficult to make appropriate use of the food in the absence of such instructions.

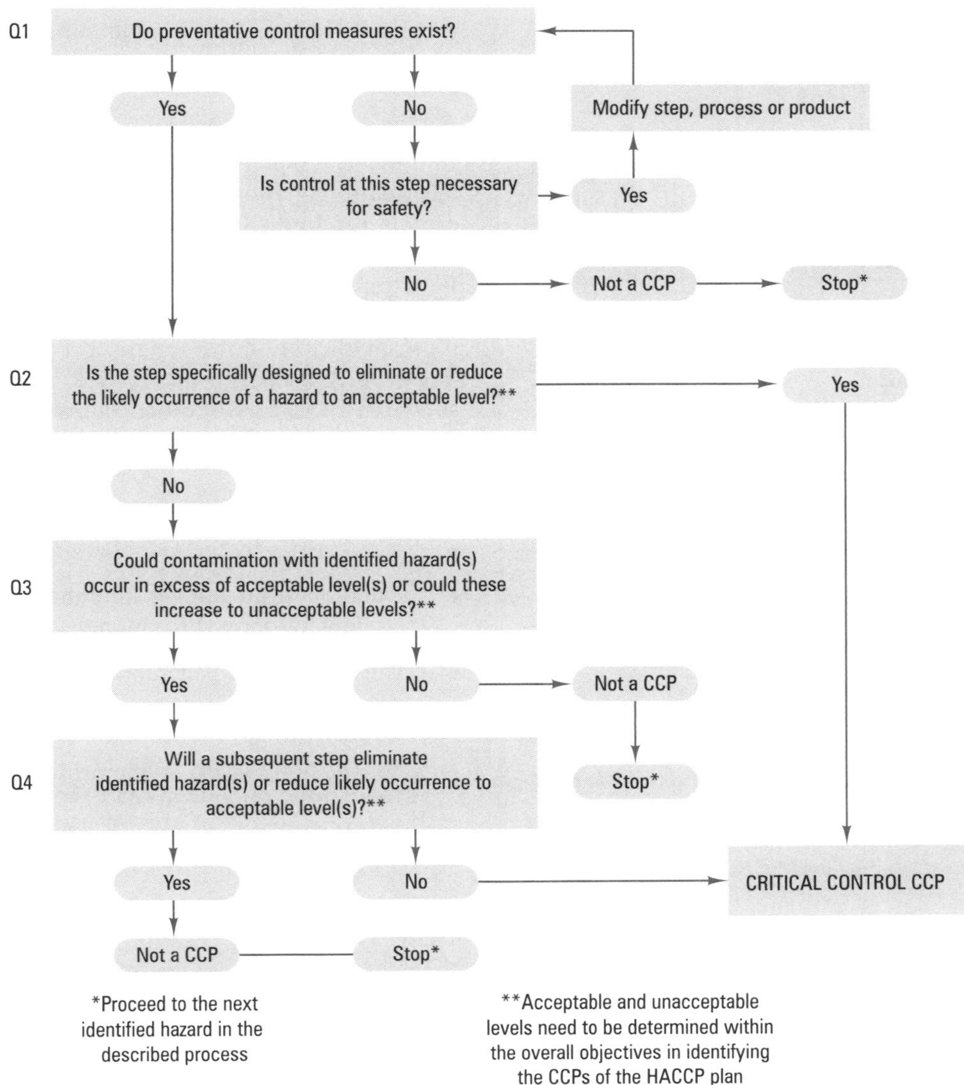

Fig. 23.2 Example of a decision tree to identify critical control points, CCPs (from FAO, 2003).

'Clean label' foods contain few E-numbers and are being driven by consumer demand for 'natural' products. E numbers are codes for food additives used in the European Union.

There is also a number of regulations and amendments relating to specific issues, such as declaration of allergens and nutrition information. It is also necessary to declare the use of sweeteners, including statements for sources of phenylalanine and polyols in excess of 10 per cent (e.g. '*excessive consumption may produce laxative effects*'). In addition, products which contain more than 150 mg of caffeine per litre are required to declare this on the label.

Allergens (p. 313) must be declared on the label, clearly stating '*contains*' followed by the allergen. Allergens which must be declared on the label include: cereals containing gluten; crustaceans; eggs; fish; peanuts; soybeans; milk (including lactose); nuts; celery; mustard; sesame seeds; sulphur dioxide and sulphites above a specified concentration; and lupins; molluscs. Manufactures often include advisory

Table 23.1 Nutrient labelling formats

Group 1 (/100 g)	
energy	kJ and kcal
protein	g
carbohydrate	g
fat	g
Group 2 (/100 g)	
energy	kJ and kcal
protein	g
carbohydrate	g
of which:	
– sugars	g
– fat	g
of which	
– saturates	g
fibre	g
sodium	g

statements such as 'may contain nuts' but this is not a legal requirement if the allergen is only likely to be present in very small amounts.

Nutrition information

In Europe, food manufacturers are only obliged to include nutrition information on food labels if they make a claim about the nutrient content e.g. *low fat*. In the USA, nutrition information should be presented on the label of all food intended for human consumption, in accordance with the Electronic Code of Federal Regulations 21, part 101.9 (http://ecfr.gpoaccess.gov/). In Europe, where nutrient information is included, it must either be shown in Group 1 format (also known informally as the 'big 4') or Group 2 format (also known as 'big 4 + little 4' or 'big 8', as shown in Table 23.1. If other claims are made about nutrients not listed in Group 2 then they must also be included.

The information for nutrient labelling can be obtained by several different means, including:

- direct analysis by the manufacturer or their representative;
- calculation from known amounts of ingredients and information on their content derived from analysis;
- calculation from generally established and accepted data, such as *McCance and Widdowson's The Composition of Foods* (FSA, 2002) (Food Labelling Regulations 1996). For example, values for energy content should be calculated using the following values: 1 g carbohydrate provides 17 kJ (4 kcal); 1 g protein provides 17 kJ (4 kcal); 1 g fat provides 37 kJ (9 kcal); 1 g ethanol provides 29 kJ (7 kcal).

Front of pack nutritional information

There is increasing use of front-of-pack (FOP) labelling to allow consumers to make more informed choices about the foods they eat. The use of FOP is not yet standardised, as some food manufacturers show guideline daily amounts (GDAs) and others use signposting such as the 'traffic light' labelling system of the UK Food Standards Agency (http://www.food.gov.uk/foodlabelling/signposting/). The GDA system (http://www.gdalabel.org.uk) is based on the amount of each nutrient per portion of the food, expressed as a percentage of daily requirements (Fig 23.3), while traffic lights are calculated per 100 g of the food, with clear guidance on the level of each nutrient in each colour band (Table 9.1). There are also separate guidelines for labelling of drinks (FSA, 2007).

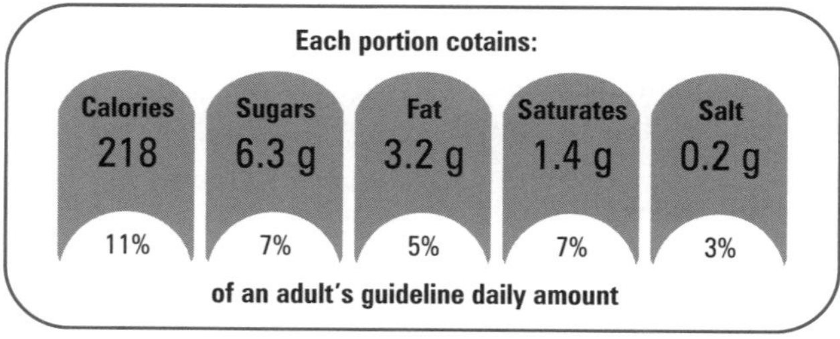

Fig. 23.3 Example of GDA (guideline daily amounts) front-of-pack label (FDF, 2009).

Fig. 23.4 Example of the Carbon Reduction label (Carbon Trust, 2008).

Currently, the inclusion of FOP information is voluntary and research is continuing to find the most suitable method of informing consumers.

Clear labelling

The *Food Labelling Regulations 1996* give clear information about how the necessary information should be presented on food labels. As well as being easily understood, the information should be legible and in a place that is clearly visible to the consumer. Voluntary guidelines by the FSA in the UK give best practice advice for clear food labelling and these should be considered if you are working on product development projects (Chapter 65) which you develop from idea to finished product. The points that need to be considered are:

- **Essential information** – all the legally required information should be on the same field of vision on the pack.
- **Font size, format contrast and print quality** should be such that the information is clear and easily read.
- **Prioritisation of information and layout** should be considered at the beginning of the design process so that the important (required) information is included.
- **Increasing the printable area on the pack** should be considered if the label is not clear.
- **Format of any date marking** should be consistent and unequivocal**.** The FSA recommend date, month, year format, e.g. 06 Jan 2011.
- **Allergen information** can be provided in a separate box in the same field of vision as the ingredients list from the ingredients. All allergens should be listed in the same box.
- **Format of nutrition information** as described above.

Carbon-reduction labelling

An increasing number of food manufacturers are incorporating 'carbon footprint' information in their food labels. The Carbon Trust's Carbon Reduction Label, shown in Fig 23.4, is currently the most widespread in use, and is supported by a Publicly Available Specification from the BSI, known as PAS 2050. This was launched in 2007 and has subsequently become a global standard for product carbon footprint calculation. The process of calculating greenhouse gas emissions is detailed in the document, and measures the amount of greenhouse gas emissions from all stages of the life of the product including production, transportation, preparation, use and disposal. The label itself highlights a brand's commitment to cut carbon / greenhouse gas emissions. Brand manufacturers must make the certified footprint measurement publically available (e.g. on their website), and also have the option of displaying the footprint measurement on the label itself. Carbon-reduction labelling is likely to become an increasingly common feature on food packaging.

Ethics

Understanding the significance of ethical practice – many professions, including dietitians, are required to follow a specific code of ethical conduct as part of their duty of care to their clients.

Ethics is defined as a system of moral principles by which people might judge 'right' from 'wrong'. Therefore it is important for you to develop an understanding of the basics of ethics and their implications to your chosen profession during your degree studies. It is also important to consider the ethical implications of your own actions in practical classes and during research projects at undergraduate level.

When planning any research (p. 129) it is always necessary to consider whether the experiments being considered have any potential impact on people or animals. All universities have ethics committees to consider the implications of any such research before it is conducted – importantly, the research can only take place once formal ethical approval has been granted. In considering whether a project can be approved, the committee will base its decision on a number of guiding principles described below. The same broad ethical principles also underpin professional codes of ethics for dietitians, nutritionists and food scientists (see also Box 21.1, p. 130).

Beneficence

The principle of beneficence means that all actions should aim to do good and be in the best interest of the individual. In research, this requires any potential risks associated with the research to be outweighed by the benefits. Ethics committees will also consider whether the welfare of the subjects is at risk by considering the competency of the investigator to undertake the procedures, and by the proposed actions themselves.

Primum non nocere – this phrase literally means 'first, do no harm', and this principle of non-maleficence is incorporated into the Hippocratic Oath, taken by medical practitioners.

Non-maleficence

Non-maleficence is defined as the principle of doing no harm and means that in considering any action it is necessary to consider all possible negative outcomes. For example, in research projects you would need to consider each step of the protocol and identify any issues or potentially harmful consequences.

Understanding cross-over trials – these studies are designed so that each subject receives both treatment and control. In its simplest form, two groups would be created, with the first receiving treatment then control, while the second would receive control then treatment.

Justice

The principle of justice ensures that everyone is treated fairly and impartially. This principle is particularly difficult to apply when resources are limited. In research projects the potential burdens and benefits of participation should be equal among all subjects. This is one reason why cross-over trials are valued in intervention trials.

Autonomy

To be autonomous each individual must have the right to make his or her own decisions. If someone is taking part in a research project or, indeed, is undergoing treatment of any kind, then he or she must be free to make his or her own choice as to whether to participate and whether to continue. The principle of autonomy is linked to the notion of informed consent and the need for an individual to be fully aware of the proposed intervention or action.

Practical aspects of research ethics – two important documents to prepare are:

1. an information sheet;
2. a consent form.

Participants should read the information sheet and confirm their agreement to be involved in the study by signing the consent form.

Informed consent

To obtain informed consent the researcher must clearly explain to the participant what is expected of him or her and exactly what the study involves. As part of this process, you must also let the person know whether there are any potential risks associated with taking part in the project. Examples of this include considering food allergies (Chapter 46) if you are asking people to taste food products, or explaining the risk of bruising if you are taking blood samples.

Box 23.1 Applying for ethics approval for your research from an ethics committee

The exact procedures and processes for applying for ethical approval will vary between organisations, but the following questions are likely to be asked (ESRC, 2010). All of the information submitted to the committee should be explained in a way that would be understandable to someone with limited scientific knowledge of the subject.

1. **Aims of the research.** You need to make it clear what you are going to achieve in the project so that the committee can judge whether your proposal is likely to achieve these aims.
2. **Scientific background of the research.** All research proposals are supported by existing research. Review of existing literature will allow you to ensure that this research has not already been done.
3. **Study design.** How do you propose to do the study?
4. **Participants.** Who will be taking part in the study and also who is not eligible to take part? How will these people be identified and how many participants do you need to achieve your aims?
5. **Vulnerable groups.** Does your research involve children or those with a learning disability or cognitive impairment?
6. **Methods of data collection.** What methods do you intend to use? The committee will consider the appropriateness of the methods to meet the aims of the project.
7. **Methods of data analysis.** It is important that you consider how you will analyse the data you collect. If you are not clear how you will do this you may need to reconsider your methods or study design.
8. **Risks and benefits to research participants or third parties.** It is generally not considered ethical to offer any kind of inducement to take part in a research project as this may be seen as coercion. Risks to participants may include physical or psychological harm.
9. **Risks to researchers.** It is important that you consider your own safety, whether you are using harmful chemicals in a laboratory or are collecting data outside of the university.
10. **Procedures for informed consent.** Participants must be fully aware of what is expected of them, what you will do with the data you collect and what will be the impact of taking part in this research. Informed consent must be collected for all aspects of the project, so if you plan to use data or samples collected at a later date you must make this clear to participants.
11. **Expected outcomes, impacts and benefits of research.** How will this research project add to existing knowledge?
12. **Dissemination of findings.** How will your results be shared with a wider audience and how will participants be able to find out the outcomes of the research? It is not usually appropriate to provide individual feedback to participants.
13. **Measures taken to ensure confidentiality, privacy and data protection.** This could involve using only numbers to identify data and ensuring that any personal information is stored appropriately, usually by your supervisor. Information would be destroyed at the end of the project.

Text references

Carbon Trust, DEFRA, BSI (2008) PAS 2050:2008 Specification for the assessment of the life cycle greenhouse gas emission of goods and services. Available: http://shop.bsigroup.com/en/Browse-by-Sector/Energy–Utilities/PAS-2050/ Last accessed: 21/12/10.

ESRC Economic & Social Research Council (2010) *Framework for Research Ethics*. Available: http://www.esrcsocietytoday.ac.uk/ESRCInfoCentre/Images/Framework%20for%20Research%20-Ethics%202010_tcm6-35811.pdf Last accessed: 21/12/10.

FAO, Food and Agriculture Organization of the United Nations (1998) *Food Quality and Safety Systems – A Training Manual on Food Hygiene and the Hazard Analysis and Critical Control Point (HACCP) System.* Available: http://www.fao.org/docrep/W8088E/w8088e00.htm#Contents Last accessed: 21/12/10.

FAO, Food and Agriculture Organization of the United Nations (2003) *Recommended International Code of Practice: General Principles of Food Hygiene CAC/RCP 1-1969, Rev. 3 (1997), Amended 1999*. Available: http://www.fao.org/docrep/005/Y1579E/y1579e02.htm Last accessed: 21/12/10.

FDF, Food and Drink Federation (2009) *GDAs Explained.* Available: http://www.gdalabel.org.uk/gda/explained.aspx Last accessed: 21/12/10.

Food Labelling (Amendment) (England) (No. 2) Regulations (2004) HMSO London.

FSA, Food Standards Agency (2002) *McCance and Widdowson's The Composition of Foods*, 6th summary edn. Royal Society of Chemistry, Cambridge.

FSA, Food Standards Agency (2007) *Front-of-pack Traffic Light Signpost Labelling Technical Guidance Issue*. Available: http://www.food.gov.uk/foodlabelling/signposting/technicalguide
Last accessed: 21/12/10.

FSA, Food Standards Agency (2008) *Food Labelling Clear Food Labelling Guidance*. Available: http://www.food.gov.uk/multimedia/pdfs/clearfoodlabelling.pdf
Last Accessed: 21/12/10.

Sources for further study

FSA, Food Standards Agency (2009) *The Food Safety Act 1990 – A Guide for Food Businesses*. Available: http://www.food.gov.uk/multimedia/pdfs/fsactguide.pdf
Last accessed: 21/12/10.

Jukes, D.J. (2007) *Foodlaw – Reading*. Available: http://www.reading.ac.uk/foodlaw/
Last accessed: 21/12/10.

Using chemicals responsibly – be considerate to others: always return storeroom chemicals promptly to the correct place. Report when supplies are getting low to the person who looks after storage/ordering. If you empty an aspirator or wash bottle, fill it up from the appropriate source.

Finding out about chemicals – *The Merck Index* (O'Neil *et al.*, 2006) and the *CRC Handbook of Chemistry and Physics* (Lide, 2008) are useful sources of information on the physical and biological properties of chemicals, including melting and boiling points, solubility, toxicity, etc. (see Fig. 24.1).

8599. Sodium Chloride. [7647-14-5] Salt; common salt. ClNa; mol wt 58.44. Cl 60.67%, Na 39.34%. NaCl. The article of commerce is also known as ***table salt***, ***rock salt*** or ***sea salt***. Occurs in nature as the mineral ***halite***. Produced by mining (rock salt), by evaporation of brine from underground salt deposits and from sea water by solar evaporation: *Faith, Keyes & Clark's Industrial Chemicals*, F. A. Lowenheim, M. K. Moran, Eds. (Wiley-Interscience, New York, 4th ed., 1975) pp 722-730. Toxicity studies: E. M. Boyd, M. N. Shanas, *Arch. Int. Pharmacodyn.* **144,** 86 (1963). Comprehensive monograph: D. W. Kaufmann, *Sodium Chloride*, ACS Monograph Series no. **145** (Reinhold, New York, 1960) 743 pp.

Cubic, white crystals, granules, or powder; colorless and transparent or translucent when in large crystals. d 2.17. The salt of commerce usually contains some calcium and magnesium chlorides which absorb moisture and make it cake. mp 804° and begins to volatilize at a little above this temp. One gram dissolves in 2.8 ml water at 25°, in 2.6 ml boiling water, in 10 ml glycerol; very slightly sol in alcohol. Its soly in water is decreased by HCl. Almost insol in concd HCl. Its aq soln is neutral. pH: 6.7-7.3. d of satd aq soln at 25° is 1.202. A 23% aq soln of sodium chloride freezes at −20.5°C (5°F). LD_{50} orally in rats: 3.75 ±0.43 g/kg (Boyd, Shanas).

Note: ***Blusalt***, a brand of sodium chloride contg trace amounts of cobalt, iodine, iron, copper, manganese, zinc is used in farm animals.

USE: Natural salt is the source of chlorine and of sodium as well as of all, or practically all, their compds, e.g., hydrochloric acid, chlorates, sodium carbonate, hydroxide, etc.; for preserving foods; manuf soap, to salt out dyes; in freezing mixtures; for dyeing and printing fabrics, glazing pottery, curing hides; metallurgy of tin and other metals.

THERAP CAT: Electrolyte replenisher; emetic; topical anti-inflammatory.

THERAP CAT (VET): Essential nutrient factor. May be given orally as emetic, stomachic, laxative or to stimulate thirst (prevention of calculi). Intravenously as isotonic solution to raise blood volume, to combat dehydration. Locally as wound irrigant, rectal douche.

Fig. 24.1 Example of typical *Merck Index* entry showing type of information given for each chemical. From *The Merck Index: An Encyclopedia of Chemicals, Drugs, and Biologicals,* Fourteenth Edition, Maryadele J. O'Neil, Patricia E. Heckelman, Cherie B. Koch, Kristin J. Roman, Eds. (Merck & Co., Inc., Whitehouse Station, NJ, USA, 2006). Reproduced with permission from *The Merck Index*, Fourteenth Edition.

Using chemicals

Safety aspects

In practical classes, the person in charge has a responsibility to inform you of any hazards associated with the use of chemicals. For routine practical procedures, a risk assessment (p. 133) will have been carried out by a member of staff and relevant safety information will be included in the practical schedule: an example is shown in Table 24.1. This information is part of basic health and safety requirements (Chapter 22, p. 133).

In project work, your first duty when using an unfamiliar chemical is to find out about its properties, especially those relating to safety. Your department must provide the relevant information to allow you to do this. If your supervisor has filled out the form, read it carefully before signing. Box 24.1 gives further advice.

KEY POINT Before you use any chemical you must find out whether safety precautions need to be taken and complete the appropriate forms confirming that you appreciate the risks involved.

Selection

Chemicals are supplied in various degrees of purity and this is always stated on the manufacturers' containers. Suppliers differ in the names given to the grades and there is no conformity in purity standards. Very pure chemicals cost more, sometimes a lot more, and should only be used if the situation demands. If you need to order a chemical, your department will have a defined procedure for doing this.

Preparing solutions

Solutions are usually prepared with respect to their molar concentrations (e.g. mmol L^{-1}, or mol m^{-3}), or mass concentrations (e.g. g L^{-1}, or kg m^{-3}):

Table 24.1 Representative risk-assessment information for a practical exercise in molecular biology, involving the isolation of DNA

Substance	Hazards	Comments
Sodium dodecyl sulphate (SDS)	Irritant Toxic	Wear gloves
Sodium hydroxide (NaOH)	Highly corrosive Severe irritant	Wear gloves
Isopropanol	Highly flammable Irritant/corrosive Potential carcinogen	No naked flames Wear gloves
Phenol	Highly toxic Causes skin burns Potential carcinogen	Use in fume hood Wear gloves
Chloroform	Volatile and toxic Irritant/corrosive Potential carcinogen	Use in fume hood Wear gloves

Box 24.1 Safe working with chemicals

You should always treat chemicals as potentially dangerous, following these general precautions:

- **Do not use any chemical until you have considered the risks involved** – for lab classes, you should carefully read all hazard and risk information provided *before* you start work. In project work, you may need to be involved in the risk-assessment process with your supervisor.
- **Wear a laboratory coat at all times** – the coat should be fully fastened and cleaned appropriately, should any chemical compound be spilled on it. Closed-toe footwear will protect your feet should any spillages occur.
- **Make sure you know where the safety apparatus is kept before you begin working** – this apparatus includes eye bath, fire extinguishers and blanket, first aid kit.
- **Wear safety glasses and gloves when working with toxic, irritant or corrosive chemicals, and for any substances where the hazards are not yet fully characterised** – make sure you understand the hazard warning signs (p. 134) along with any specific hazard-coding system used in your department. Carry out procedures with solid material in a fume cupboard.
- **Use aids such as pipette fillers to minimise the risk of contact with hazardous solutions** – these aids are further detailed on pp. 154–155.
- **Never smoke, eat, drink or chew gum in a lab where chemicals are handled** – this will minimise the risk of ingestion.
- **Label all solutions appropriately** – use the appropriate hazard warning information (see pp. 134, 150).
- **Report all spillages of chemicals/solutions** – make sure that spillages are cleaned up properly.
- **Store hazardous chemicals only in the appropriate locations** – for example, a spark-proof fridge is required for flammable liquids; acids and solvents should not be stored together.
- **Dispose of chemicals in the correct manner** – if unsure, ask a member of staff (do not assume that it is safe to use the lab waste bin or the sink for disposal).
- **Wash hands after any direct contact with chemicals or biochemical material** – always wash your hands at the end of a lab session.

Examples Using eqn [24.1], 25 g of a substance dissolved in 400 ml of water would have a mass concentration (p. 154) of $25 \div 400 = 0.0625$ g mL^{-1} ($\equiv 62.5$ mg mL^{-1} $\equiv 62.5$ g L^{-1})

Using eqn [24.1], 0.4 mol of a substance dissolved in 0.5 litres of water would have a molar concentration of $0.4 \div 0.5 = 0.8$ mol L^{-1} ($\equiv 800$ mmol L^{-1}).

Solving solubility problems – if your chemical does not dissolve after a reasonable time:

- check the limits of solubility for your compound (see *Merck Index*, O'Neil *et al.*, 2006),
- check the pH of the solution – solubility often changes with pH, e.g. you may be able to dissolve the compound in an acidic or basic solution.

both can be regarded as an amount of *substance* per unit volume of *solution*, in accordance with the relationship:

$$\text{Concentration} = \frac{\text{amount}}{\text{volume}} \qquad [24.1]$$

The most important aspect of eqn [24.1] is to recognise clearly the units involved, and to prepare the solution accordingly: for molar concentrations, you will need the relative molecular mass of the compound, so that you can determine the mass of substance required. Further advice on concentrations and interconversion of units is given on p. 161.

Box 24.2 shows the steps involved in making up a solution. The concentration you require is likely to be defined by a protocol you are following and the grade of chemical and supplier may also be specified. Success may depend on using the same source and quality, e.g. with enzyme work. To avoid waste, think carefully about the volume of solution you require, although it is always a good idea to err on the high side because you may spill some or make a mistake when dispensing it. Try to choose one of the standard volumes for vessels, as this will make measuring-out easier.

Use distilled or deionised water to make up aqueous solutions and stir to make sure all the chemical is dissolved. Magnetic stirrers are the most convenient means of doing this: carefully drop a clean magnetic stirrer bar ('flea') in the beaker, avoiding splashing; place the beaker centrally on the stirrer plate, switch on the stirrer and gradually increase the speed of stirring. When the crystals or powder have completely dissolved, switch off and

Box 24.2 How to make up an aqueous solution of known concentration from solid material

1. **Find out or decide the concentration of chemical required** and the degree of purity necessary.
2. **Decide on the volume of solution required.**
3. **Find out the relative molecular mass of the chemical (M_r).** This is the sum of the atomic (elemental) masses of the component elements and can be found on the container. If the chemical is hydrated, i.e. has water molecules associated with it, these must be included when calculating the mass required.
4. **Work out the mass of chemical that will give the concentration desired in the volume required.**

   Suppose your procedure requires you to prepare 250 ml of 0.1 mol L^{-1} NaCl.

   (a) Begin by expressing all volumes in the same units, either millilitres or litres (e.g. 250 mL as 0.25 litres).

   (b) Calculate the number of moles required from eqn [24.1]: 0.1 = amount (mol) ÷ 0.25. By rearrangement, the required number of moles is thus 0.1 × 0.25 = 0.025 mol.

   (c) Convert from mol to g by multiplying by the relative molecular mass (**M_r for** NaCl = 58.44)

   (d) Therefore, you need to make up 0.025 × 58.44 = 1.461 g to 250 mL of solution, using distilled water.

   In some instances, it may be easier to work in SI units, though you must be careful when using exponential numbers (p. 243).

   Suppose your protocol states that you need 100 ml of 10 mmol L^{-1} KCl.

   (a) Start by converting this to 100×10^{-6} m^3 of 10 mol m^{-3} KCl.

   (b) The required number of mol is thus $(100 \times 10^{-6}) \times (10) = 10^3$.

   (c) Each mol of KCl weighs 72.56 g (**M_r, the relative molecular mass).**

   (d) Therefore you need to make up 72.56×10^{-3} g = 72.56 mg KCl to 100×10^{-6} m^3 (100 mL) with distilled water.

   See Box 26.1 for additional information.
5. **Weigh out the required mass of chemical to an appropriate accuracy.** If the mass is too small to weigh to the desired degree of accuracy, consider the following options:

   (a) Make up a greater volume of solution.

   (b) Make up a stock solution that can be diluted at a later stage (see below).

   (c) Weigh the mass first, and calculate what volume to make the solution up to afterwards using eqn [24.1].
6. **Add the chemical to a beaker or conical flask then add a little less water than the final amount required.** If some of the chemical sticks to the paper, foil or weighing boat, use some of the water to wash it off.
7. **Stir and, if necessary, heat the solution to ensure all the chemical dissolves.** You can visually determine when this has happened by observing the disappearance of the crystals or powder.
8. **If required, check and adjust the pH of the solution when cool** (see p. 173).
9. **Make up the solution to the desired volume.** If the concentration needs to be accurate, use a class A volumetric flask; if a high degree of accuracy is not required, use a measuring cylinder (class B).

   (a) Pour the solution from the beaker into the measuring vessel using a funnel to avoid spillage.

   (b) Make up the volume so that the meniscus comes up to the appropriate measurement line (p. 154). For accurate work, rinse out the original vessel and use this liquid to make up the volume.
10. **Transfer the solution to a reagent bottle or a conical flask and label the vessel clearly.**

retrieve the flea with a magnet or another flea. Take care not to contaminate your solution when you do this and rinse the flea with distilled water.

'Obstinate' solutions may require heating, but do this only if you know that the chemical will not be damaged at the temperature used. Use a stirrer-heater to keep the solution mixed as you heat it. Allow the solution to cool before you measure volume or pH as these are affected by temperature.

Stock solutions

Stock solutions are valuable when making up a range of solutions containing different concentrations of a reagent or if the solutions have

Table 24.2 Use of stock solutions. Suppose you need a set of solutions 10 mL in volume containing differing concentrations of KCl, with and without reagent Q. You decide to make up a stock of KCl at twice the maximum required concentration ($50\,\text{mmol}\,\text{l}^{-1} = 50\,\text{mol}\,\text{m}^{-3}$) and a stock of reagent Q at twice its required concentration. The table shows how you might use these stocks to make up the media you require, based on eqn [24.1]. Note that the total volumes of stock you require can be calculated from the table (last column).

Stock solutions	Volume of stock required to make required solutions (ml)						Total volume of stock required (ml)
	No KCl plus Q	No KCl minus Q	15 mmol l^{-1} KCl plus Q	15 mmol l^{-1} KCl minus Q	25 mmol l^{-1} KCl plus Q	25 mmol l^{-1} KCl minus Q	
50 mmol l^{-1} KCl	0	0	3	3	5	5	16
[reagent Q] × 2	5	0	5	0	5	0	15
Water	5	10	2	7	0	5	29
Total	10	10	10	10	10	10	60

some common ingredients. They also save work if the same solution is used over a prolonged period (e.g. a nutrient solution). The stock solution is more concentrated than the final requirement and is diluted as appropriate when the final solutions are made up. The principle is best illustrated with an example (Table 24.2).

Preparing dilutions

Making a single dilution

Making a dilution – use the relationship $[C_1]V_1 = [C_2]V_2$ to determine volume or concentration (see p. 161).

You may need to dilute a stock solution to give a particular mass concentration or molar concentration. Use the following procedure:

1. Transfer an accurate volume of stock solution to a volumetric flask, using appropriate equipment (Table 25.1).
2. Make up to the calibration mark with solvent – add the last few drops from a pipette or solvent bottle, until the meniscus is level with the calibration mark.
3. Mix thoroughly, either by repeated inversion (holding the stopper firmly) or by prolonged stirring, using a magnetic stirrer. Make sure you add the magnetic flea *after* the volume adjustment step.

Removing a magnetic flea from a volumetric flask – use a strong magnet to bring the flea to the top of the flask, to avoid contamination during removal.

For routine work using dilute aqueous solutions where the highest degree of accuracy is not required, it may be acceptable to substitute test tubes or conical flasks for volumetric flasks. In such cases, you would calculate the volumes of stock solution and diluent required, with the assumption that the final volume is determined by the sum of the individual volumes of stock and diluent used (e.g. Table 24.2). Thus, a two-fold dilution would be prepared using 1 volume of stock solution and 1 volume of diluent. The dilution factor is obtained from the ratio of the initial concentration of the stock solution and the final concentration of the diluted solution. The dilution factor can be used to determine the volumes of stock and diluent required in a particular instance. For example, suppose you wanted to prepare 100 ml of a solution of NaCl at $0.2\,\text{mol}\,\text{L}^{-1}$. Using a stock solution containing $4.0\,\text{mol}\,\text{L}^{-1}$ NaCl, the dilution factor is $0.2 \div 4.0 = 0.05 = 1/20$ (a 20-fold dilution). Therefore, the amount of stock solution required is 1/20th of 100 mL = 5 mL and the amount of diluent needed is 19/20th of 100 mL = 95 mL.

Using the correct volumes for dilutions – it is important to distinguish between the volumes of the various liquids: a one-in-ten dilution is obtained using
1 volume of stock solution plus 9 volumes of diluent (1 + 9 = 10). Note that when this is shown as a ratio, it may represent the initial and final volumes (e.g. 1 : 10) or, sometimes, the volumes of stock solution and diluent (e.g. 1 : 9).

Using diluents – various liquids are used, including distilled or deionised water, salt solutions, buffers, Ringer's solution (p. 482), etc., according to the specific requirements of the procedure.

Preparing a dilution series

Dilution series are used in a wide range of procedures, including the preparation of standard curves for calibration of analytical instruments (p. 336), and in microbiology and immunoassay, where a range of dilutions of a particular sample is often required. A variety of different approaches can be used:

Linear dilution series

Here, the concentrations are separated by an equal amount, e.g. a series containing protein at 0, 0.2, 0.4, 0.6, 0.8, 1.0 $\mu g\,mL^{-1}$. Such a dilution series might be used to prepare a calibration curve for spectrophotometric assay of protein concentration (Box 53.1), or an enzyme assay (p. 406). Use $[C_1]V_1 = [C_2]V_2$ (p. 148) to determine the amount of stock solution required for each member of the series, with the volume of diluent being determined by subtraction.

Logarithmic dilution series

Here, the concentrations are separated by a constant proportion, often referred to as the step interval. This type of serial dilution is useful when a broad range of concentrations is required, e.g. for titration of biologically active substances, making a plate count of a suspension of microbes (p. 454), or when a process is logarithmically related to concentration.

The most common examples are:

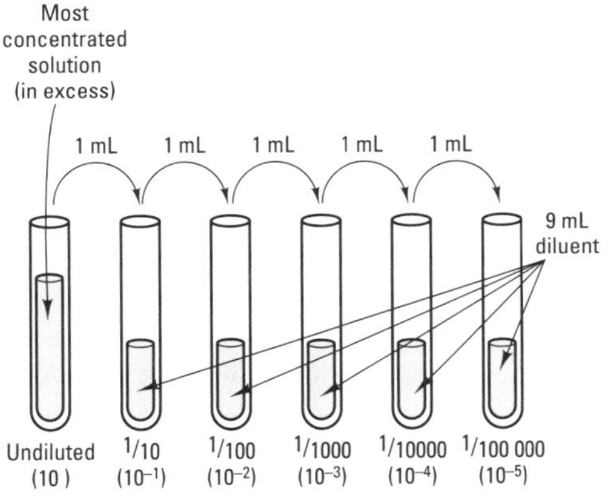

Fig. 24.2 Preparation of a dilution series. The example shown is a decimal dilution series, down to 1/100 000 (10^{-5}) of the solution in the first (left-hand) tube. Note that all solutions must be mixed thoroughly before transferring the volume to the next in the series. In microbiology and cell culture, sterile solutions and appropriate aseptic technique will be required (p. 451).

- Doubling dilutions – where each concentration is half that of the previous one (two-fold step interval, $\log_2$ dilution series). First, make up the most concentrated solution at twice the volume required. Measure out half of this volume into a vessel containing the same volume of diluent, mix thoroughly and repeat, for as many doubling dilutions as are required. The concentrations obtained will be 1/2, 1/4, 1/8, 1/16, etc., times the original (i.e. the dilutions will be two-, four-, eight- and sixteen-fold, etc.).

- Decimal dilutions – where each concentration is one-tenth that of the previous one (ten-fold step interval, $\log_{10}$ dilution series). First, make up the most concentrated solution required, with at least a 10 per cent excess. Measure out one-tenth of the volume required into a vessel containing nine times as much diluent, mix thoroughly and repeat. The concentrations obtained will be 1/10, 1/100, 1/1000, etc., times the original (i.e. dilutions of 10^{-1}, 10^{-2}, 10^{-3}, etc.). To calculate the actual concentration of solute, multiply by the appropriate dilution factor.

When preparing serial doubling or decimal dilutions, it is often easiest to add the appropriate amount of diluent to several vessels beforehand, as shown in the worked example in Fig. 24.2. When preparing a dilution series, it is essential that all volumes are dispensed accurately, e.g. using calibrated pipettors (p. 155), otherwise any inaccuracies will be compounded, leading to gross errors in the most dilute solutions.

Harmonic dilution series

Here, the concentrations in the series take the values of the reciprocals of successive whole numbers, e.g. 1, 1/2, 1/3, 1/4, 1/5, etc. The individual

Preparing a dilution series using pipettes or pipettors – use a fresh pipette or disposable tip for each dilution, to prevent carry-over of solutions. Remember to choose the appropriate pipette or pipettor for the volume you are dispensing (p. 154).

dilutions are simply achieved by a stepwise increase in the volume of diluent in successive vessels, e.g. by adding 0, 1, 2, 3, 4 and 5 times the volume of diluent to a set of test tubes, then adding a constant unit volume of stock solution to each vessel. Although there is no dilution transfer error between individual dilutions, the main disadvantage is that the series is non-linear, with a step interval that becomes progressively smaller as the series is extended.

Solutions must be thoroughly mixed before measuring out volumes for the next dilution. Use a fresh measuring vessel for each dilution to avoid contamination, or wash your vessel thoroughly between dilutions. Clearly label the vessel containing each dilution when it is made: it is easy to get confused! When deciding on the volumes required, allow for the aliquot removed when making up the next member in the series. Remember to discard any excess from the last in the series if volumes are critical.

Mixing solutions and suspensions

Various devices may be used, including:

SAFETY NOTE When using a vortex mixer with open and capped test tubes – do not vortex too vigorously or liquid will spill from the top of the tube, creating a contamination risk.

- Magnetic stirrers and fleas. Magnetic fleas come in a range of shapes and sizes, and some stirrers have integral heaters. During use, stirrer speed may increase as the instrument warms up.
- Vortex mixers. For vigorous mixing of small volumes of solution, e.g. when preparing a dilution series in test tubes. Take care when adjusting the mixing speed – if the setting is too low, the test tube will vibrate rather than creating a vortex, giving inadequate mixing. If the setting is too high, the test tube may slip from your hand.
- Orbital shakers and shaking water baths. These are used to provide controlled mixing at a particular temperature, e.g. for long-term incubation and cell-growth studies (p. 474).
- Bottle rollers. For cell-culture work, ensuring gentle, continuous mixing.

SAFETY NOTE You must always clean up any spillages of chemicals, as you are the only person who knows the risks from the spilled material.

Storing chemicals and solutions

Labile chemicals may be stored in a fridge or freezer. Take special care when using chemicals that have been stored at low temperature: the container and its contents must be warmed up to room temperature before use, otherwise water vapour will condense on the chemical. This may render any weighing you do meaningless and it could ruin the chemical. Other chemicals may need to be kept in a desiccator, especially if they are deliquescent.

KEY POINT Label all stored chemicals clearly with the following information: the chemical name (if a solution, state solute(s), concentration(s) and pH if measured), plus any relevant hazard-warning information, the date made up and your name.

Deciding on which balance to use – select a balance that weighs to an appropriate number of decimal places (p. 157). For example, you should use a top-loading balance weighing to one decimal place for less accurate work. Note that a weight of 6.4 g on such a balance may represent a true value of between 6.350 g and 6.449 g (to three decimal places).

Using balances

Electronic balances with digital readouts are now favoured over mechanical types: they are easy to read and their self-taring feature means the mass of the weighing boat or container can be subtracted automatically before weighing an object. The most common type offers accuracy down to 1 mg over the range 1 mg to 160 g, which is suitable for most biological applications.

To operate a standard self-taring balance:

1. Check that it is level, using the adjustable feet to centre the bubble in the spirit level (usually at the back of the machine). For accurate work, make sure a draught shield is on the balance.
2. Place an empty vessel in the middle of the balance pan and allow the reading to stabilise. *If the object is larger than the pan, take care that no part rests on the body of the balance or the draught shield as this will invalidate the reading.* Press the tare bar to bring the reading to zero.
3. Place the chemical or object carefully in the vessel (powdered chemicals should be dispensed with a suitably sized clean spatula). Take care to avoid spillages.
4. Allow the reading to stabilise and make a note of the value.
5. If you add excess chemical, take great care when removing it. Switch off if you need to clean any deposit accidentally left on or around the balance.

Weighing – never weigh anything directly on a balance's pan: you may contaminate it for other users. Use a weighing boat or a slip of aluminium foil. Otherwise, choose a suitable vessel such as a beaker, conical flask or aluminium tray.

Larger masses should be weighed on a top-loading balance to an appropriate degree of accuracy. Take care to note the limits for the balance: while most have devices to protect against overloading, you may damage the mechanism. In the field, spring or battery-operated balances may be preferred. Try to find a place out of the wind to use them. For extremely small masses, there are electrical balances that can weigh down to 1 μg, but these are very delicate and must be used under supervision.

Measuring length and area

When measuring linear dimensions, the device you need depends on the size of object you are measuring and the precision demanded (Table 24.3).

For many regularly shaped objects, area can be estimated from linear dimensions (see p. 242). The areas of irregular shapes can be measured with an optical measuring device or a planimeter. These have the benefits of speed and ease of use; instructions are machine-specific. A simple 'low tech' method is to trace objects on to good-quality paper or to photocopy them. If the outline is then cut around, the area can be estimated by weighing the cutout and comparing to the mass of a piece of the same paper of known area. Avoid getting moisture from the specimen onto the paper as this will affect the reading.

Table 24.3 Suitability of devices for measuring linear dimensions

Measurement device	Suitable lengths	Degree of precision
Eyepiece graticule (light microscopy)	1 μm to 10 mm	0.5 μm
Vernier calipers	1–100 mm	0.1 mm
Ruler	10 mm to 1 m	1.0 mm
Tape measure	10 mm to 30 m	1.0 mm
Optical surveying devices	1 m to 100 m	0.1 m

Measuring and controlling temperature

Heating and drying samples

Care is required when heating samples – there is a danger of fire whenever organic material is heated and a danger of scalding from heated liquids. Safety glasses should always be worn. Use a thermostatically controlled electric stirrer-heater if possible. If using a Bunsen burner, keep the flame well away from yourself and your clothing (tie back long hair). Use a non-flammable mat beneath a Bunsen to protect the bench. Switch off when no longer required. To light a Bunsen, close the air hole first, then apply a lit match or lighter. Open the air hole if you need a hotter, more concentrated flame: the hottest part of the flame is just above the apex of the blue cone in its centre.

SAFETY NOTE Heating/cooling glass vessels – take care if heating or cooling glass vessels rapidly as they may break when heat-stressed. Freezing aqueous solutions in thin-walled glass vessels is risky because ice expansion may break the glass.

Ovens and drying cabinets may be used to dry specimens or glassware. They are normally thermostatically controlled. If drying organic material for dry weight measurement, do so at about 80°C to avoid caramelising the specimen. Always state the actual temperature used as this affects results. Check that all water has been driven off by weighing until a constant mass is reached.

Cooling samples

Using thermometers – some are calibrated for use in air, others require partial immersion in liquid and others total immersion – check before use.

Fridges and freezers are used for storing stock solutions and chemicals that would either break down or become contaminated at room temperature. Dried samples should be cooled in a desiccator to ensure that water is not reabsorbed from the atmosphere. Normal fridge and freezer temperatures are about 4°C and −15°C respectively. Ice baths can be used when reactants must be kept close to 0°C. Most science departments will have a machine that provides flaked ice for use in these baths. If common salt is mixed with ice, temperatures below 0 °C can be achieved. A mixture of ethanol and solid CO_2 will provide a temperature of −72°C if required. To freeze a specimen quickly, immerse in liquid N_2 (−196°C) using tongs and wearing an apron and thick gloves, as splashes will damage your skin. Always work in a well-ventilated room.

Maintaining constant temperature

SAFETY NOTE If a mercury thermometer is broken, report the spillage, as mercury is a poison.

Thermostatically controlled temperature rooms and incubators can be used to maintain temperature at a desired level. Always check with a thermometer or thermograph that the thermostat is accurate enough for your study. To achieve a controlled temperature on a smaller scale, e.g. for an oxygen electrode, use a water bath. These usually incorporate heating elements, a circulating mechanism and a thermostat. Baths for sub-ambient temperatures have a cooling element.

Controlling atmospheric conditions

Gas composition

Example Water vapour can be removed by passing gas over dehydrated $CaCO_3$ and CO_2 may be removed by bubbling through KOH solution.

The atmosphere may be 'scrubbed' of certain gases by passing through a U-tube or Dreschel bottle containing an appropriate chemical or solution.

For accurate control of gas concentrations, use cylinders of pure gas; the contents can be mixed to give specified concentrations by controlling individual flow rates. The cylinder-head regulator (Fig. 24.3) allows you to control the pressure (and hence flow rate) of gas; adjust using the controls on the regulator or with spanners of appropriate size. Before use, ensure the regulator outlet tap is off (turn anticlockwise), then switch on at the cylinder (turn clockwise) – the cylinder dial will give you the pressure reading for the cylinder contents. Now switch on at the regulator outlet (turn clockwise) and adjust to desired pressure/flow setting. To switch off, carry out the above directions in reverse order.

Pressure

Many forms of pump are used to pressurise or provide a partial vacuum, usually to force gas or liquid movement. Each has specific instructions for use. Many laboratories are supplied with 'vacuum' (suction) and pressurised air lines that are useful for procedures such as vacuum-

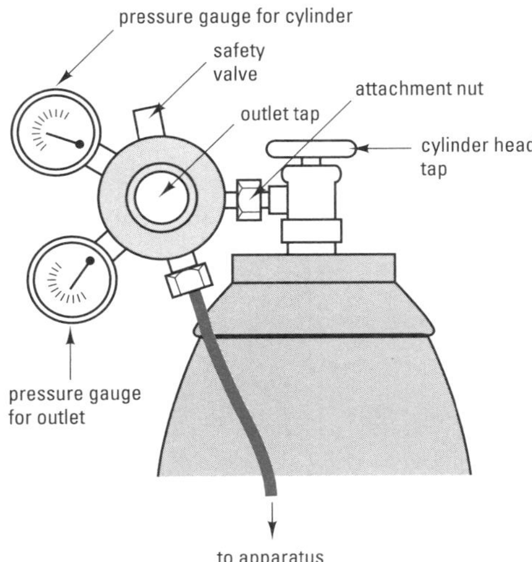

Fig. 24.3 Parts of a cylinder-head regulator. The regulator is normally attached by tightening the attachment nut clockwise; the exception is with cylinders of hydrogen, where the special regulator is tightened *anticlockwise* to avoid the chance of this potentially explosive gas being incorrectly used.

assisted filtration. Make sure you switch off the taps after use. Take special care with glass items kept at very low or high pressures. These should be contained within a metal cage to minimise the risk of injury.

Measuring time

Using a timer – always set the alarm before the critical time, so that you have adequate time to react.

Many experiments and observations need to be carefully timed. Large-faced stopclocks allow you to set and follow 'experimental time' and remove the potential difficulties in calculating this from 'real time' on a watch or clock. Some timers incorporate an alarm that you can set to warn when readings or operations must be carried out; 24-hour timers are available for controlling light and temperature regimes.

Miscellaneous methods for treating samples

Homogenising

This involves breaking up and mixing samples to give a uniform preparation. Blenders are used to homogenise animal and plant material and work best when an adequate volume of liquid is present: buffer solution may be added to samples for this purpose. Use in short bursts to avoid overheating the motor and the sample. A pestle and mortar is used for grinding up samples. Acid-washed sand grains can be added to help break up the tissues. For quantitative work with brittle samples, care must be taken not to lose material when the sample breaks into fragments.

Separation of components of mixtures and solutions

Particulate solids can be separated on the basis of size using sieves. These are available in stacking forms that fit on automatic shakers. Sieves with the largest pores are placed at the top and the assembly is shaken for a fixed time until the sample separates. Suspensions of solids in liquids may be separated out by centrifugation (see p. 383) or filtration. Various forms of filter paper are available having different porosities and purities. Vacuum-assisted filtration speeds up the process and is best carried out with a filter funnel attached to a filter flask. Filtration through pre-sterilised membranes with very small pores (e.g. the Millipore type) is an excellent method of sterilising small volumes of solution. Solvents can be removed from solutes by heating, using rotary film evaporation under low pressure and, for water, by freeze drying. The last two are especially useful for heat-labile solutes – refer to the manufacturers' specific instructions for use.

Text references

Lide, D.R. (ed.) (2008) *CRC Handbook of Chemistry and Physics*, 89th edn. Taylor and Francis, London.

O'Neil, M.J., Henckelman, P.E. *et al.* (eds.) (2006) *The Merck Index: An Encyclopedia of Chemicals, Drugs, and Biologicals*, 14th edn. Merck & Co., Inc., Whitehouse Station, New Jersey.

Sources for further study

Jack, R.C. (1995) *Basic Biochemical Laboratory Procedures and Computing*. Oxford University Press, Oxford.

Seidman, L.A. and Moore, C.J. (2000) *Basic Laboratory Methods for Biotechnology: Textbook and Laboratory Reference*. Prentice Hall, New Jersey.

25 Working with liquids

Measuring and dispensing liquids

The equipment you should choose to measure out liquids depends on the volumes being dispensed, the accuracy required and the number of times the job must be done (Table 25.1).

SAFETY NOTE Take care when using chemically hazardous liquids and solutions (flammable, corrosive, toxic, etc.) – make sure the liquid is contained properly and deal with any spillages in the correct manner (p. 146).

Reading any volumetric scale – make sure your eye is level with the bottom of the liquid's meniscus and take the reading from this point.

Table 25.1 Criteria for choosing a method for measuring out a liquid

Method	Best volume range	Accuracy	Usefulness for repetitive measurement
Pasteur pipette	0.03–2 mL	Low	Convenient
Conical flask/beaker	25–5000 mL	Very low	Convenient
Measuring cylinder	5–2000 mL	Medium	Convenient
Volumetric flask	5–2000 mL	High	Convenient
Burette	1–100 mL	High	Convenient
Glass pipette	1–100 mL	High	Convenient
Mechanical pipettor	5–1000 μL	High*	Convenient
Syringe	0.5–20 μL	Medium**	Convenient
Microsyringe	0.5–50 μL	High	Convenient
Weighing	Any (depends on accuracy of balance)	Very high	Inconvenient

* If correctly calibrated and used properly (see p. 156).
** Accuracy depends on width of barrel: large volumes are less accurate.

Certain liquids may cause problems:

- High-viscosity liquids are difficult to dispense: allow time for all the liquid to transfer.
- Organic solvents may evaporate rapidly, making measurements inaccurate: work quickly; seal containers without delay.
- Solutions prone to frothing (e.g. protein and detergent solutions) are difficult to measure and dispense: avoid forming bubbles due to overagitation; do not transfer quickly.
- Suspensions (e.g. cell cultures) may sediment: thoroughly mix them before dispensing.

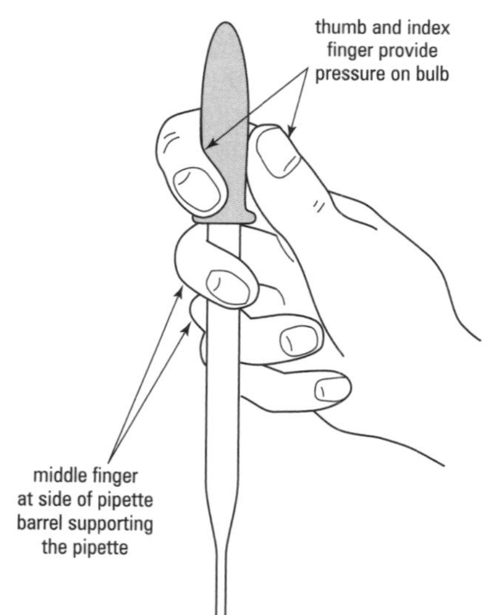

Fig. 25.1 How to hold a Pasteur pipette.

Pasteur pipettes

Hold correctly during use (Fig. 25.1) – keep the pipette vertical, with the middle fingers gripping the barrel while the thumb and index finger provide controlled pressure on the bulb. Squeeze gently to dispense individual drops. To avoid the risk of cross-contamination, take care not to draw up solution into the bulb or to lie the pipette on its side. Alternatively, use a plastic disposable 'Pastette'.

SAFETY NOTE Pasteur pipettes should be used with care for hazardous solutions: remove the tip from the solution before fully releasing pressure on the bulb – the air taken up helps prevent spillage.

Measuring cylinders and volumetric flasks

These must be used on a level surface so that the scale is horizontal; you should first fill with solution until just below the desired mark; then fill slowly (e.g. using a Pasteur pipette) until the meniscus is level with the mark. Allow time for the solution to run down the walls of the vessel.

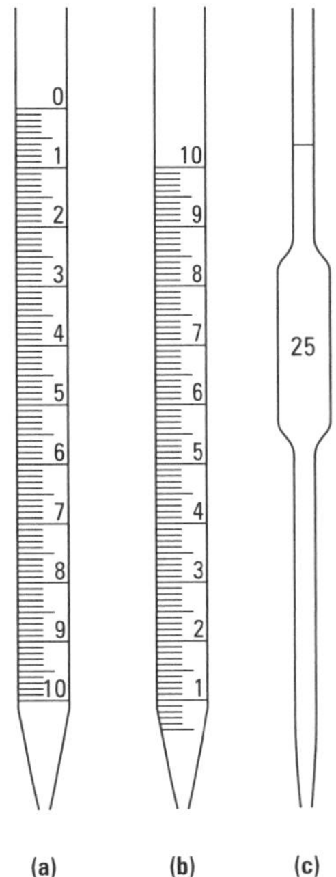

Fig. 25.2 Glass pipettes – graduated pipette, reading from zero to shoulder (a); graduated pipette, reading from maximum to tip, by gravity (b); bulb (volumetric) pipette, showing volume (calibration mark to tip, by gravity) above the bulb (c).

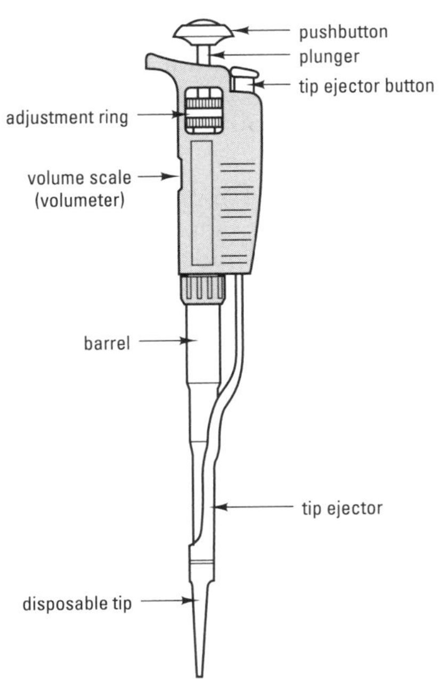

Fig. 25.3 A pipettor – the Gilson Pipetman.

Burettes

Burettes should be mounted vertically on a clamp stand – don't overtighten the clamp. First ensure the tap is closed and fill the body with solution using a funnel. Open the tap and allow some liquid to fill the tubing below the tap before first use. Take a meniscus reading, noting the value in your notebook. Dispense the solution via the tap and measure the new meniscus reading. The volume dispensed is the difference between the two readings. Titrations are usually performed on a magnetic stirrer (pp. 146, 150).

Pipettes

These come in various designs, including graduated and bulb (volumetric) pipettes (Fig. 25.2). Take care to check the volume scale before use: some empty from full volume to zero, others from zero to full volume; some scales refer to the shoulder of the tip, others to the tip either by gravity or after blowing out.

> **KEY POINT** For safety reasons, never mouth-pipette – various aids are available such as the Pi-pump. Always select the most appropriate method for the volume you need (Table 25.1).

Pipettors (autopipettors)

These come in two basic types:

- Air displacement pipettors. For routine work with dilute aqueous solutions. One of the most widely used examples is the Gilson Pipetman (Fig. 25.3). Box 25.1 gives details on its use.
- Positive displacement pipettors. For non-standard applications, including dispensing viscous, dense or volatile liquids, or certain procedures in molecular genetics, e.g. the PCR (p. 375), where an air displacement pipettor might create aerosols, leading to errors.

Air displacement and positive displacement pipettors may be:

- Fixed volume: capable of delivering a single factory-set volume.
- Adjustable: where the volume is determined by the operator across a particular range of values.
- Preset: movable between a limited number of values.
- Multi-channel: able to deliver several replicate volumes at the same time.

Whichever type you use, you must ensure that you understand the operating principles of the volume scale and the method for changing the volume delivered – some pipettors are easily misread.

A pipettor must be fitted with the correct disposable tip before use: each manufacturer produces different tips to fit particular models. Specialised tips are available for particular applications, e.g. PCR (p. 375).

Syringes

Syringes should be used by placing the tip of the needle in the solution and drawing the plunger up slowly to the required point on the scale. Check the barrel to make sure no air bubbles have been drawn up.

Box 25.1 Using a pipettor to deliver accurate, reproducible volumes of liquid

A pipettor can be used to dispense volumes with accuracy and precision, by following this step-wise procedure:

1. **Select a pipettor that operates over the appropriate range.** Most adjustable pipettors are accurate only over a particular working range and should not be used to deliver volumes below the manufacturer's specifications (minimum volume is usually 10–20 per cent of maximum value). Do not attempt to set the volume above the maximum limit, or the pipettor may be damaged.
2. **Set the volume to be delivered.** In some pipettors, you 'dial up' the required volume. Types like the Gilson Pipetman have a system where the scale (or 'volumeter') consists of three numbers, read from top to bottom of the barrel, and adjusted using the black knurled adjustment ring (Fig. 25.3). This number gives the first three digits of the volume scale and thus can only be understood by establishing the maximum volume of the Pipetman, as shown on the push-button on the end of the plunger (Fig. 25.3). The following examples illustrate the principle for two common sizes of Pipetman:

P1000 Pipetman (maximum volume 1000 μL) if you dial up	P20 Pipetman (maximum volume 20 μL) if you dial up
1 0 0	1 0 0
the volume is set at 1000 μL	the volume is set at 10.0 μL

Note: The Pipetman scale is not **a** percentage one.

3. **Fit a new disposable tip to the end of the barrel.** Make sure that it is the appropriate type for your pipettor and that it is correctly fitted. Press the tip on firmly using a slight twisting motion – if not, you will take up less than the set volume and liquid will drip from the tip during use. Tips are often supplied in boxes, for ease of use: if sterility is important, make sure you use appropriate sterile technique at all times (p. 453). *Never, ever, try to use a pipettor without its disposable tip.*
4. **Check your delivery.** Confirm that the pipettor delivers the correct volume by dispensing volumes of distilled water and weighing on a balance, assuming 1 mg = 1 μL = 1 mm^3. The value should be within 1 per cent of the selected volume. For small volumes, measure several 'squirts' together, e.g. 20 'squirts' of 5 μL = 100 mg. If the pipettor is inaccurate (p. 193) giving a biased result (e.g. delivering significantly more or less than the volume set), you can make a temporary correction by adjusting the volumeter scale down or up accordingly (the volume *delivered* is more important than the value *displayed* on the volumeter), or have the pipettor recalibrated. If the pipettor is imprecise (p. 193), delivering a variable amount of liquid each time, it may need to be serviced. After calibration, fit a clean (sterile) tip if necessary.
5. **Draw up the appropriate volume.** Holding the pipettor *vertically*, press down on the plunger/push-button until a resistance (spring-loaded stop) is met. Then place the end of the tip in the liquid. Keeping your thumb on the plunger/push-button, release the pressure slowly and evenly: watch the liquid being drawn up into the tip, to confirm that no air bubbles are present. Wait a second or so, to confirm that the liquid has been taken up, then withdraw the end of the tip from the liquid. Inexperienced users often have problems caused by drawing up the liquid too quickly/carelessly. If you accidentally draw liquid into the barrel, seek assistance from your demonstrator/supervisor as the barrel will need to be cleaned before further use.
6. **Make a quick visual check on the liquid in the tip.** Does the volume seem reasonable? (e.g. a 100 μL volume should occupy approximately half the volume of a P200 tip). The liquid will remain in the tip, without dripping, as long as the tip is fitted correctly and the pipettor is not tilted too far from a vertical position.
7. **Deliver the liquid.** Place the end of the tip against the wall of the vessel at a slight angle (10–15° from vertical) and press the plunger/push-button slowly and smoothly to the first (spring-loaded) stop. Wait a second or two, to allow any residual liquid to run down the inside of the tip, then press again to the final stop, dispensing any remaining liquid. Remove from the vessel with the plunger/push-button still depressod. Use your other hand to stabilise the pipetter during delivery, if your hand is unsteady.
8. **Eject the tip.** Press the tip ejector button if present (Fig. 25.3). If the tip is contaminated, eject directly into an appropriate container, e.g. a beaker of disinfectant, for microbiological work, or a labelled container for hazardous solutions (p. 158). For repeat delivery, fit a new tip if necessary and begin again at step 5 above. Always make sure that the tip is ejected before putting a pipettor on the bench.

Expel slowly and touch the syringe on the edge of the vessel to remove any liquid adhering to the end of the needle. Microsyringes should always be cleaned before and after use by repeatedly drawing up and expelling pure solvent. The dead space in the syringe needle can occupy up to 4 per cent of the nominal syringe volume. A way of avoiding such problems is to fill the dead space with an inert substance (e.g. silicone oil) after sampling. Alternatively, use a syringe where the plunger occupies the needle space (small volumes only).

Balances

These can be used to weigh accurately (p. 150) how much liquid you have dispensed. Convert mass to volume using the equation:

Example Using eqn [25.1], 9 g of a liquid with a density of 1.2 g mL^{-1} = 7.5 mL.

$$\text{mass/density} = \text{volume} \qquad [25.1]$$

Densities of common solvents can be found in Lide (2008). You will also need to know the liquid's temperature, as density is temperature-dependent.

Storing and holding liquids

You should clearly label all stored solutions (see p. 150), including relevant hazard information, preferably marking with hazard warning tape (p. 134). Seal vessels in an appropriate way, e.g. using a stopper or a sealing film such as Parafilm to prevent evaporation. To avoid degradation store your solution in a fridge, but allow it to reach room temperature before use.

Test tubes

These are used for colour tests, small-scale reactions, holding cultures, etc. The tube can be sterilised by heating (p. 451) and maintained in this state with a cap or cotton-wool plug. Always use a test-tube rack of the correct size for the tubes selected.

Beakers

Working with beakers and flasks – remember that volume graduations, where present, are often inaccurate and should be used only where approximations will suffice.

Beakers are used for general purposes, e.g. heating a solvent while the solute dissolves, carrying out a titration, etc.

Conical (Erlenmeyer) flasks

These are used for storage of solutions: their wide base makes them stable, while their small mouth reduces evaporation and is easily sealed.

Bottles and vials

Storing light-sensitive chemicals – use a coloured vessel or wrap aluminium foil around a clear vessel.

These are used when the solution needs to be sealed for safety, sterility or to prevent evaporation or oxidation. They usually have a screw top or ground-glass stopper to prevent evaporation and contamination. Many types are available, including 'bijou', 'McCartney', 'universal' and 'Winkler'.

Storing an aqueous solution containing organic constituents – unless this has been sterilised or is toxic, microbes will start growing, so store for short periods in a refrigerator; older solutions may not give reliable results.

Creating specialised apparatus

Glassware systems incorporating ground-glass connections such as Quickfit are useful for setting up combinations of standard glass components, e.g. for chemical reactions. In project work, you may need to adapt standard forms of glassware for a special need. A glassblowing service (often available in chemistry departments) can make special items to order.

Choosing between glass and plastic

Bear in mind the following points:

- **Reactivity.** Plastic vessels often distort at relatively low temperatures; they may be flammable, may dissolve in certain organic solvents and may be affected by prolonged exposure to ultraviolet (UV) light. Some plasticisers may leach from vessels and have been shown to have biological activity. Glass may adsorb ions and other molecules and then leach them into solutions, especially in alkaline conditions. Pyrex glass is stronger than ordinary soda glass and can withstand temperatures up to 500°C.
- **Rigidity and resilience.** Plastic vessels are not recommended where volume is critical as they may distort through time: use class A volumetric glassware for accurate work, e.g. preparing solutions (Chapter 26). Glass vessels are more easily broken than plastic, which is particularly important for centrifugation (see p. 383).
- **Opacity.** Both glass and plastic absorb light in the UV range of the electromagnetic radiation (EMR) spectrum (Table 25.2) Quartz should be used where this is important, e.g. in cuvettes for UV spectrophotometry (see p. 353).

Table 25.2 Spectral cut-off values for glass and plastics (λ_{50} = wavelength at which transmission of EMR is reduced to 50 per cent)

Material	λ_{50} (nm)
Routine glassware	340
Pyrex glass	292
Polycarbonate	396
Acrylic	342
Polyester	318
Quartz	220

Box 25.2 Safe working with glass

Many minor accidents in the laboratory are due to lack of care with glassware. You should follow these general precautions:

- **Always wear safety glasses when there is any risk of glass breakage** – e.g. when using low pressures, or heating solutions.
- **Take care when attaching tubing to glass tubes and when putting glass tubes into bungs** – always hold the tubing and glassware close together, as shown in Fig. 25.4, and wear thick gloves when appropriate.
- **Use a 'soft' Bunsen flame when heating glassware** – this avoids creating a hot spot, where cracks may start: always use tongs or special heat-resistant gloves when handling hot glassware (never use a rolled-up paper towel).
- **Do not use chipped or cracked glassware** – it may break under very slight strain and should be disposed of in the broken glassware bin.
- **Never carry large glass bottles/flasks by their necks** – support them with a hand underneath or, better still, carry them in a basket.
- **Do not force bungs too firmly into bottles** – these can be extremely difficult to remove. If you need a tight seal, use a screw-top bottle with a rubber or plastic seal.

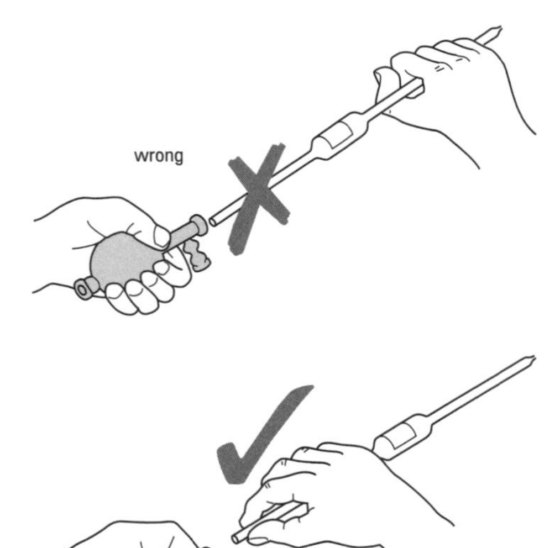

Fig. 25.4 Handling glass pipettes and tubing.

- **Dispose of broken glass thoroughly and carefully** – use disposable paper towels and wear thick gloves. Always put pieces of broken glass in the correct bin.

- **Disposability.** Plastic items may be cheap enough to make them disposable, an advantage where there is a risk of chemical or microbial contamination.

Cleaning glass and plastic

SAFETY NOTE Special cleaning of glass – for an acid wash use dilute acid, e.g. 100 mmol L^{-1} (100 mol m^{-3}) HCl. Rinse thoroughly at least three times with distilled or deionised water. Glassware that must be exceptionally clean (e.g. for a micro-nutrient study) should be washed in a chromic acid bath, but this involves toxic and corrosive chemicals and should only be used under supervision.

Take care to avoid the possibility of contamination arising from prior use of chemicals or inadequate rinsing following washing. A thorough rinse with distilled or deionised water immediately before use will remove dust and other deposits and is good practice in quantitative work, but ensure that the rinsing solution is not left in the vessel. 'Strong' basic detergents (e.g. Pyroneg) are good for solubilising acidic deposits. If there is a risk of basic deposits remaining, use an acid wash. If there is a risk of contamination from organic deposits, a rinse with Analar grade ethanol is recommended. Glassware can be disinfected by washing with a sodium hypochlorite bleach such as Chloros, with sodium metabisulphite or a blended commercial product such as Virkon – dilute as recommended before use and rinse thoroughly with sterile water after use. Alternatively, to sterilise, heat glassware to at least 121°C for 15 minutes, in an autoclave or 160°C for 3 hours in an oven.

Text references

Lide, D.R. (ed.) (2008) *CRC Handbook of Chemistry and Physics*, 89th edn. Taylor and Francis, London.

Sources for further study

Boyer, R.F. (2008) *Biochemistry Laboratory: Modern Theory and Techniques*. Pearson Education, Harlow.

Henrickson, C., Byrd, L.C. and Hunter, N.W. (2007) *General, Organic and Biochemistry: A Laboratory Manual*, 6th edn. McGraw-Hill, New York.

Seidman, L.A. and Moore, C.J. (2008) *Basic Laboratory Methods for Biotechnology: Textbook and Laboratory Reference*. 2nd edn. Benjamin Cummings, San Francisco, California.

26 Principles of solution chemistry

SAFETY NOTE Safety with solutions – many solutes and solvents used in biosciences are potentially toxic, corrosive, oxidising or flammable, and they may also be carcinogenic (see pp. 133–134). Further, there is a risk of accident involving the vessels used when preparing, storing and dispensing solutions (see Box 25.2, p. 158)

A solution is a homogeneous liquid, formed by the addition of solutes to a solvent (usually water in biological systems). The behaviour of solutions is determined by the types of solutes involved and by their proportions, relative to the solvent. Many laboratory exercises involve calculation of concentrations, e.g. when preparing an experimental solution at a particular concentration, or when expressing data in terms of solute concentration (see pp. 161–163). Make sure that you understand the basic principles set out in this chapter before you tackle such exercises.

Solutes can affect the properties of solutions in several ways, including:

Electrolytic dissociation

This occurs where individual molecules of an electrolyte dissociate to give charged particles (ions). For a strong electrolyte, e.g. NaCl, dissociation is essentially complete. In contrast, a weak electrolyte, e.g. acetic acid, will be only partly dissociated, depending upon the pH and temperature of the solution (p. 173).

Definitions

Electrolyte – a substance that dissociates, either fully or partially, in water to give two or more ions.

Relative atomic mass (A_r) – the mass of an atom relative to $^{12}C = 12$.

Relative molecular mass (M_r) – the mass of a compound's formula unit relative to $^{12}C = 12$.

Mole (of a substance) – the equivalent in mass to relative molecular mass in grams.

Osmotic effects

These are the result of solute particles lowering the effective concentration of the solvent (water). These effects are particularly relevant to biological systems since membranes are far more permeable to water than to most solutes. Water moves across biological membranes from the solution with the higher effective water concentration to that with the lower effective water concentration (osmosis).

Ideal/non-ideal behaviour

This occurs because solutions of real substances do not necessarily conform to the theoretical relationships predicted for dilute solutions of so-called ideal solutes. It is often necessary to take account of the non-ideal behaviour of real solutions, especially at high solute concentrations (see Lide, 2008 and Robinson and Stokes, 2002, for appropriate data).

Concentration

Expressing solute concentrations – you should use SI units wherever possible. However, you are likely to meet non-SI concentrations and you must be able to deal with these units too.

In SI units (p. 199), the concentration of a solute in a solution is expressed in mol m^{-3}, which is convenient for most biological purposes. The concentration of a solute is usually symbolised by square brackets, e.g. [NaCl]. Details of how to prepare a solution using SI and non-SI units are given on p. 161.

A number of alternative ways of expressing the relative amounts of solute and solvent are in general use, and you may come across these terms in your practical work or in the literature:

Molarity

Example A 1.0 molar solution of NaCl would contain 58.44 g NaCl (the relative molecular mass) per litre of solution.

This is the term used to denote molar concentration, $[C]$, expressed as moles of solute per litre volume of solution (mol L^{-1}). This non-SI term continues to find widespread usage, in part because of the familiarity of working scientists with the term, but also because laboratory glassware is

Box 26.1 Useful procedures for calculations involving molar concentrations

1. **Preparing a solution of defined molarity.** For a solute of known relative molecular mass, M_r, the following relationship can be applied:

$$[C] = \frac{\text{mass of solute/relative molecular mass}}{\text{volume of solution}} \quad [26.1]$$

So, if you wanted to make up 200 mL (0.2 L) of an aqueous solution of NaCl (M_r 58.44) at a concentration of 500 mmol L^{-1} (0.5 mol L^{-1}), you could calculate the amount of NaCl required by inserting these values into eqn [26.1]:

$$0.5 = \frac{\text{mass of solute/58.44}}{0.2}$$

which can be rearranged to

$$\text{mass of solute} = 0.5 \times 0.2 \times 58.44 = 5.844\text{ g}$$

The same relationship can be used to calculate the concentration of a solution containing a known amount of a solute, e.g. if 21.1 g of NaCl were made up to a volume of 100 mL (0.1 L), this would give

$$[\text{NaCl}] = \frac{21.1/58.44}{0.1} = 3.61\text{ mol L}^{-1}$$

2. **Dilutions and concentrations.** The following relationship is very useful if you are diluting (or concentrating) a solution:

$$[C_1]V_1 = [C_2]V_2 \quad [26.2]$$

where $[C_1]$ and $[C_2]$ are the initial and final concentrations, while V_1 and V_2 are their respective volumes: each pair must be expressed in the same units. Thus, if you wanted to dilute 200 mL of 0.5 mol L^{-1} NaCl to give a final molarity of 0.1 mol L^{-1}, then, by substitution into eqn [26.2]:

$$0.5 \times 200 = 0.1 \times V_2$$

Thus V_2 = 1 000 mL (in other words, you would have to add water to 200 mL of 0.5 mol L^{-1} NaCl to give a final volume of 1 000 mL to obtain a 0.1 mol L^{-1} solution).

3. **Interconversion.** A simple way of interconverting amounts and volumes of any particular solution is to divide the amount and volume by a factor of 10^3: thus a molar solution of a substance contains 1 mol L^{-1}, which is equivalent to 1 mmol mL^{-1}, or 1 μmol μL^{-1}, or 1 nmol nL^{-1}, etc. You may find this technique useful when calculating the amount of substance present in a small volume of solution of known concentration, e.g. to calculate the amount of NaCl present in 50 μL of a solution with a concentration (molarity) of 0.5 mol L^{-1} NaCl:
   (a) this is equivalent to 0.5 μmol μL^{-1};
   (b) therefore 50 μL will contain 50 × 0.5 μmol = 25 μmol.

Alternatively, you may prefer to convert to primary SI units, for ease of calculation (see Box 32.1).

The 'unitary method' (p. 245) is an alternative approach to these calculations.

calibrated in millilitres and litres, making the preparation of molar and millimolar solutions relatively straightforward. However, the symbols in common use for molar (M) and millimolar (mM) solutions are at odds with the SI system and many people now prefer to use mol L^{-1} and mmol L^{-1} respectively, to avoid confusion. Box 26.1 gives details of some useful approaches to calculations involving molarities.

Molality

This is used to express the concentration of solute relative to the *mass* of solvent, i.e. mol kg^{-1}. Molality is a temperature-independent means of expressing solute concentration, rarely used except when the osmotic properties of a solution are of interest (p. 163).

Example A 0.5 molal solution of NaCl would contain 58.44 × 0.5 = 29.22 g NaCl per kg of water.

Per cent composition (% w/w)

This is the solute mass (in g) per 100 g solution. The advantage of this expression is the ease with which a solution can be prepared, since it simply requires each component to be pre-weighed (for water, a volumetric measurement may be used, e.g. using a measuring cylinder) and then mixed together. Similar terms are parts per thousand (‰), i.e. mg g^{-1}, and parts per million (ppm), i.e. μg g^{-1}.

Example A 5% w/w sucrose solution contains 5 g sucrose and 95 g water (= 95 mL water, assuming a density of 1 g mL^{-1}) to give 100 g of solution.

Per cent concentration (% w/v and % v/v)

For solutes added in solid form, this is the number of grams of solute per 100 mL solution. This is more commonly used than per cent composition, since solutions can be accurately prepared by weighing out the required amount of solute and then making this up to a known volume using a volumetric flask. The equivalent expression for liquid solutes is % v/v.

The principal use of mass/mass or mass/volume terms (including $g\,L^{-1}$) is for solutes whose relative molecular mass is unknown (e.g. cellular proteins), or for mixtures of certain classes of substance (e.g. total salt in sea water). You should *never* use the per cent term without specifying how the solution was prepared, i.e. by using the qualifier w/w, w/v or v/v. For mass concentrations, it is simpler to use mass per unit volume, e.g. $mg\,L^{-1}$, $\mu g\,\mu L^{-1}$, etc.

Example A 5% w/v sucrose solution contains 5 g sucrose in 100 ml of solution. A 5% v/v glycerol solution would contain 5 mL glycerol in 100 ml of solution.

Note that when water is the solvent this is often not specified in the expression, e.g. a 20% v/v ethanol solution contains 20% ethanol made up to 100 mL of solution using water.

Parts per million concentration (ppm)

This is a non-SI weight per volume (w/v) concentration term commonly used in quantitative analysis such as flame photometry, atomic absorption spectroscopy and gas chromatography, where low concentrations of solutes are analysed. The term ppm is equivalent to the expression of concentration as $\mu g\,mL^{-1}$ (10^{-6} $g\,mL^{-1}$) and a 1.0 ppm solution of a substance will have a concentration of 1.0 $\mu g\,mL^{-1}$ (1.0×10^{-6} $g\,mL^{-1}$).

Parts per billion (ppb) is an extension of this concentration term as $ng\,mL^{-1}$ (10^{-9} $g\,mL^{-1}$) and is commonly used to express concentrations of very dilute solutions. For example, the allowable concentration of arsenic in water is 0.05 ppm, but it is more conveniently expressed as 50 ppb.

Example The concentration of a NaCl solution is stated as 3 ppm. This is equivalent to 3 $\mu g\,mL^{-1}$ (3 $mg\,L^{-1}$). The relative molecular mass of NaCl is 58.44 g mol^{-1}, so the solution has a concentration of $3 \times 10^{-6} \div 58.44$ mol mL^{-1} = 5.13×10^{-8} mol mL^{-1} = 0.0513 μmol mL^{-1} = 51.3 $\mu mol\,L^{-1}$.

Activity (a)

This is a term used to describe the *effective* concentration of a solute. In dilute solutions, solutes can be considered to behave according to ideal (thermodynamic) principles, i.e. they will have an effective concentration equivalent to the actual concentration. However, in concentrated solutions ($\geqslant 500\,mol\,m^{-3}$), the behaviour of solutes is often non-ideal, and their effective concentration (activity) will be less than the actual concentration $[C]$. The ratio between the effective concentration and the actual concentration is called the activity coefficient (γ) where

$$\gamma = \frac{a}{[C]} \qquad [26.3]$$

Equation [26.3] can be used for SI units ($mol\,m^{-3}$), molarity ($mol\,L^{-1}$) or molality ($mol\,kg^{-1}$). In all cases, γ is a dimensionless term, since a and $[C]$ are expressed in the same units. The activity coefficient of a solute is effectively unity in dilute solution, decreasing as the solute concentration increases (Table 26.1). At high concentrations of certain ionic solutes, γ may increase to become greater than unity.

Table 26.1 Activity coefficient of NaCl solutions as a function of molality. Data from Robinson and Stokes (2002).

Molality	Activity coefficient at 25°C
0.1	0.778
0.5	0.681
1.0	0.657
2.0	0.668
4.0	0.783
6.0	0.986

Example A solution of NaCl with a molality of 0.5 $mol\,kg^{-1}$ has an activity coefficient of 0.681 at 25°C and a molal activity of $0.5 \times 0.681 = 0.340$ $mol\,kg^{-1}$.

KEY POINT Activity is often the correct expression for theoretical relationships involving solute concentration (e.g. where a property of the solution is dependent on concentration). However, for most practical purposes, it is possible to use the *actual* concentration of a solute rather than the activity, since the difference between the two terms can be ignored for dilute solutions.

The particular use of the term 'water activity' is considered below, since it is based on the mole fraction of solvent, rather than the effective concentration of solute.

Equivalent mass (equivalent weight)

Equivalence and normality are outdated terms, although you may come across them in older texts. They apply to certain solutes whose reactions involve the transfer of charged ions, e.g. acids and alkalis (which may be involved in H^+ or OH^- transfer), and electrolytes (which form cations and anions that may take part in further reactions). These two terms take into account the valency of the charged solutes. Thus the equivalent mass of an ion is its relative molecular mass divided by its valency (ignoring the sign), expressed in grams per equivalent (eq) according to the relationship:

$$\text{equivalent mass} = \frac{\text{relative molecular mass}}{\text{valency}} \qquad [26.4]$$

Examples For carbonate ions (CO_3^{2-}), with a relative molecular mass of 60.00 and a valency of 2, the equivalent mass is 60.00/2 = 30.00 g eq^{-1}.
For sulphuric acid (H_2SO_4, relative molecular mass 98.08), where 2 hydrogen ions are available, the equivalent mass is 98.08/2 = 49.04 g eq^{-1}.

For acids and alkalis, the equivalent mass is the mass of substance that will provide 1 mol of either H^+ or OH^- ions in a reaction, obtained by dividing the molecular mass by the number of available ions (n), using n instead of valency as the denominator in eqn [26.4].

Normality

A 1 normal solution (1 N) is one that contains one equivalent mass of a substance per litre of solution. The general formula is:

$$\text{normality} = \frac{\text{mass of substance per litre}}{\text{equivalent mass}} \qquad [26.5]$$

Example A 0.5 N solution of sulphuric acid would contain $0.5 \times 49.04 = 24.52$ g L^{-1}.

Osmolarity

This non-SI expression is used to describe the number of moles of osmotically active solute particles per litre of solution (osmol l^{-1}). The need for such a term arises because some molecules dissociate to give more than one osmotically active particle in aqueous solution.

Example Under ideal conditions, 1 mol of NaCl dissolved in water would give 1 mol of Na^+ ions and 1 mol of Cl^- ions, equivalent to a theoretical osmolarity of 2 osmol L^{-1}.

Osmolality

This term describes the number of moles of osmotically active solute particles per unit mass of solvent (osmol kg^{-1}). For an ideal solute, the osmolality can be determined by multiplying the molality by n, the number of solute particles produced in solution (e.g. for NaCl, $n = 2$). However, for real (i.e. non-ideal) solutes, a correction factor (the osmotic coefficient, ϕ, Greek letter phi) is used:

$$\text{osmolality} = \text{molality} \times n \times \phi \qquad [26.6]$$

Example A 1.0 mol kg^{-1} solution of NaCl has an osmotic coefficient of 0.936 at 25°C and an osmolality of $1.0 \times 2 \times 0.936 = 1.872$ osmol kg^{-1}.

If necessary, the osmotic coefficients of a particular solute can be obtained from tables (e.g. Table 26.2): non-ideal behaviour means that ϕ may have values >1 at high concentrations. Alternatively, the osmolality of a solution can be measured using an osmometer.

Table 26.2 Osmotic coefficients of NaCl solutions as a function of molality. Data from Robinson and Stokes (2002).

Molality	Osmotic coefficient at 25°C
0.1	0.932
0.5	0.921
1.0	0.936
2.0	0.983
4.0	1.116
6.0	1.271

Colligative properties and their use in osmometry

Several properties vary in direct proportion to the effective number of osmotically active solute particles per unit mass of solvent and can be used

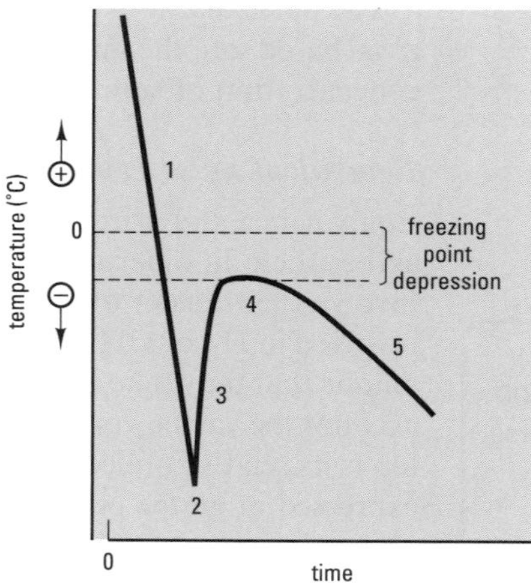

Fig. 26.1 Temperature responses of a cryoscopic osmometer. The response can be subdivided into:
1. initial supercooling;
2. initiation of crystallisation;
3. crystallisation/freezing;
4. plateau, at the freezing point;
5. slow temperature decrease.

to determine the osmolality of a solution. These colligative properties include freezing point, boiling point and vapour pressure.

Using an osmometer – it is vital that the sample holder and probe are clean, otherwise small droplets of the previous sample may be carried over, leading to inaccurate measurement.

An osmometer is an instrument that measures the osmolality of a solution, usually by determining the freezing point depression of the solution in relation to pure water, a technique known as cryoscopic osmometry. A small amount of sample is cooled rapidly and then brought to the freezing point (Fig. 26.1), which is measured by a temperature-sensitive thermistor probe calibrated in mosmol kg^{-1}. An alternative method is used in vapour pressure osmometry, which measures the relative decrease in the vapour pressure produced in the gas phase when a small sample of the solution is equilibrated within a chamber.

Osmotic properties of solutions

Several interrelated terms can be used to describe the osmotic status of a solution. In addition to osmolality, you may come across the following:

Osmotic pressure

This is based on the concept of a membrane permeable to water, but not to solute molecules. For example, if a sucrose solution is placed on one side and pure water on the other, then a passive driving force will be created and water will diffuse across the membrane into the sucrose solution, since the effective water concentration in the sucrose solution will be lower (see Fig. 26.2). The tendency for water to diffuse into the sucrose solution could be counteracted by applying a hydrostatic pressure equivalent to the passive driving force. Thus, the osmotic pressure of a solution is the excess hydrostatic pressure required to prevent the net flow of water into a vessel containing the solution. The SI unit of osmotic pressure is the pascal, Pa ($= kg\,m^{-1}\,s^{-2}$). Older sources may use atmospheres, or bars, and conversion factors are given in Box 32.1 (p. 201). Osmotic pressure and osmolality can be interconverted using the expression 1 osmol kg^{-1} = 2.479 MPa at 25°C.

Example A 1.0 mol kg^{-1} solution of NaCl at 25°C has an osmolalilty of 1.872 osmol kg^{-1} and an osmotic pressure of $1.872 \times 2.479 = 4.641$ MPa.

The use of osmotic pressure has been criticised as misleading, since a solution does not exhibit an 'osmotic pressure' unless it is placed on the other side of a selectively permeable membrane from pure water.

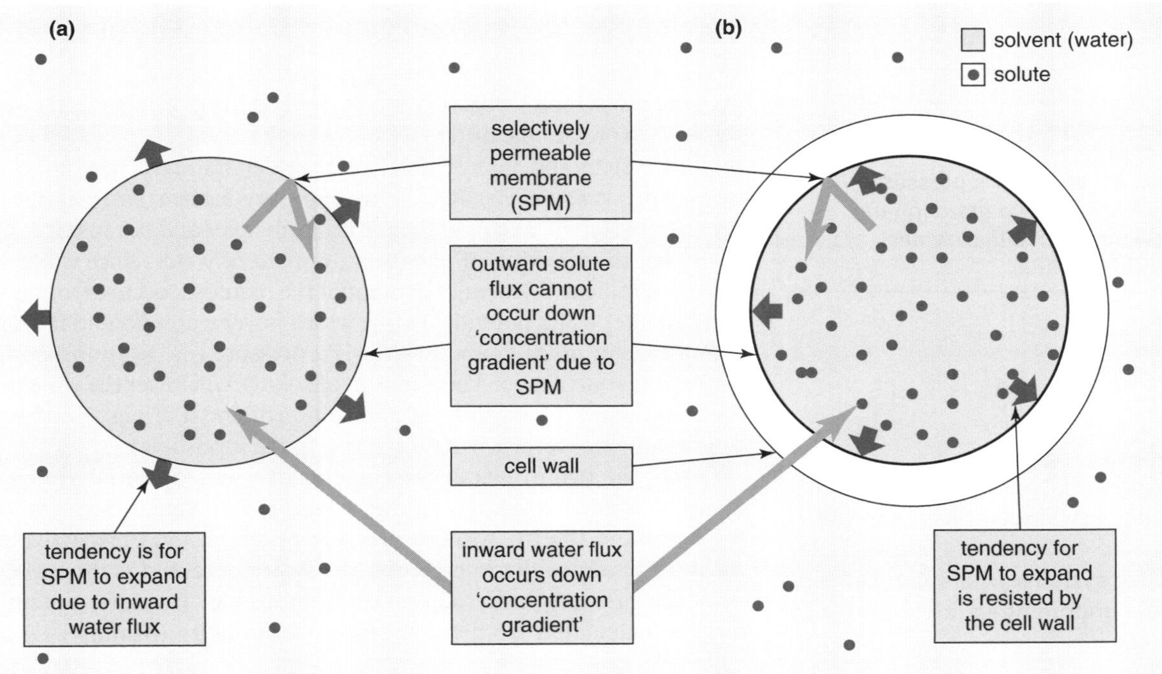

Fig. 26.2 Illustration of forces driving solvent (water) and solute movement across a selectively permeable membrane (SPM). Energetically, both solutes and solvents tend to move down their respective 'concentration gradient' (strictly, down their chemical potential gradient). However, solute molecules cannot leave the model cells illustrated because they cannot pass through the SPM. In the situation illustrated in (a), water will tend to move from outside the cell to within because the solute molecules have effectively 'diluted' the water within the cell (illustrated by the density of point shading), creating a gradient in 'concentration' and because this molecule is able to pass through the SPM. The result will be an expansion of this model cell (short arrows). The osmotic pressure is the (theoretical) pressure that would need to be applied to prevent this. If the model cell were surrounded by a cell wall, as in (b), this would resist expansion, leading to internal pressurisation (turgor pressure, p. 166).

Water activity (a_w)

This is a term often used in food microbiology to describe the osmotic behaviour of microbial cells and is also applied to foodstuffs (p. 490). It is a measure of the relative proportion of water in a solution, expressed in terms of its mole fraction, i.e. the ratio of the number of moles of water (n_w) to the total number of moles of all substances (i.e. water and solutes) in solution (n_t), taking into account the molal activity coefficient of the solvent, water (i.e. γ_w):

$$a_w = \gamma_w \frac{n_w}{n_t} \qquad [26.7]$$

The water activity of pure water is unity, decreasing as solutes are added. One disadvantage of a_w is the limited change that occurs in response to a change in solute concentration: a 1.0 mol kg^{-1} solution of NaCl has a water activity of 0.967 (Table 26.3). Water activity is considered further in the context of microbial growth within foods in Chapter 70, p. 489.

Osmolality, osmotic pressure and water activity are measurements based solely on the osmotic properties of a solution, with no regard for any other driving forces, e.g. hydrostatic and gravitational forces. In circumstances where such other forces are important, you will need to measure a variable that takes into account these aspects of water status, namely water potential.

Table 26.3 Water activity (a_w) of NaCl solutions as a function of molality. Data from Robinson and Stokes (2002).

Molality	a_w
0.1	0.997
0.5	0.984
1.0	0.967
2.0	0.932
4.0	0.852
6.0	0.760

Water potential (hydraulic potential) and its applications

Water potential, Ψ_w, is the most appropriate measure of osmotic status in many areas of bioscience. It is a term derived from the chemical potential of water. It expresses the difference between the chemical potential of water in the test system and that of pure water under standard conditions and has units of pressure (i.e. Pa). It is a more appropriate term than osmotic pressure because it is based on sound theoretical principles and because it can be used to predict the direction of passive movement of water, since water will flow down a gradient of chemical potential (i.e. osmosis occurs from a solution with a higher water potential to one with a lower water potential). A solution of pure water at 20°C and at 0.1 MPa pressure (i.e. ≈ atmospheric) has a water potential of zero. The addition of solutes will lower the water potential (i.e. make it negative), while the application of pressure, e.g. from hydrostatic or gravitational forces, will raise it (i.e. make it positive).

Examples A 1.0 mol kg^{-1} solution of NaCl has a (negative) water potential of −4.641 MPa.

Pure water at 0.2 MPa pressure (about 0.1 MPa above atmospheric pressure) has a (positive) water potential of 0.1 MPa.

Often, the two principal components of water potential are referred to as the solute potential, or osmotic potential (Ψ_s, sometimes symbolised as Ψ_π or π) and the hydrostatic pressure potential (Ψ_p) respectively. For a solution at atmospheric pressure, the water potential is due solely to the presence of osmotically active solute molecules (osmotic potential) and may be calculated from the measured osmolality (osmol kg^{-1}) at 25°C, using the relationship:

$$\Psi_w\,(\text{MPa}) = \Psi_s\,(\text{MPa}) = -2.479 \times \text{osmolality} \qquad [26.8]$$

Measuring water potential – eqn [26.9] ignores the effects of gravitational forces – for systems where gravitational effects are important an additional term is required.

For aquatic microbial cells, e.g. bacteria and fungi, equilibrated in their growth medium at atmospheric pressure, the water potential of the external medium will be equal to the cellular water potential ('isotonic') and the latter can be derived from the measured osmolality of the medium (eqn [26.8]) by osmometry (pp. 163–165). The water potential of such cells can be subdivided into two major parts, the cell solute potential (Ψ_s) and the cell turgor pressure (Ψ_p) as follows:

$$\Psi_w = \Psi_s + \Psi_p \qquad [26.9]$$

To calculate the relative contribution of the osmotic and pressure terms in eqn [26.9], an estimate of the internal osmolality is required, e.g. by measuring the freezing point depression of expressed intracellular fluid. Once you have values for Ψ_w and Ψ_s, the turgor pressure can be calculated by substitution into eqn [26.9].

The van't Hoff relationship can be used to estimate Ψ_s, by summation of the osmotic potentials due to the major solutes, determined from their concentrations, as:

$$\Psi_s = -RTn\phi[C] \qquad [26.10]$$

where RT is the product of the universal gas constant and absolute temperature (2 479 J mol^{-1} at 25°C), n and ϕ are as previously defined and $[C]$ is expressed in SI terms as mol m^{-3}.

Text references and sources for further study

Chapman, C. (1998) *Basic Chemistry for Biology*. McGraw-Hill, New York.

Lide, D.R. (ed.) (2008) *CRC Handbook of Chemistry and Physics*, 89th edn. Taylor and Francis, London.

Robinson, R.A. and Stokes, R.H. (2002) *Electrolyte Solutions*, 2nd edn. Dover Publications, New York.

27 Working with body fluids

The collection and analysis of blood, urine and saliva is fairly common in nutrition therapy and research. Biochemical tests on such samples (Chapter 45) can be used to confirm a diagnosis or to monitor a condition over time. In addition, national surveys often require samples of blood and urine to assess and monitor micronutrient levels. Collection of any body fluid must be well-planned and must follow the appropriate guidelines to ensure the safety of the individual providing the samples and of the person(s) carrying out the collection and subsequent testing.

Preparing for sample collection

Taking a blood sample – it may be necessary for a trained professional, such as a medical practitioner or phlebotomist, to carry out the procedure, particularly if you require a venous blood sample.

Ethical approval must be given by an appropriate ethics committee (Chapter 23) before taking samples of blood or other body fluids for research or use in laboratory practical classes. It is also necessary to obtain written informed consent from the participants, or if under 18, their guardian. This involves ensuring that they are fully aware of what to expect and also that clear information is given about what will be done with the results of any subsequent analysis (Chapter 23). Samples should be destroyed once the project is completed unless consent has been given for them to be used in further research.

KEY POINT Before collecting any type of sample you must first consider what tests will be done on the sample and also confirm the method of sampling that will be used.

Biological hazards and their containment

When working with body fluids it is important to understand the health and safety implications associated with this type of work. A major source of risk is the transmission of infection due to the presence of pathogenic microbes present in the samples. Pathogens are categorised into four Hazard Groups according to the severity of the diseases that they cause and the ease with which they can spread, with Hazard Group 1 being least harmful and Hazard Group 4 being severely harmful and readily transmissible (p. 453).

Example

Hazard Group 1: An organism that is unlikely to cause human disease, eg. *Lactobacillus bulgaricus*

Hazard Group 2: An organism that may cause human disease but is unlikely to spread in the community, eg. Epstein–Barr virus

Hazard Group 3: An organism that may cause severe human disease and may create a risk to the community, eg. *Mycobacterium tuberculosis*

Hazard group 4: An organism that can cause severe human disease and presents a high risk to the community, eg. Ebola virus

Most blood-borne pathogens are categorised as Hazard Groups 2 or 3 and when working with body fluids it is necessary to assume that every sample of body fluids is a potential source of blood-borne pathogens. The Health and Safety Executive (HSE) in the UK publishes an approved list of biological agents which details the level of containment required for each agent (HSE, 2004). The category of hazard generally equates to the level of containment required, although any work involving blood would normally be carried out at a minimum of containment level 2. Table 27.1 summarises the differences between containment levels 2 and 3, based on UK safety legislation. Universities would not usually have containment level 4 facilities, due to the risks involved in handling Hazard Group 4 pathogens. For most practical purposes, students will work with body fluids at containment level 2.

Table 27.1 Containment measures for health and veterinary care facilities, laboratories and animal rooms

Containment measure	Containment levels	
	2	3
Workplace to be separated from any other activities in the same building	No	Yes
Input and extract air to be filtered (HEPA* or equivalent)	No	Yes, on extract air
Access to be restricted to authorised persons only	Yes	Yes
Workplace is to be sealable to permit disinfection	No	Yes
Specified disinfection procedure required	Yes	Yes
Workplace is to be maintained at an air pressure negative to atmosphere	No	Yes
Efficient vector control, e.g. rodents and insects	Yes, for animal containment	Yes, for animal containment
Surfaces to be impervious to water and easy to clean	Yes, for bench	Yes, for bench and floor (and walls for animal containment)
Surfaces to be resistant to acids, alkalis, solvents, disinfectants	Yes, for bench	Yes, for bench and floor (and walls for animal containment)
Safe storage required for biological agents	Yes	Yes
Observation window, or alternative, required, so occupants can be seen	No	Yes
Laboratory to contain its own equipment	No	Yes, so far as is reasonably practicable
Infected material, including any animal tissues, to be handled in a biosafety cabinet or other suitable containment system	Yes, where aerosol produced	Yes, where aerosol produced
Incinerator required for disposal of animal carcases	Accessible	Accessible

*HEPA = high efficiency particulate air (COSHH, 2002)

Handwashing at containment level 2 – typically elbow-operated taps are used, to minimise the risk of transmission of infection due to touching taps with your hands.

Personal protection equipment (PPE)

This must be worn when working with potentially infectious or hazardous materials as might be the case when analysing blood samples. In containment level 2 laboratories this would normally consist of a side-fastening laboratory coat, plus safety glasses and gloves as required. It is particularly important to only wear these when working within the laboratory. Hands should be washed immediately after entering the laboratory, before putting on your lab coat. At the end of the session, once your lab coat/gloves/glasses have been removed, you must wash your hands thoroughly before leaving, to avoid transmitting pathogens to others outside the laboratory.

Urine collection

There are several different types of urine specimens and the type you use will largely depend on the analysis being carried out. It is important to know the specimen type as this can affect the interpretation of the results. All urine collection containers should be clearly labelled with the name of the participant, plus the date and time of collection.

The main types of urine sample are as follows:

Definition

Urinalysis – the analysis of urine. May including biochemical testing, and also microscopy and microbiological assay, as required.

Random specimen

This may have been collected at any time, with no specific instructions given other than to avoid touching the inside of the collection container or lid to avoid contamination. This type of specimen is relatively easy for the participant to collect but offers least information to the analyst.

First-morning specimen

This sample will be more concentrated than a typical random sample, making it suitable for urinalysis screening as it will contain higher levels of analytes.

Midstream clean-catch specimen

This sample would be used for microbiological analysis in cases of suspected urinary tract infection rather than for nutrient analysis. Measures are taken to reduce contamination due to commensal microbes. The participant should first clean the urethral opening using a towelette. They should be instructed to begin urinating into the toilet and then move the collection container into the stream of urine until the container is at least half-full. They should then remove the container from the urine stream.

24-hour urine collection

Time-based urine collections are used when testing for analytes such as creatinine, albumin, urea nitrogen, sodium and potassium. They can also be used to confirm the accuracy of dietary diaries by comparing nitrogen balance (p. 310). To confirm the completeness of a 24-hour urine collection in large-scale surveys, it is common for the subject to be given oral doses of para-aminobenzoic acid (PABA). PABA is excreted in the urine and analysis of PABA concentration can confirm that urine collection is complete with recovery of above 85 per cent accepted as a complete collection (due to differences in metabolism between individuals). Box 27.1 gives details of how to carry out 24-hour urine sampling, which you may be asked to do as part of your studies.

Urinalysis for PABA – since colorimetric methods often show interference from other compounds, HPLC (p. 362) is the preferred assay method.

Box 27.1 How to collect a 24-hour urine sample, to be verified using PABA analysis

The method described here is for collection of your own urine but you could adapt this guidance to tell a subject how to collect his or her sample.

1. **Consider which day you will collect your urine.** It may be useful to collect the 24-hour urine sample on a day when you are not planning to go out for long periods of time, as you will need to take the collecting equipment with you.
2. **Make sure you have the right collection equipment.** The collection equipment would generally consist of a plastic jug and funnel, a 5-litre bottle containing preservative, usually boric acid, and a 2-litre bottle for use away from home.
3. **Note the date and time you start the collection.** Do not collect the first urine of the day – start with the second urine.
4. **Take three PABA tablets at evenly spaced times throughout the day.** Take the first PABA tablet after you have voided the first urine. Take the other two PABA tablets at evenly spaced times during the rest of your day, usually with lunch and dinner.
5. **Remember to collect all your urine.** Urinate into a plastic jug and then pour this into the larger storage container. Collect all of your urine for the next 24 hours, including the first urine of the next morning.
6. **Repeatedly mix the urine with the preservative.** The storage container will contain a preservative, so each time you add a collection of urine, swirl the container to fully mix it with the preservative.
7. **Keep this container in a cool, preferably dark, place.**
8. **Transport the sample to the laboratory for analysis.** This should be done as soon as possible after the final urine collection – ideally early in the morning of the second day.

Understanding the difference between plasma and serum – plasma is obtained from a blood sample that has not been allowed to coagulate. Serum is from a blood sample that has coagulated, which removes fibrin and clotting factors.

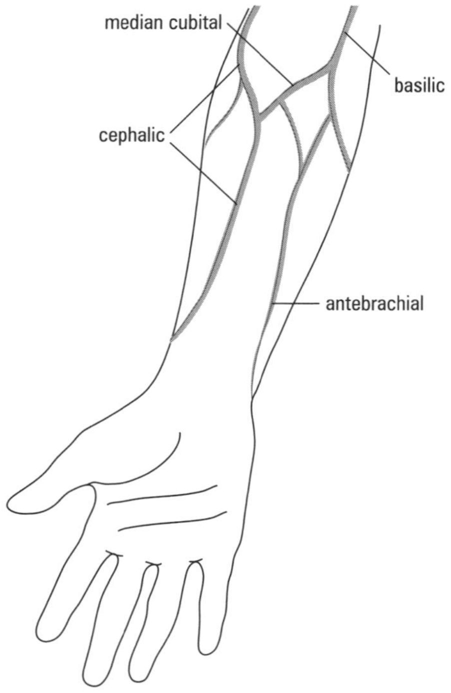

Fig. 27.1 Veins of the anterior forearm.

Blood sampling

A number of biochemical tests can be carried out on blood samples (Chapter 45). How the blood sample is collected and processed will be determined by the analysis required, since some analytes are assayed using serum, e.g. albumin analysis, while others require plasma, e.g. for measurement of iron status (see opposite).

Venous blood samples

This is the type of blood sample normally taken by a medical practitioner, nurse or phlebotomist. It should only be carried out by a trained health professional. It is generally taken from the median cubital vein (Fig. 27.1) on the inside of the elbow. Samples can either be taken with a needle and syringe or, more commonly these days, using a closed vacuum system such as a BD Vacutainer® blood collection tube, which reduces the risk of needle-stick injuries and allows multiple samples to be taken from one site. Venous blood samples would usually be taken if a large volume of blood is required.

Capillary blood samples

Capillary blood samples are suitable when a small volume of blood is required. They can be taken from a number of sites including the fingers, earlobes and heel, the latter being the site of choice for blood sampling of infants. Box 27.2 gives details on how to take a capillary blood sample from the finger.

Saliva samples

Saliva samples can be used to measure amylase levels and this is typically carried out in undergraduate practical classes using spectrophotometry (p. 357). Increasingly, saliva is also being used as a source of DNA for

Box 27.2 How to take a capillary blood sample from the finger

1. **Collect the appropriate equipment.** This includes alcohol swab, cotton wool or gauze, capillary blood tubes and a lancet for puncturing the skin. Lancets come in different depths, which will result in differences in flow rates.
2. **Ensure appropriate hygiene procedures have been followed.** Wash your hands thoroughly and wear appropriate PPE.
3. **Select the site.** The middle or ring finger would normally be selected and the puncture site would be to the side of the tip of the finger (Fig. 27.2). This is because the pad of the finger tip is likely to be more sensitive to pain during and following the procedure. Gently warming the site or getting the subject to sit with their hand down to allow gravity to increase blood flow to the area may be useful.
4. **Clean the puncture site.** You can either use fresh clean water (or an alcohol wipe, although this is not advocated in the UK). Allow the site to dry completely.
5. **Puncture the skin with the lancet.** Wipe away the first drop of blood using the cotton wool or gauze in case there is any alcohol or water left on the skin. The first flow of blood is also likely to contain excess tissue fluid.
6. **Discard the lancet into a 'sharps' container.**
7. **Collect the blood sample.** Do this by holding a capillary tube to the blood flow. Gently apply and release pressure to encourage blood flow. The volume of blood may be up to 500 μL.
8. **Stop the blood flow** by applying pressure to the puncture site with a cotton wool ball until bleeding stops. Cover with a sterile dressing.
9. **Check and confirm that the tubes are labelled correctly and then transport for analysis.**

(UCL Institute of Child Health, 2008)

Fig. 27.2 Site for puncture (shaded area) to collect capillary blood sample (BD Diagnostics, 2007).

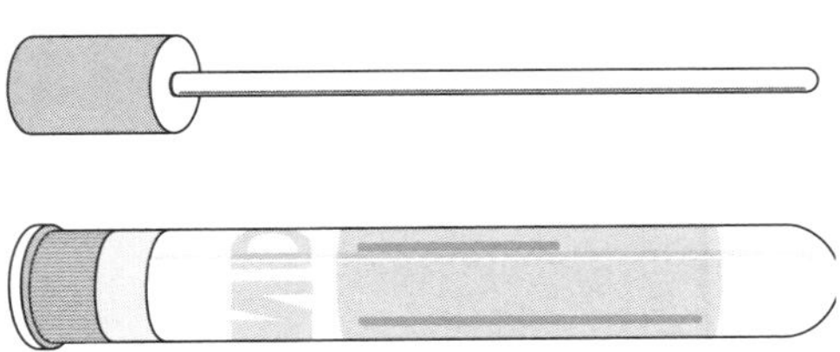

Fig. 27.3 Saliva collection device (http://www.malvernmedical.uk.com/oracol.htm).

Fig. 27.4 'Sharps' container for disposal of items such as lancets and needles. When full these are normally emptied by the technical services of your department.

genetic research and has been shown to have a higher participation rate than for blood samples (Hansen *et al.*, 2007). As saliva samples are easy and convenient to collect and are minimally invasive.

Saliva sampling

The subject should first rinse his or her mouth with water to ensure all food debris is removed. To stimulate saliva production, ask the subject to simulate a chewing action. Saliva should be collected in a sterile container, usually a 50 mL tube, although you would only expect to collect 1–2 mL of saliva. Commercially available collection kits are typically used if saliva is to be tested for circulating hormones or for peptide analysis; these kits contain absorbent material to collect the saliva rather than having the subject spit into the collection container, as this may be more acceptable to the subject (Fig. 27.3). Saliva samples can usually be frozen or refrigerated for later analysis.

Disinfection and disposal of body fluids

All body fluids must be handled very carefully from collecting the samples through to disposal, with the assumption that they may contain potentially pathogenic agents.

Disposal of equipment

All sharps, such as lancets used for capillary blood sampling, must be disposed of in an appropriate 'sharps' container immediately after use. Any sampling tubes which cannot be autoclaved (p. 452) must be disposed of in a biohazard waste bag for incineration, as should disposable tissues and any single-use protective equipment such as gloves. Nothing that has come into contact with blood or blood products should ever be disposed of via the general waste route.

Disposal of samples

Body fluid samples and specimens should either be made safe from the risk of infection either by autoclaving or by the addition of hypochlorite solution, or they should be disposed of as biohazardous waste, following the local procedures of your laboratory.

> **KEY POINT** You must make yourself aware of all local rules for the safe disposal of body fluid samples before you begin your practical work.

Dealing with spills

If a spill occurs make sure you are wearing gloves before beginning the clean-up. If there is broken glass in the spill, do not pick up the pieces of glass using your hands – use a dustpan and transfer to a sharps bin.

Soak up the spill using paper towels or ideally absorbent granules, being careful not to spread or splash the spill. Dispose of the paper towels or granules in the biohazard waste bag. Wipe the area using a concentrated solution of hypochlorite containing at least 10 000 ppm available chlorine or other disinfectant solution, e.g. Virkon®.

Text references

BD Diagnostics (2007) *Successful Specimen Collection: Fingersticks*. Available: http://www.bd.com/vacutainer/pdfs/VS7688.pdf
Last accessed: 21/12/10.

Great Britain. Health and Safety (2002) *Control of Substances Hazardous to Health Regulations 2002* (as amended). SI 2002/2677. The Stationery Office. Available: http://www.opsi.gov.uk/si/si2002/20022677.htm#1
Last accessed: 21/12/10.

Hansen, T.O., Simonsen, M.K., Neilsen, F.C., Anderson Hundrup, Y. (2007) Collection of blood, saliva, and buccal cell samples in a pilot study on the Danish nurse cohort: Comparison of the response rate and quality of genomic DNA. *Cancer Epidemiology, Biomarkers Prevention*, **16** (10), 2072–2076.

Health and Safety Executive (2004) *The Approved List of Biological Agents*. Available: http://www.hse.gov.uk/pubns/misc208.pdf
Last accessed: 21/12/10.

UCL Institute of Child Health (2008) *Clinical Guideline Capillary Blood Sampling*. Available: http://www.ich.ucl.ac.uk/clinical_information/clinical_guidelines/cpg_guideline_00136/#rationals699
Last accessed: 21/12/10.

Sources for further study

Health and Safety Executive (2005) *Biological Agents: Managing the Risks in Laboratories and Healthcare Premises*. Available: http://www.dh.gov.uk/prod_consum_dh/groups/dh_digitalassets/@dh/@ab/documents/digitalasset/dh_087496.pdf
Last accessed: 21/12/10.

National Centre for Social Research (2008) *An Assessment of Dietary Sodium Levels Among Adults (Aged 19–64) in the UK General Population in 2008, Based on Analysis of Dietary Sodium in 24 Hour Urine Collection*. Available: http://www.food.gov.uk/multimedia/pdfs/sodiumreport08.pdf
Last accessed: 21/12/10.

Skobe, C. (2004) The basics of specimen collection and handling of urine testing. Available: http://www.bd.com/vacutainer/labnotes/pdf/Volume14Number2_VS7226.pdf
Last accessed: 21/12/10.

United States Department of Labor (2008) *Regulations (Standards – 29 CFR) Bloodborne Pathogens*. Available: http://www.osha.gov/pls/oshaweb/owadisp.show_document?p_table=STANDARDS&p_id=10051
Last accessed: 21/12/10.

28 pH and buffer solutions

Definitions

Acid – a compound that acts as a proton donor in aqueous solution.

Base – a compound that acts as a proton acceptor in aqueous solution.

Conjugate pair – an acid together with its corresponding base.

Alkali – a compound that liberates hydroxyl ions when it dissociates. Since hydroxyl ions are strongly basic, this will reduce the proton concentration.

Ampholyte – a compound that can act as both an acid and a base. Water is an ampholyte since it may dissociate to give a proton and a hydroxyl ion (amphoteric behaviour).

SAFETY NOTE Safe working with strong acids or alkalis – these can be highly corrosive; rinse with plenty of water, if spilled.

Table 28.1 Effects of temperature on the ion product of water (K_w), H^+ ion concentration and pH at neutrality. Values calculated from Lide (2008).

Temp. (°C)	K_w (mol^2 L^{-2})	[H^+] at neutrality (nmol L^{-1})	pH at neutrality
0	0.11×10^{-14}	33.9	7.47
4	0.17×10^{-14}	40.7	7.39
10	0.29×10^{-14}	53.7	7.27
20	0.68×10^{-14}	83.2	7.08
25	1.01×10^{-14}	100.4	7.00
30	1.47×10^{-14}	120.2	6.92
37	2.39×10^{-14}	154.9	6.81
45	4.02×10^{-14}	199.5	6.70

Example Human blood plasma has a typical H^+ concentration of approximately 0.4×10^{-7} mol L^{-1} ($= 10^{-7.4}$ mol L^{-1}), giving a pH of 7.4.

pH is a measure of the amount of hydrogen ions (H^+) in a solution: this affects the solubility of many substances and the activity of most biological systems, from individual molecules to whole organisms. It is usual to think of aqueous solutions as containing H^+ ions (protons), although protons actually exist in their hydrated form, as hydronium ions (H_3O^+). The proton concentration of an aqueous solution [H^+] is affected by several factors:

- Ionisation (dissociation) of water, which liberates protons and hydroxyl ions in equal quantities, according to the reversible relationship:

$$H_2O \rightleftharpoons H^+ + OH^- \qquad [28.1]$$

- Dissociation of acids, according to the equation:

$$H\text{–}A \rightleftharpoons H^+ + A^- \qquad [28.2]$$

  where H–A represents the acid and A^- is the corresponding conjugate base. The dissociation of an acid in water will increase the amount of protons, reducing the amount of hydroxyl ions as water molecules are formed (eqn [28.1]). The addition of a base (usually, as its salt) to water will decrease the amount of H^+, due to the formation of the conjugate acid (eqn [28.2]).

- Dissociation of alkalis, according to the relationship:

$$X\text{–}OH \rightleftharpoons X^+ + OH^- \qquad [28.3]$$

  where X–OH represents the undissociated alkali. Since the dissociation of water is reversible (eqn [28.1]), in an aqueous solution the production of hydroxyl ions will effectively act to 'mop up' protons, lowering the proton concentration.

Many compounds act as acids, bases or alkalis: those that are almost completely ionised in solution are usually called strong acids or bases, while weak acids or bases are only slightly ionised in solution (p. 175).

In an aqueous solution, most of the water molecules are not ionised. In fact, the extent of ionisation of pure water is constant at any given temperature and is usually expressed in terms of the ion product (or ionisation constant) of water, K_w:

$$K_w = [H^+][OH^-] \qquad [28.4]$$

where [H^+] and [OH^-] represent the molar concentration (strictly, the activity) of protons and hydroxyl ions in solution, expressed as mol L^{-1}. At 25°C, the ion product of pure water (Table 28.1) is 10^{-14} mol^2 L^{-2} (i.e. 10^{-8} mol^2 m^{-6}). This means that the concentration of protons in solution will be 10^{-7} mol L^{-1} (10^{-4} mol m^{-3}), with an equivalent concentration of hydroxyl ions (eqn [28.1]). Since these values are very low and involve negative powers of 10, it is customary to use the pH scale, where:

$$pH = -\log_{10}[H^+] \qquad [28.5]$$

and [H^+] is the proton activity in mol L^{-1} (see p. 162).

KEY POINT While pH is strictly the negative logarithm (to the base 10) of H^+ activity, in practice H^+ concentration in mol L^{-1} (equivalent to kmol m^{-3} in SI terminology) is most often used in place of activity, since the two are virtually the same, given the limited dissociation of H_2O. The pH scale is not SI: nevertheless, it continues to be used widely in biological science.

Measuring pH – the pH of a neutral solution changes with temperature (Table 28.1), due to the enhanced dissociation of water with increasing temperature. This must be taken into account when measuring the pH of any solution and when interpreting your results.

The value where an equal amount of H^+ and OH^- ions are present is termed neutrality: at 25°C the pH of pure water at neutrality is 7.0. At this temperature, pH values below 7.0 are acidic while values above 7.0 are alkaline.

Always remember that the pH scale is a logarithmic one, not a linear one: a solution with a pH of 3.0 is not twice as acidic as a solution of pH 6.0, but one thousand times as acidic (i.e. contains 1000 times the amount of H^+ ions). Therefore, you may need to convert pH values into proton concentrations before you carry out mathematical manipulations (see Box 38.2, p. 254). For similar reasons, it is important that pH change is expressed in terms of the original and final pH values, rather than simply quoting the difference between the values: a pH change of 0.1 has little meaning unless the initial or final pH is known.

SAFETY NOTE Preparing a dilute acid solution using concentrated acid – always slowly add the concentrated acid to water, not the reverse, since the strongly exothermic process can trigger a violent reaction with water.

SAFETY NOTE Preparing an alkali solution – typically, the alkali will be in solid form (e.g. NaOH) and addition to water will rapidly raise the temperature of the solution: use only heat-resistant glassware, cooled with water if necessary.

Measuring pH

pH electrodes

Accurate pH measurements can be made using a pH electrode, coupled to a pH meter. The pH electrode is usually a combination electrode, comprising two separate systems: an H^+-sensitive glass electrode and a reference electrode which is unaffected by H^+ ion concentration (Fig. 28.2). When this is immersed in a solution, a pH-dependent voltage between the two electrodes can be measured using a potentiometer. In most cases, the pH electrode assembly (containing the glass and reference electrodes) is connected to a separate pH meter by a cable, although some hand-held instruments (pH probes) have the electrodes and meter within the same assembly, often using an H^+-sensitive field effect transistor in place of a glass electrode, to improve durability and portability.

Box 28.1 gives details of the steps involved in making a pH measurement with a glass pH electrode and meter.

Table 28.2 Properties of some pH indicator dyes

Dye	Acid-base colour change	Useful pH range
Thymol blue (acid)	red–yellow	1.2–6.8
Bromophenol blue	yellow–blue	1.2–6.8
Congo red	blue–red	3.0–5.2
Bromocresol green	yellow–blue	3.8–5.4
Resazurin	orange–violet	3.8–6.5
Methyl red	red–yellow	4.3–6.1
Litmus	red–blue	4.5–8.3
Bromocresol purple	yellow–purple	5.8–6.8
Bromothymol blue	yellow–blue	6.0–7.6
Neutral red	red–yellow	6.8–8.0
Phenol red	yellow–red	6.8–8.2
Thymol blue (alkaline)	yellow–blue	8.0–9.6
Phenol-phthalein	none–red	8.3–10.0

pH indicator dyes

These compounds (usually weak acids) change colour in a pH-dependent manner. They may be added in small amounts to a solution, or they can be used in paper strip form. Each indicator dye usually changes colour over a restricted pH range, typically 1–2 pH units (Table 28.2): universal indicator dyes/papers make use of a combination of individual dyes to measure a wider pH range. Dyes are not suitable for accurate pH measurement as they are affected by other components of the solution including oxidising and reducing agents and salts. However, they are useful for:

- estimating the approximate pH of a solution;
- determining a change in pH, for example at the end-point of a titration or the production of acids during bacterial metabolism (pp. 162–164);
- establishing the pH of a food (Chapter 49).

Definition

Buffer solution – one that resists a change in H^+ concentration (pH) on addition of acid or alkali.

Buffers

Rather than simply measuring the pH of a solution, you may wish to *control* the pH, e.g. in metabolic experiments, or in a growth medium for cell culture (p. 471). In fact, you should consider whether you need to control pH in any experiment involving a biological system, whether whole organisms, isolated cells, subcellular components or biomolecules. One of the most effective ways to control pH is to use a buffer solution.

A buffer solution is usually a mixture of a weak acid and its conjugate base. Added protons will be neutralised by the anionic base while a reduction in protons, e.g. due to the addition of hydroxyl ions, will be counterbalanced by dissociation of the acid (eqn [28.2]); thus the conjugate pair acts as a 'buffer' to pH change. The innate resistance of most biological fluids to pH change is due to the presence of cellular constituents that act as buffers, e.g. proteins, which have a large number of weakly acidic and basic groups in their amino acid side chains.

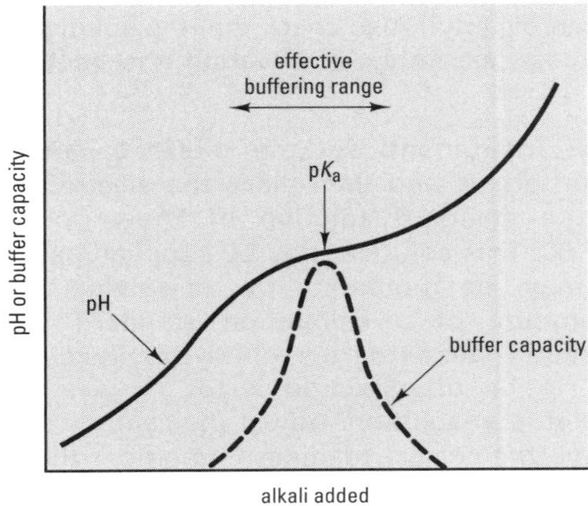

Fig. 28.1 Theoretical pH titration curve for a buffer solution. pH change is lowest and buffer capacity is greatest at the pK_a of the buffer solution.

Buffer capacity and the effects of pH

The extent of resistance to pH change is called the buffer capacity of a solution. The buffer capacity is measured experimentally at a particular pH by titration against a strong acid or alkali: the resultant curve will be strongly sigmoidal, with a plateau where the buffer capacity is greatest (Fig. 28.1). The mid-point of the plateau represents the pH where equal quantities of acid and conjugate base are present, and is given the symbol pK_a, which refers to the negative logarithm (to the base 10) of the acid dissociation constant, K_a, where

$$K_a = \frac{[H^+][A^-]}{[HA]} \quad [28.6]$$

By rearranging eqn [28.6] and taking negative logarithms, we obtain:

$$pH = pK_a + \log_{10}\frac{[A^-]}{[HA]} \quad [28.7]$$

This relationship is known as the Henderson–Hasselbalch equation and it shows that the pH will be equal to the pK_a when the ratio of conjugate base to acid is unity, since the final term in eqn [28.7] will be zero. Consequently, pK_a is an important factor in determining buffer capacity at a particular pH. In practical terms, a buffer solution will work most effectively at pH values about one unit either side of the pK_a.

Buffer selection – an ideal buffer for biological purposes would possess the following characteristics:

- impermeability to biological membranes;
- biological stability and lack of interference with metabolic and biological processes;
- lack of significant absorption of ultraviolet or visible light;
- lack of formation of insoluble complexes with cations;
- minimal effect of ionic composition or salt concentration;
- limited pH change in response to temperature.

Selecting an appropriate buffer

When selecting a buffer, you should be aware of certain limitations to their use. Citric acid and phosphate buffers readily form insoluble complexes with divalent cations, while phosphate can also act as a substrate, activator or inhibitor of certain enzymes. Both of these buffers contain biologically significant quantities of cations, e.g. Na^+ or K^+. TRIS (Table 28.3) is often toxic to biological systems: due to its high lipid solubility it can penetrate membranes, uncoupling electron transport reactions in whole cells and isolated organelles. In addition, it is markedly affected by temperature, with a ten-fold increase in H^+ concentration from 4°C to 37°C. A number of zwitterionic molecules (having both positive and negative groups) has

Box 28.1 Using a glass pH electrode and meter to measure the pH of a solution

The following procedure should be used whenever you make a pH measurement: consult the manufacturer's handbook for specific information, where necessary. Do not be tempted to miss out any of the steps detailed below, particularly those relating to the effects of temperature, or your measurements are likely to be inaccurate.

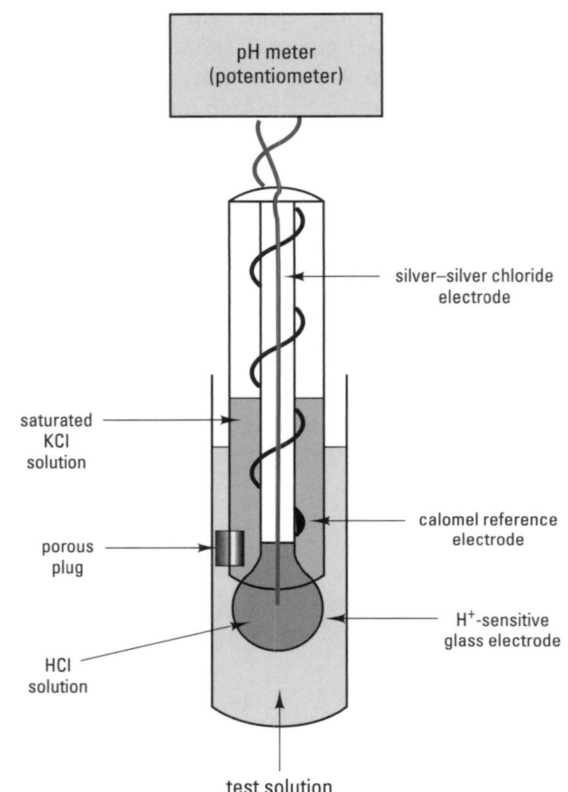

Fig. 28.2 Measurement of pH using a combination pH electrode and meter. The electrical potential difference recorded by the potentiometer is directly proportional to the pH of the test solution.

1. **Stir the test solution thoroughly before you make any measurement:** it is often best to use a magnetic stirrer. Leave the solution for sufficient time to allow equilibration at lab temperature.
2. **Record the temperature of every solution you use**, including all calibration standards and samples, since this will affect K_w, neutrality and pH.
3. **Set the temperature compensator on the meter to the appropriate value.** This control makes an allowance for the effect of temperature on the electrical potential difference recorded by the meter: it does *not* allow for the other temperature-dependent effects mentioned elsewhere. Basic instruments have no temperature compensator, and should only be used at a specified temperature, either 20°C or 25°C, otherwise they will not give an accurate measurement. More sophisticated systems have automatic temperature compensation.
4. **Rinse the electrode assembly with distilled water** and gently dab off the excess water onto a clean tissue: check for visible damage or contamination of the glass electrode (consult a member of staff if the glass is broken or dirty). Also check that the solution within the glass assembly is covering the metal electrode.
5. **Calibrate the instrument:** set the meter to 'pH' mode, if appropriate, and then place the electrode assembly in a standard solution of known pH, usually pH 7.00. This solution may be supplied as a liquid, or may be prepared by dissolving a measured amount of a calibration standard in water: calibration standards are often provided in tablet form, to be dissolved in water to give a particular volume of solution. Adjust the calibration control to give the correct reading. Remember that your calibration standards will only give the specified pH at a particular temperature, usually either 20°C or 25°C. If you are working at a different temperature, you must establish the actual pH of your calibration standards, either from the supplier or from literature information.
6. **Remove the electrode assembly from the calibration solution and rinse again with distilled water:** dab off the excess water. Basic instruments have no further calibration steps (single-point calibration), while the more refined pH meters have additional calibration procedures.

   If you are using a basic instrument, you should check that your apparatus is accurate over the appropriate pH range by measuring the pH of another standard whose pH is close to that expected for the test solution. If the standard does not give the expected reading, the instrument is not functioning correctly: consult a member of staff.

   If you are using an instrument with a slope control function, this will allow you to correct for any deviation in electrical potential from that predicted by the theoretical relationship (at 25°C, a change in pH of 1.00 unit should result in a change in electrical potential of 59.16 mV) by performing a two-point calibration. Having calibrated the

(continued)

Box 28.1 (continued)

instrument at pH 7.00, immerse in a second standard at the same temperature as that of the first standard, usually buffered to either pH 4.00 or pH 9.00, depending upon the expected pH of your samples. Adjust the slope control until the exact value of the second standard is achieved (Fig. 28.3). A pH electrode and meter calibrated using the two-point method will give accurate readings over the pH range from 3 to 11: laboratory pH electrodes are not accurate outside this range, since the theoretical relationship between electrical potential and pH is valid.

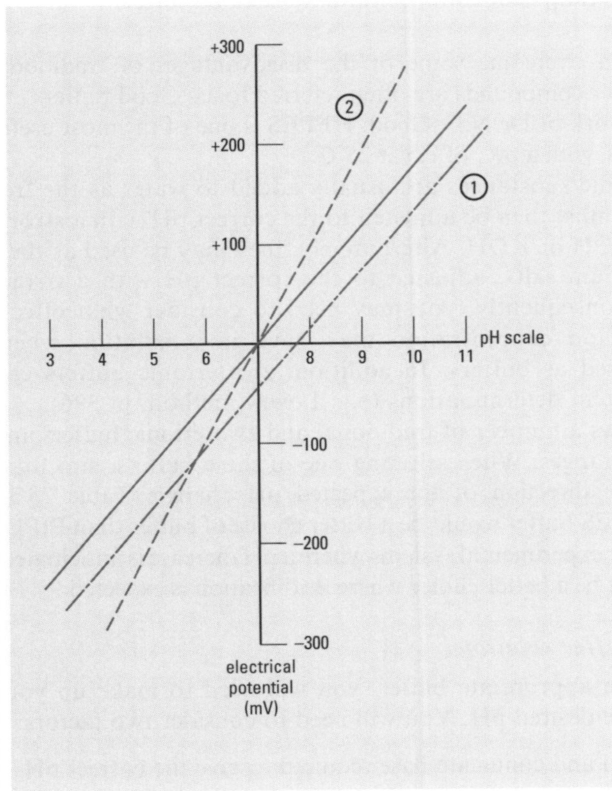

Fig. 28.3 The relationship between electrical potential and pH. The solid line shows the response of a calibrated electrode while the other plots are for instruments requiring calibration: 1 has the correct slope but incorrect isopotential point (calibration control adjustment is needed); 2 has the correct isopotential point but incorrect slope (slope control adjustment is needed).

7. **Once the instrument is calibrated, measure the pH of your solution(s)**, making sure that the electrode assembly is rinsed thoroughly between measurements. You should be particularly aware of this requirement if your solutions contain organic biological material, e.g. soil, tissue fluids, protein solutions, etc., since these may adhere to the glass electrode and affect the calibration of your instrument. If your electrode becomes contaminated during use, check with a member of staff before cleaning: avoid touching the surface of the glass electrode with abrasive material. Allow sufficient time for the pH reading to stabilise in each solution before taking a measurement: for unbuffered solutions, this may take several minutes, so do not take inaccurate pH readings due to impatience!

8. **After use, the electrode assembly must not be allowed to dry out.** Most pH electrodes should be stored in a neutral solution of KCl, either by suspending the assembly in a small beaker, or by using an electrode cap filled with the appropriate solution (typically 1.0 mol L^{-1} KCl buffered at pH 7.0). However, many labs simply use distilled water as a storage solution, leading to loss of ions from the interior of the electrode assembly. In practice, this means that pH electrodes stored in distilled water will take far longer to give a stable reading than those stored in KCl.

9. **Switch the meter to zero (where appropriate), but do not turn off the power**: pH meters give more stable readings if they are left on during normal working hours.

   Problems (and solutions) include: inaccurate and/or unstable pH readings caused by cross-contamination (rinse electrode assembly with distilled water and blot dry between measurements); development of a protein film on the surface of the electrode (soak in 1% w/v pepsin in 0.1 mol L^{-1} HCl for at least an hour); deposition of organic or inorganic contaminants on the glass bulb (use an organic solvent, such as acetone, or a solution of 0.1 mol L^{-1} disodium ethylenediamine-tetraacetic acid, respectively); drying out of the internal reference solutions (drain, flush and refill with fresh solution, then allow to equilibrate in 0.1 mol L^{-1} HCl for at least an hour); cracks or chips to the surface of the glass bulb (use a replacement electrode).

Table 28.3 pK_a values at 25°C and M_r of some acids and bases (upper section) and some large organic zwitterions (lower section) commonly used in buffer solutions. For polyprotic acids, where more than one proton may dissociate, the pK_a values are given for each ionisation step. Only the trivial acronyms of the larger molecules are provided: their full names can be obtained from the catalogues of most chemical suppliers.

Acid or base	pK_a value(s)	M_r
Acetic acid	4.8	60.1
Boric acid	9.2	61.8
Citric acid	3.1, 4.8, 5.4	191.2
Glycylglycine	3.1, 8.2	132.1
Phosphoric acid	2.1, 7.1, 12.3	98.0
Phthalic acid	2.9, 5.5	166.1
Succinic acid	4.2, 5.6	118.1
TRIS(base)*	8.3	121.1
CAPS (free acid)	10.4	221.3
CHES (free acid)	9.3	207.3
HEPES (free acid)	7.5	238.3
MES (free acid)	6.1	213.2
MOPS (free acid)	7.2	209.3
PIPES (free acid)	6.8	302.4
TAPS (free acid)	8.4	243.3
TRICINE (free acid)	8.1	179.2

*Note that this compound is hygroscopic and should be stored in a desiccator; also see text regarding its potential toxicity (p. 177).

Table 28.4 Preparation of sodium phosphate buffer solutions for use at 25°C. Prepare separate stock solutions of (a) disodium hydrogen phosphate and (b) sodium dihydrogen phosphate, both at 200 mol m^{-3}. Buffer solutions (at 100 mol m^{-3}) are then prepared at the required pH by mixing together the volume of each stock solution shown in the table, then diluting to a final volume of 100 mL using distilled or deionised water.

Required pH (at 25°C)	Volume of stock (a) Na_2HPO_4 (mL)	Volume of stock (b) NaH_2PO_4 (mL)
6.0	6.2	43.8
6.2	9.3	40.7
6.4	13.3	36.7
6.6	18.8	31.2
6.8	24.5	25.5
7.0	30.5	19.5
7.2	36.0	14.0
7.4	40.5	9.5
7.6	43.5	6.5
7.8	45.8	4.2
8.0	47.4	2.6

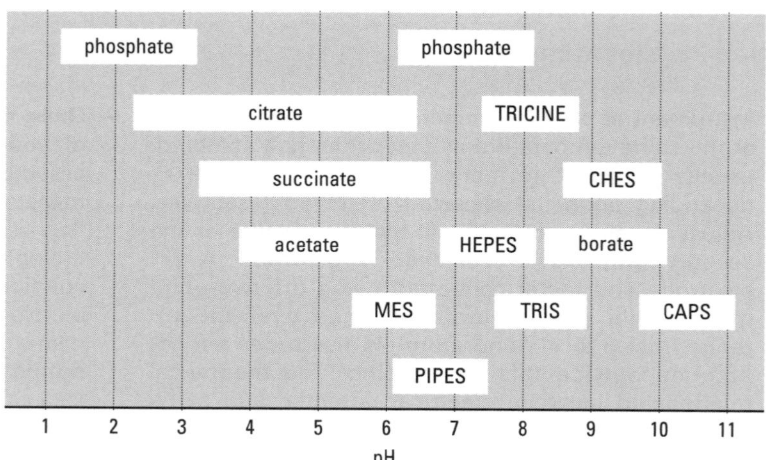

Fig. 28.4 Useful pH ranges of some commonly used buffers.

been introduced to overcome some of the disadvantages of traditional buffers. These newer compounds are often referred to as 'Good buffers', to acknowledge the work of Dr N.E. Good: HEPES is one of the most useful zwitterionic buffers, with a pK_a of 7.5 at 25°C.

These zwitterionic substances are usually added to water as the free acid: the solution must then be adjusted to the correct pH with a strong alkali, usually NaOH or KOH. Alternatively, they may be used as their sodium or potassium salts, adjusted to the correct pH with a strong acid, e.g. HCl. Consequently, you may need to consider what effects such changes in ion concentration may have in a solution where zwitterions are used as buffers. In addition, zwitterionic buffers can interfere with protein determinations (e.g. Lowry method, p. 396).

Figure 28.4 shows a number of traditional and zwitterionic buffers and their effective pH ranges. When selecting one of these buffers, aim for a pK_a that is in the direction of the expected pH change (Table 28.3). For example, HEPES buffer would be a better choice of buffer than PIPES for use at pH 7.2 for experimental systems where a pH increase is anticipated, while PIPES would be a better choice where acidification is expected.

Preparation of buffer solutions

Having selected an appropriate buffer, you will need to make up your solution to give the desired pH. You will need to consider two factors:

- The ratio of acid and conjugate base required to give the correct pH.
- The amount of buffering required; buffer capacity depends upon the absolute quantities of acid and base, as well as their relative proportions.

In most instances, buffer solutions are prepared to contain between 10 mmol L^{-1} and 200 mmol L^{-1} of the conjugate pair. Although it is possible to calculate the quantities required from first principles using the Henderson–Hasselbalch equation, there are sources that tabulate the amount of substance required to give a particular volume of solution with a specific pH value for a range of buffers (e.g. Anon., 2010a, 2010b). For traditional buffers, it is customary to mix stock solutions of acidic and basic components in the correct proportions to give the required pH (Table 28.4). For zwitterionic acids, the usual procedure is to add the compound to water, then bring the solution to the required pH by adding a specific amount of strong alkali or acid (obtained from tables). Alternatively, the required pH can be obtained by dropwise addition of alkali or acid, using a meter to check the pH, until the correct value is reached. When preparing solutions of zwitterionic buffers, the acid may be relatively insoluble. Do not wait for it to dissolve fully before adding alkali to change the pH – the addition of alkali will help bring the acid into solution (but make sure it has all dissolved before the desired pH is reached).

Remember that buffer solutions will only work effectively if they have sufficient buffering capacity to resist the change in pH expected during the course of the experiment. Thus a weak solution of HEPES (e.g. 10 mmol L^{-1}, adjusted to pH 7.0 with NaOH) will not be able to buffer the growth medium of a dense suspension of cells for more than a few minutes.

Finally, when preparing a buffer solution based on tabulated information, always confirm the pH with a pH meter before use.

Text references

Anon. (2010a) *Sigma Aldrich Buffer Reference Centre*.
Available: http://www.sigmaaldrich.com/lifescience/core-bioreagents/biological_buffers/learning_centre/buffer_referene_centre.html
Last accessed 21/12/10.

Anon. (2010b) *pH Theory and Practice: A Radiometer Analytical Guide*.
Available: http://www.radiometer-analytical.com/all_resource_centre.asp?code=112
Last accessed 21/12/10.

Lide, D.R. (ed.) (2008) *CRC Handbook of Chemistry and Physics*, 89th edn. Taylor and Francis, London.

29 Introduction to microscopy

Many features of interest in biological systems are too small to be seen by the naked eye and can only be observed with a microscope. All microscopes consist of a coordinated system of lenses arranged so that a magnified image of a specimen is seen by the viewer (Figs. 29.1 and 29.2). The main differences are the wavelengths of electromagnetic radiation used to produce the image, the nature and arrangement of the lens systems and the methods used to view the image.

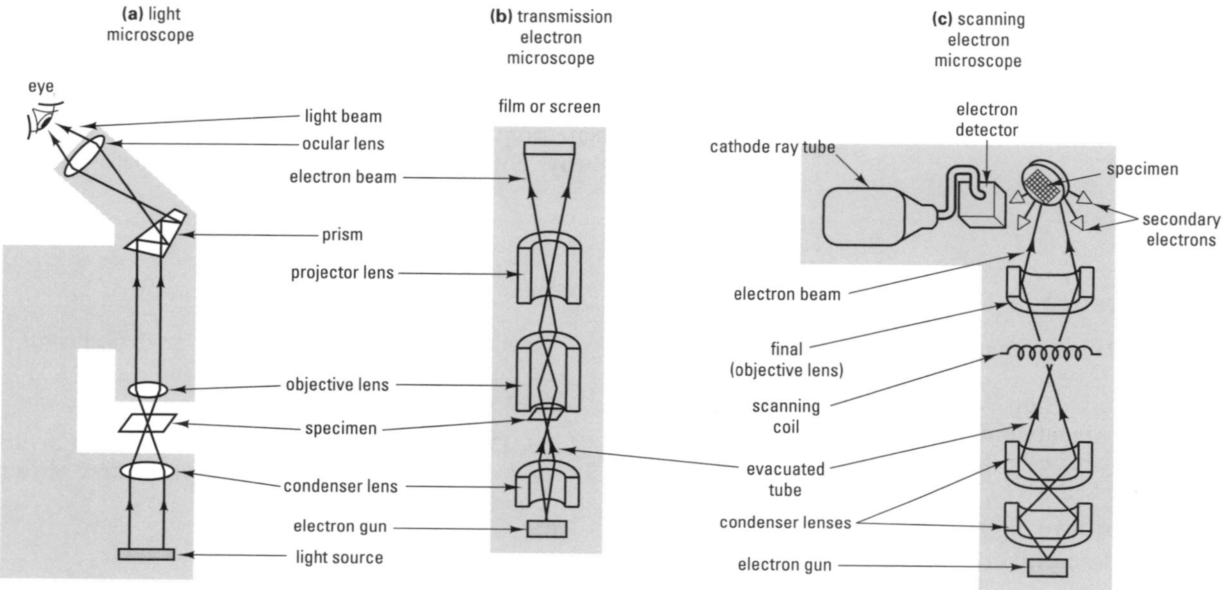

Fig. 29.1 Simplified diagrams of light and electron microscopes. Note that the electron microscopes are drawn upside-down to aid comparison with the light microscope.

Table 29.1 Comparison of microscope types. Resolution is that obtained by a skilled user. LM, light microscope; SEM, scanning electron micro-scope; TEM, transmission electron microscope

	Type of microscope		
Property	**LM**	**TEM**	**SEM**
Resolution	200 nm	1 nm	10 nm
Depth of focus	Low	Medium	High
Field of view	Good	Limited	Good
Specimen preparation (ease)	Easy	Skilled	Easy
Specimen preparation (speed)	Rapid	Slow	Quite rapid
Relative cost of instrument	Low	High	High

Microscopes allow objects to be viewed with increased resolution and contrast. Resolution is the ability to distinguish between two points on the specimen – the better the resolution, the 'sharper' the image. Resolution is affected by lens design and inversely related to the wavelength of radiation used. Contrast is the difference in intensity perceived between different parts of an image. This can be enhanced (a) by the use of stains, and (b) by adjusting microscope settings, usually at the expense of resolution.

The three main forms of microscopy are light microscopy, transmission electron microscopy (TEM) and scanning electron microscopy (SEM). Their main properties are compared in Table 29.1 and their suitability for observing cells and organelles is shown in Table 29.2.

Light microscopy

Two forms of the standard light microscope, the binocular (compound) microscope and the stereoscopic (dissecting) microscope, are described in detail in Chapter 30. These are the instruments most likely to be used in routine practical work. Figure 29.2(a) shows a typical image from a light microscope. In more advanced project work, you may use one or more of the following more sophisticated variants of light microscopy to improve image quality:

- Dark-field illumination involves a special condenser that causes reflected and diffracted light from the specimen to be seen against a dark background. The method is particularly useful for near-transparent specimens and for delicate structures like flagella. Care must be taken with the thickness of slides used – air bubbles and dust must be avoided and immersion oil must be used between the dark-field condenser and the underside of the slide.
- Ultraviolet microscopy uses short-wavelength UV light to increase resolution. Fluorescence microscopy uses radiation at UV wavelengths to make certain fluorescent substances (e.g. chlorophyll or fluorescent dyes that bind to specific cell components) emit light of visible wavelengths. Special light sources, lenses and mountants are required for UV and fluorescence microscopy, and filters must be used to prevent damage to users' eyes.
- Phase-contrast microscopy is useful for increasing contrast when viewing transparent specimens. It is superior to dark-field microscopy because a better image of the interior of specimens is obtained. Phase contrast operates by causing constructive and destructive interference effects in the image, visible as increased contrast. Adjustments must be made, using a phase telescope in place of the eyepiece, for each objective lens and a matching phase condenser, and the microscope must be set up carefully to give optimal results.
- Nomarski or Differential Interference Contrast (DIC) microscopy gives an image with a three-dimensional quality. However, the relief seen is optical rather than morphological, and care should be taken in interpreting the result. One of the advantages of the technique is the extremely limited depth of focus that results: this allows 'optical sectioning' of a specimen.
- Polarised-light microscopy can be used to reveal the presence and orientation of optically active components within specimens (e.g. cellulose fibres and starch grains; the latter displaying a characteristic 'Maltese Cross' appearance), showing them brightly against a dark background.
- Confocal microscopy allows three-dimensional views of cells or thick sections. A finely focused laser is used to create electronic images of layered horizontal 'slices', usually after fluorescent staining. Images can be viewed individually or reconstructed to provide a three-dimensional computer-generated image of the whole specimen.

(a)

(b)

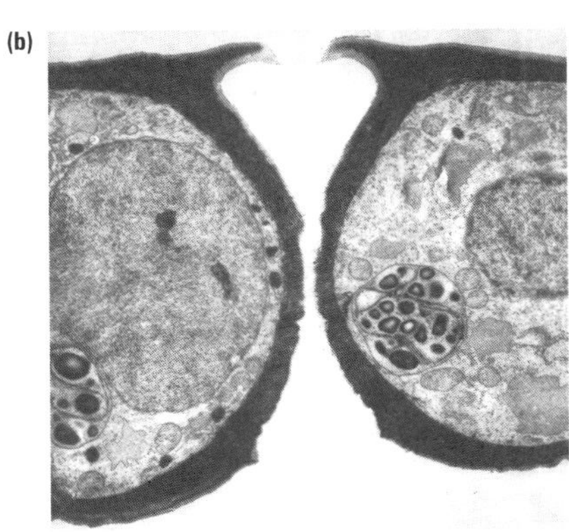

(c)

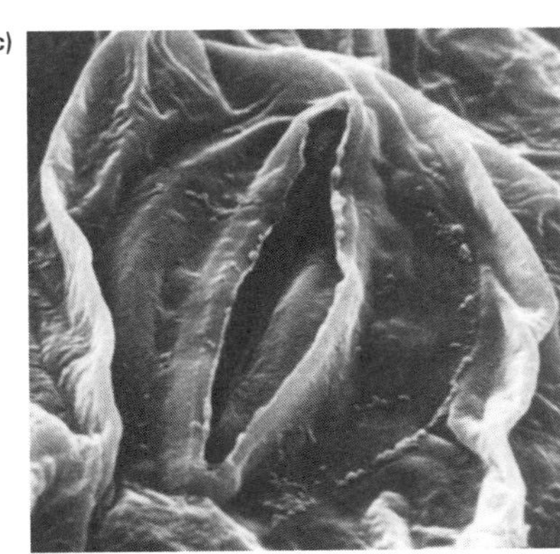

Fig. 29.2 Examples of images of a similar specimen (the stomatal complex of the plant *Commelina communis* L.) obtained using different microscopic techniques: (a) light microscopy (surface view); (b) transmission electron microscopy (transverse section through guard cell pair at mid pore); and (c) scanning electron microscopy (surface view). As an indication of scale, the width of a single guard cell is about 10 μm.

Electron microscopes

Electron microscopes offer an image resolution up to 200 times better than light microscopes (Table 29.1) because they utilise radiation of shorter wavelength in the form of an electron beam. The electrons are produced by a tungsten filament operating in a vacuum and are focused by electromagnets. TEM and SEM differ in the way in which the electron beam interacts with the specimen: in TEM, the beam passes through the specimen (Fig. 29.1(b)), while in SEM the beam is scanned across the

Table 29.2 Dimensions of some typical cells and organelles with an indication of suitable forms of microscopy for observing them. LM = light microscope; SEM = scanning electron microscope; TEM = transmission electron microscope. Column 2 data after Rubbi (1994).

Cell or organelle	Approximate diameter or width (μm)	Suitable form of microscopy
Prokaryote cell	0.15–5	LM, SEM, TEM
Eukaryote cell	10–100	LM, SEM, TEM
Fungal hypha	5–20	LM, SEM, TEM
Nucleus	5–25	LM, SEM, TEM
Mitochondrion	1–10	SEM, TEM
Chloroplast	2–8	LM, SEM, TEM
Golgi apparatus	1	SEM, TEM
Lysosome/ peroxisome	0.2–0.5	SEM, TEM
Plant cell wall	0.1–10	LM, SEM, TEM

specimen and is reflected from the surface (Fig. 29.1(c)). In both cases, the beam must fall on a fluorescent screen before the image can be seen. Permanent images ('electron micrographs') are produced after focussing the beam on photographic film (Figs. 29.2(b) and (c)).

You are unlikely to use either type of electron microscope as part of undergraduate practical work because of the time required for specimen preparation and the need for detailed training before these complex machines can be operated correctly. However, electron microscopy is extremely important in understanding cellular and subcellular structures in biological materials.

Preparative procedures

Without careful preparation of the material being studied, the structures viewed with any microscope can be rendered meaningless. Figure 29.3 summarises the processes involved for biological material, using the main types of microscopy discussed above. For details on the methods used to prepare biological specimens for light microscopy see Jones *et al.* (2007).

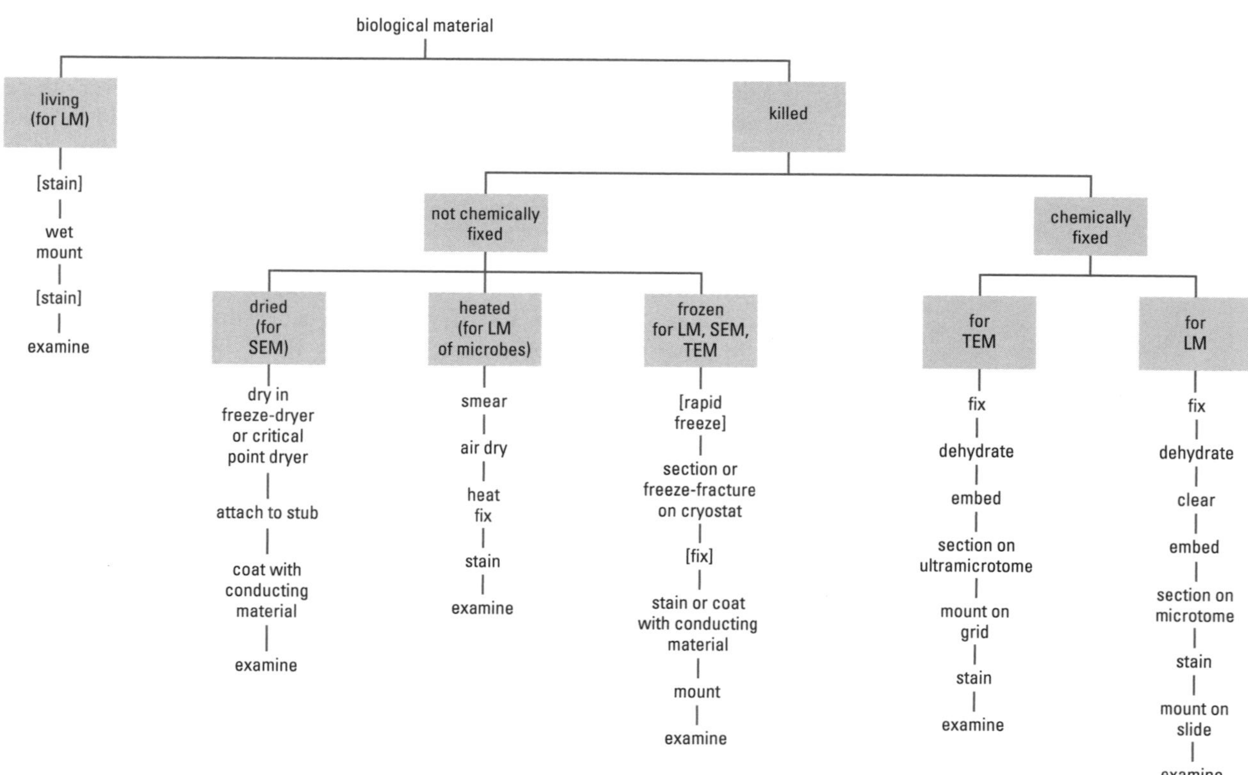

Fig. 29.3 Flowchart of procedures necessary to prepare biological material for different forms of microscopy. Steps enclosed in brackets are optional. LM, light microscope; SEM, scanning electron microscope; TEM, transmission electron microscope.

Text references

Jones, A.M., Reed, R.H. and Weyes, J.D.B. (2007) *Practical Skills in Biology*, 4th edn. Pearson Education, Harlow.

Rubbi, C.P. (1994) *Light Microscopy Essential Data*. Wiley, Chichester.

Sources for further study

Bradbury, S. (1989) *An Introduction to the Optical Microscope*. Oxford University Press, Oxford.

Davidson, M.W. and Abramowitz, M. *Molecular Expressions. Exploring the World of Optics and Microscopy*. Available: http://micro.magnet.fsu.edu/ Last accessed: 21/12/10.
[Covers many areas of basic knowledge underlying microscopy. Includes a microscopy primer.]

Jeffries, C. *Microscopy WWW Sites – by Organisation*. Available: http://www.ou.edu/research/electron/mirror/web-org.html
Last accessed: 21/12/10.
[Comprehensive set of links to microscopy websites.]

Murphy, D.B. (2001) *Fundamentals of Light Microscopy and Electronic Imaging*. Wiley-Liss, New York.

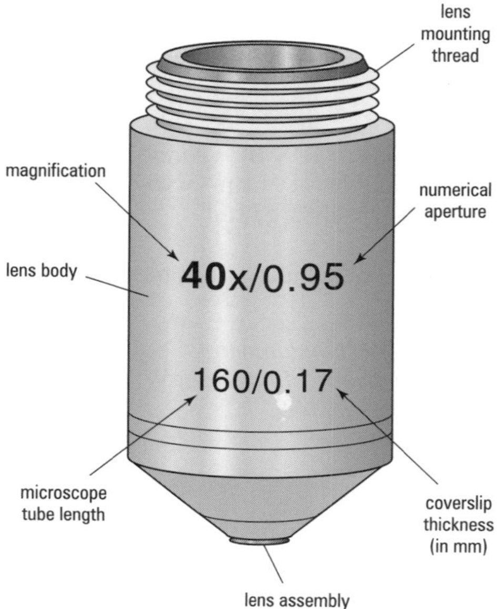

Fig. 30.1 Objective lens parameters. Most lenses are inscribed to show the details labelled above. The numerical aperture is a measure of the light-gathering power of the lens.

The light microscope is probably the most important instrument used in biology practicals and its correct use is one of the basic and essential skills of life science. A standard undergraduate binocular microscope consists of three main types of optical unit: eyepiece, objective (Fig. 30.1) and condenser. These are attached to a stand that holds the specimen on a stage (Fig. 30.2). A monocular microscope is constructed similarly but has one eyepiece lens rather than two.

Setting up a binocular light microscope

Before using any microscope, familiarise yourself with its component parts.

> **KEY POINT** Never assume that the previous person to use your microscope has left it set up correctly: apart from differences in users' eyes, the microscope needs to be properly set up for each lens combination used.

The procedures outlined below are simplified to allow you to set up microscopes like those of the Olympus CX series (Fig. 30.2). For monocular microscopes, disregard instructions for adjusting eyepiece lenses in step 5, below.

1. Place the microscope at a convenient position on the bench. Adjust your seating so that you are comfortable operating the focus and stage controls. Unwind the power cable, plug in and switch on after first ensuring that the lamp setting is at a minimum. Then adjust the lamp setting to about two-thirds of the maximum.
2. Select a low-power (e.g. ×10) objective. Make sure that the lens clicks home.
3. Set the eyepiece (ocular) lenses to your interpupillary distance; this can usually be read off a scale on the turret. You should now see a single circular field of vision. If you do not, try adjusting in either direction.
4. Put a prepared slide on the stage. Examine it first against a light source and note the position, colour and rough size of the specimen. Place the slide on the stage (coverslip up!) and, viewing from the side, position it with the stage adjustment controls so that the specimen is illuminated.
5. Focus the image of the specimen using first the coarse and then the fine focussing controls (Fig. 30.3). The image will be reversed and upside down compared with that seen by viewing the slide directly.
   (a) If both eyepiece lenses are adjustable, set your interpupillary distance on the scale on each lens. Close your left eye, look through the right eyepiece with your right eye and focus the image with the normal controls. Now close your right eye, look through the left eyepiece with your left eye and focus the image by rotating the eyepiece holder. Take a note of the setting for future use.
   (b) If only the left eyepiece is adjustable, close your left eye, look with the right eye through the static right eyepiece and focus the image with

Using binocular eyepieces – if you do not know your interpupillary distance, ask someone to measure it with a ruler. You should stare at a fixed point in the distance while the measurement is taken. Take a note of the value for future use.

Issues for spectacle and contact lens wearers – those who wear glasses can remove them for viewing, as microscope adjustments will accommodate most deficiencies in eyesight (except astigmatism). This is more comfortable and stops the spectacle lenses being scratched by the eyepiece holders. However, it may create difficulties in focussing when drawing diagrams. Those with contact lens should simply wear them as normal for viewing.

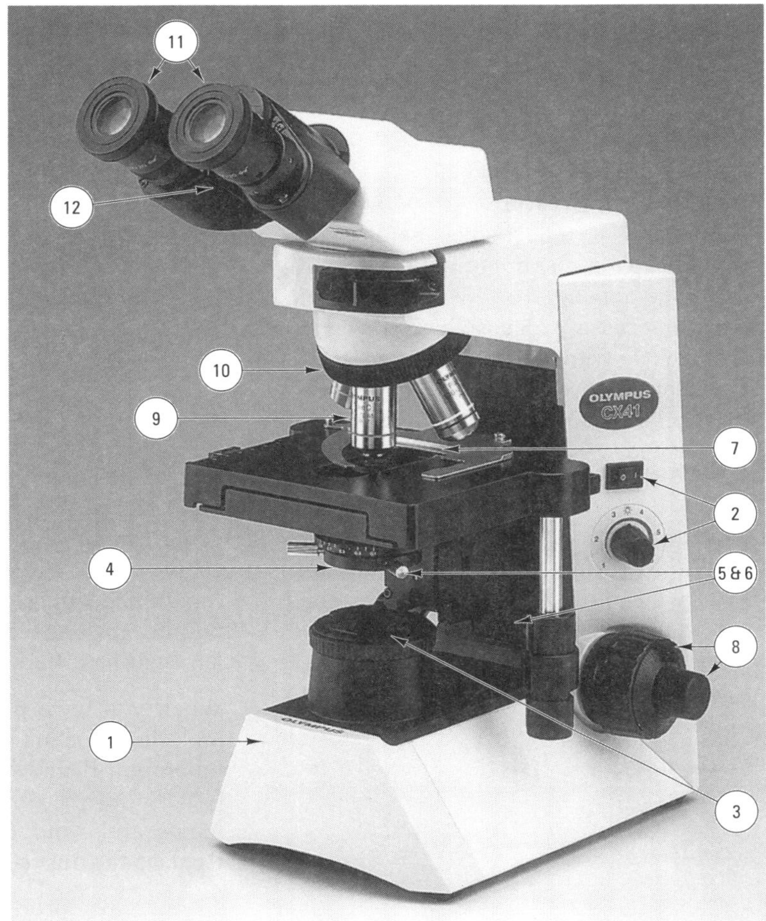

Fig. 30.2 Diagram of the Olympus binocular microscope model CX41.

- The lamp in the base of the stand (**1**) supplies light; its brightness is controlled by an on–off switch and voltage control (**2**). Never use maximum voltage or the life of the bulb will be reduced – a setting two-thirds to three-quarters of maximum should be adequate for most specimens. A field–iris diaphragm may be fitted close to the lamp to control the area of illumination (**3**).
- The condenser control focuses light from the condenser lens system (**4**) on to the specimen and projects the specimen's image on to the front lens of the objective. Correctly used, it ensures optimal resolution.
- The condenser–iris diaphragm (**5**) controls the amount of light entering and leaving the condenser; its aperture can be adjusted using the condenser–iris diaphragm lever (**6**). Use this to reduce glare and enhance image contrast by cutting down the amount of stray light reaching the objective lens.
- The specimen (normally mounted on a slide) is fitted to a mechanical stage or slide holder (**7**) using a spring mechanism. Two controls allow you to move the slide in *x* and *y* planes. Vernier scales (see p. 151) on the slide holder can be used to return to the same place on a slide. The fine and coarse focus controls (**8**) adjust the height of the stage relative to the lens systems. Take care when adjusting the focus controls to avoid hitting the lenses with the stage or slide.
- The objective lens (**9**) supplies the initial magnified image; it is the most important component of any microscope because its qualities determine resolution, depth of field and optical aberrations. The objective lenses are attached to a revolving nosepiece (**10**). Take care not to jam the longer lenses on the stage or slide as you rotate the nosepiece. You should feel a distinct click as each lens is moved into position. The magnification of each objective is written on its side; a normal complement would be ×4, ×10, ×40 and ×100 (oil immersion).
- The eyepiece lens (**11**) is used to further magnify the image from the objective and to put it in a form and position suitable for viewing. Its magnification is written on the holder (normally ×10). By twisting the holder for one or both of the eyepiece lenses you can adjust their relative heights to take account of optical differences between your eyes. The interpupillary distance scale (**12**) and adjustment knob allow compensation to be made for differences in the distance between users' pupils.

Source: From a photograph of Olympus Model CX41 binocular microscope, supplied by Olympus Microscopes, Olympus Optical Company (UK) Ltd, published courtesy of Olympus Optical Company (UK) Ltd.

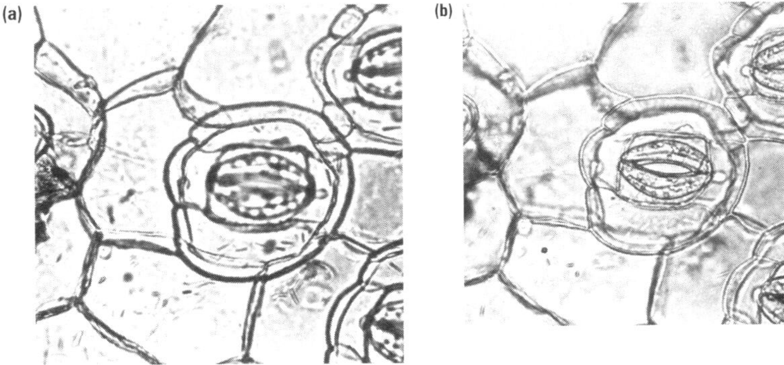

Fig. 30.3 Importance of correct focus in light microscopy. Stomatal complex of *Commelina communis* L., a specimen that is a monolayer of cells approximately 30–50 μm thick. (a) Focal plane is on 'internal' walls of the cells; (b) focal plane is on the 'external' walls and stomatal pore. The two images are different, and while it would not be possible to measure the stomatal pore in (a), it would not be possible to see the vacuolar crystals in (b). When looking at specimens, always use the fine-focus control to view different focal planes.

the normal controls. Now close your right eye, look through the left eyepiece with your left eye and focus the image by rotating the eyepiece holder. Take a note of the setting for future use.

6. Close the condenser–iris diaphragm (aperture–iris diaphragm), then open it to a position such that further opening has no effect on the brightness of the image (the 'threshold of darkening'). The edge of the diaphragm should not be in view. Turn down the lamp if it is too bright.
7. Focus the condenser. Place an opaque pointed object (the tip of a mounted needle or a sharp pencil point) on the centre of the light source. Adjust the condenser setting until both the specimen and needle tip/pencil point are in focus together. Check that the condenser–iris diaphragm is just outside the field of view.
8. For higher magnifications, swing in the relevant objective (e.g. ×40), carefully checking that there is space for it. Adjust the focus using the fine control only. If the object you wish to view is in the centre of the field with the ×10 objective, it should remain in view (magnified, of course) with the ×40. Adjust the condenser–iris diaphragm and condenser as before – the correct setting for each lens will be different.
9. When you have finished using the microscope, remove the last slide and clean the stage if necessary. Turn down the lamp setting to its minimum, then switch off. Clean the eyepiece lenses with lens tissue. Check that the objectives are clean. Unplug the microscope from the mains and wind the cable around the stand and under the stage. Replace the dust cover.

If you have problems in obtaining a satisfactory image, refer to Box 30.1; if this doesn't help, refer the problem to the class supervisor.

Adjusting a microscope with a field–iris diaphragm – adjust this before the condenser–iris diaphragm: close it until its image appears in view as a circle of light, if necessary focussing on the edge of the circle with the condenser controls and centring it with the centring screws. Now open it so the whole field is just illuminated.

High-power objectives – never remove a slide while a high-power objective lens (i.e. ×40 or ×100) is in position. Always turn back to the ×10 first. Having done this, lower the stage and remove the slide.

Box 30.1 Problems in light microscopy and possible solutions

No image; very dark image; image dark and illuminated irregularly

- Microscope not switched on (check plug and base)
- Illumination control at low setting or off
- Objective nosepiece not clicked into place over a lens
- Diaphragm closed down too much or off-centre
- Lamp failure

Image visible and focused but pale and indistinct

- Diaphragm needs to be closed down further (see Fig. 40.4)
- Condenser requires adjustment

Image blurred and cannot be focused

- Dirty objective
- Dirty slide
- Slide upside down
- Slide not completely flat on stage
- Eyepiece lenses not set up properly for user's eyes
- Fine focus at end of travel
- Oil-immersion objective in use, without oil

Dust and dirt in field of view

- Eyepiece lenses dirty
- Objective lens dirty
- Slide dirty
- Dirt on lamp glass or upper condenser lens

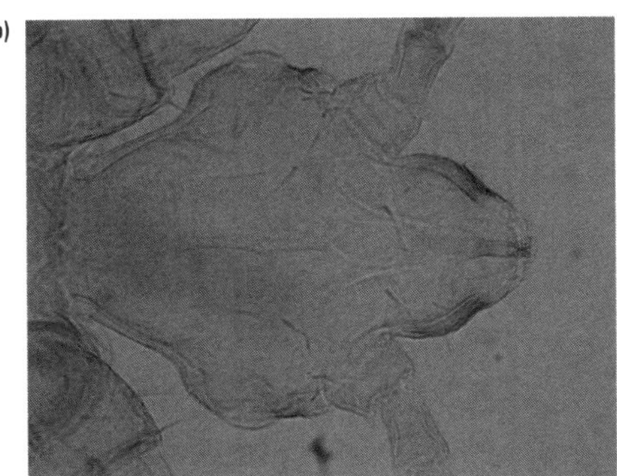

Fig. 30.4 Effect of closing the condenser–iris diaphragm on contrast. Head of human head louse, *Pediculus humanus capitus* DeGeer; (a) with condenser–iris diaphragm open; (b) with condenser–iris diaphragm closed (all other settings the same). Note difference in detail that can be seen in (b), but also that image (b) is darker – when using the condenser–iris diaphragm in this way you may need to compensate by increasing the light setting.

SAFETY NOTE Take care when moving microscopes, not only because of the cost of replacement, but also because they weigh several kilograms and could cause injury if dropped. Always carry a microscope using two hands.

Measuring specimens using a dissecting microscope – because of the low magnification, sizes can generally be estimated by comparison with a ruler placed alongside the specimen. If accurate measurements are required, eyepiece graticules can be used.

Procedure for observing transparent specimens

Some stained preparations and all colourless objects are difficult to see when the microscope is adjusted as above (Fig. 30.4). Contrast can be improved by closing down the condenser–iris diaphragm. Note that, when you do this, diffraction haloes appear around the edges of objects. These obscure the image of the true structure of the specimen and may result in loss of resolution. Nevertheless, an image with increased contrast may be easier to interpret.

Procedure for oil-immersion objectives

These provide the highest resolution of which the light microscope is capable and are used to view bacteria (Chapter 67). They must be used with immersion oil filling the space between the objective lens and the top of the slide. The oil has the same refractive index as the glass lenses, so loss of light by reflection and refraction at the glass/air interface is reduced. This increases the resolution, brightness and clarity of the image and reduces aberration. Use oil-immersion objective(s) as follows:

1. Check that the object of interest is in the field of view using, e.g., the ×10 or ×40 objective.
2. Apply a single small droplet of immersion oil to the illuminated spot on the top of the slide, having first swung the ×40 objective away. Never use too much oil: it can run off the slide and mess up the microscope.
3. Move the high-power (×100) oil-immersion objective into position carefully, checking first that there is space for it. Focus on the specimen using the fine control only. You may need a higher brightness setting.
4. Perform condenser–iris diaphragm and condenser focussing adjustments as for the other lenses.
5. When finished, clean the oil-immersion lens by gently wiping it with clean lens tissue. If the slide is a prepared one, wipe the oil off with lens tissue.

You should take great care when working with oil immersion lenses as they are the most expensive to replace. Because the working distance between the lens and coverslip is so short (less than 2 mm), it is easy to damage the lens surface by inadvertently hitting the slide or coverslip surface. You must also remember that they need oil to work properly. If working with an unfamiliar microscope, you can easily recognise oil-immersion lenses. Look for a white or black ring on the lens barrel, near the lens, or for 'oil' clearly marked on the barrel.

Care and maintenance of your microscope

Microscopes are delicate precision instruments. Handle them with care and never force any of the controls. Never touch any of the glass surfaces with anything other than a clean, dry lens tissue. Bear in mind that a replacement would be very expensive.

If moving a microscope, hold the stand above the stage with one hand and rest the base of the stand on your other hand. Always keep the

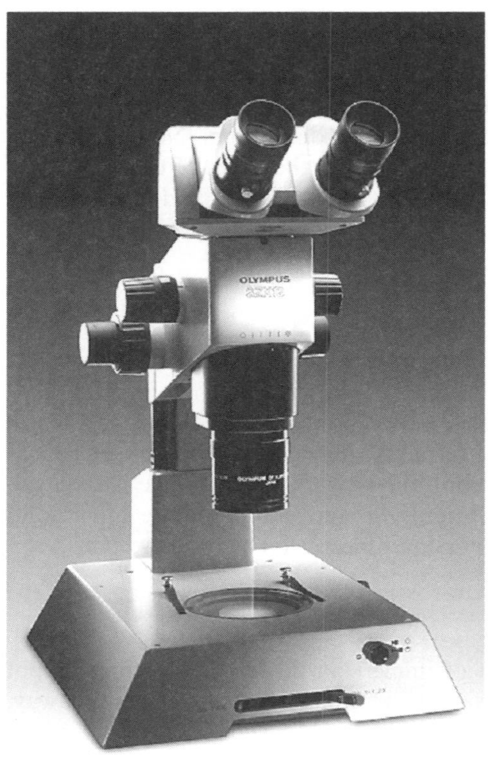

Fig. 30.5 From a photograph of Olympus Model SZX12 stereoscopic microscope, supplied by Olympus Microscopes, Olympus Optical Company (UK) Ltd, published courtesy of Olympus Optical Company (UK) Ltd.

microscope vertical (or the eyepieces may fall out). Put the microscope down gently.

Clean lenses by gently wiping with a clean, dry lens tissue. Use each piece of tissue once only. Try not to touch lenses with your fingers as oily fingerprints are difficult to clean off. Do not allow any solvent (including water) to come into contact with a lens; sea water is particularly damaging.

The stereoscopic (dissecting) microscope

The stereoscopic microscope (Fig. 30.5) is used for observations at low total magnification ($\times 4$ to $\times 50$) where a large working distance between objectives and stage is required, perhaps because the specimen is not flat or where dissecting instruments are to be used. A stereoscopic microscope essentially consists of two separate lens systems: one for each eye. Some instruments incorporate zoom objectives. The eyepiece–objective combinations are inclined at about 15° to each other and the brain resolves the compound image in three dimensions as it does for normal vision. The image is right side up and not reversed, which is ideal for dissections. Specimens are often viewed in a fresh state and need not be placed on a slide – they might be in a Petri dish or on a white tile. Illumination can be from above or below the specimen, as desired.

Most of the instructions for the binocular microscope given above apply equally well to dissecting microscopes, although the latter do not normally have adjustable condensers or diaphragms. With stereoscopic microscopes, make specially sure to adjust the eyepiece lenses to suit your eyes so that you can take full advantage of the stereoscopic effect.

Sources for further study

Bradbury, S. and Bracegirdle, B. (1998) *Introduction to Light Microscopy*. Bios Scientific Publishers, Oxford.

Davidson, M.W. and Abramowitz, M. *Nikon MicroscopyU: Introduction to Microscope Objectives.* Available: http://www.microscopyu.com/articles/optics/objectiveintro.html.
Last accessed: 21/12/10.
[Covers the optics of objective lenses.]

Olympus Microscopy Resource Centre. *Microscopy Primer* Available:
http://www.olympusmicro.com/primer/
Last accessed: 21/12/10.
[A primer on optical microscopy, including FAQs and links to further resources.]

The investigative approach

Definition

Variable – any characteristic or property that can take one of a range of values (contrast this definition with that for a **parameter**, which is a numerical constant in any particular instance).

The term data (singular = datum, or data value) refers to items of information, and you will use different types of data from a wide range of sources during your practical work. Consequently, it is important to appreciate the underlying features of data collection and measurement.

Variables

Biological variables (Fig. 31.1) can be classified as follows:

Quantitative variables

These are characteristics whose differing states can be described by means of a number. They are of two basic types:

- Continuous variables, such as mass or length; these are usually measured against a numerical scale. Theoretically, they can take any value on the measurement scale. In practice, the number of significant figures of a measurement is directly related to the precision of your measuring system; for example, dimensions measured with Vernier calipers will provide readings of greater precision than a millimetre ruler (p. 151). See Box 31.1 for practical tips on weighing foods.
- Discontinuous (discrete) variables, such as the number of teaspoons of olive oil added to a salad (see Box 31.1 for tips on measuring foods) these are always obtained by counting and therefore the data values must be whole numbers (integers). There are no intermediate values.

Working with discontinuous variables – note that while the original data values must be integers, derived data and statistical values do not have to be whole numbers. Thus, it is perfectly acceptable to express the mean number of children per family as 2.4.

Ranked variables

These provide data that can be listed in order of magnitude (i.e. ranked). A familiar example is the use of a five-point 'Hikert' scale to describe the overall taste of a food item, e.g. very bad = 1, bad = 2, neutral = 3, good = 4 and very good = 5. When such data are given numerical ranks, rather than descriptive terms, they are sometimes called 'semi-quantitative data'. Note that the difference in magnitude between ranks need not be consistent. For example, regardless of whether there was a one-year or a five-year gap between offspring in a family, their rank in order of birth would be the same (Fig. 31.1).

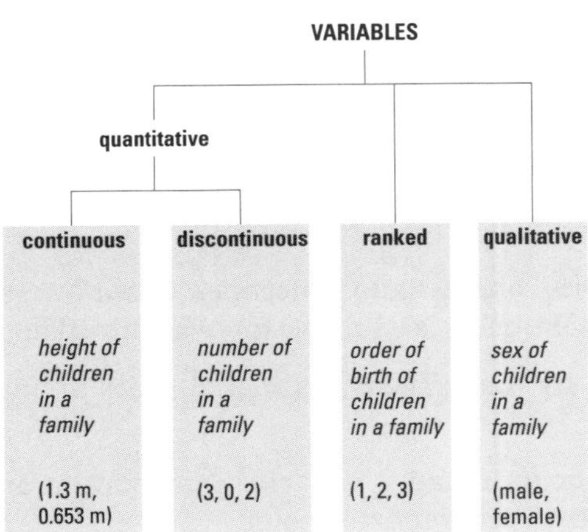

Fig. 31.1 Examples of the different types of variables as used to describe some characteristics of families.

Qualitative variables (attributes)

These are non-numerical and descriptive; they have no order of preference, and therefore are not measured on a numerical scale nor ranked in order of magnitude, but are described in terms of categories. Examples include viability (i.e. dead or alive) and shape (e.g. round, flat, elongated, etc.).

Variables may be independent or dependent. Usually, the variable under the control of the experimenter (e.g. time) is the independent variable, while the variable being measured is the dependent variable (p. 205). Sometimes it is not appropriate to describe variables in this way, and they are then referred to as interdependent variables (e.g. the length and breadth of a comcob).

Box 31.1 How to weigh and measure foods

Weighing and measuring of foods are often required in food science laboratory work, and in nutrition and dietetics to determine or manage food consumption. Weighing foods for just a week can give a person a better ability to estimate portion sizes by eye thereafter.

Weighing

1. **Consult the balance handbook** on the range of weights that can be used on a particular balance and also determine if the sensitivity of the balance is suitable for your purpose.
2. **Make sure the balance pan and surrounds are clean** before starting work.
3. **Turn the balance on** if it is not on already.
4. **Place a weigh boat, beaker, cup or other suitable clean (or sterile if required) utensil on the pan.** Allow for the weight of this measurement container in considering the maximum capacity of the balance.
5. **Tare the balance** and ensure it is on zero.
6. **Remove parts of the food not be analysed or consumed** – for example, the skin on a banana or the bone in meat.
7. **Note that foods to be cooked should be cooked before weighing** – for example, meat can lose up to 25 per cent of its weight during cooking.
8. **Cut the food into sections** of suitable size to place into the weighing container.
9. **Check that the balance is still on zero.**
10. **Place the food pieces into the weighing container** using a clean spatula, spoon or forceps.
11. **Record the weight** in your lab or record book once the balance reading has stabilised.
12. **Make sure you clean up** the balance and surrounds after completing the weighing.

Measuring

1. **Use proper metal, glass or plastic measuring cups and spoons.** Ordinary cups vary in volume and spoons used in the kitchen may differ in volume from actual measuring spoons. Liquid and dry measuring cups should only be used for their respective purposes.
2. **Ensure your measuring containers are clean and dry.**
3. **When measuring liquids, the surface of the liquid should be level with your eye.** Leave the container on the bench for stability and bend down as necessary until your eyes are level with the surface of the liquid. Or use a stand at eye level.
4. **Dry foods should be levelled off at the surface of the measuring container** using a clean knife blade or spatula edge.
5. **Clean up, dry and store** your measuring containers and equipment after use.

The majority of data values are recorded as direct measurements, readings or counts, but there is an important group, called derived (or computed), that results from calculations based on two or more data values, e.g. ratios, percentages, indices and rates.

Examples A **nominal scale** for temperature is not feasible, since the relevant descriptive terms can be ranked in order of magnitude.

An **ordinal scale** for temperature measurement might use descriptive terms, ranked in ascending order, e.g. cold = 1, cool = 2, warm = 3, hot = 4.

The **Celsius scale** is an interval scale for temperature measurement, since the arbitrary zero corresponds to the freezing point of water (0°C).

The **Kelvin scale** is a ratio scale for temperature measurement since 0 K represents a temperature of absolute zero (for information, the freezing point of water is 273.15 K on this scale).

Measurement scales

Variables may be measured on different types of scale:

- Nominal scale: this classifies objects into categories based on a descriptive characteristic. It is the only scale suitable for qualitative data.
- Ordinal scale: this classifies by rank. There is a logical order in any number scale used.
- Interval scale: this is used for quantitative variables. Numbers on an equal-unit scale are related to an arbitrary zero point.
- Ratio scale: this is similar to the interval scale, except that the zero point now represents an absence of that character (i.e. it is an absolute zero). In contrast with the interval scale, the ratio of two values is meaningful (e.g. a temperature of 200 K is twice that of 100 K).

Table 31.1 Some important features of scales of measurement

	Measurement scale			
	Nominal	**Ordinal**	**Interval**	**Ratio**
Type of variable	Qualitative (Ranked)* (Quantitative)*	Ranked (Quantitative)*	Quantitative	Quantitative
Examples	Species Sex Colour	Tenderness scale Body condition Optical assessment of colour	Fahrenheit temperature scale Date (BC/AD)	Kelvin temperature scale Weight (mass) Length Response time Most physical measurements
Mathematical properties	Identity	Identity Magnitude	Identity Magnitude Equal intervals	Identity Magnitude Equal intervals True zero point
Mathematical operations possible on data	None	Rank	Rank Addition Subtraction	Rank Addition Subtraction Multiplication Division
Typical statistics used	Only those based on frequency of counts made: contingency tables, frequency distributions, etc. Chi-square test	Non-parametric methods, sign tests. Mann–Whitney *U*-test	Almost all types of test, *t*-test, analysis of variance (ANOVA), etc. (check distribution before using, Chapter 39)	Almost all types of test, *t*-test, ANOVA, etc. (check distribution before using, Chapter 39)

*In some instances (see text for examples).

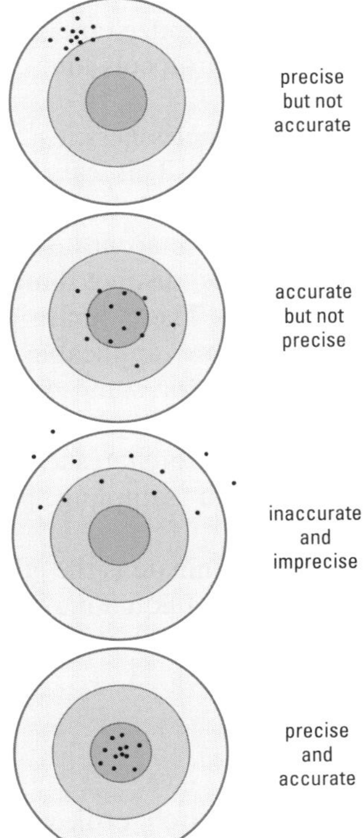

Fig. 31.2 'Target' diagrams illustrating precision and accuracy.

The measurement scale is important in determining the mathematical and statistical methods used to analyse your data. Table 31.1 presents a summary of the important properties of these scales. Note that you may be able to measure a characteristic in more than one way, or you may be able to convert data collected in one form to a different form. For instance, you might measure colour in terms of the absorption of light of a particular wavelength, e.g. using a Spectrophotometer (ratio scale), or simply as 'blue' or 'red' (nominal scale); you could find out the dates of birth of individuals (interval scale) but then use this information to rank them in order of birth (ordinal scale). Where there are no other constraints, you should use a ratio scale to measure a quantitative variable, since this will allow you to use the broadest range of mathematical and statistical procedures (Table 31.1).

Accuracy and precision

Accuracy is the closeness of a measured or derived data value to its true value, while precision is the closeness of repeated measurements to each other (Fig. 31.2). A balance with a fault in it (i.e. a bias, see below) could give precise (i.e. very repeatable) but inaccurate (i.e. untrue) results. Unless there is bias in a measuring system, precision will lead to accuracy and it is precision that is generally the most important practical consideration, if there is no reason to suspect bias. You can investigate the precision of any measuring system by repeated measurements of individual samples.

Absolute accuracy and precision are impossible to achieve, due to both the limitations of measuring systems for continuous quantitative data and the fact that you are usually working with incomplete data sets (samples, from populations). It is particularly important to avoid spurious accuracy in the presentation of results; include only those digits that the accuracy of the measuring system implies (p. 241). This type of error is common when changing units (e.g. inches to metres) and in derived data, especially when calculators give results to a large number of decimal places.

Bias (systematic error) and consistency

Bias is a systematic or non-random distortion and is one of the most troublesome difficulties in using numerical data. Biases may be associated with incorrectly calibrated instruments, e.g. a faulty pipettor, or with experimental manipulations, e.g. shrinkage during the preservation of a food sample. Bias in measurement can also be subjective, or personal, e.g. an experimenter's preconceived ideas about an 'expected' result.

Bias can be minimised by using a carefully standardised procedure, with fully calibrated instruments. You can investigate bias in 'trial runs' by measuring a single variable in several different ways, to see whether the same result is obtained.

If a personal bias is possible, 'blind' measurements should be made where the identity of individual samples is not known to the operator, e.g. using a coding system. This can be important in sensory analysis (Chapter 64).

Minimising errors – determine early in your study what the dominant errors are likely to be and concentrate your time and effort on reducing these.

Measurement error

All measurements are subject to error, but the dangers of misinterpretation are reduced by recognising and understanding the likely sources of error and by adopting appropriate protocols and calculation procedures.

A common source of measurement error is carelessness, e.g. reading a scale in the wrong direction or parallax errors. This can be reduced greatly by careful recording and may be detected by repeating the measurement. Other errors arise from faulty or inaccurate equipment, but even a perfectly functioning machine has distinct limits to the accuracy and precision of its measurements. These limits are often quoted in manufacturers' specifications and are applicable when an instrument is new; however, you should allow for some deterioration with age. Further errors are introduced when the subject being studied is open to influences outside your control. Resolving such problems requires appropriate experimental design and sampling procedures (Chapter 33).

Working with derived data – special effort should be made to reduce measurement errors because their effects can be magnified when differences, ratios, indices or rates are calculated.

One major influence virtually impossible to eliminate is the effect of the investigation itself: even putting a thermometer in a liquid may change the temperature of the liquid. The very act of measurement may give rise to a confounding variable (p. 205) as discussed in Chapter 33.

Making notes of practical work

When carrying out advanced lab work or research projects, you will need to master the important skill of managing data and observations and learn how to keep a record of your studies in a lab book.

This is important for the following reasons:

- An accurate and neat record helps when using information later, perhaps for exam purposes or when writing a report.
- It allows you to practise important skills such as scientific writing, drawing diagrams, preparing graphs and tables, and interpreting results.
- Analysing and writing up your data as you go along prevents a backlog at the end of your study time.
- You can show your work to a future employer to prove you have developed the skills necessary for writing up properly; in industry, this is vital so that others in your team can interpret and develop your work.

Understanding what's expected – especially when taking notes for a lab-based practical, pay special attention to the aims and learning objectives (pp. 22–24) of the session, as these will indicate the sorts of notes you should be taking, including content and diagrams, and the ways in which you should present these for assessment.

KEY POINT A good set of lab notes should:
- outline the purpose of your experiment or observation;
- set down all the information required to describe your materials and methods;
- record all relevant information about your results or observations and provide a visual representation of the data;
- note your immediate conclusions and suggestions for further experiments.

Collecting and recording primary data

Recording primary data – never be tempted to jot down data on scraps of paper: you are likely to lose them, or to forget what individual values mean.

Individual observations (e.g. laboratory temperature) can be recorded in the text of your notes, but tables are the most convenient way to collect large amounts of information. When preparing a table for data collection, you should:

1. Use a concise title or a numbered code for cross-referencing.
2. Decide on the number of variables to be measured and their relationship with each other, and lay out the table appropriately:
   (a) The first column of your table should show values of the independent (controlled) variable, with subsequent columns for the individual (measured) values for each replicate or sample.
   (b) If several variables are measured for the same organism or sample, each should be given a row.
   (c) In time-course studies, put the replicates as columns grouped according to treatment, with the rows relating to different times.
3. Make sure the arrangement reflects the order in which the values will be collected. Your table should be designed to make the recording process as straightforward as possible, to minimise the possibility of mistakes. For final presentation, a different arrangement may be best (Chapter 36).
4. Consider whether additional columns are required for subsequent calculations. Create a separate column for each mathematical manipulation, so the step-by-step calculations are clearly visible. Use a computer spreadsheet (pp. 69–74) if you are manipulating lots of data.
5. Use a pencil to record data so that mistakes can be easily corrected.

Designing a table for data collection – use a spreadsheet or the table-creating facility in a word processor to create your table. This will allow you to reorganise it easily if required. Make sure there is sufficient space in each column for the values – if in doubt, err on the generous side.

Identifying your notes – always put a date and time on each of your primary record sheets. You may also wish to add your name and details of the type of observation or experiment.

6. Take sufficient time to record quantitative data unambiguously – use large clear numbers, making sure that individual numerals cannot be confused.
7. Record numerical data to an appropriate number of significant figures, reflecting the accuracy and precision of your measurement. Do not round off data values, as this might affect the subsequent analysis of your data.
8. Record discrete or grouped data as a tally chart, each row showing the possible values or classes of the variable. Provided that tally marks are of consistent size and spacing, this method has the advantage of providing an 'instant' frequency distribution chart.
9. Prepare duplicated recording tables if your experiments or observations will be repeated.
10. Explain any unusual data values or observations in a footnote. Do not rely on your memory.

Recording details of project work

The recommended system is one where you make a dual record.

Primary record

Choosing a notebook for primary recording – a spiral-bound notebook is good for making a primary record – it lies conveniently open on the bench and provides a simple method of dealing with major mistakes!

The primary record is made at the bench or in the field. In this, you must concentrate on the detail of materials, methods and results. Include information that would not be used elsewhere, but that might prove useful in error tracing: for example, if you note how a solution was made up (exact volumes and weights used rather than concentration alone), this could reveal whether a miscalculation had been the cause of a rogue result. Note the origin, type and state of the chemicals and organism(s) used. Make rough diagrams to show the arrangement of replicates, equipment, etc. If you are forced to use loose paper to record data, make sure each sheet is dated and taped to your lab book, collected in a ring binder, or attached together with a treasury tag. The same applies to traces, printouts and graphs.

The basic order of the primary record should mirror that of a research report (see p. 241), including: the title and date; brief introduction; comprehensive materials and methods; the data and short conclusions.

Secondary record

Choosing a book for secondary recording – a hard-backed A4-size lined book is good because you will not lose pages. Graphs, printouts, etc. can be stuck in, as required.

You should make a secondary record concurrently or later in a bound book and it ought to be neater, in both organisation and presentation. This book will be used when discussing results with your supervisor, and when writing up a report or thesis, and may be part of your course assessment. Although these notes should retain the essential features of the primary record, they should be more concise and the emphasis should move towards analysis of the experiment. Outline the aims more carefully at the start and link the experiment to others in a series (e.g. 'Following the results of Expt D24, I decided to test whether...'). You should present data in an easily digested form, e.g. as tables of means or as summary graphs. Use appropriate statistical tests (Chapter 39) to support your analysis of the results. The choice of a bound book ensures that data are not easily lost.

Points to note

Formal aspects of keeping a record – the diary aspect of the record can be used to establish precedence (e.g. for patentable research where it can be important to 'minute' where and when an idea arose and whose it was); for error tracing (e.g. you might be able to find patterns in the work affecting the results); or even for justifying your activities to a supervisor.

Analysing data as soon as possible – always analyse and think about data immediately after collecting them as this may influence your subsequent activities.

- A graphical indication of what has happened can be particularly valuable.
- Carry out statistical analyses before moving on to the next experiment because apparent differences among treatments may not turn out to be statistically significant when tested.
- Write down any conclusions you make while analysing your data: sometimes those that seem obvious at the time of doing the work are forgotten when the time comes to write up a report or thesis.
- Note ideas for further studies as they occur to you – these may prove valuable later. Even if your experiment appears to be a failure, suggestions as to the likely causes might prove useful.

SAFETY NOTE Maintaining and consulting communal lab records – these activities may form a part of the safety requirements for working in a laboratory.

The dual method of recording deals with the inevitable untidiness of notes taken at the bench or in the field; these often have to be made rapidly, in awkward positions and in a generally complex environment. Writing a second, neater version forces you to consider again details that might have been overlooked in the primary record and provides a duplicate in case of loss or damage.

If you find it difficult to decide on the amount of detail required in materials and methods, the basic ground rule is to record enough information to allow a reasonably competent scientist to repeat your work exactly. You must tread a line between the extremes of pedantic, irrelevant detail and the omission of information essential for a proper interpretation of the data – better perhaps to err on the side of extra detail to begin with. An experienced worker can help you decide which subtle shifts in technique are important (e.g. batch numbers for an important chemical, or when a new stock solution is made up and used). Many important scientific advances have been made because of careful observation and record taking or because coincident data were recorded that did not seem of immediate value.

When creating a primary record, take care not to lose any of the information content of the data: for instance, if you only write down means and not individual values, this may affect your ability to carry out subsequent statistical analyses.

There are numerous ways to reduce the labour of keeping a record. Do not repeat materials and methods for a series of similar experiments; use devices such as 'method as for Expt B4'. A photocopy might suffice if the method is derived from a text or article (check with your supervisor). To save time, make up and copy a checklist in which details such as chemical batch numbers can be entered.

Using communal records

If working with a research team, you may need to use their communal databases. These avoid duplication of effort and ensure uniformity in techniques. You will be expected to use the databases carefully and contribute to them properly. They might include:

- a shared notebook of common techniques (e.g. how to make up media or solutions);
- a set of simplified step-by-step instructions for use of equipment. Manuals are often complex and poorly written and it may help to redraft them, incorporating any differences in procedure adopted by the group;
- an alphabetical list of suppliers of equipment and consumables (perhaps held on a card-index system);
- a list of chemicals required by the group and where they are stored;
- the risk-assessment sheets for dangerous procedures (p. 133);
- the record book detailing the use of radioisotopes and their disposal.

Sources for further study

Erikson, B.H. and Nosanchuk, T.A. (1992) *Understanding Data*, 2nd edn. Open University Press, Milton Keynes.
[A text aimed at social science students but with clear explanations of issues that are generic, including information on analysis of data.]

Friedrich, G.W. *Basic Principles of Measurement. Methods of Inquiry*. Available: http://Comminfo.rutgers.edu/~gusf/measurement.html
Last accessed: 21/12/10.
[Course notes covering diverse aspects of enquiry.]

32 SI units and their use

Dimensionless measurements – some quantities can be expressed as dimensionless ratios or logarithms (e.g. pH), and in these cases you do not need to use a qualifying unit.

When describing a measurement, you normally state both a number and a unit (e.g. 'the length is 1.85 metres'). The number expresses the ratio of the measured quantity to a fixed standard, while the unit identifies that standard measure or dimension. Clearly, a single unified system of units is essential for efficient communication of such data within the scientific community. The Système International d'Unités (SI) is the internationally ratified form of the metre-kilogram-second system of measurement and represents the accepted scientific convention for measurements of physical quantities.

Another important reason for adopting consistent units is to simplify complex calculations where you may be dealing with several measured quantities (see pp. 238 and 240). Although the rules of the SI are complex and the scale of the base units is sometimes inconvenient, to gain the full benefits of the system you should observe its conventions strictly.

The description of measurements in SI involves:

- seven base units and two supplementary units, each having a specified abbreviation or symbol (Table 32.1);
- derived units, obtained from combinations of base and supplementary units, which may also be given special symbols (Table 32.2);
- a set of prefixes to denote multiplication factors of 10^3, used for convenience to express multiples or fractions of units (Table 32.3).

Table 32.1 The base and supplementary SI units

Measured quantity	Name of SI unit	Symbol
Base units		
Length	metre	m
Mass	kilogram	kg
Amount of substance	mole	mol
Time	second	s
Electric current	ampere	A
Temperature	kelvin	K
Luminous intensity	candela	cd
Supplementary units		
Plane angle	radian	rad
Solid angle	steradian	sr

Table 32.2 Some important derived SI units

Measured quantity	Name of unit	Symbol	Definition in base units	Alternative in derived units
Energy	joule	J	$m^2\,kg\,s^{-2}$	N m
Force	newton	N	$m\,kg\,s^{-2}$	$J\,m^{-1}$
Pressure	pascal	Pa	$kg\,m^{-1}\,s^{-2}$	$N\,m^{-2}$
Power	watt	W	$m^2\,kg\,s^{-3}$	$J\,s^{-1}$
Electric charge	coulomb	C	A s	$J\,V^{-1}$
Electric potential difference	volt	V	$m^2\,kg\,A^{-1}\,s^{-3}$	$J\,C^{-1}$
Electric resistance	ohm	Ω	$m^2\,kg\,A^{-2}\,s^{-3}$	$V\,A^{-1}$
Electric conductance	siemens	S	$s^3\,A^2\,kg^{-1}\,m^{-2}$	$A\,V^{-1}$ or Ω^{-1}
Electric capacitance	farad	F	$s^4\,A^2\,kg^{-1}\,m^{-2}$	$C\,V^{-1}$
Luminous flux	lumen	lm	cd sr	
Illumination	lux	lx	$cd\,sr\,m^{-2}$	$lm\,m^{-2}$
Frequency	hertz	Hz	s^{-1}	
Radioactivity	becquerel	Bq	s^{-1}	
Enzyme activity	katal	kat	mol substrate s^{-1}	

Table 32.3 Prefixes used in the SI

Multiple	Prefix	Symbol	Multiple	Prefix	Symbol
10^{-3}	milli	m	10^3	kilo	k
10^{-6}	micro	μ	10^6	mega	M
10^{-9}	nano	n	10^9	giga	G
10^{-12}	pico	p	10^{12}	tera	T
10^{-15}	femto	f	10^{15}	peta	P
10^{-18}	atto	a	10^{18}	exa	E
10^{-21}	zepto	z	10^{21}	zetta	Z
10^{-24}	yocto	y	10^{24}	yotta	Y

Recommendations for describing measurements in SI units

Basic format

- Express each measurement as a number separated from its units by a space. If a prefix is required, no space is left between the prefix and the unit it refers to. Symbols for units are only written in their singular form and do not require full stops to show that they are abbreviated or that they are being multiplied together.

Example 10 μg is correct, while 10μg, 10 μg. and 10 μ g are incorrect. 2.6 mol is right, but 2.6 mols is wrong.

Example n stands for nano and N for newtons.

- Give symbols and prefixes appropriate upper- or lower-case initial letters as this may define their meaning. Upper-case symbols are named after persons but when written out in full they are not given initial capital letters.
- Show the decimal sign as a full point on the line. Some metric countries continue to use the comma for this purpose and you may come across this in the literature: commas should not therefore be used to separate groups of thousands. In numbers that contain many significant figures, you should separate multiples of 10^3 by spaces rather than commas.

Example 1 982 963.192 309 kg (perhaps better expressed as 1.982 963 192 309 Gg).

Compound expressions for derived units

- Take care to separate symbols in compound expressions by a space to avoid the potential for confusion with prefixes. Note, for example, that 200 m s (metre-seconds) is different from 200 ms (milliseconds).
- Express compound units by using negative powers rather than a solidus (/): for example, write mol m^{-3} rather than mol/m^3. The solidus is reserved for separating a descriptive label from its units (see p. 219).
- Use parentheses to enclose expressions being raised to a power if this avoids confusion: for example, a respiratory rate might be given in mol CO_2 (kg body mass)$^{-1}$ s^{-1}.
- Where there is a choice, select relevant (natural) combinations of derived and base units: e.g. you might choose units of Pa m^{-1} to describe a hydrostatic pressure gradient across a dialyser membrane rather than kg m^{-2} s^{-2}, even though these units are equivalent and the measurements are numerically the same.

Use of prefixes

- Use prefixes to denote multiples of 10^3 (Table 32.3) so that numbers are kept between 0.1 and 1000.
- Treat a combination of a prefix and a symbol as a single symbol. Thus, when a modified unit is raised to a power, this refers to the whole unit including the prefix.
- Avoid the prefixes deci (d) for 10^{-1}, centi (c) for 10^{-2}, deca (da) for 10 and hecto (h) for 100 as they are not strictly SI.
- Express very large or small numbers as a number between 1 and 10 multiplied by a power of 10 if they are outside the range of prefixes shown in Table 32.3.
- Do not use prefixes in the middle of derived units: they should be attached only to a unit in the numerator (the exception is in the unit for mass, kg).

Examples

10 μm is preferred to 0.000 01 m or 0.010 mm.

1 mm^2 = 10^{-6} m^2 (not one-thousandth of a square metre).

1 dm^3 (1 litre) is more properly expressed as 1×10^{-3} m^3.

The mass of a neutrino is 10^{-36} kg.

State as MW m^{-2} rather than W mm^{-2}.

KEY POINT For the foreseeable future, you will need to make conversions from other units to SI units and derived units, as much of the literature quotes data using imperial, centimetre gram second (c.g.s.) or other systems. You will need to recognise these units and find the conversion factors required. Examples relevant to the biological and food sciences are given in Box 32.1. Table 32.4 provides values of some important physical constants in SI units. Common conversion factors used in nutrition and dietetics are provided in Box 32.2.

Box 32.1 Conversion factors between some redundant units and the SI

Quantity	SI unit/symbol	Old unit/symbol	Multiply number in old unit by this factor for equivalent in SI unit*	Multiply number in SI unit by this factor for equivalent in old unit*
Area	square metre/m^2	acre	$4.046\,86 \times 10^{3}$	$0.247\,105 \times 10^{-3}$
		hectare/ha	10×10^{3}	0.1×10^{-3}
		square foot/ft^2	0.092 903	10.763 9
		square inch/in^2	645.16×10^{-9}	$1.550\,00 \times 10^{6}$
		square yard/yd^2	0.836 127	1.195 99
Angle	radian/rad	degree/°	$17.453\,2 \times 10^{-3}$	57.295 8
Energy	joule/J	erg	0.1×10^{-6}	10×10^{6}
		kilowatt hour/kWh	3.6×10^{6}	$0.277\,778 \times 10^{-6}$
		calorie/cal	4.1868	0.2388
Length	metre/m	Ångstrom/Å	0.1×10^{-9}	10×10^{9}
		foot/ft	0.304 8	3.280 84
		inch/in	25.4×10^{-3}	39.370 1
		mile	$1.609\,34 \times 10^{3}$	$0.621\,373 \times 10^{-3}$
		yard/yd	0.914 4	1.093 61
Mass	kilogram/kg	ounce/oz	$28.349\,5 \times 10^{-3}$	35.274 0
		pound/lb	0.453 592	2.204 62
		stone	6.350 29	0.157 473
		hundredweight/cwt	50.802 4	$19.684\,1 \times 10^{-3}$
		ton (UK)	$1.016\,05 \times 10^{3}$	$0.984\,203 \times 10^{-3}$
Pressure	pascal/Pa	atmosphere/atm	101 325	$9.869\,23 \times 10^{-6}$
		bar/b	100 000	10×10^{-6}
		millimetre of mercury/mmHg	133.322	$7.500\,64 \times 10^{-3}$
		torr/Torr	133.322	$7.500\,64 \times 10^{-3}$
Radioactivity	becquerel/Bq	curie/Ci	37×10^{9}	$27.027\,0 \times 10^{-12}$
Temperature	kelvin/K	centigrade (Celsius) degree/°C	°C + 273.15	K − 273.15
		Fahrenheit degree/°F	$(°F + 459.67) \times 5/9$	$(K \times 9/5) - 459.67$
Volume	cubic metre/m^3	cubic foot/ft^3	0.028 316 8	35.314 7
		cubic inch/in^3	$16.387\,1 \times 10^{-6}$	$61.023\,6 \times 10^{3}$
		cubic yard/yd^3	0.764 555	1.307 95
		UK pint/pt	$0.568\,261 \times 10^{-3}$	1 759.75
		US pint/liq pt	$0.473\,176 \times 10^{-3}$	2 113.38
		UK gallon/gal	$4.546\,09 \times 10^{-3}$	219.969
		US gallon/gal	$3.785\,41 \times 10^{-3}$	264.172

*In the case of temperature measurements, use formulae shown.

Table 32.4 Some physical constants in SI terms

Physical constant	Symbol	Value and units
Avogadro's constant	N_A	$6.022\,174 \times 10^{23}$ mol^{-1}
Boltzmann's constant	k	$1.380\,626 \times 10^{-23}$
Charge of electron	e	$1.602\,192 \times 10^{-19}$ C
Gas constant	R	8.314 43 J K^{-1} mol^{-1}
Faraday's constant	F	$9.648\,675 \times 10^{4}$ C mol^{-1}
Molar volume of ideal gas at STP	V_0	0.022 414 m^3 mol^{-1}
Speed of light *in vacuo*	c	$2.997\,924 \times 10^{8}$ m s^{-1}
Planck's constant	h	$6.626\,205 \times 10^{-34}$ J s

Box 32.2 Common conversion factors in nutrition and dietetics*

Quantity	Unit A	Unit B	Multiply quantity for Unit A by this factor for equivalent in Unit B	Multiply quantity for Unit B by this factor for equivalent in Unit A
Length	inches/in	centimetres/cm	2.54	0.394
	feet/ft	centimetres/cm	30.5	0.0328
Weight	ounces/oz	Grams/g	28.3	0.0353
	pounds/lb	kilograms/kg	0.454	2.21
Volume	fluid ounces/fl oz	millilitres/mL	29.6	0.0338
	pint/pt	millilitres/mL	473	0.00211
Unit volume	cups/c	millilitres/mL	236	0.00424
	cups/c	tablespoons/tbs	16	0.0625
	tablespoons/tbs	teaspoons/tsp	3	0.333
	tablespoons/tbs	millilitres/mL	15	0.0667
	teaspoons/tsp	millilitres/mL	5	0.200
Energy	kilocalories/kcal	kilojoules/kJ	4.19¶	0.239
Temperature	Fahrenheit	Centigrade	(°F − 32) × 0.56	(°C × 1.8) + 32

* To three significant figures for length, mass (weight), volume and energy.
¶ 4.2 is usually used as approximation for basic nutrition and dietetics energy conversions.

Expressing units of volume – in this book, we use L and mL where you would normally find equipment calibrated in that way, but use SI units where this simplifies calculations. In formal scientific writing, constructions such as $1 \times 10^{-6}\,m^3$ (= 1 mL) and $1\,mm^3$ (= 1 μL) may be used.

Expressing enzyme activity – the derived SI unit is the katal (kat), which is the amount of enzyme that will transform 1 mol of substrate in 1 s (see Chapter 62).

Some implications of SI in food science

Volume

The SI unit of volume is the cubic metre, m^3, which is rather large for practical purposes. The Litre (L) and the millilitre (mL) are technically obsolete, but are widely used and glassware is still calibrated using them. Note also that the US spelling is liter. You may find litre given the symbol L, rather than l to avoid confusion with the number 1 and the capital letter I.

Mass

The SI unit for mass is the kilogram (kg) rather than the gram (g): this is unusual because the base unit has a prefix applied.

Amount of substance

You should use the mole (mol, i.e. Avogadro's constant, see Table 32.4) to express very large numbers. The mole gives the number of atoms in the atomic mass, a convenient constant.

Concentration

The SI unit of concentration, $mol\,m^{-3}$, is quite convenient for biological systems. It is equivalent to the non-SI term 'millimolar' ($mM \equiv mmol\,L^{-1}$),

Converting between concentration units – being able to express concentrations in different units is important as this skill is frequently used when following instruc-tions and interpreting data (Chapter 26).

while 'molar' ($M \equiv mol\ L^{-1}$) becomes $kmol\ m^{-3}$. Note that the symbol M in the SI is reserved for mega and hence should not be used for concentrations. If the solvent is not specified, then it is assumed to be water (see Chapter 26).

Time

In general, use the second (s) when reporting physical quantities having a time element (e.g. give photosynthetic rates in mol $CO_2\ m^{-2}\ s^{-1}$). Hours (h), days (d) and years should be used if seconds are clearly absurd (e.g., samples were taken over a five-year period). Note, however, that you may have to convert these units to seconds when doing calculations.

Definition

STP – standard temperature and Pressure = 293.15 K and 0.101 325 MPa.

Temperature

The SI unit is the kelvin, K. The degree Celsius scale has units of the same magnitude, °C, but starts at 273.15 K, the melting point of ice at STP. Temperature is similar to time in that the Celsius scale is in widespread use, but note that conversions to K may be required for calculations. Note also that you must not use the degree sign (°) with K and that this symbol must be in upper case to avoid confusion with k for kilo; however, you *should* retain the degree sign with °C to avoid confusion with the coulomb, C.

Energy equivalents of food components:
1 g alcohol = 29 kJ or 7 kcal
1 g carbohydrate = 17 kJ or 4 kcal
1 g fat = 37 kJ or 9 kcal
1 g protein = 17 kJ or 4 kcal

Light

While the first six base units in Table 32.1 have standards of high precision, the SI base unit for luminous intensity, the candela (cd) and the derived units lm and lx (Table 32.2), are defined in 'human' terms. They are, in fact, based on the spectral responses of the eyes of 52 American GIs measured in 1923. Clearly, few organisms 'see' light in the same way as this sample of humans. Also, light sources differ in their spectral quality. For these reasons, it is better to use expressions based on energy or photon content (e.g. $W\ m^{-2}$ or mol photons $m^{-2}\ s^{-1}$) in studies other than those on human vision. Ideally you should specify the photon wavelength spectrum involved.

Sources for further study

Institute of Biology (1997) *Biological Nomenclature: Recommendations on Terms, Units and Symbols*, 2nd edn. Institute of Biology, London.

The NIST Reference on Constants, Units and Uncertainty: International System of Units (SI). Available: http://physics.nist.gov/cuu/Units/
Last accessed 21/12/10.

Pennycuick, C.J. (1988) *Conversion Factors: SI Units and Many Others: Over 2100 Conversion Factors for Biologists and Mechanical Engineers Arranged in 21 Quick-reference Tables.* University of Chicago Press, Chicago, Illinois.

Rowlett, R. *How Many? A Dictionary of Units of Measurement.* Available: http://www.unc.edu/~rowlett/units/
Last accessed 21/12/10.

33 Scientific method and experimental design

> **Definitions**
>
> There are many interpretations of the following terms. For the purposes of this chapter, the following will be used.
>
> **Paradigm** – theoretical framework so successful and well confirmed that most research is carried out within its context and doesn't challenge it – even significant diculties can be 'shelved' in favour of its retention.
>
> **Theory** – a collection of hypotheses that covers a range of natural phenomena – a 'larger-scale' idea than a hypothesis. Note that a theory may be 'hypothetical', in the sense that it is a tentative explanation.
>
> **Hypothesis** – an explanation tested in a specific experiment or by a set of observations. Tends to involve a 'small-scale' idea.
>
> **(Scientific) Law** – this concept can be summarised as an equation (law) that provides a succinct encapsulation of a system, often in the form of a mathematical relationship. The term is often used in the physical sciences (e.g. 'Beer's law', p. 353).

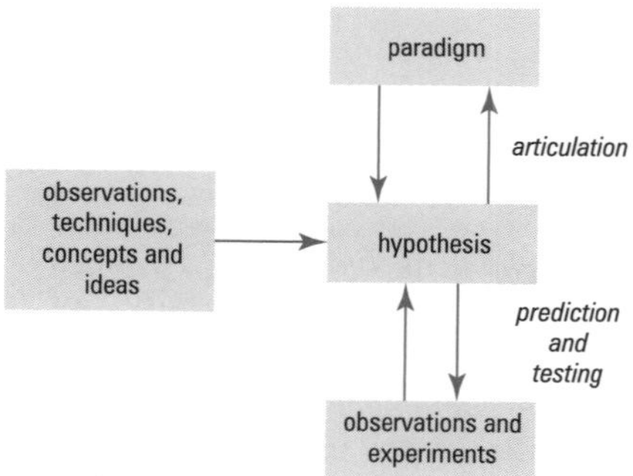

Fig. 33.1 A model of scientific method as used when testing hypotheses on a small scale. Hypotheses can arise as a result of various thought processes on the part of the scientist, and are consistent with the overlying paradigm. Each hypothesis is testable by experiment or observation, leading to its confirmation or rejection. Confirmed hypotheses act to strengthen the status of the paradigm, but rejected ones do not immediately result in the paradigm's replacement.

Food science and evidence-based nutritional science is a body of knowledge based on observation and experiment. Scientists attempt to explain life in terms of theories and hypotheses. They make predictions from these hypotheses and test them by experiment or further observations. The philosophy and sociology that underlie this process are complex topics (see e.g., Chalmers, 1999). Any brief description must involve simplifications.

Figure 33.1 models the scientific process you are most likely to be involved in – testing 'small-scale' hypotheses. These represent the sorts of explanations that can give rise to predictions that can be tested by an experiment or a series of observations. For example, you might put forward the hypothesis that drinking a glass of dark grape juice each day reduces markers for oxidative stress in people with mild to moderate hypertension, and reduces systolic and diastolic blood pressure. An experiment could be set up to test this hypothesis and the results would either confirm or falsify the hypothesis.

If confirmed, a hypothesis is retained with greater confidence. If falsified, it is either rejected outright as false, or modified and retested. Alternatively, it might be decided that the experiment was not a valid test of the hypothesis, perhaps because it was later found that some subjects were so enthused by the experiment they started drinking a range of other fruit juices during the experiment.

Nearly all scientific research deals with the testing of small-scale hypotheses. These hypotheses operate within a theoretical framework that has proven to be successful (i.e. is confirmed by many experiments and is consistently predictive). This operating model or 'paradigm' is not changed readily, and even if a result appears that seems to challenge the conventional view, would not be overturned immediately. The conflicting result would be 'shelved' until an explanation was found after further investigation. In the example used above, a relevant paradigm could be the notion that regular consumption of fruits, fruit juices and vegetables is life-enhancing.

Although changes in paradigms are rare, they are important, and the scientists who recognise them become famous. For example, a 'paradigm shift' can be said to have occurred when Darwin's ideas about Natural Selection replaced Special Creation as an explanation for the origin of species. Generally, however, results from hypothesis-testing tend to support and develop ('articulate') the paradigm, enhancing its relevance and strengthening its status. Thus, research in the area of population genetics and genetic selection studies have developed and refined Darwinism and resulted in practical outcomes.

Where do ideas for small-scale hypotheses come from? They arise from one or more thought processes on the part of a scientist:

- analogy with other systems;
- recognition of a pattern;
- recognition of departure from a pattern;
- invention of new analytical methods;
- development of a mathematical model;

- intuition;
- imagination.

Recently, it has been recognised that the process of science is not an entirely objective one. For instance, the choice of analogy that led to a new hypothesis might well be subjective, depending on past knowledge or understanding. Also, science is a social activity, where researchers put forward and defend viewpoints against those who hold an opposing view; where groups may work together towards a common goal; and where effort may depend on externally dictated financial opportunities and constraints. As with any other human activity, science is bound to involve an element of subjectivity.

Deciding whether to accept or reject a hypothesis – this is sometimes clear-cut, as in some areas of genetics, where experiments can be set up to result in a binary outcome. In many other cases, the existence of 'biological variation' means that statistical techniques need to be employed (Chapters 38 and 39).

No hypothesis can ever be rejected with certainty. Statistics allow us to quantify as vanishingly small the probability of an erroneous conclusion, but we are nevertheless left in the position of never being 100 per cent certain that we have rejected all relevant alternative hypotheses, nor 100 per cent certain that our decision to reject some alternative hypotheses was correct. However, despite these problems, experimental science has yielded and continues to yield many important findings.

> **KEY POINT** The fallibility of scientific 'facts' is essential to grasp. No explanation can ever be 100 per cent certain as it is always possible for a new alternative hypothesis to be generated. Our understanding of dietetics, nutrition and food science changes all the time as new observations and methods force old hypotheses to be retested.

The terminology of experimentation

In many experiments, the aim is to provide evidence for causality. If x causes y, we expect, repeatably, to find that a change in x results in a change in y. Hence, the ideal experiment of this kind involves measurement of y, the dependent (measured) variable, at one or more values of x, the independent variable, and subsequent demonstration of some relationship between them. Experiments therefore involve comparisons of the results of treatments – changes in the independent variable as applied to an experimental subject. The change is engineered by the experimenter under controlled conditions.

Subjects given the same treatment are known as replicates (they may be called plots). A block is a grouping of replicates or plots. The blocks are contained in a field, i.e. the whole area (or time) available for the experiment (Fig. 33.2). These terms originated from the statistical analysis of agricultural experiments, but they are now used for all areas of biology.

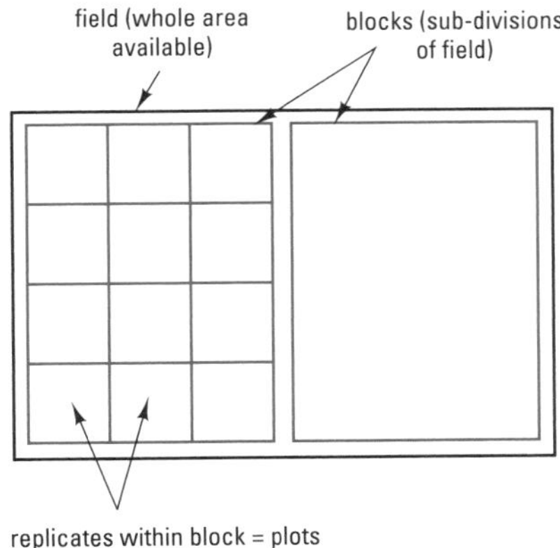

Fig. 33.2 Terminology and physical arrangement of elements in an experiment. Each block should contain the complete range of treatments (treatments may be replicated more than once in each block).

Why you need to control variables in experiments

Interpretation of experiments is seldom clear-cut because uncontrolled variables always change when treatments are given.

Confounding variables

These increase or decrease systematically as the independent variable increases or decreases. Their effects are known as systematic variation. This form of variation can be disentangled from that caused directly by treatments by incorporating appropriate controls in the experiment. A control is really just another treatment where a potentially confounding variable is adjusted so that its effects, if any, can be taken into account. The results from a control may therefore allow an alternative hypothesis to be rejected. There are often many potential controls for any experiment.

The consequence of systematic variation is that you can never be certain that the treatment, and the treatment alone, has caused an observed result. By careful design, you can, however, 'minimise the uncertainty' involved in your conclusion. Methods available include:

Example Suppose you wish to investigate the effect of a metal ion on the growth of a culture. If you add the metal as a salt to the culture and then measure the growth, you will immediately introduce at least two confounding variables, compared with a control that has no salt added. Firstly, you will introduce an anion that may also affect growth in its own right, or in combination with the metal ion; secondly, you will alter the osmotic potential of the medium (see p. 166). Both of these effects could be tested using appropriate controls.

- Ensuring, through experimental design, that the independent variable is the only major factor that changes in any treatment.
- Incorporating appropriate controls to show that potential confounding variables have little or no effect.
- Selecting experimental subjects randomly to cancel out systematic variation arising from biased selection.
- Matching or pairing individuals among treatments so that differences in response due to their initial status are eliminated.
- Arranging subjects and treatments randomly so that responses to systematic differences in conditions do not influence the results.
- Ensuring that experimental conditions are uniform so that responses to systematic differences in conditions are minimised. When attempting this, beware 'edge effects' where subjects on the periphery of the layout receive substantially different conditions from those in the centre, for example, in a 96-well microtitre plate assay.

Reducing edge effects – one way to do this is to incorporate a 'buffer zone' of untreated subjects around the experiment proper.

Nuisance variables

These are uncontrolled variables that cause differences in the value of y independently of the value of x, resulting in random variation. Experimental biology is characterised by the high number of nuisance variables that are found and their relatively great influence on results: biological data tend to have large errors. To reduce and assess the consequences of nuisance variables:

- incorporate replicates to allow random variation to be quantified;
- choose subjects that are as similar as possible;
- control random fluctuations in environmental conditions.

Constraints on experimental design

Box 33.1 outlines the important stages in designing an experiment. At an early stage, you should find out how resources may constrain the design. For example, limits may be set by availability of subjects, cost of treatment, availability of a chemical or bench space. Logistics may be a factor (e.g. time taken to record or analyse data).

Evaluating design constraints – a good way to do this is by processing an individual subject through the experimental procedures – a 'preliminary run' can help to identify potential difficulties.

Your equipment or facilities may affect design because you cannot regulate conditions as well as you might desire. For example, you may be unable to ensure that temperature and lighting are equal over an experiment laid out in a glasshouse or you may have to accept a great deal of initial variability if your subjects are human volunteers. This problem is especially acute for epidemiological studies.

Box 33.1 Checklist for designing and performing an experiment

1. **Preliminaries**
   (a) **Read background material** and decide on a subject area to investigate.
   (b) **Formulate a simple hypothesis to test.** It is preferable to have a clear answer to one question than to be uncertain about several questions.
   (c) **Decide which dependent variable you are going to measure and how:** is it relevant to the problem? Can you measure it accurately, precisely and without bias?
   (d) **Think about and plan the statistical analysis of your results.** Will this affect your design?
2. **Designing**
   (a) **Find out the limitations on your resources.**
   (b) **Choose treatments that alter the minimum of confounding variables.**
   (c) **Incorporate as many effective controls as possible.**
   (d) **Keep the number of replicates as high as is feasible.**
   (e) **Ensure that the same number of replicates is present in each treatment.**
   (f) **Use effective randomisation and blocking arrangements.**
3. **Planning**
   (a) **List all the materials you will need.** Order any chemicals and make up solutions; grow, collect or breed the experimental subjects you require; check equipment is available.
   (b) **Organise space and/or time** in which to do the experiment.
   (c) **Account for the time taken to apply treatments and record results.** Make out a timesheet if things will be hectic.
4. **Carrying out the experiment**
   (a) **Record the results and make careful notes of everything you do.** Make additional observations to those planned if interesting things happen.
   (b) **Repeat experiment** if time and resources allow.
5. **Analysing**
   (a) **Graph data as soon as possible** (during the experiment if you can). This will allow you to visualise what has happened and make adjustments to the design (e.g. timing of measurements).
   (b) **Carry out the planned statistical analysis.**
   (c) **Jot down conclusions and new hypotheses** arising from the experiment.

Use of replicates

Replicate results show how variable the response is within treatments. They allow you to compare the differences among treatments in the context of the variability within treatments – you can do this via statistical tests such as analysis of variance (Chapter 39). Larger sample sizes tend to increase the precision of estimates of parameters and increase the chances of showing a significant difference between treatments if one exists. For statistical reasons (weighting, ease of calculation, fitting data to certain tests), it is often best to keep the number of replicates even. Remember that the degree of independence of replicates is important: subsamples cannot act as replicate samples – they tell you about variability in the measurement method but not in the quantity being measured.

Deciding the number of replicates in each treatment – try to:

- maximise the number of replicates in each treatment;
- make the number of replicates even.

If the total number of replicates available for an experiment is limited by resources, you may need to compromise between the number of treatments and the number of replicates per treatment. Statistics can help here, for it is possible to work out the minimum number of replicates you would need to show a certain difference between pairs of means (say 10 per cent) at a specified level of significance (say $P = 0.05$). For this, you need to obtain a prior estimate of variability within treatments (see Sokal and Rohlf, 1994).

Randomisation of treatments

The two aspects of randomisation you must consider are:

- positioning of treatments within experimental blocks;
- allocation of treatments to the experimental subjects.

For relatively simple experiments, you can adopt a completely randomised design; here, the position and treatment assigned to any subject is defined randomly. You can draw lots, use a random number generator on a calculator, or use the random number tables that can be found in most books of statistical tables (see Box 33.2).

A completely randomised layout has the advantage of simplicity but cannot show how confounding variables alter in space or time. This information can be obtained if you use a blocked design in which the degree of randomisation is restricted. Here, the experimental space or time is divided into blocks, each of which accommodates the complete set of treatments (Fig. 33.2). When analysed appropriately, the results for the blocks can be compared to test for differences in the confounding variables and these effects can be separated out from the effects of the treatments. The size and shape (or timing) of the block you choose is important: besides being able to accommodate the number of replicates desired, the suspected confounding variable should be relatively uniform within the block.

Box 33.2 How to use random number tables to assign subjects to positions and treatments

This is one method of many that could be used. It requires two sets of *n* random numbers – where *n* is the total number of subjects used.

1. **Number the subjects in any arbitrary order** but in such a way that you know which is which (i.e. mark or tag them).
2. **Decide how treatments will be assigned**, e.g. first five subjects selected, treatment A; second five – treatment B, etc.
3. **Use the first set of random numbers in the sequence obtained to identify subjects and allocate them to treatment groups** in order of selection as decided in (2).
4. **Map the positions for subjects in the block or field. Assign numbers to these positions using the second set of random numbers**, working through the positions in some arbitrary order, e.g. top left to bottom right.
5. **Match the original numbers given to subjects with the position numbers.**

To obtain a sequence of random numbers:

1. **Decide on the range of random numbers you need.**
2. **Decide how you wish to sample the random number tables** (e.g. row by row and top to bottom) and your starting point.
3. **Moving in the selected manner, read the sequence of numbers until you come to a group that fits your needs** (e.g. in the sequence 978186, 18 represents a number between 1 and 20). Write this down and continue sampling until you get a new number. If a number is repeated, ignore it. Small numbers need to have the appropriate number of zeros preceding (e.g. 5 = 05 for a range in the tens, 21 = 021 for a range in the hundreds).
4. **When you come to the last number required, you don't need to sample any more:** simply write it down.

Example: You find the following random number sequence in a table and wish to select numbers between 1 and 10 from it.

9059146823 4862925166 1063260345
1277423810 9948040676 6430247598
8357945137 2490145183 5946242208
6588812379 2325701558 3260726568

Working left to right and top to bottom, the order of numbers found is 5, 10, 3, 9, 4, 6, 2, 1, 8, 7 as indicated by the coloured type. If the table is sampled by working row by row right to left from bottom to top, the order is 6, 10, 7, 2, 9, 3, 4, 8, 1, 5.

> **Example** In food production research, if you knew that soil type varied in a graded fashion across a field, you might arrange blocks to be long thin rectangles at right angles to the gradient to ensure conditions within the block were as even as possible.

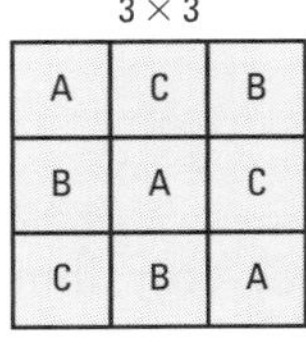

4 × 4

A	C	B	D
D	B	C	A
C	D	A	B
B	A	D	C

Fig. 33.3 Examples of Latin square arrangements for 3 and 4 treatments. Letters indicate treatments; the number of possible arrangements for each size of square increases greatly as the size increases.

A Latin square is a method of placing treatments so that they appear in a balanced fashion within a square block or field. Treatments appear once in each column and row (see Fig. 33.3), so the effects of confounding variables can be 'cancelled out' in two directions at right angles to each other. This is effective if there is a smooth gradient in some confounding variable over the field. It is less useful if the variable has a patchy distribution, where a randomised block design might be better.

Latin square designs are useful in serial experiments where different treatments are given to the same subjects in a sequence (e.g. Fig. 33.4). A disadvantage of Latin squares is the fact that the number of plots is equal to the number of replicates, so increases in the number of replicates can only be made by the use of further Latin squares.

Pairing and matching subjects

The paired comparison is a special case of blocking used to reduce systematic variation when there are two treatments. Examples of its use are:

- 'Before and after' comparison. Here, the pairing removes variability arising from the initial state of the subjects, e.g. weight gain of mice on a diet, where the weight gain may depend on the initial weight.
- Application of a treatment and control to parts of the same subject or to closely related subjects. This allows comparison without complications arising from different origin of subjects, e.g. food additive or placebo given to sibling rats, virus-containing or control solution swabbed on left or right halves of a leaf.
- Application of treatment and control under shared conditions. This allows comparison without complications arising from different environments of subjects, e.g. rats in a cage, plants in a pot.

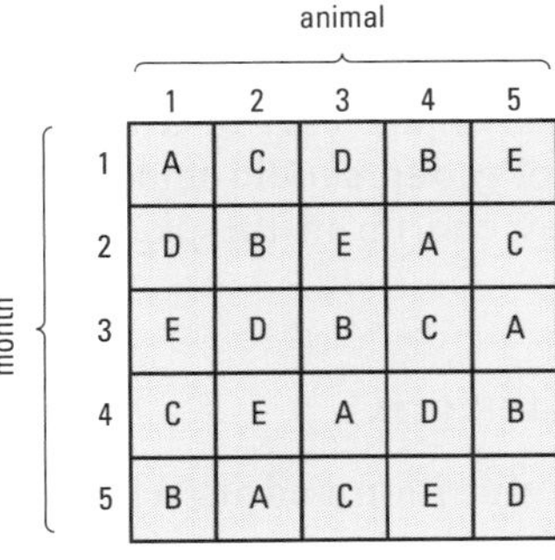

Fig. 33.4 Example of how to use a Latin square design to arrange sequential treatments. The experimenter wishes to test the effect of drugs A–E on weight gain, but only has five animals available. Each animal is fed on control diet for the first three weeks of each month, then on control diet plus drug for the last week. Weights are taken at start and finish of each treatment. Each animal receives all treatments.

Matched samples represent a restriction on randomisation where you make a balanced selection of subjects for treatments on the basis of some attribute or attributes that may influence results, e.g. age, sex, prior history. The effect of matching should be to 'cancel out' the unwanted source(s) of variation. Disadvantages include the subjective element in choice of character(s) to be balanced, inexact matching of quantitative characteristics, the time matching takes and possible wastage of unmatched subjects.

When analysed statistically, both paired comparisons and matched samples can show up differences between treatments that might otherwise be rejected on the basis of a fully randomised design, but note that the statistical analysis may be different.

Multifactorial experiments

The simplest experiments are those in which one treatment (factor) is applied at a time to the subjects. This approach is likely to give clear-cut answers, but it could be criticised for lacking realism. In particular, it cannot take account of interactions among two or more conditions that are likely to occur in real life. A multifactorial experiment (Fig. 33.5) is an attempt to do this; the interactions among treatments can be analysed by specialised statistics.

Definition

Interaction – where the effect of treatments given together is greater or less than the sum of their individual effects.

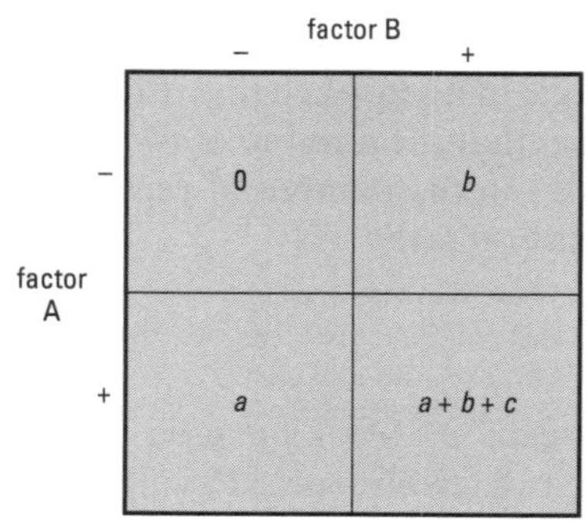

Fig. 33.5 Design of a simple multifactorial experiment. Factors A and B have effects *a* and *b* when applied alone. When both are applied together, the effect is denoted by $a + b + c$.

- If $c = 0$, there is no interaction (e.g. $2 + 2 + c = 4$).
- If *c* is positive, there is a positive interaction (synergism) between A and B (e.g. $2 + 2 + c = 5$).
- If *c* is negative, there is a negative interaction (antagonism) between A and B (e.g. $2 + 2 + c = 3$).

Reporting correctly – it is good practice to report how many times your experiments were repeated (in materials and methods); in the results section, you should add a statement saying that the illustrated experiment is representative.

Multifactorial experiments are economical on resources because of 'hidden replication'. This arises when two or more treatments are given to a subject because the result acts statistically as a replicate for each treatment. Choice of relevant treatments to combine is important in multifactorial experiments; for instance, an interaction may be present at certain concentrations of a chemical but not at others (perhaps because the response is saturated). It is also important that the measurement scale for the response is consistent, otherwise spurious interactions may occur. Beware when planning a multifactorial experiment that the numbers of replicates do not get out of hand: you may have to restrict the treatments to 'plus' or 'minus' the factor of interest (as in Fig. 33.5).

Repetition of experiments

Even if your experiment is well designed and analysed, only limited conclusions can be made. Firstly, what you can say is valid for a particular place and time, with a particular investigator, experimental subject and method of applying treatments. Secondly, if your results were significant at the 5 per cent level of probability (p. 259), there is still an approximately one-in-twenty chance that the results did arise by chance. To guard against these possibilities, it is important that experiments are repeated. Ideally, this would be done by an independent scientist with independent materials. However, it makes sense to repeat work yourself so that you can have full confidence in your conclusions. Many scientists recommend that experiments are done three times in total, but this may not be possible in undergraduate work.

Permits and approvals

Experiments involving animals or human subjects (including surveys) require the appropriate ethical considerations and permits as enacted in legislations and organisational policies (Chapter 23). A particular point to note is that people who are considering taking part in a double-blind experiment containing a placebo control group should understand they may not receive any benefit should they end up in the placebo control group.

Nutritional epidemiology study designs

An increasing focus is being placed on nutrition and its relationship to either chronic diseases or wellness. Nutritional researchers need to investigate the effects of nutrition and various diets as scientifically as possible to produce evidence-based data for public health education. Various studies are possible based on their relevance to the situation (Freudenheim, 1993; Langseth,1996; Khoury *et al.*, 2004). They each have their strengths and weaknesses. For a summary of study designs see Figure 33.6. For dietary assessment methods used in nutritional epidemiology see Chapter 41.

Placebo-controlled double-blind clinical trials

Placebo-controlled double-blind experiments can be conducted where a single nutrient or supplement is being studied, e.g. beta carotene could be

Definitions

Placebo – a control treatment with a similar appearance to the experimental treatment, but without the specific activity, e.g. a 'sugar pill'.

Double-blind – experiment in which both participants and investigators remain unaware during the experiment of which subjects are receiving the treatment and which subjects are receiving the placebo, with the aim of totally objective results and the avoidance of bias.

Nutritional epidemiology – Greek *epi* = among; *demos* = people; *logos* = word or study; the study of nutritional determinants of disease and wellness in human populations.

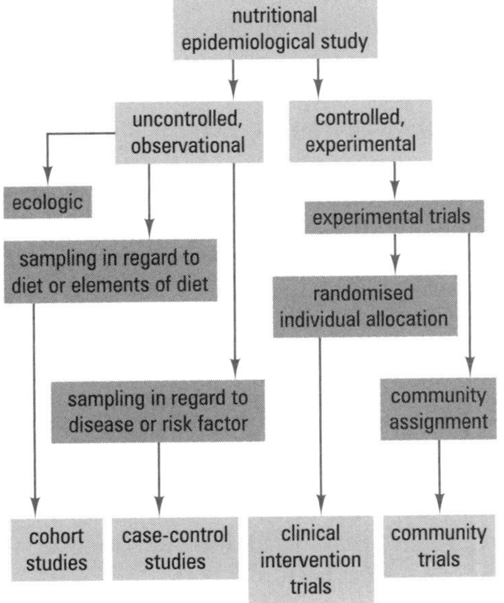

Fig. 33.6 Basic nutritional epidemiology study methods.

supplied in an opaque capsule for the treated group, while an identical capsule containing no beta carotene would be given to the control group. As the study is double-blind, the capsules would be given to the participants by a qualified third party who would keep the group identities from the participants, principal investigators and data analysts. Participants are usually assigned randomly to groups, which helps to minimise confounding factors (pp. 205-6). Because the end point (disease manifestation) may be lengthy for chronic diseases, often an intermediate point is used, such as changes to a marker for a risk factor (e.g. plasma LDL cholesterol). In contrast to cohort and case-control studies (see below), a clinical trial usually only concentrates on a single focused hypothesis, e.g. 'Vitamin B6 supplementation reduces homocysteine concentrations in the blood.'

A placebo-controlled double-blind study is less easy to design in nutritional experiments, such as studying the effects of the inclusion of 30 g of whole walnuts in the diet each day. The investigators could be blinded to the treatment groups, but it would be very difficult to disguise the whole walnuts from participants eating them or to produce a placebo that looks and tastes like a walnut. In a long-term nutritional study involving significant dietary changes, compliance may also become an issue.

Ecologic studies

Ecologic studies are mostly used as an explorative tool. The unit of observation is usually a group of people in a particular geographic region. A comparison of diseases amongst different groups in different regions is made in relationship to determinations of foods eaten and related nutrient intakes. An example was the initial determination that people of the Mediterranean region had a lower incidence of cardiovascular disease than groups in other areas, even though the diet included oils and red wine. Such studies are helpful for generating hypotheses. One limitation is that it is not clear whether the people with/without the disease are the ones consuming the dietary factor, speculated to be related to the condition. There could also be inaccuracies in the regional data on food consumption and data may also vary in its accuracy for different groups.

Cross-sectional studies

A marker for nutritional status (e.g. body mass index, p. 295) is evaluated in individuals to look for concurrent association with a disease or disease risk factor (e.g. hypertension). Other factors such as activity level, which may impact on disease risk, are also recorded and factored into the study. A limitation in cross-sectional studies is the measurement of the risk factor and nutritional marker at the same time; it is not possible to determine which came first.

Case-control studies

A group of people with a disease is compared to a group without the disease, e.g. in regard to their diets. The study design is fixed for the disease, then the current/previous diet is compared between the controls and cases. Other variables that could influence the disease, for example smoking, need to be accounted for in the survey and analysis. This type of study can be efficient and cost-effective, in that subjects that already have the disease are studied.

Limitations include that it may be difficult to determine the time course for events, e.g. if blood concentrations of vitamin C are different for cases and controls it is not known whether the difference is caused by the diet or the disease. There is also the possibility of subject selection bias and recall bias for information on diet prior to disease.

Class exercise – design an experiment to test if Chamomile tea has relaxing properties.

Cohort studies

A population is identified and their dietary consumption and additional relevant exposure factors are surveyed. The population is then studied to determine, over time, those subjects that develop disease. Study of exposure to various factors prior to the development of disease allows time-related inferences to be made, and reduces the possibility of recall bias. People with preclinical conditions and cases diagnosed close to the survey or interview time are best excluded from the analysis. A limitation is in rare diseases where a large group of subjects are needed over a long time period.

Nutritional genomics studies

Genome-wide association studies (GWAS), through large epidemiology studies, aim to relate novel genetic variants with diseases such as cancer, diabetes and circulatory disorders. Genetic variations between the DNA sequences of individuals are characterised and then analysed for their possible relationship to biological and physical trait markers, including body mass index, plasma cholesterol concentrations, plasma glucose concentrations, blood pressure and bone density. Non-genetic factors such as diet and activity level are then studied for any interaction with the genetic factors. Potentially, studies on the wide-ranging effects of the genetic differences between individuals may lead to more individually targeted diets for preventing or ameliorating chronic diseases. A limitation of this type of study is the complexity when several or many different genes may contribute to a particular trait, e.g. in obesity (Qi and Cho, 2008).

Text references

Chalmers, A.F. (1999) *What is this Thing called Science?*, 3rd edn. Open University Press, Buckingham.

Freudenheim, J.L. (1993) A Review of study designs and methods of dietary assessment in nutritional epidemiology of chronic disease. *Journal of Nutrition*, **123**, 401–405.

Khoury, M.J., Millikan, R., Little, J. and Gwinn, M. (2004) The emergence of epidemiology in the genomics age, *International Journal of Epidemiology*, **33**, 936–944. Available: http://ije.oxfordjournals.org/cgi/reprint/33/5/936 Last accessed: 21/12/10.

Langseth, L. (1996) *Nutritional Epidemiology: Possibilities and Limitations.* ISLI Europe Concise Monograph Series, International Life Sciences Institute. Brussells. Available: http://europe.ilsi.org/file/Ilsiepid.pdf Last accessed: 21/12/10.

Qi, L. and Cho, A. (2008) Gene-environment interaction and obesity. *Nutrition Reviews*, **66**, 684–694.

Sokal, R.R. and Rohlf, F.J. (1994) *Biometry*, 3rd edn. W.H. Freeman, San Francisco, California.

Sources for further study

Heath, D. (1995) *An Introduction to Experimental Design and Statistics for Biology.* UCL Press, London.

Quinn, G.P. and Keough, M.J. (2002) *Experimental Design and Data Analysis for Biologists.* Cambridge University Press, Cambridge.

34 Research project work

Research projects are an important component of the final-year syllabus for many degree programmes across the sciences, while shorter projects may also be carried out during courses in earlier years. Project work can be extremely rewarding, although it does present a number of challenges. The assessment of your project is likely to contribute significantly to your degree grade, so all aspects of this work should be approached in a thorough manner.

Obtaining ethical approval – if any aspect of your project involves work with human or animal subjects, then you must obtain the necessary ethical clearance before you begin; consult your Department's ethical committee for details.

Deciding on a topic to study

Assuming you have a choice, this important decision should be researched carefully. Make appointments to visit possible supervisors and ask them for advice on topics that you find interesting. Use library texts and research papers to obtain further background information. Perhaps the most important criterion is whether the topic will sustain your interest over the whole period of the project. Other things to look for include:

- **Opportunities to learn new skills.** Ideally, you should attempt to gain experience and skills that you might be able to 'sell' to a potential employer.
- **Ease of obtaining valid results.** An ideal project provides a means to obtain 'guaranteed' data for your report, but also the chance to extend knowledge by doing genuinely novel research.
- **Assistance.** What help will be available to you during the project? A busy lab with many research students might provide a supportive environment should your potential supervisor be too busy to meet you often; on the other hand, a smaller lab may provide the opportunity for more personal interaction with your supervisor.
- **Impact.** Your project may result in publishable data: discuss this with your prospective supervisor.

The Internet as an information source – since many university departments have home pages on the World Wide Web, searches using relevant keywords may indicate where research in your area is currently being carried out. Academics usually respond positively to emailed questions about their area of expertise.

Asking around – one of the best sources of information about supervisors, laboratories and projects is past students. Some of the post-graduates in your department may be products of your own system and they could provide an alternative source of advice.

Planning your work

As with any lengthy exercise, planning is required to make the best use of the time allocated (p. 10). This is true on a daily basis as well as over the entire period of the project. It is especially important not to underestimate the time it will take to write and produce your thesis (see below). If you wish to benefit from feedback given by your supervisor, you should aim to have drafts in his/her hands in good time. Since a large proportion of marks will be allocated to the report, you should not rush its production.

If your department requires you to write an interim report, look on this as an opportunity to clarify your thoughts and get some of the time-consuming preparative work out of the way. If not, you should set your own deadlines for producing drafts of the introduction, materials and methods section, etc.

Liaising with your supervisor(s) – this is essential if your work is to proceed efficiently. Specific meetings may be timetabled, e.g. to discuss a term's progress, review your work plan or consider a draft introduction. Most supervisors also have an 'open-door' policy, allowing you to air current problems. Prepare well for all meetings: have a list of questions ready before the meeting; provide results in an easily digestible form (but take your lab notebook along); be clear about your future plans for work.

> **KEY POINT** Project work can be very time-consuming at times. Try not to neglect other aspects of your course – make sure your lecture notes are up-to-date and collect relevant supporting information as you go along.

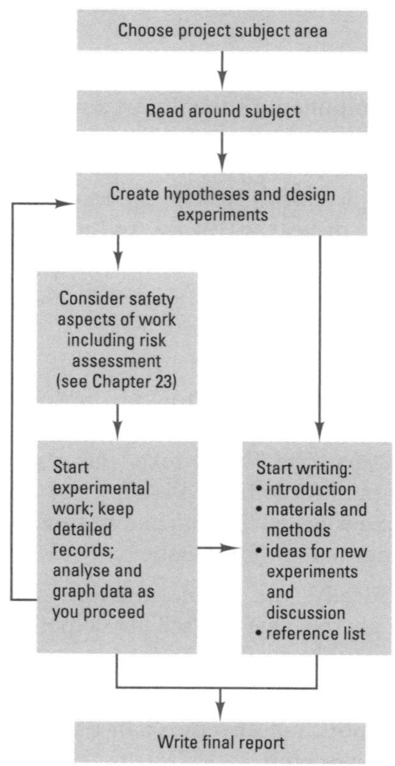

Fig. 34.1 Flowchart showing a recommended sequence of events in carrying out an undergraduate research project.

SAFETY NOTE It is essential that you follow all the safety rules applying to the laboratory or field site. Make sure you know all relevant procedures – normally there will be prominent warnings about these. If in doubt, ask.

Getting started

Figure 34.1 is a flowchart illustrating how a project might proceed; at the start, don't spend too long reading the literature and working out a lengthy programme of research. Get stuck in and do an experiment. There's no substitute for 'getting your hands dirty' for stimulating new ideas:

- even a 'failed' experiment will provide some useful information that may allow you to create a new or modified hypothesis;
- pilot experiments may point out deficiencies in experimental technique that will need to be rectified;
- the experience will help you create a realistic plan of work.

Designing experiments

Design of experiments is covered in Chapter 33. Avoid being too ambitious at the start of your work. It is generally best to work with a simple hypothesis and design your experiments or sampling around this. A small pilot experiment or test sample will highlight potential stumbling blocks including resource limitations, whether in materials or time, or both.

Working in a laboratory environment

During your time as a project student, you are effectively a guest in your supervisor's laboratory.

- Be considerate – keep your 'area' tidy and offer to do your share of lab duties such as calibrating the pH meter, replenishing stock solutions, distilled water, etc., maintaining cultures, tending plants or animals.
- Use instruments carefully – they could be worth more than you'd think. Careless use may invalidate calibration settings and ruin other people's work as well as your own.
- Do your homework on techniques you intend to use – there's less chance of making costly mistakes if you have a good background understanding of the methods you will be using.
- Always seek advice if you are unsure of what you are doing.

Keeping notes and analysing your results

Tidy record keeping is often associated with good research, and you should follow the advice and hints given in Chapter 31. Try to keep copies of all files relating to your project. As you obtain results, you should always calculate, analyse and graph data as soon as you can (see Fig. 34.1). This can reveal aspects that may not be obvious in numerical or readout form. Don't be worried by negative results – these can sometimes be as useful as positive results if they allow you to eliminate hypotheses – and don't be too dispirited if things do not work first time. Thomas Edison's maxim 'Genius is one per cent inspiration and ninety-nine per cent perspiration' certainly applies to research work.

Writing your project report

The structure of scientific reports is dealt with in Chapter 17. The following advice concerns methods of accumulating relevant information.

Brushing up on IT skills – word processors and spreadsheets are extremely useful when producing a thesis. Chapters 11 and 12 detail key features of these programs. You might benefit from attending courses on the relevant programs or studying manuals or texts so that you can use them more efficiently.

Introduction

This is a big piece of writing that can be very time-consuming. Therefore, the more work you can do on it early on, the better. You should allocate some time at the start for library work (without neglecting field or bench work), so that you can build up a database of references (Chapter 8). Although photocopying can be expensive, you will find it valuable to have copies of key reviews and references handy when writing away from the library. Discuss proposals for content and structure with your supervisor to make sure your effort is relevant. Leave space at the end for a section on aims and objectives. This is important to orient readers (including assessors), but you may prefer to finalise the content after the results have been analysed.

Materials and methods

You should note as many details as possible *when doing the experiment or making observations*. Don't rely on your memory or hope that the information will still be available when you come to write up. Even if it is, chasing these details can waste valuable time.

Using drawings and photographs – these can provide valuable records of samples or experimental set-ups and can be useful in your report. Plan ahead and do the relevant work at the time of carrying out your research rather than afterwards.

Results

Show your supervisor graphed and tabulated versions of your data promptly. These can easily be produced using a spreadsheet (p. 69), but you should seek your supervisor's advice on whether the design and print quality is appropriate to be included in your report. You may wish to access a specialist graphics program to produce publishable-quality graphs and charts: allow some time for learning its idiosyncrasies. If you are producing a poster for assessment (Chapter 13), be sure to mock up the design well in advance. Similarly, think ahead about your needs for any seminar or poster you will present.

Discussion

Because this comes at the end of your report, and some parts can only be written after you have all the results in place, the temptation is to leave the discussion to last. This means that it might be rushed – not a good idea because of the weight attached by assessors to your analysis of data and thoughts about future experiments. It will help greatly if you keep notes of aims, conclusions and ideas for future work as you go along (Fig. 34.1). Another useful tip is to make notes of comparable data and conclusions from the literature as you read papers and reviews.

Acknowledgements

Make a special place in your notebook for noting all those who have helped you carry out the work, for use when writing this section of the report.

References

Because of the complex formats involved (p. 46), these can be tricky to type. To save time, process them in batches as you go along, or consider using bibliographic software, such as Microsoft EndNote®.

KEY POINT Make sure you are absolutely certain about the deadline for submitting your report and try to submit a few days before it. If you leave things until the last moment, you may find that access to printers, photocopiers and binding machines is difficult.

Sources for further study

Luck, M. (2008) *Student Research Projects: Guidance on Practice in the Biosciences*. HEA Centre for Bioscience, York.

Marshall, P. (1997) *Research Methods: How to Design and Conduct a Successful Project*. How To Books, Plymouth.

Robson, C. (2006) *How to do a Research Project: A Guide for Undergraduate Students*. Wiley, New York.

Analysis and presentation of data

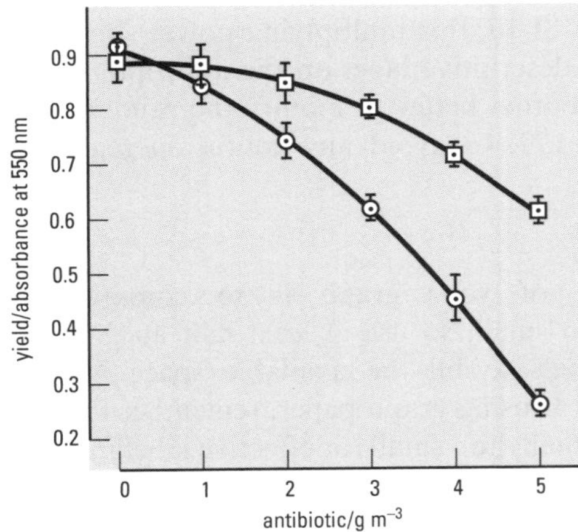

Fig. 35.1 Effect of antibiotic on yield of two bacterial isolates: ○, sensitive isolate; □, resistant isolate. Vertical bars show standard errors ($n = 6$).

Selecting a title – it is a common fault to use titles that are grammatically incorrect: a widely applicable format is to state the relationship between the dependent and independent variables within the title, e.g. 'The relationship between enzyme activity and external pH'. Do not start the title with 'A graph to show . . .'.

Remembering which axis is which – a way of remembering the orientation of the x axis is that x is a 'cross', and it runs 'across' the page (horizontal axis) while y is the first letter of yacht, with a large vertical mast (vertical axis).

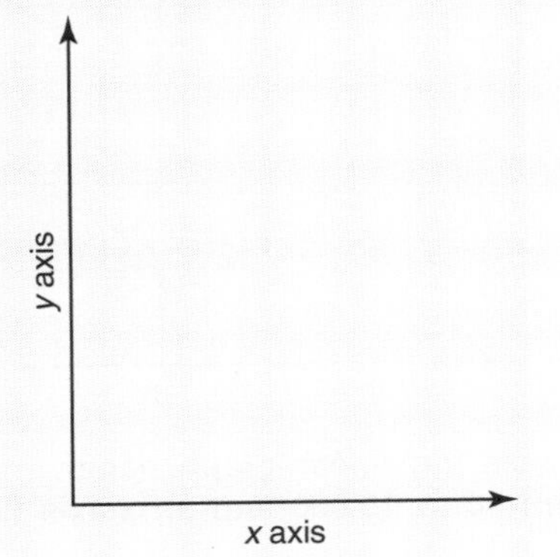

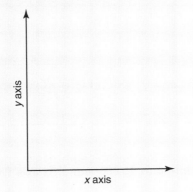

Graphs can be used to show detailed results in an abbreviated form, displaying the maximum amount of information in the minimum space. Graphs and tables present findings in different ways. A graph (figure) gives a visual impression of the content and meaning of your results, while a table provides an accurate numerical record of data values. You must decide whether a graph should be used, e.g. to illustrate a pronounced trend or relationship, or whether a table (Chapter 36) is more appropriate.

A well-constructed graph will combine simplicity, accuracy and clarity. Planning of graphs is needed at the earliest stage in any write-up as your accompanying text will need to be structured so that each graph delivers the appropriate message. Therefore, it is best to decide on the final form for each of your graphs before you write your text. The text, diagrams, graphs and tables in a laboratory write-up or project report should be complementary, each contributing to the overall message. In a formal scientific communication it is rarely necessary to repeat the same data in more than one place (e.g. as a table and as a graph). However, graphical representation of data collected earlier in tabular format may be applicable in laboratory practical reports.

Practical aspects of graph drawing

The following comments apply to graphs drawn for laboratory reports. Figures for publication, or similar formal presentation, are usually prepared according to specific guidelines provided by the publisher/organiser.

> KEY POINT Graphs should be self-contained – they should include all material necessary to convey the appropriate message without reference to the text. Every graph must have a concise explanatory title to establish the content. If several graphs are used, they should be numbered, so they can be quoted in the text.

- Consider the layout and scale of the axes carefully. Most graphs are used to illustrate the relationship between two variables (x and y) and have two axes at right angles (e.g. Fig. 35.1). The horizontal axis is known as the abscissa (x axis) and the vertical axis as the ordinate (y axis).
- The axis assigned to each variable must be chosen carefully. Usually the x axis is used for the independent variable (e.g. treatment) while the dependent variable (e.g. biological response) is plotted on the y axis (p. 205). When neither variable is determined by the other, or where the variables are interdependent, the axes may be plotted either way around.
- Each axis must have a descriptive label showing what is represented, together with the appropriate units of measurement, separated from the descriptive label by a solidus or 'slash' (/), as in Fig. 35.1, or by brackets, as in Fig. 35.2.
- Each axis must have a scale with reference marks ('ticks') on the axis to show clearly the location of all numbers used.
- A figure legend should be used to provide explanatory detail, including a key to the symbols used for each data set.

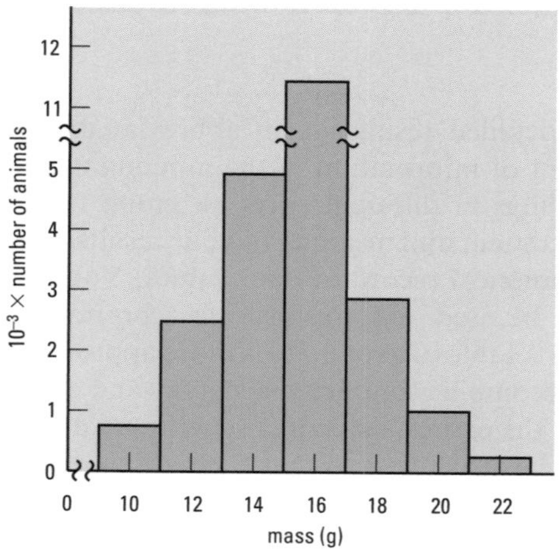

Fig. 35.2 Frequency distribution of masses for a sample of animals (sample size 24 085); the size class interval is 2 g.

Example For a data set where the smallest number on the log axis is 12 and the largest number is 9000, three-cycle log–linear graph paper would be used, covering the range 10–10 000 (Fig 35.3).

Handling very large or very small numbers

To simplify presentation when your experimental data consist of either very large or very small numbers, the plotted values may be the measured numbers multiplied by a power of 10: this multiplying power should be written immediately before the descriptive label on the appropriate axis (as in Fig. 35.2). However, it is often better to modify the primary unit with an appropriate prefix (p. 199) to avoid any confusion regarding negative powers of 10.

Size

Remember that the purpose of your graph is to communicate information. It must not be too small, so use at least half an A4 page and design your axes and labels to fill the available space without overcrowding any adjacent text. If using graph paper, remember that the white space around the grid is usually too small for effective labelling. The shape of a graph is determined by your choice of scale for the x and y axes which, in turn, is governed by your experimental data. It may be inappropriate to start the axes at zero (e.g. Fig. 35.1). In such instances, it is particularly important to show the scale clearly, with scale breaks where necessary, so the graph does not mislead. Note that Fig. 35.1 is drawn with 'floating axes' (i.e. the x and y axes do not meet in the lower left-hand corner), while Fig. 35.2 has clear scale breaks on both x and y axes.

Graph paper

In addition to conventional linear (squared) graph paper, you may need the following:

- Probability graph paper. This is useful when one axis is a probability scale (e.g. p. 263).
- Log–linear graph paper. This is appropriate when one of the scales shows a logarithmic progression, e.g. the exponential growth of cells in liquid culture (p. 473). Log–linear paper is defined by the number of logarithmic divisions (usually termed 'cycles') covered (e.g. Fig. 35.3), so make sure you use a paper with the appropriate number of cycles

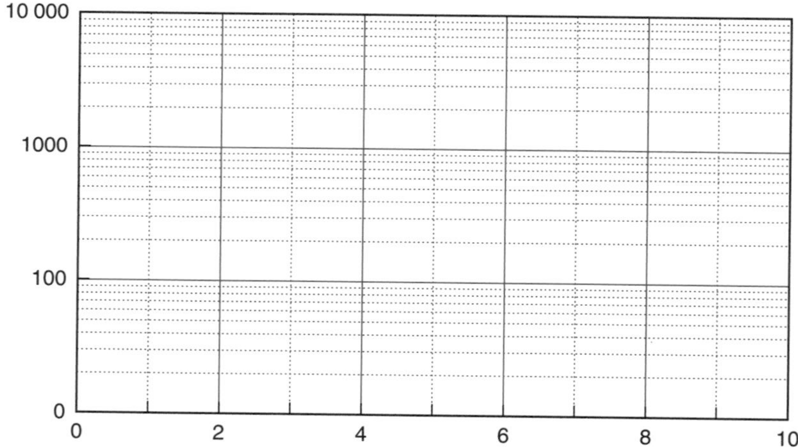

Fig. 35.3 Representation of three-cycle log–linear graph paper, marked up to show a y-axis (log) scale from 10 to 10 000 and an x-axis (linear) scale from 0 to 10.

for your data. An alternative approach is to plot the log-transformed values on 'normal' graph paper.

- Log–log graph paper. This is appropriate when both scales show a logarithmic progression.

Choosing between a histogram and a bar chart – use a histogram for continuous quantitative variables and a bar chart for discrete variables (see Chapter 31 for details of these types of measurement scale).

Types of graph

Different graphical forms may be used for different purposes, including:

- Plotted curves – used for data where the relationship between two variables can be represented as a continuum (e.g. Fig. 35.4).
- Scatter diagrams – used to visualise the relationship between individual data values for two interdependent variables (e.g. Fig. 35.5) often as a preliminary part of a correlation analysis (p. 267).
- Three-dimensional graphs show the interrelationships of three variables, often one dependent and two independent (e.g. Fig. 35.6). A contour diagram is an alternative method of representing such data.
- Histograms represent frequency distributions of continuous variables (e.g. Fig. 35.7). An alternative is the tally chart (p. 196).
- Frequency polygons emphasise the form of a frequency distribution by joining the coordinates with straight lines, in contrast to a histogram. This is particularly useful when plotting two or more frequency distributions on the same graph (e.g. Fig. 35.8).
- Bar charts represent frequency distributions of a discrete qualitative or quantitative variable (e.g. Fig. 35.9). An alternative representation is the line chart (Fig. 39.3, p. 262).
- Pie charts illustrate portions of a whole (e.g. Fig. 35.10).
- Pictographs give a pictorial representation of data (e.g. Fig. 35.11).

Using computers to produce graphs – never allow a computer program to dictate size, shape and other aspects of a graph: find out how to alter scales, labels, axes, etc., and make appropriate selections (see Box 35.2 for Microsoft Excel). Draw curves freehand if the program only has the capacity to join the individual points by straight lines.

The plotted curve

This is the commonest form of graphical representation used in biology. The key features are outlined below and given in checklist form in Box 35.1, while Box 35.2 gives advice on using Microsoft Excel.

Data points

Each data point must be shown accurately, so that any reader can determine the exact values of x and y. In addition, the results of each treatment must be readily identifiable. A useful technique is to use a dot for each data point, surrounded by a hollow symbol for each treatment (see Fig. 35.1). An alternative is to use symbols only (Fig. 35.8), although the coordinates of each point are defined less accurately. Use the same symbol for the same entity if it occurs in several graphs, and provide a key to all symbols.

Choosing graphical symbols – plotted curves are usually drawn using a standard set of symbols: ●, ○, ■, □, ▲, △, ◆, ◇. By convention, paired symbols ('closed' and 'open') are often used to represent 'plus' (treatment) and 'minus' (control) treatments.

Statistical measures

If you are plotting average values for several replicates and if you have the necessary statistical knowledge, you can calculate the standard error (p. 252, or the 95 per cent confidence limits (p. 267) for each mean value and show these on your graph as a series of vertical bars (see Fig. 35.1).

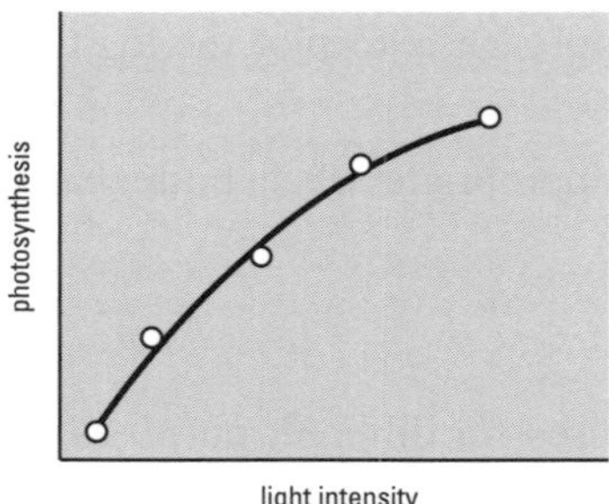

Fig. 35.4 Plotted curve: the rate of photosynthesis as a function of light intensity.

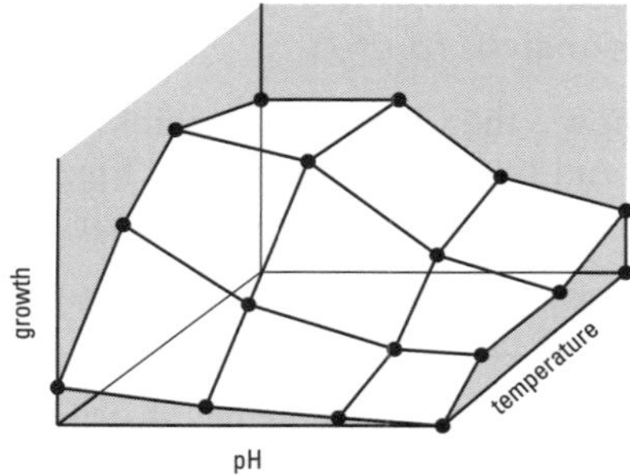

Fig. 35.6 Three-dimensional graph: growth of an organism as a function of temperature and pH.

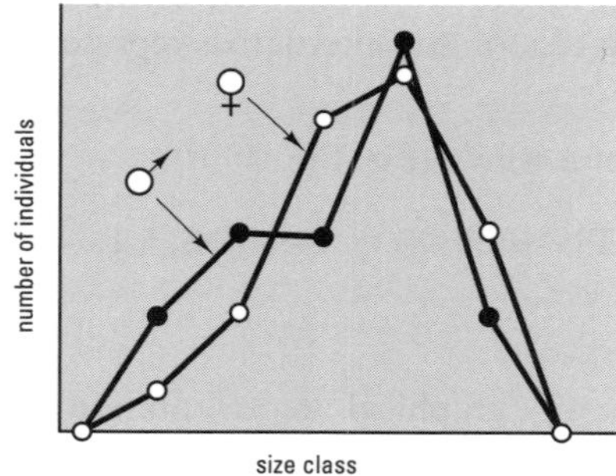

Fig. 35.8 Frequency polygon: frequency distributions of male and female animals according to size.

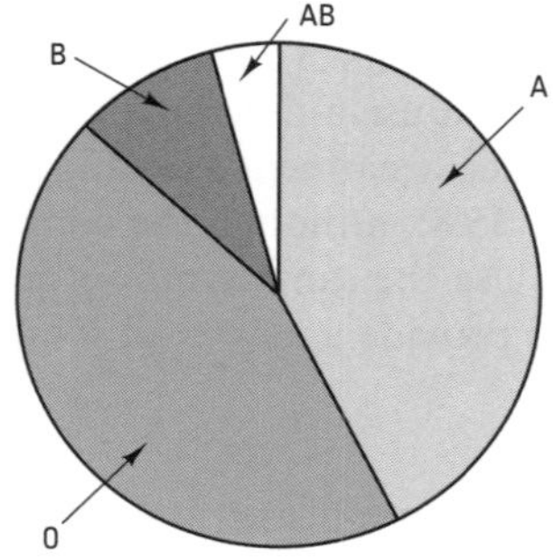

Fig. 35.10 Pie chart: relative abundance of human blood groups.

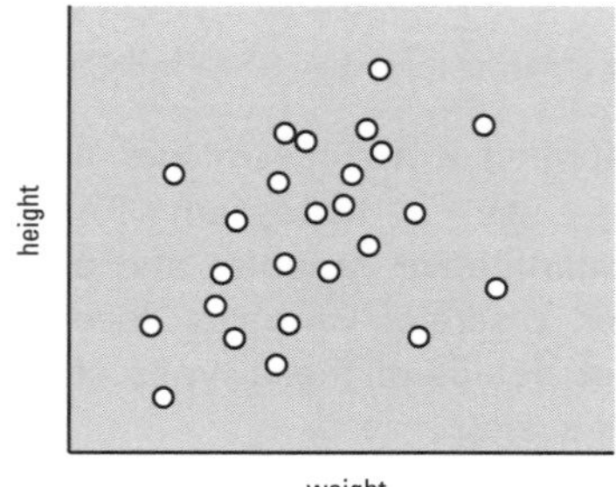

Fig. 35.5 Scatter diagram: height and weight (mass) of individual people in a sample.

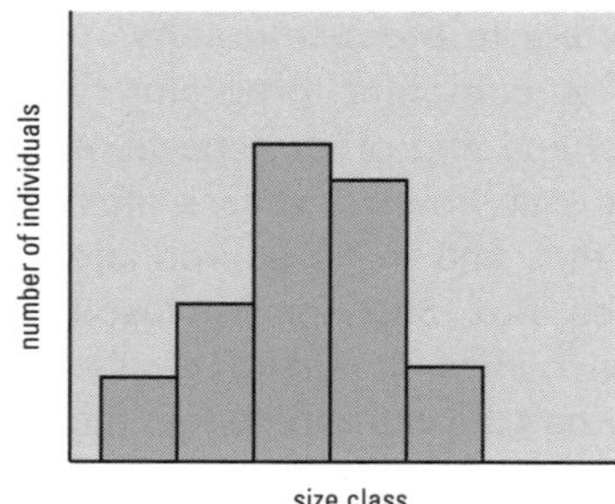

Fig. 35.7 Histogram: the number of individuals within different size classes.

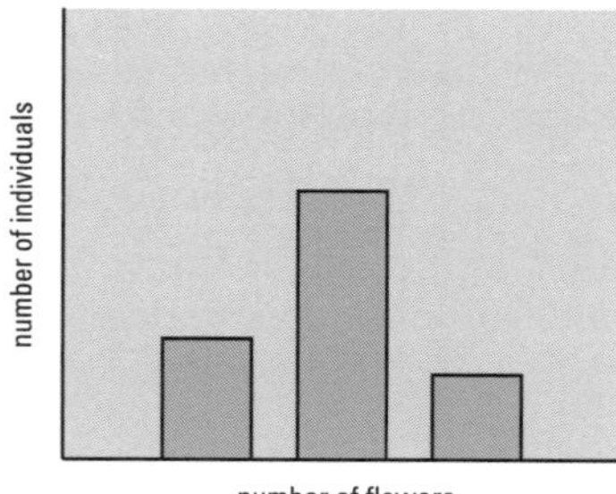

Fig. 35.9 Bar chart: number of flowers per plant.

Fig. 35.11 Pictograph: consumption of desserts over a one-week period.

Box 35.1 Checklist for the stages in drawing a graph

The following sequence can be used whenever you need to construct a plotted curve: it will need to be modified for other types of graph.

1. **Collect all of the data values and statistical values** (in tabular form, where appropriate).
2. **Decide on the most suitable form of presentation**: this may include transformation to convert data to linear form.
3. **Choose a concise descriptive title**, together with a reference (figure) number and date, where necessary.
4. **Determine which variable is to be plotted on the *x* axis and which on the *y* axis.**
5. **Select appropriate scales for both axes** and make sure that the numbers and their location (scale marks) are clearly shown, together with any scale breaks.
6. **Decide on appropriate descriptive labels for both axes**, with SI units of measurement, where appropriate.
7. **Choose the symbols for each set of data points** and decide on the best means of representation for statistical values.
8. **Plot the points** to show the coordinates of each value with appropriate symbols.
9. **Draw a trend line for each set of points.** Use a see-through ruler, so you can draw the line to have an equal number of points on either side of it.
10. **Write a figure legend**, to include a key that identifies all symbols and statistical values and any descriptive footnotes.

Box 35.2 How to create and amend graphs within a spreadsheet (Microsoft Excel 2007) for use in coursework reports and dissertations

Microsoft Excel can be used to create graphs of reasonable quality, as long as you know how to amend the default settings so that your graph meets the formal standards required for practical and project reports. As with a hand-drawn graph, the basic stages in graph drawing (Box 35.1) still apply. The following instructions explain how to produce an X–Y graph (plotted curve, p. 221), bar graph (p. 222), pie graph and histogram using Excel 2007, where all types of graphs are termed 'charts'. Earlier versions use broadly similar commands, although not always found in the same locations within the software.

Producing an *x–y* graph (*Scatter* chart in Excel)

1. **Create the appropriate type of graph for your data.** Enter the numeric values for your *x* variable data in the cells of a single column and the equivalent values for the *y* variable in the adjacent cells of the next column to the right. Then click on the whole data array (highlight the appropriate cells by clicking and holding down the left mouse button and dragging the cursor across the cells so that all values are included). Then click on the *Insert* tab at the top of the sheet, and select (left-click) *Scatter* chart from the options provided in the upper ribbon. Note that you should never use the *Line* chart option, as it is based on an *x* axis that is not quantitative, so all *x* values will appear as equally spaced categories, rather than having a true scale). Select the first option from the *Scatter* menu (*Scatter with only Markers*). Once selected, this will produce an embedded scatter chart of the type shown in Fig. 35.12(a). The line is then added later, as described below.
2. **Change the default settings to improve the appearance of your graph.** Consider each element of the image in turn, including the overall size, height and width of the graph (resize by clicking and dragging one of the 'sizing handles' around the edge of the chart). The graph shown in Fig. 35.12(b) was produced by altering the default settings, typically by moving the cursor over the feature and then clicking the right mouse button to reveal an additional menu of editing and formatting options. (Note that the example given below is for illustrative purposes only, and should not necessarily be regarded as prescriptive.)

***Example for an x-y* graph** (compare Fig. 35.12(a) with Fig. 35.12(b)):

- Unnecessary legend box on the right-hand side can be removed using the *Delete* option (right-click).
- Chart border can be removed using the *Format Chart Area* function (available by right-clicking within the chart area).
- Gridlines can either be removed, using the *Delete* function, or, if desired (as in Fig. 35.12(b)), changed by clicking on each axis and using the *Add Minor Gridlines* and *Format Gridlines* options to alter the *Color* and *Style* of the gridlines to make them more like those of conventional graph paper.

(continued)

Box 35.2 (continued)

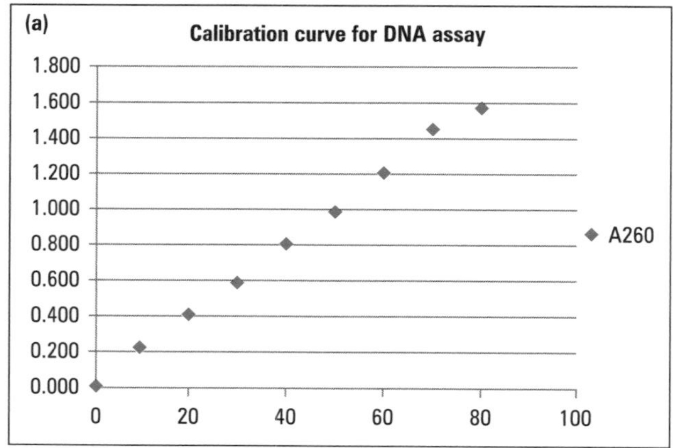

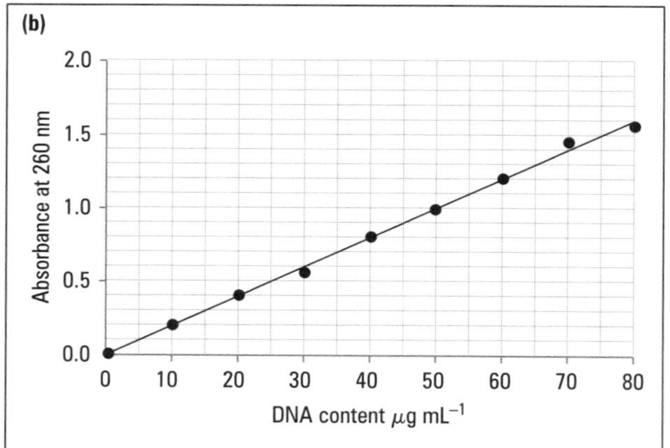

Fig. 1 Calibration curve for DNA assay. Performed using an A100X spectrophotometer (on 01.04.07). Values shown are averages of triplicate measurements.

Fig. 35.12 Examples of a plotted curve produced in Microsoft Excel using (a) default settings and (b) modified (improved) settings.

- x and y axes can be reformatted by selecting each in turn, and using the *Format Axis* menu options to select appropriate scales for major and minor units, line colour, style of tick marks, etc. Remember that it is better to use a figure legend in Word, rather than the *Chart Title* option within Excel.
- x- and y-axis labels can be added by selecting the *Layout* tab, then *Axis Titles*, then *Primary Horizontal Axis Title* and *Primary Vertical Axis Title*, which will produce a text box beside each axis into which can be typed the axis label and any corresponding units. This can be then changed from the default font using the options on the *Home* tab.
- Data x- and y-axis point style can be changed by right-clicking any data point and following the *Format Data Series* options to choose appropriate styles, colours and fill for the data markers.
- A straight line of best fit can be added by selecting any data point and using the *Add Trendline* option to choose a *Linear* line type with appropriate colour and style (explore other options within the *Format, Layout* and *Design* tabs at the top of the worksheet).

Producing a bar graph (*Column* chart in Excel)

1. **Create the appropriate type of graph for your data.** Enter the category names (for the x axis) in one column and the numeric values (for the y axis) in the next column. Select (highlight) the data array, then click the *Insert* tab, and choose *Column* chart from the options provided. For a standard bar graph, select the first option from the *Column* menu (*Clustered Column*). Once selected, this will produce an embedded bar graph of the type shown in Fig. 35.13(a).
2. **Change the default settings to improve the appearance of your graph.** The bar graph shown in Fig. 35.13(b) was produced by selecting each feature and altering the default settings, as detailed below (illustrative example).

Example for a bar graph (compare Fig. 35.13(a) with Fig. 35.13(b)):

- Unnecessary legend box on the right-hand side can be removed using the *Delete* option.
- Chart border can be removed using the *Format Chart Area* function (available by right-clicking within chart area).
- Gridlines can either be removed, using the *Delete* function, or changed by selecting the gridlines and using the *Format Gridlines* option to alter the *Color* and *Style*.
- y axis can be reformatted by selecting the axis, then using the *Format Axis* menu options to select appropriate scales, tick marks, line colour, etc. Note that the x axis should already contain category labels from the spreadsheet cells (modify the original cells to update the spreadsheet, if necessary).

(continued)

Box 35.2 (continued)

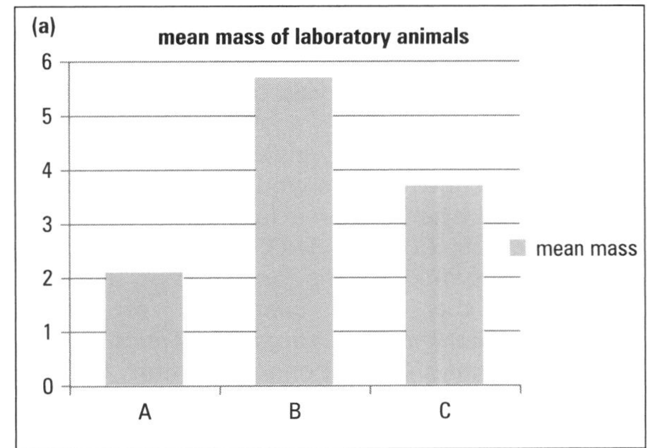

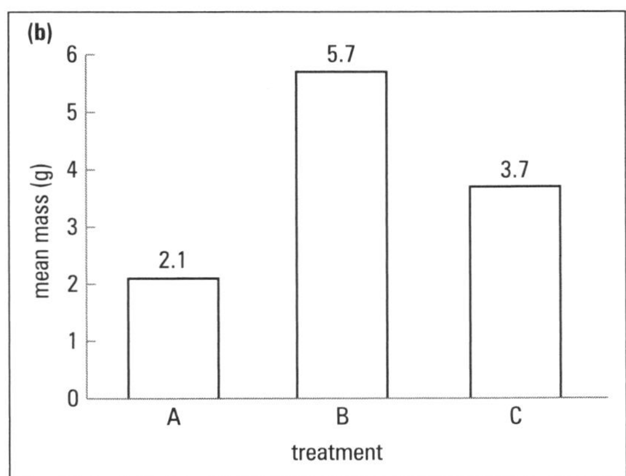

Fig. 2 Bar chart for mean mass of laboratory animals fed three different diets. Key: A = fruits, B = cereals, C = vegetables.

Fig. 35.13 Examples of a bar chart produced in Microsoft Excel using (a) default settings and (b) modified (improved) settings.

- *x*- and *y*-axis labels can be added by selecting the *Layout* tab, then *Axis Titles*, then *Primary Horizontal Axis Title* and *Primary Vertical Axis Title* , as detailed for the plotted curve (p. 222).
- Bar colour can be modified using the *Format Data Series* menu, and selecting appropriate *Fill* and *Border* colours, e.g. white and black respectively in Fig. 35.13(b).
- Individual *y*-data values can be shown using the *Add Data Labels* option (other options and adjustments can be made using the *Format*, *Layout* and *Design* tabs at the top of the worksheet).

Note that for all types of graph, it is better not to use the *Chart Title* option within Excel, which places the title at the top of the chart (as in Fig. 35.13(a)), but to copy and paste your untitled graph into a word-processed document, such as a Microsoft Word file (gives details of the procedure), and then type a formal figure legend below the graph, as in Figs 35.12(b) and 35.13(b). However, once your graph is embedded into a Word file, it is generally best not to make further amendments – you should go back to the original Excel file, make the required changes and then repeat the copy–paste procedure to reinsert the graph into the Word file.

Producing a pie graph (*Pie* chart in Excel)

1. **Create the appropriate type of graph for your data.** Enter the category names for each part of the pie chart in one column, and the corresponding numbers (counts, fractions or percentages) in the next column. Select (highlight) the data array, then click the *Insert* tab, and choose the *Pie* chart from the options provided. For a standard pie graph, select the first option from the menu (*Pie*). Once selected, this will produce an embedded pie graph.
2. **Change the default settings to improve the appearance of your graph.** For example, you can show the data values (*Data labels*), adjust colours and shading (e.g. switch from multi-colour to shades of grey), remove the chart border, etc., as required (an illustrative example is shown as Fig. 35.14).

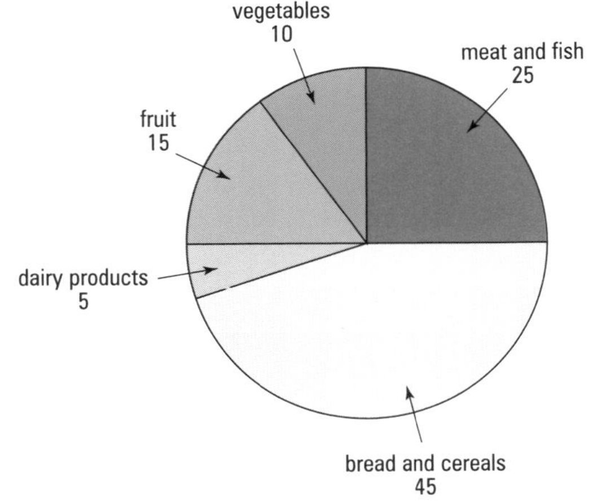

Fig. 35.14 Food consumed in a one-week period.

(continued)

Box 35.2 (continued)

Producing a histogram

The histogram function in Excel 2007 requires a little more effort to master, compared with other chart types. Essentially, a histogram is a graphical display of frequencies (counts) for a continuous quantitative variable, where the data values are grouped into classes. It is possible to select the upper limit for each class into which you want the data to be grouped (these are termed 'bin range values' in Excel 2007). An alternative approach is to let the software select the class intervals (bin range values) for you: Excel selects evenly distributed bins between the minimum and maximum values. However, this is often less effective than selecting your own class intervals.

To use the Excel histogram function, you will need to make sure that the *Analysis ToolPak* is loaded (check under the *Data* tab and, if not loaded, select via the *Office Button* > *Excel Options* > *Add-ins* > *Manage Excel Add-ins* > *Analysis ToolPak*. The *Histogram* function should then be available in the *Analysis ToolPak* option on the *Data* tab/ribbon.

The following steps outline the procedure used to create the histograms shown in Fig. 35.15 for the table of data below (length, in mm, of 24 fruits from a single plant.

7.2	6.5	7.1	8.5	6.6	7.2
7.0	7.3	8.6	9.1	7.5	8.3
7.1	5.7	7.3	7.6	6.9	7.1
8.3	7.6	5.4	8.6	7.9	8.0

1. **Enter the raw data values** e.g. as a single column of numbers, or as an array, as above.
2. **Decide on the class intervals to be used.** Base your choice on the number of data points and the maximum and minimum values (for a small data set such as that shown above, you can do this by visual examination, whereas for a large data set use the Excel functions *COUNT*, *MAX* and *MIN* (find these under the Σ symbol (*More Functions*) on the *Editing* section on the *Home* tab/ribbon, or use the *Descriptive Statistics* > *Summary statistics* option of the *Data Analysis* component on the *Data* tab/ribbon). A typical histogram would have 4–10 classes, depending on the level of discrimination required. Enter the upper limit for each class (*bin range values*) in ascending order in a separate array of cells, e.g. in a column close to the data values (in the above example, 6, 7, 8 and 9 were chosen – the few data values above the final bin value will be shown on the histogram as a group labelled 'more').
3. **Select the histogram function, then input your data and class interval values.** From the *Data* tab/ribbon, select *Data Analysis* > *Histogram*. A new window will open: input your data into the *Input Range* box (highlight the appropriate cells by clicking on the first data value and dragging to the final data value while holding down the left mouse button). Next, type the *Bin Range* values into the appropriate box (if this is left empty, Excel will select default bin range values). Most of the remaining boxes can be left empty, although you must click the last box to get a *Chart Output*, otherwise the software will give

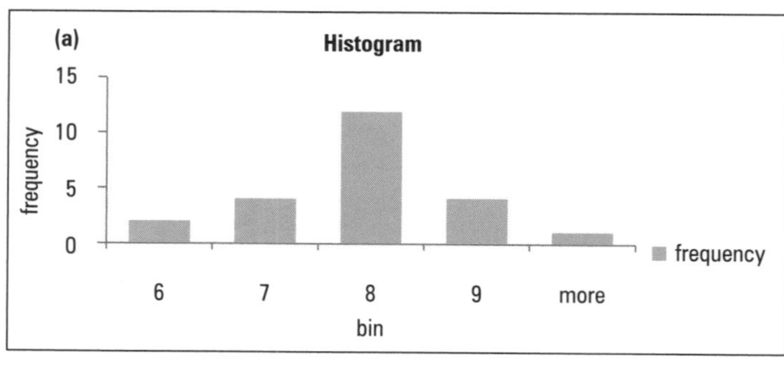

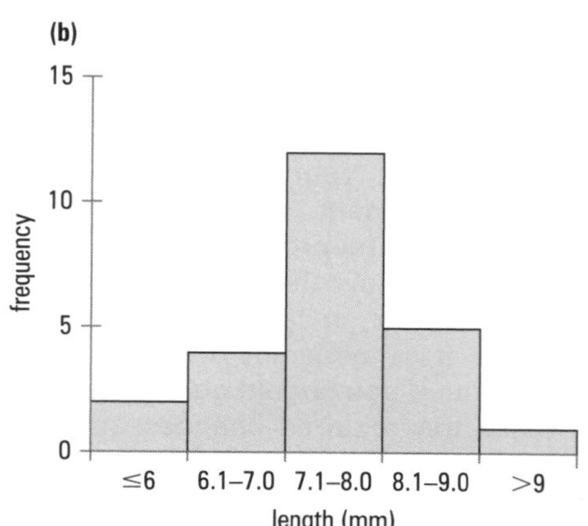

Fig. 35.15 Examples of histogram output from Microsoft Excel using (a) default settings and (b) modified (improved) settings.

(continued)

Box 35.2 (continued)

the numerical counts for each group, without drawing a histogram. Click *OK* and entries will be created within a new worksheet, showing the upper limits of each group (in a column labelled *Frequency*), plus a poorly constructed chart based on Excel 2007 default settings, as shown in Fig. 35.15(a) (note that the default output is a bar chart, rather than a histogram, since there are gaps between the groups).

Example for a histogram (compare Fig. 35.15(a) with Fig. 35.15(b)):

- Chart can be resized to increase height, using the 'sizing handles' at the edges of the chart and border line around graph can be removed using *Format Chart Area > Border Color > No line* (the menu is available by right-clicking within chart area).
- Title and unnecessary legend box can be removed using the *Delete* option.
- Axis scales can be reformatted using the *Format Axis* menu options (e.g. scales, tick marks, line colour).
- *x*-axis labels (class intervals) can be amended by typing directly into the cells containing the bin range values.
- Axis titles can be changed by typing directly into the axis title box (double-click to access).
- Bar colour can be changed (e.g. to grey, with black outline) using the *Format Data Series > Fill* and *Border Color* options.
- Bar chart converted to correct histogram format (no gaps between bars) using the *Format Data Series > Series Options*, setting *Gap Width* to 0%.
- Figure legend can be added below the figure in Microsoft Word, following copying and pasting of the Excel histogram into a Word file.

Importing an Excel 2007 chart into a Word 2007 document

One problem encountered with Microsoft Office 2007 products (but not with earlier versions, e.g. Office 1997–2003) is that the standard *Cut Paste* procedure gives a poor-quality figure, with a grainy appearance and fuzzy lines/text; similar problems occur using the *Insert* tab in Word 2007. The simplest approach is to follow the procedure below:

1. Select your Excel 2007 chart: right-click outside the chart itself, near to the edge, then choose the *Copy* option from the drop-down menu.
2. Open your Word 2007 file, go to the *Home* tab/ribbon and select *Paste special > Microsoft Office Word Document Object* from the *Clipboard* options (on the left-hand side of the ribbon).
3. This should give a graph with the same crisp axis/line/text formatting as the original chart in Excel 2007.

The alternative approach is to use Excel 2007 to print the entire graph (chart) as a single sheet, and then add this to the print-out from your word-processed document. However, the disadvantage with this approach is that you cannot produce a professional looking figure legend below your graph.

Make it clear in the legend whether the bars refer to standard errors or 95 per cent confidence limits and quote the value of n (the number of replicates per data point). Another approach is to add a least significant difference bar (p. 264) to the graph.

Interpolation

Once you have plotted each point, you must decide whether to link them by straight lines or a smoothed curve. Each of these techniques conveys a different message to your reader. Joining the points by straight lines may seem the simplest option, but may give the impression that errors are very low or non-existent and that the relationship between the variables is complex. Joining points by straight lines is appropriate in certain graphs involving time sequences (e.g. the number of animals at a particular

Adding error bars to Microsoft Excel graphs – you can do this as follows (for Excel 2007):

1. On your graph, right-click on one of the data points of the series to which you want to add error bars.
2. Select *Layout* tab > *Error bars* > *More error bar options* and use either the *Y Error Bars* tab (vertical error bars) or the *X Error Bars* tab (horizontal error bars):
   - select the type of error bar that you want under *Display* (e.g. *Both*);
   - select the method to be used to input the error values under *Error amount*; typically, you will want to select *Custom* > *Specify*, to input pre-calculated numerical values. Then, in the *Plus* and *Minus* boxes, either specify the range of cells in your worksheet where the error values are located, or enter the relevant numbers directly, separated by commas and click 'ok'.

Conveying the correct message – the golden rule is: 'always draw the simplest line that fits the data reasonably well and is biologically reasonable'.

Extrapolating plotted curves – try to avoid the need to extrapolate by better experimental design.

Drawing a histogram – each datum is represented by a column with an area proportional to the magnitude of y: in most cases, you should use columns of equal width, so that the height of each column is then directly proportional to y. Shading or stippling may be used to identify individual columns, according to your needs.

site each year), or for repeat measurements where measurement error can be assumed to be minimal (e.g. recording a patient's temperature in a hospital, to emphasise any variation from one time point to the next). However, in most plotted curves the best straight line or curved line should be drawn (according to appropriate mathematical or statistical models, or by eye), to highlight the relationship between the variables – after all, your choice of a plotted curve implies that such a relationship exists. Do not worry if some of your points do not lie on the line: this is caused by errors of measurement and by biological variation. Most curves drawn by eye should have an equal number of points lying on either side of the line. You may be guided by 95 per cent confidence limits, in which case your curve should pass within these limits wherever possible.

Curved lines can be drawn using a flexible curve, a set of French curves, or freehand. In the latter case, turn your paper so that you can draw the curve in a single, sweeping stroke by a pivoting movement at the elbow (for larger curves) or wrist (for smaller ones). Do not try to force your hand to make complex, unnatural movements, as the resulting line will not be smooth.

Extrapolation

Be wary of extrapolation beyond the upper or lower limit of your measured values. This is rarely justifiable and may lead to serious errors. Whenever extrapolation is used, a dotted line ensures that the reader is aware of the uncertainty involved. Any assumptions behind an extrapolated curve should also be stated clearly in your text.

The histogram

Whereas a plotted curve assumes a continuous relationship between the variables by interpolating between individual data points, a histogram involves no such assumptions. Histograms are also used to represent frequency distributions (p. 227), where the y axis shows the number of times a particular value of x was obtained (e.g. Fig. 35.2). As in a plotted curve, the x axis represents a continuous quantitative variable that can take any value within a given range (e.g. plant height), so the scale must be broken down into discrete classes and the scale marks on the x axis should show either the mid-points (mid-values) of each class, or the boundaries between the classes.

The columns are contiguous (adjacent to each other) in a histogram, in contrast to a bar chart, where the columns are separate because the x axis of a bar chart represents discrete values.

Interpreting graphs

The process of analysing a graph can be split into five phases:

1. Consider the context. Look at the graph in relation to the aims of the study in which it was reported. Why were the observations made? What hypothesis was the experiment set up to test? This information can usually be found in the introduction or results section of a report. Also relevant are the general methods used to obtain the results. This might be obvious from the figure title and legend, or from the materials and methods section.

Examining graphs – do not be tempted to look at the data displayed within a graph before you have considered its context, read the legend and decided the scale of each axis.

2. Recognise the graph form and examine the axes. Firstly, what kind of graph is presented (e.g. histogram, plotted curve)? You should be able to recognise the main types summarised on pp. 221–222 and their uses. Next, what do the axes measure? You should check what quantity has been measured in each case and what units are used.
3. Look closely at the scale of each axis. What is the starting point and what is the highest value measured? For the x axis, this will let you know the scope of the treatments or observations (e.g. whether they lasted for five minutes or 20 years; whether a concentration span was two-fold or 50-fold). For each axis, it is especially important to note whether the values start at zero; if not, then the differences between any treatments shown may be magnified by the scale chosen (see Box 35.3).
4. Examine the symbols and curves. Information will be provided in the key or legend to allow you to determine what these refer to. If you have made your own photocopy of the figure, it may be appropriate to note this directly on it. You can now assess what appears to have happened. If, say, two conditions have been observed while a variable is altered, when exactly do they differ from each other; by how much; and for how long?
5. Evaluate errors and statistics. It is important to take account of variability in the data. For example, if mean values are presented, the underlying errors may be large, meaning that any difference between two treatments or observations at a given x value could simply have arisen by chance. Thinking about the descriptive statistics used (Chapter 38) will allow you to determine whether apparent differences could be significant in both statistical and biological senses.

Understanding graphs within scientific papers – the legend should be a succinct summary of the key information required to interpret the figure without further reference to the main text. This is a useful approach when 'skimming' a paper for relevant information (p. 18).

Sometimes graphs are used to mislead. This may be unwitting, as in an unconscious favouring of a 'pet' hypothesis of the author. Graphs may be used to 'sell' a product in the field of advertising or to favouring of a viewpoint as, perhaps, in politics. Experience in drawing and interpreting graphs will help you spot these flawed presentations, and understanding how graphs can be erroneously presented (Box 35.3) will help you avoid the same pitfalls.

Box 35.3 How graphs can misrepresent and mislead

1. **The 'volume' or 'area' deception** – this is mainly found in histogram or bar chart presentations where the size of a symbol is used to represent the measured variable. For example, the amount of food additive used in different years might be represented on a chart by different sizes of a chemical drum, with the y axis (height of drum) representing the amount of waste. However, if the symbol retains its *shape* for all heights as in Fig. 35.16(a), its *volume* will increase as a cubic function of the height, rather than in direct proportion. To the casual observer, a two-fold increase may look like an eight-fold one, and so on. Strictly, the *height* of the symbol should be the measure used to represent the variable, with no change in symbol width, as in Fig. 35.16(b).

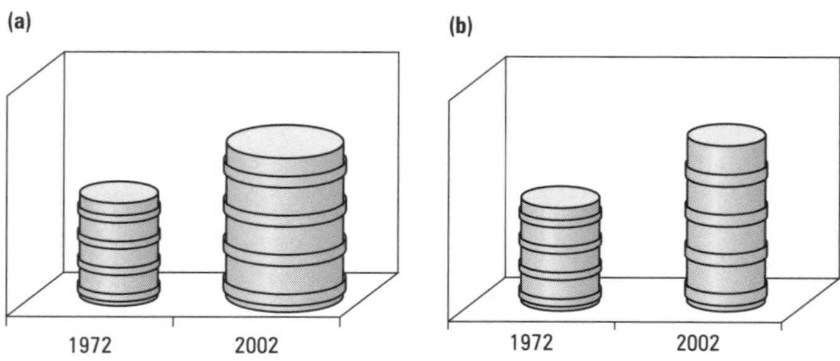

Fig. 35.16 Increase in food additive use between 1972 and 2002.

2. **Effects of a non-zero axis** – a non-zero axis acts to emphasise the differences between measures by reducing the range of values covered by the axis. For example, in Fig. 35.17(a), it looks as if there are large differences in mass between males and females; however, if the scale is adjusted to run from zero, then it can be seen that the differences are not large as a proportion of the overall mass. Always scrutinise the scale values carefully when interpreting any graph.

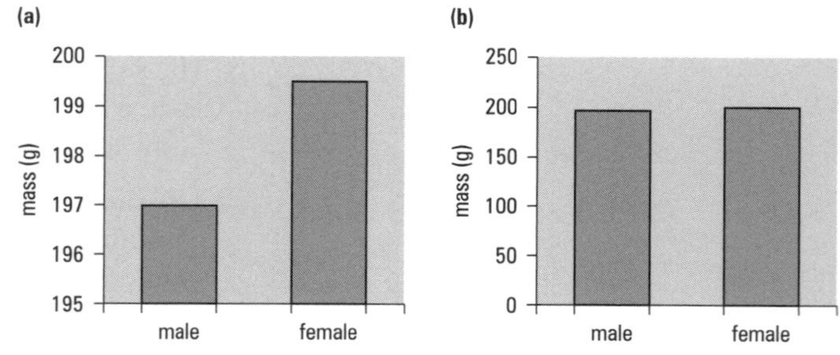

Fig. 35.17 Average mass (weight) of males and females in test group.

3. **Use of a relative rather than absolute scale** – this is similar to the above, in that data compared using relative scales (e.g. percentage or ratio) can give the wrong impression if the denominator is not the same in all cases. In Fig. 35.18(a), two treatments are shown as equal in *relative* effect, both resulting in 50 per cent relative response compared (say) to the respective controls. However, if treatment A is 50 per cent of a control value of 200 and treatment B is 50 per cent of a control value of 500, then the actual difference in *absolute* response would have been masked, as shown by Fig. 35.18(b).

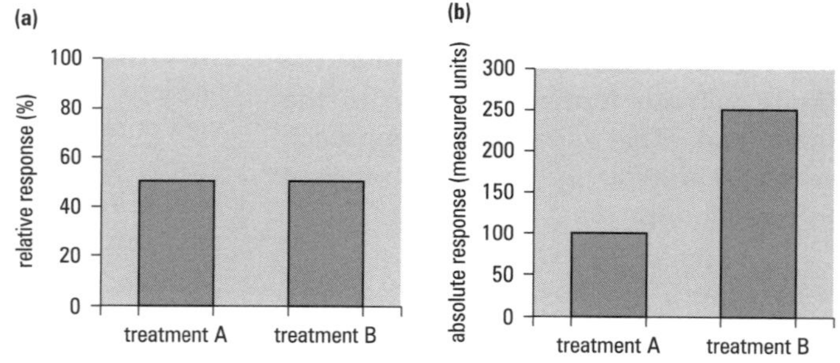

Fig. 35.18 Responses to treatments A and B.

4. **Effects of a non-linear scale** – when interpreting graphs with non-linear (e.g. logarithmic) scales, you may interpret any changes on an imagined linear scale. For example, the pH scale is logarithmic, and linear changes on this scale mean less in terms of absolute H^+ concentration at high (alkaline) pH than they do at low (acidic) pH. In Fig. 35.19(a), the cell density in two media is compared on a logarithmic scale, while in Fig. 35.19(b), the same data are graphed on a linear scale. Note, also, that the log y axis scale in Fig. 35.19(a) cannot be shown to zero, because there is no logarithm for 0.

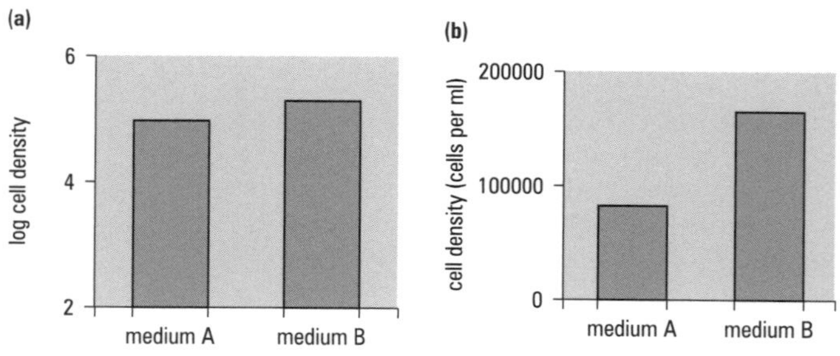

Fig. 35.19 Effect of different media on cell density.

(continued)

Box 35.3 (continued)

5. **Unwarranted extrapolation** – a graph may be extrapolated to indicate what would happen if a trend continued, as in Fig. 35.20(a). However, this can only be done under certain assumptions (e.g. that certain factors will remain constant or that relationships will hold under new conditions). There may be no guarantee that this will actually be the case. Figure 35.20(b) illustrates other possible outcomes if the experiment were to be repeated with higher values for the x axis.

6. **Failure to account for data point error** – this misrepresentation involves curves that are overly complex in relation to the scatter in the underlying data. When interpreting graphs with complex curves, consider the errors involved in the data values. It is probably unlikely that the curve would pass through all the data points unless the errors were very small. Figure 35.21(a) illustrates a curve that appears to assume zero error and is thus overly complex, while Fig. 35.21(b) shows a curve that takes possible errors of the points into account.

7. **Failure to reject outlying points** – this is a special case of the previous example. There may be many reasons for outlying data, from genuine mistakes to statistical 'freaks'. If a curve is drawn through such points on a graph, it indicates that the point carries equal weight with the other points, when in fact, it should probably be ignored. To assess this, consider the accuracy of the measurement, the number and position of adjacent points, and any special factors that might be involved on a one-off basis. Figure 35.22(a) shows a curve where an outlier (arrowed) has perhaps been given undue weight when showing the presumed relationship. If there is good reason to think that the point should be ignored, then the curve shown in Fig. 35.22(b) would probably be more valid.

8. **Inappropriate fitted line** – here, the mathematical function chosen to represent a trend in the data might be inappropriate. A straight line might be fitted to the data, when a curve would be more correct, or vice versa. These cases can be difficult to assess. You need to consider the theoretical validity of the model used to generate the curve (this is not always stated clearly). For example, if a straight line is fitted to the points, the implicit underlying model states that one factor varies in direct relation to another, when the true situation may be more complex. In Fig. 35.23(a), the relationship has been shown as a linear relationship, whereas an exponential relationship, as shown in Fig. 35.23(b), could be more correct.

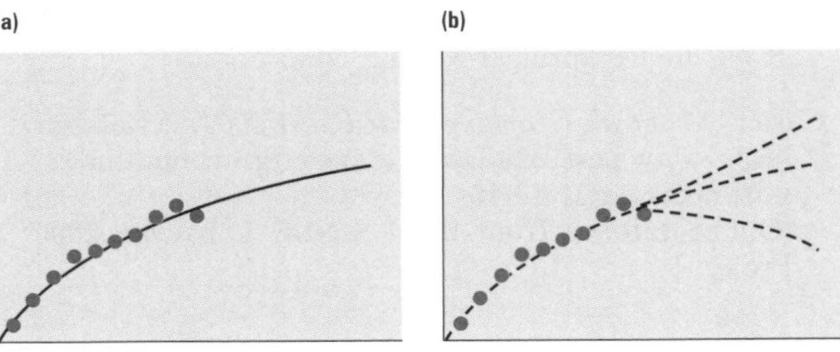

Fig. 35.20 Extrapolation of data under different assumptions.

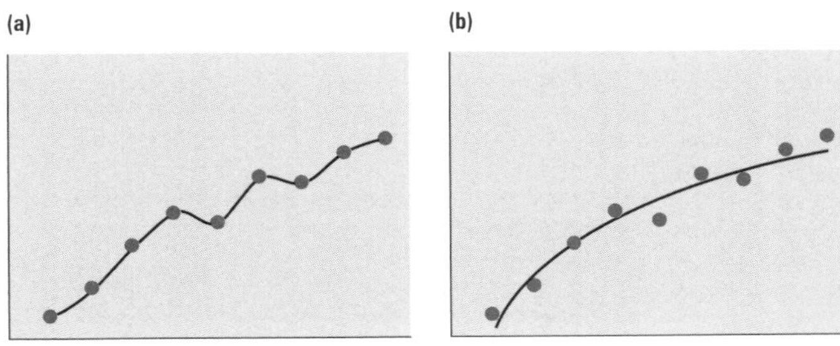

Fig. 35.21 Fitted curves under different assumptions of data error.

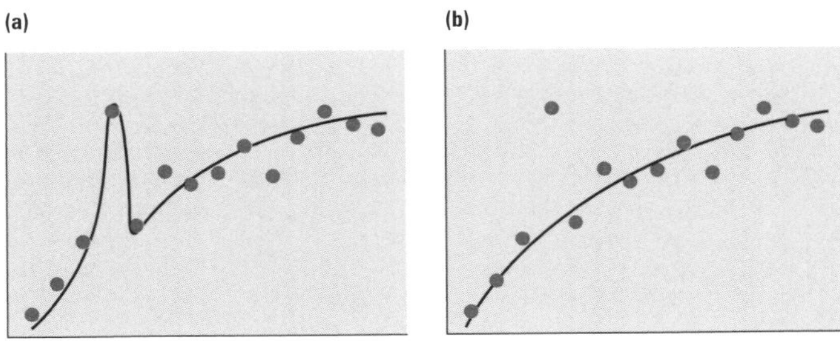

Fig. 35.22 Curves with and without outlier taken into account.

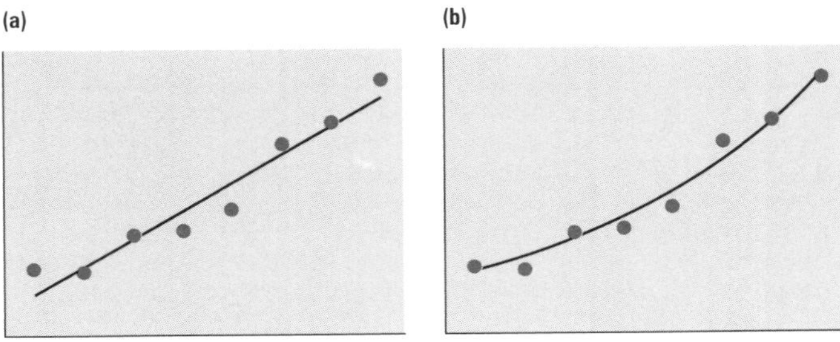

Fig. 35.23 Different mathematical model used to represent trends in data.

Sources for further study

Briscoe, M.H. (1996) *Preparing Scientific Illustrations: A Guide to Better Posters, Presentations and Publications*. Springer-Verlag, New York.

Carter, M. *et al. Graphing with Excel 2003*. Available: http://www.ncsu.edu/labwrite/res/gt/gt-menu.html Last accessed 21/12/10.
[Online tutorial from the US-NSF LabWrite 2000 Project.]

Institute of Biology (2000) *Biological Nomenclature: Recommendations on Terms, Units and Symbols*. Institute of Biology, London.
[Includes a section on presentation of data.]

Robbins, N.B. (2005) *Creating More Effective Graphs*. Wiley, New York.

36 Presenting data in tables

A table is often the most appropriate way to present numerical data in a concise, accurate and structured form. Assignments and project reports should contain tables that have been designed to condense and display results in a meaningful way and to aid numerical comparison. The preparation of tables for recording primary data is discussed on p. 195.

Decide whether you need a table, or whether a graph is more appropriate. Histograms and plotted curves can be used to give a visual impression of the relationships within your data (pp. 222–228). On the other hand, a table gives you the opportunity to make detailed numerical comparisons.

KEY POINT Always remember that the primary purpose of your table is to communicate information and allow appropriate comparison, not simply to put down the results on paper.

Preparation of tables

Title

Constructing titles – take care over titles as it is a common mistake in student practical reports to present tables without titles or to misconstruct the title.

Every table must have a brief descriptive title. If several tables are used, number them consecutively so they can be quoted in your text. The titles within a report should be compared with one another, making sure they are logical and consistent, and that they describe accurately the numerical data contained within them.

Structure

Saving space in tables – you may be able to omit a column of control data if your results can be expressed as percentages of the corresponding control values.

Display the components of each table in a way that will help the reader understand your data and grasp the significance of your results. Organise the columns so that each category of like numbers or attributes is listed vertically, while each horizontal row shows a different experimental treatment, organism, sampling site, etc. (as in Table 36.1). Where appropriate, put control values near the beginning of the table. Columns that need to be compared should be set out alongside each other. Use rulings to subdivide your table appropriately, but avoid cluttering it up with too many lines.

Table 36.1 Characteristics of selected microbes

			Intracellular carbohydrate	
Division	**Species**	**Optimum [NaCl]* (mol m^{-3})**	**Identity**	**Quantity† (nmol (g dry wt)$^{-1}$)**
Chlorophyta	*Scenedesmus quadruplicatum*	340	Sucrose	49.7
	Chlorella emersonii	780	Sucrose	102.3
	Dunaliella salina	4700	Glycerol	910.7
Cyanobacteria	*Microcystis aeruginosa*	<20‡	None	0.0
	Anabaena variabilis	320	Sucrose	64.2
	Rivularia atra	380	Trehalose	ND

* Determined after 28-day incubation in modified Von Stosch medium.
† Individual samples, analysed by gas–liquid chromatography.
‡ Poor growth in all media with added NaCl (minimum NaCl concentration 5 mol m^{-3}).
ND: sample lost – no quantitative data.

Examples If you measured the width of a fungal hypha to the nearest one-tenth of a micrometre, quote the value in the form '52.6 μm'.
Quote the width of a fungal hypha as 52.6 μm, rather than 0.000 052 6 m or 52.6 10^{-6} m.

Saving further space in tables – in some instances a footnote can be used to replace a whole column of repetitive data.

Using spreadsheets and word-processing packages – these can be used to prepare high-quality versions of tables for project work (Box 36.2).

Headings and sub-headings

These should identify each set of data and show the units of measurement, where necessary. Make sure that each column is wide enough for the headings and for the longest data value.

Numerical data

Within the table, do not quote values to more significant figures than necessary, as this will imply spurious accuracy (pp. 239, 241). By careful choice of appropriate units for each column you should aim to present numerical data within the range 0 to 1 000. As with graphs, it is less ambiguous to use derived SI units, with the appropriate prefixes, in the headings of columns and rows, rather than quoting multiplying factors as powers of 10. Alternatively, include exponents in the main body of the table (see Table 28.1), to avoid any possible confusion regarding the use of negative powers of 10.

Other notations

Avoid using dashes in numerical tables, as their meaning is unclear; enter a zero reading as '0' and use 'NT' not tested or 'ND' if no data value was obtained, with a footnote to explain each abbreviation. Other footnotes, identified by asterisks, superscripts or other symbols in the table, may be used to provide relevant experimental detail (if not given in the text) and an explanation of column headings and individual results, where appropriate. Footnotes should be as condensed as possible. Table 36.1 provides examples.

Statistics

In tables where the dispersion of each data set is shown by an appropriate statistical parameter, you must state whether this is the (sample) standard deviation, the standard error (of the mean) or the 95 per cent confidence limits and you must give the value of n (the number of replicates). Other descriptive statistics should be quoted with similar detail, and hypothesis-testing statistics should be quoted along with the value of P (the probability). Details of any test used should be given in the legend or in a footnote.

Text

Sometimes a table can be a useful way of presenting textual information in a condensed form (see example on p. 237).

When you have finished compiling your tabulated data, carefully double-check each numerical entry against the original information, to ensure that the final version of your table is free from transcriptional errors. Box 36.1 gives a checklist for the major elements of constructing a table.

Box 36.1 Checklist for preparing a table

Every table should have the following components:

1. **A title**, plus a reference number and date where necessary.
2. **Headings for each column and row**, with appropriate units of measurement.
3. **Data values**, quoted to the nearest significant figure and with statistical parameters, according to your requirements.
4. **Footnotes** to explain abbreviations, modifications and individual details.
5. **Rulings to emphasise groupings** and distinguish items from each other.

Box 36.2 How to use a word processor (Microsoft Word 2007) or a spreadsheet (Microsoft Excel 2007) to create a table for use in coursework reports and dissertations

Creating tables with Microsoft Word

Word-processed tables are suitable for text-intensive or number-intensive tables, although in the second case entering data can be laborious. When working in this way, the natural way to proceed is to create the 'shell' of the table, add the data, then carry out final formatting on the table.

1. **Move the cursor to the desired position in your document.** This is where you expect the top left corner of your table to appear. Go to the *Insert* tab, then choose *Table*.

2. **Select the appropriate number of columns and rows.** Don't forget to add rows and columns for headings. As default, a full-width table will appear, with single rulings for all cell boundaries, with all columns of equal width and all rows of equal height.

   Example of a 4 x 3 table:

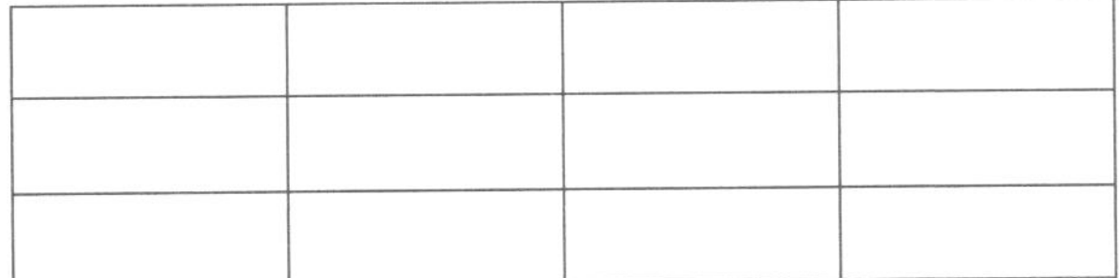

3. **Customise the columns.** By placing the cursor over the vertical rulings then 'dragging', you can adjust their width to suit your heading text entries, which should now be added.

Heading 1	Heading 2	Heading 3	Heading 4

4. **Work through the table adding the data**. Entries can be numbers or text.

Heading 1	Heading 2	Heading 3	Heading 4
xx	xx	xx	xx
xx	xx	xx	

5. **Make further adjustments to column and row widths to suit**. For example, if text fills several rows within a cell, consider increasing the column width, and if a column contains only single or double digit numbers, consider shrinking its width. To combine cells, first highlight them, then right-click and select *Merge Cells*. You may wish to reposition text within a cell using the formatting commands on the *Home* tab.

Heading 1	Heading 2	Heading 3	Heading 4
xx	xx	xx	xx
	xx	xx	xx

6. **Finally, remove selected borders to cells.** One way to do this is using the table borders options accessed from the *Design* tab > *borders* option on the toolbar, so that your table looks like the examples shown in this chapter.

7. **Add a table title.** This should be positioned *above* the table (contrast this with a figure title and legend p. 219).

   Final version of the table:

Table xx. Some data

Heading 1	Heading 2^a	Heading 3	Heading 4
Aaa	xx	xx	xx
	xx	xx	xx

^{a}An example of a footnote.

Creating tables with Microsoft Excel

Tables derived from spreadsheets are effective when you have lots of numerical data, especially when these are stored or created using the spreadsheet itself. When working in this way, you can design the table as part of an output or summary section of the spreadsheet, add explanatory headings, format, then possibly export to a word processor when complete.

1. **Design the output or summary section.** Plan this as if it were a table, including adding text headings within cells.

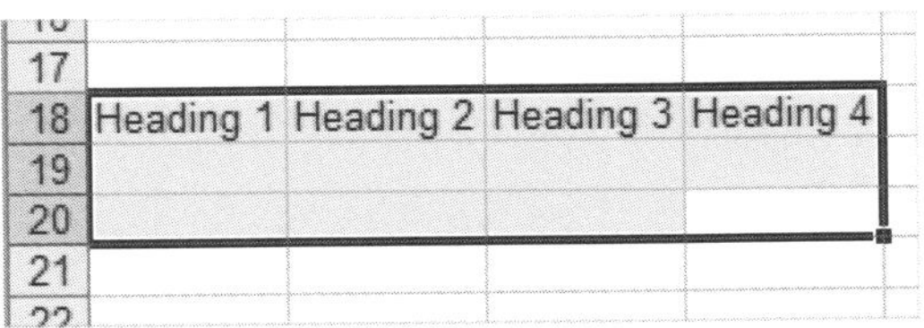

(continued)

Box 36.2 (continued)

2. **Insert appropriate formulae within cells to produce data.** If necessary, formulae should draw on the other parts of the spreadsheet.

17				
18	Heading 1	Heading 2	Heading 3	Heading 4
19	Aaa	=A1	=C3*5	=SDEV(A1:A12)
20	Bbb	=A2	=F45/G12	=SDEV(B1:B12)
21				
22				

3. **Format the cells.** This is important to control the number of decimal places presented (*Cells* tab > *Format* > *Format cells* > *Number*).

4. **Adjust column width to suit.** You can do this via the column headings, by placing the cursor over the rulings between columns then 'dragging'.

17				
18	Heading 1	Heading 2	Heading 3	Heading 4
19	Aaa	=A1	=C3*5	=SDEV(A1:A12)
20	Bbb	=A2	=F45/G12	=SDEV(B1:B12)
21				
22				

5. **Add rulings as appropriate.** Use the borders menu within the *Format cells* function as described above.

17				
18	Heading 1	Heading 2	Heading 3	Heading 4
19	Aaa	=A1	=C3*5	=SDEV(A1:A12)
20	Bbb	=A2	=F45/G12	=SDEV(B1:B12)
21				

6. **Add 'real' data values to the spreadsheet.** This should result in the summary values within the table being filled. Check that these are presented with the appropriate number of significant figures (p. 241).

7. **The table can now be copied and pasted to a Word document.** If you wish to link the spreadsheet and the word-processed document so that the latter is updated whenever changes are made to the spreadsheet values, then click *Paste* on the *Home* tab. Click *Paste special*, and then select *Paste link and Microsoft Office Excel Worksheet object.* Alternatively select *Insert tab* > *Table* > *Excel spreadsheet* and an Excel spreadsheet will open within the Word document. Click outside of the spreadsheet when you have finished formatting the table to continue in the Word document. The contents of the table can be accessed and changed by double-clicking within the table.

Sources for further study

Kirkup, L. (1994) *Experimental Methods: An Introduction to the Analysis and Presentation of Data.* Wiley, New York.

Simmonds, D. and Reynolds, L. (1994) *Data Presentation and Visual Literacy in Medicine and Science.* Butterworth-Heinemann, London.

Willis, J. (2004) *Data Analysis and Presentation Skills.* Wiley, New York.

37 Hints for solving numerical problems

The following boxes give advice on dealing with numerical procedures:

- Box 24.2: preparing solutions
- Box 26.1: molar concentrations
- Box 68.1: cell-counting chambers
- Box 68.2: plate (colony) counts
- Box 50.1: calibration curves

Biology often requires a numerical or statistical approach. Not only is mathematical modelling an important aid to understanding, but computations are often needed to turn raw data into meaningful information or to compare them with other data sets. Moreover, calculations are part of laboratory routine, perhaps required for making up solutions of known concentration (see p. 147 and below) or for the calibration of a microscope (see p. 180). In research, 'trial' calculations can reveal what input data are required and where errors in their measurement might be amplified in the final result (see p. 194).

KEY POINT If you find numerical work difficult, practice at problem-solving is especially important.

Practising at problem-solving:

- demystifies the procedures involved, which are normally just the elementary mathematical operations of addition, subtraction, multiplication and division (Table 37.1);
- allows you to gain confidence so that you don't become confused when confronted with an unfamiliar or apparently complex form of problem;
- helps you recognise the various forms a problem can take as, for instance, in crossing experiments in classical genetics.

Table 37.1 Sets of numbers and operations

Sets of numbers	
Whole numbers:	0, 1, 2, 3, ...
Natural numbers:	1, 2, 3, ...
Integers:	... −3, −2, −1, 0, 1, 2, 3, ...
Real numbers:	integers and anything between (e.g. −5, 4.376, 3/16, π, $\sqrt{5}$)
Prime numbers:	subset of natural numbers divisible by 1 and themselves only (i.e. 2, 3, 5, 7, 11, 13, ...)
Rational numbers:	p/q where p (integer) and q (natural) have no common factor (e.g. 3/4)
Fractions:	p/q where p is an integer and q is natural (e.g. −6/8)
Irrational numbers:	real numbers with no exact value (e.g. π)
Infinity:	(symbol ∞) is larger than any number (technically not a number as it does not obey the laws of algebra)
Operations and symbols	
Basic operators:	+, −, × and ÷ will not need explanation; however, / may substitute for ÷, * may substitute for × or this operator may be omitted
Powers:	a^n, i.e. 'a to the power n', means a multiplied by itself n times (e.g. $a^2 = a \times a$ = 'a squared', $a^3 = a \times a \times a$ = 'a cubed'). n is said to be the index or exponent. Note $a^0 = 1$ and $a^1 = a$
Logarithms:	the common logarithm (log) of any number x is the power to which 10 would have to be raised to give x (i.e. the log of 100 is 2; $10^2 = 100$); the antilog of x is 10^x. Note that there is no log for 0, so take this into account when drawing log axes by breaking the axis. Natural or Napierian logarithms (ln) use the base e (= 2.718 28 ...) instead of 10
Reciprocals:	the reciprocal of a real number a is $1/a$ ($a \neq 0$)
Relational operators:	$a > b$ means 'a is greater (more positive) than b', $<$ means less than, $\leqslant$ means less-than-or-equal-to and $\geqslant$ means greater-than-or-equal-to
Proportionality:	$a \propto b$ means 'a is proportional to b' (i.e. $a = kb$, where k is a constant). If $a \propto 1/b$, a is inversely proportional to b ($a = k/b$)
Sums:	Σx_i is shorthand for the sum of all x values from $i = 0$ to $i = n$ (more correctly the range of the sum is specified under the symbol)
Moduli:	$\|x\|$ signifies modulus of x, i.e. its absolute value (e.g. $\|4\| = \|-4\| = 4$)
Factorials:	$x!$ signifies factorial x, the product of all integers from 1 to x (e.g. $3! = 6$). Note $0! = 1! = 1$

Steps in tackling a numerical problem

Tracing errors in mathematical problems – this is always easier when all the stages in a calculation are laid out clearly.

Using a spreadsheet for numerical problems – this may be very useful in repetitive work or for 'what if?' case studies (see Chapter 11).

The step-by-step approach outlined below may not be the fastest method of arriving at an answer, but most mistakes occur where steps are missing, combined or not made obvious, so a logical approach is often better.

Have the right tools ready

Scientific calculators (p. 131) greatly simplify the numerical part of problem-solving. However, the seeming infallibility of the calculator may lead you to accept an absurd result that could have arisen because of faulty key-pressing or faulty logic. Make sure you know how to use all the features on your calculator, especially how the memory works; how to introduce a constant multiplier or divider; and how to obtain an exponent (note that the 'exp' button on most calculators gives you 10^x, not 1^x or y^x; so 1×10^6 would be entered as [1] [exp] [6], *not* [10] [exp] [6]).

Approach the problem thoughtfully

If the individual steps have been laid out on a worksheet, the 'tactics' will already have been decided. It is more difficult when you have to adopt a strategy on your own, especially if the problem is presented as a story and it isn't obvious which equations or rules need to be applied.

- Read the problem carefully as the text may give clues as to how it should be tackled. Be certain of what is required as an answer before starting.
- Analyse what kind of problem it is, which effectively means deciding which equation(s) or approach will be applicable. If this is not obvious, consider the dimensions/units of the information available and think how they could be fitted to a relevant formula. In examinations, a favourite ploy of examiners is to present a problem such that the familiar form of an equation must be rearranged (see Table 37.2 and Box 37.1). Another is to make you use two or more equations in series (see Box 37.2). If you are unsure whether a recalled formula is correct, a dimensional analysis can help: write in all the units for the variables and make sure that they cancel out to give the expected answer.
- Check that you have, or can derive, all of the information required to use your chosen equation(s). It is unusual but not unknown for examiners to supply redundant information. So, if you decide not to use some of the information given, be sure that you know why you do not require it.
- Decide in what format and units the answer should be presented. This is sometimes suggested to you. If the problem requires many changes in the prefixes to units, it is a good idea to convert all data to base SI units (multiplied by a power of 10) at the outset.
- If a problem appears complex, break it down into component parts.

Table 37.2 Simple algebra – rules for manipulating

If $a = b + c$, then $b = a - c$ and $c = a - b$
If $a = b \times c$, then $b = a \div c$ and $c = a \div b$
If $a = b^c$, then $b = a^{1/c}$ and $c = \log a \div \log b$
$a^{1/n} = \sqrt[n]{a}$
$a^{-n} = 1 \div a^n$
$a^b \times a^c = a^{(b+c)}$ and $a^b \div a^c = a^{(b-c)}$
$(a^b)^c = a^{(b \times c)}$
$a \times b = \text{antilog}(\log a + \log b)$

Present your answer clearly

Presenting calculations in assessed work – always show the steps in your calculations; most markers will only penalise a mistake once and part marks will be given if the remaining operations are performed correctly. This can only be done if those operations are visible. You should also remember to check whether your final answer seems reasonable.

The way you present your answer obviously needs to fit the individual problem. The example shown in Box 37.2 has been chosen to illustrate several important points, but this format would not fit all situations. Guidelines for presenting an answer include:

Box 37.1 Example of using the algebraic rules of Table 37.2

Problem: if $a = (b - c) \div (d + e^n)$, find e

1. Multiply both sides by $(d + e^n)$; formula becomes: $a(d + e^n) = (b - c)$
2. Divide both sides by a; formula becomes: $$d + e^n = \frac{b - c}{a}$$
3. Subtract d from both sides; formula becomes: $e^n = \frac{b - c}{a} - d$
4. Raise each side to the power $1/n$; formula becomes: $e = \left\{\frac{b - c}{a} - d\right\}^{1/n}$

(a) Make your assumptions explicit. Most mathematical models of biological phenomena require that certain criteria are met before they can be legitimately applied (e.g. 'assuming the tissue is homogeneous ...'), while some approaches involve approximations that should be clearly stated (e.g. 'to estimate the mouse's skin area, its body was approximated to a cylinder with radius x and height y ...').

(b) Explain your strategy for answering, perhaps giving the applicable formula or definitions that suit the approach to be taken. Give details of what the symbols mean (and their units) at this point.

(c) Rearrange the formula to the required form with the desired unknown on the left-hand side (see Table 37.2).

(d) Substitute the relevant values into the right-hand side of the formula, using the units and prefixes as given (it may be convenient to convert values to SI beforehand). Convert prefixes to appropriate powers of 10 as soon as possible.

Units – never write any answer without its unit(s) unless it is truly dimensionless.

(e) Convert to the desired units step by step, i.e. taking each variable in turn.

(f) When you have the answer in the desired units, rewrite the left-hand side and underline the answer for emphasis. Make sure that the result is presented to an appropriate number of significant figures (see below).

Check your answer

Having written out your answer, you should check it methodically, answering the following questions:

- Is the answer realistic? You should be alerted to an error if a number is absurdly large or small. In repeated calculations, a result standing out from others in the same series should be double-checked.
- Do the units make sense and match up with the answer required? Don't, for example, present a volume in units of m^2.
- Do you get the same answer if you recalculate in a different way? If you have time, recalculate the answer using a different 'route', entering the numbers into your calculator in a different form and/or carrying out the operations in a different order.

Rounding off to a specific number of significant figures – do not round off numbers until you arrive at the final answer or you will introduce errors into the calculation.

Rounding: decimal places and significant figures

In many instances, the answer you produce as a result of a calculation will include more figures than is justified by the accuracy and precision of the original data. Sometimes you will be asked to produce an answer to a

Box 37.2 Model answer to a typical biological problem

Problem

Estimate the total length and surface area of the fibrous roots on a maize seedling from measurements of their total fresh weight and mean diameter. Give your answers in m and cm^2 respectively to four significant figures.

Measurements

Fresh weight[a] = 5.00 g, mean diameter[b] = 0.5 mm.

Answer

Assumptions: (1) the roots are cylinders with constant radius[c] and the 'ends' have negligible area; (2) the root system has a density of 1 000 kg m^{-3} (i.e. that of water[d]).

Strategy: from assumption (1), the applicable equations are those concerned with the volume and surface area of a cylinder (Table 37.3), namely:

$$V = \pi r^2 h \qquad [37.1]$$
$$A = 2\pi r h \text{ (ignoring ends)} \qquad [37.2]$$

where V is volume (m^3), A is surface area (m^2), $\pi \approx 3.141\,59$, h is height (m) and r is radius (m). The total length of the root system is given by h and its surface area by A. We can find h by rearranging eqn [37.1] and then substitute its value in eqn [37.2] to get A.

To calculate total root length: rearranging eqn [37.1], we have $h = V/\pi r^2$. From measurements[e], $r = 0.25\text{ mm} = 0.25 \times 10^{-3}\text{ m}$.

From density = weight/volume,

$$V = \text{fresh weight/density}$$
$$= 5\text{ g}/1000\text{ kg m}^{-3}$$
$$= 0.005\text{ kg}/1000\text{ kg m}^{-3}$$
$$= 5 \times 10^{-6}\text{ m}^3$$

Total root length,

$$h = V/\pi r^2$$
$$5 \times 10^{-6}\text{ m}^3/3.141\,59 \times (0.25 \times 10^{-3}\text{ m})^2$$

$\therefore$ Total root length = 25.46 m

To calculate surface area of roots: substituting value for h obtained above into eqn [37.2], we have:

Root surface area

$$= 2 \times 3.141\,59 \times 0.25 \times 10^{-3}\text{ m} \times 25.46\text{ m}$$
$$= 0.04\text{ m}^2$$
$$= 0.04 \times 10^4\text{ cm}^2$$
(there being $100 \times 100 = 10^4$ cm^2 per m^2)

$\therefore$ Root surface area = 400.0 cm^2

Notes

(a) The fresh weight of roots would normally be obtained by washing the roots free of soil, blotting them dry and weighing.

(b) In a real answer you might show the replicate measurements giving rise to the mean diameter.

(c) In reality, the roots will differ considerably in diameter and each root will not have a constant diameter throughout its length.

(d) This will not be wildly inaccurate as about 95 per cent of the fresh weight will be water, but the volume could also be estimated from water displacement measurements.

(e) Note conversion of measurements into base SI units at this stage and on line 3 of the root volume calculation. Forgetting to halve diameter measurements where radii are required is a common error.

specified number of decimal places or significant figures and other times you will be expected to decide for yourself what would be appropriate.

KEY POINT Do not simply accept the numerical answer from a calculator or spreadsheet, without considering whether you need to modify this to give an appropriate number of significant figures or decimal places.

Rounding to n *decimal places*

This is relatively easy to do.

1. Look at the number to the right of the *n*th decimal place.
2. If this is less than five, simply 'cut off' all numbers to the right of the *n*th decimal place to produce the answer (i.e. round down).

Examples
The number 4.123 correct to two decimal places is 4.12
The number 4.126 correct to two decimal places is 4.13
The number 4.1251 correct to two decimal places is 4.13
The number 4.1250 correct to two decimal places is 4.12
The number 4.1350 correct to two decimal places is 4.14
The number 99.99 correct to one decimal place is 100.0.

Examples
The number of significant figures in 194 is 3
The number of significant figures in 2305 is 4
The number of significant figures in 0.003482 is 4
The number of significant figures in 210×10^8 is 3 (21×10^9 would be 2).

Examples
The number of significant figures in 3051.93 is 6
To five significant figures, this number is 3051.9
To four significant figures, this number is 3052
To three significant figures, this number is 3050
To two significant figures, this number is 3100
To one significant figure, this number is 3000
3051.93 to the nearest 10 is 3050
3051.93 to the nearest 100 is 3100
Note that in this last case you must include the zeros before the decimal point to indicate the scale of the number (even if the decimal point is not shown). For a number less than 1, the same would apply to the zeros before the decimal point. For example, 0.00305193 to three significant figures is 0.00305. Alternatively, use scientific notation (in this case, 3.05×10^{-3}).

3. If the number is greater than five, 'cut off' all numbers to the right of the *n*th decimal place and add one to the *n*th decimal place to produce the answer (i.e. round up).
4. If the number is 5, then look at further numbers to the right to determine whether to round up or not.
5. If the number is *exactly* 5 and there are no further numbers to the right, then round to the nearest even number. *Note:* When considering a large number of calculations, this procedure will not affect the overall mean value. Some rounding systems do the opposite to this (i.e. round to the nearest odd number), while others always round up where the number is exactly 5 (which *will* affect the mean). Take advice from your tutor and stick to one system throughout a series of calculations.

Whenever you see any number quoted, you should assume that the last digit has been rounded. For example, in the number 22.4, the '.4' is assumed to be rounded and the calculated value may have been between 22.35 and 22.45.

Quoting to n *significant figures*

The number of significant figures indicates the degree of approximation in the number. For most cases, it is given by counting all the figures except zeros that occur at the beginning or end of the number. Zeros *within* the number are always counted as significant. The number of significant figures in a number like 200 is ambiguous and could be one, two or three; if you wish to specify clearly, then quote as, e.g., 2×10^2 (one significant figure), 2.0×10^2 (two significant figures), etc. to avoid spurious accuracy (pp. 193, 234). When quoting a number to a specified number of significant figures, use the same rules as for rounding to a specified number of decimal places, but do not forget to keep zeros before or after the decimal point. The same principle is used if you are asked to quote a number to the 'nearest 10', 'nearest 100', etc.

When deciding for yourself how many significant figures to use, adopt the following rules of thumb:

- Always round *after* you have done a calculation. Use *all* significant figures available in the measured data during a calculation.
- If adding or subtracting with measured data, then quote the answer to the number of decimal places in the data value with the least number of decimal places (e.g. $32.1 - 45.67 + 35.6201 = 22.1$, because 32.1 has one decimal place).
- If multiplying or dividing with measured data, keep as many significant figures as are in the number with the least number of significant places (e.g. $34901 \div 3445 \times 1.3410344 = 13.59$, because 3445 has four significant figures).
- For the purposes of significant figures, assume 'constants' (e.g. number of mm in a metre) have an infinite number of significant figures.

Some reminders of basic mathematics

Errors in calculations sometimes appear because of faults in mathematics rather than computational errors. For reference purposes, Tables 37.1–37.3

Table 37.3 Geometry and trigonometry – analysing shapes

Shape/object	Diagram	Perimeter	Area
Two-dimensional shapes			
Square	x, x	$4x$	x^2
Rectangle	x, y	$2(x + y)$	xy
Circle	r	$2\pi r$	πr^2
Ellipse	a, b	$\pi[1.5(a + b) - \sqrt{a} * b]$ (approx.)	πab
Triangle (general)	x, y, h, z	$x + y + z$	$0.5zh$
(right-angled)	r, θ, y, x	$x + y + r$ $\sin\theta = y/r$, $\cos\theta = x/r$, $\tan\theta = y/x$; $r^2 = x^2 + y^2$	$0.5xy$
Shape/object	**Diagram**	**Surface area**	**Volume**
Three-dimensional shapes			
Cube	x, x, x	$6x^2$	x^3
Cuboid	x, y, z	$2xy + 2xz + 2yz$	xyz
Sphere	r	$4\pi r^2$	$4\pi r^3/3$
Ellipsoid	a, b	no simple formula	πrab
Cylinder	h, r	$2\pi rh + 2\pi r^2$	$\pi r^2 h$
Cone and pyramid	L, B, P	$0.5PL + B$	$BL/3$

Key: x, y, z = sides a, b = half minimum and maximum axes; r = radius or hypotenuse; h = height; B = base area; L = perpendicular height; P = perimeter of base.

give some basic mathematical principles that may be useful. Eason *et al.* (1992) or Stephenson (2003) should be consulted for more advanced/specific needs.

Examples
1/8 as a percentage is $1 \div 8 \times 100 = 100 \div 8 = 12.5\%$
0.602 as a percentage is $0.602 \times 100 = 60.2\%$.

Percentages and proportions

A percentage is just a fraction expressed in terms of hundredths, indicated by putting the percentage sign (%) after the number of hundredths. So 35% simply means 35 hundredths. To convert a fraction to a percentage, just multiply the fraction by 100. When the fraction is in decimal form, multiplying by 100 to obtain a percentage is easily achieved just by moving the decimal point two places to the right.

Examples
190% as a decimal fraction is $190 \div 100 = 1.9$
5/2 as a percentage is $5 \div 2 \times 100 = 250\%$.

To convert a percentage to a fraction, just remember that, since a percentage is a fraction multiplied by 100, the fraction is the percentage divided by 100. For example: 42% = 42/100 = 0.42. In this example, since we are dealing with a decimal fraction, the division by 100 is just a matter of moving the decimal point two places to the left (42% could be written as 42.0%). Percentages greater than 100% represent fractions greater than 1. Percentages less than 1 may cause confusion. For example, 0.5% means half of one per cent (decimal fraction 0.005) and must not be confused with 50% (which is the decimal fraction 0.5).

Example
A population falls from 4 million to 3.85 million. What is the percentage change? The decrease in numbers is $4 - 3.85 = -0.15$ million. The fractional decrease is $-0.15 \div 4 = -0.0375$ and we multiply by 100 to get the percentage change = minus 3.75%.

To find a percentage of a given number, just express the percentage as a decimal fraction and multiply the given number. For example: 35% of 500 is given by $0.35 \times 500 = 175$. To find the percentage change in a quantity, work out the difference (= value 'after' – value 'before'), and divide this difference by the original value to give the fractional change, then multiply by 100.

Exponents

Exponential notation is an alternative way of expressing numbers in the form a^n ('a to the power n'), where a is multiplied by itself n times. The number a is called the base and the number n the exponent (or power or index). The exponent need not be a whole number, and it can be negative if the number being expressed is less than 1. See Table 37.2 for other mathematical relationships involving exponents.

Example $2^3 = 2 \times 2 \times 2 = 8$.

Scientific notation

In scientific notation, also known as 'standard form', the base is 10 and the exponent a whole number. To express numbers that are not whole powers of 10, the form $c \times 10^n$ is used, where the coefficient c is normally between 1 and 10. Scientific notation is valuable when you are using very large numbers and wish to avoid suggesting spurious accuracy. Thus if you write 123 000, this may suggest that you know the number to ±0.5, whereas 1.23×10^5 might give a truer indication of measurement accuracy (i.e. implied to be ±500 in this case). Engineering notation is similar, but treats numbers as powers of 10 in groups of 3, i.e. $c \times 10^0$, 10^3, 10^6, 10^9, etc. This corresponds to the SI system of prefixes (p. 199).

Example Avogadro's number, ≈ 602 352 000 000 000 000 000 000, is more conveniently expressed as $6.023\,52 \times 10^{23}$.

A useful property of powers when expressed to the same base is that when multiplying two numbers together, you simply add the powers, while if dividing, you subtract the powers. Thus, suppose you counted 8 bacteria in a known volume of a 10^{-7} dilution (see p. 149), there would be 8×10^7 in the same volume of undiluted solution; if you now dilute this 500-fold (5×10^2), then the number present in the same volume would be $8/5 \times 10^{(7-2)} = 1.6 \times 10^5 = 160\,000$.

Examples The logarithm to the base 10 ($\log_{10}$) of 1000 is 3, since $10^3 = 1000$. The logarithm to the base e ($\log_e$ or ln) of 1000 is 6.907755 (to six decimal places).

Examples (use to check the correct use of your own calculator)
102963 as a $\log_{10} = 5.012681$ (to six decimal places)
$10^{5.012681} = 102962.96$
(Note loss of accuracy due to loss of decimal places.)
102963 as a natural logarithm (ln) = 11.542125 (to six decimal places), thus $2.718282^{11.542125} = 102963$.

Logarithms

When a number is expressed as a logarithm, this refers to the power n that the base number a must be raised to give that number. Any base could be used, but the two most common are 10, when the power is referred to as $\log_{10}$ or simply log, and the constant e (2.718282), used for mathematical convenience in certain situations, when the power is referred to as $\log_e$ or ln (natural logarithm). Note that (a) logs need not be whole numbers; (b) there is no log value for the number zero; and (c) that $\log_{10} = 0$ for the number 1.

To obtain logs, you will need to use the log key on your calculator, or special log tables (now largely redundant). To convert back (antilog), use

- the $\boxed{10^x}$ key, with x = log value;
- the $\boxed{\text{inverse}}$ then the $\boxed{\text{log}}$ key; or
- the $\boxed{y^x}$ key, with $y = 10$ and x = log value.

If you have used log tables, you will find complementary antilogarithm tables to do this.

There are many uses of logarithms in biology, including pH ($= -\log[H^+]$), where $[H^+]$ is expressed in mol l^{-1} (see p. 173); the exponential growth of micro-organisms, where if log(cell number) is plotted against time, a straight-line relationship is obtained; and allometric studies of growth and development, where if data are plotted on log axes, a series of straight-line relationships may be found.

Linear functions and straight lines

One of the most straightforward and widely used relationships between two variables x and y is that represented by a straight-line graph, where the corresponding mathematical function is known as the equation of a straight line, where:

$$y = a + bx \qquad [37.3]$$

In this relationship, a represents the intercept of the line on the y (vertical) axis, i.e. where $x = 0$, and therefore $bx = 0$, while b is equivalent to the slope (gradient) of the line, i.e. the change in y for a change in x of 1. The constants a and b are sometimes given alternative symbols, but the mathematics remains unchanged, e.g. in the equivalent expression for the slope of a straight line, $y = mx + c$. Figure 37.1 shows what happens when these two constants are changed, in terms of the resultant straight lines.

The two main applications of the straight-line relationship are:

1. Function fitting. Here, you determine the mathematical form of the function, i.e. you estimate the constants a and b from a data set for x and y, either by drawing a straight line by eye (p. 228) and then working out the slope and y intercept, or by using linear regression (p. 269) to obtain the most probable values for both constants. When putting a straight line of best fit by eye on a hand-drawn graph, note the following:
   - Always use a *transparent* ruler, so you can see data points on either side of the line.
   - For a data series where the points do not fit a perfect straight line, try to have an equal number of points on either side of the line, as

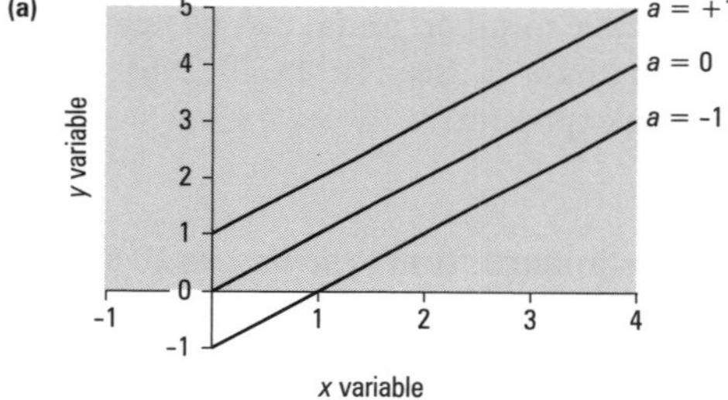

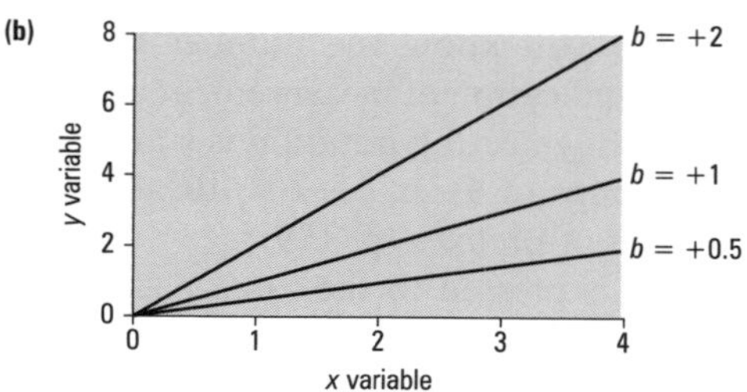

Fig. 37.1. Straight-line relationships ($y = a + bx$), showing the effects of (a) changing the intercept at constant slope, and (b) changing the slope at constant intercept.

in Fig. 37.2(a), and try to minimise the average distance of these points from the line.

- Once you have drawn the line of best fit use this line, rather than your data values, in all subsequent procedures (e.g. in a calibration curve, Chapter 50).
- Tangents drawn to a curve give the slope (gradient) at a particular point, e.g. in an enzyme reaction progress curve, Chapter 62. These are best drawn by bringing your ruler up to the curve at the exact point where you wish to estimate the slope and then trying to make the two angles immediately on either side of this point approximately the same, by eye (Fig. 37.2(b)).
- Once you have drawn the straight line or tangent, choose two points reasonably far apart at either end of your line and then draw construction lines to represent the change in y and the change in x between these two points: make sure that your construction lines are perpendicular to each other. Determine the slope as the change in y divided by the change in x (Fig. 37.2).

2. Prediction. Where a and b are known, or have been estimated, you can use eqn [37.3] to predict any value of y for a specified value of x, e.g. during exponential growth of a cell culture (p. 474), where $\log_{10}$ cell number (y) increases as a linear function of time (x): note that in this example the dependent variable has been transformed to give a linear relationship. You will need to rearrange eqn [37.3] in cases where a prediction of x is required for a particular value of y (e.g. in calibration curves, p. 336, or bioassays, pp. 477–9), as follows:

$$x = (y - a) \div b \qquad [37.4]$$

This equation can also be used to determine the intercept on the x (horizontal) axis, i.e. where $y = 0$.

Examples
Using eqn [37.3], the predicted value for y for a linear function where $a = 2$ and $b = 0.5$, where $x = 8$ is:
$y = 2 + (0.5 \times 8) = 6$.
Using eqn [64.4], the predicted value for x for a linear function where $a = 1.5$ and $b = 2.5$, where $y = -8.5$ is:
$x = (-8.5 - 1.5) \div 2.5 = -4$.
Using eqn [37.4] the predicted x intercept for a linear function where $a = 0.8$ and $b = 3.2$ is:
$x = (0 - 0.8) \div 3.2 = -0.25$.

Hints for some typical problems

Calculations involving proportions or ratios

The 'unitary method' is a useful way of approaching calculations involving proportions or ratios, such as those required when making up solutions from stocks (see also Chapter 26) or as a subsidiary part of longer calculations.

1. If given a value for a multiple, work out the corresponding value for a single item or unit.
2. Use this 'unitary value' to calculate the required new value.

Example A lab schedule states that 5 g of a compound with a relative molecular mass of 220 are dissolved in 400 mL of solvent. For writing up your materials and methods, you wish to express this as mol L^{-1}.

1. If there are 5 g in 400 mL, then there are 5 ÷ 400 g in 1 mL.
2. Hence, 1000 mL will contain 5 ÷ 400 × 1000 g = 12.5 g.
3. 12.5 g = 12.5 ÷ 220 mol = 0.0568 mol, so [solution] = 56.8 mmol L^{-1} (= 56.8 mol m^{-3}).

Calculations involving series

Series (used in, e.g., dilutions, see also pp. 148–50) can be of three main forms:

1. Arithmetic, where the *difference* between two successive numbers in the series is a constant, e.g. 2, 4, 6, 8, 10, . . .
2. Geometric, where the *ratio* between two successive numbers in the series is a constant, e.g. 1, 10, 100, 1000, 10 000, . . .

Examples For a geometric dilution series involving ten-fold dilution steps, calculation of concentrations is straightforward, e.g. two serial decimal dilutions (= 100-fold dilution) of a solution of NaCl of 250 mmol L^{-1} will produce a dilute solution of 250 ÷ 100 = 2.5 mmol L^{-1}. Similarly, for an arithmetic dilution series, divide by the overall dilution to give the final concentration, e.g. a 16-fold dilution of a solution of NaCl of 200 mg mL^{-1} will produce a dilute solution of 200 ÷ 16 = 12.5 mg mL^{-1}.

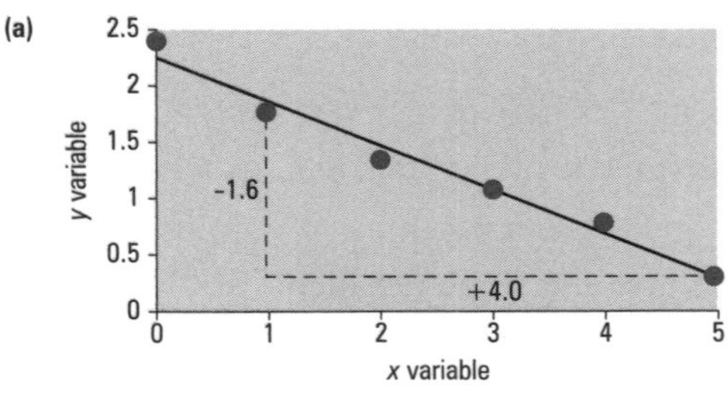

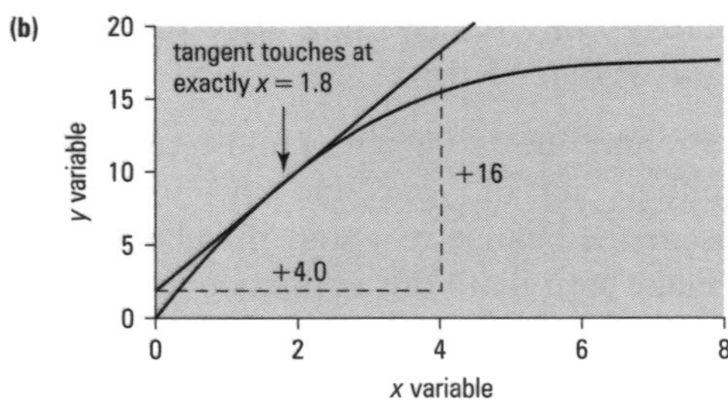

Fig. 37.2. Drawing straight lines. (a) Simple linear relationship, giving a straight line with an intersect of 2.3 and a slope of −1.6 ÷ 4.0 = 0.4. (b) Tangent drawn to a curve at x = 1.8, giving a slope of 16 ÷ 4 = 4.

3. Harmonic, where the values are reciprocals of successive whole numbers, e.g. $1, \frac{1}{2}, \frac{1}{3}, \frac{1}{4}, \ldots$

Note that the logs of the numbers in a geometric series will form an arithmetic series (e.g. 0, 1, 2, 3, 4,... in the above case). Thus, if a quantity y varies with a quantity x such that the rate of change in y is proportional to the value of y (i.e. it varies in an exponential manner), a semi-log plot of such data will form a straight line. This form of relationship is relevant for exponentially growing cell cultures and radioactive decay.

Statistical calculations

The need for long, complex calculations in statistics has largely been removed because of the widespread use of spreadsheets with statistical functions (Chapter 11) and more specialised programs such as SPSS. It is, however, important to understand the principles behind what you are trying to do (see Chapters 38 and 39) and interpret the program's output correctly, either using the 'help' function or a reference manual.

Text references

Eason, G., Coles, C.W. and Gettinby, G. (1992) *Mathematics and Statistics for the Bio-Sciences*. Ellis Horwood, Chichester.

Stephenson, F.H. (2003) *Calculations in Molecular Biology and Biotechnology*. Academic Press, London.

Sources for further study

Anon. *S.O.S. Mathematics*. Available: http://www.sosmath.com/
Last accessed 21/12/10.
[A basic Web-based guide with very wide coverage.]

Britton, N.F. (2005) *Essential Mathematical Biology*. Springer. New York.
[Describes a range of mathematical applications in population dynamics, epidemiology, genetics, biochemistry and medicine.]

Cann, A.J. (2002) *Maths from Scratch for Biologists*. Wiley, Chichester.
[Deals with basic manipulations, formulae, units, molarity, logs and exponents as well as basic statistical procedures.]

Causton, D.R. (1992) *A Biologist's Basic Mathematics*. Cambridge University Press, Cambridge.

Forster, P.C. (2003) *Easy Mathematics for Biologists*. Taylor and Francis, London.

Harris, M., Taylor, G. and Taylor, J. (2005) *Catch Up Maths and Stats for the Life and Medical Sciences*. Scion, Bloxham.
[Covers a range of basic mathematical operations and statistical procedures.]

Koehler, K.R. *College Physics for Students of Biology and Chemistry*. Available: http://www.rwc.uc.edu/koehler/biophys/text.html.
Last accessed 21/12/10.
[A 'hypertextbook' written for first-year undergraduates. Assumes that you have a working knowledge of algebra.]

Lawler, G. (2007) *Understanding Maths. Basic Mathematics Explained*. Studymates, Abergele.

38 Descriptive statistics and survey

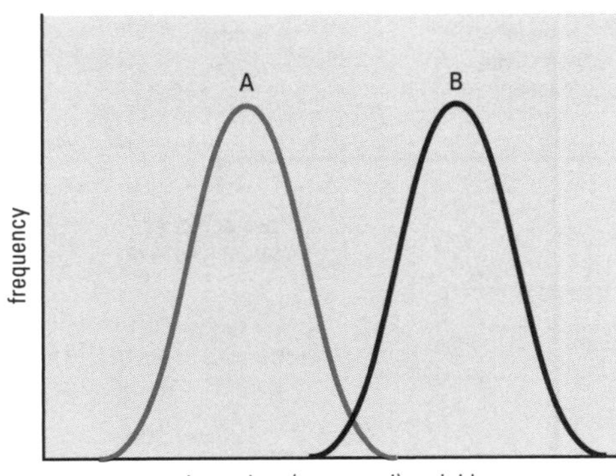

Fig. 38.1 Two distributions with different locations but the same dispersion. The data set labelled B could have been obtained by adding a constant to each datum in the data set labelled A.

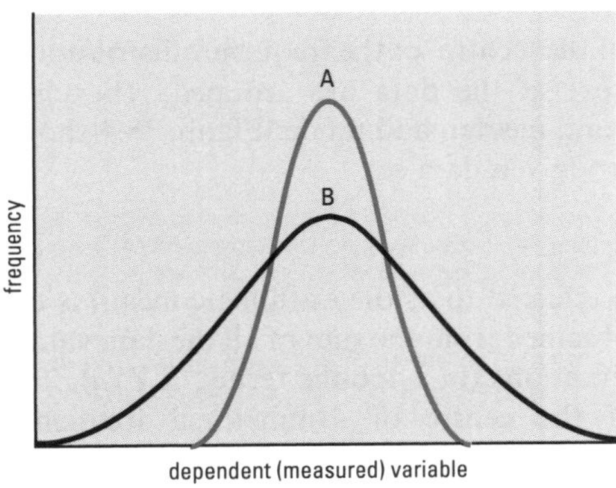

Fig. 38.2 Two distributions with different dispersions but the same location. The data set labelled A covers a relatively narrow range of values of the dependent (measured) variable, while that labelled B covers a wider range.

Example Box 38.1 shows a set of data and the calculated values of the measures of location, dispersion and shape for which methods of calculation are outlined here. Check your understanding by calculating the statistics yourself and confirming that you arrive at the same answers.

As scientists, and particularly in the field of life sciences, you will appreciate that the data which is generated, either experimentally or through observation, will exhibit a degree of variability. This is due to either experimental variation, as in the case of replicated experiments, or it is due to biological variability between humans (or animals). For these reasons it is necessary to be able to appropriately handle and manipulate the raw data which are generated, for example, so that they can be presented in the form of tables/figures or to carry out appropriate descriptive/summary statistics. You can use these statistics to:

- condense a large data set for presentation in figures or tables;
- provide estimates of parameters of the frequency distribution of the population being sampled (p. 260).

KEY POINT The appropriate descriptive statistics to choose depend on both the type of data, i.e. quantitative, ranked or qualitative, and the nature of the underlying population frequency distribution.

If you have no clear theoretical grounds for assuming what the underlying frequency distribution is like, graph one or more sample frequency distributions, ideally with a sample size >100. In smaller projects you can also base your decisions on information from published research.

The methods used to calculate descriptive statistics depend on whether data have been grouped into classes. You should use the original data set if it is still available, because grouping into classes loses information and accuracy. However, large data sets may make calculations unwieldy and are best handled using computer programs.

Three important features of a frequency distribution that can be summarised by descriptive statistics are:

- the sample's location, i.e. its position along a given dimension representing the dependent (measured) variable (Fig. 38.1);
- the dispersion of the data, i.e. how spread out the values are (Fig. 38.2);
- the shape of the distribution, i.e. whether symmetrical, skewed, U-shaped, etc. (Fig. 38.3).

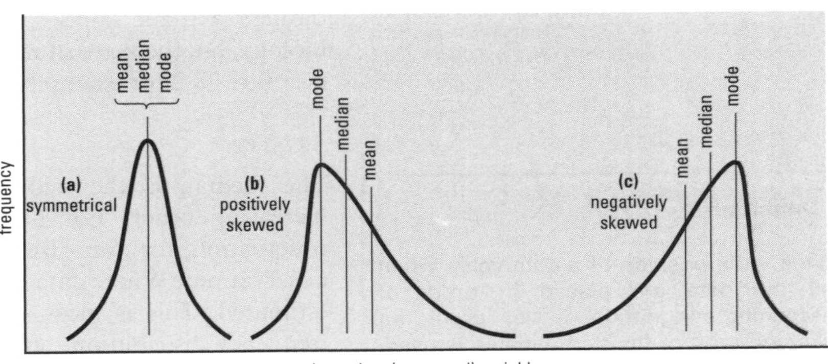

Fig. 38.3 Symmetrical and skewed frequency distributions, showing relative positions of mean, median and mode.

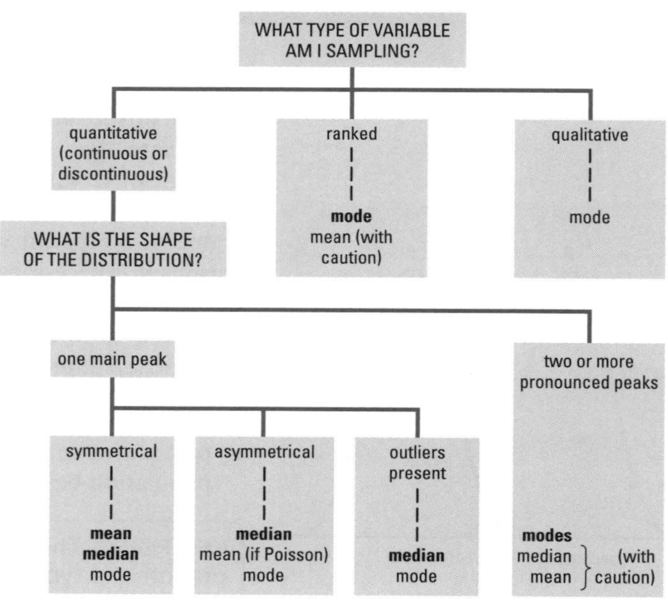

Fig. 38.4 Choosing a statistic for characterising a distribution's location. Statistics written in bold are the preferred option(s).

Use of symbols – Y is used in Chapters 38 and 39 to signify the dependent variable in statistical calculations (following the example of Sokal and Rohlf, 1994, Heath, 1995 and Wardlaw, 2000). Note, however, that some authors use X or x in analogous formulae and many calculators refer to, e.g., $\bar{x}$, Σx^2, etc., for their statistical functions.

Measuring location

Here, the objective is to pinpoint the 'centre' of the frequency distribution, i.e. the value about which most of the data are grouped. The chief measures of location are the mean, median and mode. Figure 38.4 shows how to choose among these for a given data set.

Mean

The mean (denoted $\bar{Y}$ and also referred to as the arithmetic mean) is the average value of the data. It is obtained from the sum of all the data values divided by the number of observations (in symbolic terms, $\Sigma\, Y/n$). The mean is a good measure of the centre of symmetrical frequency distributions. It uses all of the numerical values of the sample and therefore incorporates all of the information content of the data. However, the value of a mean is greatly affected by the presence of outliers (extreme values). The arithmetic mean is a widely used statistic in biology, but there are situations when you should be careful about using it (see Box 38.2 for examples).

Median

Definition

Rank – the position of a data value when all the data are placed in order of ascending magnitude. If ties occur, an average rank of the tied variates is used. Thus, the rank of the datum 6 in the sequence 1,3,5,6,8,8,10 is 4; the rank of each datum with value 8 is 5.5.

The median is the mid-point of the observations when ranked in increasing order. For odd-sized samples, the median is the middle observation; for even-sized samples it is the mean of the middle pair of observations. Where data are grouped into classes, the median can only be estimated. This is most simply done from a graph of the cumulative frequency distribution, but can also be worked out by assuming the data to be evenly spread within the class. The median may represent the location of the main body of data better than the mean when the distribution is asymmetric or when there are outliers in the sample.

Box 38.1 Descriptive statistics for an illustrative sample of data

Value (Y)	Frequency (f)	Cumulative frequency	fY	fY^2
1	0	0	0	0
2	1	1	2	4
3	2	3	6	18
4	3	6	12	48
5	8	14	40	200
6	5	19	30	180
7	2	21	14	98
8	0	21	0	0
Totals	$21 = \Sigma f\ (=n)$		$104 = \Sigma fY$	$548 = \Sigma fY^2$

In this example, for simplicity and ease of calculation, integer values of Y are used. In many practical exercises, where continuous variables are measured to several significant figures and where the number of data values is small, giving frequencies of 1 for most of the values of Y, it may be simpler to omit the column dealing with frequency and list all the individual values of Y and Y^2 in the appropriate columns. To gauge the underlying frequency distribution of such data sets, you would need to group individual data into broader classes (e.g. all values between 1.0 and 1.9, all values between 2.0 and 2.9, etc.) and then draw a histogram (p. 228). Calculation of certain statistics for data sets that have been grouped in this way (e.g. median, quartiles, extremes) can be tricky and a statistical text should be consulted.

Statistic	Value*	How calculated
Mean	4.95	$\Sigma fY/n$, i.e. 104/21
Median	5	Value of the $(n+1)/2$ variate, i.e. the value ranked $(21+1)/2 = 11$th (obtained from the cumulative frequency column)
Mode	5	The most common value (Y value with highest frequency)
Upper quartile	6	The upper quartile is between the 16th and 17th values, i.e. the value exceeded by 25% of the data values
Lower quartile	4	The lower quartile is between the 5th and 6th values, i.e. the value exceeded by 75% of the data values
Semi-interquartile range	1.0	Half the difference between the upper and lower quartiles, i.e. $(6-4)/2$
Upper extreme	7	Highest Y value in data set
Lower extreme	2	Lowest Y value in data set
Range	5	Difference between upper and lower extremes
Variance (s^2)	1.65	$s^2 = \dfrac{\Sigma fY^2 - (\Sigma fY)^2/n}{n-1} = \dfrac{548 - (104)^2/21}{20}$
Standard deviation (s)	1.28	$\sqrt{s^2}$
Standard error (SE)	0.280	$s/\sqrt{n}$
95% confidence limits	4.36 – 5.54	$\bar{Y} \mp t_{0.05}[20] \times \text{SE}$, (where $t_{0.05}[20] = 2.09$, Table 66.2)
Coefficient of variation (CoV)	25.9%	$100s/\bar{Y}$

*Rounded to three significant figures (see p. 241), except when it is an exact number.

> **Definition**
>
> **An outlier** – any datum that has a value much smaller or bigger than most of the data.

Mode

The mode is the most common value in the sample. The mode is easily found from a tabulated frequency distribution as the most frequent value. If data have been grouped into classes then the term modal class is used for the class containing most values. The mode provides a rapidly and easily found estimate of sample location and is unaffected by outliers. However, the mode is affected by chance variation in the shape of a sample's distribution and it may lie distant from the obvious centre of the distribution.

The mean, median and mode have the same units as the variable under discussion. However, whether these statistics of location have the same or similar values for a given frequency distribution depends on the symmetry and shape of the distribution. If it is near-symmetrical with a single peak, all three will be very similar; if it is skewed or has more than one peak, their values will differ to a greater degree (see Fig. 38.3).

> **Describing the location of qualitative data** – the mode is the only statistic that is suitable for this task. For example, 'the modal (most frequent) eye colour was brown'.

Measuring dispersion

Here, the objective is to quantify the spread of the data about the centre of the distribution. Figure 38.5 indicates how to decide which measure of dispersion to use.

Range

> **Example** In a sample of data with values 3, 7, 15, 8, 5, 10 and 4, the range is 12 (i.e. the difference between the highest value, 15, and the lowest value, 3).

The range is the difference between the largest and smallest data values in the sample (the extremes) and has the same units as the measured variable. The range is easy to determine, but is greatly affected by outliers. Its value may also depend on sample size: in general, the larger this is, the greater

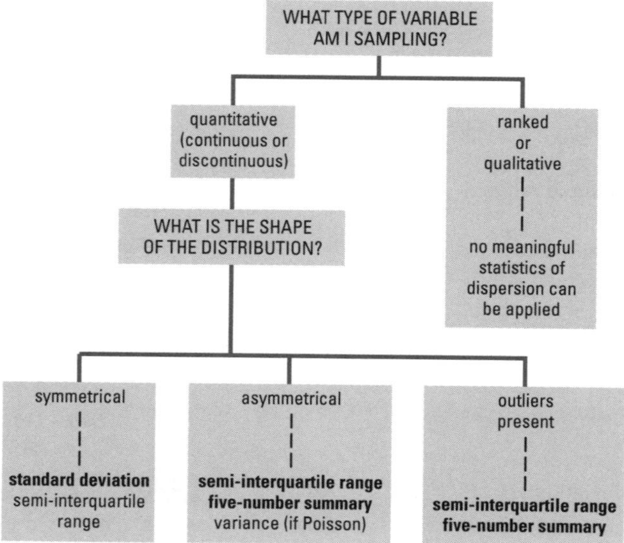

Fig. 38.5 Choosing a statistic for characterising a distribution's dispersion. Statistics written in bold are the preferred option(s). Note that you should match statistics describing dispersion with those you have used to describe location, i.e. standard deviation with mean, semi-interquartile range with median.

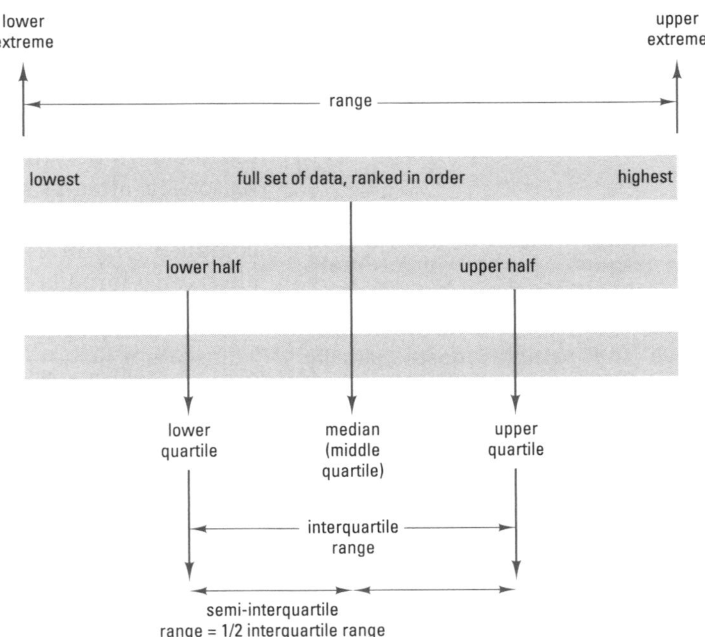

Fig. 38.6 Illustration of median, quartiles, range and semi-interquartile range.

will be the range. These features make the range a poor measure of dispersion for many practical purposes.

Semi-interquartile range

The semi-interquartile range is an appropriate measure of dispersion when a median is the appropriate statistic to describe location. For this, you need to determine the first and third quartiles, i.e. the medians for those data values ranked below and above the median of the whole data set (see Fig. 38.6). To calculate a semi-interquartile range for a data set:

1. Rank the observations in ascending order.
2. Find the values of the first and third quartiles.
3. Subtract the value of the first quartile from the value of the third.
4. Halve this number.

For data grouped in classes, the semi-interquartile range can only be estimated. Another disadvantage is that it takes no account of the shape of the distribution at its edges. This objection can be countered by using the so-called 'five-number summary' of a data set, which consists of the three quartiles and the two extreme values; this can be presented on graphs as a box and whisker plot (see Fig. 38.7) and is particularly useful for summarising skewed frequency distributions. The corresponding 'six-number summary' includes the sample's size.

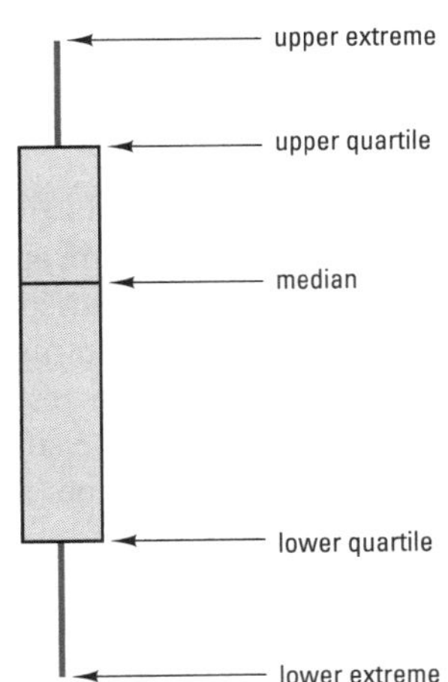

Fig. 38.7 A box and whisker plot, showing the 'five-number summary' of a sample as it might be used on a graph.

Variance and standard deviation

For symmetrical frequency distributions, an ideal measure of dispersion would take into account each value's deviation from the mean and provide a measure of the average deviation from the mean. Two such statistics are the sample variance, which is the sum of squares ($\Sigma(Y - \bar{Y})^2$) divided by $n - 1$ (where n is the sample size), and the sample standard deviation, which is the positive square root of the sample variance.

The variance (s^2) has units that are the square of the original units, while the standard deviation (s or SD) is expressed in the original units, one reason s is often preferred as a measure of dispersion. Calculating s or s^2 longhand is a tedious job and is best done with the help of a calculator or computer. If you do not have a calculator that calculates s for you, an alternative formula that simplifies calculations is:

$$s = +\sqrt{\frac{\Sigma Y^2 - (\Sigma Y)^2/n}{n - 1}} \qquad [38.1]$$

To calculate s using a calculator:

1. Obtain ΣY, square it, divide by n and store in memory.
2. Square Y values, obtain ΣY^2, subtract memory value from this.
3. Divide this answer by $n - 1$.
4. Take the positive square root of this value.

Take care to retain significant figures, or errors in the final value of s will result. If continuous data have been grouped into classes, the class mid-values or their squares must be multiplied by the appropriate frequencies before summation (see example in Box 38.1). When data values are large, longhand calculations can be simplified by coding the data, e.g. by subtracting a constant from each datum, and decoding when the simplified calculations are complete (see Sokal and Rohlf, 1994).

Using a calculator for statistics – make sure you understand how to enter individual data values and which keys will give the sample mean (usually shown as $\bar{X}$ or $\bar{x}$) and sample standard deviation (often shown as σ_{n-1}). In general, you should not use the population standard deviation (usually shown as σ_n).

Example Consider two methods of bioassay for a toxin in fresh water. For a given standard, Method A gives a mean result of = 50 'response units' with s = 8, while Method B gives a mean result of = 160 'response units' with s = 18. Which bioassay gives the more reproducible results? The answer can be found by calculating the CoV values, which are 16 per cent and 11.25 per cent respectively. Hence, Method B is the more precise, even though the absolute value of s is larger.

Coefficient of variation

The coefficient of variation (CoV) is a dimensionless measure of variability relative to location that expresses the sample standard deviation, usually as a percentage of the sample mean, i.e.

$$\text{CoV} = 100s/\bar{Y}\,(\%) \qquad [38.2]$$

This statistic is useful when comparing the relative dispersion of data sets with widely differing means or where different units have been used for the same or similar quantities. Table 38.1 gives a list of common symbols and abbreviations.

A useful application of the CoV is to compare different analytical methods or procedures, so that you can decide which involves the least proportional error – create a standard stock solution, then base your comparison on the results from several subsamples analysed by each method. You may find it useful to use the CoV to compare the precision of your own results with those of a manufacturer, e.g. for an autopippettor (p. 155). The smaller the CoV, the more precise (repeatable) is the apparatus or technique (note: this does not mean that it is necessarily more *accurate*, see p. 193).

Table 38.1 Commonly used symbols (descriptive statistics)

Term (also known as)	Symbol or abbreviations used
Sample mean (arithmetic mean, average)	$\bar{x}$ (x bar)[1] or $\bar{y}$ (y bar)
Population mean	μ (mu)
Standard error (standard error of the sample mean)	SE or SEM
Sample standard deviation	s
Population standard deviation	σ (sigma)
Sample variance	s^2
Sum	Σ (capital sigma)
Count (sample size)	n
Coefficient of variation	CoV or CV

[1]The letter used will depend on the letter used to describe the data set.

Measuring the precision of the sample mean as an estimate of the true value using the standard error

Most practical exercises are based on a limited number of individual data values (a sample) that are used to make inferences about the population from which they were drawn. For example, the haemoglobin content might be measured in blood samples from 100 adult females and used as an estimate of the adult female haemoglobin content, with the sample mean ($\bar{Y}$) and sample standard deviation (s) providing estimates of the true values of the underlying population mean (μ) and the population standard deviation (σ). The reliability of the sample mean as an estimate of the true (population) mean can be assessed by calculating a statistic termed the standard error of the sample mean (often abbreviated to standard error or SE), from:

$$\mathrm{SE} = s/\sqrt{n} \qquad [38.3]$$

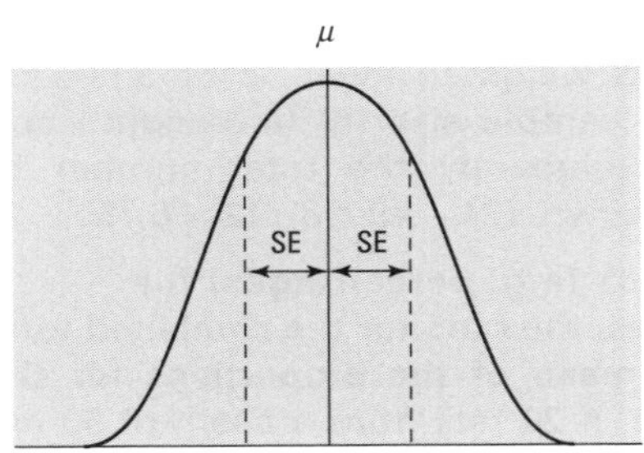

Fig. 38.8 Frequency distribution of sample means around the population mean (μ). Note that SE is equivalent to the standard deviation of the sample means, for sample size = *n*.

Strictly, the standard error is an estimate of the dispersion of repeated sample means around the true (population) value: if several samples were taken, each with the same number of data values (n), then their means would cluster around the population mean (μ) with a standard deviation equal to SE, as shown in Fig. 38.8. Therefore, the *smaller* the SE, the more reliable the sample mean is likely to be as an estimate of the true value, since the underlying frequency distribution would be more tightly clustered around μ. At a practical level, eqn [38.3] shows that SE is directly affected by the dispersion of individual data values within the sample, as represented by the sample standard deviation (s). Perhaps more importantly, SE is inversely related to the *square root* of the number of data values (n). Therefore, if you wanted to increase the precision of a sample mean by a factor of 2 (i.e. to reduce SE by half), you would have to increase n by a factor of 2^2 (i.e. four-fold).

Summary descriptive statistics for the sample mean are often quoted as $\bar{Y} \pm \mathrm{SE}\,(n)$, with the SE being given to one significant figure more than the mean. For example, summary statistics for the sample mean and standard error for the data shown in Box 38.1 would be quoted as 4.95 ± 0.280 ($n = 21$).You can use such information to carry out a t-test between two sample means (Box 39.1); the SE is also useful because it allows calculation of confidence limits for the sample mean (p. 267).

Describing the 'shape' of frequency distributions

Frequency distributions may differ in the following characteristics:

- number of peaks;
- skewness or asymmetry;
- kurtosis or pointedness.

The shape of a frequency distribution of a small sample is affected by chance variation and may not be a fair reflection of the underlying population frequency distribution: check this by comparing repeated samples from the same population or by increasing the sample size. If the original shape were due to random events, it should not appear consistently in repeated samples and should become less obvious as sample size increases.

Genuinely bimodal or polymodal distributions may result from the combination of two or more unimodal distributions, indicating that more than one underlying population is being sampled (Fig. 38.9). An example of a bimodal distribution is the height of adult humans (females and males combined).

Box 38.2 Three examples where simple arithmetic means are inappropriate

Mean	n
6	4
7	7
8	1

1. **If means of samples are themselves meaned, an error can arise if the samples are of different size.** For example, the arithmetic mean of the means in the table shown left is 7, but this does not take account of the different 'reliabilities' of each mean due to their sample sizes. The correct weighted mean is obtained by multiplying each mean by its sample size (n) (a 'weight') and dividing the sum of these values by the total number of observations, i.e. in the case shown, $(24 + 49 + 8)/12 = 6.75$.
2. **When making a mean of ratios (e.g. percentages) for several groups of different sizes, the ratio for the combined total of all the groups is not the mean of the proportions for the individual groups.** For example, if 20 rats from a batch of 50 are male, this implies 40 per cent are male. If 60 rats from a batch of 120 are male, this implies 50 per cent are male. The mean percentage of males $(50 + 40)/2 = 45\%$ is *not* the percentage of males in the two groups combined, because there are $20 + 60 = 80$ males in a total of 170 rats = 47.1% approx.
3. **If the measurement scale is not linear, arithmetic means may give a false value.** For example, if three media had pH values 6, 7 and 8, the appropriate mean pH is not 7 because the pH scale is logarithmic. The definition of pH is $-\log_{10}[H^+]$, where $[H^+]$ is expressed in mol L^{-1} ('molar'); therefore, to obtain the true mean, convert data into $[H^+]$ values (i.e. put them on a linear scale) by calculating $10^{(-\text{pH value})}$ as shown. Now calculate the mean of these values and convert the answer back into pH units. Thus, the appropriate answer is pH 6.43 rather than 7. Note that a similar procedure is necessary when calculating statistics of dispersion in such cases, so you will find these almost certainly asymmetric about the mean.

pH value	$[H^+]$ (mol L^{-1})
6	1×10^{-6}
7	1×10^{-7}
8	1×10^{-8}
mean	3.7×10^{-7}
$-\log_{10}$ mean	6.43

Mean values of log-transformed data are often termed geometric means – they are sometimes used in microbiology and in cell culture studies, where log-transformed values for cell density counts are averaged and plotted, rather than using the raw data values. The use of geometric means in such circumstances serves to reduce the effects of outliers on the mean.

A distribution is skewed if it is not symmetrical, a symptom being that the mean, median and mode are not equal (Fig. 38.3). Positive skewness is where the longer 'tail' of the distribution occurs for higher values of the measured variable; negative skewness where the longer tail occurs for

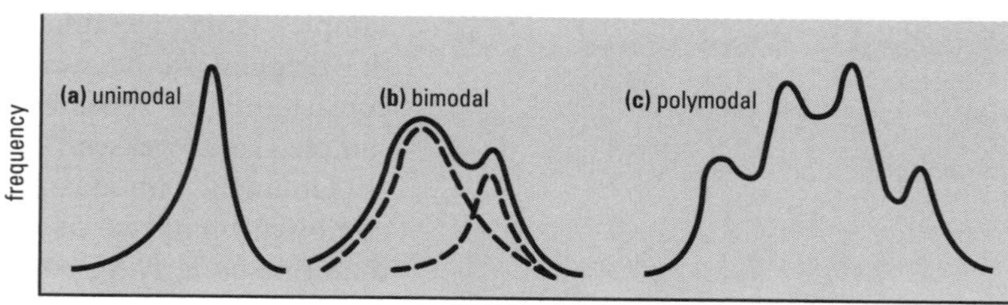

Fig. 38.9 Frequency distributions with different numbers of peaks. A unimodal distribution (a) may be symmetrical or asymmetrical. The dotted lines in (b) indicate how a bimodal distribution could arise from a combination of two underlying unimodal distributions. Note here how the term 'bimodal' is applied to any distribution with two major peaks – their frequencies do not have to be exactly the same.

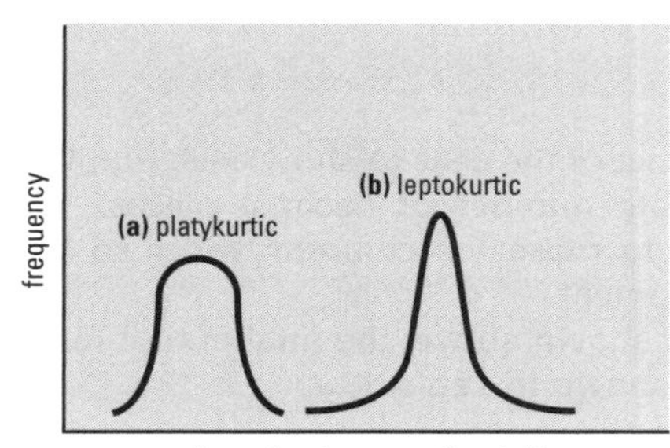

Fig. 38.10 Examples of the two types of kurtosis.

lower values. Some biological examples of characteristics distributed in a skewed fashion are volumes of plant protoplasts, insulin levels in human plasma and bacterial colony counts.

Kurtosis is the name given to the 'pointedness' of a frequency distribution. A platykurtic frequency distribution is one with a flattened peak, while a leptokurtic frequency distribution is one with a pointed peak (Fig. 38.10). While descriptive terms can be used, based on visual observation of the shape and direction of skew, the degree of skewness and kurtosis can be quantified and statistical tests exist to test the 'significance' of observed values (see Sokal and Rohlf, 1994), but the calculations required are complex and best done with the aid of a computer.

Box 38.3 How to use a spreadsheet (Microsoft Excel 2007) to calculate descriptive statistics

Method 1: Using spreadsheet functions to generate the required statistics.
Suppose you had obtained the following set of data, stored within an array (block of columns and rows) of cells (A2:L6) within a spreadsheet:

	A	B	C	D	E	F	G	H	I	J	K	L
1	My data set											
2	4	4	3	3	5	4	3	7	7	3	5	3
3	6	2	9	7	3	4	5	6	6	9	4	8
4	5	3	2	5	4	5	7	2	8	3	6	3
5	11	3	5	2	4	3	7	8	4	4	4	3
6	3	6	8	5	6	4	3	4	3	6	10	5

The following functions (*formulas* tab) could be used to extract descriptive statistics from this data set:

Descriptive statistic	Example of use of function[a,b]	Result for the above data set
Sample size *n*	=COUNT((A2:L6)	60
Mean	=AVERAGE(A2:L6)[c]	4.9
Median	=MEDIAN(A2:L6)	4.0
Mode	=MODE(A2:L6)	3
Upper quartile	=QUARTILE(A2:L6,3)[d]	6.0
Lower quartile	=QUARTILE(A2:L6,1)	3.0
Semi-interquartile range	=QUARTILE(A2:L6,3)-QUARTILE(A2:L6,1)	3.0
Upper extreme	=QUARTILE(A2:L6,4) *or* =MAX(A2:L6)	11
Lower extreme	=QUARTILE(A2:L6,0) *or* =MIN(A2:L6)	2
Range	=MAX(A2:L6)- MIN(A2:L6)[e]	9.0
Variance	=VAR(A2:L6)	4.464
Standard deviation	=STDEV(A2:L6)	2.113
Standard error	=STDEV(A2:L6)/(SQRT(COUNT(A2:L6)))[f]	0.273
Coefficient of variation	=100*STDEV(A2:L6)/AVERAGE(A2:L6)	43.12%

Notes:
[a] Typically, in an appropriate cell, you would use *Formulas > Insert function > COUNT*, then select the input range and press return.
[b] Other descriptive statistics can be calculated – these mirror those shown in Box 3.1, but for this specific data set.
[c] There is no function termed 'MEAN' in Microsoft Excel.
[d] The first argument within the brackets relates to the array of data, the second relates to the quartile required (consult the *Help* feature for further information).
[e] There is no direct 'RANGE' function in Microsoft Excel.
[f] There is no direct 'STANDARD ERROR' function in Microsoft Excel. The SQRT function returns a square root and the COUNT function determines the number of filled data cells in the array.

(continued)

Box 38.3 (continued)

Method 2: Using the *Data* tab, then > *Data Analysis* option within the *Analysis* group

This can automatically generate a table of descriptive statistics for the data array selected, although the data must be presented as a single row or column. This option might need to be installed for your network or personal computer before it is available to you (in the latter case use the *Office* button, then select *Excel options* > *Add-Ins* > *Analysis ToolPak* from the menu – consult the *Help* feature for details). Having entered or rearranged your data into a row or column, the steps involved are as follows:

1. Select *Data* > *Data Analysis.*
2. From the *Data Analysis box,* select *Descriptive Statistics.*
3. Input your data location into the *Input Range* (left-click and hold down to highlight the column of data).
4. From the menu options, select *Summary Statistics* and *Confidence Level for Mean: 95%.*
5. When you click *OK* you should get a new worksheet, with descriptive statistics and confidence limits shown. Alternatively, at step 3, you can select an area of your current worksheet as a data output range (select an area away from any existing content as these cells would otherwise be overwritten by the descriptive statistics output table).
6. Change the format of the cells to show each number to an appropriate number of decimal places. You may also wish to make the columns wider so you can read their content.
7. For the data set shown above, the final output table should look as shown in Table 38.2.

Table 38.2 Descriptive statistics for a data set.

Column1[a,b]	
Mean	4.9
Standard error	0.27
Median	4.0
Mode	3
Standard deviation	2.113
Sample variance	4.464
Kurtosis	0.22
Skewness	0.86
Range	9.00
Minimum	2.0
Maximum	11.0
Sum	294
Count	60
Confidence level (95.0%)	0.55

Notes:
[a] These descriptive statistics are specified (and are automatically presented in this order) – any others required can be generated using Method 1.
[b] A more descriptive heading can be added if desired – this is the default.

Using computers to calculate descriptive statistics

There are many specialist statistical packages (e.g. SPSS) that can be used to simplify the process of calculation of statistics. Note that correct interpretation of the output requires an understanding of the terminology used and the underlying process of calculation, and this may best be obtained by working through one or more examples by hand before using these tools. Spreadsheets offer increasingly sophisticated statistical analysis functions, some examples of which are provided in Box 38.3 for Microsoft Excel 2007.

Definition

Epidemiology – the observational study of factors which affect health and illness incidence in populations.

The use of surveys in food science and nutrition

Surveys are used for a number of purposes. Large-scale surveys are conducted to try to identify links between dietary variables and disease, or at least with risk factors associated with a disease. This epidemiological data can be used to make comparisons between different populations or within the same population but over time.

Smaller-scale surveys are done to measure opinion or preferences of food, such as in taste panels in product development. Food diaries can be regarded as a type of survey (Chapter 41).

Dietary surveys

Large-scale dietary surveys are an important source of information about the diets of population. Many countries carry out national surveys of the diets of the population, the results of which are an important resource in developing dietary guidelines. They are also used to inform food and nutrition policy. UK examples are the National Diet & Nutrition Survey and the Expenditure and Food Survey.

Planning a survey

> **Definition**
>
> **Quantitative data** – numerical data, e.g. grams of saturated fat consumed per day.
>
> **Qualitative data** – data that are non-numerical, e.g. focus group discussion (see Chapter 31).

The key questions to ask yourself are:

- **What are you trying to find out?** When planning a survey it is important to consider what it is you are trying to find out. This may seem obvious but if you are not clear about exactly what you want to know then it is very likely that your results will be equally unclear. This is part of the process of formulating your hypothesis (Chapter 39).
- **Who is the target population?** In planning a survey it is also necessary to know who will be the target audience. This may a specific age range or a group of people with a particular set of disease risk factors. The challenge is to identify an appropriate sample so that you achieve the goals you have set. You will also need to consider how many people you will need to ask in order to be as sure as you can be that the findings you present are an accurate reflection of the habits or opinions of the population. Sample bias is virtually impossible to avoid but being aware of potential bias in your sample will go some way to overcoming the problem at the planning stage. An example of bias would be only questioning females and assuming the results represent the entire population.
- **How are you going to collect the data?** The process of data collection will dictate the format of the questions and will depend on the target population and the purpose of the survey. Surveys can be conducted by face-to-face interviews, telephone, Internet or mail.
- **What type of questions are you going to ask?** When you are designing your survey or questionnaire think about how you will analyse the results. If you think about your analysis at this stage it will make your task much easier in the long term. If the data you generate are not numerical (quantitative) how will you analyse and describe this in your report?

A common mistake is to ask too many questions which are not relevant to what you are actually trying to find out; ask yourself why you are asking every question and discard that question if you cannot justify it. Including questions you don't need increases the time it takes to complete the survey and this might reduce the number of people who are willing to take part.

Pilot your survey

A pilot study is where you 'test drive' your methods. This process should include all stages, from recruiting to analysis. This will ensure that you are using the appropriate method for your requirements and will also confirm that the instructions are clear. Although your pilot study will only be done on a few people you should analyse the data so that you are confident you can present the data in tables and graphs.

Ethical considerations

Although not directly related to analysis, if you are doing any research which involves people you need to consider the impact their participation may have (Chapter 23). Careful consideration of how much personal information you need is also important.

Coding the survey

Coding is the process of transforming your survey responses into a format which you can analyse. Each question you ask will have a number but each response should also be given a number or code. If the data are not quantitative you could code it in binary form so that Excel or a statistics software package can be used in the analysis, e.g. if the question asks whether the respondent is male or female convert to 0 for male, 1 for female. Often, in questionnaires or surveys you will see numbers in subscript beside the boxes you have been asked to tick; this is coding. It is important to remember that these codes should be used appropriately in your analysis so in the above male or female example, you could calculate how many respondents were male and how many were females, but obviously it would not be appropriate to calculate a mean score of the ones and zeroes. Once the data are coded and entered into a spreadsheet you can use descriptive statistics to describe the data and further statistical analysis to test your hypothesis.

Text references and sources for further study

Heath, D. (1995) *An Introduction to Experimental Design and Statistics for Biology*. UCL Press, London.

Schmuller, J. (2009) *Statistical Analysis with Excel for Dummies*. Wiley, Hoboken, New Jersey.

Sokal, R.R. and Rohlf, F.J. (1994) *Biometry*, 3rd edn. W.H. Freeman, San Francisco, California.

Sokal, R.R. and Rohlf, F.J. (2009) *Introduction to Biostatistics*. Dover Publications, Mineola, New York.

Wardlaw, A.C. (2000) *Practical Statistics for Experimental Biologists*, 2nd edn. Wiley, New York.

This chapter outlines the philosophy of hypothesis-testing statistics, indicates the steps to be taken when choosing a test, and discusses features and assumptions of some important tests. For details of the mechanics of tests, consult appropriate texts (e.g. Sokal and Rohlf, 1994). Most tests are now available in statistical packages for computers (see p. 79) and many in spreadsheets (p. 72).

To carry out a statistical test:

1. Decide what it is you wish to test (create a null hypothesis H_0 and its alternative H_1).
2. Determine whether your data fit a standard distribution pattern.
3. Select a test and apply it to your data.

Setting up a null hypothesis

Hypothesis-testing statistics are used to compare the properties of samples either with other samples or with some theory about them. For instance, you may be interested in whether two samples can be regarded as having different means, whether the counts of an organism in different quadrats can be regarded as randomly distributed, or whether property A of an organism is linearly related to property B.

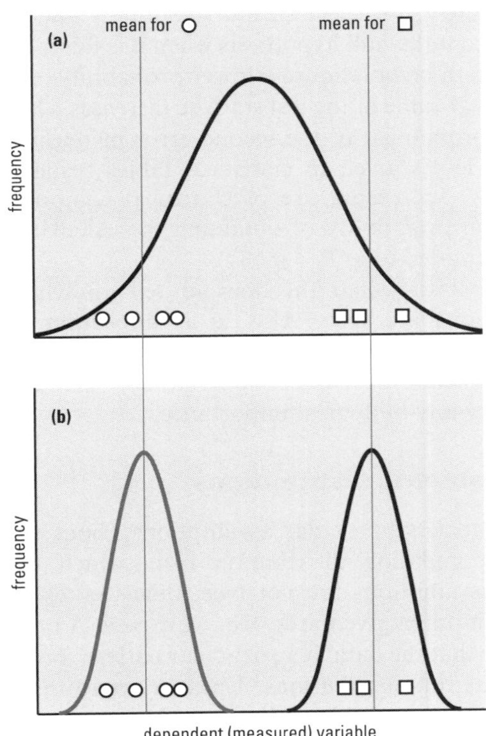

Fig. 39.1 Two explanations for the difference between two means. In case (a) the two samples happen by chance to have come from opposite ends of the same frequency distribution, i.e. there is no true difference between the samples. In case (b) the two samples come from different frequency distributions, i.e. there is a true difference between the samples. In both cases, the means of the two samples are the same.

> **KEY POINT** You can't use statistics to prove any hypothesis, but they can be used to assess how likely it is to be wrong.

Statistical testing operates in what at first seems a rather perverse manner. Suppose you think a treatment has an effect. The theory you actually test is that it has no effect; the test tells you how improbable your data would be if this theory were true. This 'no effect' theory is the null hypothesis. If your data are very improbable under the null hypothesis, then you may suppose it to be wrong, and this would support your original idea (the 'alternative hypothesis'). The concept can be illustrated by an example. Suppose two groups of subjects were treated in different ways, and you observed a difference in the mean value of the measured variable for the two groups. Can this be regarded as a 'true' difference? As Fig. 39.1 shows, it could have arisen in two ways:

- Because of the way the subjects were allocated to treatments, i.e. all the subjects liable to have high values might, by chance, have been assigned to one group and those with low values to the other (Fig. 39.1(a)).
- Because of a genuine effect of the treatments, i.e. each group came from a distinct frequency distribution (Fig. 39.1(b)).

A statistical test will indicate the probabilities of these options. The null hypothesis states that the two groups come from the same population (i.e. the treatment effects are negligible in the context of random variation). To test this, you calculate a test statistic from the data, and compare it with tabulated critical values giving the probability of obtaining the observed or a more extreme result by chance (see Boxes 39.1 and 39.2). This probability is sometimes called the significance of the test.

Note that you must take into account the degrees of freedom (d.f.) when looking up critical values of most test statistics. The d.f. is related to the size(s) of the samples studied; formulae for calculating it depend on the test being used. Biologists normally use two-tailed tests, i.e. we have no expectation beforehand that the treatment will have a positive or negative effect compared with the control (in a one-tailed test we expect one particular treatment to be bigger than the other). Be sure to use critical values for the correct type of test.

Definition

Modulus – the absolute value of a number, e.g. modulus $-3.385 = 3.385$.

By convention, the critical probability for rejecting the null hypothesis is 5 per cent (i.e. $P = 0.05$). This means we reject the null hypothesis if the observed result would have come up by chance a maximum of one time in 20. If the modulus of the test statistic is less than or equal to the tabulated critical value for $P = 0.05$, then we accept the null hypothesis and the result is said to be 'not significant' (NS for short). If the modulus of the test statistic is greater than the tabulated value for $P = 0.05$, then we reject the null hypothesis in favour of the alternative hypothesis that the treatments had different effects and the result is 'statistically significant'.

Quoting significance – the convention for quoting significance levels in text, tables and figures is as follows:

$P > 0.05$ = 'not significant' (or NS)

$P \leqslant 0.05$ = 'significant' (or *)

$P \leqslant 0.01$ = 'highly significant' (or **)

$P \leqslant 0.001$ = 'very highly significant' (or ***)

Thus, you might refer to a difference in means as being 'highly significant ($P \leqslant 0.01$)'. For this reason, the word 'significant' in its everyday meaning of 'important' or 'notable' should be used with care in scientific writing.

Two types of error are possible when making a conclusion on the basis of a statistical test. The first occurs if you reject the null hypothesis when it is true and the second if you accept the null hypothesis when it is false. To limit the chance of the first type of error, choose a lower probability, e.g. $P = 0.01$, but note that the critical value of the test statistic increases when you do this and results in the probability of the second error increasing. The conventional significance levels given in statistical tables (usually 0.05, 0.01, 0.001) are arbitrary. Increasing use of statistical computer programs now allows the actual probability of obtaining the calculated value of the test statistic to be quoted (e.g. $P = 0.037$).

Note that if the null hypothesis is rejected, this does not tell you which of many alternative explanations are true. Also, it is important to distinguish between statistical significance and biological relevance: identifying a statistically significant difference between two samples doesn't mean that this will carry any biological importance.

Comparing data with parametric distributions

A parametric test is one that makes particular assumptions about the mathematical nature of the population distribution from which the samples were taken. If these assumptions are not true, then the test is obviously invalid, even though it might give the answer we expect. A non-parametric test does not assume that the data fit a particular pattern, but it may assume some things about the distributions. Used in appropriate circumstances, parametric tests are better able to distinguish between true but marginal differences between samples than their non-parametric equivalents (i.e. they have greater 'power').

Choosing between parametric and non-parametric tests – always plot your data graphically when determining whether they are suitable for parametric tests as this may save a lot of unnecessary effort later.

The distribution pattern of a set of data values may be biologically relevant, but it is also of practical importance because it defines the type of statistical tests that can be used. The properties of the main distribution types found in biology are given below with both rules of thumb and more rigorous tests for deciding whether data fit these distributions.

Binomial distributions

These apply to samples of any size from populations when data values occur independently in only two mutually exclusive classes (e.g. type A

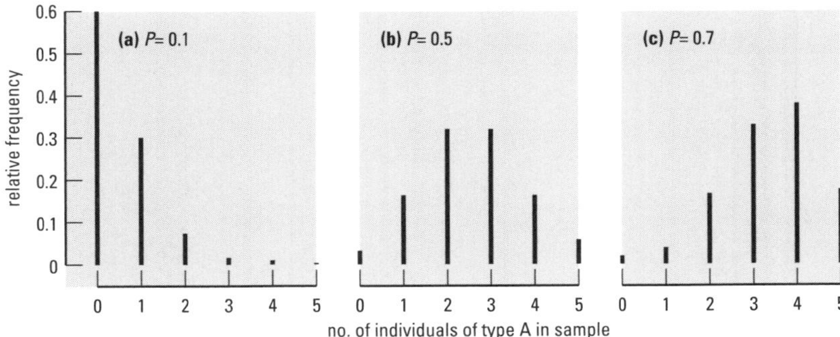

Fig. 39.2 Examples of binomial frequency distributions with different probabilities. The distributions show the expected frequency of obtaining n individuals of type A in a sample of 5. Here P is the probability of an individual being type A rather than type B.

or type B). They describe the probability of finding the different possible combinations of the attribute for a specified sample size k (e.g. out of 10 specimens, what is the chance of 8 being type A?). If p is the probability of the attribute being of type A and q the probability of it being type B, then the expected mean sample number of type A is kp and the standard deviation is $\sqrt{kpq}$. Expected frequencies can be calculated using mathematical expressions (see Sokal and Rohlf, 1994). Examples of the shapes of some binomial distributions are shown in Fig. 39.2. Note that they are symmetrical in shape for the special case $p = q = 0.5$ and the greater the disparity between p and q, the more skewed the distribution.

Some biological examples of data likely to be distributed in binomial fashion are: possession of two alleles for seed-coat morphology (e.g. smooth and wrinkly); whether an organism is infected with a microbe or not; whether an animal is male or female. Binomial distributions are particularly useful for predicting gene segregation in Mendelian genetics and can be used for testing whether combinations of events have occurred more frequently than predicted (e.g. more siblings being of the same sex than expected). To establish whether a set of data is distributed in binomial fashion: calculate expected frequencies from probability values obtained from theory or observation, then test against observed frequencies using a χ^2 test or a G test (see Sokal and Rohlf, 1994).

Tendency towards the normal distribution – under certain conditions, binomial and Poisson distributions can be treated as normally distributed:

- where samples from a binomial distribution are large (i.e. > 15) and p and q are close to 0.5;
- for Poisson distributions, if the number of counts recorded in each outcome is greater than about 15.

Poisson distributions

These apply to discrete characteristics that can assume low whole-number values, such as counts of events occurring in area, volume or time. The events should be 'rare' in that the mean number observed should be a small proportion of the total that could possibly be found. Also, finding one count should not influence the probability of finding another. The shape of Poisson distributions is described by only one parameter, the mean number of events observed, and has the special characteristic that the variance is equal to the mean. The shape has a pronounced positive skewness at low mean counts, but becomes more and more symmetrical as the mean number of counts increases (Fig. 39.3).

Some examples of characteristics distributed in a Poisson fashion are: number of plants in a quadrat; number of microbes per unit volume of

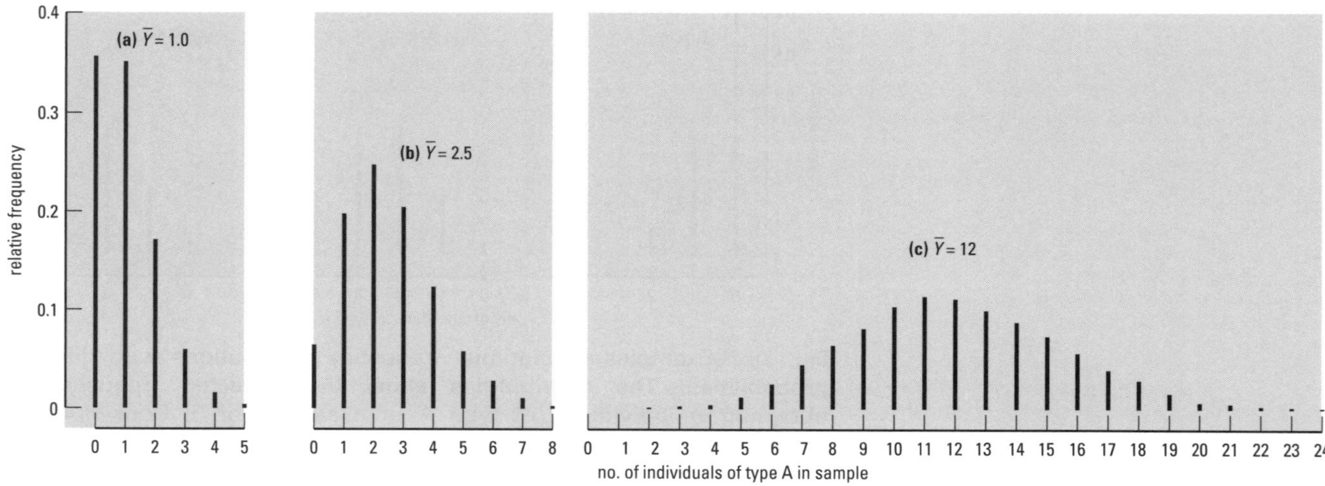

Fig. 39.3 Examples of Poisson frequency distributions differing in mean. The distributions are shown as line charts because the independent variable (events per sample) is discrete.

medium; number of animals parasitised per unit time; number of radioactive disintegrations per unit time. One of the main uses for the Poisson distribution is to quantify errors in count data such as estimates of cell densities in dilute suspensions. To decide whether data are Poisson distributed:

> **Definition**
>
> **Coefficient of dispersion** = $s^2/\bar{Y}$. This is an alternative measure of dispersion to the coefficient of variation (p. 252).

- Use the rule of thumb that if the coefficient of dispersion ≈ 1, the distribution is likely to be Poisson.
- Calculate 'expected' frequencies from the equation for the Poisson distribution and compare with actual values using a χ^2 test or a *G*-test.

It is sometimes of interest to show that data are *not* distributed in a Poisson fashion, e.g. the distribution of parasite larvae in hosts. If $s^2/\bar{Y} > 1$, the data are 'clumped' and occur together more than would be expected by chance; if $s^2/\bar{Y} < 1$, the data are 'repulsed' and occur together less frequently than would be expected by chance.

Normal distributions (Gaussian distributions)

These occur when random events act to produce variability in a continuous characteristic (quantitative variable). This situation occurs frequently in biology, so normal distributions are very useful and much used. The bell-like shape of normal distributions is specified by the population mean and standard deviation (Fig. 39.4): it is symmetrical and configured such that 68.27 per cent of the data will lie within ± 1 standard deviation of the mean, 95.45 per cent within ± 2 standard deviations of the mean, and 99.73 per cent within ± 3 standard deviations of the mean.

Some biological examples of data likely to be distributed in a normal fashion are: fresh weight of plants of the same age; linear dimensions of bacterial cells; height of either adult female or male humans. To check whether data come from a normal distribution, you can:

- Use the rule of thumb that the distribution should be symmetrical and that nearly all the data should fall within $\pm 3s$ of the mean and about two-thirds within $\pm 1s$ of the mean.

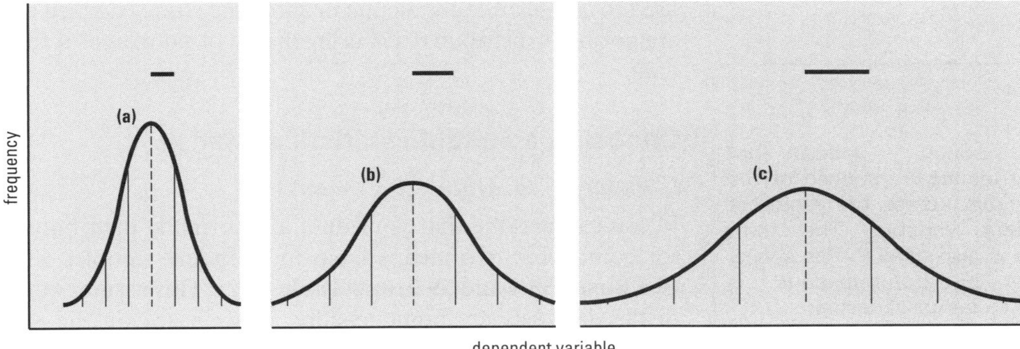

Fig. 39.4 Examples of normal frequency distributions differing in mean and standard deviation. The horizontal bars represent population standard deviations for the curves, increasing from (a) to (c). Vertical dashed lines are population means, while vertical solid lines show positions of values ±1, 2 and 3 standard deviations from the means.

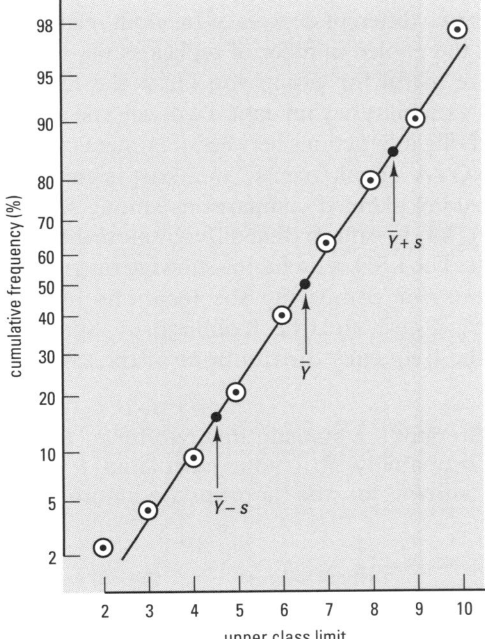

Fig. 39.5 Example of a normal probability plot. The plotted points are from a small data set where the mean $\bar{Y} = 6.93$ and the standard deviation $s = 1.895$. Note that values corresponding to 0% and 100% cumulative frequency cannot be used. The straight line is that predicted for a normal distribution with $\bar{Y} = 6.93$ and $s = 1.895$. This is plotted by calculating the expected positions of points for $\bar{Y} \pm s$. Since 68.3% of the distribution falls within these bounds, the relevant points on the cumulative frequency scale are $50 \pm 34.15\%$; thus this line was drawn using the points (4.495, 15.85) and (8.285, 84.15) as indicated on the plot.

- Plot the distribution on normal probability graph paper. If the distribution is normal, the data will tend to follow a straight line (see Fig. 39.5). Deviations from linearity reveal skewness and/or kurtosis (see p. 253), the significance of which can be tested statistically (see Sokal and Rohlf, 1994).
- Use a suitable statistical computer program to generate predicted normal curves from the $\bar{Y}$ and s values of your sample(s). These can be compared visually with the actual distribution of data and can be used to give 'expected' values for a χ^2 test or a G-test.

The wide availability of tests based on the normal distribution and their relative simplicity means you may wish to transform your data to make them more like a normal distribution. Table 39.1 provides transformations that can be applied. The transformed data should be tested for normality as described above before proceeding – do not forget that you may need to check that transformed variances are homogeneous for certain tests (see below).

A very important theorem in statistics, the Central Limit Theorem, states that as sample size increases, the distribution of a series of means from any frequency distribution will become normally distributed. This fact can be

Table 39.1 Suggested transformations altering different types of frequency distributions to the normal type. To use, modify data by the formula shown; then examine effects with the tests described on pp. 260–263.

Type of data; distribution suspected	Suggested transformation(s)
Proportions (including percentages); binomial	arcsine $\sqrt{x}$ (also called the angular transformation)
Scores; Poisson	$\sqrt{x}$ or $\sqrt{(x + 1/2)}$ if zero values present
Measurements; negatively skewed	x^2, x^3, x^4, etc. (in order of increasing strength)
Measurements; positively skewed	$1/\sqrt{x}$, $\sqrt{x}$, $\ln x$, $1/x$ (in order of increasing strength)

used to devise an experimental or sampling strategy which ensures that data are normally distributed, i.e. using means of samples as if they were primary data.

Definition

Homogeneous variance – uniform (but not necessarily identical) variance of the dependent variable across the range of the independent variable. The term homoscedastic is also used in this sense. The opposite of homogeneous is heterogeneous (= hereoscedastic).

Understanding 'degrees of freedom' – this depends on the number of values in the data set analysed, and the method of calculation depends on the statistical test being used. It relates to the number of observations that are free to vary before the remaining quantities for a data set can be determined.

Checking the assumptions of a test – always acquaint yourself with the assumptions of a test. If necessary, test them before using the test.

Choosing a suitable statistical test

Comparing location (e.g. means)

If you can assume that your data are normally distributed, the main test for comparing two means from independent samples is Student's t-test (see Boxes 39.1 and 39.2, and Table 39.2). This assumes that the variances of the data sets are homogeneous. Tests based on the t-distribution are also available for comparing means of paired data or for comparing a sample mean with a chosen value.

When comparing means of two or more samples, analysis of variance (ANOVA) is a very useful technique. This method also assumes data are normally distributed and that the variances of the samples are homogeneous. The samples must also be independent (e.g. not subsamples). The test statistic calculated is denoted F and it has two different degrees of freedom related to the number of means tested and the pooled number of replicates per mean. The nested types of ANOVA are useful for letting you know the relative importance of different sources of variability in your data. Two-way and multi-way ANOVAs are useful for studying interactions between treatments.

For data satisfying the ANOVA requirements, the least significant difference (LSD) is useful for making planned comparisons among several means (see Sokal and Rohlf, 1994). Any two means that differ by more than the LSD will be significantly different. The LSD is useful for showing on graphs.

The chief non-parametric tests for comparing the locations of two samples are the Mann–Whitney U-test and the Kolmogorov–Smirnov test. The former assumes that the frequency distributions of the samples

Table 39.2 Critical values of Student's t statistic (for two-tailed tests). Reject the null hypothesis at probability P if your calculated t value equals or exceeds the value shown for the appropriate degrees of freedom $= (n_1 - 1) + (n_2 - 1)$.

Degrees of freedom	Critical values for $P = 0.05$	Critical values for $P = 0.01$	Critical values for $P = 0.001$
1	12.71	63.66	636.62
2	4.30	9.92	31.60
3	3.18	5.84	12.94
4	2.78	4.60	8.61
5	2.57	4.03	6.86
6	2.45	3.71	5.96
7	2.36	3.50	5.40
8	2.31	3.36	5.04
9	2.26	3.25	4.78
10	2.23	3.17	4.59
12	2.18	3.06	4.32
14	2.14	2.98	4.14
16	2.12	2.92	4.02
20	2.09	2.85	3.85
25	2.06	2.79	3.72
30	2.04	2.75	3.65
40	2.02	2.70	3.55
60	2.00	2.66	3.46
120	1.98	2.62	3.37
∞	1.96	2.58	3.29

Box 39.1 How to carry out a *t*-test

The *t*-test was devised by a statistician who used the pen name 'Student', so you may see it referred to as Student's *t*-test. It is used when you wish to decide whether two samples come from the same population or from different ones (Fig. 39.1). The samples might have been obtained by observation or by applying two different treatments to an originally homogeneous population (Chapter 33).

The null hypothesis is that the two groups can be represented as samples from the same overlying population (Fig. 39.1(a)). If, as a result of the test, you accept this hypothesis, you can say that there is no significant difference between the group means.

The alternative hypothesis is that the two groups come from different populations (Fig. 39.1(b)). By rejecting the null hypothesis as a result of the test, you can accept the alternative hypothesis and say that there is a significant difference between the sample means, or, if an experiment were carried out, that the two treatments affected the samples differently.

How can you decide between these two hypotheses? On the basis of certain assumptions (see below), and some relatively simple calculations, you can work out the probability that the samples came from the same population. If this probability is very low, then you can reasonably reject the null hypothesis in favour of the alternative hypothesis and, if it is high, you will accept the null hypothesis.

To find out the probability that the observed difference between sample means arose by chance, you must first calculate a '*t* value' for the two samples in question. Some computer programs provide this probability as part of the output, otherwise you can look up statistical tables (e.g. Table 39.2). These tables show 'critical values' – the borders between probability levels. If your value of *t* equals or exceeds the critical value for probability *P*, you can reject the null hypothesis at this probability ('level of significance').

Note that:

- for a given difference in the means of the two samples, the value of *t* will get larger the smaller the scatter within each data set; and
- for a given scatter of the data, the value of *t* will get larger, the greater the difference between the means.

So, at what probability should you reject the null hypothesis? Normally, the threshold is arbitrarily set at 5 per cent – you quite often see descriptions such as 'the sample means were significantly different ($P < 0.05$)'. At this 'significance level' there is still up to a 5 per cent chance of the *t* value arising by chance, so about one in 20 times, on average, the conclusion will be wrong. If *P* turns out to be lower, then this kind of error is much less likely.

Tabulated probability levels are generally given for 5, 1 and 0.1 per cent significance levels (see Table 39.2). Note that this table is designed for 'two-tailed' tests, i.e. where the treatment or sampling strategy could have resulted in either an increase or a decrease in the measured values. These are the most likely situations you will deal with in biology.

Examine Table 39.2 and note the following:

- The larger the size of the samples (i.e. the greater the 'degrees of freedom'), the smaller *t* needs to be to exceed the critical value at a given significance level.
- The lower the probability, the greater *t* needs to be to exceed the critical value.

The mechanics of the test

A calculator that can work out means and standard deviations is helpful.

1. **Work out the sample means $\bar{Y}_1$ and $\bar{Y}_2$ and calculate the difference between them.**
2. **Work out the sample standard deviations s_1 and s_2.** (NB if your calculator offers a choice, chose the '$n-1$' option for calculating s – see p. 252).
3. **Work out the sample standard errors $SE_1 = s_1/\sqrt{n_1}$ and $SE_2 = s_2/\sqrt{n_2}$; now square each, add the squares together, then take the positive square root of this** (n_1 and n_2 are the respective sample sizes, which may, or may not, not be equal).
4. **Calculate *t*** from the formula:

$$t = \frac{\bar{Y}_1 - \bar{Y}_2}{\sqrt{((SE_1)^2) + ((SE_2)^2)}} \qquad [39.1]$$

   The value of *t* can be negative or positive, depending on the values of the means; this does not matter and you should compare the modulus (absolute value) of *t* with the values in tables.
5. **Work out the degrees of freedom** $= (n_1 - 1) + (n_2 - 1)$.
6. **Compare the *t* value with the appropriate critical value (see e.g. Table 39.2) and decide on the significance of your findings (see p. 260).**

Box 39.2 provides a worked example – use this to check that you understand the above procedures.

Assumptions that must be met before using the test

The most important assumptions are:

- The two samples are independent and randomly drawn (or, if not, drawn in a way that does not create bias). The test assumes that the samples are quite large.
- The underlying distribution of each sample is normal. This can be tested with a special statistical test, but a rule of thumb is that a frequency distribution of the data should be (a) symmetrical about the mean and (b) nearly all of the data should be within 3 standard deviations of the mean and about two-thirds within 1 standard deviation of the mean (see p. 262).
- The two samples should have uniform variances. This again can be tested (by an *F*-test), but may be obvious from inspection of the two standard deviations.

Box 39.2 Worked example of a *t*-test

Suppose the following data were obtained in an experiment (the units are not relevant):

Control: 6.6, 5.5, 6.8, 5.8, 6.1, 5.9
Treatment: 6.3, 7.2, 6.5, 7.1, 7.5, 7.3

Using the steps outlined in Box 39.1, the following values are obtained (denoting control with subscript 1, treatment with subscript 2):

1. $\bar{Y}_1 = 6.1167$; $\bar{Y}_2 = 6.9833$: difference between means $= \bar{Y}_1 - \bar{Y}_2 = -0.8666$
2. $s_1 = 0.49565$; $s_2 = 0.47504$
3. $SE_1 = 0.49565/2.44949 = 0.202348$
   $SE_2 = 0.47504/2.44949 = 0.193934$
4. $t = \dfrac{-0.8666}{\sqrt{(0.202348^2 + 0.193934^2)}} = \dfrac{-0.8666}{0.280277} = -3.09$
5. d.f. $= (5 + 5) = 10$
6. Looking at Table 39.2, we see that the modulus of this t value exceeds the tabulated value for $P = 0.05$ at 10 degrees of freedom ($= 2.23$). We therefore reject the null hypothesis, and conclude that the means are different at the 5 per cent level of significance. If the modulus of t had been $\leqslant 2.23$, we would have accepted the null hypothesis. If the modulus of t had been > 3.17, we could have concluded that the means are different at the 1 per cent level of significance.

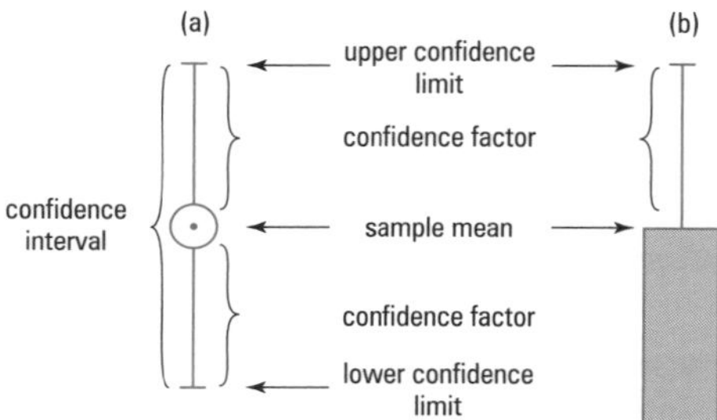

Fig. 39.6 Graphical representation of confidence limits as 'error bars' for (a) a sample mean in a plotted curve, where both upper and lower limits are shown; and (b) a sample mean in a histogram, where, by convention, only the upper value is shown. For data that are assumed to be symmetrically distributed, such representations are often used in preference to the 'box and whisker' plot shown on p. 251. Note that SE is an alternative way of representing sample imprecision/error (e.g. Fig. 35.1).

Confidence limits for statistics other than the mean – consult an advanced statistical text (e.g. Sokal and Rohlf, 1994) if you wish to indicate the reliability of estimates of, e.g., population variances.

are similar, whereas the latter makes no such assumption. In both cases the sample's size must be $\geqslant 4$ and for the Kolmogorov–Smirnov test the samples must have equal sizes. In the Kolmogorov–Smirnov test, significant differences found with the test could be due to differences in location or shape of the distribution, or both.

Suitable non-parametric comparisons of location for paired data (sample size $\geqslant 6$) include Wilcoxon's signed rank test, which is used for quantitative data and assumes that the distributions have similar shape. Dixon and Mood's sign test can be used for paired data scores where one variable is recorded as 'greater than' or 'better than' the other.

Non-parametric comparisons of location for three or more samples include the Kruskal–Wallis H-test. Here, the number of samples is without limit and they can be unequal in size, but again the underlying distributions are assumed to be similar. The Friedman S-test operates with a maximum of five samples and data must conform to a randomised block design. The underlying distributions of the samples are assumed to be similar.

Comparing dispersions (e.g. variances)

If you wish to compare the variances of two sets of data that are normally distributed, use the F-test. For comparing more than two samples, it may be sufficient to use the F_{max}-test, on the highest and lowest variances. The Scheffé–Box (log-anova) test is recommended for testing the significance of differences between several variances. Non-parametric tests exist but are not widely available: you may need to transform the data and use a test based on the normal distribution.

Determining whether frequency observations fit theoretical expectation

The χ^2 test is useful for tests of 'goodness of fit', e.g. comparing expected and observed progeny frequencies in genetical experiments or comparing observed frequency distributions with some theoretical function. One limitation is that simple formulae for calculating χ^2 assume that no expected number is less than 5. The G-test ($2I$-test) is used in similar circumstances.

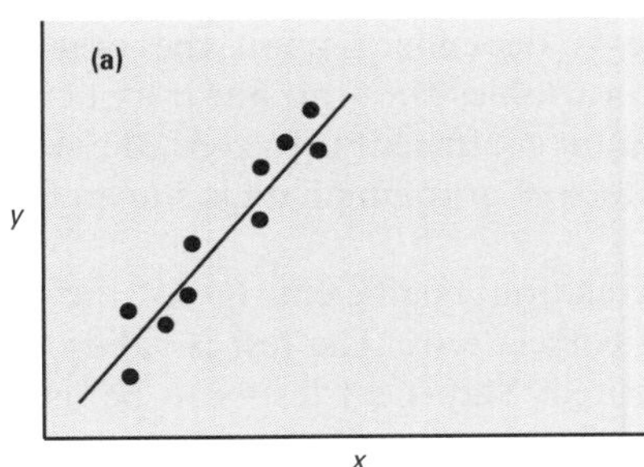

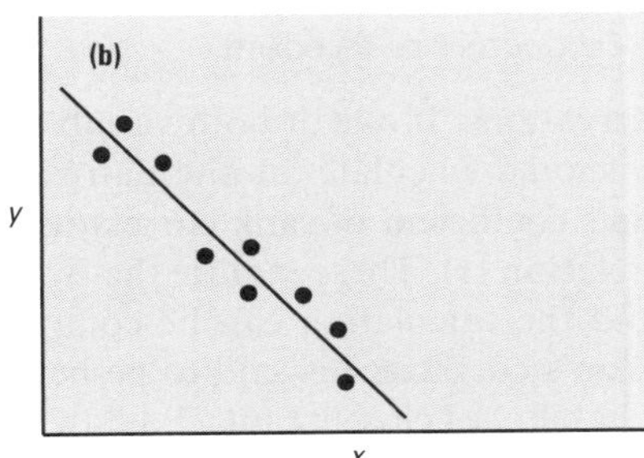

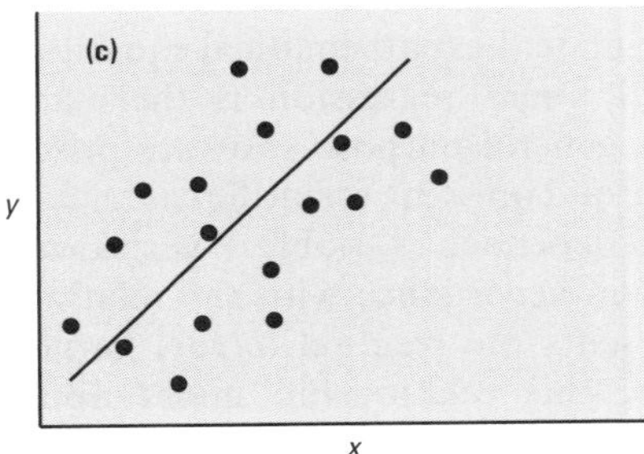

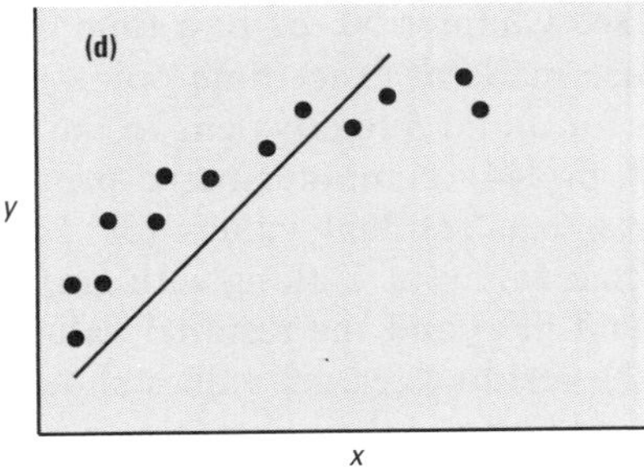

Fig. 39.7 Examples of correlation. The linear regression line is shown. In (a) and (b), the correlation between x and y is good: for (a) there is a positive correlation and the correlation coefficient, r, would be close to 1; for (b) there is a negative correlation and the correlation coefficient would be close to -1. In (c) there is a weak positive correlation and r would be close to 0. In (d) the correlation coefficient may be quite large, but the choice of linear regression is clearly inappropriate.

Comparing proportion data

When comparing proportions between two small groups (e.g. whether 3/10 is significantly different from 5/10), you can use probability tables such as those of Finney *et al.* (1963) or calculate probabilities from formulae; however, this can be tedious for large sample sizes. Certain proportions can be transformed so that their distribution becomes normal.

Placing confidence limits on an estimate of a population parameter

On many occasions, a sample statistic is used to provide an estimate of a population parameter, and it is often useful to indicate the reliability of such an estimate. This can be done by putting confidence limits on the sample statistic, i.e. by specifying an interval around the statistic within which you are confident that the true value (the population parameter) is likely to fall, at a specified level of probability. The most common application is to place confidence limits on the mean of a sample taken from a population of normally distributed data values. In practice, you determine a confidence factor for a particular level of probability that is added to and subtracted from the sample mean ($\bar{Y}$) to give the upper confidence limit and lower confidence limit respectively. These are calculated as:

$$\begin{aligned} &\bar{Y} + (t_{P[n-1]} \times \text{SE}) \text{ for the upper limit and} \\ &\bar{Y} - (t_{P[n-1]} \times \text{SE}) \text{ for the lower limit} \end{aligned} \qquad [39.2]$$

where $t_{P[n-1]}$ is the tabulated critical value of Student's t statistic for a two-tailed test with $n - 1$ degrees of freedom at a specified probability level (P) and SE is the standard error of the sample mean (p. 252). The 95 per cent confidence limits (i.e. $P = 0.05$) tells you that, on average, 95 times out of 100 the interval between the upper and lower limits will contain the true (population) value. Confidence limits are often shown as 'error bars' for individual sample means plotted in graphical form. Figure 39.6 illustrates how this is applied to plotted curves and histograms (note that this can be carried out for data series within a Microsoft Excel graph (chart) using the *Format data series* and *Y error bars* commands).

Correlation and regression

These methods are used when testing the relationship between data values for two variables. Correlation is used to measure the extent to which changes in the two sets of data values occur together in a linear manner. If one variable can be assumed to be dependent upon the other (i.e. a change in X causes a particular change in Y), then regression techniques can be used to provide a mathematical description of the underlying relationship between the variables, e.g. to find a line of best fit for a data series. If there is no *a priori* reason to assume dependency, then correlation methods alone are appropriate.

A correlation coefficient measures the strength of the linear relationship between two variables, but does not describe the relationship. The coefficient is expressed as a number between -1 and $+1$: a positive coefficient indicates a direct relationship, where the two variables change in the same direction, while a negative coefficient indicates an inverse relationship, where one variable decreases as the other increases (Fig. 39.7). The nearer the coefficient is to -1 or $+1$, the stronger the linear relationship between the variables, i.e. the less 'scatter' there would be about a straight line of best fit (note that this does *not* imply that one variable is dependent upon the other). A coefficient of 0 implies that the two variables show no linear association and therefore the closer the correlation coefficient is to zero, the weaker the linear relationship. The importance of graphing data is shown by the case illustrated in Fig. 39.7(d).

Pearson's product moment correlation coefficient (r) is the most commonly used statistic for testing correlations. The test is valid only if both variables are normally distributed. Statistical tests can be used to decide whether the correlation is significant (e.g. using a one-sample t-test to see whether r is significantly different from zero, based on the equation:

$$t = r \div \sqrt{[(1 - r^2) \div (n - 2)]} \text{ at } n - 2 \text{ degrees of freedom,} \qquad [39.3]$$

where n is the number of paired observations. If one or both variables are not normally distributed, then you should calculate an alternative non-parametric coeffcient, e.g. Spearman's coefficient of rank correlation (r_s) or Kendall's coefficient of rank correlation (τ). These require the two sets of data to be ranked separately, and the calculation can be complex if there are tied (equal) ranks. Spearman's coefficient is said to be better if there is any uncertainty about the reliability of closely ranked data values.

Using more advanced types of regression – these include:

- Model II linear regression, which applies to situations where a dependent variable Y varies with an independent variable X, and where both variables may have error terms associated with them.
- Multiple regression, which applies when there is a relationship between a dependent variable and two or more independent variables.
- Non-linear regression, which extends the principles of linear regression to a wide range of functions. Technically, this method is more appropriate than transforming data to allow linear regression.

Advanced statistics books should be consulted for details of these methods, which may be offered by some statistical computer programs.

If underlying theory or empirical graphical analysis indicate a linear relationship between a dependent and an independent variable, then linear regression can be used to estimate the mathematical equation that links the two variables. Model I linear regression is the standard approach, and is available within general-purpose software programs such as Microsoft Excel (Box 39.3), and on some scientific calculators. It is suitable for experiments where a dependent variable Y varies with an *error-free* independent variable X in accordance with the relationship $Y = a + bX + e_Y$, where e_Y represents the residual (error) variability in the Y variable. For example, this relationship might apply in a laboratory procedure where you have carefully controlled the independent variable and the X values can be assumed to have zero error (e.g. in a calibration curve, see Chapter 50, or in a time course experiment where measurements are made at exact time points). The regression analysis gives estimates for a and b (equivalent to the slope and intercept of the line of best fit, p. 244): computer-based programs usually provide additional features, e.g. residual values for Y (e_Y), estimated errors for a and b, predicted values of Y along with graphical plots of the line of best fit (the trend line) and the residual values. In order for the model to be valid, the residual (error) values should be normally distributed around the trend line and their variance should be uniform (homogeneous), i.e. there should be a similar scatter of data points around the trend line along the x axis (independent variable).

If the relationship is not linear, try a transformation. For example, this is commonly done in analysis of enzyme kinetics. However, you should be aware that the transformation of data to give a straight line can lead to errors when carrying out linear regression analysis: take care to ensure that (a) the assumptions listed in the previous paragraph are valid for the transformed data set and (b) the data points are evenly distributed throughout the range of the independent variable. If these criteria cannot be met, non-linear regression may be a better approach, but for this you will require a suitable computer program, e.g. GraphPad Prism.

Example If a regression analysis gives a value for r^2 of 0.75 (i.e. $r = 0.84$), then 75% of the variance in Y can be explained by the trend line, with $1 - r^2 = 0.25$ (25%) remaining as unexplained (residual) variation.

The strength of the relationship between Y and X in Model I linear regression is best estimated by the coefficient of determination (r^2 or R^2),

Box 39.3 Using a spreadsheet (Microsoft Excel 2007) to calculate hypothesis-testing statistics

Presented below are three examples of the use of Microsoft Excel to investigate hypotheses about specific data sets. In each case, there is a brief description of the problem; a table showing the data analysed; an outline of the Microsoft Excel commands used to carry out the analysis and an annotated table of results from the spreadsheet.

Example 1: A *t*-test

As part of a project, a student applied a chemical treatment to a series of flasks containing fungal cultures with nutrient solution. An otherwise similar set of control flasks received no chemical treatment. After three weeks' growth, she measured the wet mass of the filtered cultures:

Wet mass of samples (g)

Replicate	1	2	3	4	5	6	7	8	Mean	Variance
Treated with ZH52	2.342	2.256	2.521	2.523	2.943	2.481	2.601	2.449	2.515	0.042
Control	2.658	2.791	2.731	2.402	3.041	2.668	2.823	2.509	2.703	0.038

The student proposed the null hypothesis that there was no difference between the two means and tested this using a *t*-test, as she had evidence from other studies that the fungal masses of replicate flasks were normally distributed. She also established, by calculation, that the assumption that the populations had homogeneous variances was likely to be valid. Using the *Data* tab > *Data Analysis* > *t-Test: Two-Sample Assuming Equal Variance* option, with *Hypothesized Mean Difference* = 0 and *Alpha* (= *P*) = 0.05, and adjusting the number of significant figures displayed, the following table was obtained:

t-test: Two-sample assuming equal variances

	Variable 1	Variable 2
Mean	2.515	2.703
Variance	0.042	0.038
Observations	8	8
Pooled variance	0.040	
Hypothesised mean difference	0	
d.f.	14	
t Stat	−1.881	
P(T <= t) one-tail	0.040	
t Critical one-tail	1.761	
P(T <= t) two-tail	0.081	
t Critical two-tail	2.145	

The value of *t* obtained was −1.881 (row 7, '*t* stat') and the probability of obtaining this value for a two-tailed test (row 10) was 0.081 (or 8.1%), so the student was able to accept the null hypothesis and conclude that ZH52 had no significant effect on fungal growth in these circumstances.

Example 2: An ANOVA test

A biochemist made six replicate measurements of four different batches (A–D) of alcohol dehydrogenanse, obtaining the following data:

Alcohol dehydrogenase activity (U l^{-1})

Batch/ Replicate	1	2	3	4	5	6	Mean	Variance
A	0.562	0.541	0.576	0.545	0.542	0.551	0.552833	0.000189
B	0.531	0.557	0.537	0.521	0.559	0.538	0.540500	0.000221
C	0.572	0.568	0.551	0.549	0.564	0.559	0.560500	0.000085
D	0.532	0.548	0.541	0.538	0.547	0.536	0.540333	0.000039

The biochemist wanted to know whether the observed differences were statistically significant, so he carried out an ANOVA test, assuming the samples were normally distributed and the variances in the three populations were homogeneous. Using the *Data* tab > *Data Analysis* > *Anova: Single Factor* option, with *Alpha* (=*P*) =0.05, and adjusting the number of significant figures displayed, the following table was obtained:

ANOVA: Single factor

SUMMARY

Groups	*Count*	*Sum*	*Average*	*Variance*
A	6	3.317	0.552833	0.000189
B	6	3.243	0.5405	0.000221
C	6	3.363	0.5605	8.51E-05
D	6	3.242	0.540333	3.95E-05

ANOVA

Source of variation	SS	d.f.	MS	*F*	*P-value*	*F* crit
Between groups	0.001761	3	0.000587	4.397856	0.015669	3.098391
Within groups	0.002669	20	0.000133			
Total	0.00443	23				

The *F*-value calculated was 4.397856. This comfortably exceeds the stated critical value (F_{crit}) of 3.098391, and the probability of obtaining this result by chance (*P*-value) was calculated as 0.015669 (1.57% to three significant figures); hence the biochemist was able to reject the null hypothesis and conclude that there was a significant difference in average enzyme activity between the four batches, since $P < 0.05$. Such a finding might lead on to an investigation into why there was batch variation, e.g. had they been stored differently?

(continued)

Box 39.3 (continued)

Example 3: Testing the significance of a correlation

A researcher wanted to know whether his observations of earthworm casts on the surface of closely mown grass were related to how wet the soil was. He took weekly measurements of precipitation using a rain gauge and counted the mean number of casts per m^2 taking mean results from nine quadrats per weekly observation:

Observation	Precipitation in previous week (mm)	Mean density of casts (m^{-2})
1	11	4.4
2	1	3.1
3	0	2.3
4	5	4.6
5	8	4.5
6	2	3.3
7	4	3.5
8	15	6.4

The researcher used the Microsoft Excel function PEARSON(array1, array2), available through the *Formulas* tab to obtain a value of +0.927 857 674 for Pearson's product moment correlation coefficient *r*, specifying the precipitation data as array1 and the cast density as array2. He then used a spreadsheet to calculate the *t* statistic (p. 265) for this *r* value, using eqn. [39.3]. The value of *t* was 7.037, with six degrees of freedom. The critical value from tables (e.g. Table 39.2) at $P = 0.001$ is 5.96, so he concluded that there was a very highly significant positive correlation between his two sets of observations. The investigator moved on from this observation and next investigated the effect of artificial hosepipe rainfall in a sheltered grass plot, to test whether there was a causal relationship involved.

which is equivalent to the square of the Pearson correlation coefficient. The coefficient of determination varies between 0 and +1 and provides a measure of the goodness of fit of the *Y* data to the regression line: the closer the value is to 1, the better the fit. In effect, r^2 represents the fraction of the variance in *Y* that can be accounted for by the regression equation. Conversely, if you subtract this value from 1, you will obtain the residual (error) component, i.e. the fraction of the variance in *Y* that cannot be explained by the line of best fit. Multiplying the values by 100 allows you to express these fractions in percentage terms.

Using computers to calculate hypothesis-testing statistics

As with the calculation of descriptive statistics (p. 256), specialist statistical packages such as SPSS can be used to simplify the calculation of hypothesis-testing statistics. The correct use of the software and interpretation of the output requires an understanding of relevant terminology and of the fundamental principles governing the test, which is probably best obtained by working through one or more examples by hand before using these tools (e.g. Box 39.2). Spreadsheets offer increasingly sophisticated statistical analysis functions, three examples of which are provided in Box 39.3.

Text references and sources for further study

Ennos, R. (2007) *Statistical and Data Handling Skills in Biology*, 3rd edn. Pearson Education, Harlow.

Finney, D.J., Latscha, R., Bennett, B.M. and Hsu, P. (1963) *Tables for Testing Significance in a 2 × 2 Table*. Cambridge University Press, Cambridge.

Schmuller, J. (2005) *Statistical Analysis with Excel for Dummies*. Wiley, Hoboken, New Jersey.

Sokal, R.R. and Rohlf, F.J. (1994) *Biometry*, 3rd edn. W.H. Freeman, San Francisco, California.

Dietary assessment and intervention

40 Nutritional recommendations and guidelines

Definitions

Nutrition – the scientific study of foods, including nutrients, energy and other food components, and how they act within the body, organs and tissues.

Carbohydrates – compounds composed of carbon, hydrogen and oxygen in the ratio $(CH_2O)_n$ that form the main source of energy in our diet; examples include sugars from fruit and starch from potatoes (p. 275).

Proteins – molecules composed of amino acids linked together in a particular order specified by a DNA sequence (p. 394). From foods, they act as a source of amino acids for protein synthesis in the body and as a source of energy.

Lipids – highly reduced components with the highest energy availability of any foodstuff, they are soluble in organic solvents but not water-soluble; principally fats and oils and including fatty acids, phospholipids and cholesterol (p. 413).

Vitamins – organic food components required in very small amounts for growth and for maintaining good health. They often act as co-enzymes in metabolic reactions.

Minerals – inorganic elements; some minerals are essential nutrients needed in mg quantities, such as calcium and phosphorous. Others are essential trace minerals, needed in μg quantities.

Using bulleted lists in nutrition guidelines – these are used in preference to numbered recommendations because each is considered equally important.

Being able to determine how to eat a nutritious, healthy and balanced diet on a daily basis requires an understanding of the fundamental principles of good nutrition. Nutrition guidelines and their approach may vary from country to country, but the basic principles of good nutrition, based on research, are still applicable. The World Health Organisation (WHO) recommendations are used as a basis for the development of national nutrition programmes established around local circumstances and the availability of particular types of foods. As a newcomer to the subject, a good starting point is to compare your own diet with the recommendations and guidelines outlined in this chapter.

KEY POINT A well-planned, balanced and varied diet can go a long way towards helping prevent or correct many chronic diseases.

World Health Organisation (WHO) guidelines

The WHO suggests that unhealthful diets and lack of physical activity are associated with the development of a number of chronic diseases in modern society, e.g. type II diabetes mellitus and cardiovascular diseases. WHO guidelines are based on international research, reviewed by scientific experts, and provide information on nutrient intake that can help prevent such diseases. Their recommendations include:

- That individuals and populations achieve energy balance and a healthy weight. A good starting point is to limit the overall kilojoule intake from fats, replacing saturated fats in the diet with unsaturated fats, and avoiding harmful trans-fatty acids (Figs 40.1 and 40.2).
- The intake of free sugars should be minimised, while maximising the intake of fruits, vegetables, legumes, wholegrain cereals and nuts.
- Salt consumption should be limited; where salt is used, it should be iodised.

The WHO asserts that the improvement of dietary habits is a society issue rather than simply an individual problem, therefore requiring a population-based, multi-disciplinary, multi-sectoral and culturally relevant approach (WHO, 2009).

The Global Strategy for Infant and Young Child Feeding (WHO, 2003), jointly developed by WHO and UNICEF, recommends that infants under six months should be exclusively fed breast milk to achieve optimal growth, development and health. After six months they also need to be fed sufficient, safe, nutrient-rich complementary food and drink, while continuing breastfeeding for up to two years or more. Care with feeding for the period up to 18–24 months of age is critical for preventing malnutrition, which can lead to lifelong consequences. Inappropriate feeding can result in risk factors for later disease, could lead to poor school performance, reduced productivity, impaired mental function and social development, and may lead to chronic disease. Cow's milk should not be fed to children until they are at least one year old, since it can cause intestinal disturbance and may place stress on the infant's kidneys.

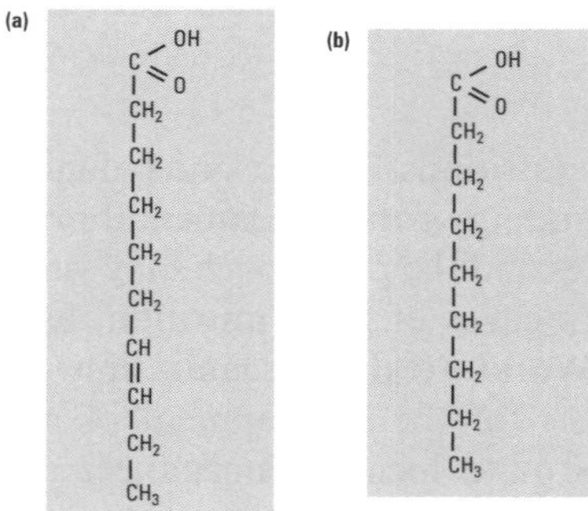

Fig. 40.1 **(a)** Unsaturated fatty acid. These have one (monounsaturated, shown above) or more than one (polyunsaturated) double bonds between the carbon atoms. Unsaturated fatty acids, particularly the monounsaturated variety, are so-called 'good' fats because they lower the concentration of cholesterol in the blood. They are found in plant foods. Olive oil, avocado and macadamia nuts contain principally monounsaturated fats. **(b)** Saturated fatty acid. These do not have any double bonds between the carbons in the chain. Saturated fatty acids are so-called 'bad' fats as they increase the concentration of cholesterol in the blood.

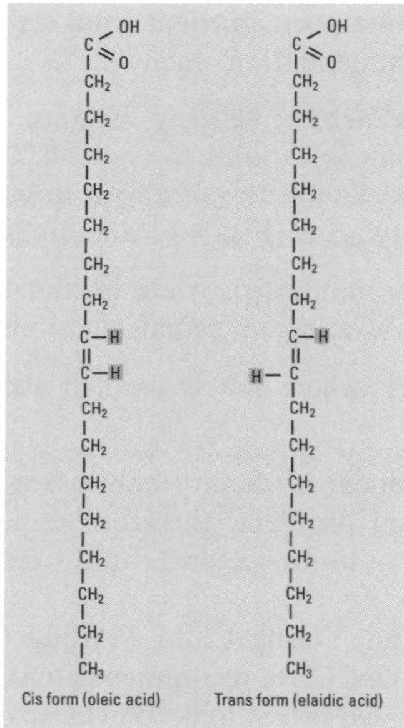

Fig. 40.2 *Cis* and *Trans* unsaturated fatty acids. These have hydrogen (H) molecules in the *Cis* orientation around a double bond, as shown on the left. Trans-fatty acid hydrogens are in the *Trans* orientation around the double bond as illustrated on the right, and have been shown to be 'bad' fats because they increase blood cholesterol concentrations. They are often produced by hydrogenation of un-saturated fats to make them more solid and are also produced during high temperature frying.

KEY POINT Parents should be encouraged to instill healthy nutrition and physical activity as a lifetime habit for their children. Habits developed as children often remain in place through adulthood and studies show that obese children often become obese adults.

UK (eatwell plate) guidelines

The 'eatwell plate' (Fig. 40.3) is used in the United Kingdom to provide a pictorial guide to the types and proportions of foods to eat from each of the main food groups in order to achieve a balanced diet. It shows how much of your food should come proportionally from each food group. The broad recommendations from the eatwell plate approach are:

- Try to get the balance right over time such as a whole day or week. It is not necessary to get the balance exactly right at every meal, e.g. if you cannot avoid eating something quite fatty, then for the next meal(s) balance this out with nutritious low-fat foods.
- Eat plenty of fruits and vegetables, and plenty of breads, rice, potatoes, pasta and other starchy foods (wholegrain where available).
- Eat some meat, fish, eggs, beans, and milk and dairy products.
- Only consume small amounts of foods and drinks that are high in fat or sugar.

Note: additional information is available through the UK Food Standards Agency website for different stages of the life cycle (http://www.eatwell.gov.uk/agesandstages/).

The eatwell plate
Use the eatwell plate to help you get the balance right. It shows how much of what you eat should come from each food group.
FOOD STANDARDS AGENCY
food.gov.uk
Fruit and vegetables
Bread, rice, potatoes, pasta and other starchy foods
Meat, fish, eggs, beans and other non-dairy sources of protein
Foods and drinks high in fat and/or sugar
Milk and dairy foods

Fig. 40.3 The 'eatwell plate'.

Fig. 40.4 'MyPyramid'.

USA (MyPyramid) guidelines

The United States Department of Agriculture uses a pictorial guide termed 'MyPyramid' (Fig. 40.4) to enable people to gauge the relative proportions of foods to eat from each of the main food groups, and for physical activity recommendations. Current recommendations are based on 2005 dietary guidelines and new guidelines are planned for release in 2011.

The MyPyramid website (http://www.mypyramid.gov) allows you to input data for age, gender, height, weight and activity level for interactive guidance on menu planning. However, it uses the imperial measurement system, so if you are working in metric measurement you should use the table on p. 202 to convert from centimetres to feet and inches and from kilograms to pounds. MyPyramid also allows you to interactively assess and track and store your diet and activity levels. Additional MyPyramids are available for nutrition for children and for nutrition in pregnancy.

MyPyramid recommendations for food groups include the following:

- For the grain group, half of your grain products intake should be wholegrain.
- For vegetables, focus on variety, including more dark green and orange vegetables, for example broccoli, spinach, carrots and sweet potatoes.
- For the fruit group, aim for variety, including fresh, frozen, canned or dried fruits, but moderate the consumption of fruit juices.
- For milk products, these are recommended as a good source of calcium. Choose low-fat or fat-free varieties of milk, yogurt and other dairy products. (Note: low-fat products are *not* recommended for infants.) Lactose-free dairy products or calcium-fortified foods and drinks are suggested for those with lactose intolerance.

Appreciating the benefits of eating oily fish – salmon, trout, mackerel and other oily fish contain beneficial essential omega 3 unsaturated fatty acids and are recommended to be eaten twice a week. (An omega 3 fatty acid is one that has a final carbon=carbon double bond in the bond between the third and fourth carbon from the methyl end of the fatty acid, see Fig. 40.1a.)

Understanding dietary sources of sugar – some people use honey, rather than sugar, to sweeten foods, believing that it is more nutritious. However, honey is no more nutritious than purified cane sugar and is an energy-dense food.

- For the meat and beans group, focus on eating lean or low-fat meats and poultry and cook by baking, broiling or grilling to reduce fat. Also vary protein intake by including more fish, beans, peas, nuts and seeds.
- For fats and oils (the smallest component on MyPyramid), principal sources should be oils from fish, nuts and vegetables, while consumption of solid fats such as butter, margarine, shortening and lard should be limited.

The person climbing the stairs on the MyPyramid image (Fig. 40.4) indicates the need for physical activity as part of a healthy lifestyle. It is recommended to undertake physical activity for at least 30 minutes most days of the week and to stay within daily kilojoule requirements for energy consumption. However, it is further suggested that 60 minutes a day of exercise may be needed to prevent weight gain with 60–90 minutes required per day for sustained weight loss. The recommendation for children and teenagers is 60 minutes of physical activity every day or on most days.

People should be encouraged to read the nutrition labels on food products to curb their saturated fat, trans fat and salt (sodium) intake and to think carefully about marketing/labelling, choosing food and beverages low in added sugars to avoid consumption of energy-rich components with few, if any, nutrients (e.g. a yogurt labelled 99 per cent fat-free may have hidden added sugar).

Understanding national guidelines – serving sizes differ in guidelines from different countries and regions, so the relevant values should be used for a particular set of guidelines. Likewise, the serving sizes on packaged foods may differ from those in national guidelines.

Australian guidelines

Australia uses the Australian Guide to Healthy Eating (Fig. 40.5) from the National Health and Medical Research Council's (NHMRC) *Food for Health – Dietary Guidelines for Australians* (http://www.nhmrc.gov.au/) to determine healthy and balanced food choices. The 'plate' provides an

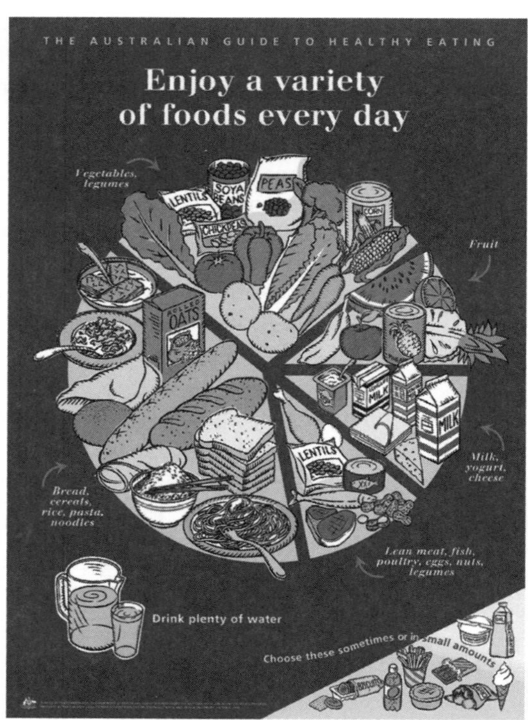

Fig. 40.5 Australian Guide to Healthy Eating.

indication of the relative amounts of foods from each class to be consumed by adults. Additional nutrition guides are available on the NHMRC nutrition website for different stages of the life cycle.

The following dietary guidelines summarise the recommendations for adults, which are broadly similar to those in the UK and USA:

- Eat a wide variety of nutritious foods including plenty of vegetables, legumes, fruits and cereals, preferably wholegrain.
- Include lean meat, fish, poultry and reduced-fat milk and dairy products.
- Limit your intake of total fat and saturated fat, salt, sugars and foods containing sugars, and limit alcohol consumption if you choose to drink.
- Be physically active and eat as required by your energy needs to prevent weight gain.
- Drink plenty of water.
- Take care of food by preparing and storing it safely (p. 134).

The basic principles for children and adolescents are similar to those above for adults except:

- Reduced-fat milks are not suitable for children under two years of age, because of a young child's high energy requirements. However, reduced-fat varieties should be encouraged for older children and adolescents, depending on individual energy needs.
- Saturated and total fat intake should be moderated; however, low-fat diets are not suitable for infants.
- Water should be chosen as a drink, in preference to sugary soft drinks.

Understanding dietary fibre – wholegrain products, vegetables and legumes are high in dietary fibre, which is important for healthy gastrointestinal function and mobility. Some types of fibre, such as that found in oats, indirectly reduce blood cholesterol concentration.

Design a balanced, varied menu – use the UK and USA guidelines outlined on p. 274 and p. 275 to produce a menu for yourself: check your menu against the examples given at http://examplemypyramid and www.eatwell/menus.

Comparing national guidelines and recommendations

While the basic principles remain the same for different sets of guidelines, the details vary from one country to another, e.g. in the Australian guideline the daily protein (meat or alternatives) consumption for a healthy adult male is 65–100 g, which is low compared with the USA guidelines, at around 170 g (6oz). The US MyPyramid guidelines suggest that half of grain products consumed should be wholegrain, whereas the Australian recommendations suggest that all grains consumed be preferably wholegrain, with the latter being probably more beneficial, based on research. You might also consider whether national agricultural self-interest may be reflected in guidelines that recommend the consumption of certain foods. For example, in the Australian guidelines, men should have 6–12 servings a day of cereals, one serving being equivalent to two slices of bread (12–24 slices per day, in total).

Most national nutrition guidelines do not include specific references to the many processed, packaged and tinned foods and drinks, although the Australian guidelines factor in limited snack foods. Freshly prepared foods are generally recommended. However, it is also generally recommended that people should read the labels on processed, packaged and canned foods because many contain added sodium and/or added sugar (p. 276), and this will need to be considered in a balanced diet, and as part of any dietary assessment (Chapter 41). Some processed foods may also have reduced magnesium, potassium and vitamin content.

For other parts of the world, illustrations used in food guides differ considerably, from pyramids to pagodas to plates to rainbows (Table 40.1

Table 40.1 Pyramids, plates and guidelines by country/region

Asia: http://www.oldwayspt.org/asian_pyramid.html
Australia: http://www.nhmrc.gov.au/publications/subjects/nutrition.htm
Canada: http://www.hc-sc.gc.ca/fn-an/food-guide-aliment/index-eng.php
China: http://www.fao.org/ag/agn/nutrition/education_guidelines_chn_en.stm
Latin America: http://www.oldwayspt.org/latin_pyramid.html
Mediterranean: http://www.oldwayspt.org/med_pyramid.html
Philippines: http://www.fnri.dost.gov.ph/index.php?option=com_content&task=view&id=1275&Itemid=162
Singapore: http://www.hpb.gov.sg/hpb/default.asp?pg_id=865&aid=316
Sweden: http://www.slv.se/upload/nfa/documents/food_nutrition/The%20Food%20Circle.pdf
United Kingdom: http://www.eatwell.gov.uk/healthydiet/eatwellplate/
USA: http://www.mypyramid.gov
WHO: http://www.who.int/dietphysicalactivity/diet/en/index.html

gives relevant weblinks). However, the recommendations in these various food guides all emphasise eating relatively large helpings of grains, vegetables and fruits, and only moderate amounts of meat and dairy products.

Text references

Food Standards Agency, UK. *Eat Well, Be Well.* Available: http://www.eatwell.gov.uk Last accessed: 20/12/10.

MyPyramid. *MyPyramid, Steps to a Healthier You.* USDA. Available: http://www.mypyramid.gov/ Last accessed: 20/12/10.

NHMRC. *Nutrition & Diet Publications.* Available: http://www.nhmrc.gov.au/publications/subjects/nutrition.htm Last accessed: 20/12/10.

US Department of Agriculture (2005) *Dietary Guidelines.* Available: http://www.cnpp.usda.gov/Publications/DietaryGuidelines/2005/2005DGPolicyDocument.pdf Last accessed: 20/12/10.

WHO (2003) *Global Strategy for Infant and Young Child Feeding.* Available: http://amro.who.int/english/ad/fch/ca/GSIYCF_infantfeeding_eng.pdf Last accessed: 20/12/10.

WHO (2009) *Global Strategy on Diet, Physical Activity and Health.* Available: http://www.who.int/dietphysicalactivity/diet/en/index.html Last accessed: 20/12/10.

Sources for further study

Food Standards Agency, UK. *Ages and Stages* (Nutrition Guidelines Through the Lifecycle). Available: http://www.eatwell.gov.uk/agesandstages/ Last accessed: 20/12/10.

Food Standards Agency, UK. *Keeping Food Safe.* Available: http://www.eatwell.gov.uk/keepingfoodsafe/ Last accessed: 20/12/10.

Thomson, J. and Manore, M. (2007) *Nutrition for Life.* Pearson Benjamin Cummings, San Francisco, California.

41 Dietary assessment and analysis

The assessment of a person's diet is a fundamental aspect of nutritional practice. It is also one of the most challenging practical skills to develop, due to the many factors that influence what people eat, and to the variety of methods available.

There are two main questions to consider before undertaking dietary assessment:

1. Who are you going to collect the information from?
2. What are you going to do with the information once you have it?

Developing your skills in dietary analysis – it is useful to have tried all of the methods of dietary assessment yourself, so that you fully understand and appreciate what you are asking individuals to do.

The first question is important as you need to be aware of any possible limitations on the method imposed by your clients/participants. For example, if you are collecting dietary information from children you would probably need to use a different method from that used with adults. Similarly, within a heterogeneous adult population you might need to consider the level of literacy and numeracy of participants as this may affect your choice of method. The second question will have an impact on your decision as, broadly speaking, the more detailed and accurate the information you require, the more detailed, and therefore probably time-consuming, the method.

KEY POINT All methods of dietary assessment have their strengths and weaknesses and you need to fully consider these before starting work.

All dietary assessment relies on an individual's ability to record their food intake, either by recall or recording at the time of consumption. As there is a whole range of social, psychological and physical factors that influence what people eat, people may not always give an honest and accurate account of their food consumption. This may be influenced by the reason for their participation in the process. For example, you might expect that if someone has volunteered to take part in a research project that requires them to provide details of their diet then they would be likely to be honest about their intake. You might also hope that people who need their diet analysed for specific health reasons would be motivated enough to give honest and accurate information, although this is not always the case. Consequently, you need to make sure that individuals understand what is required of them and also why it is important that the information that they provide is accurate. In recording methods, another major issue is that people may, perhaps subconsciously, change their diet as a result of being asked to record it (an example of measurement bias, p. 194). This may be because of their heightened awareness of what they are eating, or to make the process of recording easier.

Abbreviations used in dietary analysis –

AI	adequate intake
AMDR	acceptable macronutrient distribution range
BMI	body mass index
DRI	dietary reference intakes (USA)
DRV	dietary reference value
EAR	estimated average requirement
EER	estimated energy requirement (USA)
LRNI	lower reference nutrient intake
NRV	nutrient reference values
PAL	physical activity level
RDA	recommended daily amount (EU labelling standards) or recommended dietary allowance (USA)
RDI	recommended dietary intake (Australia)
RNI	reference nutrient intake
UL	upper level of intake

Dietary assessment methods

24-hour recall

This is a retrospective method of dietary assessment, where the individual or the interviewer records all food and drinks consumed over the previous

Box 41.1 How to carry out 24-hour recall of food intake

Typically, this will be carried out by questioning a respondent, in the following sequence:

1. **Prepare a record sheet with appropriate sections to record the information you require.** This would normally include the day/date as this may be important in analysing how usual or average the intake is. There are likely to be columns for the time of each eating event, a description of the food and brand name, a description of the portion size or estimated weight and the cooking method. It may be useful to include an example of a completed entry at the top of the sheet, either for the subject to refer to, or to act as a reminder for the interviewer.
2. **List all the food and drinks that they consumed in the previous 24 hours.** Either allow the respondent to quickly list what they ate and then go back through it more slowly with them or take them through their day sequentially, prompting as you go. Include portion sizes in household measures, e.g. dessertspoonful.
3. **Provide additional information for each food item.** Prompt the respondent for more details, so that you have information on cooking methods, leftovers, added foods (e.g. sugar added to hot drinks) and brand names.
4. **Review and finalise the list.** Double check that no foods have been missed by questioning the respondent, e.g. 'Did you not have anything to drink with your lunch?' or 'Did you eat anything when cooking dinner for the children?'

Carrying out 24-hour recall – one way of reducing some of the errors associated with portion size is to use food portion images with the individual.

24 hours. The method is also useful for collecting information on a person's usual intake of food, where you might ask whether the food they consumed over the past day is what they would usually consume. Portion sizes are typically recorded in household measures, e.g. cups, spoons, etc.

This quick and simple method is useful when trying to identify aspects of the diet which might need further investigation. This could be related to eating patterns, e.g. snacking habits or types of foods consumed, such as 'fast food'. However, in practice, detailed nutrient analysis of the information could be considered a starting point for further investigation, given the potential levels of error involved in data collection. The main drawback of the 24-hour recall method is that it relies on memory and you may need to prompt the individual to recall all foods consumed.

Food frequency questionnaire (FFQ)

This method comprises a list of foods and drinks and requires the subject to indicate how often he or she consumes each item from the list (Fig 41.1). The frequency classes used to group the data will depend on the aims of the dietary assessment, but typically would range from 'more than once a day' through 'once a week' to 'never'. The advantage of this method is that it allows identification of foods which may be eaten infrequently and might therefore not be picked up using shorter-term dietary assessment such as 24-hour recall. An FFQ is a relatively quick method of obtaining dietary information, but the large number of foods included in a typical questionnaire may be daunting for some people. This method is useful for identifying general patterns of food intake but is not generally suitable for measuring nutrient intake, as the data collected are not generally detailed enough. The example in Fig. 41.1 indicates that this person consumes between 15 and 25 portions of these fruits per week so FFQ is useful to categorise high, medium or low consumers of particular types of foods, e.g. fruit or vegetables.

FFQs may be used to verify other methods of dietary assessment and vice versa. The foods listed in the FFQ used should be appropriate for the population under investigation, because if the questionnaire does not

Please estimate your average food use as best you can, and please answer every question - do not leave **ANY** lines blank. **Please put a tick (✓) on every line.**

FOODS & AMOUNTS	AVERAGE CONSUMPTION LAST YEAR								
4. FRUIT ***(1 fruit or medium serving)*** *For very seasonal fruits such as strawberries, please estimate your average use when the fruit is in season									
	Never or less than once/ month	1-3 per month	Once a week	2-4 per week	5-6 per week	Once a day	2-3 per day	4-5 per day	6+ per day
Apples	1	2	3	4	✓ 5	6	7	8	9
Pears	1	✓ 2	3	4	5	6	7	8	9
Oranges, satsumas, mandarins, tangerines, clementines	1	2	3	4	✓ 5	6	7	8	9
Grapefruit	✓ 1	2	3	4	5	6	7	8	9
Bananas	1	2	3	✓ 4	5	6	7	8	9
Grapes	1	2	3	✓ 4	5	6	7	8	9
Melon	✓ 1	2	3	4	5	6	7	8	9
*Peaches, plums, apricots, nectarines	1	2	3	✓ 4	5	6	7	8	9
*Strawberries, raspberries, kiwi fruit	1	2	✓ 3	4	5	6	7	8	9
Tinned fruit	✓ 1	2	3	4	5	6	7	8	9
Dried fruit, e.g. raisins, prunes, figs	✓ 1	2	3	4	5	6	7	8	9
	Never or less than once/ month	1-3 per month	Once a week	2-4 per week	5-6 per week	Once a day	2-3 per day	4-5 per day	6+ per day

Fig. 41.1 Example of a completed food frequency questionnaire.

contain the appropriate foods for that population then it would result in under-estimation of food intake. For example, if the population you are questioning eat a lot of cauliflower but you have not included this on the list then you will get an incorrect measure of their vegetable intake.

Food diary

Recording portion sizes in a food diary – portion sizes are generally recorded in household, descriptive measures in an unweighed (estimated) food diary, while a weighed food diary requires all foods and drinks consumed to be weighed and recorded. The latter requires greater commitment by the participant, but provides more robust data.

Where more specific or accurate nutrient intake is required, then a food diary is a useful method, as it requires the subject to record everything he or she consumes over a number of days, usually between three and seven. It also has the benefit of being a recording method and therefore does not rely on memory. The length of time the diary is kept will depend on the requirements of the analysis, but will have an impact on the level of engagement required from the participant. If you require a very accurate analysis of a person's nutrient intake then you should consider a seven-day record as this will allow for daily variation, particularly over weekends, when intake is likely to be different from weekdays. If fewer than seven days are recorded then at least one of the days should be at the weekend. You should remember that the level of accuracy of the information recorded will depend on the motivation, honesty, literacy and numeracy skills of the subject.

Participants should complete as much information about each foodstuff as possible, including methods of cooking and brand names, where appropriate, as well as details of any foods not eaten (leftovers). The recording format is similar to that of the 24-hour recall method and, again, it is useful to include an example entry to show the level of detail you require.

To ensure the accuracy of the information provided it is good practice to discuss the content of the diary with the subject very shortly after completion, although this will depend on the purpose of the dietary assessment. During this process the interviewer can prompt the subject to confirm the details of the diary. This should help to ensure clarity of content and also provides a check that there are no omissions, as people will often forget to enter details such as what type of milk or how many spoonfuls of sugar they have in their coffee. This process is particularly important in unweighed food diaries, where you can use standardised pictures of portion sizes (such as in *A Photographic Atlas of Food Portion Sizes*, Nelson *et al.*, 1997) to estimate quantities, which will add to the accuracy of any analysis.

Using mobile phones in dietary assessment – some of the more innovative projects have used daily text messages to prompt participants to complete their food diaries.

Innovative dietary assessment methods

Several research projects have evaluated the use of technology to enhance or replace more traditional dietary assessment methods. For example, digital cameras and mobile phones can be used to takes pictures of food before it is consumed. There is also a number of PDA (personal digital assistant) software packages that may be used, although there are cost-implications associated with their use. However, these methods may be easier for the participant to use and so may increase compliance without compromising the accuracy of the data.

Analysis of dietary assessment

Once a record of food intake has been obtained, the next stage is to analyse the data. How you do this analysis will depend on the resources you have available to you, but in all cases you will be able to calculate the intake of nutrient you are particularly interested in. The key issue with nutrient analysis is that the accuracy and therefore validity of the analysis will only be as good as the initial data collection, so if the assessment is based on poor or vague information, then the outcome of the subsequent nutrient analysis will be similarly flawed. Whether you do the analysis manually using food composition tables or use appropriate software, you will need to use composition data relevant to the country in which you are working, as the nutrient contents of similar foods can be different in different countries.

Food composition tables

Using commercial dietary analysis software packages – these include Microdiet®, Compeat®, Nutmeg®, WinDiet® and Hamilton Grant®.

Food composition tables are the source of the majority of information required for nutrient analysis. These are lists of thousands of foods and the amounts of nutrients per 100 g. In the UK these composition tables are available from the National Nutrient Databank, which is currently maintained and managed by the Food Standards Agency. There is now a rolling programme of food analysis to ensure that the data are as up-to-date as possible and reflect changing production, processing and preparation methods. The printed version in the UK is known as *McCance and Widdowson's The Composition of Foods* (Food Standards

Table 41.1 Nutrient content of different types of chicken per 100 g

	Energy (kJ)	Protein (g)	Fat (g)
Roasted meat, average	742	27.3	7.5
Roasted dark meat	819	24.4	10.9
Roasted light meat	645	30.2	3.6

Food Standards Agency (2002)

Agency, 2002a). The information in Table 41.1 shows some of the different types of chicken meat which may all be described as 'roast chicken' in a food diary, so you can see it is important to know whether it is dark or light meat as the nutrient composition will be different for each. In the USA the United States Department of Agriculture (USDA) publishes the National Nutrient Database online (USDA, 2009) so it is possible to access nutrient information for food products very easily.

Nutrient analysis software

The data from food composition tables have been used in a range of dietary analysis and recipe analysis software packages. Your university will probably subscribe to a single product, but the principles for using them are the same for all.

Using food tables and nutrient analysis software

Since the information provided in food tables or analysis software is given per 100 g of food, you will need to know how much of each food someone has eaten to be able to calculate his or her daily nutrient intake. Consider the different methods of dietary assessment outlined above to evaluate how accurate the information you enter into the analysis package might be.

Figure 41.2a shows an example of a food diary which would be difficult to analyse because not only are the food products not very well detailed

(a)

Meal	Food	Portion size
Breakfast	cereal	bowlful
	milk	splash
Snack	chocolate bar	
Lunch	salad	bought
	coke	can
Snack	apple	average
Dinner	fish potatoes salad water	whole 3 bowl glass
Snack	ice cream	one

(b)

Meal	Food	Portion size
Breakfast	Kellogg's Cornflakes	4 heaped tablespoons
	Semi-skimmed milk	½ cup
Snack	Snickers bar black coffee, instant	standard size mug
Lunch	Greek salad: lettuce feta cheese tomato raw red onion olive oil	 ½ bag matchbox size 6 cherry ¼ 2 tablespoons
	can of diet coke	330ml
Snack	green apple	large
Dinner	grilled salmon (no skin) new potatoes, boiled, lettuce cucumber low-fat French dressing sparkling water	supermarket portion 3 egg sized ¼ bag 6cm 1 tablespoon 500ml
Snack	Häagen Dazs vanilla ice cream	2 scoops

Fig. 41.2 Example of a food diary with insufficient (a) and sufficient (b) detail.

Box 41.2 How to analyse a food diary

1. **Create a separate file for each respondent.** If you are using any form of software, remember to give an appropriate file name and save your work regularly.
2. **Search for the first food listed in the food diary.** Each food will probably have a number of types, e.g. bread – is it white, brown or wholemeal?
3. **Choose the appropriate cooking method.** Has the food been grilled, baked, roasted, etc.?
4. **Find and substitute an alternative food, if necessary.** If you can't find an exact match, search using a different term or choose an appropriate alternative. Make a note on the diary if you have had to use a substitution.
5. **Enter ingredients of composite foods (if you know them).** Some composite foods are included in food composition tables but if not, try to source an appropriate recipe and use values for the component ingredients (consider whether the respondent has given enough information for you to do this accurately). Most processed foods give nutrient information on the tables, which can also be entered if necessary.
6. **Enter the portion size,** once you have identified the appropriate food product. If you do not have the exact portion size use average portion size information provided in the software or consult portion size books (Food Standards Agency, 2002b).
7. **Make a note of the code number of common foods.** In composition tables each food will have been assigned a unique code number. For foods which are likely to have been listed more than once, e.g. milk, it may save time to use the code number to search the database.
8. **Enter the meal code (if available).** This will allow you to add another aspect to your analysis, although it is not a requirement.
9. **Work methodically through the food diary.** Tick off each food as you go so that you don't lose your place.
10. **Calculate average daily nutrient intakes.** Divide the nutrient totals by the number of days you are analysing. This is sometimes called a divisor. For example, if you have a completed three-day food diary then you would divide the nutrient totals by 3.

but the portion sizes are also very vague. This shows how important it is to collect accurate and detailed information from the person. In contrast the example shown in Fig. 41.2(b) provides sufficient detail to allow accurate analysis.

Comparing nutrient intake to reference standards

Once you have calculated the nutrient intake or composition of the food then software packages have a number of options. If you have analysed a recipe, you could calculate the nutrients per 100 g of the product for use on food labels (see pp. 54–56 for legislative requirements for food labels). More usually you will want to compare the results against dietary standards, to begin to make a judgement on the appropriateness of the dietary choices of the individual and the adequacy of their nutrient intake.

Dietary reference values (DRV)

DRV are a series of estimates of adequate intakes of nutrients for a normal healthy population. The terminology for dietary reference values differ slightly among countries although the principles are largely the same. To set these values a full review of all of the scientific data available is carried out and these values are periodically reviewed in light of developing findings.

The reference values differ for different sexes, ages and stages of life, and include values for pregnancy and lactation. DRVs were established with the assumption that nutrient requirements follow a normal frequency distribution (a bell-shaped curve, see Fig. 41.3). They are also based on average body weights and average physical activity levels, therefore the

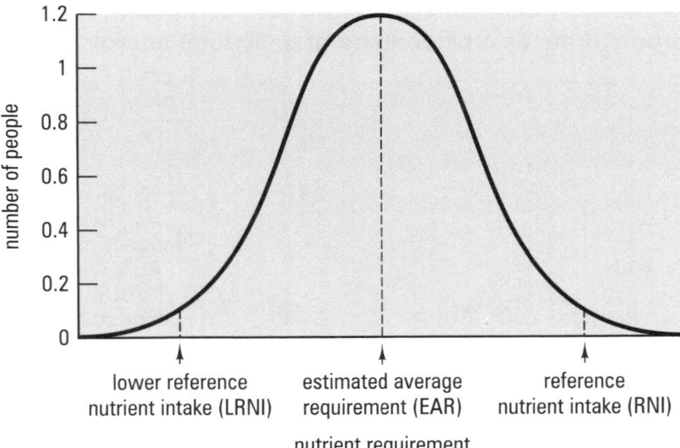

Fig. 41.3 Normal distribution of nutrient requirements (Department of Health, 1991).

more information you have about the person you are analysing the more accurate you can be with your evaluation and recommendations.

DRVs in the UK are given as three values:

- The reference nutrient intake (RNI) would be sufficient to meet the needs of the majority of the population.
- Estimated average requirement (EAR) is the amount required to meet the nutrient needs of half of the population and is the value against which energy intake is normally compared.
- Finally there is the lower reference nutrient intake (LRNI) value, which represents the amount of a nutrient that would only meet the needs of a 2.5 per cent of the population. Table 41.2 gives a summary

Table 41.2 Dietary reference values for males and females in UK (19–50 y) (Department of Health, 1991), USA/Canada (19–30 y) (Food and Nutrition Board Institute of Medicine of the National Academies, 2005) and Australia/NZ (19–30 y) (National Health and Medical Research Council, 2006)

	UK RNI* 19–50 years		USA/Canada RDA* 19–30 years		Australia/NZ RDI*19–30 years	
	male	female	male	female	male	female
Energy MJ (kcal)	10.6 (2550)[a]	8.10 (1940)[a]	12.2 (2900)[b]	9.2 (2200)[c]	10.3[e]	7.7[f]
Protein (g)	55.5	45	63	50	64	46
Thiamin (mg)	1.0	0.8	1.2	1.1	1.2	1.1
Riboflavin (mg)	1.3	1.1	1.3	1.1	1.3	1.1
Niacin (mg)	17	13	16	14	16	14
Vitamin B6 (mg)	1.4	1.2	1.3	1.3	1.3	1.3
Vitamin B12 (mg)	1.5	1.5	2.4[d]	2.4[d]	2.4	2.4
Folate (μg)	200	200	400	400	400	400
Vitamin C (mg)	40	40	90	75	45	45
Vitamin A (mg)	700	600	900	700	900	700
Vitamin D (mg)	ND	ND	5[d]	5[d]	5[d]	5[d]
Calcium (mg)	700	700	1000[d]	1000[d]	1000	1000
Phosphorus (mg)	550	550	700	700	1000	1000
Magnesium (mg)	300	270	400	310	400	310
Sodium (mg)	1600	1600	1500[d]	1500[d]	2300[g]	2300[g]
Potassium (mg)	3500	3500	4700[d]	4700[d]	3800[d]	3800[d]
Chloride (mg)	2500	2500	2300[d]	2300[d]	ND	ND
Iron (mg)	8.7	14.8	8	18	8	18
Zinc (mg)	9.5	7.0	11	8	14	8
Copper (mg)	1.2	1.2	0.9	0.9	1.7[h]	1.2[h]
Selenium (μg)	75	60	55	55	70	60
Iodine (μg)	140	140	150	150	150	150

* Unless otherwise stated: [a]EAR; [b]AI; [c]EER based on age 30 y, BMI 24.99 kg/m^2, height 1.8 m, low active; [d]EER based on age 30 y, BMI 24.99 kg/m^2, height 1.65 m, low active; [e]based on EER age 19–30 y, height 1.8 m, weight 71.3 kg, PAL 1.4; [f]based on EER age 19–30 y, height 1.6 m, weight 56.3 kg, PAL 1.4; [g]UL; [h]AI; ND = no data.

Table 41.3 Dietary reference values (UK) and acceptable macronutrient distribution ranges (USA and Australia/NZ) for fat and carbohydrate as a percentage of daily total energy intake

	DRV/AMDR % total energy intake		
	UK	**USA**	**Australia/NZ**
Total fat	33	20–35	20–35
Saturated fat	10	As low as possible while consuming a nutritionally adequate diet	No more than 10
Trans-fatty acids	2	As low as possible while consuming a nutritionally adequate diet	
Polyunsaturated fatty acids (PUFA)	6 –10	Not stated	Not stated
n-6 PUFA	0.2	5–10	4–10
n-3 PUFA	1.0	0.6–1.2	0.4–1
Monounsaturated fatty acids	12	Not stated	Not stated
Total carbohydrate	47	45–65	45–65
Non-milk extrinsic sugars	10	25	Not stated
Intrinsic, milk sugars and starch	37	Not stated	Not stated

of dietary reference values in different countries. Values vary between countries as the values are based on data collected from the different populations of each country and therefore it is important to use the appropriate values for the given population you are studying.

Correct use of DRVs – For UK populations it is important to consider whether these values are *maximum* values (in the case of dietary fat) or *minimum* values (for carbohydrate). Other countries have identified ranges for these values rather than a single figure.

DRVs have also been established to provide guidance on the amount of energy the average person should consume from fat and carbohydrate, expressed as percentage of either total energy intake (this includes food and alcohol consumption) or food energy intake (if the person does not consume alcohol). These values allow you to evaluate whether the proportions of macronutrients are appropriate in the diet. Table 41.3 gives details of DRV and acceptable macronutrient distribution ranges for macronutrients.

Using dietary reference values

DRVs are used most effectively when comparing intakes of a population, rather than an individual, so if you have surveyed a number of people from a homogenous population (e.g. of the same sex, age, activity level) and you find that their average intake is around the RNI (or RDA in USA) level then you can be fairly confident that there is little risk of them not meeting their requirement for that nutrient. Conversely, if the average intake is at the LRNI level then you would be concerned that this was insufficient to meet their requirements. If you are using DRVs to assess the diet of an individual, then you should be more cautious when making judgements, without doing further dietary or biochemical analysis to test for deficiencies.

KEY POINT DRVs should be used as guidelines not as definitive goals for individuals.

Box 41.3 gives an example of dietary analysis, showing the factors that you might need to take into consideration.

Box 41.3 Dietary analysis – an example

Mr Smith has completed a three-day food diary which you have analysed, and from your analysis you conclude that he consumes an average of 10.6 MJd^{-1}. As Mr Smith is 30 years old you will need to choose the appropriate DRV for his sex and age. In this case the EAR for energy in the UK is 10.6 MJ/d (see Table 41.2), suggesting that his energy intake is ideal. However, you must consider that this EAR is based on an average body weight of 74 kg and a physical activity level of 1.4. If Mr Smith weighs more than this and/or has a higher activity level then his average intake may not be adequate to meet his requirements. Although weight and physical activity may affect other nutrient requirements less than they do for energy, you must still be cautious in the interpretation of DRVs for micronutrients. So, if you calculate that Mr Smith has an average iron intake of 4.7 mgd^{-1} you can see that this is at the LRNI level. However, you cannot conclude that his dietary intake is inadequate as he may have a low individual requirement for iron and therefore his dietary intake may be meeting his requirements. In this case you could do further dietary analysis over a longer period of time to establish whether this was indeed a long-term pattern of intake and also investigate other signs or and symptoms of potential iron deficiency (see p. 290).

Text references and sources for further study

Department of Health (1991) *Dietary Reference Values for Food Energy and Nutrients for the United Kingdom. Reports on Public Health and Medical Subjects No. 41.* HM Stationery Office, London.

Food and Nutrition Board Institute of Medicine of the National Academies (2005) *Dietary Reference Intakes for Energy, Carbohydrate, Fiber, Fat, Fatty Acids, Cholesterol, Protein, and Amino Acids.* The National Academies Press, Washington, DC. Available: http://books.nap.edu/openbook.php?record_id=10490&page=R1
Last accessed: 21/12/10.

Food Standards Agency (2002a) *McCance and Widdowson's The Composition of Foods,* 6th summary edn. Royal Society of Chemistry, Cambridge.

Food Standards Agency (2002b) *Food Portion Sizes,* 3rd edn. HMSO, London.

National Health and Medical Research Council (2006) *Nutrient Reference Values for Australia and New Zealand.* Commonwealth of Australia. Available: http://www.nhmrc.gov.au/_files_nhmrc/file/publications/synopses/n35.pdf
Last accessed 21/12/10.

Nelson, M., Atkinson, M. and Meyer, J. (1997) *A Photographic Atlas of Food Portion Sizes.* MAFF Publications, London.

United States Department of Agriculture (2009) *National Nutrient Database for Standard Reference, Release 22 Nutrient Data Laboratory Home Page.* Available: http://www.ars.usda.gov/ba/bhnrc/ndl
Last accessed 21/12/10.

42 Physical examination

> **Definitions**
>
> **Nutritional assessment** – the measurement of indicators of dietary status and nutrition-related health status to identify the possible occurrence, nature and extent of impaired nutritional status (US Department of Health and Human Services, 2000)
>
> **Signs** – objective physical manifestations, usually observed by the practitioner.
>
> **Symptoms** – subjective manifestations experienced and described by the person to the practitioner.

Physical examination should be viewed as part of an overall assessment of an individual's nutritional status. Observations from a physical assessment would normally be used in conjunction with anthropometry (Chapter 43), dietary assessment (Chapter 41) and, where appropriate, biochemical analysis (Chapter 45). Methods of integrating all of this information are described in Chapter 47 as no single measure can be seen as diagnostic in nutrition-related conditions.

Physical examination can provide information in two main ways:

1. It can aid the diagnosis of a condition, in combination with other information.
2. It can be used to track changes in a condition over time, starting from baseline measurement and observations to gauge improvement or deterioration during treatment.

> **KEY POINT** While signs and symptoms noted during a physical examination cannot provide enough information on their own to diagnose nutrition-related conditions, they can provide an indication of possible problems that can then be followed up.

The main limitation of a physical examination is the non-specificity of the signs. Most nutrient deficiencies do not occur in isolation and therefore there are likely to be other deficiencies, resulting in a range of signs and symptoms.

> **Developing your skills** – it is important to be aware of your personal scope of practice; if you note abnormal findings in you practical classes, you should bring these to the attention of your lab supervisor

Although signs are, by definition, objective, there may be inconsistencies between examiners. This is more likely to occur if an observation needs to be graded. For example, in the subjective global assessment (Chapter 47), physical observations, including ankle oedema, are rated from normal to severe. To ensure consistency it is necessary to practice making these observations by repeating the same examination during training, and by comparing your observations with other, more experienced, examiners. If the presence of signs are just recorded as positive or negative, i.e. present or not, then the possibility inconsistencies is reduced.

In some cases the same physical sign in different people may not suggest the same condition. For example, the normal ageing processes may result in signs which could be attributed to malnutrition in younger people, but which do not necessarily indicate this in later life.

If the physical examination suggests possible deficiencies or malnutrition, specific biochemical tests can be carried out; however, it is possible to start treatment to improve the diet of the person even before definite diagnosis is available, as improved dietary intake is very unlikely to have any detrimental effect on the person even if the biochemical analysis fails to indicate a specific condition.

Condition of the hair			
Shiny		Dull/dry	
Healthy scalp	✓	Colour changes	
		Corkscrew	
		Easily plucked	
Condition of the lips			
Smooth		Cheilosis	
Moist	✓	Angular stomatitis	
Condition of the nails			
Smooth	✓	Koilonychia	
Pink	✓	Transverse ridging	
		Brittle	

Fig. 42.1 Extract from record sheet for signs relevant in nutrition-focused physical assessment.

Definition

Oedema (UK)/edema (US) – excess fluid accumulation in intercellular spaces of body tissues. It may accompany malnutrition, infection or may be a side-effect of medication. Oedema can also occur after strenuous physical activities, high salt intake or during pregnancy.

Performing a physical examination

The process of examination is also useful in getting to know the individual and should act as a way of putting the individual at ease. The following points outline the key aspects:

- **Use appropriate hygiene precautions**. Before conducting an examination always wash your hands thoroughly. If you are using equipment, ensure that this is also cleaned before use. In some clinical settings, where there is a risk from infection, it may be necessary to wear protective clothing such as gloves.
- **Avoid being impersonal**. Always introduce yourself to the person and explain the purpose and process of the examination. That way, he or she is more likely to provide you with open and honest responses to questions during the examination if he or she is not wondering what your reasons are for asking the questions.
- **Ensure the individual's comfort and privacy.** When carrying out the examination, ensure that the physical environment allows the person to feel comfortable and there is no risk of interruption, for example by using a separate room with a closed door and 'do not disturb' sign. If it is necessary for the person to remove any items of clothing, allow him or her to remain covered except for the part of the body that you are examining. Be sensitive to any possible traits of the person such as gender, religion, sexual orientation or colour that might influence your interaction with them to your responses, to your actions or questions.
- **Keep appropriate notes.** Use a record sheet to note your observations (Fig 42.1). Although there are no universally standard record sheets, many institutions have their own versions and you could prepare your own. This approach serves at least three purposes:

  1. it provides an *aide memoire* to ensure no part of the examination is missed;
  2. it allows comparisons to be made over time, so that changes are readily identified;
  3. it facilitates the communication of key signs and symptoms between health professionals.

Components of a nutrition-focused physical examination

This type of detailed examination would be carried out if initial nutritional screening, using tools such as MUST and MNA (Chapter 47), indicates that the person is at risk of impaired nutritional status. As a student you are likely to use these methods in theoretical case-study assessments or in practical classes to help develop your skills. Some of the signs which are described below are unlikely to be observed in the general population in developed countries, although may be relevant in clinical settings.

KEY POINT If the person you are examining is wearing make-up you may need to ask for it to be removed so that you can make accurate observations about his or her skin, eyes, lips or nails.

Overall Physical appearance

It is possible to make a number of general observations about an individual before a more thorough examination is carried out. These observations can provide an initial indication about a person's nutritional health. Overall body size can be noted from initial impressions and this may be useful for checking that your BMI calculations are correct. It is also possible to observe whether or not a person's clothes are loose and ill-fitting, suggesting weight loss or, conversely, their clothes may appear tight, suggesting recent weight gain. Tight clothes or rings and/or swollen ankles may also indicate fluid retention (oedema). In older adults weight loss may also be apparent from ill-fitting dentures.

Mobility

Observing the general movement of the individual can act as a starting point for assessment of any musculoskeletal signs of nutritional problems. These will also need to be considered in the context of any other underlying medical conditions.

If the subject is in good health, you would expect to see a good upright posture and no signs of pain when he or she is walking. If movement proves difficult or painful, then either muscle wasting or protein energy malnutrition may be suspected. If the person reports muscle pain as a symptom, this may suggest thiamin deficiency.

In children, vitamin D deficiency results in rickets, which is characterised by bowed legs and delayed walking.

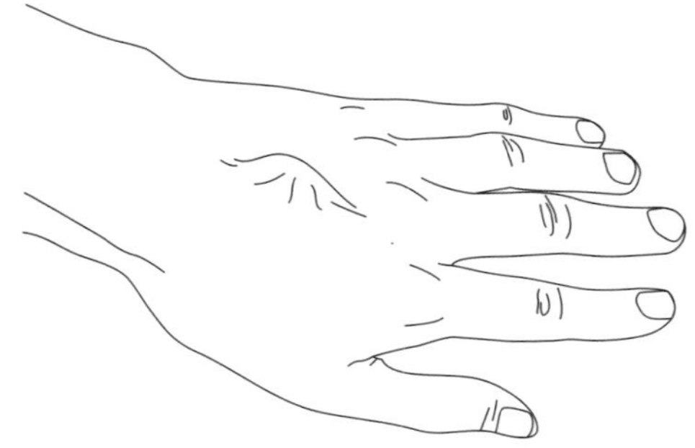

Fig. 42.2 Drawing of dehydrated skin, showing 'tenting' when the skin on the back of the hand is gently pinched.

Mood

Observations of the general behaviour of the individual may also provide important information. If the person is listless, lethargic and shows poor concentration this may be symptomatic of inadequate nutrition. If there are signs of dementia and disorientation this may be indicative of a B vitamin deficiency, although only in very advanced deficiency states, and rarely seen in developed countries. Dehydration may result in confusion or apathy.

Skin

Skin should normally look smooth with a good, even colour. It should be slightly moist. Signs of possible malnutrition include:

- Bruising or bleeding under the skin (*purpura*) can be linked to vitamin C deficiency. Other signs of vitamin C deficiency in the skin are poor wound healing.
- Susceptibility to bruising suggests vitamin K deficiency. If the skin is dry and flaky, or shows signs of scaling, it may indicate a deficiency of essential fatty acids, zinc or vitamin A.
- Pale skin may indicate anaemia. This may be caused by deficiency of iron, vitamin B12 or folate.
- Reduced elasticity of the skin is linked to dehydration. This can be observed by gently pinching the skin and releasing; if it remains in the pinched position and does not return immediately to its original state, this is known as 'tenting' (see Fig. 42.2).

Some of the signs listed previously will also be observed if the individual has poor skin care, and they are also linked to ageing. Some

Fig. 42.3 Flag sign of protein malnutrition showing a band of pigmentation loss.

> **Definitions**
>
> **Bitot's spots** – oval or triangular spots which occur on the conjunctiva.
>
> **Xerophthalmia** – literally 'dry eye' (in Greek); lack of tears usually causes significant damage to the cornea over time. It is a major cause of blindness in children in Africa and South America.

Fig. 42.4 Examination of eye membranes (Sheth *et al.*, 1997).

medications may also result in alterations to the texture and appearance of the skin. For example, long-term use of inhaled high-dose corticosteroids for conditions such as asthma may be associated with increased evidence of skin bruising (Dahl, 2006).

Hair

In healthy people the hair will be shiny and firm in the scalp. Signs of possible malnutrition include:

- Hair that can be easily plucked from the head with no pain, or which is falling out, may indicate an inadequate protein intake. Protein deficiency may also result in thin hair, although this is also observed in zinc and biotin deficiencies. In rare cases, intermittent protein deficiency can also result in loss of pigmentation of the hair, giving rise to a so-called 'flag sign', which is a band of colour loss or colour change (Fig. 42.3).
- Corkscrew-shaped hair and unemerged coiled hairs may suggest a deficiency of vitamin C.

Other non-nutritional factors which may result in these signs include ageing and overprocessing of the hair such as excessive bleaching.

Eyes

The eyes should be shiny and clear with pink membranes and rapid pupillary adjustment to bright light. Signs of possible malnutrition include:

- Redness at the corners of the eyes may be due to deficiencies of B vitamins.
- Pale membranes may suggest iron deficiency. This can be checked by gently pulling down the lower eyelid and observing the colour of the conjunctiva (Fig. 42.4), the rim of the eye.
- Dry eyes (xerosis), which can lead to ulceration of the cornea. This can be caused by deficiency of vitamin A. This deficiency can also cause the development of Bitot's spots. Collectively, these signs are the clinical features of xerophthalmia.

Some of these signs may develop as a result of allergies or other eye disorders. Dry eyes may also occur as part of the normal ageing process.

Lips

The lips should be smooth and moist. Signs of possible malnutrition include:

- Dry and cracked lips with sores at the corners of the mouth may indicate deficiency of B vitamins, riboflavin, B6 or niacin.
- Dry, cracked or sore lips may also be caused by dehydration.

Similar signs can be due to weather conditions such as cold, wind or sun.

Definition

Koilonychia – the word koilonychia is from the Greek words *koilos* for hollow, and *onyx* for nail.

Fig. 42.5 Transverse ridges on nails which may indicate periods of protein malnutrition.

Dentition and dietary deficiencies – although the teeth may give limited indication of deficiencies it is worth noting any tooth loss or ill-fitting dentures as this could contribute to inadequate dietary intake and consequently impact on nutrient intake.

Tongue and mouth

The tongue should be pink and not show signs of swellings. The subject should report having normal taste functions and there should be no bleeding from the gums. Signs of possible malnutrition include:

- A bright red tongue may be linked to a deficiency of B vitamins, e.g. niacin, or a magenta tongue in the case of riboflavin deficiency. Other symptoms of B vitamin deficiency are glossitis, which is a swelling of the tongue and atrophic lingual papillae (also known as slick tongue).
- Signs of bleeding and swollen gums may be observed in cases of vitamin C deficiency.

Some medications may affect the colour and general appearance of the tongue, for example some antihistamines can cause dry mouth. Periodontal disease and generally poor oral hygiene will also give rise to similar signs to those described above.

Nails

Healthy nails are smooth and pink. Signs of possible malnutrition include:

- Spoon-shaped nails which have lost the usual convex contour and become concave may be seen in iron deficiency. This condition is known as koilonychia.
- Transverse (horizontal) ridging of the nails (Fig. 42.5) is associated with protein energy malnutrition. These are also known as Beau's ridges and reflect periods of interruption in the growth of the nail and will be evident on all the nails.

Ridging of the nails may also be caused by skin conditions such as eczema or psoriasis, or by physical damage to the nail bed. Vertical ridging of the nails is a normal sign of ageing. Koilonychias may be hereditary.

Functional assessment

Assessment of physical ability may be required in certain circumstances, such as for elderly residents in care homes, where protein energy malnutrition or generalised poor nutrition may be suspected. Loss of physical ability to carry out everyday functions may be caused by malnutrition, or conversely malnutrition may be as a result of loss of function. Either way, it is important to establish the subject's ability to perform these functions and so evaluate their care regime.

As an example, this could involve a series of tests which measure hand grip or exercise tolerance. It is also possible to measure physical ability by appraising the ability of a person to carry out routine daily activities such as dressing, bathing, feeding and transferring from their bed or chair. A tool commonly used to assess function in independently living older adults is the Lawton Instrumental Activities of Daily Living (IADL) scale (Lawton and Brody, 1969). This questionnaire includes activities such as shopping, housekeeping, laundry and using the telephone to measure physical ability.

Text references

Dahl, R. (2006) Systemic side effects of inhaled corticosteroids in patients with asthma. *Respiratory Medicine,* **100**, 1307–1317.

Lawton, M.P. and Brody, E.M. (1969) Assessment of older people: Self-maintaining and instrumental activities of daily living. *Gerontologist*, **9**(3) 179–186.

Sheth T.A., Choudry N.K., Bowes, M. and Detsky A.S. (1997) The relation of conjunctival pallor to the presence of anemia. *Journal of General Internal Medicine*, **12**(2), 102–106.

US Department of Health and Human Services (2000) *Healthy People 2010: Understanding and Improving Health*. Available: http://www.healthypeople.gov/document/pdf/uih/2010uih.pdf
Last accessed 21/12/10.

Sources for further study

Fuhrman, M.P. (2009) Nutrition-focused physical assessment. In Charney, P. and Malone, A.M. (eds) *ADA Pocket Guide to Nutrition Assessment,* 2nd edn. American Dietetic Association, Chicago, Illinois.

Gibson, R.S. (2005) *Principles of Nutritional Assessment,* 2nd edn. Oxford University Press, New York.

Jarvis, C. (2008) *Physical Examination & Health Assessment*, 5th edn. Saunders Elsevier, St Loius, Illinois.

Lee, D.L. and Nieman, D.C. (2009) *Nutritional Assessment,* 4th edn. McGraw Hill Higher Education, New York.

Rolfes, R.R., Pinna, K. and Whitney, E. (2006) *Understanding Normal and Clinical Nutrition*, 8th edn. Wadsworth-Cengage, California.

43 Anthropometric and body composition measurements

> **Definition**
>
> **Anthropometry** – measurement of the size, weight and proportional composition of the human body using non-invasive techniques.

Anthropometry is used to obtain information on a person's physical body stature and composition, to determine health/health risk status. Variables measured include (i) fat and lean muscle composition and (ii) body fat distribution. Each anthropometric procedure has its strengths and weaknesses, and often a combination of two or more methods is used for a more complete analysis.

Height and length

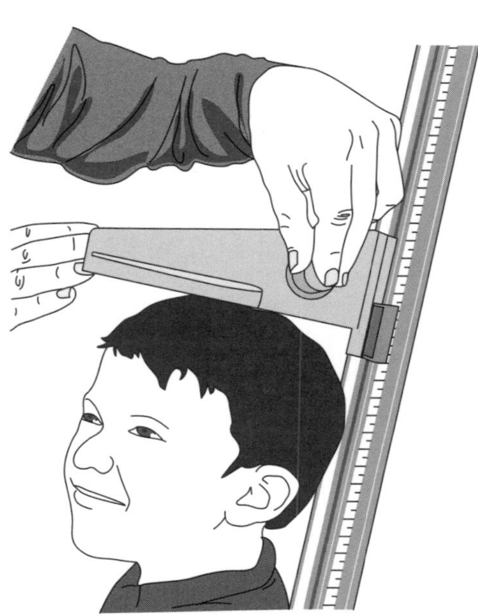

Fig. 43.1 Measuring height with a stadiometer.

Height should be measured without footwear, with feet together and heels, buttocks, shoulders and back of the head against a wall, measuring board or stadiometer, with the person standing straight and looking directly forward. The top end of the ear and outer edge of the eye should be in a line parallel to the floor (the Frankfort plane). From the top of the stadiometer a horizontal bar is then lowered to lie flat on top of the head (Fig. 43.1) and the height read from the scale and recorded. Height should be measured to the nearest 0.5 cm.

If a height board or stadiometer is not available, a simple but effective means of measuring height can be set up by attaching the zero end of a tape measure to a wall, perpendicular with the floor, and taping the measure again at about 2 metres straight up the wall. A flat ruler is placed horizontally back-to-front over the person's head to reach the wall and the measurement is then read and recorded.

The body length of infants can be measured using a horizontal measuring board with a movable foot board and a fixed head board. Two people are usually required for this measurement, one to gently hold the infant's head against the head board, the other to carefully straighten the infant's legs and move the foot board to the bottom of the feet. Record lengths to the nearest 0.5 cm. If a measuring board is not available, the infant can be placed on a section of cardboard and a mark can be made below the top of the lying infant's head and at the bottom of the heel and then the distance between can be measured after carefully removing the infant. Once the infant has learned to stand, standing measurements can be implemented.

Body mass

> **Avoiding unreliability in self-reported measures** – always try to measure someone's weight or height yourself rather than asking him or her for details, because self-reported weights are often lower than their true value, while self-reported heights are often higher.

Body mass is the quantity of matter in a person's body. It is determined by the measurement of weight, which is the force that the matter brings to bear on the scales due to gravity. Beam balances or electronic scales are used for accurate mass measurements for adults and children. For infants, special scales have platforms for the infant to sit or lie on.

When using scales or balances, you should always ensure that they are regularly calibrated as indicated in the manufacturer's guidelines and that they are tested for accuracy, e.g. using a series of known weights. Ensure that the scale is reading zero before use.

Footwear should not be worn. The person should stand still on the platform, with his or her weight equally distributed between both feet, without touching or leaning on anything. Record measurements to the nearest 0.1 kg. Also record the time of day the measurement was taken, as results can vary by 1–2 kg during a 24-hour period.

Where people are carrying out their own body mass measurements at home they should weigh themselves at roughly the same time each day (e.g. before breakfast) naked or in the same amount of clothing, following voiding.

Body mass index (BMI)

The BMI provides an indication of mass relative to height, and is used to determine whether individuals have an appropriate body mass for their height. It is calculated according to the equation:

$$\text{BMI} = \text{body mass in kg}/(\text{height in m})^2 \qquad [43.1]$$

Example – calculation of BMI

Body mass = 71 kg
Height = 153 cm (1.53 m)
BMI = $71/1.53^2$
= 71/2.34
= 30.3
= obese category

BMI categories and ranges have been defined by the WHO, which uses them as an international standard for comparisons between countries and regions (Table 43.1).

Table 43.1 BMI values and categories for adults (kg/m^2)

Underweight	<18.5
Normal	18.5–24.9
Overweight	25.0–29.9
Obese	30.0–39.9
Morbidly obese	>40.0

WHO (2009).

Being underweight according to BMI may indicate under-nutrition, malnutrition, serious illness, or the need for further enquiries into the possibility of anorexia nervosa and related conditions. Being underweight may reduce immune system function and increase the risk of infectious diseases such as respiratory infections, cancer and osteoporosis. Anorexia nervosa is a serious condition related to the person under-eating due to a distorted self-body image. If not corrected it may eventually lead to organ and multiple organ failure.

Being overweight or obese are moderate and high risk factors respectively, for type II diabetes, hypertension, atherosclerosis, stroke, cardiovascular diseases and some cancers.

The weaknesses of BMI include:

- it does not give any real indication of body composition;
- it does not indicate where the fat is located on the body;
- it can vary between different ethnic groups;
- it can overestimate body fat in very muscular people;
- it can underestimate body fat in very thin people;
- it may overestimate body fat in shorter people.

Waist circumference

This measurement provides a good correlation with total abdominal fat and has been shown an independent indicator of morbidity. Abdominal fat has been shown to be a higher risk factor than gluteal (buttocks and hips) fat with regard to type II diabetes, hypertension and cardiovascular diseases (Table 43.2).

The waist circumference is the circumference of the abdomen at its narrowest point between the lower costal rib (tenth rib – lowest of the ribs that articulate with the sternum) border and the top of the iliac crest (hip bone), perpendicular to the long axis of the trunk (Fig. 43.2). The subject's arms are folded across the thorax. The measurement is taken at the end of a normal breath. If there is no obvious narrowing, the measurement is taken at the mid-point between the lower costal (tenth rib) and the iliac crest. The tape should be wrapped snug around the waist but should not compress the skin. The tape must be parallel to the ground. The measurement should be recorded to the nearest centimetre.

Table 43.2 Risks associated with high waist circumference

Classification	Waist circumference – men (cm)	Waist circumference – women (cm)	Risk of metabolic complications
Neither overweight nor obese	<94	<80	None
Abdominally overweight but not obese	≥ 94 and <102	≥ 80 and <88	Increased
Abdominally obese	≥ 102	≥ 88	Significantly increased

ABHI (2008).

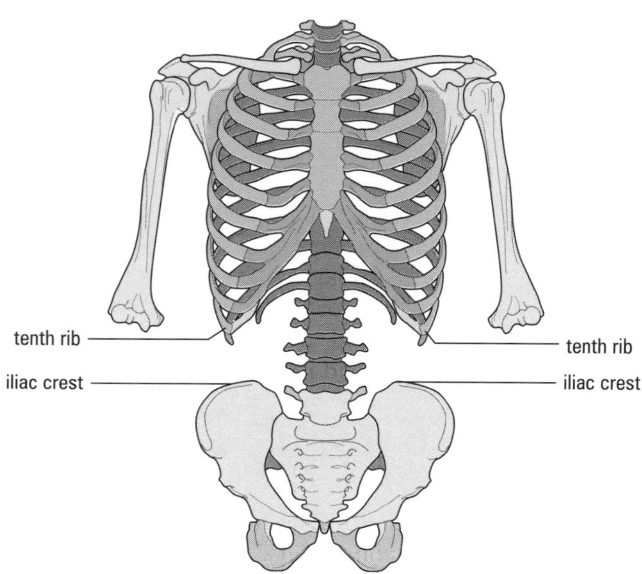

Fig. 43.2 Landmarks for determining the site for waist circumference measurement.

Hip circumference

Hip circumference is used in conjunction with waist circumference to calculate the waist to hip ratio, which gives an additional indication of fat distribution.

Hip circumference is the circumference of the buttocks at the level of the greatest posterior protuberance, perpendicular to the long axis of the trunk. The subject's feet must be together. The tape must be parallel to the ground. Record the measurement to the nearest centimetre.

Waist to hip ratio (WHR)

This ratio differentiates between abdominal obesity (apple shaped) versus hip and buttocks obesity (pear shaped):

$$\text{WHR} = \text{waist circumference/hip circumference} \qquad [43.2]$$

A WHR of > 1.0 for men and > 0.8 for women indicates abdominal obesity with increased risks for type II diabetes and cardiovascular diseases.

SAFETY NOTE Measurement by bioelectrical impedance introduces a low-level electrical signal into the body to determine the electrical resistance of tissues, therefore people with heart pacemakers and similar devices should not use this unless their doctor approves.

Prepare for accurate skinfold measurements –

- Measurements must be taken directly on the skin and not through clothes.
- Find the correct anatomical features to ascertain precise points for the measurements.
- Undertake lots of practice using the calipers.
- Repeat the measurement at least three times at each site and note your consistency. Is it improving with practice? How does your variability compare with others?
- For repeat assessments take the measurements at the same time of day. Early morning before activity and exercise is a good time for a more consistent hydration state.
- Do not take measurements straight after exercise because of a resulting shift in body fluids.

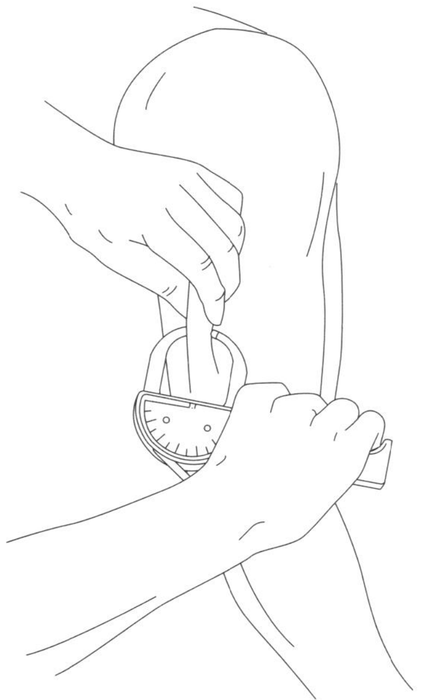

Fig. 43.3 Talking a mid-triceps skinfold measurement.

Measuring body composition using bioelectrical impedance

Basic scales that also measure body fat percentage using bioelectrical impedance are readily available and are relatively inexpensive. They allow for input of sex, age and activity level and provide a result in the context of the appropriate healthy range. Shoes and socks need to be removed so that the soles of the feet are exposed to the electrical signal. Feet should be dry, to ensure the effective flow of current through the body.

Degree of hydration can affect body fat composition readings. For consistent readings a person should use the body fat monitor at the same time each day, under the same conditions each time.

There are also more expensive professional body composition analysers that are more sensitive and accurate, and used both clinically and for research. They are usually programmed to give body mass, % fat, fat mass, muscle mass, bone mass, total body water, basal metabolic rate and BMI.

A specialised body fat analyser is required for athletes due to different hydration levels and composition of muscle tissue, which can cause overestimatation of body fat content when using a standard adult model.

Skinfold measurement

Skinfold measures rely on the understanding that the fat thickness under the skin correlates with the amount of fat throughout the body. About half of the body's adipose tissue is subcutaneous. Use of at least four measurement sites gives a more representative result. The advantages of skinfold measurements are that they relatively fast, simple and inexpensive to undertake.

1. **Consult the directions for the calipers model you are using** – or await demonstration by your instructor as skinfold calipers may vary in their mechanism of operation.
2. **All measurements should be taken on the right-hand side of the body** – unless the subject has a medical condition at the site (ISAK, 2007). Arm measurements should be taken with the arm hanging loosely by the side.
3. **Take the measurements at four sites**. Common skinfold measurement sites include mid-triceps (Fig. 43.3), mid-biceps (Fig. 43.4, subscapular (Fig. 43.5) and supra-iliac (Fig. 43.6).
4. **Grasp the tissue firmly between thumb and forefinger and gently pull away from the underlying muscle**. If you are new to this and not sure where the muscle is have the person tense the muscle so you can identify the layer above it. However, the arm should be relaxed when taking the actual measurements.
5. **Place the calipers where the skinfold sides are parallel**, usually about 1 cm below the operator's fingers. The skinfold should be held for the full duration that the calipers are in contact with the skin. Three measurements should be taken, with intervals of at least 10 seconds between them, and the readings averaged.
6. **Take each reading within 2 to 3 seconds** – or the calipers may over-compress the tissue.

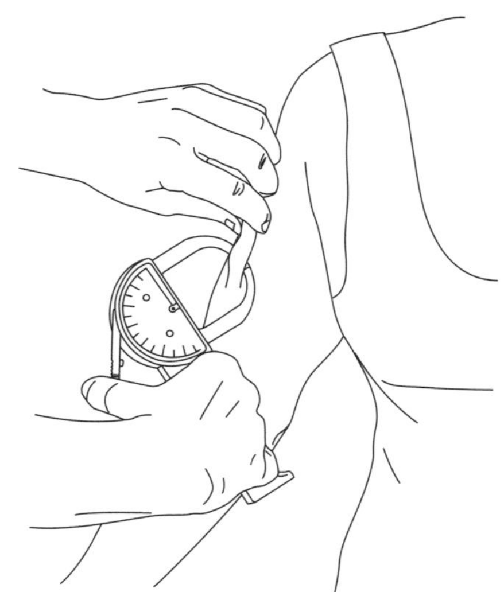

Fig. 43.4 Taking a mid-biceps skinfold measurement.

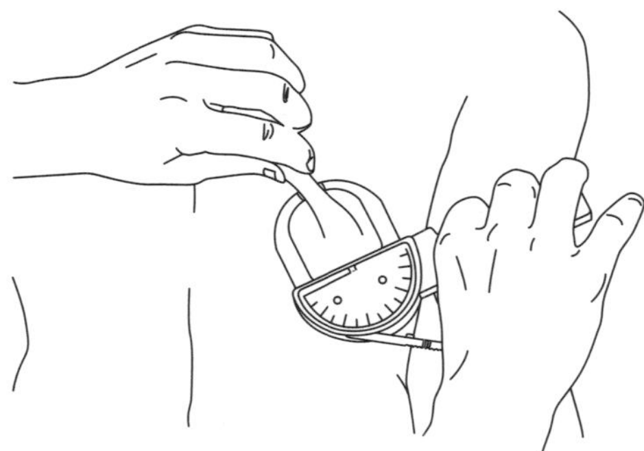

Fig. 43.5 Taking a subscapular skinfold measurement.

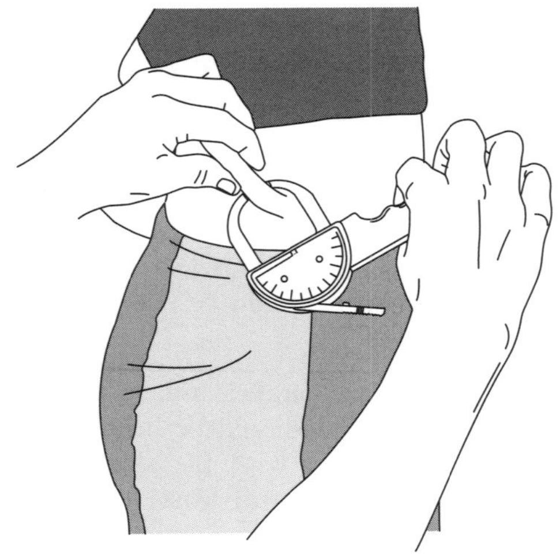

Fig. 43.6 Taking a supra-iliac skinfold measurement.

7. **Refer to the calipers handbook or information provided by your instructor** for the formulas and tables used for converting the sum of the four skinfold measurements covered below into percentage body fat.

Mid-triceps

1. **A vertical fold is taken** on the back of the upper arm mid-way between the shoulder and the elbow (Fig. 43.3).
2. **Take the measurements on the right arm**. This should be hanging loosely by the side.
3. **Measure the upper arm length and mark the mid-point**. Length is measured from the acromion process on the scapula, (bony tip of shoulder) to the olecranon (approximately the elbow joint) on the mid-line of the back surface of the upper arm (over the triceps muscle). The mid-point is then marked.
4. **Produce the skinfold lengthwise on the arm, at the mark.**
5. **Take the skinfold measurement**, as covered above.

Mid-biceps

Measure the same as for the triceps but over the biceps muscle on the front of the arm (Fig. 43.4).

Subscapular

This site is just beneath the inferior (lower) border of the scapula (shoulder blade). The fold should be 45° to the horizontal (Fig. 43.5).

Supra-iliac

This is measured about 2 cm above the iliac crest (hip bone) at the midaxillary line. The skinfold axis is made diagonally (Fig. 43.6).

Lean mass measurements

Lean mass is the mass of a person's muscles, bones, connective tissue and organs, excluding fat. It can be estimated from the mid-arm circumference in conjunction with the triceps skinfold.

Mid-arm circumference (MAC)

1. **Ensure the subject is seated.** The non-dominant arm should be hanging loosely by his or her side.
2. **Wrap a measuring tape around the arm at the midpoint between the tip of the elbow and bony tip of the shoulder**. (The same point as used for the triceps skinfold.)
3. **Read and record the circumference measurement in mm** (see Fig. 43.7).

Mid-arm muscle circumference (MAMC)

$$\text{MAMC} = \text{MAC (mm)} - (\pi \times \text{triceps skinfold in mm}) \quad [43.3]$$

(Note: $\pi = 3.142$)

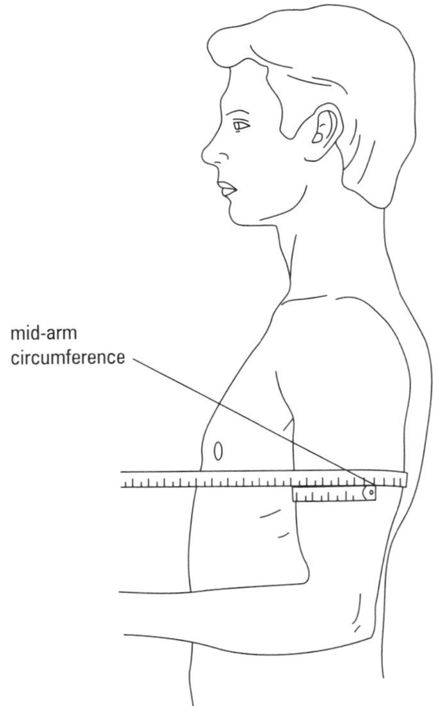

Fig. 43.7 Measuring mid-arm circumference.

Arm muscle area (AMA)

$$\text{AMA (cm}^2) = \frac{(\text{MAC (cm)} - [\pi \times \text{triceps skinfold in cm}])^2}{4 \times \pi} \quad [43.4]$$

AMA (cm^2) corrected for sex:

$$\text{cAMA} = \text{AMA} - 6.5 \text{ (women)} \quad [43.5]$$

$$\text{cAMA} = \text{AMA} - 10.0 \text{ (men)} \quad [43.6]$$

Muscle mass (kg)

$$\text{Muscle mass (kg)} = \text{height (cm)} \times [0.0264 + (0.0029 \times \text{cAMA})] \quad [43.7]$$

The mean muscle mass for healthy adults are around 17 kg for women and 30 kg for men. Lean muscle mass declines as people age but good nutrition and aerobic exercise combined with resistance exercise or strength training can help preserve it. Lean muscle mass changes over time can be used to monitor the effects on muscle of changes in nutrition and exercise.

Overestimation or underestimation of skinfolds can result in large errors in muscle mass determinations, so operators need to be well trained.

Monitoring body mass changes

Body mass changes reflect changes in nutritional status, exercise patterns and illnesses over time, and can be used to monitor these.

Percentage mass change (%MC)

$$\%\text{MC} = (\text{usual body mass} - \text{current body mass}) \times 100/\text{usual body mass} \quad [43.8]$$

The significance of percentage body mass change depends on the time period involved. For example, a body mass loss of 1–2% over a week, >5% over a month, >7.5% over 3 months, >10% over 6 months and >20% over a year are significant losses (Chapter 47).

Table 43.3 %IBM and %UBM ranges and indicators

%IBM	%UBM	Nutrition status
>120	–	Obese
110–120	–	Overweight
90–109	–	Adequate
80–89	85–95	Mildly underweight
70–79	75–84	Moderately underweight
<70	<75	Severely underweight

Percentage of ideal body mass (%IBM)

$$\%\text{IBM} = \text{actual body mass} \times 100/\text{ideal body mass} \quad [43.9]$$
(use body mass [weight] for height tables)

A %IBM greater than 120 generally indicates obesity; a %IBM under 90 indicates possible malnutrition (Table 43.3).

Percentage of usual body mass (%UBM)

$$\%\text{UBM} = \text{actual body mass} \times 100/\text{usual body mass} \quad [43.10]$$

This is a useful index to monitor body mass changes following surgery or during an illness and convalescence, when loss of body mass often occurs (Table 43.3).

Fig. 43.8 Hydrostatic weighing cradle suspended from scales over pool.

Hydrostatic weighing (hydrodensitometry)

Underwater weighing and water displacement provide for a measure of body density from which percentage of body fat can be estimated. Hydrodensitometry is considered one of the most accurate anthropometric procedures for estimating percentage body fat. A specialised weighing cradle and tank, containing a submergible seat suspended from a scale is used. A swimming pool may also be used for the tank (Fig. 43.8). The body mass in air of the person is first recorded (p. 294). Then the body mass in water of the person is recorded. The difference between wet and dry mass is indicative of the volume of water displaced by the person's body. The same minimal clothing should be worn for both air and water measurements, e.g. a swimsuit.

For taking the mass in water, the person sits in the seat, expels air from their lungs by breathing out fully and is lowered into the tank until his or her whole body is underwater. The measurement is taken with the person sitting still. This procedure should be repeated at least four times and until a consistent reading is obtained. For the most exact result, hydrostatic weighing should be done in the morning after voiding and before breakfast to minimise the amounts of non-body mass in the system.

To calculate percentage body fat, first calculate body density using the relationship:

$$D_b = M_a/([M_a - M_w]/D_w) - (RV + 100) \quad [43.11]$$

where:

D_b = body density (g mL^{-1})

M_a = body mass in air (g)

M_w = body mass in water (g)

D_w = density of water (approx. 1 g mL^{-1}, varies with temperature (Table 43.4);

RV = residual lung volume (mL). This can be measured using methods called nitrogen washout, helium dilution or oxygen dilution and should be done while immersed in the water. More commonly, the estimation formulae below can be used:

$$\text{Males: } RV = (0.017 \times \text{age}) + (0.06858 \times \text{height [inches]}) - 3.447 \quad [43.12]$$

$$\text{Females: } RV = (0.009 \times \text{age}) + (0.08128 \times \text{height [inches]}) - 3.9 \quad [43.13]$$

(note 1 inch = 2.54 cm)

100 = estimate of gases in the gastrointestinal tract (mL)

The percentage of body fat is then usually estimated using the equation of Brozek *et al.* (1963):

$$\%\text{ body fat} = (497.1/D_b) - 451.9 \quad [43.14]$$

Table 43.4 Density of water at different temperatures

Temp (°C)	Density (g mL^{-1})
23.0	0.997 541 2
24.0	0.997 299 4
25.0	0.997 048 0
26.0	0.996 787 0
27.0	0.996 516 6
28.0	0.996 237 1
29.0	0.995 948 6
30.0	0.995 651 1
31.0	0.995 345 0
32.0	0.995 030 2
33.0	0.994 734 0
34.0	0.994 707 1
35.0	0.994 035 9

Text references

ABHI (2008) *Waist Measurement Fact Sheet*. Australian Better Health Initiative, Australian Government. Available: http://www.health.gov.au/internet/abhi/publishing.nsf/Content/factsheet-waist-measurement
Last accessed: 21/12/10.

Brozek. J., Grande, F., Anderson, J. and Keys, A. (1963) Densitometric analysis of body composition: Revision of some quantitative assumptions. *Annals of The New York Academy of Science*, **110**, 113–140.

ISAK (2007) International Society for the Advancement of Kinathropometry. Available: http://www.isakonline.com/
Last accessed: 21/12/10.

Stocks, J. and Quanjer, Ph.H. (1995) Reference values for residual volume, functional residual capacity and total lung capacity. *European Respiratory Journal*, **8**, 492–506. Available: http://erj.ersjournals.com/cgi/reprint/8/3/492
Last accessed: 21/12/10.

WHO (2009) *BMI Classification*. World Health Organisation. http://apps.who.int/bmi/index.jsp?introPage=intro_3.html
Last accessed: 21/12/10.

Sources for further study

Heymsfield, S., Lohman, T., Wang, Z.M. and Going, S. (2005) *Human Body Composition*, Human Kinetics, Champaign, Illinois.

Heyward, V. and Wagner, D. (2004) *Applied Body Composition Assessment*. Human Kinetics, Champaign, Illinois.

Lee, R.D. and Nieman, D.C. (2007) *Nutritional Assessment*. McGraw Hill, New York.

44 Estimating energy requirements

Abbreviations used when estimating dietary energy requirements

BEE	basal energy expenditure
BMR	basal metabolic rate
DIT	dietary induced thermogenesis
PAL	physical activity level
REE	resting energy expenditure
RMR	resting metabolic rate
TEE	total energy expenditure

Definitions

Dietary-induced thermogenesis – the amount of energy that a person uses in the process of digestion and absorption of food.

Relative proportions of total energy expenditure (TEE) – depending on a person's lifestyle, this is usually 45%–70% due to basal metabolic rate, 20–45% due to physical activity and 10% due to dietary-induced thermogenesis.

The ability to calculate the energy requirements of a person is fundamental in nutrition and dietetic practice. It forms the basis of most regimes, whether preventative or as treatment. In practice, energy requirements are based on measurement or calculation of total energy expenditure (TEE). If an individual is neither gaining nor losing weight then their energy intake will equal their energy expenditure and they are described as having achieved energy balance.

Energy expenditure can be measured by direct or indirect calorimetry (Chapter 51), but because of the cost and time required to carry out these methods, energy requirements are usually estimated using predictive equations which have been derived from calorimetry data.

In healthy people, total energy expenditure (TEE) is made up of three main components:

1. Basal Metabolic Rate (BMR);
2. thermic effect of eating or dietary-induced thermogenesis (DIT);
3. physical activity.

If a person has additional energy requirements due to illness, trauma or surgery, for example, then a 'stress factor' is also used in the calculation (see below).

Estimating basal metabolic rate (BMR)

BMR, sometimes referred to as basal energy expenditure (BEE), is the largest component of energy expenditure as it accounts for up to 70% of the total daily amount. It is the amount of energy that is required for an individual to maintain the vital functions of their internal organs and maintain body temperature. BMR should be measured using the following standardised conditions:

- ensure that the subject has fasted for 12 hours;
- he or she should not have performed strenuous physical activity the day before;
- he or she should be fully rested;
- the subject must be fully awake and lie completely still;
- the measurement should take place in a thermoneutral environment (22–26°C);
- the subject should be free from emotional stress and there should be no physical stimulus.

If these strict conditions are met then accurate measurements of basal metabolism will be recorded, as heat production (as measured by calorimetry) will not be affected by processes which require energy, such as digestion or body temperature regulation. If one or more of these strict criteria are not met then the measurement is generally referred to as resting metabolic rate (RMR) or resting energy expenditure (REE).

Comparison of different methods of predicting basal metabolic rate:

Subject: 40-year-old female, weight 60 kg, height 1.7 m

1. Estimation based on Schofield equations:
   BMR = 8.3W + 846
   BMR = 8.3 × 60 + 846
   BMR = 1344 kcal/24 hours
2. Estimation based on Harris–Benedict equations:
   BEE = 655.1 + 9.6W +1.8H − 4.7A
   BEE = 655.1 + (9.6×60) + (1.8 × 170) − (4.7×40)
   BEE = 655.1 + 576 + 306 − 188
   BEE = 1349 kcal/24 hours
3. Estimation based on Mifflin–St Joer equations:
   RMR = 10W + 6.25H − 5A − 161
   RMR = (10 ×60) + (6.25 × 170) − (5 ×40) −161
   RMR = 600 + 1062.5 − 200 − 161
   RMR = 1301.5 kcal/24 hours
4. Estimation based on Owen equations:
   RMR = 795 + 7.18W
   RMR = 795 + (7.18 × 60)
   RMR = 795 + 430.8
   RMR = 1226 kcal/24 hours

Using predictive equations to estimate energy requirements

Estimating energy requirements in individuals begins with calculating their BMR. The most commonly used and most convenient method for doing this is to use prediction equations. There is a number of prediction equations used in practice, which differ slightly because they are derived from different data sets. When choosing which equation to use, check that the original subjects used in the data set correspond to the people you are evaluating as this will increase the accuracy of the estimation.

The following equations are the main ones used in Europe and the USA:

- **Schofield equations (Table 44.1).** These are based on a data set of 1300 10–17-year-olds, 3500 men and 1200 women (Schofield, 1985). They are sometimes also referred to as 'the WHO equations', although they are in fact an extension of work published by the Food and Agriculture Organisation and the World Health Organisation (FAO/WHO/UNU, 1985). To use the equations it is necessary to know the sex, age and body weight (in kg) of the subject. These equations have been used to calculate dietary reference values for energy in the UK (Department of Health, 1991).
- **Harris–Benedict equations (Table 44.2).** These were developed from a data set of healthy Americans (136 male, 103 women) and are still widely used in clinical settings, particularly in the USA. The equations differ from Schofield's equations in that they also include height as a parameter.

Table 44.1 Schofield equations for estimating BMR

Age range (years)	BMR MJ/day	kcal/day
Males		
10–17	0.074W + 2.754	17.7W + 657
18–29	0.063W + 2.896	15.1W + 692
30–59	0.048W + 3.653	11.5W + 873
60–74	0.0499W + 2.930	11.9W + 700
75+	0.0350W + 3.434	8.4W + 821
Females		
10–17	0.056W+ 2.898	13.4W + 692
18–29	0.062W + 2.036	14.8W + 487
30–59	0.034W + 3.538	8.3W + 846
60–74	0.0386W + 2.875	9.2W + 687
75+	0.0410W + 2.610	9.8W + 624

W = weight in kg.

Table 44.2 Harris–Benedict equations for estimating BEE

BEE (kcal/24 hours)	
Males	Females
66.5 +13.8W + 5.0H − 6.8A	655.1 + 9.6W +1.8H − 4.7A

W = weight (kg); H = height (cm); A = age (years).

Table 44.3 Mifflin–St Joer equations for estimating RMR

RMR (kcal/24 hours) Males	Females
10W + 6.25H − 5A + 5	10W + 6.25H − 5A − 161

W = weight (kg); H = height (cm); A = age (years).

Table 44.4 Owen equations for estimating RMR

RMR (kcal/24 hours) Males	Females
879 + 10.2W	795 + 7.18W

W = weight (kg).

- **Mifflin–St Joer equations (Table 44.3).** These were published in 1990 and are endorsed by the American Dietetic Association. They are based on a sample of almost 500 individuals ranging from normal weight to obese.
- **Owen equations (Table 44.4)**. These are derived from a small data set of 60 people of normal weight (44 women, 16 men). They are not suitable for use when predicting resting energy expenditure for obese people.

Physical activity level (PAL)

Once you have estimated BMR the next step is to factor in the energy required for physical activity. This is the most variable component of TEE as each activity has a different energy 'cost', so activities which are more strenuous, such as running, require more energy than less strenuous activities, such as walking.

There is a number of ways of measuring and estimating PAL. Using the equipment such as an accelerometer to directly measure physical activity of an individual can achieve very accurate data. An alternative method to measure activity is to ask the person to keep a physical activity diary, which provides information on activity, usually over a period of seven days and reported in 15-minute blocks. Activities can then be categorised according to energy cost as defined by FAO/WHO/UNU (1985) or Department of Health (1991). This is known as the physical activity ratio (PAR) for each activity which can range from 1.0 for when someone is asleep to 7.0 for strenuous exercise activities. This shows that it takes seven times more energy to play certain sports than it does to lay quietly (as when directly measuring BMR). Table 44.5 details the calculations for energy requirements based on information about physical activity patterns.

Although the process of detailed physical activity diaries can give accurate values for physical activity, it is very time consuming for the individual. A quicker and more convenient way to estimate PAL is to base the estimation on occupational and non-occupational activity levels using a matrix as in Table 44.6. You only need limited information about the individual to use the matrix. Using the example of the office clerk detailed

> **Example**
> Classification of occupational activity
>
> - **Light**: administrative and managerial workers, sales representatives, clerical workers.
> - **Moderate**: service workers, domestic helpers, students, transport workers.
> - **Moderate/heavy**: labourers, agricultural workers, some constructions workers.

Table 44.5 Example calculation of energy requirements taking account of different periods of physical activity. Data are based on figures from FAO/WHO/UNU (1985). These figures estimate the energy requirement of a male office clerk who takes part in some sporting activity. The details used were: age 25; weight 65 kg; basal metabolic rate (BMR) 1680 kcal (6.96 MJ)/24 h.

Activity	Energy use relative to BMR (PAR)	Time taken (h)	Proportion of 24 h day	Energy used (kcal [MJ])
In bed, sleeping	1.0	8.0	0.333	560 [2.34]
Occupational activities	1.7	6.0	0.250	710 [2.97]
Non-occupational activities:				
• household tasks, such as cooking, cleaning etc and walking to work	3.0	2.0	0.083	420 [1.76]
• Aerobic exercise, such as swimming or jogging	6.0	0.33	0.014	140 [0.58]
Remainder (no information given)	1.4	7.77	0.320	750 [3.14]
Total	**1.54**	**24**	**1.0**	**2580 [10.79]**

Table 44.6 BMR multiples for different levels of occupational and non-occupational activity

Non-occupational activity	Occupational activity					
	Light		Moderate		Moderate/heavy	
	M	F	M	F	M	F
Non-active	1.4	1.4	1.6	1.5	1.7	1.5
Moderately active	1.5	1.5	1.7	1.6	1.8	1.6
Very active	1.6	1.6	1.8	1.7	1.9	1.7

Department of Health (1991).

in Table 44.5, he would be described as having light occupational activity and moderately active non-occupational activity, therefore his energy requirements would be calculated by multiplying his estimated BMR by the PAL value of 1.5. This factorial method also takes DIT into account.

Additional factors affecting energy requirements

When estimating the energy requirements of healthy individuals, it is appropriate to use the predictive equations and PAL factors. If, however, the estimate is for a patient with a clinical condition this will have to be adjusted for. These are sometime described as 'stress factors'. Stress would need to be considered if, in the last week, the patient had any of the following:

- surgery
- injury

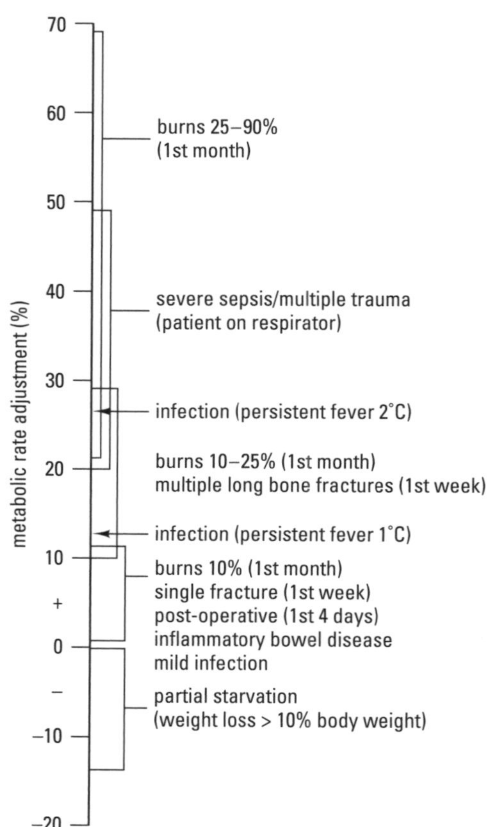

Fig. 44.1 Nomogram, showing guidelines for adjusting BMR for level of stress to calculate energy requirements (Elia, 1990).

- inflammation
- infection.

Energy requirements can be calculated from BMR and adjustments based on a nomogram giving guidelines for a range of conditions (Fig. 44.1).

In clinical settings the following factors should be applied:

- bedbound, immobile: BMR + 10% (PAL equivalent 1.1);
- bedbound, mobile or sitting: BMR + 15–20% (PAL equivalent 1.15–1.2);
- mobile, on ward: BMR + 25% (PAL equivalent 1.25).

Using estimations of energy requirements

Knowledge of energy requirements is fundamental in nutritional and dietetic practice. As an undergraduate you are likely to use these types of calculations in case-study exercises or in assessment of your own or classmates' energy requirements. In professional practice, calculated requirements would be used to advise on dietary intakes in interventions such as weight loss and treatment regimes in clinical settings.

Given the individual differences in energy requirements, the predictive equations and physical activity factors can only provide estimates of requirements so the person would be monitored regularly so that advice or intakes could be adjusted if necessary. To illustrate this point consider a weight loss regime. To achieve weight loss the person would need to consume less energy than he or she requires for BMR and physical activity. If weight loss does not occur, and assuming the person is following the diet plan, then the energy intake would need to be reduced further. As weight loss occurs that person will require less energy (as can be seen from the fact that BMR prediction calculations use body weight to estimate requirements). Each time the person's weight plateaus, indicating energy balance, repeat calculations should be carried out to estimate the new energy requirements until the target weight is reached.

Predictive equations can also be useful when assessing the accuracy of a dietary assessment, particularly with self-reported methods (Chapter 41). If there is discrepancy between predicted and reported intakes, with no change to body mass it may be necessary to repeat the dietary assessment over a longer period of time.

Text references

Department of Health (1991) *Dietary Reference Values for Food Energy and Nutrients for the United Kingdom. Reports on Public Health and Medical Subjects No. 41*. HMSO, London.

Elia, M. (1990) Artificial nutritional support. *Medicine International*, **82**, 3392–3396.

FAO/WHO/UNU (1985) *Energy and Protein Requirements: Report of a Joint FAO/WHO/UNU Expert Consultation*. WHO Technical Report Series No. 724. Geneva. Available: http://www.fao.org/docrep/003/AA040E/AA040E06.htm#ref5 Last accessed: 21/12/10.

Schofield, W. (1985) Predicting basal metabolic rate, new standards and reviews of previous work. *Human Nutrition Clinical Nutrition*, **39**, S5–S91.

Sources for further study

FAO (2004) *Human Energy Requirements. Report of the Joint FAO/WHO/UNU Expert Consultation.* Available: http://www.fao.org/docrep/007/Y5686E/y5686e00.htm#Contents
Last accessed: 21/12/10.

Ferrie, S. and Ward, M. (2007) Back to basics: Estimating energy requirements for adult hospital patients. *Nutrition Dietetics*, **64**, 192–199.

Frankenfield, D., Roth-Yousey, L., Compher, C. (2005) Comparison of predictive equations for resting metabolic rate in healthy nonobese and obese adults: A systematic review. *Journal of the American Dietetic Association*, **105**, 775–789.

Henry, C.J.K. (2007) Basal metabolic rate studies in humans: Measurement and development of new equations. *Public Health Nutrition*, **8**(7A), 1133–1152.

Thomas, B. and Bishop, J. (2007) *Manual of Dietetic Practice,* 4th edn. Blackwell, Oxford.

45 Assaying biochemical markers of malnutrition

Definitions

Biochemical markers – measurable and quantifiable biological parameters (for example, specific protein concentration) which serve as indices for health-related and physiology-related assessments, such as disease risk.

Half-life – period of time required for 50 per cent of a particular protein to be catabolised (degraded). The degradation of proteins is one of the ways the body regulates specific protein levels.

Body pool – total body amount found in both the circulation and in tissues.

Oncotic pressure – osmotic pressure exerted by proteins in blood plasma that draws fluid into the capillaries.

Nutritional markers from blood samples are used to determine undernutrition and malnutrition. They include several proteins whose concentration is reduced in the blood during poor nutrition. Biochemical analyses of the levels of specific metabolites in blood are not necessarily good indicators of nutritional status when used in isolation. However, when considered carefully in conjunction with nutritional assessment (Chapter 41), medical history and any current clinical findings they can be of great importance.

KEY POINT When interpreted correctly, biochemical markers can provide an objective measure of a person's nutritional status.

Serum proteins

These provide an overall indication of nutritional status especially with regard to possible protein–energy malnutrition (PEM) and recovery from this condition. Protein markers are chosen according to their amount in the body and their biological half-life in the blood. Those proteins with a relatively smaller amount in the body and shorter half-life are best for assessing early malnutrition because they better reflect more recent conditions. Proteins such as albumin with a relatively long half-life and significant body pool are used to evaluate nutritional status over a longer period of time.

Albumin

This protein functions as a transport molecule for hydrophobic molecules in the circulation and also in maintaining blood volume by regulating oncotic pressure. It is relatively resistant to degradation in the body. Albumin has a half-life in the body of 20 days, which is relatively long, compared with other protein markers. Albumin also represents 55–60 per cent of total serum proteins (Table 45.1), with a large body pool. Low serum albumin indicates long-term protein depletion and possibly serious malnutrition. Levels increase slowly during nutritional therapy (appropriate changes to diet/protein supplements). Limitations in nutritional interpretation of albumin levels include oedema (fluid retention), hypermetabolic states, eclampsia (convulsions associated with pregnancy-induced hypertension) and liver and renal diseases which also reduce albumin levels. It may also be increased in any condition causing dehydration, e.g. diarrhoea, vomiting, hyperthermia, infection, type II diabetes mellitus, in older people who have lost their ability to effectively sense thirst, redistribution caused by inflammation, and dilution from any infused fluids. It should also be noted that even in severe starvation serum albumin may remain at normal levels, particularly if there is accompanying dehydration.

Table 45.1 Marker serum reference ranges for adults.

Marker	Reference range (g/L)
Albumin	35–52
Transferrin	M: 2.15–3.65 F: 2.50–3.80
Transthyretin	0.18–0.45
Retinol-binding protein	0.03–0.06
C-reactive protein	< 0.01

Wu (2006); Pagana and Pagana (2006).
M = male; F = female.

KEY POINT Many biochemical markers are affected by other conditions, including a range of diseases states and medications. Consequently, it is important to consider these markers in the broader context of nutritional assessment (Chapter 47) and the overall health of the individual.

Example (case study)
A 56-year-old male trauma patient has been slow in recovering his energy levels. Anthropometric measurements (p. 294) suggest he is underweight and may have PEM. Routine clinical blood tests by his physician are within usual reference ranges, including iron status, lymphocyte count, and liver and renal function tests. Further laboratory tests are ordered to help evaluate his nutritional status, including several protein markers for nutrition status and also CRP to indicate the presence of any inflammation.

Results
- albumin 32 g/L
- transthyretin 0.15 g/L
- transferrin 1.40 g/L
- CRP 0.0044 g/L

The CRP level indicates no significant inflammation.
It is concluded, in conjunction with anthropometric data and diet history, that the patient is moderately undernourished. Implementation of a nutrition care plan (p. 321) is recommended including adequate protein and energy sources and reassess regularly. Educate patient regarding improved eating habits.

Definitions

Proinflammatory cytokines – regulatory proteins produced by activated immune system cells which amplify inflammatory reactions.

C-reactive protein (CRP) – protein produced by the liver and found in rapidly increasing concentrations in the blood following inflammation.

Methods for analysis of serum albumin include dye-binding-spectroscopy (p. 397), immunoturbidimetry and immunonephelometry.

Dye-binding-spectroscopy methods rely on dyes that specifically bind to serum albumin. Bromocresol purple binds to albumin at a pH of 5.2. Absorbance is measured at 605 nm. With this dye there is minimal non-specific binding, and it will not bind to non-human albumins. Therefore, human albumin standards must be used.

Immunoturbidimetric quantification is based upon the reaction of a protein-specific antiserum with the specific protein in the sample. The decrease in light transmission is then recorded. Immunoturbidimetric assays can be quantified using a standard spectrophotometer or clinical biochemistry autoanalyser. An advantage of immunoturbidimetry over immunodiffusion (p. 348) is its relatively short reaction time of 2–10 minutes.

Immunonephelometry is also based upon optical detection of specific antigen–antibody complexes. However, in this case the generated antigen–antibody complexes are quantified by measuring side-scattered light, using a nephelometer. These assays can also be automated and are therefore suitable for the routine measurement of protein in a large number of samples. Immunonephelometry requires specialised analysers that do only a limited range of tests.

Transferrin

This protein binds and transports iron (as Fe^{3+}) from tissues to the liver for storage. Transferrin has a biological half-life of about eight days. It has lower serum concentrations compared with albumin (Table 45.1) and about one-third is bound to iron. Very low levels may indicate severe PEM. Response to nutritional therapy is relatively slow. Hypermetabolic states and liver and renal diseases reduce transferrin levels while pregnancy, hepatitis, iron deficiency and blood loss increase levels. Therefore medical condition and iron status need to be carefully considered before making an interpretation of the concentration of this protein as a nutritional marker. Serum transferrin is routinely assayed by immunonephelometry.

*Transthyretin (***formerly named prealbumin***)*

The function of transthyretin is to transport thyroid hormones T3 and T4, and retinol (together with retinol binding protein). Its synthesis in the liver is highly dependent on the availability of sufficient protein and energy stores. The half-life of transthyretin in the body is two days and there is only a relatively low body pool and serum concentration (Table 45.1). Because of these factors it shows early responses to nutritional deficits and recovery. However, inflammation, liver dysfunction, cancer, cystic fibrosis, hypothyroidism, female oral contraceptives, haemodialysis and chronic illnesses decrease transthyretin levels, while high levels of corticosteroids, chronic renal failure and Hodgkin's disease increase levels.

Synthesis of transthyretin is repressed by pro-inflammatory cytokines and transthyretin levels fluctuate with variations in inflammatory status. Therefore plasma markers for pro-inflamatory cytokines, such as C-reactive protein (CRP), must be simultaneously monitored and accounted for in any changes to transthyretin. CRP serum concentrations can increase more than a thousand-fold with severe inflammation.

Plasma transthyretin should be measured every three days and day-to-day changes are more relevant to nutritional status than absolute values (Cynober, 2009). Methods of assay for transthyretin include radial immunodifusion (p. 348), immunonephelometry and immunoturbidimetry.

Retinol-binding protein (RBP)

This protein circulates bound to transthyretin as a macromolecular complex. Around 90 per cent of RBP is saturated with retinol. Retinol is transported from the liver to target tissues. Specific receptors for retinol/retinol-RBP are found on target cells. RBP shows a rapid response to protein changes and short-term changes in nutritional status. However, levels are lowered in vitamin A deficiency, hypermetabolism, liver disease, hyperthyroidism, cystic fibrosis and zinc deficiency while chronic renal failure increases levels. Another practical limitation is that it is also the most expensive nutritional marker to assay. Assays can be performed by ELISA (p. 350), radial immunodiffusion (p. 348) or immunonephelometry.

Understanding the interpretation of serum RBP – this protein has a biological half-life of 12 hours, which is the shortest half-life of the major nutritional protein markers. Since it also has a very small body pool and a low blood serum concentration (Table 45.1), this makes it one of the most sensitive nutritional markers.

Nitrogen balance

Malnutrition results in a state of negative nitrogen balance when basic intake of protein is insufficient to cover losses, and/or when proteins are broken down at an increased rate in a hypermetabolic state, which often accompanies serious illness. Appropriate nutritional support with increased dietary or infused protein usually leads to rapid improvements in nitrogen balance.

Renal disease can also affect nitrogen balance due to reduced renal clearance of urea. Blood urea nitrogen, urinary urea nitrogen and plasma and urinary creatinine are laboratory tests used for monitoring kidney function. If renal failure is diagnosed by a medical practitioner, a renal diet will need to be implemented to avoid accumulation of an excess of nitrogenous wastes in the body.

Immunological markers

Malnutrition reduces immune function. Lymphocyte count, as determined by differential white blood cell counting using microscopy (p. 181) or an electronic particle counter (p. 474), has been used to assess this (1000–4000 lymphocyte cells/μL being a typical acceptable adult range). However, lymphocyte count should also be considered in regard to known or possible viral infections, in which the count may be increased.

Skin sensitivity tests have been used, placing an antigen under the skin and noting the response (see Chapter 46 for further details). However, results may not be sufficiently sensitive to reflect the efficacy of nutritional support during recovery. Also there have been reports of scarring at the injection site.

Table 45.2 Analytical methods used for plasma/serum vitamin assays

Vitamin	Methods	Page ref.
A	fluorimetry HPLC HPLC-MS	 362
B6	Enzymatic assay radioimmunoassay HPLC	 350 362
B12	Competitive binding protein (CPB) radioassay radioimmunoassay HPLC	 350 362
C	colorimetry	354
D	radioimmunoassay CPB radioassay	 350
E	HPLC	362
K	HPLC	362

Micronutrient assays

Vitamin and mineral status are assessed in blood serum or plasma as needed, based on indications from the physical examination and reported symptoms (Chapter 42). However, blood levels of vitamins and minerals

Table 45.3 Analytical methods used for plasma/serum mineral assays

Mineral	Method	Page ref.
Sodium Potassium Chloride	Measured simultaneously by flame emission photometry	357
Calcium	Spectrophotometric	353
Magnesium	AAS	358
Phosphorous	Colourimetric	354

Combining foods to aid iron uptake – tomato slices (containing natural vitamin C) on a wholemeal sandwich will help absorption of iron from the bread; MFP in meat eaten with vegetables will assist iron from the vegetables to be absorbed.

are affected by many variables and results must be considered along with the other nutritional assessment data. Some analysis methods for determining vitamin and mineral plasma/serum concentrations are shown in Tables 45.2 and 45.3.

Calcium is the most abundant mineral in the human body. However blood calcium levels may not reflect true calcium nutritional status. Blood calcium levels are normal in people with osteoporosis, as the bones act as a calcium reservoir to supply the blood. Consequently, bone density scans usually provide a better assessment of calcium status, along with dietary history of calcium intake.

Serum iron determination is a measure of the quantity of iron bound to the iron-carrying protein transferrin. This may be measured using colourimetry (p. 354) or AAS (p. 358). Iron deficiency may lead to anaemia, which is determined by measuring blood haemoglobin concentration, to see if it is below the reference range (adults: male 11.6–31.3 μmol/L; female 9.0–30.4 μmol/L). However, it is important to note that symptoms of iron deficiency can arise, particularly in children and adolescents, before a deficiency can be detected in blood tests. Symptoms which might suggest a dietary assessment of iron intake include behavioural problems, such as impaired cognitive function, and reduced physical energy and activity. In treating iron deficiency it is useful to note that the consumption of foods containing vitamin C along with those containing iron can aid iron absorption. Also **m**eat, **f**ish and **p**oultry contain an unidentified factor (called the MFP factor) that assists in absorption of non-haem iron from plant sources.

Text references

Cynober, L. (2009) Basics in clinical nutrition: Some laboratory measures of response to nutrition in research and clinical studies. *European e-journal of Clinical Nutrition and Metabolism*, **4**, e226–228.

Pagana, K.D. and Pagana, T.J. (2006) *Mosby's Manual of Diagnostic and Laboratory Tests.* Mosby, Elsevier, St Louis, Missouri.

Wu, A.H.B. (ed.) (2006) *Tietz Clinical Guide to Laboratory Tests.* WB Saunders, Elsevier, St Louis, Missouri.

Resources for further study

Anon. (1973) Laboratory assessment of nutritional status. *American Journal of Public Health*, **63**, 11 Suppl, 28–37. Available: http://www.ncbi.nlm.nih.gov/pmc/articles/PMC1775329/ Last accessed: 21/12/10.

Chernecky, C.C. and Berger, B.J. (2008) *Laboratory Tests and Diagnostic Procedures.* WB Saunders, Elsevier, St Louis, Missouri.

Pagana, K.D. and Pagana, T.J. (2007) *Mosby's Diagnostic and Laboratory Test Reference.* Mosby, Elsevier, St Louis, Missouri.

Sauberlich, H.E. (1999) *Laboratory Tests for the Assessment of Nutritional Status.* CRC Press, Boca Raton, Florida.

Vaughn, G. (1999) *Understanding and Evaluating Common Laboratory Tests.* Prentice Hall, Upper Saddle River, New Jersy.

46 Testing for food allergies and intolerances

> **Definitions**
>
> **Food allergy** – an exaggerated immune response in which antibodies are produced against a specific food.
>
> **Food intolerance** – a range of reproducible adverse responses to a specific food or food ingredient.
>
> **Anaphylactic shock** – an extreme reaction to an antigen, with loss of blood pressure and tissue swelling, leading to life-threatening airway, breathing and circulation problems.

Food allergies occur in about 8 per cent of children, and are usually associated with digestion and absorption of the protein content in foods. Food allergies affect about 3 per cent of adults. The incidence of allergies is increasing in Western societies. For example, in the Australian Capital Territory, the incidence of allergies to peanuts has doubled in the last ten years and similar increases in peanut allergies have also been documented in the UK and USA (Mullins *et al.*, 2009). In some instances food allergies can be life-threatening and in others they can cause physical disturbance and medical conditions. Therefore nutritionists will likely encounter in clients both sufferers of acute and masked food allergies as well as food intolerances.

This chapter outlines the general principles involved in food allergy testing. While you are unlikely to gain practical experience of allergy testing within your degree programme, it is important to develop an understanding of these principles in your training, so that you aware of how allergies and intolerances are dealt with.

Acute severe allergy

Typical allergy testing involves exposure of the subject to the suspect allergen in a 'challenge' test (p. 315). However, if a child or adult has previously suffered a very obvious and severe allergic reaction to a food, such as shortness of breath or difficulty swallowing, then the allergen should not be tested by 'challenge', and the food should be totally avoided. If a severe reaction occurs, prompt treatment with adrenaline is necessary, followed by medical observation for at least four hours. The known allergy sufferer would be referred to a physician for a prescription of adrenalin. This comes in a ready-to-use pen syringe, to be administered by a carer or self-administered.

> **SAFETY NOTE** children with a severe food allergy should always have an adrenaline pen syringe available, and schoolteachers and other carers should be made aware of how to use it.

Peanuts and tree nuts are the main cause of the potentially fatal food allergy reaction known as anaphylactic shock, although other foods can be responsible. Carefuly scrutiny of food labels is essential. However, possible contamination also poses a major risk, e.g. traces of peanut in a plain chocolate bar.

Food allergy

> **Understanding food allergies in infants and children** – these are most commonly caused by cows' milk (not recommended for infants, p. 273), soy products, eggs, fish, peanuts, tree nuts or wheat. However, people can have allergies to almost any type of food.

Food allergies are more common in young children whose digestive system is still developing. Allergy symptoms are mostly induced by immunoglobulin E (IgE) antibody responses to one or more proteins in food. Delayed reactions may involve IgG production and may be more difficult to relate to a specific food because of chronic (long-term) symptoms, although this is somewhat controversial.

Allergy responses can be almost immediate, or delayed for up to several hours after eating the food. Symptoms include:

- tightness, swelling and/or itching of the tongue, lips and throat, which may be accompanied by coughing and hoarse voice;
- tightness in the chest, difficulty breathing (note similarity here to asthma);

Understanding the diagnosis of food allergies and intolerances – in recognising food allergies and intolerances it is important to appreciate that similar symptoms can be caused by airborne allergens or chemicals (e.g. pollen or natural salicylates in scented products) or by contact with allergens (e.g. poison ivy), or by reactions to environmental chemicals (e.g. cleaning fluids). Other medical conditions can also cause similar symptoms, including metabolic disorders, pancreatic insufficiency, food poisoning and some infections.

Investigating undiagnosed conditions – where a person has one or more chronic symptoms for which no medical diagnosis has been determined, it may well be worth investigating the possibility of food allergy or intolerance, even if only to rule it out.

- nasal congestion, sneezing, itchy and/or watery eyes;
- nausea, vomiting, abdominal cramps and diarrhoea;
- urticaria (hives), an itchy skin rash;
- swelling of the face, trunk, limbs;
- eczema in chronic unrecognised exposure.

Food intolerance

Food intolerances are believed to be more common than food allergies and may result from individual pharmacological responses to foods. These responses do not involve the immune system, but are a reaction to the chemicals present in foods, which can be both natural and/or artificial. There are three classes of natural chemicals linked to food intolerance – salicylates, amines and glutamates. Individuals can have different levels of sensitivity to one, two or all three of these chemicals. Foods and food additives linked with intolerance include chocolate, yeast, red wine, aged cheeses, berries, tomato, lactose, monosodium glutamate (MSG) and many artificial preservatives and colours. Food intolerances may affect the following systems:

- **nervous system**, e.g. headache, hyperactivity and irritability in children;
- **skin**, e.g. rashes;
- **gastrointestinal system**, e.g. nausea;
- **respiratory system**, e.g. coughing.

Note that the symptoms of food intolerances are much broader than those for food allergies and can also be caused by a range of other conditions. The methods described below for diet history, food record and elimination diet apply equally to the determination of food intolerances and food allergies. Food intolerance symptoms can mimic the symptoms of an allergy, but the immune system is not playing a role, therefore IgE based testing is not going to diagnose the intolerance.

Identifying a food allergen

Diet history

A medical, medication and diet history (p. 279) will be taken at the first consultation. If the patient is a child the guardian(s) should also be present.

The person would normally be asked a series of questions, to provide information on the condition, such as:

- **What are the symptoms of your suspected allergic reaction?** (This can provide diagnostic information.)
- **Are you aware of reactions to any particular foods?** (This relates to previous experience – establishing a pattern.)
- **How soon did your reaction occur after eating the food?** (This can provide insight into the likelihood of a link between the suspect food and the symptoms.)

Daily Food Record
Date: 24 Jan
Breakfast 3 strawberries, glass of milk, wholegrain tuna sandwich, Reaction: 30 minutes later, chest tight
Lunch apple, wholemeal roast beef and salad sandwich, white coffee Reaction: none
Dinner chicken burger and chips, orange juice Reaction: none
Snacks ice cream and walnuts (10 am), carrot cake, coffee (3 pm), left-over tuna on cracker(9 pm) Reaction: difficulty breathing shortly after tuna and cracker

Fig. 46.1 Example of a daily food record, noting possible allergy symptoms.

Elimination diets are also used to test for food intolerance – an intolerance can be quite complex if the individual responds to a chemical rather than a particular food, e.g. if it is thought the person is reacting to amines, then all foods containing amines need to be excluded, e.g. cheese, chocolate, tomatoes, bananas and broccoli, etc.

- **Have you tried antihistamines and, if so, did they help?** (This can indicate whether IgE is involved.)
- **Did anyone else who ate the same food have a similar reaction?** (This may point to food poisoning or infection.)
- **How much of the food did you eat?** (This can provide information relating to severity.)
- **Did you eat other foods along with the suspect food?** (These may slow digestion and allergic reaction, or may indicate an alternative source of the allergy.)
- **In the case of seafoods, was the food well cooked?** (Thorough cooking can sometimes inactivate allergens.)

Food record

If the diet history is unclear, the subject would often be recommended to keep a written food record (p. 281), initially perhaps for two weeks, to document every food consumed along with any reactions that follow. This may show more clearly whether or not there is a consistent response to a particular food. Figure 46.1 shows an example, for a single day.

Elimination diet

Dieticians, in conjunction with the principal healthcare provider, can test for a suspected allergy-causing food by temporary removal from the diet. This will involve careful checking of food labels for hidden traces, e.g. wheat flour is present in many packaged foods. If the symptoms clear up after removing the food, this is a good indicator that it is causing a problem. If it is a minor allergy the food can be reintroduced, under appropriate supervision, to see if the symptoms recur, for confirmation. If it is a severe allergy the food should not be reintroduced.

A related approach taken when it is not clear what food may be causing an allergy, is to remove all of the main allergy-causing foods from the diet for 3–6 weeks, see if the problem clears up, then reintroduce one food at a time and watch for symptoms. For food intolerance, it is often the case that small amounts of the food can be tolerated, and this can be tested following elimination. However, with allergies, usually even small amounts of the food can cause a reaction, and therefore complete avoidance is the best approach.

KEY POINT With elimination diets, particularly in children, it is extremely important that low-allergenic substitutes of similar nutritional value be substituted for the eliminated food(s) to avoid malnutrition (for example, rice cereals instead of wheat cereals; calcium-fortified soy milk instead of cow's milk; legumes instead of nuts; meat and poultry dishes instead of eggs or fish).

Skin testing

If the diet history, food record or elimination diet indicates a potential reaction to a specific food, an allergy specialist can confirm the finding with a skin prick test or scratch test. A negative control (allergen-free solution) should also be used with each set of tests, as well as a positive

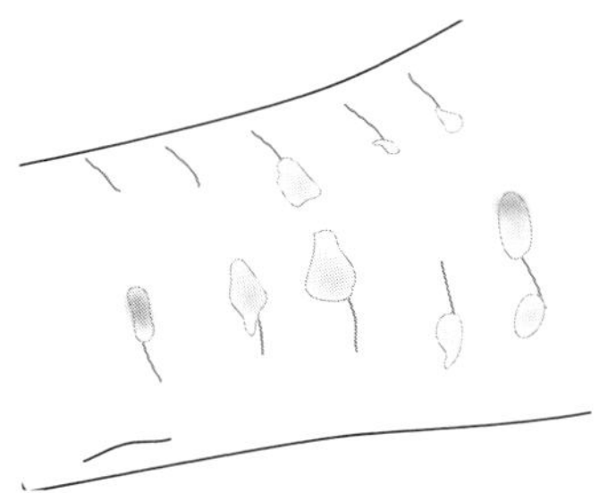

Fig. 46.2 Skin allergy test on forearm showing pen line markings of testing sites and inflammatory swelling responses.

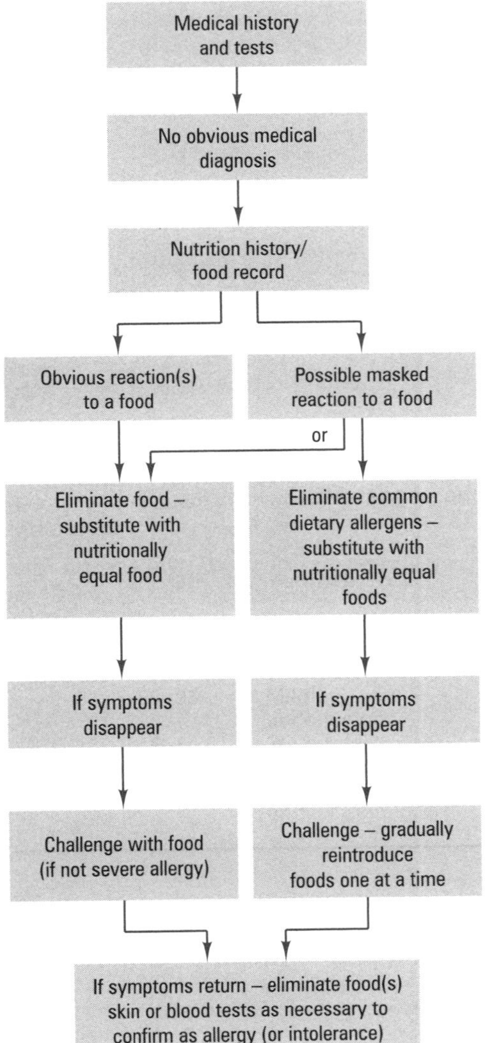

Fig. 46.3 Flow chart of procedures for testing for food allergy and intolerance.

control (histamine solution). Antihistamines should not be taken by the patient for at least the preceding 48 hours and withdrawn only under medical supervision if a prescribed medication. Before the test the skin is cleaned with alcohol. The sites of the tests are then marked with ink using a surgical marker (Fig. 46.2). There are two main variations:

1. With a scratch test, the practitioner applies an extract of food to the skin of the lower forearm then applies a small scratch to the site with a sterile needle.
2. A skin prick test involves inserting a sterile lancet or needle just below the patient's lower arm skin surface then applying a small amount of food extract under the skin. Alternatively, in the 'prick through drop' test, the lancet is used to prick the skin through a surface droplet of food extract.

Any excess extract can then be carefully blotted from the skin so that it does not run into adjacent testing sites. A positive reaction takes up to 15 minutes, showing as swelling and/or redness at the site due to a reaction of specific IgE antibodies on the skin's mast cells with the extract (Fig. 46.2). Following recording of the results, ink marks and extracts are then cleaned from the skin with alcohol, and the medical practitioner may then apply a mild cortisone cream if there is itching. Some allergists recommend the patient stays at the clinic for a further 30 minutes, to be sure there are no unexpected side-effects.

A practical advantage of skin testing is that common airborne and contact allergens such as pollens and dust mites can be tested at the same time as foods. A disadvantage of skin testing for food allergens is that a positive reaction does not always indicate the patient is allergic to the food when eaten. So diet history needs to be considered along with the skin test result.

Blood testing

Immunological blood tests can also be done to confirm IgE food allergies, and these are a better choice than skin tests in cases of previous severe allergic reactions, or if the patient has extensive eczema. These include screening for specific IgEs with the original radioallergosorbent test (RAST), the improved, more sensitive commercial CAP-RAST test, and/or ELISA (p. 350). Additionally, commercial ELISA kits can be used to screen for allergy-induced IgG antibodies.

Figure 46.3 gives an overview of all above testing procedures, as a flow chart showing the major steps and decisions taken.

Double-blind oral food challenge

This test follows a similar process to that used in double-blind experimental trials (p. 210). Foods to be tested as possible allergens are placed in capsules or masked in other foods known not to be allergenic. These are consumed sequentially, about every 60 minutes, along with randomly allocated placebos, and the subject's responses recorded. The tester is also 'blinded', as the capsules/food mixes are prepared by another qualified person. This method is useful to eliminate possible psychological factors influencing patient responses, or the subconscious bias of a recorder.

Text references

Mullins, R.J., Dear, K.B.G. and Tang, M.L.K. (2009) Characteristics of childhood peanut allergy in the Australian Capital Territory, 1995 to 2007. *Journal of Allergy and Clinical Immunology*, **123**, 689–693.

Sources for further study

Brostoff, J. and Gamlin, L. (2000) *Food Allergies and Food Intolerance: The Complete Guide to Their Identification and Treatment*. Healing Arts Press, Rochester.

Food Standards Agency, UK. *Food Allergy and Intolerance*. Available: http://www.eatwell.gov.uk/healthissues/foodintolerance/ Last accessed: 21/12/10.

Jackson, F.W. (2003) *Food Allergy*. International Life Sciences Institute, Brussels. Available: http://europe.ilsi.org/NR/rdonlyres/655C8D9B-909F-4BDE-92E1-244893B2961C/0/CMFoodAllergyv2.pdf Last accessed: 21/12/10.

National Institute of Allergy and Infectious Diseases (2007) *Food Allergy, an Overview*. Available: http://www3.niaid.nih.gov/topics/foodAllergy/PDF/foodallergy.pdf Last accessed: 21/12/10.

Swain, A., Soutter, V. and Loblay, R. (2009) *RPAH Elimination Diet Handbook*. Allergy Unit, Royal Prince Alfred Hospital, Australia. Available: http://www.sswahs.nsw.gov.au/rpa/allergy/ Last accessed: 21/12/10.

47 Integrating nutritional assessment data

Assessing Nutritional status and risk of under nutrition – the range of methods take into account some easily assessed characteristics, such as:

- Short-term weight changes
- Medium-term weight changes
- Body Mass Index (BMI)
- Illness history
- Physical characteristics
- Patient mobility

Nutritional assessment is made up from a number of different elements and no one measurement can provide the whole picture. It is necessary to consider a number of variables and observations so that an accurate evaluation of a person's nutritional status can be made. The process of becoming skilled in nutritional assessment is developed over time. As an undergraduate you are likely to work through the process when working on case studies or you may have direct experience while on placement. Once qualified, you will have guidance from suitably trained and appropriately experienced people.

The main tools, either for use in initial screening for risk of malnutrition or for full assessment, are described below.

KEY POINT There are a number of methods available and it is important to choose the appropriate one. This will depend on the subject group, the setting and also local guidelines.

Subjective global assessment (SGA)

This is a standard questionnaire which was developed by Detsky *et al.* (1987) and is still widely used in a hospital setting. It can be used to predict the risk of post-treatment complications such as poor wound healing and infections, particularly in surgical patients. It is based on a patient's history and aspects of physical examination. The main elements of the questionnaire are detailed below but there are often minor changes made to the format in different organisations. The information recorded on the form comes from both direct questioning of the subject and from his or her medical notes. In some cases it may also come from other health professionals or carers if the subject was not able to give the information.

Example Calculation of percentage weight loss if the body mass six months ago was 65.5 kg and current body mass is 61.3 kg:
(current mass − mass 6 months ago) ÷ mass 6 months ago × 100
Percent loss = (65.5 − 61.3) ÷ 65.5 × 100 = 6.4%

Body mass (body weight) change

The first element of the patient history looks at the amount and the pattern of body mass loss by noting the weight changes over the past 6 months and in the past two weeks. From these measurements it is easy to calculate the percentage loss (see example opposite). A weight loss of <5 per cent is considered small; 5–10 per cent would be considered potentially significant and above 10 per cent is definitely significant.

Weight loss in the past 2 weeks should be noted and you would also question the patient about his or her maximum weight to compare that with his or her current weight. These two questions allow you to build a picture of the pattern of weight loss, i.e. whether it has been sudden or over a long period of time.

Changes in dietary intake

If the subject reports that their dietary intake is different from his or her usual intake six months ago you should question when this change took place. You will also need to ask what type of change occurred. Some examples of the questions you could ask are:

- Do you eat more or less frequently than you did six months ago?
- Has the amount you eat changed?
- Are there foods which you used to eat but don't anymore?

You should try to find out the reasons for any changes to identify whether there are problems with food purchasing or preparation, swallowing or changes in taste perceptions, for example.

Gastrointestinal function

Any symptoms of gastrointestinal disturbance which have lasted more than two weeks would be noted as significant. Diarrhoea or vomiting for one or two days would not be.

Functional capacity

The next step of the assessment is a description of whether the subject's ability to conduct activities of daily living is affected. The types of activities which you would focus on depend on whether he or she has been living independently or, particularly in the case of older adults, has been living in a care home. The grading goes from 'no dysfunction' when he or she has full capacity to perform all activities to 'severe dysfunction' such as being confined to bed. If functional capacity has been affected you should try to find out how long this has been the case.

Definitions

Functional capacity – the capability to perform routine tasks and activities.

Oedema/edema – accumulation of fluid between the cells. Sometimes described as 'fluid retention'.

Ascites – fluid in the abdominal cavity.

Physical characteristics

The elements of physical examination included in the SGA are as follows:

- **Loss of subcutaneous fat** – check shoulders, triceps, chest and hands for evidence of loose skin or loss of appearance of 'fullness' (you may wish to look at your own hands for a healthy comparison). Older adults may have signs of loose skin which do not indicate malnutrition, so the presence of this feature on its own is not an indicator.
- **Muscle wasting** – this is best assessed by looking at the deltoid muscles at the sides of the shoulders and the quadriceps at the front of the thigh. It is difficult to make this type of assessment until you have assessed people who are well-nourished and people who are malnourished, so you should initially take advice from more experienced colleagues or supervisors.
- **Ankle oedema or ascites** – these would be assessed as absent, mild, moderate or severe. If present, the assessment of loss of body mass becomes less relevant as any loss may be masked by the accumulation of fluid.

SGA rating

The final step of the SGA is assigning a rating which is a subjective opinion of their overall nutritional status. The assessor categorises the individual as follows:

A. **Well-nourished** – patients would have gained weight or had no significant weight loss, no issues with their dietary intake, no gastrointestinal (GI) symptoms which would lead to nutritional problems, and no physical or functional impairment.

B. **Moderately malnourished** – patients would usually present with at least 5 per cent weight loss, reduced dietary intake, and mild to moderate loss of subcutaneous fat and muscle. Moderately malnourished people may not show any loss of function or physical symptoms at this stage.

C. **Severely malnourished** – patients would have continuing weight loss of more than 10 per cent, poor dietary intake, and severe fat and muscle loss and subsequent loss of function (Detsky *et al.*, 1994).

The National Institute of Health and Clinical Excellence (NICE) in the UK recommends that all people who are admitted to hospitals or care homes are weighed and measured.

The SGA is generally carried out when a patient is about to undergo treatment. It can then be repeated to monitor progress over a defined period of time. After this time people would be reclassified as well-nourished (rated A) if they had increased their body mass and their dietary intake had improved even if their body weight loss was still between 5 per cent and 10 per cent.

Malnutrition universal screening tool (MUST)

MUST has been developed to assess malnutrition in adults and is widely used both in the community and in clinical settings, and includes management guidelines which clearly define what should be done once the final score is calculated.

Scores are given for:

- **current Body Mass Index (BMI) kg/m^2** – if BMI is above 20 a score of 0 is recorded; if between 18.5 and 20 score 1; and if less than 18.5 score 2; (a BMI of less than 18.5 is considered underweight);
- **unplanned weight loss in the previous three–six months** – score 0 if body mass loss is under 5 per cent; a score of 1 if it is between 5 per cent and 10 per cent; and a score of 2 if the loss is greater than 10 per cent (see example above to calculate the percentage weight loss);
- **acute disease effect** – a score of 2 if the person is acutely ill or there has been or is likely to be no nutritional intake for more than five days.

Scores for each of the three sections are added together and a final score calculated. This final score indicates the patient's overall risk of malnutrition, as follows:

- a score of 0 indicates low risk;
- a score of 1 indicates medium risk;
- a score of 2 or more indicates high risk.

Management guidelines based on MUST score

The course of action which should be taken once a MUST score is calculated depends on the situation in which it is being used, for example whether in the community, in a care home or in a hospital.

If the overall risk of malnutrition is low, then routine clinical care would be continued and repeated screening would be carried out annually for those in the community, monthly for people in care homes and weekly for hospitalised patients. If the screening suggested a medium risk of malnutrition (a total MUST score of 1) then repeat screening would be more frequent and dietary intake may be documented and evaluated. If dietary intake is adequate there is generally no clinical concern at this stage. If the dietary intake is inadequate the patient may require dietetic referral. High-risk patients with a MUST score of 2 or more would be treated by referring to a dietitian or nutritional support team with a view to increasing and improving dietary intake. Monitoring and review of the care plan would be frequent: weekly in hospital and monthly in care homes or in the community.

For detailed further information on MUST, including alternative measurements for calculating BMI if it is not possible to measure height, and for BMI and weight loss tables see BAPEN (2006).

Indirect measures of height – use length of ulna and refer to tables provided by BAPEN (2006) to estimate height or use demi span or knee height.

Understanding the terminology
Nutritional risk screening – this is generally referred to as NRS 2002 to indicate the most recent version.

ESPEN – the European Society of Parenteral and Enteral Nutrition is now known as the European Society for Clinical Nutrition and Metabolism but still uses the acronym, ESPEN.

Defining nutritional risk – it is necessary to take into account both present nutritional status and risk of impairment to present status. The latter may be affected by the patient's clinical condition, which may result in increased stress-related metabolism (Kondrup *et al.,* 2003).

Nutritional risk screening (NRS)

The NRS was developed for use with hospital patients to detect existing under nutrition and to measure the risk of developing under nutrition. It was validated by ESPEN and is recommended by them (Kondrup *et al.,* 2003). You will normally gain experience of this tool during placements and when working through case studies.

NRS 2002 is a two-part process. The first part (initial screening) has four questions:

1. Is the person's BMI less than 20.5 kg/m^2?
2. Has the person lost weight within the last three months?
3. Has the person had a reduced dietary intake in the last week?
4. Is the person severely ill?

If the answer to any of these initial questions is yes, then the second part of the screening process is carried out. If all answers are no, then the initial screening procedure is repeated on a weekly basis.

The second part of the screening scores both the level of impaired nutritional status and the severity of disease. There are detailed criteria for each score, which range from 0 to 3 for each of the two components, the scoring criteria for which are detailed in Table 47.1 (Kondrup *et al.,* 2003). The two scores are then added and age-adjusted by adding 1 to the score if the person is aged 70 years or above. If the final score is 3 or more then the patient is considered nutritionally at risk and a nutritional care plan would be implemented. If the score is less than 3 the patient would be screened on a weekly basis.

This method has been shown to be useful to a range of health professionals. However, it has been criticised for not giving clear guidance on how to estimate BMI if height or weight cannot be measured, for example, because the person is unable to stand up straight (Stratton *et al.,* 2006). It is appropriate to use the instructions for alternative measurements provided with the MUST in other screening tools.

Table 47.1 Scoring criteria in the final screening of NRS 2002 (Kondrup *et al.,* 2003)

Impaired nutritional status		**Severity of disease (approximately equal to increase in requirements)**	
Absent: score 0	Normal nutritional status	Absent: Score 0	Normal nutritional requirements
Mild: score 1	Weight loss >5% in 3 months or food intake below 50–75% of normal requirements in preceding week	Mild: score 1	Hip fracture; chromic patient, in particular those with acute complications; cirrhosis; chronic obstructive pulmonary disease; chronic haemodialysis, diabetes, oncology
Moderate: score 2	Weight loss >5% in 2 months or BMI 18.5–20.5 kg/m^2 and impaired general condition or food intake 25–60% of normal requirements in preceding week	Moderate: score 2	Major abdominal surgery; stroke; severe pneumonia, haematologic malignancy
Severe: score 3	Weight loss >5% in 1 month (>15% in 3 months) or BMI < 18.5 kg/m^2 and impaired general condition or food intake 0–25% of normal requirement in preceding week	Severe: score 3	Head injury; bone marrow transplantation; intensive care patients

Table 47.2 MNA scoring system

Category	Scores and descriptions
A	0 = severe decrease 1 = moderate decrease 2 = no decrease
B	0 = weight loss >3 kg 1 = does not know 2 = weight loss > 1 kg <3 kg 3 = no weight loss
C	0 = bed or chair bound 1 = able to get out of bed/chair but does not 2 = goes out
D	0 = yes 2 = no
E	0 = severe dementia or depression 1 = mild dementia 2 = no psychological problems
F	0 = <19 1 = 19 to < 21 2 = 21 to < 2 3 = ≥23

Evaluating integrated nutritional assessments – A comprehensive review of these tools by Green and Watson (2005) gives details of each method and also an evaluation of the validity, reliability and sensitivity of the different methods.

Mini nutritional assessment (MNA®)

This has been validated for use with elderly people and is appropriate for use in hospital, care homes and in the community (Nestlé Nutrition Institute, 2009). The MNA® is split into two parts; the initial screening and the full assessment. Initial screening comprises six questions and gives a possible maximum score of 14. This part has also been validated as a standalone version known as the MNA® Short Form.

The six questions in the initial screening/MNA® short form are:

A. **Has food intake declined over the past three months?**

B. **Has there been weight loss in the previous three months?**

C. **How mobile is the patient?**

D. **Have there been any psychological stress or acute illness in the past three months?**

E. **What are the patient's neuropsychological problems, such as depression or dementia?**

F. **What is the patient's current BMI kg/m^2?**

The scoring system is shown in Table 47.2. If the score is 12 or more then the person is not considered to be at risk and no further assessment is required. If score is less than 12 the second part of the assessment should be carried out. This involves a more detailed evaluation of the person's dietary intakes and health status by asking a further 12 questions (refer to Nestlé Nutrition Institute, 2009, for details). The score for this section is added to the screening score to give a total malnutrition indicator score, with a possible maximum of 30. If the total score is between 17 and 23.5 it indicates that the person is at risk of malnutrition. If the score is less than 17 points the subject would be considered malnourished. In both cases he or she would be referred to the dietitian and a nutritional care plan would be implemented.

The MNA® takes about ten minutes to complete and can be carried out by most health professionals following minimal training.

Text references

BAPEN (2006) *Malnutrition Universal Screening Tool.* Available: http://www.bapen.org.uk/must_tool.html Last accessed: 21/12/10.

Detsky, A.S., McLaughlin, J.R., Baker, J.P. *et al.* (1987) What is subjective global assessment of nutritional status? *Journal of Parenteral and Enteral Nutrition*, **11**, 8–13.

Detsky, A.S., Smalley, P.S. and Chang, J. (1994) Is this patient malnourished? *Journal of the American Medical Association*, **271**(1), 54–58.

Green, S.M. and Watson, R. (2005) Nutritional screening and assessment tools for use by nurses: Literature review. *Journal of Advanced Nursing*, **50** (1), 69–83.

Kondrup, J., Allison, S.P., Elia, M., Vellus, B. and Plauth, M. (2003) ESPEN guidelines for nutrition screening 2002. *Clinical Nutrition*, **22**(4), 415–421. Available: http://www.espen.org/documents/Screening.pdf
Last accessed: 21/12/10.

Nestlé Nutrition Institute (2009) *MNA Mini Nutritional Assessment Overview*. Available: http://www.mna-elderly.com
Last accessed: 21/12/10.

Stratton, R.J., King, C.L., Stroud, M.A., Jackson, A.A. and Elia, M. (2006) 'Malnutrition Universal Screening Tool' predicts mortality and length of hospital stay in acutely ill elderly. *British Journal of Clinical Nutrition*, **95**, 325–330.

Sources for further study

Gibson, R.S. (2005) *Principles of Nutritional Assessment*, 2nd edn. Oxford University Press, New York.

Lee, D.L. and Nieman, D.C. (2007) *Nutritional Assessment*, 4th edn. McGraw Hill Higher Education, New York.

48 Sports nutrition

An understanding of sports nutrition first requires knowledge of basic nutrition principles (Chapter 40) as the foundation. This basic foundation can then be built upon to meet the specific nutritional challenges found in different types of sports and physical activities. Some goals of sports nutrition include optimisation of diet for specific sports, achieving maximal performance, and assisting recovery after sports and related injuries. While this is a specialist area of nutrition, you are likely to work with sports persons within your degree programme, so it is useful to have an appreciation of the fundamental principles.

Dietary matters that sports persons might enquire about –

- Unable to eat enough
- Limited time/opportunities for eating
- Suppressed appetite
- Media reports
- Promoted new sports diets/supplements
- Weight loss or gain
- Eating while travelling
- Eating and drinking to hasten recovery
- Eating and drinking to enhance performance.

Energy

Adequate energy intake is essential for very active people to prevent:

- poor physical performance;
- increased vulnerability to illnesses;
- loss of muscle tissues;
- inadequate recovery following exercise.

> KEY POINT When working with sports persons, a medical, dietary and exercise history should be taken and advice should be tailored to the person's individual requirements.

Some broad guidelines for adult energy intake are:

- **General fitness** (30–40 minutes a day three days a week)
  — 105–150 kJ (25–35 kcal) $kg^{-1}day^{-1}$
- **Moderate exercise** (60 to 120 minutes per day, five to six times a week)
  — 210–340 kJ (50–80 kcal) $kg^{-1}day^{-1}$
- **Endurance training** (>120 minutes per day)
  — 420–840 kJ (100–200 kcal) kg^{-1} day^{-1}

Regularly checking body weight on a reliable set of weighing scales is a simple yet effective way for athletes to monitor the adequacy of their energy intake related to activity. Weight loss equals inadequate intake.

Understanding the effects of dehydration during exercise – Loss of 1 per cent body weight of water = increased feelings of effort.

Loss of 2 per cent body weight of water = decreased performance.

Loss of 4 per cent body weight of water = heat exhaustion, heat stroke, possible death.

Water – a key nutrient

If a person feels thirsty, then dehydration is already present. So advise sports and exercise enthusiasts to drink between 150 and 200 mL of plain water just before starting exercise and 125 to 250 mL every 15 minutes during exercise, depending upon ambient temperature, humidity and personal sweating rate (PSR).

> KEY POINT Dehydration leads to poor performance in sports.

Example (PSR) –

1. Pre-exercise weight = 72.64 kg
2. Post-exercise weight = 71.94 kg
3. Weight lost = 0.70 kg = 700 grams ≈ 700 mL water
4. Water drunk = 250 mL
5. Fluid replacement requirements = 700 + 250 = approximately 950 mL/hour

It is best when possible to drink water every 15–20 minutes during exercise – in this case about 320 mL/20 min.

Determining personal sweating rate (PSR)

Calculating PSR, while useful for most active sports, can be particularly important in water sports such as swimming, where the amount of water lost as sweat will not be obvious.

The following step-by-step procedure can be followed:

1. **Record the subject's weight in kg**. Ensure the person is well hydrated, have him or her do a short warm up, then weigh him/herself undressed.
2. **Have the subject exercise for 60 minutes** at the same intensity expected to be faced in forthcoming training or competition. If possible this should also be arranged for a period of expected similar ambient conditions of the forthcoming event.
3. **Measure and record the amount of water consumed during the hour of exercise.** This is usually recorded in millilitres (mL).
4. **Re-weigh the subject**. Stop the person after the hour and have him or her remove his or her sweat-soaked clothes and shoes, towel him/herself thoroughly and weigh him/herself naked again.
5. **Subtract the post-exercise weight from the pre-exercise weight**. If working in kg convert this to grams, which approximates to mL of water lost.
6. **Add to this figure the number of mL of water drunk during the hour**, to determine the approximate hourly fluid requirements for this individual in order to compensate for sweating.

Carbohydrates

These should be the principal energy source in the diet. Starchy carbohydrates are important for athletes to build up glycogen (Fig. 48.1) stores in the muscles and liver to sustain a high level of

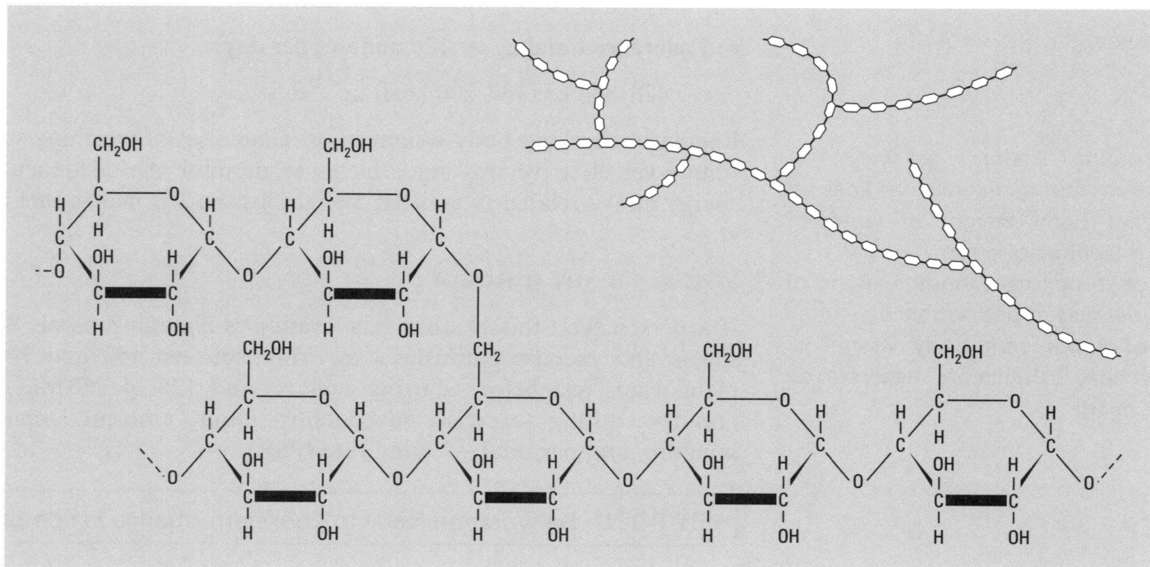

Fig. 48.1 Branched chain structure of glycogen consisting of linked glucose molecules.

activity. If glycogen supplies run low, tiredness and decreased performance can result. Eating carbohydrate before and after exercise is essential to maximise and restore glycogen levels, respectively.

Carbohydrate loading

Carbohydrate loading can significantly increase pre-exercise muscle and liver glycogen stores. It is particlarly useful for endurance athletes to increase performance during exercise of 90 minutes duration or longer. Studies to date show carbohydrate loading may be more effective in males than females.

The following example demonstrates carbohydrate loading for a marathon:

Taking care with carbohydrate loading – nutritionists should ensure, through client medical history and blood glucose tests, that carbohydrate loading does not adversely affect blood glucose levels, particularly in people with diabetes.

- 4–6 days before, eat 5 g carbohydrate kg (body weight)$^{-1}$day^{-1}.
- 1–3 days before 10 g kg^{-1} day^{-1}.
- The night before the race focus on eating starchy carbohydrates (for example, rice, pasta). In this instance wholegrains are not recommended (slow to digest and feeling full quickly).
- Advise athletes to avoid eating something new or unusual the night before the race or for breakfast in the morning in case it results in a gastrointestinal upset.
- Breakfast should be two–three hours before the race and should be rich in light carbohydrates, such as cereal with skimmed milk, toast and jam, pancakes, banana, fruit juice or sports drink.
- In conjunction with the carbohydrate loading detailed above, scale down the training before the event to allow glycogen to accumulate, e.g. 90-minutes training six days before the race, reducing to 40 minutes on days five and four, then 20 minutes on days three and two, with no training the day before the event.

Protein

High-quality proteins to recommend for sports persons – include lean red meats, lean chicken with skin removed, reduced fat dairy products/skim milk powder, eggs/egg white, fish, nuts and seeds.

Protein is required for muscle growth and repair. People who are very active and train a lot require more protein than those that do not. However, most people in the developed world already eat more than enough protein, even those doing very active sports and body building. If a person is following the basic nutrition guidelines (Chapter 40) and is eating a varied, healthy diet, based on current evidence there is little to be gained by taking protein supplements.

However, before a nutritionist would make a recommendation to a client, they should check the nutritional guidelines for protein foods for their region in case they are set lower than in other regions, e.g. in Australia the general recommended protein intake is about 0.75–1.0 g kg (body weight)$^{-1}$ day^{-1}, which is reasonable for a basic exercise programme. However, for adolescents, endurance training, body building and high-intensity exercise, it is suggested to increase intake to about 1.5–2.0 g kg (body weight)$^{-1}$day^{-1}. For a 70-kg person this would still only be 105–140 g of protein foods a day.

Advice for nutrition-related problems –

Lack of appetite/weight loss

- Smaller more frequent meals
- Nutrient and energy dense foods, e.g dried fruits.

Weight gain

- Is it muscle or fat? Use bioelectrical impedance scales (p. 297) to check.

Gastrointestinal upsets

- Avoid large heavy meals before exercise
- Do not eat gas-inducing food such as beans before exercise
- Avoid fruit as fructose has been linked to gut upsets.

Stitch or abdominal cramp

- Do not eat big meals before exercising
- Limit fat and fibre intake before exercise
- Avoid concentrated sports drinks, cordials and soft drinks.

General advice on pre-event meal –

- Consume two–four hours before event
- Small meal, e.g. < 4000 kJ
- Meal high in carbohydrates
- Meal lower in fat and protein
- Avoid spicy, irritating and gas-inducing foods.

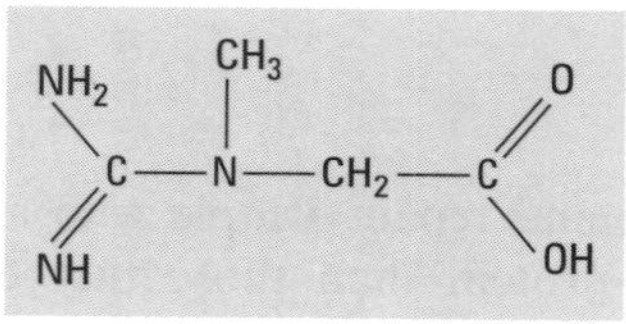

Fig. 48.2 Structure of creatine, which supplies energy in muscle cells.

Fats

Advice for amounts of fats in the diet for sports nutrition is similar to that found in basic nutrition guidelines for non-athletes (Chapter 36), and also depends upon the person's individual body condition. Generally, advise limiting fats in the diet overall, and the use of primarily monounsaturated and polyunsaturated sources.

Vitamins and minerals

A varied and balanced diet should usually provide all the vitamins and minerals needed for an athlete. Vitamin and mineral supplementation is usually required only in special circumstances (e.g. people on a vegetarian diet or those who avoid a whole food group such as milk and dairy). If necessary, a multivitamin-mineral tablet that provides 100 per cent of the RDA (p. 279) should provide all nutrients needed, but this will not replace the benefits of a nutritious diet. Strict vegetarians may also require a calcium supplement. 'Megadoses' of vitamins are not recommended and may do more harm than good. Iron supplements should not be taken unless medical tests show more iron is needed. If iron levels are adequate, taking more will not improve sports performance and can cause negative side effects. It is usually recommended that endurance athletes have their haemoglobin and serum ferritin levels checked annually.

Other supplements

Creatine

The metabolite creatine (Fig. 48.2) is naturally found in meat and fish. About half the creatine used in the body comes directly from our diet while about half is also synthesised in the body from amino acids. Creatine supplies energy in muscle cells. Supplementation with creatine monohydrate for high-intensity strength training and in sports necessitating high-intensity intermittent bursts of strength and speed, appears to be beneficial to performance and strength. Creatine monohydrate has not been demonstrated to improve performance in endurance sports. Example doses are: loading for competition 20–30 g per day (5 × 6 g doses) for a week, and training maintenance 10–15 g per day. Short-term studies have not shown any serious ill effects. However, possible long-term effects of chronic use of creatine are not known.

HMB (hydroxymethylbutyrate)

This is a metabolite of the amino acid leucine, and is found naturally occurring in citrus fruits. Preliminary research suggests that HMB may suppress muscle protein breakdown and increase strength. No harmful effects have yet been noted. An example dose of HMB is 1 g three times a day.

Glutamine

The amino acid glutamine may be helpful to boost the immune system when a person is unwell. Glutamine may also reduce muscle protein breakdown in athletes. However, a healthy athlete on a suitable diet with adequate protein intake will be already taking in sufficient glutamine. Recent research from the University of Western Australia shows that endurance exercise increases plasma concentrations of glutamine, apparently as a natural training adaptation (Kargotich *et al.*,

Retrospective designer drug testing – new drugs can become available before testing methods are available. Sports clients who indicate they take such drugs should be cautioned that samples from sports athletes are now being retained and tested retrospectively when new tests are available.

2007). No toxicity has been shown to glutamine over several weeks when taking it at recommended doses (e.g. 0.1 to 0.3 g glutamine kg (body weight)$^{-1}$ day^{-1}).

Controlled substances and harmful supplements

Steroid precursors

Nutritionists should warn clients against using steroid precursors such as dehydroepiandrosterone (DHEA) and androstenedione. Studies show no or little benefit on strength or muscle growth. However, long-term studies have suggested that use of these may be linked to liver damage, alteration of secondary sexual characteristics, and may be associated with cancers of the prostate and uterus.

Assessing the value of a new, improved supplement – if a new athletic supplement or promoted diet sounds too good to be true it usually is. Look for independent scientific evidence rather than at advertised pictures of big muscles.

Anabolic steroids

These are steroid hormones related to testosterone. Possible adverse affects include hypertension, increase in the LDL to HDL cholesterol ratio, increased risk of cardiovascular disease and high doses can cause liver damage. Psychiatric effects can include increased aggressive behaviour, violence and mania.

Ephedra (Ma Huang)

This is a Chinese medicinal herb form of the stimulant ephedrine. It has been banned following more than 80 confirmed deaths associated with the cardiovascular system, including in some professional athletes.

Overseas travel and food – sports persons should check whether foods can be taken through customs.

Travelling

Sports persons travelling away from home to a sporting activity should not rely on fast food take-aways and transport catering facilities, where healthy choice foods may be limited. Whenever possible they should plan their meals ahead of time and take suitable foods with them. They could contact the venue's organisers to see if special foods will be catered for there. Examples of foods that sports persons could take travelling include sandwiches, fresh or dried fruit, rice and oatcakes, nuts, cereal bar, yogurt, and for drinks: bottled water, sports drinks, smoothies and fruit juices. When eating out, choose the healthy options from the menu.

Text reference

Kargotich, S., Keast, D., Goodman, C., *et al.* (2007) Monitoring 6 weeks of progressive endurance training with plasma glutamine. *International Journal of Sports Medicine*, **28**, 211–216.

Sources for further study

Burke, L. (2007) *Practical Sports Nutrition.* Human Kinetics, Champaign, Illinois.

Food Standards Agency, UK. *Food for Sport.* Available: http://www.eatwell.gov.uk/healthydiet/foodforsport/?lang=en Last accessed: 21/12/10.

Williams, M. (2005) Dietary supplements and sports performance: amino acids. *Journal of the International Society of Sports Nutrition*, **2**, 63–67. Available: http://www.pubmedcentral.nih.gov/articlerender.fcgi?artid=2129148 Last accessed: 21/12/10.

Wolinsky, I. and Driskell, J.A. (eds) (2007) *Sports Nutrition – Energy Metabolism and Exercise.* CRC Press, Boca Raton, Florida.

Analytical techniques in food science

49 Basic physico-chemical techniques for food analysis

Definitions

Organoleptic – relating to the sensory properties of an item, namely taste, smell, touch, sight and/or sound.

Rheology – scientific study of the flow and deformation of matter.

Foods are analysed for various reasons, including: (i) quality assurance; (ii) determination of nutritive value and 'shelf life'; (iii) compliance with legal requirements, including labelling (p. 138); (iv) determination of properties, e.g. rheological characteristics; and (v) detection of contaminants. This chapter deals with the fundamental procedures used to provide information on the key physical and chemical properties of food. Evaluation of organoleptic properties of food is covered in Chapter 64 (p. 434).

KEY POINT Physico-chemical analysis provides information on the overall structure, composition and properties of food, rather than details of specific biochemical constituents (see Chapters 58–63).

Understanding moisture content – this term is preferred to 'water content' since widely used methods such as drying and distillation do not strictly measure total water. Consequently, the methods used in the food industry are highly standardised, to provide consistent measurements of moisture under specified conditions (e.g. AOAC, 2009).

Moisture

The water content of a foodstuff is of interest because of its effect on quality and stability, e.g. low-moisture foods may discourage the growth of many microbes, due to their low water activity (p. 490).

Typically, the moisture content is expressed as a percentage of total mass, based on the following equation:

$$\% \text{ moisture} = (\text{mass of water}/\text{total mass}) \times 100 \quad [49.1]$$

While the total mass of a food sample is simply determined by weighing, the mass of water is usually measured by one of the following methods:

Alternatives to conventional forced draft drying ovens – microwave ovens and infra-red lamps can be used to provide rapid drying, although care is required to obtain reproducible and reliable measurements. Another approach is to use a desiccator (p. 150).

- **Drying.** Here, the moisture content is determined from the difference in the mass (weight) of a sample of food before and after drying, typically in a forced draft oven. The relationship is usually expressed as:

  $$\% \text{ moisture} = (\text{initial mass} - \text{mass after drying})/\text{initial mass} \times 100 \quad [49.2]$$

  Drying is typically carried out at 100°C, or at slightly lower temperatures under partial vacuum. Samples are kept in this regime until they reach a constant weight, with the assumption that all water is now removed. While drying is relatively straightforward and inexpensive, it is also slow and may destroy some food components. This approach also makes the assumption that the loss in mass is equivalent to the total water content, which ignores (i) retention of any remaining 'bound' water, (ii) loss of any other volatile substances, and (iii) chemical changes, such as hydrolysis of complex sugars.

- **Distillation.** In contrast to drying, where the moisture content is determined indirectly from the change in mass of a food sample [eqn 49.2], this is a direct method, using eqn [49.1]. A known mass (weight) of food is heated in an immiscible solvent (e.g. xylene) and the evaporated water is condensed and collected in a graduated tube, e.g. in a Dean and Stark distillation apparatus (Fig 49.1). The process is

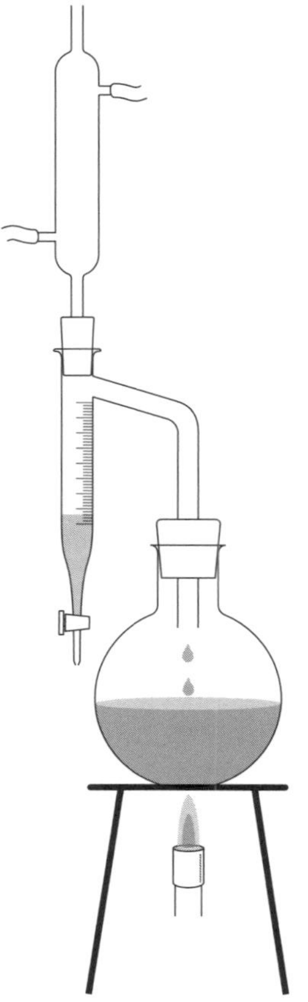

Fig. 49.1 Dean and Stark distillation apparatus.

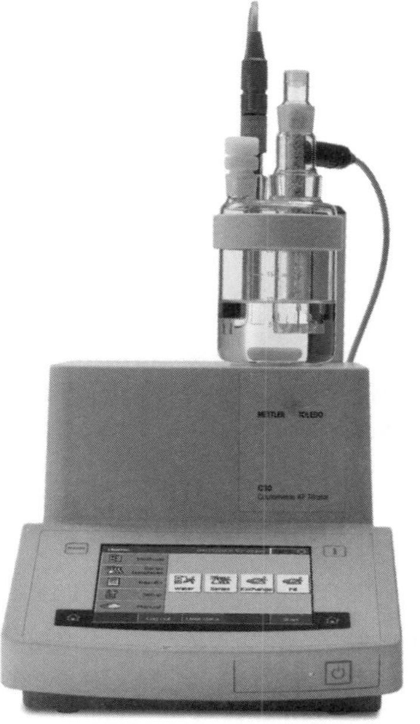

Fig. 49.2 METTLER TOLEDO's C20 Compact Coulometric Karl Fischer titrator.

stopped when no more water condenses in the collection tube. As with oven drying, distillation is time-consuming and destructive, since the foodstuff cannot be used for further analysis.

- **Chemical reaction.** The most important method here is the Karl Fischer titration method, which is based on the following overall reaction:

$$I_2 + SO_2 + H_2O \quad 2HI + H_2SO_4 \qquad [49.3]$$

Thus iodine and water are consumed in a 1:1 ratio. Since I_2 is strongly coloured and HI essentially colourless, the simplest approach is to titrate a food sample using a suitable Karl Fischer reagent containing a known concentration of I_2, with the end point being determined by the appearance of dark red-brown I_2. The water content can then be determined from the I_2 concentration of the Karl Fischer reagent and the volume used to reach the end-point. Semi-automated Karl Fischer titrators based on coulometric (electrolytic) determination of the titration end-point are now used in the food industry and in some university laboratories (Fig 49.2). Since it is based on a specific chemical reaction, this approach avoids some of the limitations of the indirect methods described above. However, a major practical problem is access of the Karl Fischer reagent to the water within a food sample – this is usually aided by homogenising the sample and/or solvent extraction, e.g. using methanol.

Total solids

This is typically determined as the % mass remaining after subtraction of the % moisture content. Because of the potential errors associated with measurement of water content described above, it is important that assays are carried out under standardised conditions, often specified within standard procedures (e.g. AOAC, 2009).

Ash

This term is used to describe the inorganic fraction remaining after all organic matter has been removed, e.g. by combustion or oxidation. This is usually assayed by one of two approaches:

1. **Dry ashing.** Using a muffle furnace at 500–600°C to remove water by evaporation and organic matter by incineration. Food samples are weighed in porcelain crucibles, kept in the muffle furnace for 12–18 hours, then reweighed. Ash content is typically expressed either as a percentage of the dry mass or the total mass. The advantages of dry ashing are that it is relatively safe (apart from the risk of burns due to the high temperatures used) and that it requires no reagents.
2. **Wet ashingn.** This uses strong acids to fully oxidise the organic matter in a foodstuff to CO_2. Nitric, sulfuric and perchloric acids are used in various combinations for different types of food samples. Wet ashing must be carried out in a fume hood and should follow the procedure carefully, with appropriate safety precautions. It is mostly used as a preliminary step in analysis of specific minerals, e.g. by atomic spectroscopy (p. 357) or using an ion-selective electrode.

Sub-dividing the dry ash fraction of a foodstuff – water-soluble and water-insoluble fractions can be determined by boiling the residue in water, drying and re-weighing, giving the water-insoluble fraction directly and the water-soluble fraction by subtraction.

pH and titratable acidity

pH

The pH of a food and its surrounding liquid can provide information on its overall quality, especially in relation to processing and storage, since low pH foods inhibit growth and multiplication of many microbes (p. 495). Basic theory and practical measurement is covered in Chapter 28. At a practical level, problems are encountered with solid foods such as cheeses and meats, where conventional glass electrodes (p. 174) are easily broken, and their surfaces may become coated with a protein film. Purpose-designed pH electrodes for use with foods are robust, for easy use and cleaning, typically with a narrow pointed tip to penetrate the food material. An alternative approach is to homogenise the food sample (p. 380) and then use a conventional pH electrode, although this may also cause pH changes in the food, e.g. due to enzyme activity in the tissue being homogenised. An alternative approach is to use a solid-state ion-selective field effect transistor (ISFET) electrode which avoids the use of glass; manufacturers' websites give specific details for individual products.

Definitions

pH – the negative logarithm of the hydrogen ion (proton) concentration or, to be more exact, the hydronium ion activity (p. 173).

Titratable acidity – a practical measure of the total acidity in foods and drinks.

Titratable acidity

This is dependent upon the concentration and type of acids in a food, and influences sourness/tartness to a greater extent than pH alone. It is determined by titrating a measured amount of food or drink against a solution of a strong base of known concentration (typically 0.1 mol L^{-1} NaOH) and measuring the pH to determine the end-point in terms of the amount of base required to neutralise all of the acid. This end-point can be established either by (i) colorimetric analysis using phenolphthalein, which changes from colourless to deep pink at pH 8.2, or (ii) conventional pH measurement to produce a titration curve (Fig. 49.3) for coloured foods/drinks that are unsuitable for assay using phenolphthalein. A typical practical exercise would be assaying grape juice or wine, with the overall acidity typically expressed in terms of tartaric acid (g L^{-1}) where 1 mL of 0.1 mol L^{-1} NaOH titrant is equivalent to 7.5 mg tartaric acid.

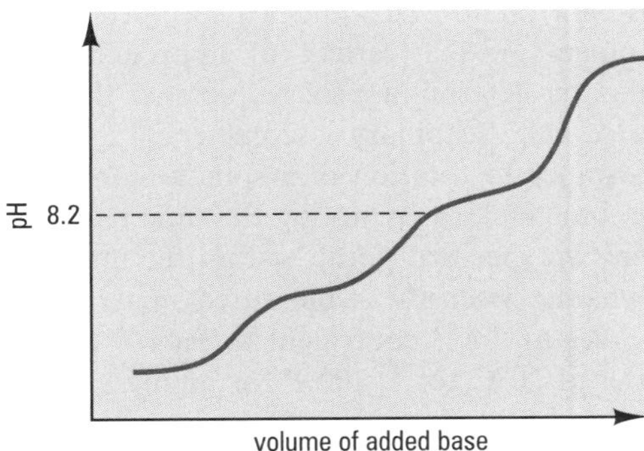

Fig. 49.3 Titration curve for an acidic food sample against NaOH.

Colour analysis

This can be considered from two main perspectives, namely:

1. **Visual analysis** – textual descriptions of the colour of foods (e.g. 'light orange') are difficult to standardise, due to variations in lighting conditions and to intrinsic differences in colour perception between individual people. Consequently, most methods are based on comparison and matching of food samples against appropriate colour standards under standardised illumination. A similar comparative approach is also taken for other visual aspects of food, e.g. marbling of beef is assessed by comparison with photographic standards (for more details, see Hutchings, 1999). More recent developments include the use of computer-based digital image analysis to provide a more objective system that is independent of human visual judgement (e.g. Quevedo *et al.*, 2009). Chapter 64 covers other practical aspects of sensory analysis and organoleptic techniques.

2. **Pigment analysis** – extracted from food samples, typically using an appropriate solvent. Total extracted pigment can be assayed by a

Examples (colour standards)
Commission Internationale de l'Eclairage (CIE colour system http://cie.co.at/)
Hunter laboratories (http://www.hunterlab.com/)

suitable means, e.g. spectrophotmetry (p. 353). Alternatively, individual pigments can be analysed by additional processing, e.g. chromatographic separation, e.g. using TLC (p. 360) or HPLC (p. 362) and subsequent spectroscopic analysis (p. 353) to identify and quantify the constituents.

Rheology and texture

Definitions

Elasticity – the tendency of a material to return to its original shape after stretching or compression.

Viscosity – the resistance of a material to flow when subjected to an applied force.

The behaviour of many foods lies somewhere between that of a perfect solid and a perfect liquid, and an understanding of their texture, consistency and responses to applied force is important in preparation (e.g. making bread), commercial production (e.g. flow of food materials on a production line) and consumption (e.g. 'mouthfeel', p. 434) of food.

Rheology is typically considered under two main headings:

1. **Viscosity.** Some materials have a relatively constant viscosity that is independent of the applied force. These are termed 'Newtonian fluids', e.g. water, wine and honey (sugar syrup). Many other materials change their viscosity in response to applied force and are termed 'non-Newtonian fluids'. Most foods decrease their viscosity when subjected to applied forces such as mixing, shaking, kneading or chewing; examples of shear-thinning fluids include fruit puree, melted chocolate and tomato ketchup. In contrast, shear-thickening fluids increase their viscosity with increasing applied force, e.g. corn starch in water. Viscosity is typically measured using a commercial viscometer under carefully controlled conditions, to provide comparative data on the apparent viscosity of non-Newtonian fluids under a specified set of conditions. Analytical instruments use a variety of approaches to measure resistance to flow, e.g. by determining the volumetric flow of the material within a calibrated tube (capillary viscometers; Fig. 49.4 and 'consistometers'), the resistance to rotation of a spindle immersed in the material (rotational viscometers), or by noting the time required for a steel ball to fall between two specified points within the material (falling ball viscometers). Dynamic viscosity is measured in terms of pressure and time (poise, $P = \text{kg m}^{-1}\text{ s}^{-1}$ equivalent to Pa s^{-1} in SI), varying from 0.001 P for water at 20°C to > 100 P for highly viscous corn syrups. Instruments are calibrated using standards of known viscosity.

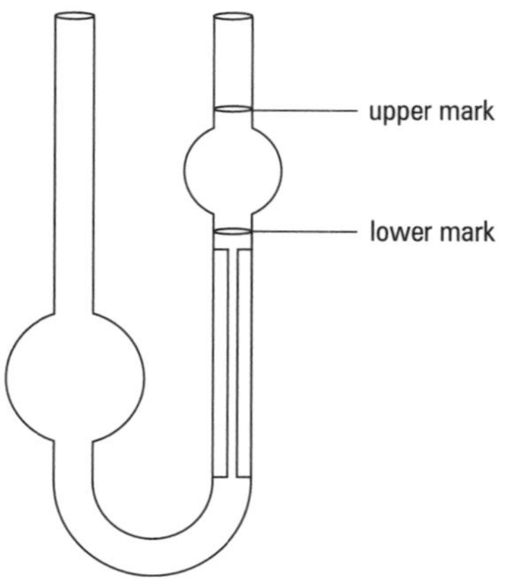

Fig. 49.4 Ostwald viscometer. The time taken for the meniscus of a food material to flow from the upper mark to the lower mark is used to measure viscosity.

2. **Elasticity.** A variety of commercial machines (sometimes collectively known as 'universal testing machines') have been manufactured to apply forces (strain) to foods and measure the resulting force (stress) together with any deformation of the material, usually in terms of elongation or compression and any subsequent recovery on removal of the applied force. For further details of the underlying principles and the methods involved, including those for determining the modulus of elasticity, see Bourne (2002).

Some foods exhibit behaviour intermediate between an elastic solid and a viscous fluid – these are termed viscoelastic materials, e.g. peanut butter and cream and their behaviour is more complex (see McClements, 1999).

Other aspects of food texture that can be analysed by method-specific equipment include:

Sensory evaluation of food texture – despite the use of mechanical instruments to measure specific physical characteristics under controlled laboratory conditions, this is probably still best assessed by trained sensory evaluators who can process all of the organoleptic attributes of a food and decide on its overall quality, e.g. in comparative taste tests (see Chapter 64).

- **Penetration** – provides information on the force required for a probe to puncture the surface of a food.
- **Bending** – where the deflection resulting from the application of a force is measured.
- **Shearing/cutting** – measures the ease with which food material can be cut by a knife or cutting wire in response to specific applied forces.
- **Crispness/brittleness** – for example, the 'snap test' used to determine the force required to break a biscuit. Other versions measure the tension required to break a sample under extension (e.g. stretched between two points).
- **Plasticity** – the ability of a material to be moulded once the elastic limit has been exceeded.

Rosenthal (1999) provides further details of the various techniques of materials science used to assess food texture.

Text references

AOAC (2009) Homepage. Available: http://www.aoac.org/
Last accessed: 21/12/10.

Bourne, M.C. (2002) *Food Texture and Viscosity: Concept and Measurement*, 2nd edn. Academic Press, London.

Hutchings, J.B. (1999) *Food Color and Appearance*. Springer, New York.

McClements, D.J. (1999) *Food Emulsions: Principles, Practice and Techniques*. CRC Press, Boca Raton, Florida.

Quevedo, R.A., Aguilera, J.A. and Pedreschi, F. (2009) Color of salmon fillets by computer vision and sensory panel. *Food and Bioprocess Technology*. Available: http://www.springerlink.com/content/653w7054m870rx41/fulltext.pdf?page=1
Last accessed: 21/12/10.

Rosental (1999) *Food Texture: Measurement and Perception*. Springer, New York.

Sources for further study

Food and Agriculture Organisation of the United Nations. *Food Energy – Methods of Analysis and Conversion Factors*. Available: http://www.fao.org/docrep/006/y5022e/y5022e00.HTM
Last accessed: 21/12/10.

Nielsen, S.S. (2003) *Food Analysis: Laboratory Manual*, 3rd edn. Springer, New York.

Pomeranz, Y. and Meloan, C.E. (2002) *Food Analysis: Theory and Practice*, 3rd edn. Springer, New York.

Potter, N.N. and Hotchkiss J.H. (1999) *Food Science*, 5th edn. Springer, New York.

50 Calibration and its application to quantitative analysis

There are many instances where it is necessary to measure the quantity of a test substance using a calibrated procedure. You are most likely to encounter this approach in one or more of the following practical exercises:

Understanding quantitative measurement – Chapter 31 contains details of the basic principles of valid measurement, while Chapters 52–56 deal with some of the specific analytical techniques used in biology. The use of internal standards is covered on p. 366.

- Quantitative spectrophotometric assay of biomolecules (Chapter 53).
- Flame or atomic absorption spectroscopic analysis of metal ions in biological solutions (p. 357).
- Using a chromogenic or fluorogenic substrate to determine the activity of an enzyme (p. 408).
- Quantitative chromatographic analysis, e.g. GC or HPLC (Chapter 54).
- Using a bioassay system to quantify a test substance: examples include immunodiffusion (Chapter 52) and radioimmunoassay.

KEY POINT In most instances, calibration involves the establishment of a relationship between the measured response (the 'signal') and one or more 'standards' containing a known amount of substance.

Calibrating laboratory apparatus – this is important in relation to validation of equipment, e.g. when determining the accuracy and precision of a pipettor by the weighing method: see p. 156.

In some instances, you can measure a signal due to an inherent property of the substance, e.g. the absorption of UV light by nucleic acids (p. 426), whereas in other cases you will need to react it with another substance to see the result (e.g. molecular weight measurements of DNA fragments after electrophoresis, visualised using ethidium bromide, p. 425), or to produce a measurable response (e.g. the reaction of cupric ions and peptide bonds in the Biuret assay for proteins, p. 396).

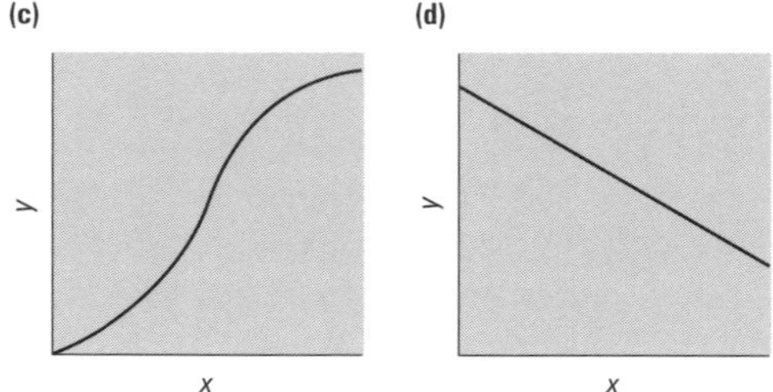

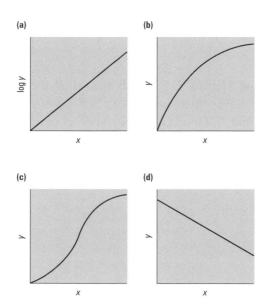

Fig. 50.1 Calibration curves: (a) log–linear; (b) curvilinear; (c) sigmoid, or S-shaped; (d) inverse.

The different types of calibration curve

By preparing a set of solutions (termed 'standards'), each containing either (i) a known *amount* or (ii) a specific *concentration* of the substance, and then measuring the response of each standard solution, the underlying relationship can be established in graphical form as a 'calibration curve', or 'standard curve'. This can then be used to determine either (i) the amount or (ii) the concentration of the substance in one or more test samples. Alternatively, the response can be expressed solely in mathematical terms: an example of this approach is the determination of chlorophyll pigments in plant extracts by measuring absorption at particular wavelengths, and then applying a formula based on previous (published) measurements for purified pigments (p. 356).

There are various types of standard curve: in the simplest cases, the relationship between signal and substance will be linear, or nearly so, and the calibration will be represented best by a straight-line graph (see Box 50.1). In some instances (Fig. 50.1(a)), you will need to transform either the x values or y values (Chapter 61), in order to produce a linear graph (e.g. in radial immunodiffusion bioassay, where the y values are squared, p. 349. In other instances, the straight-line relationship may only hold up to a certain value (the 'linear dynamic range') and beyond this point the graph may curve (e.g. in quantitative spectrophotometry, the

Box 50.1 The stages involved in preparing and using a calibration curve

1. **Decide on an appropriate test method** – for example, in a project, you may need to research the best approach to the analysis of a particular metabolite in your biological material.

2. **Select either (a) amount or (b) concentration, and an appropriate range and number of standards** – in practical classes, this may be given in your schedule, along with detailed instructions on how to make up the standard solutions. In other cases, you may be expected to work this out from first principles (Chapters 24–26 give worked examples) – aim to have evenly spaced values along the *x* axis.

3. **Prepare your standards very carefully** – due attention to detail is required: for example, you should ensure that you check the calibration of pipettors beforehand, using the weighing method (p. 156). Don't forget the 'zero standard' plus any other controls required, e.g. to test for interference due to other chemical substances. Your standards should cover the range of values expected in your test samples.

4. **Assay the standards and the unknown (test) samples** – preferably all at the same time, to avoid introducing error due to changes in the sensitivity or drift in the zero setting of the instrument with time. It is a good idea to measure all of your standard solutions at the outset, and then measure your test solutions, checking that the 'zero standard' and 'top standard' give the same values after, say, every six test measurements. If the re-measured standards do not fall within a reasonable margin of the previous value, then you will have to go back and recalibrate the instrument, and repeat the last six test measurements. If your test samples lie outside the range of your standards, you may need to repeat the assay using diluted test samples (extrapolation of your curve may not be valid, see p. 228).

5. **Draw the standard curve, or determine the underlying relationship** – Fig. 50.2 gives an example of a typical linear calibration curve, where the spectrophotometric absorbance of a series of standard solutions is related to the amount of substance. When using a spreadsheet (Box 50.2) or graphic package (Chapter 12), it is often appropriate to use Model I linear regression (p. 268) to produce a linear trend line (also termed the 'line of best fit') and you can then quote the value of r^2, which is a measure of the 'fit' of the measurements to the line (see p. 268). However, you should take care not to use a linear plot when the underlying relationship is clearly non-linear and you must consider whether the assumptions of the regression analysis are valid.

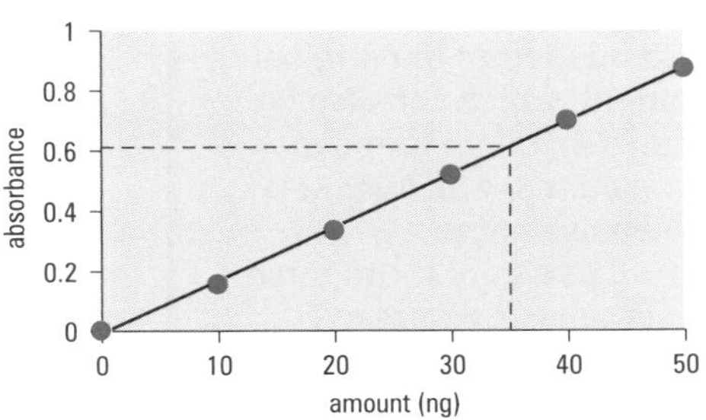

Fig. 50.2 Typical calibration curve for spectrophotometric analysis.

6. **Determine the amount or concentration in each unknown sample** – either by reading the appropriate value from the calibration curve, or by using the underlying mathematical relationship, i.e. $y = a + bx$ (p. 244). Make sure you draw any horizontal and vertical construction lines very carefully – students often lose marks unnecessarily by submitting poorly drawn construction lines within practical reports.

7. **Correct for dilution or concentration, where appropriate** – for example, if you diluted each test sample by ten-fold, then you would need to multiply by 10 to determine the value for the undiluted test sample. As another example, if you assayed 0.2 mL of test sample, you would need to multiply the value obtained from the calibration curve by 5, to give the value per mL.

8. **Quote your test results to an appropriate number of significant figures** – this should reflect the accuracy of the method used (see pp. 239–241), not the size of your calculator's display.

Example A test tube cntaining 8 mL of water plus 2 mL of 1 % w/v NaCl (M_r) = 58.44) would have a *mass concentration* of 0.2 % w/v NaCl (a five-fold dilution of the original NaCl solution), which can also be expressed as 2 gL^{-1}; in terms of *molar concentration* (p. 154), this would be equivalent to 2 ÷ 58.44 = 0.0342 mol L^{-1}(to three signifiant figures). Expressed in terms of the <l amount of NaCl test tuble, this would be 0.02 g in *mass*, or 0.02 ÷ 58.44 = 3.42 x 10^{-4} mol (342 μmol in *moles*.

Beer–Lambert relationship often becomes invalid at high absorbance, giving a curve, Fig. 50.1(b)). Some calibration curves are sigmoid (Fig. 50.1(c)). Finally, the signal may *decrease* in response to an increase in the substance (Fig. 50.1(d)), e.g. radioimmunoassay, where an inverse sigmoid calibration curve is obtained. In some practical classes, you may be told that the relationship is expected to be linear, curvilinear, or whatever, while in others you may be expected to decide the form of the standard curve as part of the exercise.

Practical considerations

Amount or concentration?

This first step is often the most confusing for new students. It is vital that you understand the difference between *amount* of substance (e.g. mg, ng, etc.), and *concentration* (the amount of substance per unit volume, e.g. mmol l^{-1}, mol m^{-3}, % w/v, etc.) before you begin your practical work.

KEY POINT Essentially, you have to choose whether to work in terms of either (i) the total amount of substance in your assay vessel (e.g. test tube or cuvette) or (ii) the final concentration of the substance in your assay vessel, which is independent of the volume used.

The interconversion of amount and concentration is covered in more detail on p. 146. Either way, this is usually plotted on the x (horizontal) axis and the measured response on the y (vertical) axis.

Organisations providing national/ international standards – these include:

- Laboratory of the Government Chemist (http://www.lgc.co.uk/), for UK standards and European Reference Materials.
- National Institute for Science and Technology (http://www.cstl.nist.gov/), for biochemical/chemical standards in the USA.
- Institute for Reference Materials and Measurements (http://www.irmm.jrc.be/), for European standards, including BCR and ERM materials.
- OIE Biological Standards Commission (http://www.oie.int/bsc/eng/en_bsc.htm), for international animal standards.

Choice of standards

In your early practical classes, you may be provided with a stock solution (p. 147), from which you then have to prepare a specified number of standard solutions. In such cases, you will need to understand how to use dilutions to achieve the required amounts or concentrations (p. 147). In later work and projects, you may need to prepare your standards from chemical reagents in solid form, where the important considerations are purity and solubility (p. 145). For professional analysis (e.g. in forensic science or clinical biochemistry), it is often important to be able to trace the original standard or stock solution back to national or international standards or to certified reference materials.

How many standards are required?

This may be given in your practical schedule, or you may have to decide what is appropriate (e.g. in project work and research). If the form and working range of the standard curve is known in advance, this may influence your choice – for example, linear calibration curves can be established with fewer standards than curvilinear relationships. In some instances, analytical instruments can be calibrated using a single standard solution, often termed a ‘calibrator’. Replication of each standard solution is a good idea, since it will give you some information on the variability involved in preparing and assaying the standards. Consider whether you should plot mean values on your standard curve, or whether it is better to plot the individual values (if one value appears to be well off the line, you have made an error, and you may need to check and repeat).

Plotting a standard curve – do not force your calibration line to pass through zero if it clearly does not. There is no reason to assume that the zero value is any more accurate than any other reading you have made.

Box 50.2 How to use a spreadsheet (Microsoft Excel 2007) to produce a linear regression plot

Two approaches are possible: either using the *Trendline* feature or using the regression analysis function within the data analysis tool pack. Both make the assumption that the criteria for Model I linear regression are met (see p. 268). In the example shown below, the following simple data set has been used:

Amount (ng)	Absorbance
0	0.00
10	0.19
20	0.37
30	0.56
40	0.63
50	0.78

Using the *Trendline* feature

This quick method provides a line of best fit on an Excel chart and can also provide a set of equation values for predictive purposes.

1. **Create a graph (*chart*) of your data.** Enter the data in two columns within your spreadsheet, select the data array (highlight using left mouse button) and then, using the *Insert* tab, select *Scatter* without any plotted lines.
2. **Add a trend line.** Right-click on any of the data points on your graph, and select the *Add Trendline* menu. Choose the *Linear* trend-line option, but do not click *OK* at this stage. Rather, from the menu, select: (i) *Display equation on chart* and (ii) *Display R-squared value on chart*. Now click OK. The equation (shown in the form $y = bx + a$) gives the slope and intercept of the line of best fit, while the R-squared value (coefficient of determination, p. 268) gives the proportional fit to the line (the closer this value is to 1, the better the fit of the data to the trend line).
3. **Modify the graph to improve its effectiveness.** For a graph that is to be used elsewhere (e.g. in a lab write-up or project report), adjust the display to remove the default background and gridlines and change the symbol shape (see Box 35.2, p. 223 for more advice and examples). For example, right-click on the trend line and use the *Format Trendline* > *Line style* menu to adjust the *Width* of the line to make it thinner, or to create a dashed line. Drag and move the equation panel if you would like to alter its location on the chart, or delete it, having noted the values. Figure 50.3 shows a calibration curve produced in this way for the data presented above.

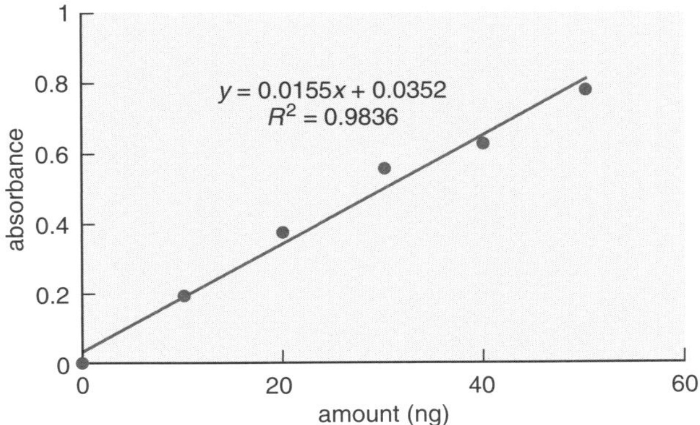

Fig. 50.3 Calibration curve showing line of best fit and details of linear regression equation.

4. **Use the regression equation to estimate unknown (test) samples.** By rearranging the equation for a straight line and substituting a particular *y*-value, you can predict the amount/concentration of substance (*x*-value) in a test sample. This is more precise than simply reading the values from the graph using construction lines (Box 50.1). If you are carrying out multiple calculations, the appropriate equation, $x = (y - a)/b$, can be entered into a spreadsheet, for convenience.

Using the Regression Analysis tool

This requires the 'data analysis tool pack' to be loaded beforehand (see Box 38.3, p. 256) and provides summary output that contains details of slope, intercept and coefficient of determination along with an analysis of variance (ANOVA) table (Chapter 39 gives further details of these aspects).

Dealing with interfering substances – one approach is to use the method of 'standard additions', where the standards all contain a fixed additional amount of the sample (see e.g. Dean, 1997). Internal standards (p. 366) can also be used to detect such problems.

Preparing your standards

It is extremely important to take the greatest care to measure out all chemicals and liquids very accurately, to achieve the best possible standard curve. The grade of volumetric flask used and temperature of the solution also affect accuracy (grade A apparatus is best). You may also consider what other additives might be required in your standard solutions. For example, do your test samples have high levels of potentially interfering substances, and should these also be added to your standards? Also consider what controls and blank solutions to prepare.

KEY POINT The validity of your standard curve depends upon careful preparation of standards, especially in relation to accurate dispensing of the volumes of any stock solution and diluting liquid – the results for your test samples can only be as good as your standard curve.

Preparing the calibration curve and determinining the amount of the unknown (test) sample(s)

This is described in stepwise fashion within Box 50.1. Check you understand the requirements of graph drawing, especially in relation to plotted curves (p. 336) and the mathematics of straight-line graphs (p. 244). Spreadsheet programs such as Microsoft Excel can be used to produce a regression line for a straight-line calibration plot (pp. 369–370). Examples of how to do this are provided in Box 50.2.

Text reference

Dean, J.D. (1997) *Atomic Absorption and Plasma Spectroscopy*, 2nd edn. Wiley, Chichester.
[Chapter 1 deals with calibration, and covers the principle of standard additions.]

Sources for further study

Brown, P.J. (1994) *Measurement, Regression, and Calibration*. Clarendon Press, Oxford.
[Covers advanced methods, including curve fitting and multivariate methods.]

Mark, H. (1991) *Principles and Practice of Spectroscopic Calibration*. Wiley, New York.

Miller, J.N. and Miller, J.C. (2005) *Statistics and Chemometrics for Analytical Chemistry*, 5th edn. Prentice Hall, Harlow.
[Gives detailed coverage of calibration methods and the validity of analytical measurements.]

51 Indirect calorimetry

Definitions

Direct calorimetry – measurement of the amount of heat produced, for example by a person in a known amount of time, or from a food item when it is fully oxidised.

Indirect calorimetry – method of estimating energy expenditure by measuring volumes of oxygen consumed and with some equipment, carbon dioxide produced over a given time period.

Thermic effects of food – increase in energy expenditure above resting metabolic rate caused by the body's work of processing food for storage and use.

Respiratory quotient – the ratio of the amount (in molar or volume terms) of carbon dioxide expired to the amount of oxygen consumed during the oxidation of food.

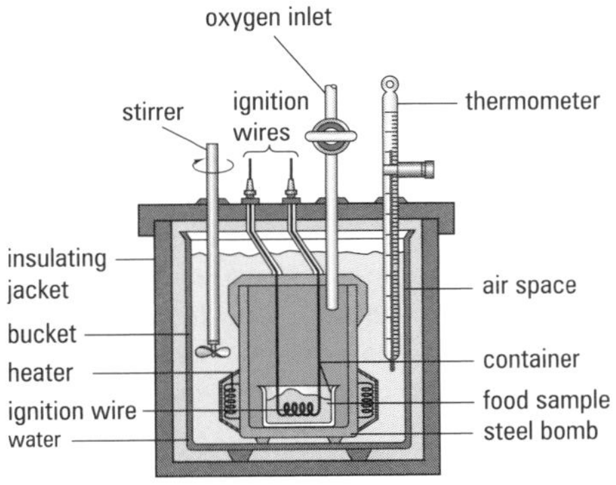

Fig. 51.1 Diagram of a bomb calorimeter.

Calorimetry is the science of measuring of the heat absorbed or evolved during chemical reactions and changes of state. In relation to dietetics, it often refers to an individual's energy use while in food science it is the measure of the amount of energy in food. A person's total energy expenditure can be estimated in a number of ways, including by calculation, e.g. using Schofield equations (p. 303), or by measurement, using direct or indirect calorimetry. In practice, direct methods are either prohibitively expensive or technically very challenging as they measure the direct heat loss of a subject while they are contained in a purpose-designed chamber. A more commonly used method is indirect calorimetry. This can be used to measure basal and resting metabolic rates, the thermic effects of food and also the energy expended by a human subject during physical activity. In contrast, direct calorimetry is often used in the laboratory to estimate the energy content of foods, e.g. using a bomb calorimeter.

Direct calorimetric analysis of foods

Oxygen bomb calorimeters are standard instruments used in the laboratory to measure the calorific value of foods to provide an estimate of their energy content. This energy content information is required for packaged food labels (p. 138), and also may be used in food science research.

A typical bomb calorimeter (Fig. 51.1) comprises a stainless steel container (the 'bomb') which is able to withstand several atmospheres of gas pressure. Box 51.1 describes how to use a bomb calorimeter, but as each instrument is slightly different you should always refer to the specific instructions. In practical classes or projects the equipment should only be used under supervision by a member of staff.

Principles of indirect calorimetry

Indirect calorimetry approaches work on the principle that all of the energy that is expended by the body is lost as heat. Rather than measuring this heat loss directly (as in direct calorimetry), heat energy, expressed in kJ or kcal, is calculated from respiratory gas analysis. When nutrients in food are oxidised, the process consumes oxygen, and produces carbon dioxide and water. In the case of protein, nitrogenous products such as urea are also generated, although in practice this is not generally used in calculations as it requires urinary excretion to be analysed.

KEY POINT Indirect calorimetry calculates energy expenditure from information on the amount of oxygen consumed and (sometimes) the amount of carbon dioxide generated in a known time period.

Methods of indirect calorimetry

Various types of equipment can be used to measure the amount of oxygen consumed by a subject. The 'classic' method of collection is the Douglas bag, which collects expired air up to a volume of 100 Litres (Fig. 51.2). The

Box 51.1 Procedure for measuring the energy contents of a food sample using a bomb calorimeter with a worked example of the calculations

1. **Accurately weigh the sample of food.** The sample should be approximately 1 g. If the sample is a powder it will need to be compacted into a pellet to ensure complete combustion.
2. **Place the sample within the container in contact with an ignition or fuse wire.** If a disposable ignition fuse wire is used it should also be weighed before being inserted.
3. **Tighten the bomb cap.** This should be done tightly but *only by hand* to avoid over-tightening.
4. **Connect the oxygen tank to the inlet valve.** Make sure this pipe is securely attached.
5. **Slowly fill to approximately 20 atmospheres of pressure** (or as indicated in the instruction manual) through the inlet valve, then seal.
6. **Attach the electrode wires to the bomb head.**
7. **Place the bomb into a known amount of water (e.g. 2000 mL) in the calorimeter vessel (water bath).** To prevent heat loss this water bath is surrounded by thermal insulation or a water jacket that needs to be temperature-controlled to be at the same temperature as the water surrounding the bomb during the measurement process. Some models make this adjustment in temperature of the water jacket automatically.
8. **Press the firing button and hold for a maximum of five seconds.** This switches on an electric current that ignites the sample. The combustion of the food generates heat which raises the temperature of the calorimeter and the water surrounding the bomb.
9. **Measure the temperature changes using a thermometer capable of measuring to 0.01°C** (in some instruments this is in-built). Readings should be taken every 30 seconds for two minutes before ignition, and then every minute until the temperature has reached a plateau (usually five–eight minutes after ignition). If the instrument uses a water jacket without automatic temperature adjustment, maintain the jacket and calorimeter vessel temperatures to within 0.03°C by adding hot water slowly to the jacket.
10. **Retrieve any remaining fuse wire, weigh it and subtract this mass from the initial wire mass.** Part of the fuse wire combusts adding to the heat generated, and this needs to be subtracted from the total. The amount of energy per gram of fuse wire is usually listed on the spool.
11. **To relate the temperature increase to energy values requires use of a calorimeter constant (kJ g^{-1} °C^{-1}).** This is obtained for a particular bomb by combusting a standard sample of known mass (for example 1 g) with a known energy content, therefore obtaining the number of kJ per gram per degree temperature rise for this bomb. The energy value of the sample is then calculated as in eqn. [51.1]:

$$\text{kJ g}^{-1}\text{ (sample)} = \Delta T\text{ (°C) (sample)} \times \text{calorimeter constant (kJ g}^{-1}\text{ °C}^{-1}\text{)/mass (g)} \qquad [51.1]$$

Example

Calculate the energy value of 1 g of sugar if ΔT of the sample is 1.65 while 1 g of standard benzoic acid with an energy value of 26.44 kJ raises the temperature by 2.59°C. This would give a calorimeter constant of 10.21 kJ g^{-1} °C^{-1}. Using eqn [51.1], kJ g^{-1} (sample) = $1.65 \times 10.21/1 = 16.85$.

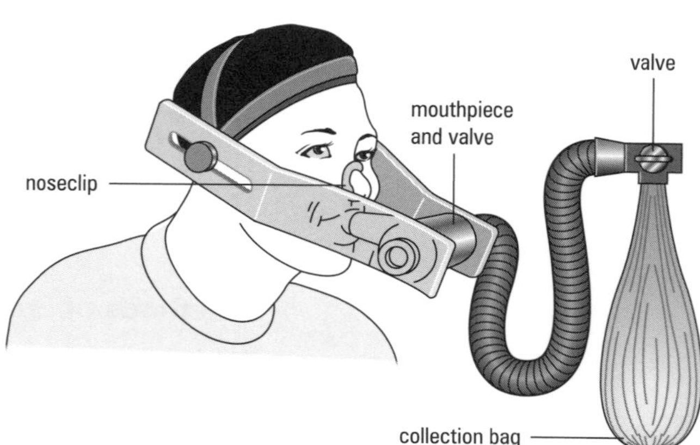

Fig. 51.2 Douglas bag apparatus for respiratory gas analysis.

Understanding the composition of air – dry atmospheric air contains 20.95% oxygen, 0.04% carbon dioxide, 78.09% nitrogen and 0.93% argon.

subject wears a noseclip and breathes though a mouthpiece with a one-way valve. The subject should be allowed to become used to breathing through the equipment before the expired air is collected. This ensures that potential factors which might affect oxidation, such as stress, are reduced. Because of the limited volume of expired air which can be collected, this method can only be used for short experiments (up to 15 minutes). The volume of air is measured and the oxygen and carbon dioxide concentrations are then analysed, typically using an infra-red gas analyser.

The amount of oxygen consumed can be calculated from the difference between its concentration in the inspired (atmospheric) air and expired air, and the volume of air expired. This method is suitable for calculating the basal or resting metabolic rate and the thermic effects of food, but because of the limitations to movement it imposes, it cannot be used during activity.

A method using more portable equipment, which can be used during static exercise such as when using a cycle ergometer, is the Kofrani–Michaelis respirometer, also known as KM or Max Planck respirometer. This equipment is still in widespread use although no longer manufactured. Newer, portable versions that work on the same principles include Oxylog®, Cosmed Quark PFT™ (Fig. 51.3) and Reevue™. The subject wears a face mask or mouthpiece, but the respirometer measures the volume and oxygen concentration of expired air as it is produced. This allows for longer measurement times than can be achieved using the Douglas bag. Some of these newer calorimeters use a different equation for calculating energy expenditure that does not require CO_2 expiration measurements.

Ventilated hood systems such as Deltatrac® II avoid the discomfort associated with the use of clips, mouthpieces or face masks and are suitable in a hospital setting (Fig. 51.4). They require the subject to lie or sit at rest and air flow is controlled in the restricted space in which the subject breathes. They are suitable for intensive-care patients and for babies.

Whole-body indirect calorimetry chambers are specialised equipment which would only be found in dedicated units. They are suitable for long-term studies, up to 14 days, therefore measures of energy expenditure during a normal lifestyle, albeit a relatively sedentary one, are possible.

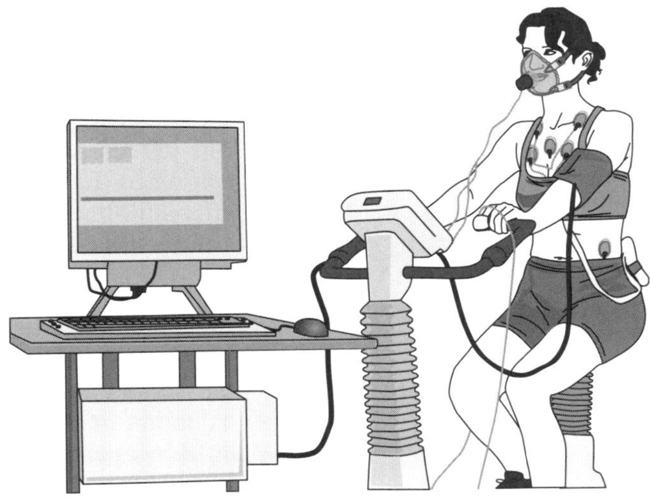

Fig. 51.3 Use of the Cosmed Quark PFT™ indirect calorimeter in conjunction with a cycle ergometer.

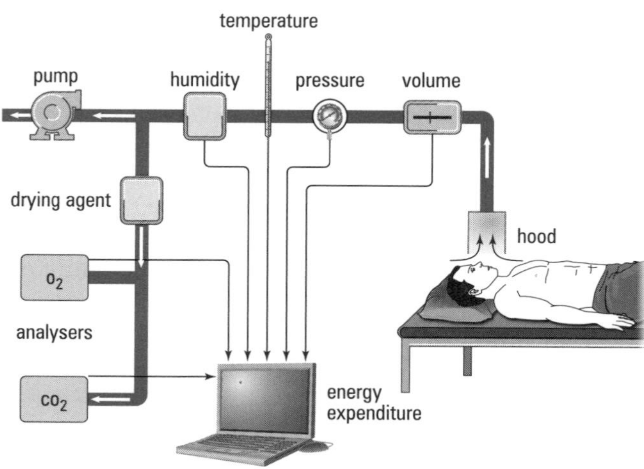

Fig. 51.4 Ventilated hood respirometer.

Calculating energy expenditure from respiratory gas analysis

It is possible to calculate energy expenditure by measuring the amount of oxygen which has been used and carbon dioxide produced. The type of equipment you use will influence the exact way that energy expenditure is calculated but will be based on Weir's modified equation according to manufacturer's instructions eqn [51.2] or for the models that do not measure CO_2 production. The volumes of gases must be adjusted to those applying at standard temperature and pressure using the relationship $P_1V_1/T_1=P_2V_2/T_2$, where, in standard conditions, P_2 is 101.3 kPa (1 atmosphere), T_2 is 273.15 K and V_2 is the adjusted volume.

$$EE = ([VO_2 \times 16.318] + [VCO_2 \times 4.602]) \times 1440 \qquad [51.2]$$

where:
EE = energy expenditure (kJ/day)
VO_2 = volume of oxygen consumed (Litres per minute)
VCO_2 = volume of CO_2 produced (Litres per minute)
1440 = minutes per day

Respiratory quotient (RQ) and its application to indirect calorimetry

Example
RQ of palmitic acid:
$CH_3(CH_2)_{14}COOH + 23O_2 \rightarrow 16CO_2 + 16H_2O$
Therefore, RQ = $16CO_2$:$23O_2$ = 0.7

The RQ represents the molar ratio of carbon dioxide produced to oxygen consumed (eqn [51.3]). Instead of moles, volumes of gases can be used after adjustment to standard temperature and pressure. The RQ provides a further indication of nutritional status.

$$RQ = \text{moles } CO_2 \text{ produced} / \text{moles } O_2 \text{ consumed} \qquad [51.3]$$

The RQ depends upon the combination of macronutrients being used as the respiratory substrates. It is different for each macronutrient as oxidation of each source requires different amounts of oxygen. These differences can be expressed in terms of the energy equivalence of each Litre of oxygen consumed.

Examples The energy equivalence of nutrients has been calculated as follows (Brockway, 1987)

Protein: 19.48 kJ/L O_2

Fat: 19.61 kJ/L O_2

Ethanol: 20.33 kJ/L O_2

Carbohydrate: 21.12 kJ/L O_2

The respiratory quotient is another way of expressing the differences in respiratory gas exchange for each macronutrient. In the case of carbohydrate the RQ is equal to 1.0 because if one mole of glucose is fully oxidised then 6 moles of oxygen are consumed and 6 moles of carbon dioxide are produced, as shown in eqn [51.4].

$$C_6H_{12}O_6 + 6O_2 = 6CO_2 + 6H_2O \quad [51.4]$$

Fat has an RQ of 0.7 and protein somewhere in between these two values, but usually quoted as 0.8. RQ tables are available to calculate the energy equivalent of oxygen if RQ is calculated (Zuntz, 1901, in McArdle *et al.*, 2000). As humans tend to be oxidising a mixture of nutrients (fuels) at any one time the generally accepted value is 20.3 kJ/L oxygen, assuming appropriate proportions of each nutrient is consumed, i.e. 50 per cent energy from carbohydrate, 35 per cent energy from fat, 15 per cent energy from protein.

If RQ $>$ 1.0 this is indicative of the over consumption of energy-containing foods and fat synthesis, while RQ $<$ 0.8 may suggest poor nutritional intake as RQ tables show that an RQ of 0.75 indicates that 84 per cent energy is being derived from fat.

Factors affecting the accuracy of indirect calorimetry

Definition

A **thermoneutral environment** maintains body temperature with a minimal requirement of energy. The subject will feel neither hot nor cold.

Where indirect calometry is to be used to calculate basal metabolic rate (BMR), it is necessary for the subject to be in a post-absorptive state (having fasted for 12 hours) and for the measurements to take place in thermoneutral (22–26°C) conditions. No stimulants such as nicotine, coffee or tea should have been taken in the past 12 hours, and no heavy physical activity done in the previous 24 hours. There should be no external stimuli (e.g. music, television, reading) and the subject should be lying still and rested mentally and physically, but not asleep. If it is impractical to meet these requirements the calorimetry result is termed resting metabolic rate (RMR) or the equivalent term resting energy expenditure (REE). For further details, see Chapter 44.

It is necessary to ensure that the equipment is properly maintained and correctly calibrated before the measurements are recorded, as indicated in the instrument's handbook. Some newer model indirect calorimeters such as the Reevue™ automatically calibrate when turned on and results of the calibration test are displayed on the screen. All connections should be checked and particular attention should be paid to any potential gas leaks which would invalidate the data collected.

Understanding the effects of injury or illness on measurements – RMR may be increased due to metabolic stress, e.g. fever or increased release of stress hormones due to injury, leading to a hypermetabolic state. Measuring RMR provides an indication of the degree of hypermetabolism and subsequent nutritional requirements, particularly with regard to protein.

In a clinical setting it would be necessary to take into account the age, physical condition and treatment regime of the patient when calculating energy expenditure, e.g. acute respiratory infection causes an increase in RMR, followed by a return to more normal RMR levels following successful antibiotic treatment; many elderly people tend to have reduced lean body mass and a resulting reduced energy metabolism.

KEY POINT Energy expenditure results should not be considered on their own, but within an integrated framework with anthropometric measurements and nutritional and medical assessments (see Chapter 47).

Text reference

Brockway, J.M. (1987) Derivation of formulae used to calculate energy expenditure in man. *Human Nutrition: Clinical Nutrition,* **41C**, 463–471.

McArdle, W.D., Katch, F.I., and Katch V.L. (2000) *Essentials of Exercise Physiology*. Lippencott Williams & Wilkins, Philadelphia, Pennsylvania.

Sources for further study

da Rocha, E.E., Alves, V.G. and da Fonseca, R.B. (2006) Indirect calorimetry: methodology, instruments and clinical application. *Current Opinion in Clinical Nutrition & Metabolic Care*, **9**, 247–256.

Haugen, H.A., Chan, L.N. and Li, F. (2007) Indirect calorimetry: A practical guide for clinicians. *Nutrition in Clinical Practice*, **22**, 377–388.

Kaletun, G. (2009) *Calorimetry in Food Processing – Analysis and Design of Food Systems*. Wiley-Blackwell, Hoboken, New Jersy.

Webb, G.P. (2002) *Nutrition, A Health Promotion Approach*, 2nd edn. Arnold, London.

Weekes, C.E. (2007) Controversies in the determination of energy requirements. *Proceedings of the Nutrition Society*, **66**, 367–377.

Weir, J.B. de V. (1949) New methods for calculating metabolic rate with special reference to protein metabolism. *Journal of Physiology*, **109**, 1–9.

52 Immunological methods

Definitions

Antibody – a protein produced in response to an antigen (an *anti*body-*gen*erating foreign macromolecule).

Epitope – a site on the antigen that determines its interaction with a particular antibody.

Hapten – a substance that contains at least one epitope, but is too small to induce antibody formation unless it is linked to a macromolecule.

Ligand – a molecule or chemical group that binds to a particular site on another molecule.

Antibodies are important components of the immune system, which protect animals against certain diseases (see Delves *et al.*, 2006). They are produced by B lymphocytes in response to foreign macromolecules (antigens). A particular antibody will bind to a site on a specific antigen, forming an antigen–antibody complex (immune complex). Immunological assays use the specificity of this interaction for:

- identifying biomolecules or whole cells;
- quantifying a particular substance.

In food science, immunological methods are used for:

- detection of particular microbes in foods, e.g. viruses and bacteria;
- identification and assay of specific food components, e.g. microbial toxins in foods, or herbicide/pesticide residues;
- identification of bacteria at the sub-species level, e.g. serotyping (p. 467) of *Salmonella enterica* and *Escherichia coli*.

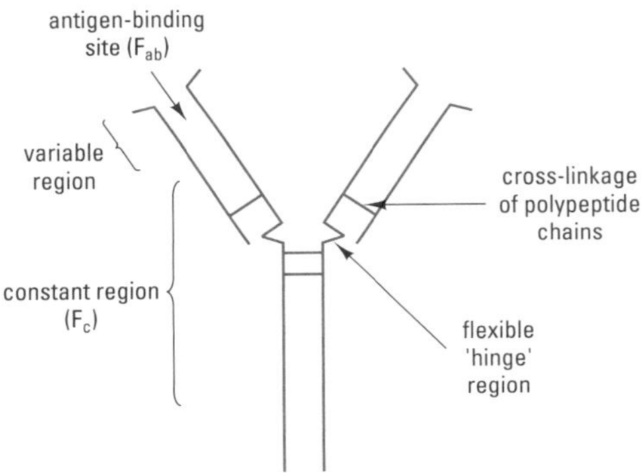

Fig. 52.1 Diagrammatic representation of IgG (antibody).

Antibody structure

An antibody is a complex globular protein, or immunoglobulin (Ig). Although there are several types, IgG is the major soluble antibody in vertebrates and is used in most immunological assays. Its main features are:

- **Shape:** IgG is a Y-shaped molecule (Fig. 52.1), with two antigen-binding sites.
- **Specificity:** variation in amino acid composition at the antigen-binding sites explains the specificity of the antigen–antibody interaction.
- **Flexibility:** each IgG molecule can interact with epitopes that are different distances apart, including those on different antigen molecules.
- **Labelling:** regions other than the antigen-binding sites can be labelled, e.g. using a radioisotope or enzyme (p. 350).

KEY POINT The presence of two antigen-binding sites on a single flexible antibody molecule is relevant to many immunological assays, especially the agglutination and precipitation reactions.

Producing polyclonal antibodies – in the UK, this is controlled by government regulations, since it involves vertebrate animals: personnel must be licensed by the Home Office and must operate in accordance with the *Animals Scientific Procedures Act (1986)* and with the *Code of Practice for the Housing and Care of Animals (2005)*.

Antibody production

Polyclonal antibodies

These are commonly used at undergraduate level. They are produced by repeated injection of antigen into a laboratory animal. After a suitable period (three–four weeks) blood is removed and allowed to clot, leaving a liquid phase (polyclonal antiserum) containing many different IgG antibodies, resulting in:

- cross-reaction with other antigens or haptens;
- batch variation, as individual animals produce slightly different antibodies in response to the same antigen;
- non-specificity, as the antiserum will contain many other antibodies.

Standardisation of polyclonal antisera therefore is difficult. You may need to assess the amount of cross-reaction, interbatch variation or non-specific binding using appropriate controls, assayed at the same time as the test samples.

Monoclonal antibodies

These are specific to a single epitope and are produced from individual clones of cells (hybridomas), grown using cell culture techniques (p. 474). Such cultures provide a stable source of antibodies of known, uniform specificity. Although monoclonal antibodies are likely to be used increasingly in future years, polyclonal antisera are currently employed for many routine immunological assays.

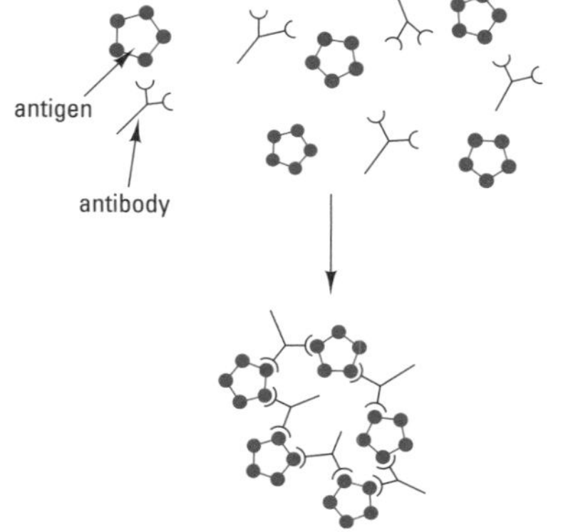

Fig. 52.2 Formation of an antigen–antibody complex.

Agglutination tests

When antibodies interact with a suspension of a particulate antigen, e.g. cells or latex particles, the formation of immune complexes (Fig. 52.2) causes visible clumping, termed agglutination. A positive haemagglutination reaction gives an even 'carpet' of red cells over the base of the tube, while a negative reaction gives a tightly packed 'button' of red cells at the bottom of the tube. Agglutination tests are used in several ways:

- **Microbial identification:** at the species or subspecies level (serotyping), e.g. mixing an unknown bacterium with the appropriate antiserum will cause the cells to agglutinate.
- **Latex agglutination using bound antigens:** by coating soluble (non-particulate) antigens on to microscopic latex spheres, their reaction with a particular antibody can be visualised.
- **Latex agglutination using bound antibodies:** antibodies can be bound to latex microspheres, leaving their antigen-binding sites free to react with soluble antigen.
- **Haemagglutination:** red blood cells can be used as agglutinating particles. However, in some instances, such reactions do not involve antibody interactions (e.g. some animal viruses may haemagglutinate unmodified red blood cells).

Precipitin tests

Immune complexes of antibodies and soluble antigens (or haptens) usually settle out of solution as a visible precipitate: this is termed a precipitin test, or precipitation test. The formation of visible immune complexes in agglutination and precipitation reactions only occurs if antibody and antigen are present in an optimal ratio (Fig. 52.3). It is important to appreciate the shape of this curve: cross-linkage is maximal in the zone of equivalence, decreasing if either component is present in excess. The quantitative precipitin test can be used to measure the antibody content of a solution. Visual assessment of precipitation reactions forms the basis of several other techniques, described below.

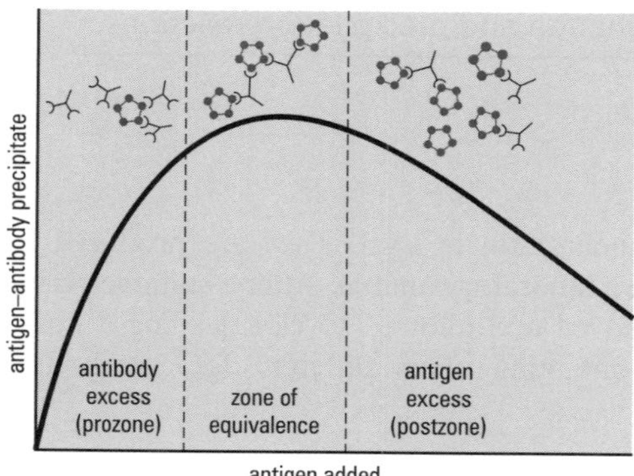

Fig. 52.3 Precipitation curve for an antigen titrated against a fixed amount of antibody.

Immunodiffusion assays

These techniques are easier to perform and interpret than the quantitative precipitin test. Precipitation of antibody and antigen occurs within an agarose gel, giving a visible line corresponding to the zone of equivalence (Fig. 52.3). Details of the main techniques are given in Box 52.1.

Box 52.1 How to carry out immunodiffusion assays

The two most widespread approaches are (i) single radial immunodiffusion (Mancini technique) and (ii) double-diffusion immunoassay (Ouchterlony technique).

Single radial immunodiffusion (RID)

This is used to quantify the amount of antigen in a test solution, as follows:

1. **Prepare an agarose gel** (1.5% w/v), containing a fixed amount of antibody: allow to set on a glass slide or plate, on a level surface.
2. **Cut several circular wells in the gel**. These should be of a fixed size between 2 and 4 mm in diameter (see Fig. 52.4(a)). Cut your wells carefully. They should have straight sides and the agarose must not be torn or lifted from the glass plate. All wells should be filled to the top, with a flat meniscus, to ensure identical diffusion characteristics. Non-circular precipitin rings, resulting from poor technique, should not be included in your analysis.
3. **Add a known amount of the antigen or test solution to each well.**
4. **Incubate on a level surface at room temperature in a moist chamber:** diffusion of antigen into the gel produces a precipitin ring. This is usually measured after two–seven days, depending on the molecular mass of the antigen.
5. **Examine the plates against a black background** (with side illumination), or stain using a protein dye (e.g. Coomassie blue).
6. **Measure the diameter of the precipitin ring,** e.g. using Vernier calipers.

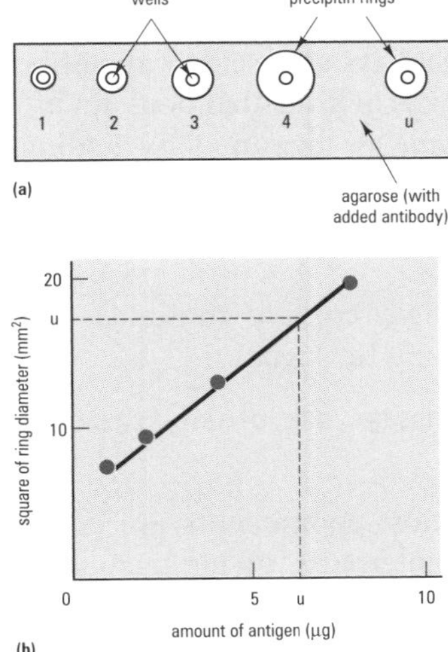

Fig. 52.4 Single radial immunodiffusion (RID). (a) Assay: four standards are shown (wells 1 to 4, each one double the strength of the previous standard), and an unknown (u), run at the same time, (b) calibration curve. The unknown contains 6.25 μg of antigen. Note the non-zero intercept of the calibration curve, corresponding to the square of the well diameter: do not force such calibration lines through the origin.

7. **Prepare a calibration curve** (Chapter 50) from the samples containing known amounts of antigen (Fig. 52.4(b)): the squared diameter of the precipitin ring is directly proportional to the amount of antigen in the well.
8. **Quantify the amount of antigen in your test solutions**, using a calibration curve prepared from standards assayed at the same time.

Double-diffusion immunoassay (Ouchterlony technique)
This technique is widely used to detect particular antigens in a test solution, or to look for cross-reaction between different antigens.

1. **Prepare an agarose gel** (1.5% w/v) on a level glass slide or plate: allow to set.
2. **Cut several circular wells in the gel.**
3. **Add test solutions of antigen or polyclonal antiserum to adjacent wells.** Both solutions diffuse outwards, forming visible precipitin lines where antigen and corresponding antibody are present in optimal ratio (Fig. 52.5).

The various reactions between antigen and antiserum are:

- **Identity**: two wells containing the same antigen, or antigens with identical epitopes, will give a fused precipitin line (identical interaction between the antiserum and the test antigens, Fig. 52.5(a)).
- **Non-identity**: where the antiserum contains antibodies to two different antigens, each with its own distinct epitopes, giving two precipitin lines that intersect without any interaction (no cross-reaction, Fig. 52.5(b)).
- **Partial identity**: where two antigens have at least one epitope in common, but where other epitopes are present, giving a fused precipitin line with a spur (cross-reaction, Fig. 52.5(c)).

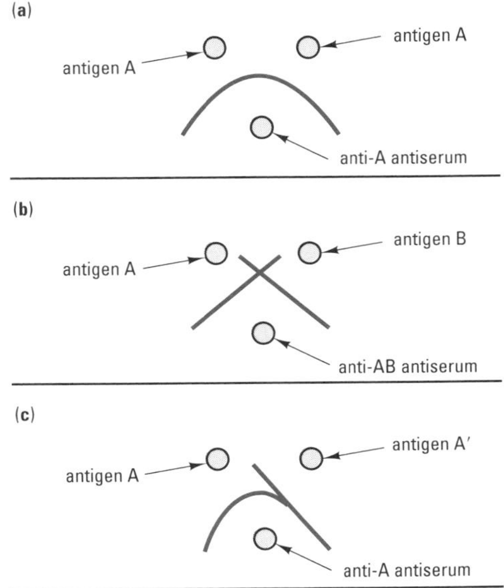

Fig. 52.5 Precipitin reactions in double-diffusion immunoassay: (a) identity; (b) non-identity; (c) partial identity.

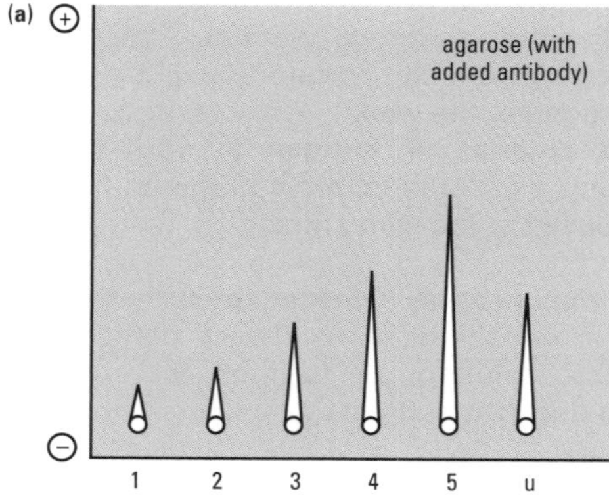

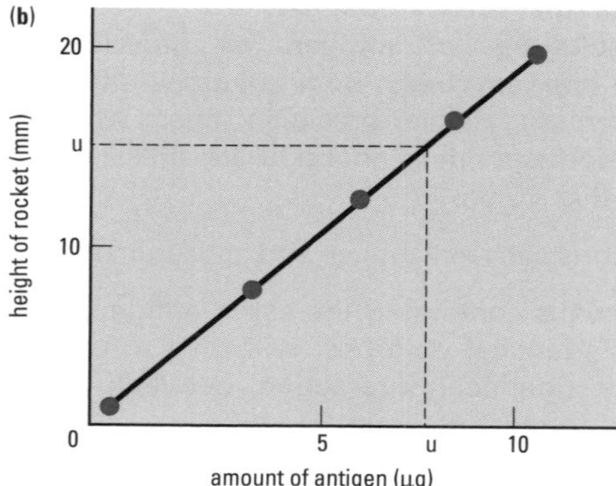

Fig. 52.6 Laurell rocket immunoelectrophoresis. (a) Assay: precipitin rockets are formed by electrophoresis of five standards of increasing concentration (wells 1 to 5) and an unknown (u). (b) Calibration curve: the unknown sample contains 7.7 µg of antigen.

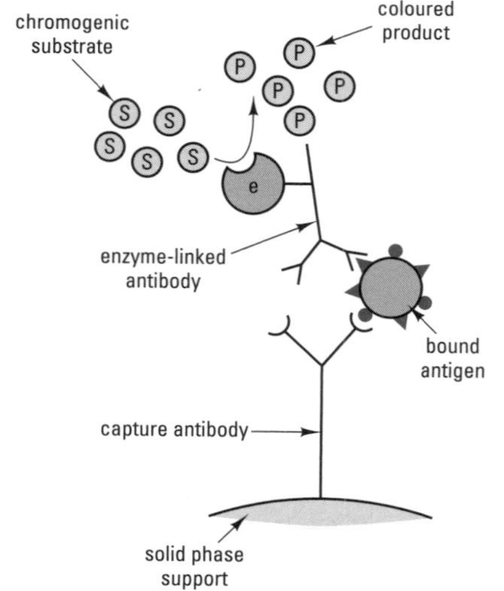

Fig. 52.7 Double-antibody sandwich ELISA.

Immunoelectrophoretic assays

These methods combine the precipitin reaction with electrophoretic migration, providing sensitive, rapid assays with increased separation and resolution.

Crossover electrophoresis (countercurrent electrophoresis)

Similar to the Ouchterlony technique, since antigen and antibody are in separate wells. However, the movement of antigen and antibody towards each other is driven by a voltage gradient: most antigens migrate towards the anode, while IgG migrates towards the cathode. This method is faster and more sensitive than double immunodiffusion, taking 15–20 minutes to reach completion.

Quantitative immunoelectrophoresis (Laurell rocket immunoelectrophoresis)

Similar to RID, as the antibody is incorporated into an agarose plate while the antigen is placed in a well. However, a voltage gradient moves the antigen into the gel, usually towards the anode, while the antibody moves towards the cathode, giving a sharply peaked, rocket-shaped precipitin line, once equivalence is reached (within 2–10 hours). The height of each rocket shape at equivalence is directly proportional to the amount of antigen added to each well. A calibration curve for samples containing known amount of antigen can be used to quantify the amount of antigen present in test samples (Fig. 52.6).

Enzyme immunoassays (EIA)

These techniques are also known as enzyme-linked immunosorbent assays (ELISA). They combine the specificity of the antibody–antigen interaction with the sensitivity of enzyme assays using either an antibody or an antigen conjugated (linked) to an enzyme at a site that does not affect the activity of either component. The enzyme is measured by adding an appropriate chromogenic substrate, which yields a coloured product (Box 52.2). Enzymes offer the following advantages over radioisotopic labels:

- **Increased sensitivity:** a single enzyme molecule can produce many product molecules, amplifying the signal.
- **Simplified assay:** enzyme assays are usually easier than radioisotope immunoassays.
- **Improved stability of reagents:** components are generally more stable than their radiolabelled counterparts, giving them a longer shelf life.
- **Automation is straightforward:** using disposable microtitre plates and an optical scanner.

The principal techniques are:

Double-antibody sandwich ELISA

This is used to detect specific antigens, involving a three-component complex between a capture antibody linked to a solid support, the antigen, and a second, enzyme-linked antibody (Fig. 52.7). This can be used to

Box 52.2 How to perform an ELISA assay

While the following example is for a sandwich (capture) ELISA assay in microplate format, the same general principles apply to the other types of ELISA:

1. **Prepare the apparatus.** Switch on all equipment required:
   (i) The microplate reader – used to measure the absorbance of the solution in each well: set the reader to the required wavelength.
   (ii) The microplate washer (where used) – each well must be washed at various stages during the procedure. When using an automated washer, first check that the wash bottle contains sufficient diluent and then test using an old microplate, to check that all wells are being washed correctly. Where required, use a wire needle to clean any blocked wash delivery tubes and repeat. For manual washing, use a wash bottle or multichannel pipettor – make sure you fill each well and empty out all of the wash solution at the end of each wash stage.
   (iii) The computer – this will contain the software required to label the wells, draw the calibration curve and calculate the results for test samples: fill out the ELISA template with details of the assays to be carried out.
2. **Prepare the various solutions to be analysed.** These include:
   (i) Test samples – make sure that each sample is identified with a code that enables you to record what each test well contains.
   (ii) Calibrators/standards – including 'cutoff' calibrators and known positive standards.
   (iii) Controls – positive and negative controls and blanks.
3. **Coat the wells with the capture antibody.** Typically 100 μl of a solution of the appropriate monoclonal or polyclonal antibody is added to each well and microplates are then incubated overnight at 4 °C to allow binding to the well.
4. **Wash the wells.** Transfer the microplate to the washer (or wash manually) – wash six times to remove excess coating (capture) antibody. The final rinse should be programmed so that the washer leaves the wells empty of diluent.
5. **Add blocking solution to each well.** Typically 100 μl of an inert protein solution (bovine serum albumin) is added to each well to block any free binding sites on the well. Microplates are incubated at room temperature for 30 minutes and then washed, as in step 4.
6. **Add test samples, calibrators and controls to wells.** Typically 100 μL of appropriately diluted test sample, control, etc. is added to each well. Microplates are incubated at room temperature for 90 minutes and then washed as step 4.
7. **Add the detection antibody to each well.** Add 100 μl of monoclonal or polyclonal detection antibody labelled with a suitable enzyme (e.g. horseradish peroxidise HRP) to each well. The microplate is then incubated at room temperature for 30–60 min and then washed, as in step 4.
8. **Add chromogenic substrate to each well.** For example, with HRP-labelled antigen add 100 μL of a standard solution of tetramethylbenzidine (TMB) and hydrogen peroxide to each well and re-incubate in darkness for 30 minutes, to allow colour development. The TMB is oxidised in the presence of hydrogen peroxide to produce a blue colour.

   SAFETY NOTE TMB and hydrogen peroxide are harmful by inhalation – use a fume hood.

9. **Stop the reactions.** For example, by adding 100 μl of 2 mol L^{-1} sulphuric acid to denature the enzyme. The colour of the oxidised TMB will change from blue to yellow as a result of the pH shift. While the human eye can readily distinguish different shades of blue, it is more difficult to visually assess different shades of yellow, once the reaction has been stopped.
10. **Measure the absorbance of each sample/calibrator/control well.** Transfer the microplate to the reader and assay at an appropriate wavelength: for TMB, use 450 nm.
11. **Interpret the results.** Check that the absorbance values of calibrators and control are within the required range. Then, for each sample, either record the absolute value (convert to concentration or amount e.g. using a calibration curve, pp. 336–338) or record as 'positive' or 'negative' (based on values for 'cutoff' calibrators), as appropriate.

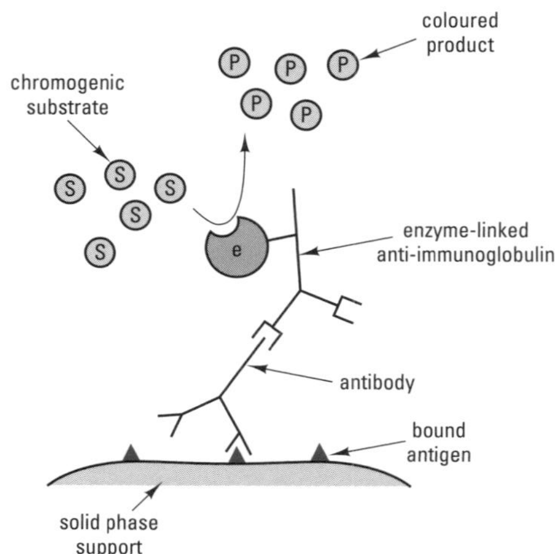

Fig. 52.8 Indirect ELISA.

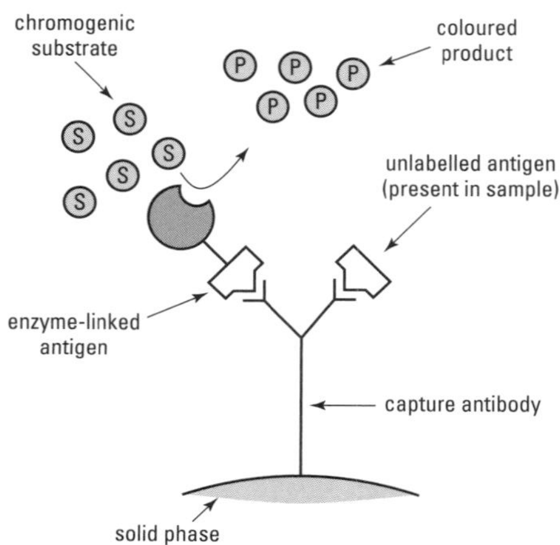

Fig. 52.9 Competitive ELISA.

detect a particular antigen, e.g. a virus in a food sample, or to quantify the amount of antigen.

Indirect ELISA

This is used for antibody detection, with a specific antigen attached to a solid support. When the appropriate antibody is added, it binds to the antigen and will not be washed away during rinsing. Bound antibody is then detected using an enzyme-linked anti-immunoglobulin, e.g. a rabbit IgG antibody raised against human IgG (Fig. 52.8). One advantage of the indirect assay is that a single enzyme-linked anti-immunoglobulin can be used to detect several different antibodies, since the specificity is provided by the bound antigen.

Competitive ELISA

Here, any antigen present in a test sample competes with added enzyme-labelled antigen for a limited number of binding sites on the capture antibody (Fig. 52.9). Most commercial systems use 96-well microplates (12 columns by 8 rows), where each well is coated with the appropriate antibody. Following addition of known volumes of (i) sample and (ii) enzyme-labelled antigen, the plates are incubated (typically, up to one hour), then washed thoroughly to remove all unbound material. Bound enzyme-labelled antigen is then detected using a suitable substrate. Quantitative results can be obtained by measuring the absorbance of each well using a spectrophotometric microplate reader: the absorbance at a particular wavelength is *inversely* proportional to the amount of antigen present in the test sample. Calibration standards (p. 338) are required to convert the readings to an amount or concentration. Alternatively, the test can be carried out in positive/negative format. Box 52.2 gives practical details for a sandwich-type ELISA.

Text references and sources for further study

Delves, P.J., Martin, S., Burton, D. and Roitt, I. (2006) *Roitt's Essential Immunology*, 11th edn. Blackwell, Oxford.

Gosling, J.R.G. (2000) *Immunoassays: A Practical Approach*. Oxford University Press, Oxford.

Hay, F.C., Westwood, O.M.R. and Nelson, P.N. (2002) *Practical Immunology*. Blackwell, Oxford.

53 Spectroscopic techniques

Using spectroscopy – this technique is valuable for:

- tentatively identifying compounds, by determining their absorption or emission spectra;
- quantifying substances, either singly or in the presence of other compounds, by measuring the signal strength at an appropriate wavelength;
- determining molecular structure;
- following reactions, by measuring the disappearance of a substance, or the appearance of a product as a function of time.

The absorption and emission of electromagnetic radiation of specific energy (wavelength) is a characteristic feature of many molecules, involving the movement of electrons between different energy states, in accordance with the laws of quantum mechanics. Spectroscopic techniques are used to measure and interpret such interactions between molecules and radiation.

UV/visible spectrophotometry

This is a widely used technique for measuring the absorption of radiation in the visible and UV regions of the spectrum. A spectrophotometer is an instrument designed to allow precise measurement at a particular wavelength, while a colorimeter is a simpler instrument, using filters to measure broader wavebands (e.g. light in the green, red or blue regions of the visible spectrum).

Principles of light absorption

Two fundamental principles govern the absorption of light by a solution:

- The absorption of light passing through a solution is exponentially related to the number of molecules of the absorbing solute, i.e. the solute concentration $[C]$.
- The absorption of light passing through a solution is exponentially related to the length of the absorbing solution, l.

These principles are combined in the Beer–Lambert relationship (sometimes referred to simply as 'Beer's law'), usually expressed in terms of absorbance (A) – the logarithm of the ratio of incident light (I_0) to emergent light (I):

$$A = \varepsilon l [C] \qquad [53.1]$$

where A is absorbance, ε is a constant for the absorbing substance at a specific wavelength, termed the absorptivity, or absorption coefficient, and $[C]$ is expressed either as mol L^{-1} or g L^{-1} (see p. 160) and l is given in cm. This relationship is extremely useful, since most spectrophotometers are constructed to give a direct measurement of absorbance, sometimes also termed extinction (E), of a solution (older texts may use the outdated term optical density, OD).

Definitions

Absorbance (A) – this is given by:

$A = \log_{10}(I_0/I)$, usually shown as A_x where 'x' is the wavelength in nanometres.

Transmittance (T) – this is usually expressed as a percentage often at a particular wavelength, T_x, where

$T_x = (I/I_0) \times 100$ (%).

KEY POINT The Beer–Lambert relationship states that there is a direct linear relationship between the concentration of a substance in a solution, [C], and the absorbance of that solution, A.

Example For incident light (I_0) = 1.00 and emergent light (I) = 0.16 (expressed in relative terms) then $A = \log_{10}$ (1.00 ÷ 0.16) = $\log_{10}$ 6.25 = 0.796 (to three significant figures). The corresponding transmittance, T = (0.16 ÷ 1.00) ×100 = 16%.

Absorbance at a particular wavelength is often shown as a subscript, e.g. A_{550} represents the absorbance at 550 nm. The proportion of light passing through the solution is known as the transmittance (T), and is calculated as the ratio of the emergent and incident light intensities.

Some instruments have two scales:

- an exponential scale from zero to infinity, measuring absorbance;
- a linear scale from 0 to 100, measuring (per cent) transmittance.

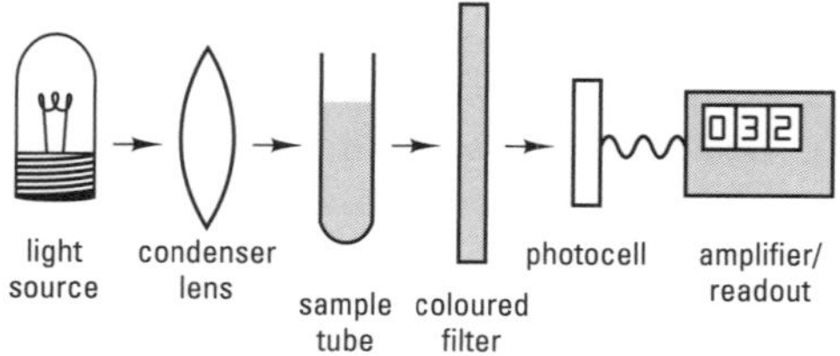

Fig. 53.1 Components of a colorimeter.

SAFETY NOTE Take care not to spill water when using a colorimeter or spectrophotometer, due to the risk of electric shock (switch off at mains and seek assistance if this should happen).

For most practical purposes, the Beer–Lambert relationship (eqn [53.1]) applies and you should use the absorbance scale.

Colorimeter

This can be used with solutions where the test substance is highly coloured and present as the major constituent, e.g. haemoglobin in blood, or where a substance is assayed by adding a reagent that gives a coloured product (a chromophore), e.g. amino acid assay using ninhydrin reagent. Quantification of a particular substance requires a calibration curve, constructed using known amounts of the compound measured at the same time as the test samples, rather than using the Beer–Lambert relationship.

The light source is usually a tungsten filament bulb, focused by a condenser lens to give a parallel beam of light that passes through a glass sample tube or cuvette containing the solution, then through a coloured filter to a photocell detector, which develops an electrical potential in direct proportion to the intensity of the light falling on it (Fig. 53.1). The signal from the photocell is then amplified and passed to a galvanometer, or digital readout, calibrated on a logarithmic scale (see Box 53.1).

The broad bandwidth of most filters means that colorimetry cannot be used to identify a particular compound, nor to distinguish between two compounds with closely related absorption characteristics, e.g. in mixed solution. The photocells used in colorimeters have coefficients of variation (p. 252) of around 0.5 per cent, so they are not suitable for work requiring a high degree of precision. In the simplest instruments, the logarithmic measurement scale has arbitrary units, adjusted via sensitivity/scale zero controls, and values obtained on one instrument will not be directly comparable with other instruments, or with the same instrument on

Box 53.1 How to use a colorimeter

1. **Switch on and stabilise** allowing at least five minutes for the lamp to warm up before use.
2. **Choose a filter that is complementary** to the colour of the substance to be measured. Thus haemoglobin is red because it absorbs light within the blue/green regions of the spectrum and should be measured using a blue filter. Similarly, a blue substance should be measured using a red filter.
3. **Set the scale zero** using an appropriate solution blank.
4. **Adjust the sensitivity** to give a reasonable deflection on the galvanometer/readout device, e.g. for a standard solution containing a known amount of the test substance. You would normally adjust the sensitivity so that your highest standard gave a reading close to the maximum on the readout device.
5. **Analyse your samples and standard solutions,** making sure you take all the measurements you need in a single 'run', otherwise they will not be directly comparable.
6. **Use the same tube in the same orientation** in the sample holder to improve precision, since individual tubes may differ in their light-absorbing characteristics, wall thickness, etc.
7. **Rinse the tube between samples,** making sure that the rinsing solution does not dilute the next sample and that the outside of the tube is dry before it is loaded into the instrument.
8. **Make frequent checks on reproducibility** by repeat measurements of the same solution (e.g. after every six to eight samples). You should also prepare test and standard solutions in duplicate.
9. **Plot a calibration curve for your standard solutions** and draw the line of best fit through the points. Do not worry if it is a curve, rather than a straight line, or if it does not pass through the origin. Do not extrapolate the calibration curve. If a sample has a reading greater than the highest standard it should be diluted and reassayed. Chapter 50 gives further advice on preparing and using calibration curves.

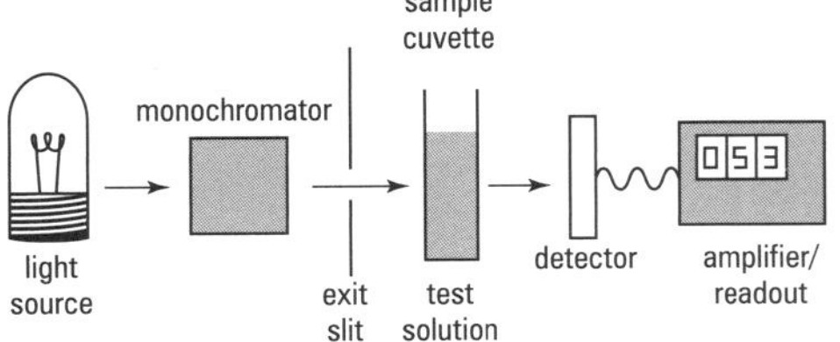

Fig. 53.2 Components of a UV/visible spectrophotometer.

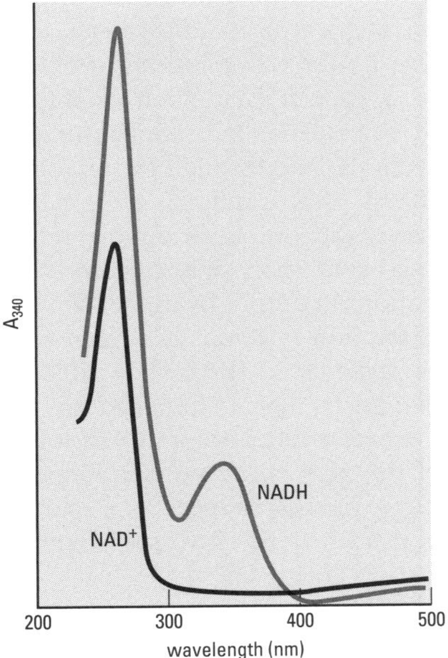

Fig. 53.3 Absorption spectra of nicotinamide adenine dinucleotide in oxidised (NAD^+) and reduced (NADH) form. Note the 340 nm absorption peak, used for quantitative work (p. 360).

Using plastic disposable cuvettes – these are adequate for work in the near-UV region, e.g. for enzyme studies using nicotinamide coenzymes, at 340 nm, as well as the visible range. Table 25.2 (p. 155) gives details of the spectral cut-off wavelengths of different materials.

Measuring low absorbances – for accurate readings with dilute solutions, use a cuvette with a longer optical path length (e.g. 5 cm, rather than 1 cm).

different settings. A colorimeter is not suitable for quantitative work at a particular wavelength.

UV/visible spectrophotometer

The principal components of a UV/visible spectrophotometer are shown in Fig. 53.2. High-intensity tungsten bulbs are used as the light source in basic instruments, capable of operating in the visible region (i.e. 400–700 nm). Deuterium lamps are used for UV spectrophotometry (200–400 nm); these lamps are fitted with quartz envelopes, since glass does not transmit UV radiation.

A major improvement over the simple colorimeter is the use of a diffraction grating to produce a parallel beam of monochromatic light from the (polychromatic) light source. In practice the light emerging from such a monochromator does not have a single wavelength, but is a narrow band of wavelengths. This bandwidth is an important characteristic, since it determines the wavelengths used in absorption measurements – the bandwidth of basic spectrophotometers is around 5–10 nm, while research instruments have bandwidths of less than 1 nm.

Bandwidth is affected by the width of the exit slit (the slit width), since the bandwidth will be reduced by decreasing the slit width. To obtain accurate data at a particular wavelength setting, the narrowest possible slit width should be used. However, decreasing the slit width also reduces the amount of light reaching the detector, decreasing the signal-to-noise ratio. The extent to which the slit width can be reduced depends upon the sensitivity and stability of the detection/amplification system and the presence of stray light.

Most UV/visible spectrophotometers are designed to take cuvettes with an optical path length of 1 cm. Disposable plastic cuvettes are suitable for routine work in the visible range using aqueous and alcohol-based solvents, and glass cuvettes are useful for other organic solvents. Glass cuvettes are manufactured to more exacting standards, so use optically matched glass cuvettes for accurate work, especially at low absorbances (<0.1), where any differences in the optical properties of cuvettes for reference and test samples will be pronounced. Glass and plastic absorb UV light, so quartz cuvettes must be used at wavelengths below 300 nm.

KEY POINT Before taking a measurement, make sure that cuvettes are clean, unscratched, dry on the outside, filled to the correct level and located in the correct position in their sample holders.

Proteins and nucleic acids in biological samples can accumulate on the inside faces of glass/quartz cuvettes, so remove any deposits using acetone on a cotton bud, or soak overnight in 1 mol L^{-1} nitric acid. Corrosive and hazardous solutions must be used in cuvettes with tightly fitting lids, to prevent damage to the instrument and to reduce the risk of accidental spillage.

Basic instruments use photocells similar to those used in colorimeters or photodiode detectors. In many cases, a different photocell must be used at wavelengths above and below 550–600 nm, due to differences in the sensitivity of such detectors over the visible waveband. The detectors used in more sophisticated instruments give increased sensitivity and stability when compared with photocells.

Measuring high absorbances in colorimetric analysis – if any final solution has an absorbance that is too high to be read with accuracy on your spectrophotometer (i.e. A > 1), it is not appropriate to dilute the solution so that it can be measured. This dilutes both the sample molecules and the colour reagents to an equal extent. Instead, you should dilute the original sample and re-assay.

Examples in quantitative analysis
The molar absorptivity of NADH is 6.22×10^3 L mol^{-1} cm^{-1} at 340 nm. For a test solution giving an absorbance of 0.21 in a cuvette with a light path of 5 mm, using eqn [53.1] this is equal to a concentration of:

$0.21 = 6.22 \times 10^3 \times 0.5 \times [C]$
$[C] = 0.0000675$ mol L^{-1}
(or 67.5 μmol L^{-1}).

The specific absorptivity (10 g L^{-1}) of double-stranded DNA is 200 at 260 nm, therefore a solution containing 1 g L^{-1} will have an absorbance of 200/10 = 20. For a DNA solution, giving an absorbance of 0.35 in a cuvette with a light path of 1.0 cm, using eqn [59.1] this is equal to a concentration of:

$0.35 = 20 \times 1.0\,[C]$
$[C] = 0.0175$ g L^{-1}
(equivalent to 17.5 μg mL^{-1}).

Chlorophylls a and b in vascular plants and green algae can be assayed in 90% v/v acetone/water by measuring the absorbance of the mixed solution at two wavelengths, according to the formulae:

Chlorophyll a (mg L^{-1}) = $11.93\,A_{664} - 1.93\,A_{647}$

Chlorophyll b (mg L^{-1}) = $20.36\,A_{647} - 5.5\,A_{664}$

Note: different equations are required for other solvents.

SAFETY NOTE In atomic spectroscopy, the use of high-pressure gas cylinders can be particularly hazardous. Always consult a member of staff before using such apparatus.

Digital displays are increasingly used in preference to needle-type meters, as they are not prone to parallax errors and misreading of the scale. Some digital instruments can be calibrated to give a direct readout of the concentration of the test substance.

KEY POINT Basic spectrophotometers are most accurate within the absorbance range from 0.00 to 1.00 and your standard and test solutions should be prepared to give readings within this range.

Types of UV/visible spectrophotometer

Basic instruments are single-beam spectrophotometers in which there is only one light path. The instrument is set to zero absorbance using a blank solution, which is then replaced by the test solution, to obtain an absorbance reading. An alternative approach is used in double-beam spectrophotometers, where the light beam from the monochromator is split into two separate beams, one beam passing through the test solution and the other through a reference blank. Absorbance is then measured by an electronic circuit that compares the output from the reference (blank) and sample cuvettes. Double-beam spectrophotometry reduces measurement errors caused by fluctuations in output from the light source or changes in the sensitivity of the detection system, since reference and test solutions are measured at the same time (Box 53.2). Recording spectrophotometers are double-beam instruments, designed to record either the difference in absorbance between reference and test solutions across a predetermined waveband to give an absorption spectrum (Fig. 53.3), or to record the change in absorbance at a particular wavelength as a function of time (e.g. in an enzyme assay, see Chapter 62).

Quantitative spectrophotometric analysis

A single substance in solution can be quantified using the Beer–Lambert relationship (eqn [53.1]), provided its absorptivity is known at a particular wavelength (usually the absorption maximum for the substance, since this will give the greatest sensitivity). The molar absorptivity is the absorbance given by a solution with a concentration of 1 mol L^{-1} (= 1 kmol m^{-3}) of the compound in a light path of 1 cm. The appropriate value may be available from tabulated spectral data (e.g. Anon., 1963), or it can be determined experimentally by measuring the absorbance of known concentrations of the substance (Box 53.2) and plotting a standard curve (Chapter 50). This should confirm that the relationship is linear over the desired concentration range and the slope of the line will give the molar absorptivity.

The specific absorptivity is the absorbance given by a solution containing 10 g L^{-1} (i.e. 1% w/v) of the compound in a light path of 1 cm. This is useful for substances of unknown molecular weight, e.g. proteins or nucleic acids, where the amount of substance in solution is expressed in terms of its mass, rather than as a molar concentration. For use in eqn [53.1], the specific absorptivity should be divided by 10 to give the solute concentration in g L^{-1}.

This simple approach cannot be used for complex samples e.g. food homogenates. In such cases, it may be possible to estimate the amount of each substance by measuring the absorbance at several wavelengths, e.g. protein estimation in the presence of nucleic acids. Further details on

Box 53.2 How to use a spectrophotometer

1. **Switch on and select the correct lamp** for your measurements (e.g. deuterium for UV, tungsten for visible light).
2. **Allow at least 15 min for the lamp to warm up** and for the instrument to stabilise before use.
3. **Select the appropriate wavelength:** on older instruments a dial is used to adjust the monochromator, whereas newer machines have microprocessor-controlled wavelength selection.
4. **Select the appropriate detector**: some instruments choose the correct detector automatically (on the basis of the specified wavelength), and others have manual selection.
5. **Choose the correct slit width** (if available): this may be specified in the protocol you are following, or may be chosen on the manufacturer's recommendations.
6. **Insert appropriate reference blank(s)**: single-beam instruments use a single cuvette, whereas double-beam instruments use two cuvettes (a matched pair for accurate work). The reference blanks should match the test solution in all respects apart from the substance under test, i.e. they should contain all reagents apart from this substance. *Make sure that the cuvettes are positioned correctly, with their polished (transparent) faces in the light path, and check that they are accurately located in the cuvette holder(s).*
7. **Check/adjust the 0% transmittance**: most instruments have a control that allows you to zero the detector output in the absence of any light (so-called dark current correction). Some microprocessor-controlled instruments carry out this step automatically.
8. **Set the absorbance reading to zero**: usually via a dial, or digital readout.
9. **Analyse your samples**: replace the appropriate reference blank with a test sample, allow the absorbance reading to stabilise (5–10 s) and read the absorbance value from the meter/readout device. For absorbance readings greater than 1 (i.e. <10% transmittance), the signal-to-noise ratio is too low for accurate results. Your analysis may require a calibration curve (Chapter 50), or you may be able to use the Beer–Lambert relationship (eqn [53.1]) to determine the concentration of test substance in your samples.
10. **Check the scale zero at regular intervals** using a reference blank, e.g. after every 10 samples.
11. **Check the reproducibility of the instrument**: measure the absorbance of a single solution several times during your analysis. It should give the same value.

Problems (and solutions): inaccurate/unstable readings are most often due to incorrect use of cuvettes, e.g. dirt, fingerprints or test solution on outside of cuvette (wipe the polished faces using a soft tissue before insertion into the cuvette holder), condensation (if cold solutions aren't allowed to reach room temperature before use), air bubbles (which scatter light and increase the absorbance; tap gently to remove), insufficient solution (causing refraction of light at the meniscus), particulate material in the solution (centrifuge before use, where necessary) or incorrect positioning in light path (locate in correct position).

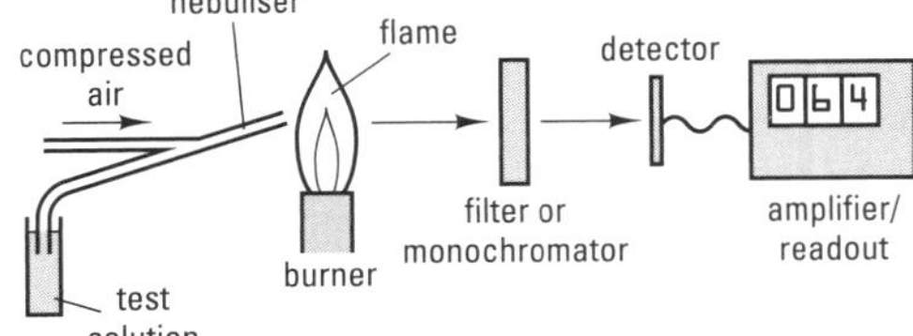

Fig. 53.4 Components of a flame photometer.

SAFETY NOTE When carrying out acid digestion, always work within a fume hood and wear gloves and safety glasses throughout the procedure. Rinse any spillages with a large volume of water.

spectrophotometric methods of determining the amount of protein in an aqueous sample are given in Chapter 58.

Atomic spectroscopy

Atoms of certain metals will absorb and emit radiation of specific wavelengths when heated in a flame, in direct proportion to the number of atoms present. Atomic spectrophotometric techniques measure the absorption or emission of particular wavelengths of UV and visible light, to identify and quantify such metals.

Flame atomic-emission spectrophotometry (or flame photometry)

The principal components of a flame photometer are shown in Fig. 53.4. A liquid sample is converted into an aerosol in a nebuliser (atomiser) before

Box 53.3 How to use a flame photometer

1. **Switch on the instrument and allow it to stabilise.** Light the flame and then wait for at least five minutes before analysing your solutions.
2. **Check for impurities in your reagents.** For example, if you are measuring K^+ in an acid digest of a food sample, you should check the K^+ content of a reagent blank (a solution containing everything except the food, prepared in exactly the same way as the samples). Once converted to concentration (see below) you should then subtract this value from the concentration determined for each of your sample solutions, to obtain the true K^+ content.
3. **Quantify your samples using a calibration curve** (p. 336). Calibration standards should cover the expected concentration range for the test solutions – your calibration curve may be non-linear, especially at higher concentrations (e.g. above 1 mmoL l^{-1}, i.e. 1 mol m^{-3} in SI units).
4. **Assay all solutions in duplicate,** so that repeatability can be assessed (if duplicate readings do not agree, repeat until you are satisfied that the measurements are reliable).
5. **Check your calibration.** Make repeated measurements of a standard solution of known concentration after every six or seven samples, to confirm that the instrument calibration is still valid.
6. **Consider the possibility of interference.** Other metal atoms may emit light that is detected by the photocell, since the filters cover a wider waveband than the emission wavelength of a particular element. This can be a serious problem if you are trying to measure low concentrations of a particular metal in the presence of high concentrations of other metals (e.g. Na^+ in sea water), or if there are other substances present that form complexes with the test metal, suppressing the signal (e.g. phosphate).

being introduced into the flame, where a small proportion (typically less than 1 in 10 000) of the atoms will be raised to a higher energy level, releasing this energy as light of a particular wavelength, which is passed through a filter to a photocell detector. Flame photometry is used to measure the alkali metal ions K^+, Na^+ and Ca^{++} in fluids e.g. beverages or food homogenates. Box 53.3 gives details on how to use a flame photometer.

Plotting calibration curves in quantitative analysis – do not force your calibration line to pass through zero if it clearly does not. There is *no* reason to assume that the zero value is any more accurate than any other reading you have made (see also Chapter 50).

Atomic-absorption spectrophotometry (or flame absorption spectrophotometry)

This technique is applicable to a broad range of metal ions, including those of Pb, Cu, Zn, etc. It relies on the absorption of light of a specific wavelength by atoms dispersed in a flame. The appropriate wavelength is provided by a cathode lamp, coated with the element to be analysed, focused through the flame and on to the detector. When the sample (e.g. an acid digest of a particular food) is introduced into the flame, it will decrease the light detected in direct proportion to the amount of metal present. Practical advantages over flame photometry include improved sensitivity, increased precision and decreased interference. Newer variants of this method include flameless atomic absorption spectrophotometry and atomic fluorescence spectrophotometry, both of which are more sensitive than the flame-absorption technique. For more advanced spectroscopic techniques, See Reed *et al.* (2007).

Text references and sources for further study

Anon. (1963) *Tables of Spectrophotometric Absorption Data for Compounds used for the Colorimetric Detection of Elements (International Union of Pure and Applied Chemistry)*. Butterworth-Heinemann, London.

Gore, M.G. (ed.) (2000) *Spectrophotometry and Spectrofluorimetry: A Practical Approach,* 2nd edn. Oxford University Press, Oxford.

Nielsen, S.S. (2010) *Food Analysis*, 4th edn. Springer, Berlin.

Reed, R.H., Holmes, D.H., Weyers, J.D.B. and Jones, A.M. (2007) *Practical Skills in Biomolecular Sciences,* 3rd edn. Prentice Hall, Harlow.

Wilson, K. and Walker, J. (eds.) (2010) *Principles and Techniques of Biochemistry and Molecular Biology,* 7th edn. Cambridge University Press, Cambridge. [Also covers a range of other topics, including chromatography and centrifugation.]

54 Chromatography

Making compromises in chromatography – the process is often a three-way compromise between:

1. separation of analytes;
2. time taken for analysis;
3. volume of eluent.

Thus, if you have a large sample volume and want to achieve a good separation of a mixture of analytes, the time taken for the chromatography will be lengthy.

SAFETY NOTE The solvents used as the mobile phases of chromatographic systems are often toxic and may produce noxious fumes – where necessary, work in a fume hood.

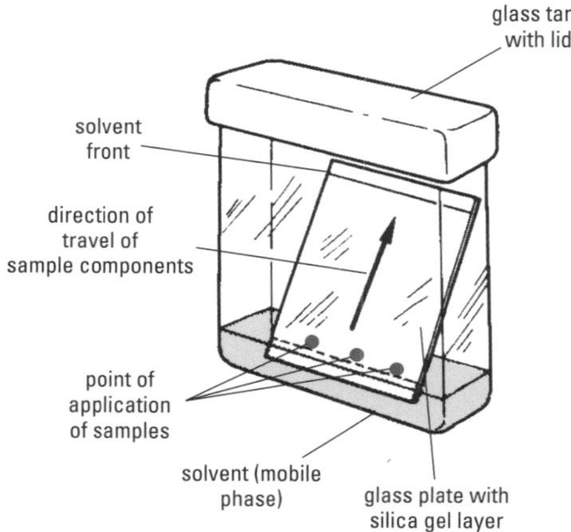

Fig. 54.1 Components of a TLC system.

Using a TLC system – it is essential that you allow the solvent to pre-equilibrate in the chromatography tank for at least two hours before use, to saturate the atmosphere with vapour. Deliver drops of sample with a blunt-ended microsyringe. Make sure you know exactly where each sample is applied, so that R_F values can be calculated.

Chromatography is used to separate the individual constituents within a sample (e.g. a food homogenate) on the basis of differences in their physical characteristics, e.g. molecular size, shape, charge, volatility, solubility and/or adsorptivity. The essential components of a chromatographic system are:

- A stationary phase, either a solid, a gel or an immobilised liquid, held by a support matrix.
- A chromatographic bed: the stationary phase may be packed into a glass or metal column, spread as a thin layer on a sheet of glass or plastic, or adsorbed on cellulose fibres (paper).
- A mobile phase, either a liquid or a gas that acts as a solvent, carrying the sample through the stationary phase and eluting from the chromatographic bed.
- A delivery system to pass the mobile phase through the chromatographic bed.
- A detection system to monitor the test substances.

Individual substances interact with the stationary phase to different extents as they are carried through the system, enabling separation to be achieved.

KEY POINT In a chromatographic system, those substances that interact strongly with the stationary phase will be retarded to the greatest extent while those that show little interaction will pass through with minimal delay, leading to differences in distances travelled or elution times.

Types of chromatographic system

Chromatographic systems can be categorised according to the form of the chromatographic bed, the nature of the mobile and stationary phases and the method of separation.

Thin-layer chromatography (TLC) and paper chromatography

Here, you apply the sample as a single spot near one end of the sheet or plate, by microsyringe or microcapillary. This sheet is allowed to dry fully, then it is transferred to a glass tank containing a shallow layer of solvent (Fig. 54.1). Remove the sheet when the solvent front has travelled across 80–90 per cent of its length.

You can express movement of an individual substance in terms of its relative frontal mobility, or R_F value, where:

$$R_F = \frac{\text{distance moved by substance}}{\text{distance moved by solvent}} \quad [54.1]$$

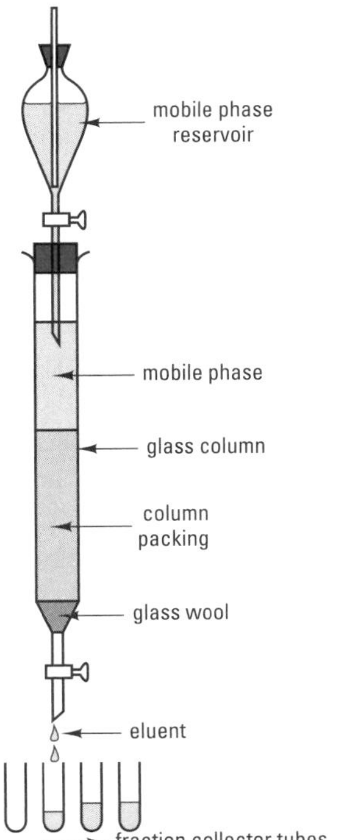

Fig. 54.2 Equipment for column chromatography (gravity-feed system).

Alternatively, you may express movement with respect to a standard of known mobility, as R_X, where:

$$R_X = \frac{\text{distance moved by test substance}}{\text{distance moved by standard}} \qquad [54.2]$$

The R_F (or R_X) value is a constant for a particular substance and solvent system (under standard conditions) and closely reflects the partitioning of the substance between the stationary and mobile phases. Tabulated values are available for a range of biomolecules and solvents (e.g. Stahl, 1969). However, you should analyse one or more reference compounds on the same sheet/plate as your unknown sample, to check their R_F values.

Column chromatography

Here, you pack a glass column with the appropriate stationary phase and equilibrate the mobile phase by passage through the column, either by gravity (Fig. 54.2), or using a low-pressure peristaltic pump. You can then introduce the sample to the top of the column, to form a discrete band of material. This is then flushed through the column by the mobile phase. If the individual substances have different rates of migration, they will separate within the column, eluting at different times as the mobile phase travels through the column.

You can detect eluted substances by collecting the mobile phase as it elutes from the column in a series of tubes (discontinuous monitoring), either manually or with an automatic fraction collector. Fractions of 2–5 per cent of the column bed volume are usually collected and analysed, e.g. by chemical assay. You can now construct an elution profile (or chromatogram) by plotting the amount of substance against either time, elution volume or fraction number, which should give a symmetrical peak for each substance (Fig. 54.3).

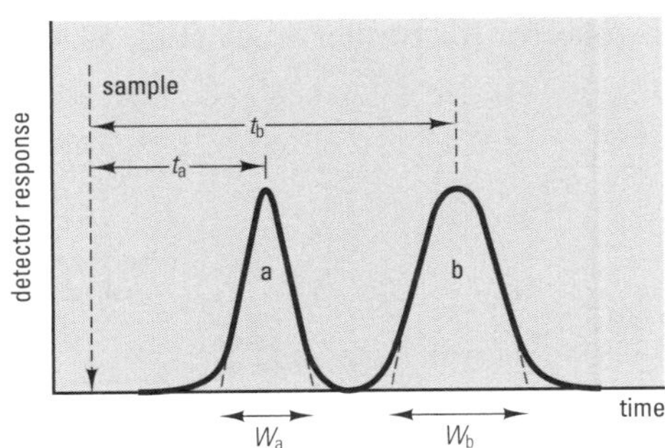

Fig. 54.3 Peak characteristics in a chromatographic separation, i.e. a chromatogram.

You can express the migration of a particular substance at a given flow rate in terms of its retention time (t), or elution volume (V_e). The separation efficiency of a column is measured by its ability to distinguish between two similar substances, assessed in terms of:

- selectivity (α), measured using the following equation, which takes into account the retention times of the two peaks (i.e. t_a and t_b), plus the column dead time (t_0):

$$\alpha = \frac{t_b - t_0}{t_a - t_0} \qquad [54.3]$$

  The column dead time is the time it takes for an unretained compound to pass through the column without interacting with the stationary phase.

- resolution (R), quantified in terms of the retention time and the base width (W) of each peak:

$$R = \frac{2(t_a - t_b)}{W_a + W_b} \qquad [54.4]$$

The subscripts a and b in eqns [54.3] and [54.4] refer to substances a and b respectively (Fig. 54.3). For most practical purposes, R values of 1 or more are satisfactory, corresponding to 98 per cent peak separation for symmetrical peaks.

Using HPLC – HPLC is a versatile form of chromatography, used with a wide variety of stationary and mobile phases, to separate individual compounds of a particular class of molecules on the basis of size, polarity, solubility or adsorption characteristics.

> **Learning from experience** – if you are unable to separate your test substance(s) using a particular method, do not regard this as a failure, but instead think about what this tells you about either the substance(s) or your sample.

> **Applications of high-performance liquid chromatography** – the speed and sensitivity of HPLC makes this the method of choice for the separation of many small molecules of biological interest, normally using reverse-phase partition chromatography (p. 363).

> **Applications of gas chromatography** – GC is used to separate volatile, non-polar compounds: substances with polar groups must be converted to less polar derivatives prior to analysis, to prevent adsorption on the column, leading to poor resolution and peak tailing.

High-performance liquid chromatography (HPLC)

Column chromatography originally used large 'soft' stationary phases that required low-pressure flow of the mobile phase to avoid compression; separations were usually time-consuming and of low resolution ('low performance'). Subsequently, the production of small, incompressible, homogeneous particulate support materials and high-pressure pumps with reliable, steady flow rates have enabled high-performance systems to be developed. These systems operate at pressures up to 10 MPa, forcing the mobile phase through the column at a high flow rate to give rapid separation with reduced band broadening, due to smaller particle size.

HPLC columns are usually made of stainless steel, and all components, valves, etc., are manufactured from materials that can withstand the high pressures involved. The two main solvent-delivery systems are:

- Isocratic separation: a single solvent (or solvent mixture) is used throughout the analysis.
- Gradient elution separation: the composition of the mobile phase is altered using a microprocessor-controlled gradient programmer, which mixes appropriate amounts of two different substances to produce the required gradient.

Most HPLC systems are linked to a continuous monitoring detector of high sensitivity, e.g. proteins may be detected spectrophotometrically by monitoring the absorbance of the eluent at 280 nm as it passes through a flow cell (cuvette). Other detectors can be used to measure changes in fluorescence, refractive index, ionisation, radioactivity, etc. The detector is linked to a recorder or microcomputer to produce an elution profile. Most detection systems are non-destructive, which means that you can collect eluent with an automatic fraction collector for further study (Fig. 54.4).

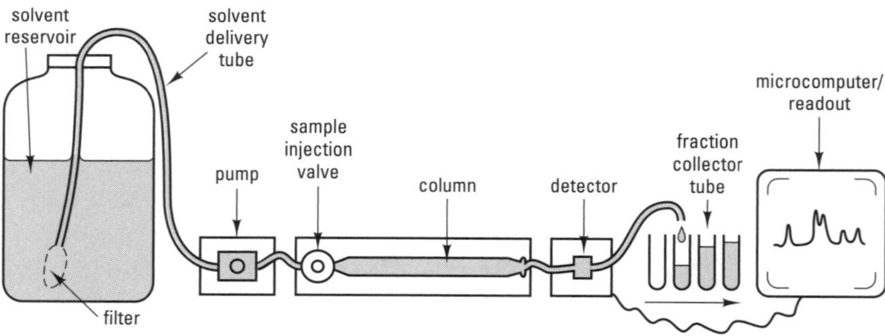

Fig. 54.4 Components of an HPLC system.

Separation of biological macromolecules (especially proteins and nucleic acids) usually requires 'biocompatible' systems in which stainless steel components are replaced by titanium, glass or fluoroplastics, using lower pressures to avoid denaturation, e.g. the GE-Pharmacia FPLC system. Such separations are carried out using ion-exchange, gel-permeation and/or hydrophobic-interaction chromatography (p. 365).

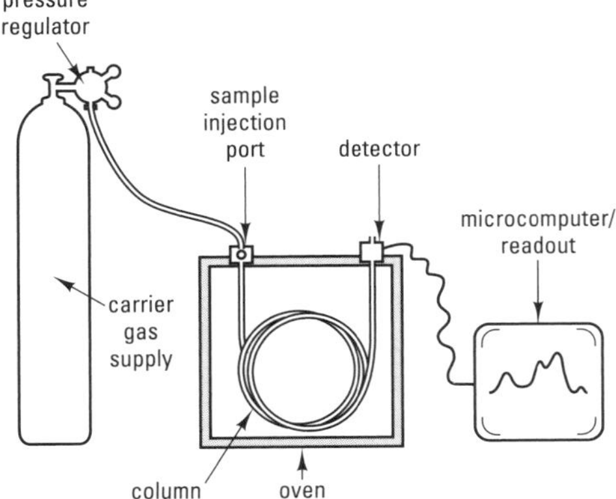

Fig. 54.5 Components of a GC system.

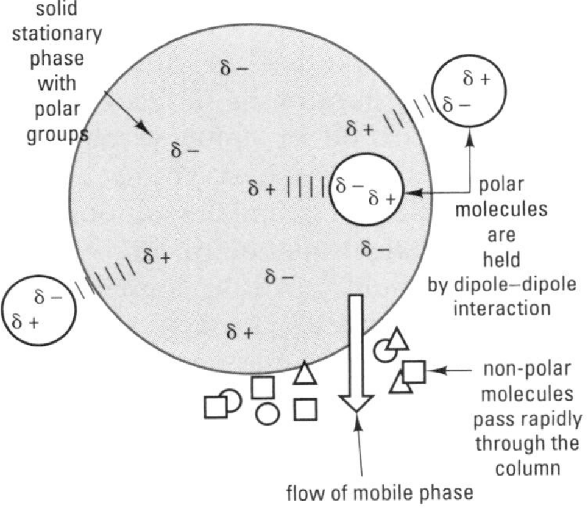

Fig. 54.6 Adsorption chromatography (polar stationary phase).

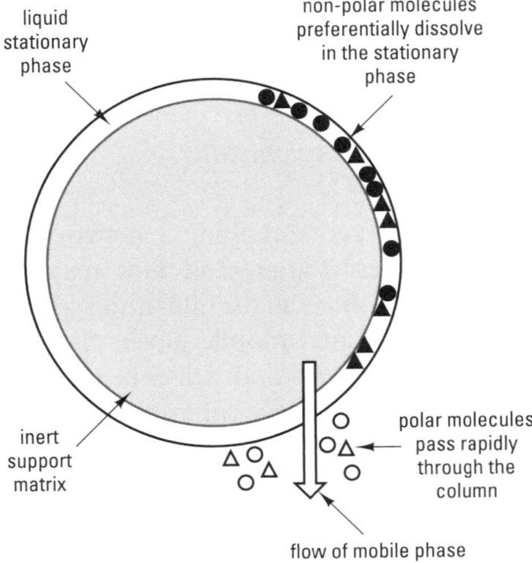

Fig. 54.7 Liquid–liquid partition chromatography, e.g. reverse-phase HPLC.

Gas chromatography (GC)

Modern GC uses capillary chromatography columns (internal diameter 0.1–0.5 mm) up to 50 m in length (Fig. 54.5). The stationary phase is generally a cross-linked silicone polymer, coated as a thin film on the inner wall of the capillary: at normal operating temperatures, this behaves in a similar manner to a liquid film, but is far more robust. The mobile phase ('carrier gas') is usually nitrogen or helium. Selective separation is achieved as a result of the differential partitioning of individual compounds between the carrier gas and silicone polymer phases. The separation of most biomolecules is influenced by the temperature of the column, which may be constant during the analysis ('isothermal' – usually 50–250°C) or, more commonly, may increase in a pre-programmed manner (e.g. from 50°C to 250°C at 10°C per min). Samples are injected on to the 'top' of the column, through an injection port containing a gas-tight septum. The output from the column can be monitored by:

- Flame ionisation: the outflow gas is passed through a flame where any organic compounds will be ionised and subsequently detected by an electrode mounted near the flame tip.
- Electron capture: using a beta-emitting radioisotope as the means of ionisation. This is capable of detecting extremely small amounts (pmol) of electrophilic compounds.
- Spectrometry: including mass spectrometry (GC-MS) and infrared spectrometry (GC-IR).
- Thermal conductivity: changes in the composition of the gas at the outflow alter the resistance of a platinum wire.

GC can only be used directly with samples capable of volatilisation at the operating temperature of the column, e.g. short-chain fatty acids. Other substances need to be chemically modified to produce more volatile compounds, e.g. long-chain fatty acids are usually analysed as methyl esters while monosaccharides are converted to trimethylsilyl derivatives.

Separation methods

Adsorption chromatography

This is a form of solid–liquid chromatography. The stationary phase is a porous, finely divided solid that adsorbs molecules of the test substance on its surface due to dipole–dipole interactions, hydrogen bonding and/or van der Waals interactions (Fig. 54.6). The range of adsorbents is limited, e.g. polystyrene-based resins (for non-polar molecules), silica, aluminium oxide and calcium phosphate (for polar molecules). Most adsorbents must be activated by heating to 110–120°C before use, since their adsorptive capacity is decreased by bound water. Adsorption chromatography can be carried out in column or thin-layer form, using a range of organic solvents.

Partition chromatography

This is based on the partitioning of a substance between two liquid phases, in this instance the stationary and mobile phases. Substances that are more soluble in the mobile phase will pass rapidly through the system, whereas those that favour the stationary phase will be retarded (Fig. 54.7).

Selecting a separation method – it is often best to select a technique that involves direct interaction between the substance(s) and the stationary phase, (e.g. ion-exchange, affinity or hydrophobic interaction chromatography), due to the increased capacity and resolution compared with other methods, (e.g. partition or gel-permeation chromatography).

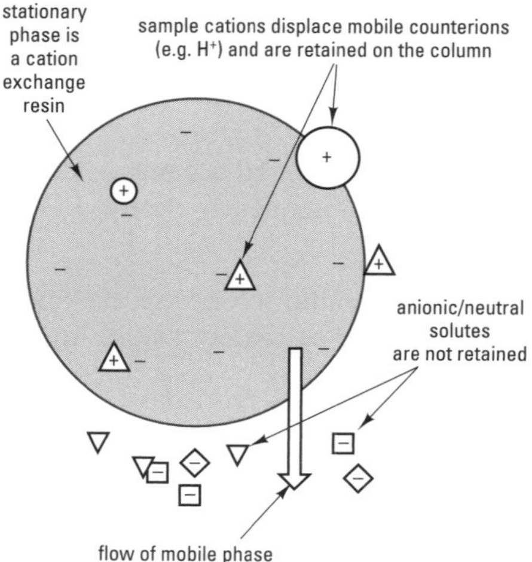

Fig. 54.8 Ion-exchange chromatography (cation exchanger).

Using a gel-permeation system – keep your sample volume as small as possible, as low resolution means that the sample will be diluted during passage through the column.

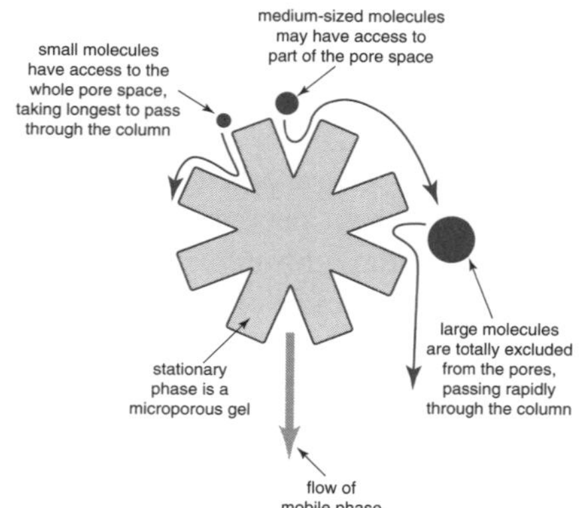

Fig. 54.9 Gel-permeation chromatography.

In normal phase partition chromatography the stationary phase is a polar solvent, usually water, supported by a solid matrix (e.g. cellulose fibres in paper chromatography) and the mobile phase is an immiscible, non-polar organic solvent. For reverse-phase partition chromatography the stationary phase is a non-polar solvent (e.g. a C_{18} hydrocarbon, such as octadecylsilane) that is chemically bonded to a porous support matrix (e.g. silica), while the mobile phase can be chosen from a wide range of polar solvents, usually water or an aqueous buffered solution containing one or more organic solvents, e.g. acetonitrile. Solutes interact with the stationary phase through non-polar interactions and so the *least* polar solutes elute last from the column. Solute retention and separation are controlled by changing the composition of the mobile phase (e.g. % v/v acetonitrile). Reverse-phase high-performance liquid chromatography (RP–HPLC) can separate a range of non-polar, polar and ionic biomolecules, including peptides, proteins, oligosaccharides and vitamins.

Ion-exchange chromatography

Here, separations are carried out using a column packed with a porous matrix that has a large number of ionised groups on its surfaces, i.e. an ion-exchange resin. The groups may be cation or anion exchangers, depending upon their affinity for positive or negative ions. The net charge of a resin depends on the pK_a of the ionisable groups and the solution pH, in accordance with the Henderson–Hasselbalch equation (p. 177).

For most practical applications, you should select the ion-exchange resin and buffer pH so that the test substances are strongly bound by electrostatic attraction to the ion-exchange resin on passage through the system, while the other components of the sample are rapidly eluted (Fig. 54.8). You can then elute the bound ions by changing the pH (which will alter the affinity of the test substance for the resin) or by raising the salt concentration of the mobile phase (to displace the bound ions); for instance, if you pass a solution of increasing concentration through the system, the weakly bound ions will elute first while the strongly bound ions will be elute at a higher concentration.

Ion-exchange chromatography can be used to separate mixtures of a wide range of ions, including amino acids, peptides, proteins and nucleotides. Electrophoresis (Chapter 55) is an alternative means of separating charged molecules, as described on p. 373 for DNA.

Gel-permeation chromatography (molecular-exclusion chromatography)

Here, the stationary phase is a cross-linked gel containing a network of minute pores and channels of various sizes. Large molecules may be completely excluded from these pores, passing through the interstitial spaces and eluting rapidly from the column in the liquid mobile phase. Smaller molecules will penetrate the gel matrix to an extent that will depend upon their size and shape, retarding their progress through the column (Fig. 54.9). Substances will therefore be eluted from a gel-permeation column in decreasing order of their molecular size.

Cross-linked dextrans (e.g. Sephadex), agarose (e.g. Sepharose) and polyacrylamide (e.g. Bio-Gel) are used to separate mixtures of macromolecules, particularly enzymes, antibodies and other globular proteins. Calibration of a gel-filtration column using molecules of similar

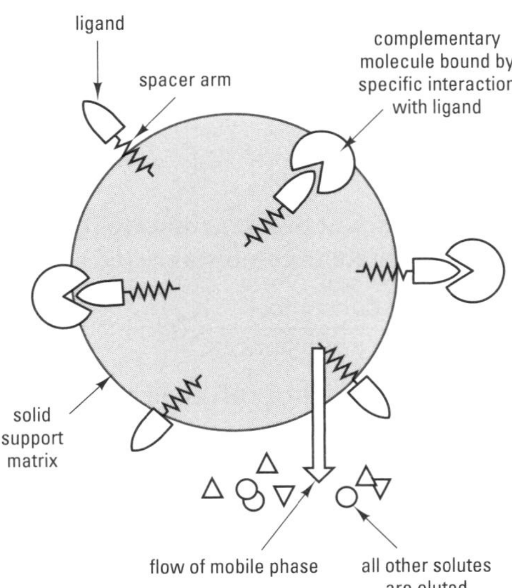

Fig. 54.10 Affinity chromatography.

Elution of substances from an affinity system – make sure that your elution conditions do not affect the interaction between the ligand and the stationary phase, or you may elute the ligand from the column.

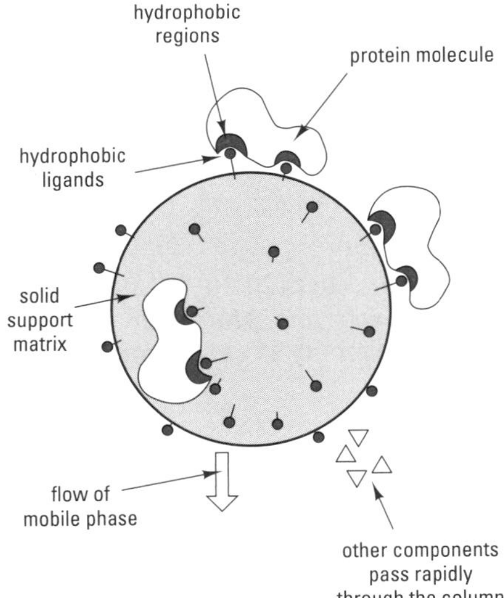

Fig. 54.11 Hydrophobic interaction chromatography. Hydrophobic interactions between li-gands and hydrophobic amino acid residues release 'structured' water, making the inter-actions thermodynamically favourable.

shape and known molecular mass enables the molecular mass of other components to be estimated, since a plot of elution volume (V_e) against $\log_{10}$ molecular mass is approximately linear. A further application of gel-permeation chromatography is for separating low-molecular-mass and high-molecular-mass components, e.g. desalting a protein extract using a Sephadex G-25 column is faster and more efficient than dialysis.

Affinity chromatography

This is a highly specific form of adsorption chromatography, where a particular binding molecule (or ligand) is covalently attached to a solid matrix in such a way that the affinity of the ligand for its complementary molecule is unchanged. The immobilised ligand is then packed into a chromatography column where it will selectively adsorb the complementary molecule, preventing its passage through the system (Fig. 54.10). Once the sample has been applied to the column and the contaminating substances washed through with buffer, the purified complementary molecule can be eluted, e.g. by changing the pH or salt concentration, or by adding other substances with a greater affinity for the ligand. Providing a suitable ligand is available, affinity chromatography can be used for single-step purification of small quantities of a particular molecule in the presence of large amounts of contaminating substances. Examples of ligands include:

- triazine dyes, for protein purification;
- enzyme substrates and co-factors, for certain enzymes;
- antibodies, for specific antigens (see p. 347);
- single-stranded oligonucleotides, for complementary nucleic acids, e.g. mRNA, or particular single-stranded DNA sequences;
- lectins, for specific monosaccharide subunits.

Hydrophobic-interaction chromatography

This technique is used to separate proteins. The stationary phase consists of a non-polar ligand group (e.g. an octyl or phenyl group) bound to a support matrix (Fig. 54.11). This binds proteins differentially, according to the number and position of hydrophobic surface groups. The principle is similar to reverse-phase HPLC (Fig. 54.7), but samples are analysed under non-denaturing conditions, to retain biological activity.

Quantitative analysis

Most detectors and chemical assay systems give a linear response with increasing amounts of the test substance over a given 'working range' of concentration. Alternative ways of converting the measured response to an amount of substance are:

- **External standardisation:** this is applicable where the sample volume is sufficiently precise to give reproducible results (e.g. HPLC, column chromatography). Peak areas (or heights) of known amounts of the substance give a calibration factor or calibration curve, used to calculate the amount of test substance in the sample (Chapter 50).

Problems with peaks – non-symmetrical peaks may result from column overloading, co-elution of solutes, poor packing of the stationary phase, or interactions between the substance and the support material.

Using external standardisation – samples and standards should be analysed more than once, to confirm the reproducibility of the technique.

Using an internal standard – you should add an internal standard to the sample at the first stage in the extraction procedure, so that any loss or degradation of test substance during purification is accompanied by an equivalent change in the internal standard.

- **Internal standardisation:** where you add a known amount of a reference substance (not originally present in the sample) to the sample, to give an additional peak in the elution profile. You determine the response of the detector to the test and reference substances by analysing a standard containing known amounts of both substances, to provide a response factor (r), where:

$$r = \frac{\text{peak area (or height) of test substance}}{\text{peak area (or height) of reference substance}} \quad [54.5]$$

Use this response factor to quantify the amount of test substance (Q_t) in a sample containing a known amount of the reference substance (Q_r), from:

$$Q_t = \frac{\text{peak area (or height) of test substance}}{\text{peak area (or height) of reference substance}} \times (Q_r \div r) \quad [54.6]$$

Internal standardisation should be the method of choice wherever possible, since it is unaffected by small variations in sample volume (e.g. for GC microsyringe injection). The internal standard should be chemically similar to the test substance(s) and must give a peak that is distinct from all others in the sample. An additional advantage of an internal standard that is chemically related to the test substance is that it may show up problems due to changes in detector response, etc. A disadvantage is that it may be difficult to find the space to fit an internal standard peak into a complex chromatogram.

Text reference

Stahl, E. (1969) *Thin Layer Chromatography – A Laboratory Handbook*, 2nd edn. Springer-Verlag, Berlin.

Sources for further study

Cazes, J. (2005) *Encyclopedia of Chromatography*. CRC Press, Boca Raton, Florida.

McMaster, M. (2006) *HPLC: A Practical User's Guide*. Wiley, New York.

Snyder, L.R., Kurkland, J.J. and Dolan, J.W. (2009) *Introduction to Modern Chromatography*. Wiley, New York.

Wilson, K. and Walker, J. (eds) (2010) *Principles and Techniques of Biochemistry and Molecular Biology*, 7th edn. Cambridge University Press, Cambridge.

55 Electrophoresis

Definitions

Anode – positive electrode (negatively charged ions which move to this electrode are termed anions).

Cathode – negative electrode (positively charged ions which move to this electrode are termed cations).

Electrophoresis is used to separate individual charged molecules (ionic substances) in a sample on the basis of their differential movement in an electric field. The essential components of an electrophoretic system (see Fig. 55.1) are:

- **two electrodes**, between which a voltage can be applied using a microprocessor-controlled 'power pack' that provides control over voltage (V) and current (A) at constant power (W) for a specified time;
- **a buffer (salt) solution**, through which the current flows and in which the ionic substances move at a controlled constant pH;
- **a supporting medium**, which limits diffusion and thermal convection, leading to enhanced and consistent separation of the charged molecules;
- **a visualisation/detection system**, to view the results following electrophoresis.

KEY POINT Biomolecules move at different rates within an electrophoretic system, leading to their separation on the basis of their size, shape and/or charge, depending on the conditions used.

Different buffer systems are used in electrophoresis, according to the pH requirements of the procedure. Buffers are usually prepared as standard mixtures, including a chelating agent such as ethylene diamine tetra-acetic acid (EDTA, and are often referred to by their acronyms, for example TBE (for Tris/boric acid/ EDTA), or TAE (for Tris/acetic acid/EDTA).

Types of electrophoretic system

These can be characterised according to the supporting medium used. The main types that you are likely to come across are detailed below.

Cellulose acetate

Here, the support medium has a large and uniform pore structure that enables unrestricted movement of large biomolecules. Strips of cellulose acetate are often used for the rapid separation of plasma proteins in biochemistry practicals, allowing the procedure to take less than an hour.

Handling cellulose acetate electrophoresis strips – these are fragile and must be handled carefully – avoid touching the flat surfaces with your fingers.

Agarose

This is a purified polymer obtained from seaweed. It is prepared in gel format by mixing powder with electrophoresis buffer at concentrations of 0.5–3.0 per cent, boiled until the mixture becomes clear, then poured onto a glass plate and allowed to cool until it forms a gel. The size of pores within an agarose gel depends on the concentration of agarose used: high concentrations of agarose produce gels with small pores, which means that large biomolecules are slowed in their movement through such gels, leading to 'molecular sieving'. This is the basis of the electrophoretic separation of nucleic acids (p. 425).

Preparing agarose gels – agarose 'sets' at around 36–38°C, so gels must be allowed to cool fully after pouring, or they will be distorted/damaged.

SAFETY NOTE Preparing gels for PAGE–acrylamide and bisacrylamide are extremely potent neurotoxins, so you must wear plastic gloves when handling solutions containing these reagents, and when handling polyacrylamide.

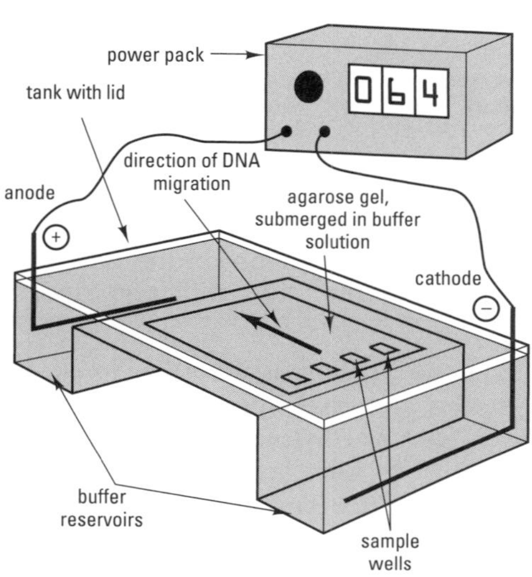

Fig. 55.1 Components of an electrophoresis system.

Polyacrylamide

Polyacrylamide gel electrophoresis (PAGE) has a major role in the analysis of proteins. The gel is formed by polymerising acrylamide monomers into long cross-linked chains using bisacrylamide under standard conditions to give a gel with a pore size that can be 'tailored' to suit the size of biomolecule to be separated.

PAGE is especially useful for protein separation, where it is used in two main ways:

- **Non-dissociating conditions**, where proteins will retain their native three-dimensional configuration.
- **Dissociating conditions**, where the addition of the detergent sodium dodecyl sulphate (SDS) to the buffer causes proteins to denature, producing linear polypeptides with a uniform negative charge per unit length of polypeptide. Since SDS-dissociated polypeptides will have identical charge densities, when they are subjected to PAGE they will then migrate according to their size. This not only give effective separation, but additionaly enables the molecular mass of each polypeptide to be determined, by comparing its mobility to polypeptide standards of known mass, run under the same conditions (Fig. 55.1)

Applications of electrophoresis

Agarose gel electrophoresis of nucleic acids

DNA and RNA carry a net negative charge, due to the phosphate groups in the polymer. As a result, they move towards the anode in electrophoretic systems. Typically, the support medium is agarose, used at low concentration (e.g. 0.3 per cent) for large nucleic acid fragments and at higher concentration (e.g. 0.8 per cent) for smaller fragments, while very small fragments are best separated using a polyacrylamide gel. Box 55.1 gives advice on running agarose gel electrophoresis of nucleic acids.

Polyacrylamide gel electrophoresis of proteins

PAGE plays a major role in protein anaylsis, both for one-dimensional and two-dimensional separations. In addition to the choice of using SDS (dissociating conditions) or not, there is a number of additional choices available for PAGE, including:

- **Whether to use rod gels or slab gels.** The former are useful for individual large samples, e.g. where separation of a protein is required in large amounts, while the latter can be used for multiple samples, e.g. 20 or more individual samples can be separated under identical conditions.
- **Whether to use a continuous or discontinuous buffer system.** Discontinuous systems have different buffers in the gel compared with the reservoirs – they are more time-consuming to prepare but they give better separation and resolution with dilute samples of large volume.
- **What pH to use.** The pH is not critical for SDS-PAGE, since SDS-treated polypeptides are negatively charged over a wide pH range.

Box 55.1 How to carry out agarose gel electrophoresis of nucleic acids

1. **Prepare the gel**. Typically, a small volume (10–20 mL) of buffer plus agarose is heated gently until the powder dissolves – take care not to overheat, or it will boil over. Nowadays, gels are often cast with a small amount of a non-toxic visualising dye, such as SYBR®Safe.
2. **Prepare the samples**. A small amount of sucrose or glycerol is usually added, to increase the density of the sample. A water-soluble anionic 'tracking' dye (e.g. bromophenol blue or xylene cyanol) is also added to each sample, so that migration can be followed visually.
3. **Load the samples onto the gel.** Individual samples are added to the pre-formed wells using a pipettor (the sample should be retained within the well due to its higher density, compared with the buffer solution). The volume of sample added to each well is small – typically less than 25 μL, so a very steady hand and careful dispensing are needed to pipette each sample accurately.
4. **Load the nucleic acid markers onto the gel.** Typically, these standards of know size are added to the first and last wells of the gel: after electrophoresis, the relative positions of the bands of known size can be used to prepare a calibration curve (p. 336), usually by plotting $\log_{10}$ size (length) against distance travelled).
5. **Carry out (run) the electrophoresis.** Nucleic acid separation is usually carried out at 100–150 V for 30–60 min (see manufacturer's instructions for specific details, according to which 'power pack' you are using): the gel should be run until the 'tracking' dye has migrated across 80 per cent of the gel.
6. **Examine the result.** If you have used a visualising dye within the agarose, then you simply transfer the gel to a UV transilluminator and look for 'bands' of fluorescence corresponding to each nucleic acid fragment. **Safety note – always wear suitable UV-filtering safety glasses when working with UV radiation, to protect your eyes.** Use a digital camera to photograph your gel. Alternatively, a dedicated image capture system can be used, e.g. GelDoc®.
7. **Extract any nucleic acid bands of interest.** If a particular band is required for further study, the piece of gel containing that band can be cut from the gel using either a clean scalpel or a specialised gel band cutter.

However, for non-dissociating conditions, the pH can be critical, particularly when the biological activity of proteins is to be retained, e.g. in enzyme separations.

- **What gel concentration to use.** For separation of unknown mixtures, a gradient gel can be used (e.g. 5–20 per cent monomer).

Box 55.2 gives advice on running standard SDS-PAGE in slab gel format for protein separation and size determination. Further practical details for electrophoresis of nucleic acid and proteins are given in Westermeier (2005).

Advanced electrophoretic methods

A number of advanced techniques is available that give very high resolution with very small amounts of sample material. These are more likely to be used in project work, e.g. for isolation and characterisation of individual protein constituents in blood/tissues.

Isoelectric focusing (IEF)

Definition

Isoelectric point – the pH at which a biomolecule (e.g. a protein) carries no net charge.

In contrast with standard electrophoresis, which is carried out at constant pH using a buffer system, IEF is carried out using a pH gradient across the gel. When a sample containing a mixture of proteins is applied to this gradient, the individual proteins will migrate to their isoelectric point on the pH gradient. This enables the proteins to be separated according to their isoelectric points, which is, in turn, a function of the relative proportions of different amino acids present.

Box 55.2 How to carry out SDS-PAGE for protein separation

1. **Prepare the gel.** Nowadays, many laboratories use pre-cast gels, bought from a manufacturer (e.g. Bio-Rad®). If you are preparing your own gel, you will need to follow the protocol very carefully. Typically the correct proportions of acrylamide, bisacrylamide and SDS are mixed together and degassed, under vacuum. Then ammonium persulfate and tetramethylethylenediamine (TEMED) are added to trigger the polymerisation. Once the latter two constituents are added, the gel should be poured immediately into the casting tray (including the well former 'comb').
2. **Prepare the samples.** The protein sample is mixed with a buffer solution containing SDS (to bind to the dissociated proteins), plus dithiothreitol (to cleave disulphide bonds in the proteins) and a 'tracker' dye (e.g. bromophenol blue), then heated to $\geq 60°C$, to disrupt the tertiary structure and 'linearise' the polypeptide chains.
3. **Load the samples onto the gel.** Individual samples are added to the wells using a pipettor (p. 155). The volume of sample added to each well is typically less than 100 μL, so a very steady hand and careful dispensing are needed to accurately pipette each sample (steady the pipettor using your other hand if this helps). To optimise the separation of proteins, the volume added should be kept as small as possible.
4. **Load the molecular mass standards.** Nowadays, many labs use 'rainbow' markers, with a wide range of proteins of known molecular mass, each of which is stained a different colour, to facilitate estimation of the molecular mass of unknown proteins, by visual comparison.
5. **Run the electrophoresis.** The gel is positioned with the well/samples closest to the cathode (negative electrode), since they will move towards the anode during electrophoresis as a result of their negative charge. Protein separation is typically carried out at 80–100 V for one–two hours (see manufacturer's instructions for specific details, according to which 'power pack' you are using): the gel should be run until the 'tracking' dye has migrated across 80 per cent of the gel). Higher voltages give faster separation, but poorer separation of protein 'bands'.
6. **Fix and stain the gel.** One of the most widely used approaches is Coomassie Brilliant Blue staining. Typically, this involves immersing the gel in Coomassie Blue for one hour then destaining overright in a methanol/acetic acid solution. A safer, water-based alternative can be bought from a manufacturer. (e.g. Bio-Rad Bio-Safe Coomassie Stain), which allows the gel to be destained in water overright. After destaining, separated proteins are visible as blue bands against an unstained background. For higher resolution, a silver stain can be used instead of Coomassie Brilliant Blue.
7. **Examine the results.** After destaining, separated proteins are visible as blue bands against an unstained background (Fig. 55.2a). The position of these bands can be compared with the molecular mass standards to determine the size of each band in the test sample. For greater accuracy, the distance moved by the molecular mass standards can be plotted against $\log_{10}$ relative molecular mass and the resulting relationship (Fig. 55.2b) used to determine the size of the unknown bands.

(a)

1 2 3 4 5 6 7 8

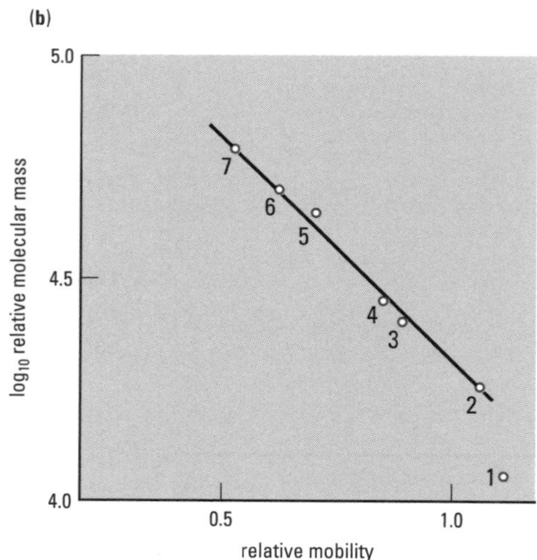

Fig. 55.2 Determination of relative molecular mass (M_r) of proteins by SDS-PAGE: (a) gel samples: 1, cytochrome c; 2, myoglobin; 3, γ-globulin; 4, carbonic anhydrase; 5, ovalbumin; 6, albumin; 7, transferrin; 8, mixture of samples 1–7 (photo courtesy of Pharmacia Biotech). (b) plot of log M_r against distance travelled through gel.

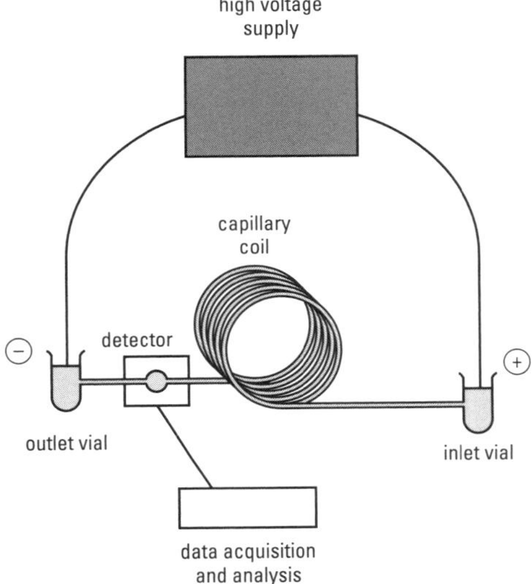

Fig. 55.3 Capillary electrophoresis.

Two-dimensional electrophoresis

The most commonly used version of this technique involves separating a mixture of proteins in one dimension using IEF, followed by separation by molecular mass using denaturing SDS-PAGE. This technique allows up to 1000 proteins to be separated from a single sample. Anaylsis of the complex patterns that result from two-dimensional electrophoresis requires computer-aided gel scanners to acquire and process data from the gel image. These systems can compare, match and quantify protein 'spots' from several gels; slight variation in the patterns found in different runs are accounted for using internal reference 'landmark' proteins, which are either added as standard proteins or are proteins known to be present in all samples.

Capillary electrophoresis (CE)

This technique, which is also known as capillary zone electrophoresis (CZE), combines the high resolving power of electrophoresis with the speed and versatility of high-performance liquid chromatography (p. 362). The technique largely overcomes the poor resolution problems encountered when carrying out electrophoresis without a supporting medium by using a capillary tube with a high surface area to volume ratio. A further advantage is that extremely small sample volumes can be analysed (e.g. 5–10 nL). The versatility of CE is demonstrated by its use in the separation of a range of biomolecules, including amino acids, proteins, nucleic acids, vitamins and drugs.

The components of a typical CE set-up are shown in Fig. 55.2. The capillary (internal diameter 25–50 μm) is made of fused silica and externally coated with a polymer for added mechanical strength–a gap in the polymer provides a 'window' for detection purposes. Samples are injected into the capillary and the components then move through the capillary at different rates, depending on their interaction with the negatively charged surface of the silica capillary, which means that they leave the other end of the capillary as discrete 'peaks', often detected using UV/visible absorption spectrophotometry. Several advanced variants are used for more specialised work; these include micellar electrokinetic capillary chromatography (MECC), capillary gel electrophoresis (CGE) and capillary isoelectric focussing (CIEF).

Text reference

Westermeier, R. (2005) *Electrophoresis in Practice: A Guide to Methods and Applications of DNA and Proteins Separations*, 4th edn. Wiley-VCH, Weinheim.

Sources for further study

American Electrophoresis Society. Homepage. Available: http://www.aesociety.org/
Last accessed: 21/12/10.
[Includes details on electrophoretic techniques including IEF 2D electrophoresis and CE.]

British Society for Proteome Research. Homepage. Available: http://www.bspr.org/
Last accessed 21/12/10.
[Includes links to a wide range of web resources on electrophoresis.]

Guttmann, A. (2009) *Capillary Gel Electrophoresis and Related Microseparation Techniques*. Elsevier, Amsterdam.

Walker, J.M (2009) *The Protein Protocols Handbook*, 3rd edn. Humana Press, New Jersey.

56 Molecular biology techniques

Deoxyribonucleic acid (DNA) is the genetic material of all cellular organisms. The sequential arrangement of nucleotide subunits represents a genetic code for the synthesis of specific cellular proteins or functional RNA molecules. A portion of DNA that encodes the information for a single polypeptide, protein or polyribonucleotide is usually referred to as a gene while the entire genetic information of an organism is known as its genome.

Recent advances in the manipulation of nucleic acids in the laboratory have increased our understanding of the structure and function of genes at the molecular level. Additionally, these techniques can be used to alter the genome of an organism (genetic engineering or gene cloning), e.g. to create a bacterium capable of synthesising a foreign protein such as a recombinant protein, or a genetically engineered food, e.g. a herbicide-resistant crop. In food science these techniques can be used to detect and identify the nucleic acids of particular microbes in food. These molecular techniques are also important in research and development of functional foods by investigating interactions in metabolic processes at the molecular level.

Conforming to regulations – in the UK, the *Genetically Modified Organisms (Contained Use) (Amendment) Regulations (2005)* provide the regulatory framework for all research procedures involving the genetic modification of organisms.

KEY POINT Although genetic manipulation must be carried out under strict containment, in accordance with appropriate legislation, the procedures involved in the isolation, amplification, recombination and cloning of DNA are often used at undergraduate level, to illustrate the general features of the techniques.

Definitions

Plasmids – circular molecules of DNA that are capable of autonomous replication. They can be isolated, manipulated and then reintroduced into bacterial cells.

Cosmids – hybrid plasmid vectors containing the cos (cohesive) sites from phage λ, enabling *in-vitro* packaging into phage capsids. Useful for cloning large segments of DNA, typically up to 50 kb.

Transformation – in bacteria: stable incorporation of external DNA, e.g. a plasmid. In eukaryotes: the conversion of a cell culture of finite life to a continuous (immortal) cell line (also occurs in cancer).

Transfection – in bacteria: uptake of viral nucleic acid. In eukaryotes: uptake of any naked, foreign DNA.

There is an increasing amount of genomic information available for food and nutritional scientists (Chapter 10) and an understanding of the basic techniques used in molecular biology is required in order to fully utilise this knowledge.

Basic principles

Gene cloning involves several steps:

1. **Isolation of the DNA sequence (gene) of interest** from the genome of an organism. This usually involves DNA purification followed by either enzymic digestion, mechanical fragmentation or PCR amplification, to liberate the target DNA sequence.
2. **Creation of an artificial recombinant DNA molecule** (sometimes referred to as rDNA), by inserting the gene into a DNA molecule capable of replicating in a host cell, i.e. a 'cloning vector'. Suitable cloning vectors for bacterial cells include plasmids, bacteriophages (bacterial viruses) and cosmids.
3. **Introduction of the recombinant DNA molecule into a suitable host**, e.g. *Escherichia coli* (transformation when a plasmid is used, or transfection for recombinant viral nucleic acid).
4. **Selection and growth of the transformed (or transfected) cell**, using the techniques of cell culture (p. 474). Since a single transformed host cell

Preparing glassware – all glassware for DNA purification must be siliconised, to prevent adsorption of DNA. All glass and plastic items must be sterilised before use.

Extracting DNA – many laboratories now use small-scale chromatographic columns for routine extraction of nucleic acids. These are available in kit form, e.g. Qiagen miniprep columns.

SAFETY NOTE Working with solutions in molecular biology – note that phenol is toxic and corrosive and chloroform is potentially carcinogenic (p. 160). Take appropriate safety precautions (e.g. wear gloves, use a fume hood where available).

Using ultrapure sterile water in molecular biology – to avoid contamination, prepare all solutions using pre-sterilised (autoclaved) deionised purified water (e.g. MilliQ water).

Maximising recovery of DNA – these large molecules are easily damaged by mechanical forces, e.g. vigorous shaking or stirring during extraction. In addition all glassware must be scrupulously cleaned and gloves must be worn, to prevent DNase contamination of solutions.

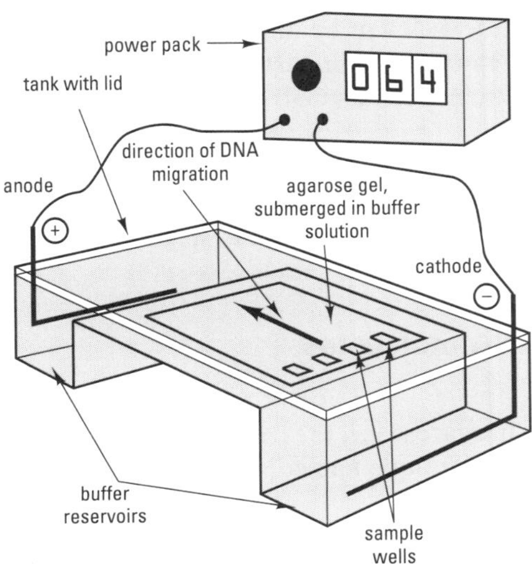

Fig. 56.1 Agarose gel electrophoresis of DNA.

can be grown to give a clone of genetically identical cells, each carrying the gene of interest, the technique is often referred to as 'gene cloning', or molecular cloning.

Extraction and purification of DNA

Specific details of the steps involved in the isolation of DNA vary, depending upon the source material. However, the following sequence shows the principal stages in the purification of plasmid DNA from bacterial cells.

1. **Cell wall digestion:** incubation of bacteria in a lysozyme solution will remove the peptidoglycan cell wall. This is often carried out under isotonic conditions, to stop the cells from bursting open and releasing chromosomal DNA. Note that Gram-negative bacteria are relatively insensitive to lysozyme, requiring additional treatment to allow the enzyme to reach the cell wall layer, e.g. osmotic shock, or incubation with a chelating agent, e.g. ethylenediaminetetra-acetic acid (EDTA). The latter treatment will also inactivate any bacterial deoxyribonucleases (DNases) in the solution, preventing enzymic degradation of plasmid DNA during extraction.
2. **Lysis using strong alkali (NaOH) and a detergent**, e.g. sodium dodecyl sulphate (SDS), to solubilise the cellular membranes and partially denature the proteins. Neutralisation of this solution (e.g. using potassium acetate) causes the chromosomal DNA to aggregate as an insoluble mass, leaving the plasmid DNA in solution.
3. **Removal of other macromolecules**, particularly RNA and proteins by, for example, enzymic digestion using ribonuclease and proteinase. Additional chemical purification steps give further increases in purity, e.g. proteins can be removed by mixing the extract with water-saturated phenol (50% v/v), or a phenol/chloroform mixture. On centrifugation, the DNA remains in the upper aqueous layer, while the proteins partition into the lower organic layer. Repeated cycles of phenol/chloroform extraction can be used to minimise the carry-over of these macromolecules. Additional purification can be obtained using isopycnic density gradient centrifugation in CsCl (p. 385).
4. **Precipitation of DNA** using around 70% v/v ethanol : water (produced by adding two volumes of 95% v/v ethanol to one volume of aqueous extract), followed by centrifugation, to recover the DNA pellet. Further rinsing with 70% v/v ethanol : water will remove any salt contamination from the previous stages. The extracted DNA can then be redissolved in buffer solution and frozen for future use.

Separation of DNA using agarose gel electrophoresis

Electrophoresis is the term used to describe the movement of ions in an applied electrical field, as shown for DNA in Fig. 56.1. DNA molecules are negatively charged, migrating through an agarose gel towards the anode at a rate that is dependent upon molecular size – smaller, compact DNA molecules can pass through the sieve-like agarose matrix more easily than large, extended fragments. Electrophoresis of plasmid DNA is usually carried out using a submerged agarose gel. The amount of agarose is adjusted, depending on the size of the DNA molecules to be separated,

Working with small volumes in molecular biology – use a pipettor of appropriate size (e.g. P2 or P20 Pipetman, p. 155) with a fine tip. For very small volumes, pre-wet the tip before delivering the required volume.

Using DNA fragment-size markers – for accurate determination of fragment size (length) your standards must have the same conformation as the DNA in your sample, i.e. linear DNA standards for linear (restriction) fragments and closed circular standards for plasmid DNA.

SAFETY NOTE When using an electrophoresis tank, ensure that the top cover is in position, to prevent evaporation of buffer solution and to reduce the possibility of electric shock.

e.g. 0.3% w/v agarose is used for large fragments (>20 000 bases) while 0.8% is used for smaller fragments. Note the following:

- Individual samples are added to preformed wells using a pipettor. The volume of sample added to each well is usually less than 25 μL so a steady hand and careful dispensing are needed to pipette each sample.
- The density of the samples is usually increased by adding a small amount of glycerol or sucrose, so that each sample is retained within the appropriate well.
- A water-soluble anionic tracking dye (e.g. bromophenol blue) is also added to each sample, so that migration can be followed visually.
- DNA fragment-size markers ('molecular weight' standards) are added to one or more wells. After electrophoresis, the relative position of bands of known size (length) can be used to prepare a calibration curve (usually, by plotting $\log_{10}$ of the molecular weight of each band against the distance travelled).
- The gel should be run until the tracking dye has migrated across 80 per cent of the gel (see manufacturer's instructions for appropriate voltages/times).
- After electrophoresis, the bands of DNA can be visualised by soaking the gel for around five minutes in ethidium bromide, which binds to DNA by intercalation between the paired nucleotides of the double helix. Ethidium bromide is carcinogenic so always use gloves when handling stained gels and make sure you do not spill any staining solution, or use a safer alternative such as SYBR Safe.
- Under UV light, bands of DNA are visible due to the intense orange–red fluorescence of the ethidium bromide. The limit of detection using this method is around 10 ng DNA per band. The migration of each band from the well can be measured using a ruler. Alternatively, a photograph can be taken, using a digital camera and adaptor or a dedicated system such as GelDoc.
- If a particular band is required for further study (e.g. a plasmid), the piece of gel containing that band is cut from the gel using a scalpel and the DNA extracted, e.g. using a silica membrane 'spin column'.

Identification of specific DNA molecules using Southern blotting

After electrophoretic separation in agarose gel, the fragments of DNA can be immobilised on a filter using a technique named after E.M. Southern:

SAFETY NOTE When working with any source of UV radiation, always wear suitable plastic safety glasses or goggles to protect your eyes.

1. The fragments are first denatured using concentrated alkali, giving single-stranded DNA (ssDNA); this is necessary to allow hybridisation with 'probe' DNA after blotting.

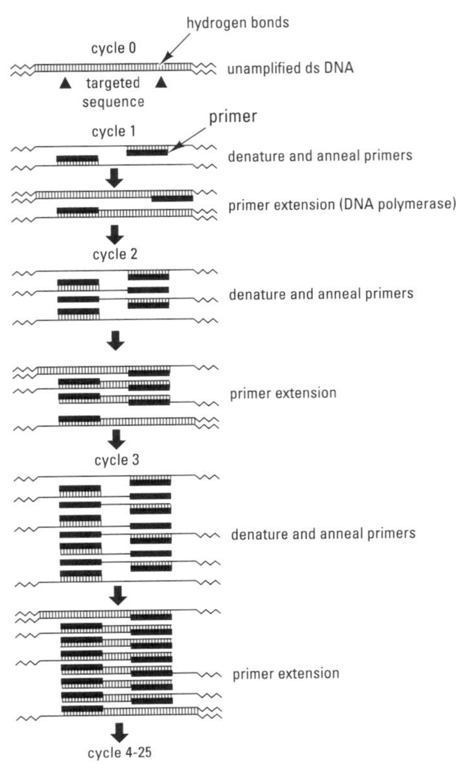

Fig. 56.2 The polymerase chain reaction (PCR).

2. A nitrocellulose filter is then placed directly on to the gel, followed by several layers of absorbent paper. The DNA is 'blotted' on to the filter as the buffer solution is pulled through.
3. Specific sequences can be identified by incubation with labelled complementary probes of single stranded DNA, which will hybridise with a particular sequence, followed by visualisation, often using an enzyme-based system.

DNA amplification using the polymerase chain reaction (PCR)

The polymerase chain reaction is particularly useful for amplifying small amounts of DNA, e.g. from a virus in a clinical specimen, or from a small tissue sample in forensic science. The three basic steps are:

1. **Denaturation of double-stranded DNA** by heating to 94–98°C, to separate the individual DNA strands.
2. **Annealing of oligonucleotide primers** (short, synthetic DNA sequences that will hybridise at a specific position on the target DNA, when the temperature is reduced to 37–65°C.
3. **Primer extension** by a thermostable DNA polymerase (e.g. *Taq* polymerase from *Thermus aquaticus*) at 72°C.

Repeated cycling generates an exponentially increasing number of DNA fragments corresponding to the sequence between the two primers (see Fig. 56.2). Box 56.1 gives further practical details.

> **KEY POINT** PCR is so sensitive that one of the main problems associated with the technique is contamination with 'foreign' DNA. Great care is required to avoid sample contamination during *in vitro* amplification. It is good practice to carry out PCR amplification in a designated laboratory, to minimise the risk of contamination.

Assaying nucleic acids in solution – double-stranded DNA at 50 μg mL^{-1} has an absorbance of 1 at 260 nm, and the same absorbance is obtained for single-stranded DNA at 33 μg mL^{-1} and (single-stranded) RNA at 40 μg mL^{-1}. These values can be used to convert absorbance of a test solution to amount of nucleic acid; e.g. a solution of RNA giving $A_{260} = 0.65$ would contain $40 \times 0.65 = 26$ μg mL^{-1} RNA.

DNA assay

The relative amount of DNA in an agarose gel can be quantified by visual comparison of stained bands with 'marker' bands of known quantity. A more accurate method is to use band densitometry, e.g. using the GelDoc system. A simple approach to quantifying the amount of nucleic acid in an aqueous solution is to measure the absorbance of the solution at 260 nm using a spectrophotometer, as detailed on p. 356. Note that the A_{260} value applies to purified DNA, whereas a plasmid DNA extract prepared using the protocol on p. 372 will contain a substantial amount of contaminating RNA, with similar absorption characteristics to DNA. Any contaminating protein would also invalidate the calculation; the protein can be detected by measuring the absorbance of the solution at 280 nm. Purified DNA has a value for A_{260}/A_{280} of around 1.8 and contaminating protein will give a lower ratio. If your solution gives a ratio substantially lower than this (<1.7), you should repeat the later steps of your extraction procedure (p. 373).

Box 56.1 How to carry out the polymerase chain reaction (PCR)

The protocol below shows the main stages in a typical PCR: note that temperatures, incubation times and the number of cycles will vary with the particular application.

1. **Check you have the required apparatus and reagents to hand**, including (i) thermal cycler; (ii) template DNA (at least 50 ng μL^{-1}); (iii) stock solutions of deoxyribonucleoside triphosphates (dNTPs; at 5 mmol L^{-1} each); (iv) thermostable DNA polymerase (5 U μL^{-1}); (v) primers (e.g. at 30 pmol μL^{-1}); (vi) stock buffer solution, e.g. containing 100 mmol L^{-1} TRIS (pH 8.4), 500 mmol L^{-1} KCl, 15 mmol L^{-1} $MgCl_2$, 1% w/v gelatine, 1% v/v Triton X0100 (the stock buffer is often termed 10x PCR buffer).
2. **Prepare a reaction mixture for your template DNA**: for example, a mixture containing (i) 1.0 μL of template DNA; (ii) 2.5 μL of stock buffer solution; (iii) 1.0 μL of each of the two primers; 1.0 μL of each of the four dNTPs; (iv) 0.1 μL of DNA polymerase; (v) 15.4 μL of sterile ultrapure water, to give a total reaction volume of 25.0 μL.
3. **Prepare appropriate positive and negative controls**: a positive control is a PCR template that is known to work under the reaction conditions used. The most commonly used negative control is the reaction mixture minus the template DNA, though negative controls can be set up lacking any one of the reaction components.
4. **Cycle in the thermal cycler**: for example, an initial period of 5 min at 94°C, followed by 30 cycles of 94°C for one minute (denaturation), then 50°C for one minute (primer annealing), then 72°C for one minute (chain extension).
5. **Assess the effectiveness of the PCR**: for example, by agarose gel electrophoresis (Fig. 56.1) and ethidium bromide (or SYBR Safe) staining.

To avoid contamination in PCR:

- Use a laminar-flow cabinet (p. 453) as a dedicated workstation for PCR use, located in a separate lab from that used to carry out other DNA manipulation. For very sensitive work, use a short-wavelength UV light inside the workstation for 20–30 minutes before starting work, to degrade any contaminating DNA. Make sure that you do not expose your skin or eyes to the short-wavelength UVC source.
- Keep exclusive supplies of pipettors, tips, tubes and reagents for sample preparation, reactions and product analysis.
- Autoclave all buffer solutions, pipette tips and tubes.
- Wear disposable gloves at all times and change them frequently.

Troubleshooting – common problems include:

- Lack of a PCR product (one or more components of the reaction mixture may have been left out; annealing temperature may be too high; denaturation temperature may be too low).
- Too many bands present (one or more components of the reaction mixture may be present in excess; primers may not be specific enough to the target sequence; annealing temperature may be too low).
- Bands corresponding to primer-dimers present (the primers may show partial complementarity – check sequences; primer concentration may be too high).

Enzymic cleavage and ligation of DNA

Type II restriction endonucleases (commonly called restriction enzymes) recognise and cleave a particular sequence of double-stranded DNA (usually, four or six nucleotide pairs), known as the restriction site. Each enzyme is given a code name derived from the name of the bacterium from which it is isolated, e.g. *Hin* dIII was the third restriction enzyme isolated from *Haemophilus influenzae* strain Rd (Fig. 56.3). Most enzymes cleave each strand at a different position, producing short, single-stranded regions known as cohesive ends, or 'sticky ends'. A few enzymes cleave DNA to give blunt-ended fragments. Restriction enzymes that cleave DNA to give 'sticky ends' are widely used in genetic engineering, since two DNA molecules cut with the same restriction enzyme will have complementary single-stranded

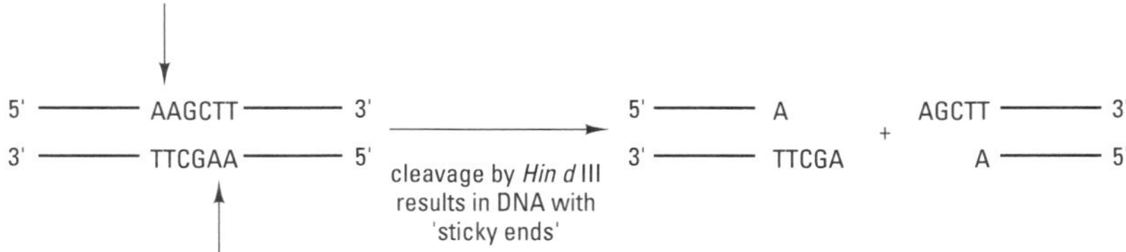

Fig. 56.3 Recognition site for the restriction enzyme *Hin* dIII. This is the conventional representation of double-stranded DNA, showing the individual bases, where A is adenine, C cytosine, G guanine and T thymine. The cleavage site on each strand is shown by an arrow.

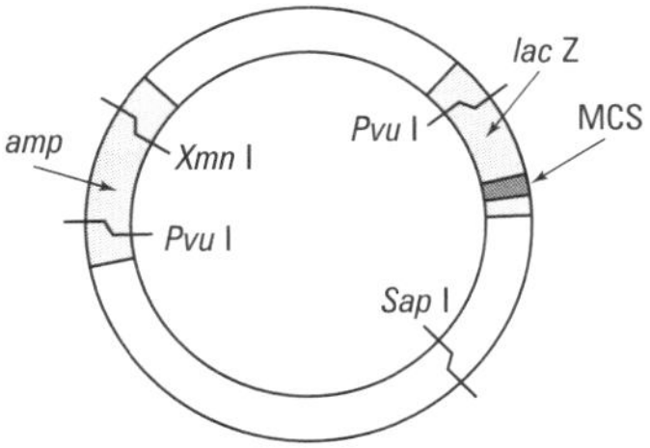

Fig. 56.4 Restriction map of the plasmid pUC19. The position of some individual restriction sites is shown together with the genes for ampicillin resistance (*bla*) and β-galactosidase (*lac Z*). MCS = multiple cloning site for 40 restriction enzymes within the *lac Z* gene.

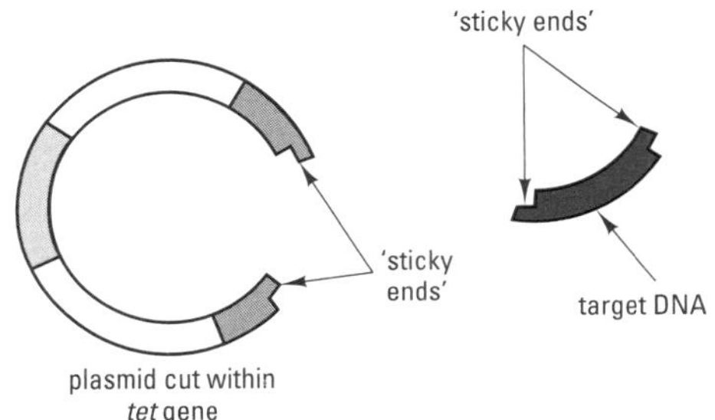

Fig. 56.5 Restriction of plasmid and foreign DNA e.g. with *Hin* dIII.

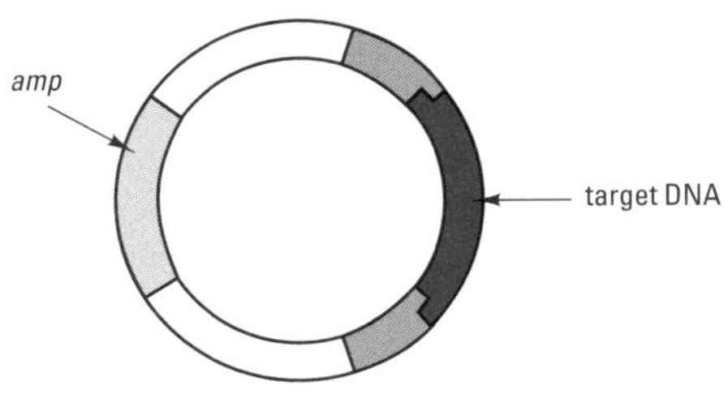

Fig. 56.6 Annealing and ligation of plasmid and target DNA to give a recombinant plasmid that confers resistance to ampicillin. As target DNA has been inserted within the *lac Z* gene, the gene is now discontinuous and inactive.

regions, allowing them to anneal (base pair), due to the formation of hydrogen bonds between individual bases within this region.

Restriction enzymes have two important applications in molecular genetics:

- Mapping. A DNA molecule can be cleaved into several restriction fragments whose number, size and orientation relative to one another can be determined using agarose gel electrophoresis. The position of individual restriction sites can be used to create a restriction enzyme map for a particular molecule, e.g. a plasmid such as pUC19 (Fig. 56.4).
- Genetic engineering. Two restriction fragments cut using the same enzyme (Fig. 56.5) and annealed by complementary base pairing can be permanently joined together using another bacterial enzyme (DNA ligase), which forms covalent bonds between the annealed strands, creating the recombinant molecule (Fig. 56.6). When the two molecules involved are the cloning vector and the target DNA, the size of the recombinant plasmid can be predicted (e.g. a plasmid of 4500 base pairs, plus a target DNA fragment of 2500 base pairs will give a recombinant molecule of 7000 base pairs), allowing separation by electrophoresis. Note that most of the plasmids used in genetic engineering code for two or more easily detectable marker genes (e.g. antibiotic resistance), with single restriction sites on each plasmid (Fig. 56.4).

Transformation/transfection of a suitable host cell

Once a recombinant plasmid has been produced *in vitro*, it must be introduced into a suitable host cell (e.g. *E. coli*). Several procedures can be used:

- Pretreatment with $CaCl_2$ at low temperature: actively growing cells are incubated with hypotonic $CaCl_2$ at 4°C for around 30 minutes, followed by a brief heat shock (e.g. 42°C for two minutes). The low-temperature incubation allows DNA to adhere to the cells, while the heat shock promotes DNA uptake.
- Electroporation: cells or protoplasts are subjected to electric shock treatment (typically, $>1\,\text{kV cm}^{-1}$) for short periods ($<10\,\text{ms}$).
- For animal and plant cells a range of techniques can be used, including electroporation, various microinjection treatments, either using a microsyringe or microprojectiles ('biolistics'), or cationic lipids

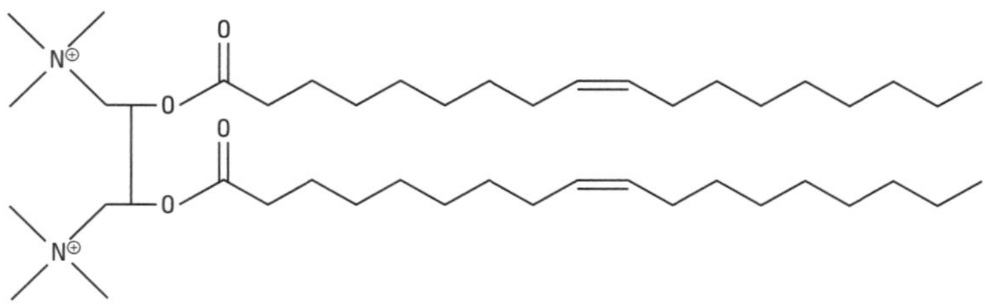

Fig. 56.7 General structure of a synthetic cationic lipid, e.g. Tfx.

(Fig. 56.7), which form lipid–DNA complexes that fuse with cell membranes and allow DNA to enter the cell.

These treatments mostly cause a temporary increase in membrane permeability, leading to the uptake of plasmid DNA from the external medium. Such systems are often inefficient, with fewer than 0.1 per cent of all cells showing stable transformation. However, this is not usually a significant problem, since a single viable transformant can be grown to give a large number of identical cells, using standard microbiological plating techniques (p. 454).

To maximise the transformation efficiency of *E. coli* in the heat shock/$CaCl_2$ procedure, use the minimum volume of solution in a thin-walled glass tube, so that the cells experience a rapid change in temperature.

Selection and detection of transformants

As the efficiency of transformation is often very low, many of the plasmids used in genetic manipulation code for antibiotic resistance and other selectable traits, e.g. pUC19 (Fig. 56.4) carries the gene for the β-lactamase gene (*bla*) and also the gene for the enzyme β-galactosidase (*lac Z*). These genes act as markers in a single-step selection procedure. Expression of *bla* confers resistance to ampicillin, which can be demonstrated by including this antibiotic in the growth medium, thus selecting for transformants (non-transformed cells would be unable to grow on a medium containing ampicillin). The other gene is used as a marker to distinguish transformants that have acquired the recombinant plasmid from those that have taken up the original plasmid, as follows: an intact *lac Z* gene can be detected using a combination of (i) an inducer (e.g. isothiopropylgalactoside, IPTG) plus (ii) a suitable chromogenic substrate, e.g. 5-bromo-4-chloro-3-indolyl-β-D-galactoside (XGAL), which is broken down to release a blue-green indigo derivative. Thus a colony of cells containing the native plasmid would express *lac Z* and would appear blue, while the equivalent recombinant colony would have a discontinuous *lac Z* gene (Fig. 56.5) and would appear off-white, due to insertional inactivation of this gene. The multiple cloning site ('polylinker' region) within the *lac Z* gene enables use of any one of a number of different restriction enzymes, depending on the sites present in the target DNA sequence. Another useful feature of pUC19 and similar plasmids is that they are present in high copy number within a transformed cell, with several thousand identical copies per cell, giving improved yield of plasmid DNA on extraction.

Recognising transformants – after plating bacteria onto medium containing ampicillin, you may notice a few small 'feeder' or 'satellite' colonies surrounding a single larger (transformant) colony. The feeder colonies are derived from non-transformed cells that survive due to the breakdown of antibiotic in the medium around the transformant colony, and they should not be selected for subculture.

Other plasmids make use of different markers; for example the luciferase gene can be detected by bioluminescence in the presence of the substrate luciferin, or the green fluorescent protein (GFP) from the jellyfish *Aequorea victoria*. More specialised vectors are available for specialised functions, such as M13mp19, a single-stranded phage vector for DNA sequencing.

Chapter 10, p. 58 provides an overview of Web-based resources for bioinformatics, while Reed *et al.* (2007) give further practical information on additional procedures in molecular biology (including DNA sequencing, probes and hybridisation, and various forms of PCR including real-time PCR), bioinformatics (databases and tools) and proteomics (studying the expressed proteins in an organism).

Text reference

Reed, R.H., Holmes, D., Weyers, J.D.B. and Jones, A.M. (2007) *Practical Skills in Biomolecular Sciences*, 3rd edn. Prentice Hall, Harlow.

Sources for further study

Brown, T.A. (2006) *Gene Cloning and DNA Analysis*, 5th edn. Blackwell, Oxford.

Hardin, C., Edwards, J., Riell, A. et al. (2001) *Cloning, Gene Expression, and Protein Purification: Experimental Procedures and Process Rationale.* Oxford University Press, Oxford.
[A practical handbook and laboratory manual, with basic experimental protocols together with underlying theoretical principles.]

McPherson, M.J. and Møller, S.G. (2006) *PCR: The Basics*, 2nd edn. Taylor & Francis, London.

Russell, P. (2009) *iGenetics: A Molecular Approach*, 3rd edn. Pearson Education, Harlow.

Twyman, R.H. and Primrose, S.B. (2004) *Principles of Gene Manipulation and Genomics*, 7th edn. Blackwell, Oxford.

57 Homogenisation and centrifugation

Most biomolecules within food samples must be isolated from their source material in order to be studied in any detail. Unless the biomolecule is already a component of an aqueous medium (e.g. plasma, tissue exudate), the first step will be to disrupt the structure of the cells or tissues. Following disruption, the *in vitro* environment of the disrupted tissue will be very different from that of the cell or tissues, and it is important that the integrity of the biomolecule is preserved as far as possible during the isolation procedure.

Definitions

Disruption – a process involving structural damage to cells and tissues, to an extent where the biomolecule or complex of interest is released.

Homogenisation – a process where cells and tissues are broken into fragments small enough to create a uniform, stable emulsion (a homogenate).

Homogeniser – a general term for any equipment used to disrupt or homogenise cells and tissues.

KEY POINT Disruption may be achieved by chemical, physical or mechanical procedures – the rigour of the technique(s) required will depend on the intracellular location of the molecule and the nature of the source material.

Homogenising media for food samples

The solution used for homogenisation serves several purposes, since it acts as a solvent or suspension medium for the released components; it serves as a cooling medium (since many biomolecules are denatured by heat); and it contains various reagents that may help to preserve the biological integrity of components. A typical medium will contain:

- **Buffer.** This replaces the intracellular buffer systems and is needed to prevent pH changes that might denature proteins. TRIS and phosphate buffers are often used for 'physiological' pH values (7.0–8.0); alternatives include various zwitterionic buffers (p. 178).
- **Inorganic salts.** The intracellular ionic strength is quite high, so KCl and NaCl are often included to maintain the ionic strength of the homogenate. However, the total concentration of inorganic salts should be kept below 100 mmol L^{-1}, to avoid thickening of the homogenate due to solubilisation of structural proteins.
- **EDTA.** This chelates divalent cations and removes metal ions (e.g. Cu^{2+}, Pb^{2+}, Hg^{2+}) that inactivate proteins by binding to thiol groups. In addition, it removes Ca^{2+} which could activate certain proteases, nucleases and lipases in the homogenate.
- **Sucrose.** This can be used to prevent osmotic lysis of organelles (e.g. mitochondria, lysosomes), and stabilises proteins from hydrophobic intracellular environments by reducing the polarity of the aqueous medium.
- **Mg^{2+}.** This helps to preserve the integrity of membrane systems by counteracting the fixed negative charges of membrane phospholipids.
- **Protease inhibitors** (e.g. phenylmethanesulphonylfluoride, PMSF; L-*trans*-epoxysuccinylleucylamido-(4-guanodino)-butane, E-64; leupeptin). These protect solubilised proteins from digestion by intracellular proteases, mainly released from lysosomes on disruption of the cell. Lysosomal proteases have acid pH optima, another reason for maintaining the pH of the medium close to neutrality.

Using chelating agents – EDTA (ethylenediaminetetra-acetic acid) and EGTA (ethylenebis(oxethylenenitrilo)-tetraacetic acid). If Mg^{2+} is an important component of the medium, use EGTA, which does not chelate Mg^{2+}.

Extracting proteins from plants – carry out all post-homogenisation procedures as quickly as possible, because plant cell vacuoles may contain phenols that will inactivate proteins when released.

- **Reducing agents** (e.g. 2-mercaptoethanol, dithiothreitol, cysteine at $\approx 1\,\text{mmol}\,\text{L}^{-1}$). These reagents prevent oxidation of certain proteins, particularly those with free thiol groups that may be oxidised to disulphide bonds when released from the cell under aerobic conditions.
- **Detergents** (e.g. Triton X-100, SDS). These cause dissociation of proteins and lipoproteins from the cell membrane, aiding the release of membrane-bound and intracellular components.

Definitions

Isotonic – a medium with the same water potential as the cells or tissues.

Hypertonic – a medium with a more negative water potential, compared with the cells or medium.

Hypotonic – a medium with a less negative water potential, compared with the cells or medium.

Methods of disrupting tissues and food samples

Prior to disruption, animal tissues will need to be freed of any visible fat deposits and connective components, and then randomly sliced with fine scissors or a scalpel. Any fibrous and vascular tissue should be removed from plant material. Disruption can be achieved by mechanical and non-mechanical means: the principal applications of various methods are described in Table 57.1 for the major cell and tissue types.

Non-mechanical methods

- **Osmotic shock:** cells are first placed in a hypertonic solution of high osmolality (p. 163), e.g. 20% w/v sucrose, leading to loss of water. On dilution of this solution (e.g. by addition of water or transfer to a hypotonic solution), the cells will burst due to water influx. This is only effective for wall-less cells.

Table 57.1 Summary of techniques for the disruption of tissues and cells – note that safety glasses should be worn for all procedures

Technique	Suitability	Comments
Non-mechanical methods		
Osmotic shock	Animal soft tissues Some plant cells	Small scale only
Freeze/thaw	Animal soft tissues Some bacteria	Time consuming; small scale; closed system – suitable for pathogens with appropriate safety measures; some enzymes are cold-labile
Lytic enzymes, e.g. lipases; proteases pectinase; cellulase	Animal cells Plant cells	Mild and selective; small scale; expensive; enzymes must be removed once lysis is complete
Lysozyme	Some bacteria	Gram-negative bacteria must be pretreated with EDTA. Suitable for some organisms resistant to mechanical disruption.
Mechanical methods		
Pestle and mortar + abrasives	Tough tissues	Not suitable for delicate tissues
Ball mills + glass beads	Bacteria and fungi	May cause organelle damage in eukaryotes
Blenders and rotor-stators	Plant and animal tissues	Ineffective for microbes
Homogenisers (glass and Teflon)	Soft, delicate tissues e.g. white blood cells, liver	Glass may shatter – wear safety glasses during use
Solid extrusion (Hughes press)	Tough plant material; bacteria; yeasts	Small scale
Liquid extrusion (French pressure cell)	Microbial cells	Small scale
Ultrasonication	Microbial cells	Cooling required; small scale; may cause damage to organelles, especially in eukaryotic cells

- **Freezing and thawing:** causing leakage of intracellular material, following cell wall and membrane damage and internal disruption due to ice crystal formation.
- **Lytic enzymes:** damaging the cell wall and/or plasma membrane. Cells can then be disrupted by osmotic shock or gentle mechanical treatment.

Avoiding protein denaturation during homogenisation – excessive frothing of the homogenate indicates denaturation of proteins (think of whipping egg whites for meringue).

Mechanical methods

All mechanical procedures for cell disruption generate heat, and this may denature proteins. Therefore it is very important to cool the starting material, the homogenising medium, and, if possible, the homogeniser itself (to ≈4°C). The homogenisation should be carried out in short bursts, and the homogenate should be cooled in an ice bath between each burst. Cooling will also reduce the activity of any degradative enzymes in the homogenate. Ideally, carry out homogenisation in a walk-in cooler (a 'cold room') typically at 4–10°C.

Equipment commonly used includes:

- **Mixers and blenders.** These are similar to domestic liquidisers, with a static vessel and rotating blades. The Waring blender is widely used in food service: it has a stainless steel vessel that will stay cool if pre-chilled. The vessel and blades are designed to maximise turbulence, both disrupting and homogenising tissues and cells.
- **Ball mills** (e.g. Retch mixer mill, Mickle mill). These devices contain glass beads that vibrate and collide with each other and with tissues/cells, leading to disruption.
- **Liquid extrusion devices** (e.g. French pressure cell). Cells are forced from a vessel to the outside, through a very narrow orifice at high pressures (≈100 MPa). The resulting pressure changes are a powerful means of disrupting cells.
- **Solid extrusion** (e.g. Hughes press). Here, a frozen cell paste is forced through a narrow orifice, where the shear forces and the abrasive properties of the ice crystals cause cell disruption.
- **Rotor-stators** (e.g. Ultra-turrax homogeniser). These have a rotor (a set of stainless steel blades) and a stator (a slotted stainless steel cylinder) at the tip of a stainless steel shaft, immersed in the homogenising medium: the arrangement is illustrated in Figure 57.1. The high speed of the rotor blades causes material in the homogenising fluid to be sucked into the dispersing head, where it is pressed radially through the slots in the stator. Along with the cutting action of the rotor blades, the material is subjected to very high shear and thrust and the resulting turbulence in the gap between rotor and stator gives effective mixing. The vigour of the homogenisation process can be altered by varying the rotor speed setting. Various sizes of rotor-stator are available, with typical diameters in the range 8–65 mm: the smaller sizes are particularly useful for small-scale preparations.

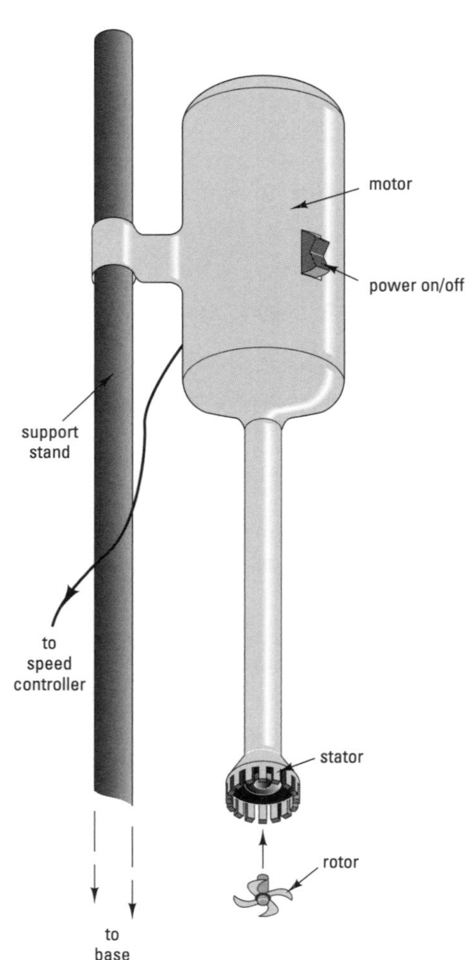

Fig. 57.1 Components of a rotor-stator homogeniser.

- **Sonicators.** Ultrasonic waves are transmitted to an aqueous suspension of cells via a metal probe. The ultrasound creates bubbles within the liquid and these produce shock waves when they collapse. Successful disruption depends on the correct choice of power and incubation time, together with pH, temperature and ionic strength of the suspension medium, often obtained by trial and error. You can reduce the effects of heating during ultrasonication by using short 'bursts' of power (10–30 s), with rests of 30–60 s in between, and by keeping your cell suspension on ice during disruption. An ultrasonic water bath provides a more gentle means of disrupting certain types of cells, e.g. some bacterial and animal cells.
- **Homogenisers.** These involve the reciprocating movement of a ground glass or Teflon pestle within a glass tube (Fig. 57.2). Cells are forced against the walls of the tube, releasing their contents. For glass pestles, the tubes also have ground glass homogenising surfaces and may have an overflow chamber. The homogeniser can either be hand operated (e.g. Dounce), or motorised (e.g. Potter-Elvejham). The clearance between the pestle and the tube (range 0.05–0.5 mm) must be chosen to suit the particular application.

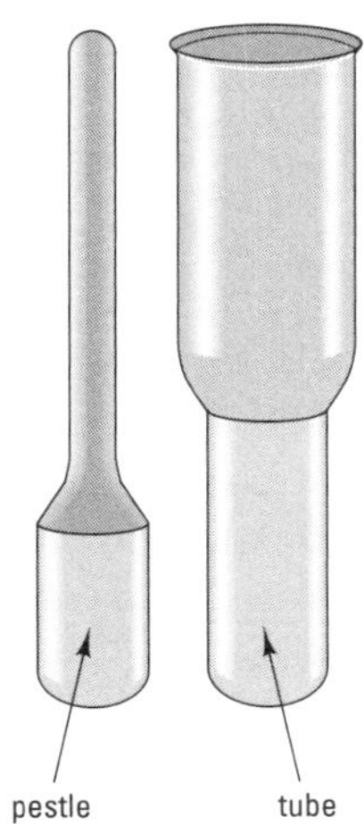

Fig. 57.2 Ground-glass homogeniser.

Centrifugation

Particles suspended in a liquid will move at a rate that depends on:

- **the applied force** – particles in a liquid within a gravitational field, e.g. a stationary test tube, will move in response to the Earth's gravity;
- **the density difference between the particles and the liquid** – particles less dense than the liquid will float upwards while particles denser than the liquid will sink;
- **the size and shape of the particles;**
- **the viscosity of the medium.**

For most particles in a homogenised food sample the rate of flotation or sedimentation in response to the Earth's gravity is too slow to be of practical use in separation.

KEY POINT A centrifuge is an instrument designed to produce a centrifugal force far greater than the Earth's gravity, by spinning the sample about a central axis (Fig. 57.3). Particles of different size, shape or density will thereby sediment at different rates, depending on the speed of rotation and their distance from the central axis.

Working in SI units – to convert RCF to acceleration in SI units, multiply by $9.80\,m\,s^{-2}$: e.g. an RCF of 290 is equivalent to an acceleration of $290 \times 9.80 = 2842\,m\,s^{-1}$.

How to calculate centrifugal acceleration

The acceleration of a centrifuge is usually expressed as a multiple of the acceleration due to gravity ($g = 9.80\,m\,s^{-2}$), termed the relative centrifugal field (RCF, or 'g value'). The RCF depends on the speed of the rotor (n, in

Table 57.2 Relationship between speed (r.p.m.) and acceleration (relative centrifugal field, RCF) for a typical bench centrifuge with an average radius of rotation, r_{av} = 115 mm.

r.p.m.	RCF*
500	30
1000	130
1500	290
2000	510
2500	800
3000	1160
3500	1570
4000	2060
4500	2600
5000	3210
5500	3890
6000	4630

*RCF values rounded to nearest 10.

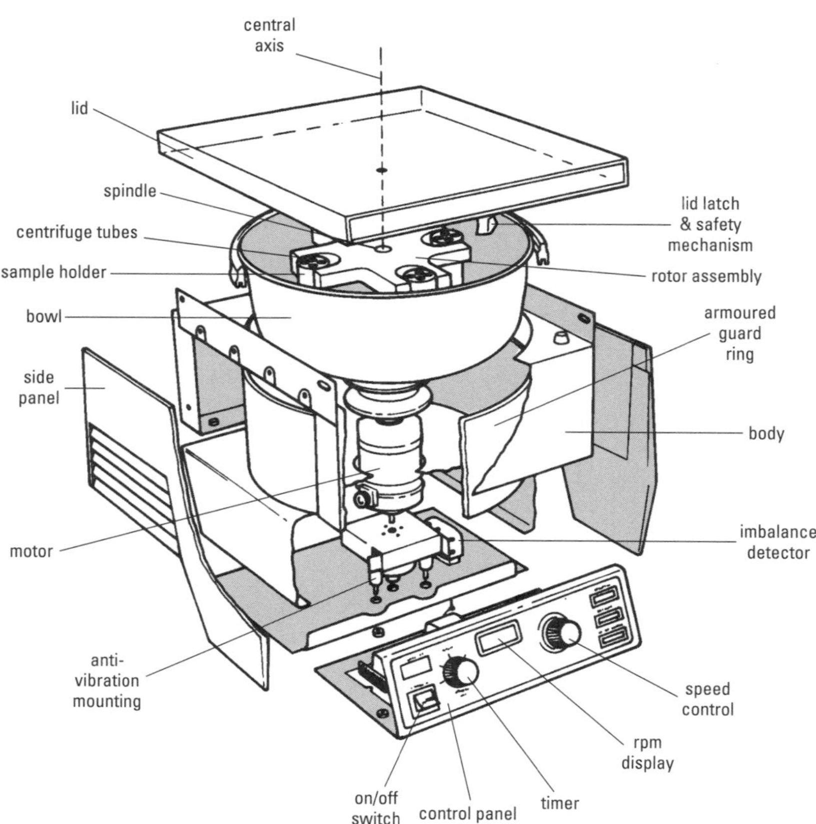

Fig. 57.3 Principal components of a low-speed bench centrifuge, from diagram of low-speed centrifuge *model MSE Centaur 2*, supplied by Fisher Scientific UK Ltd, reproduced by kind permission of Fisher Scientific UK Ltd.

revolutions per minute, r.p.m.) and the radius of rotation (r, in mm) where:

$$\text{RCF} = 1.118\,r\left(\frac{n}{1\,000}\right)^2 \qquad [57.1]$$

This relationship can be rearranged, to calculate the speed (r.p.m.) for specific values of r and RCF:

$$n = 945.7\sqrt{\left(\frac{\text{RCF}}{r}\right)} \qquad [57.2]$$

However, you should note that RCF is not uniform within a centrifuge tube: it is highest near the outside of the rotor (r_{max}) and lowest near the central axis (r_{min}). In practice, it is customary to report the RCF calculated from the average radius of rotation (r_{av}), as shown in Fig. 57.7. It is also worth noting that RCF varies relative to the *square* of the speed: thus the RCF will be doubled by an increase in speed of approximately 41 per cent (Table 57.2).

Examples Suppose you wanted to calculate the RCF of a bench centrifuge with a rotor of r_{av} = 95 mm running at a speed of 3000 r.p.m. Using eqn [57.1] the RCF would be:
$1.118 \times 95 \times (3)^2 = 956\,g$.

You might wish to calculate the speed (r.p.m.) required to produce a relative centrifugal field of 2000 g using a rotor of r_{av} = 85 mm. Using eqn [57.2] the speed would be:
$945.7\sqrt{(2000 \div 85)} = 4587$ r.p.m.

Centrifugal separation methods

Differential sedimentation (pelleting)

By centrifuging a mixed suspension of particles at a specific RCF for a particular time, the mixture will be separated into a pellet and a supernatant

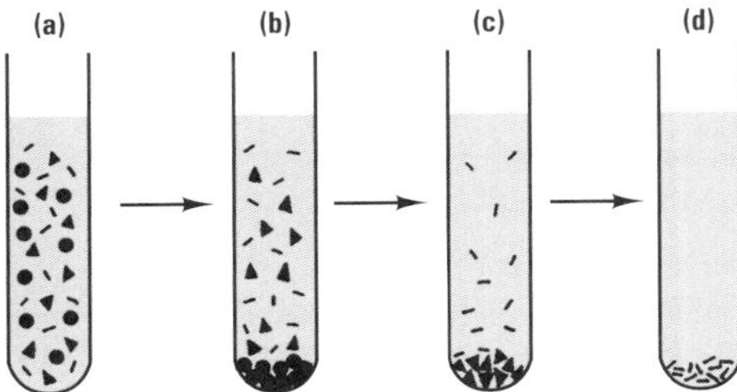

Fig. 57.4 Differential sedimentation. (a) Before centrifugation, the tube contains a mixed suspension of large, medium and small particles of similar density. (b) After low-speed centrifugation, the pellet is predominantly composed of the largest particles. (c) Further high-speed centrifugation of the supernatant will give a second pellet, predominantly composed of medium-sized particles. (d) A final ultracentrifugation step pellets the remaining small particles. Note that all of the pellets apart from the final ones will have some degree of cross-contamination.

Working with silicone oil – the density of silicone oil is temperature-sensitive, so work in a location with a known, stable temperature or the technique may fail.

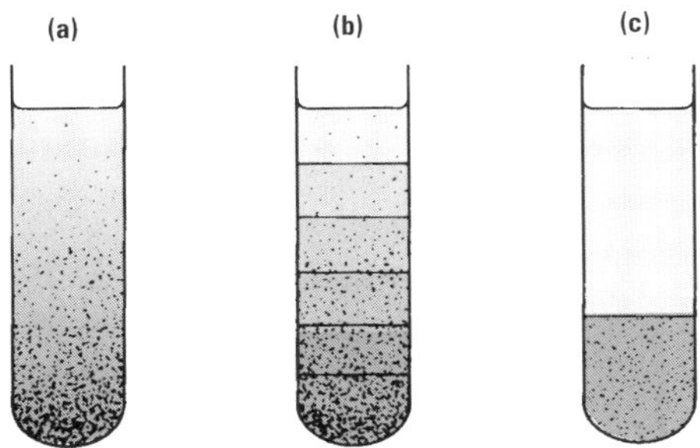

Fig. 57.5 Density gradients. (a) A continuous (linear) density gradient. (b) A discontinuous (stepwise) density gradient, formed by layering solutions of decreasing density on top of each other. (c) A single-step density barrier, designed to allow selective sedimentation of one type of particle.

(Fig. 57.4). The successive pelleting of a suspension by spinning for a fixed time at increasing RCF is widely used to separate organelles from cell homogenates. The same principle applies when cells are harvested from a liquid medium.

Density-gradient centrifugation

The following techniques use a density gradient, a solution that increases in density from the top to the bottom of a centrifuge tube (Fig. 57.5).

- Rate-zonal centrifugation. By layering a sample on to a shallow pre-formed density gradient, followed by centrifugation, the larger particles will move faster through the gradient than the smaller ones, forming several distinct zones (bands). This method is time-dependent, and centrifugation *must* be stopped before any band reaches the bottom of the tube (Fig. 57.6).
- Isopycnic centrifugation. This technique separates particles on the basis of their buoyant density. Several substances form density gradients during centrifugation (e.g. sucrose, CsCl, Ficoll, Percoll, Nycodenz). The sample is mixed with the appropriate substance and then centrifuged – particles form bands where their density corresponds to that of the medium (Fig. 57.6). This method requires a steep gradient and sufficient time to allow gradient formation and particle redistribution, but is unaffected by further centrifugation.

Bands within a density gradient can be sampled using a fine Pasteur pipette, or a syringe with a long, fine needle. Alternatively, the tube may be punctured and the contents (fractions) collected dropwise in several tubes.

Density-barrier centrifugation

A single-step density barrier (Fig. 57.5(c)) can be used to separate cells from their surrounding fluid, e.g. using a layer of silicone oil adjusted to the correct density using dinonyl phthalate. Blood cell types can be separated using a density barrier of e.g. Ficoll.

Types of centrifuge and their uses

Low-speed centrifuges

These are bench-top instruments for routine use, with a maximum speed of 3000–6000 r.p.m. and RCF up to 6000 *g* (Fig. 57.3). They are used to harvest cells, and coarse precipitates (e.g. antibody–antigen complexes, p. 347). Most modern machines also have a sensor that detects any imbalance when the rotor is spinning and cuts off the power supply (Fig. 57.3). However, some of the older models do not, and must be switched off as soon as any vibration is noticed, to prevent damage to the rotor or harm to the operator. Box 57.1 gives details of operation for a low-speed centrifuge.

Microcentrifuges (microfuges)

These are bench-top machines, capable of rapid acceleration up to 12 000 r.p.m. and 10 000 g. They are used to sediment small sample volumes (up to 1.5 mL) of larger particles (e.g. cells, precipitates) over

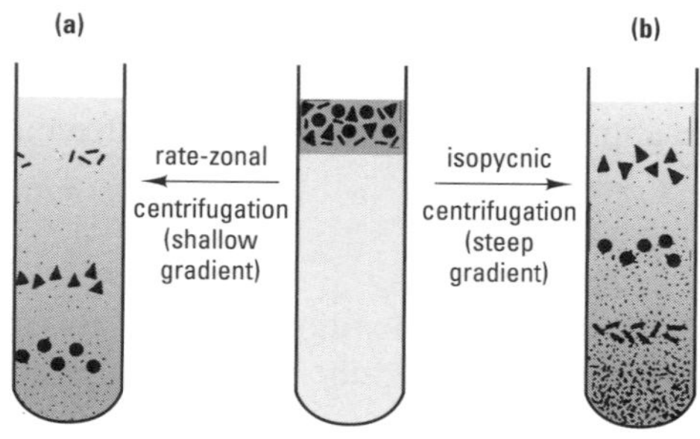

Fig. 57.6 Density-gradient centrifugation. The central tube shows the position of the sample prior to centrifugation, as a layer on top of the density gradient medium. Note that particles sediment on the basis of size during rate-zonal centrifugation (a), but form bands in order of their densities during isopycnic centrifugation (b). ●, large particles, intermediate density; ▲, medium-sized particles, low density; ▬ small particles, high density.

Recording usage of high-speed centrifuges and ultracentrifuges – most departments have a log book (for samples/speeds/times): make sure you record these details, as the information is important for servicing and replacement of rotors.

SAFETY NOTE When changing a rotor, make sure that you carry it properly (don't knock/drop it), that you fit it correctly (don't cross-thread it, and tighten to the correct setting using a torque wrench) and that you store it correctly (clean it after use and don't leave it lying around).

SAFETY NOTE When working with centrifuge tubes never be tempted to use a tube or bottle that was not designed to fit the machine you are using (e.g. a general-purpose glass test tube, or a screw-capped bottle), or you may damage the centrifuge and cause an accident.

short timescales (typically, 0.5–15 minutes). They are particularly useful for the rapid separation of cells from a liquid medium, e.g. silicone oil microcentrifugation.

Continuous-flow centrifuges

Useful for harvesting large volumes of cells from their growth medium. During centrifugation, the particles are sedimented as the liquid flows through the rotor.

High-speed centrifuges

These are usually larger, free-standing instruments with a maximum speed of up to 25 000 r.p.m. and RCF up to 60 000 *g*. They are used for microbial cells, and protein precipitates. They often have a refrigeration system to keep the rotor cool at high speed. You would normally use such instruments only under direct supervision.

Ultracentrifuges

These are the most powerful machines, having maximum speeds in excess of 30 000 r.p.m. and RCF up to 60 0000 *g*, with sophisticated refrigeration and vacuum systems. They are used to separate biological macromolecules in food samples. You would not normally use an ultracentrifuge, though your samples may be run by a member of staff.

Rotors

Many centrifuges can be used with tubes of different size and capacity, either by changing the rotor, or by using a single rotor with different buckets/adaptors.

- Swing-out rotors: sample tubes are placed in buckets that pivot as the rotor accelerates (Fig. 57.7(a)). Swing-out rotors are used on many low-speed centrifuges: their major drawback is their extended path length and the resuspension of pellets due to currents created during deceleration.
- Fixed-angle rotors: used in many high-speed centrifuges and microcentrifuges (Fig. 57.7(b)). With their shorter path length, fixed rotors are more effective at pelleting particles than swing-out rotors.
- Vertical-tube rotors: used for isopycnic density-gradient centrifugation in high-speed centrifuges and ultracentrifuges (Fig. 57.7(c)). They cannot be used to harvest particles in suspension as a pellet is not formed.

Centrifuge tubes

These are manufactured in a range of sizes (from 1.5 mL up to 1000 mL) and materials. The following aspects may influence your choice:

- **Capacity.** This is obviously governed by the volume of your sample. Note that centrifuge tubes must be completely full for certain applications, e.g. for high-speed work.

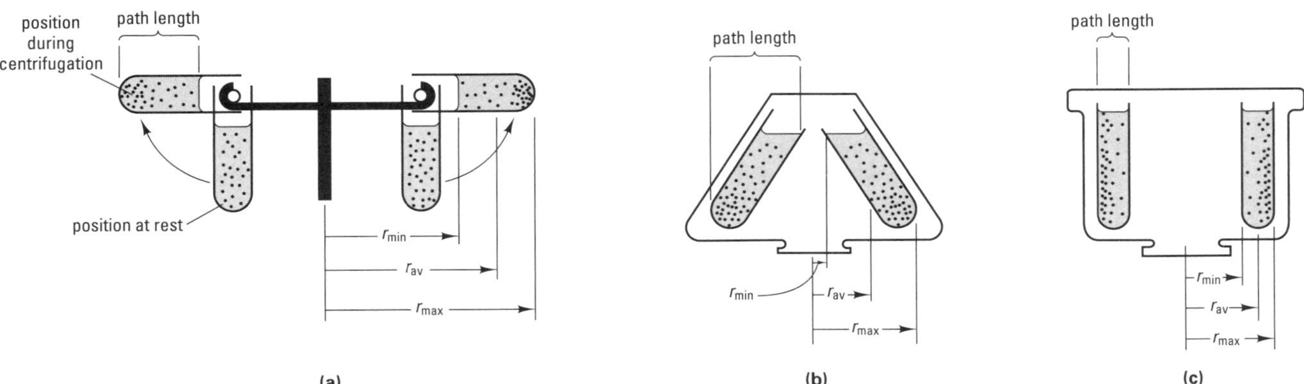

Fig. 57.7 Rotors: (a) swing-out rotor; (b) fixed-angle rotor; (c) vertical-tube rotor.

- **Shape.** Conical-bottomed centrifuge tubes retain pellets more effectively than round-bottomed tubes, while the latter may be more useful for density-gradient work.
- **Maximum centrifugal force.** Detailed information is supplied by the manufacturers. Standard Pyrex glass tubes can only be used at low centrifugal force (up to 2000 *g*).
- **Caps.** Most fixed-angle and vertical-tube rotors require tubes to be capped, to prevent leakage during use and to provide support to the tube during centrifugation. For low-speed centrifugation, caps must be used for any hazardous samples. Make sure you use the correct caps for your tubes.

Using microcentrifuge tubes – the integral push-on caps of microcentrifuge tubes must be correctly pushed home before use or they may come off during centrifugation.

- **Solvent resistance.** Glass tubes are inert, polycarbonate tubes are particularly sensitive to organic solvents (e.g. ethanol, acetone), while polypropylene tubes are more resistant. See manufacturer's guidelines for detailed information.
- **Sterilisation.** Disposable plastic centrifuge tubes are often supplied in sterile form. Glass and polypropylene tubes can be repeatedly sterilised. Cellulose ester tubes should *not* be autoclaved. Repeated autoclaving of polycarbonate tubes may lead to cracking/stress damage.
- **Opacity.** Glass and polycarbonate tubes are clear, while polypropylene tubes are more opaque.
- **Ability to be pierced.** If you intend to harvest your sample by puncturing the tube wall, cellulose acetate and polypropylene tubes are readily punctured using a syringe needle.

SAFETY NOTE Never balance centrifuge tubes 'by eye' – use a balance. Note that a 35 mL tube full of liquid at an RCF of 3000 g has an effective weight greater than a large adult human.

SAFETY NOTE It is vital that you balance your loaded centrifuge tubes before use. As a general rule, balance all sample tubes to within 1 per cent or better, using a toppan balance or scales. Place balanced tubes opposite each other.

Balancing the rotor

For the safe use of centrifuges, the rotor must be balanced during use (Box 57.1), or the spindle and rotor assembly may be damaged permanently; in severe cases, the rotor may fail and cause a serious accident.

Safe practice

Given their speed of rotation and the extremely high forces generated, centrifuges have the potential to be extremely dangerous, if used incorrectly. You should only use a particular centrifuge if you fully

Box 57.1 How to use a low-speed bench centrifuge

1. **Choose the appropriate tube size and material for your application**, with caps where necessary. Most low-speed machines have four-place or six-place rotors – use the correct number of samples to *fill* the rotor assembly whenever possible.
2. **Fill the containers to the appropriate level**: do not overfill, or the sample may spill during centrifugation.
3. **It is vital that the rotor is balanced during use.** Therefore, *identical* tubes must be prepared, to be placed opposite each other in the rotor assembly. This is particularly important for density-gradient samples, or for samples containing materials of widely differing densities, e.g. soil samples, since the density profile of the tube will change during a run. However, for low-speed work using small amounts of particulate matter in aqueous solution, it is sufficient to counterbalance a sample with a second tube filled with water, or a saline solution of similar density to the sample.
4. **Balance each pair of sample tubes** (plus the corresponding caps, where necessary) to within 0.1 g using a top-pan balance; add liquid dropwise to the lighter tube, until the desired weight is reached. Alternatively, use a set of scales. For small sample volumes (up to 10 ml) added to disposable, lightweight plastic tubes, accurate pipetting of your solution may be sufficient for low-speed use.
5. **For centrifuges with swing-out rotors**, check that each holder/bucket is correctly positioned in its locating slots on the rotor and that it is able to swing freely. All buckets must be in position on a swing-out rotor, even if they do not contain sample tubes – buckets are an integral part of the rotor assembly.
6. **Load the sample tubes into the centrifuge.** Make sure that the outside of the centrifuge tubes, the sample holders and sample chambers are dry: any liquid present will cause an imbalance during centrifugation, in addition to the corrosive damage it may cause to the rotor. For sample holders where rubber cushions are provided, make sure that these are correctly located. Balanced tubes must be placed opposite each other – use a simple code if necessary, to prevent mix-ups.
7. **Bring the centrifuge up to operating speed** by gentle acceleration. Do not exceed the maximum speed for the rotor and tubes used.
8. **If the centrifuge vibrates at any time during use, switch off** and find the source of the problem.
9. **Once the rotor has stopped spinning, release the lid and remove all tubes.** If any sample has spilled, make sure you clean it up thoroughly using a non-corrosive disinfectant, e.g. Virkon, so that it is ready for the next user.
10. **Close the lid (to prevent the entry of dust) and return all controls to zero.**

understand the operating principles: if unsure, check with a member of staff. For safety reasons, all centrifuges are manufactured with an armoured casing that should contain any fragments in cases of rotor failure. Machines usually have a safety lock to prevent the motor from being switched on unless the lid is closed and to stop the lid from being opened while the rotor is moving. Do not be tempted to use older machines without a safety lock, or centrifuges where the locking mechanism is damaged/inoperative. Be particularly careful to make sure that hair and clothing are kept well away from moving parts.

Sources for further study

Boyer, R.F. (2005) *Biochemistry Laboratory: Modern Theory and Techniques.* Benjamin Cummings, San Francisco, California.
[Also covers other topics, including chromatography and spectroscopy.]

Graham, J. (2001) *Biological Centrifugation.* Bios Scientific Publishers, Oxford.

Analysing food components and properties

58 Analysis of biomolecules in food: fundamental principles

> **Definitions**
>
> **Accuracy** – the closeness of an individual measurement, or a mean value based on a number of measurements, to the true value.
>
> **Concentration range** – the range of values from the detection limit to the upper concentration at which the technique becomes inaccurate or imprecise.
>
> **Detection limit** – the minimum concentration of an analyte that can be detected at a particular confidence level.
>
> **Drift** – 'baseline' movement in a particular direction: drift can be a problem between analyses (e.g. using a spectrophotometer for colorimetric analysis), or for a single analysis (e.g. when separating biomolecules by chromatography).
>
> **Noise** – random fluctuations in a continuously monitored signal.
>
> **Precision** – the extent of mutual agreement between replicate data values for an individual sample.
>
> **Quality assurance** – procedures to monitor, document and audit a process against a set of criteria, usually defined externally (e.g. ISO 9000), to confirm that the process meets the specified requirements.
>
> **Quality control** – procedures designed to ensure that the performance of a process meets a particular standard, e.g. analysis of a test sample containing a known amount of a substance, to confirm that the measurement falls within an acceptable range of the true value.
>
> **Replicate** – repeated measurement.
>
> **Selectivity** – the extent to which a method is free from interference due to other substances in the sample.
>
> **Sensitivity** – the ability to discriminate between small differences in analyte concentration.
>
> **Validation** – the process whereby the accuracy and precision of a particular analytical method are checked in relation to specific standards, using an appropriate reference material containing a known amount of analyte.

Biochemical analysis involves the characterisation of biological components within food and biological samples using appropriate laboratory techniques. Most analytical methods rely on one or more chemical or physical properties of the test substance (the analyte) for detection and/or measurement. There are two principal approaches:

1. **Qualitative analysis** – where a sample is assayed to determine whether a biomolecule is present or absent. As an example, a blood sample might be analysed for a particular drug or a specific antibody (p. 347), or a bacterial cell might be probed for a nucleic acid sequence (p. 374).
2. **Quantitative analysis** – where the quantity of a particular biomolecule in a sample is determined, either as an amount (e.g. as g, or mol) or in terms of its concentration in the sample (e.g. as $g\,L^{-1}$, or $mol\,m^{-3}$). For example, a blood sample might be analysed to determine its pH ($-\log_{10}$ [H^+]), alcohol concentration in mg mL^{-1}, or glucose concentration in $mmol\,L^{-1}$. Skoog *et al.* (2006) give details of methods.

Your choice of approach will be determined by the purpose of the investigation and by the level of accuracy and precision required. Many of the basic quantitative methods rely on chemical reactions of the analyte and involve assumptions about the nature of the test substance and the lack of interfering compounds in the sample: such assumptions are unlikely to be wholly valid at all times. If you need to make more exacting measurements of a particular analyte, it may be necessary to separate it from the other components in the sample, e.g. using chromatography (Chapter 54), centrifugation (Chapter 57), or electrophoresis (Chapter 55), and then identify the separated components, e.g. using spectroscopic methods (Chapter 53). However, each stage in the separation and purification procedure may introduce further errors and/or loss of sample, as described for protein purification in Chapter 59.

> **KEY POINT** In general, you should aim to use the simplest procedure that satisfies the purpose of your investigation – there is little value in using a complex, time-consuming or costly analytical procedure to answer a simple problem where a high degree of accuracy is not required.

Most of the routine methods based on chemical analysis are destructive, since the analyte is usually converted to another substance which is then assayed, e.g. in colorimetric assays of the major types of biomolecules (Table 58.1). In contrast, many of the analytical methods based on physical properties are non-destructive (e.g. the intrinsic absorption and emission of electromagnetic radiation in spectrophotometry, p. 355, or nuclear magnetic resonance techniques). Non-destructive methods are often preferred, as they allow the further characterisation of a particular sample. Most biological methods are destructive: bioassays are often sensitive to interference and require validation.

Evaluating a new method – results from a novel technique can be compared with an established 'standardised' technique by measuring the same set of samples by each method and analysing the results by correlation (p. 267).

Table 58.1 Some examples of colorimetric assays

Analyte	Reagent/wavelength
Amino acids	ninhydrin/540 nm
Proteins	biuret/520 nm
	Folin–Ciocalteau/600 nm
Carbohydrates	anthrone/625 nm
Reducing sugars	dinitrosalicylate/540 nm
DNA	diphenylamine/600 nm
RNA	orcinol/660 nm

Interpreting results from 'spiked' samples – remember that such procedures tell you nothing about the extraction efficiency of biomolecules from a particular food sample, e.g. during homogenisation.

Criteria for the selection of a particular analytical method:

- the required level of accuracy and precision;
- the number of samples to be analysed;
- the amount of each sample available for analysis;
- the physical form of the samples;
- the expected concentration range of the analyte in the samples;
- the sensitivity and detection limit of the technique;
- the likelihood of interfering substances;
- the speed of analysis;
- the ease and convenience of the procedure;
- the skill required by the operator;
- the cost and availability of the equipment.

Validity and quantitative analysis

Before using a particular procedure, you should consider its possible limitations in terms of:

- **measurement errors, and their likely magnitude:** these might include processing errors (e.g. in preparing solutions and making dilutions), instrumental errors (e.g. a pH electrode that has not been set up correctly), calibration errors (e.g. converting a digital readout analyte concentration) and errors due to the presence of interfering substances;
- **sampling errors:** these may occur if the material used for analysis is not representative, e.g. due to differences between the food items used in the sampling procedure.

Replication will allow you to make quantitative estimates of several potential sources of error: for example, repeated measurements of the same sample can provide information on the precision of the analytical method, e.g. by calculating the coefficient of variation (p. 252), while measurements of several different samples can provide information on biological and sample variability, e.g. by calculating the standard deviation (Chapter 38). Analyses of different sets of samples at different times (e.g. on different days) can provide information on 'between batch' variability, as opposed to 'within batch' variability (based on a single set of analytical data).

The reliability of a particular method can be assessed by measuring 'standards' (sometimes termed 'controls'). These are often prepared in the laboratory by adding a known amount of analyte to a real sample (this is often termed 'spiking' a sample), or by preparing an artificial sample containing a known amount of analyte along with other relevant components (e.g. the major sample constituents and possible interfering substances). In many instances, several standards (including a 'blank' or 'zero') are assayed to construct a 'standard curve', which is then used to convert sample measurements to amounts of analyte (see Chapter 50 for details). Such standard curves form the basis of many routine laboratory assays: while hand-drawn linear calibration curves are sufficient for basic assays, more complex curves are often fitted to a particular mathematical function using a computer program (e.g. bioassays). Standards can also be used to check the calibration of a particular method: a mean value based on repeated measurements of an individual 'standard' can be compared with the true value using a modified t-test (p. 265), in which there is only one standard error term, i.e. that associated with the measured values.

Validation of a particular method can be important in certain circumstances, e.g. in a government food analysis laboratory, where particular results can have important implications. Such laboratories operate strict validation procedures, including: (i) adherence to standard operating procedures for each analytical method; (ii) calibration of assays using certified reference materials containing known amount of analyte and traceable to a national reference laboratory; (iii) effective systems for internal quality control and external quality assurance; (iv) detailed record-keeping, covering all aspects of the analysis and recording of results. Although such rigour is not required for routine analysis in your undergraduate work, the general principles of standardisation, calibration, assessment of performance and record-keeping are equally valid for all analytical work.

Text reference

Skoog, D.A., Holler, F.J. and Crouch, S.R. (2006) *Instru-mental Analysis Principles*. Brooks-Cole, Pacific Grove, CA.

Sources for further study

Center for Drug Evaluation and Research. *Guidance for Industry Bioanalytical Method Validation*. Available: http://www.fda.gov/CDER/GUIDANCE/4252fnl.htm
Last accessed: 21/12/10.
[US FDA guidance/details of procedures.]

Evans, G. (2003) *Handbook of Bioanalysis and Drug Metabolism*. CRC Press, Boca Raton, Florida.

International Union of Biochemistry and Molecular Biology. *Homepage*. Available: http://www.iubmb.unibe.ch/
Last accessed: 21/12/10.
[Promotes international standardisation of methods, nomenclature and symbols in biomolecular analysis.]

Laboratory of the Government Chemist. *Valid Analytical Measurement Homepage*. Available: http://www.vam.org.uk
Last accessed: 21/12/10.

Venn, R.F. (2000) *Principles and Practice of Bioanalysis*. Taylor and Francis, London.

Wilson, K. and Walker, J. (eds) (2000) *Principles and Techniques of Practical Biochemistry*, 5th edn. Cambridge University Press, Cambridge.

59 Assaying proteins, amino acids and enzymes

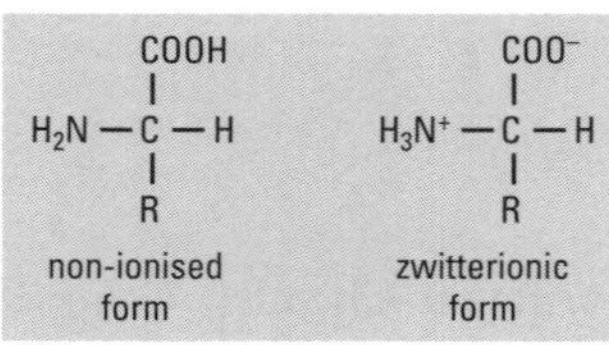

Fig. 59.1 Structure of α-amino acids. R = side chain (see Table 59.1 on p. 399 for examples).

Proteins are polymers formed by the linkage of α-amino acids (Fig. 59.1) to create polypeptide chains. Some of them act as biological catalysts (enzymes). Due to the broad range of structural and functional roles of proteins in biological systems, they are major components of many different types of foods. Protein-rich foods include meats, fish, milk, eggs, nuts, beans and peas. During digestion, proteins are broken down into their constituent amino acids or into short chains (peptides), which are then absorbed and utilised to form new human proteins.

Because foods are derived from biological tissues, they contain a diverse range of proteins of different types, in terms of their sizes. From a practical perspective, there are three major aspects to consider:

1. **Measurement of total protein content**, e.g. as part of an assessment of macronutrient composition. Such information also forms part of the labelling requirements of food, as detailed in Chapter 23 (p. 138).

2. **Characterisation of protein composition**, which involves the fractionation, separation and identification of individual constituent proteins.

3. **Analysis of amino acid composition**, particularly in relation to the essential amino acids, which cannot be synthesised by the body, for which there are dietary guidelines. This first requires the proteins to be hydrolysed to monomeric amino acids, followed by chromatographic separation (e.g. HPLC, p. 362) and quantitative detection of the individual components.

The assay of enzyme activity is covered as the final section of this chapter, pp. 499–503.

Remembering the names of the essential amino acids – use the mnemonic 'Private Tim Hall' (PVT TIM HALL), which will give you the first letter of each of the ten amino acids, namely: phenylalanine, valine, tryptophan, threonine, isoleucine, methionine, histidine, arginine, lysine and leucine (note that histidine and arginine are essential for children but not for adults).

Measuring total protein content

Kjeldahl standard assay

This is one of the international standard methods used to measure the protein content of food. The assay involves the following major steps:

1. **Digestion**. A known amount of food is added to a mixture of concentrated sulphuric acid, potassium sulphate and a catalyst (e.g. copper sulfate). The overall result is to hydrolyse the polypeptides to their constituent amino acids and then liberate the nitrogen from these in the form of $(NH_4)_2SO_4$.

2. **Neutralisation and distillation**. $(NH_4)_2SO_4$ is neutralised with NaOH to liberate gaseous ammonia (NH_3), which is then captured by distillation into a solution of boric acid, forming ammonium borate according to the reaction:

$$NH_3 + H_3BO_3 \rightarrow NH_4H_2BO_3 \qquad [59.1]$$

3. **Titration**. The amount of ammonium borate formed is assayed by titration of the borate anion against HCl, as follows:

$$H_2BO_3^- + H^+ \rightarrow H_3BO_3 \qquad [59.2]$$

SAFETY NOTE The Kjeldahl assay involves the use of hot, concentrated acids and toxic vapours. Always use a fume hood and follow any specific guidance for the specific procedure used, e.g. cool liquids before dilution and/or titration.

The end point of this reaction is determined using a suitable pH indicator, e.g methyl red (p. 174).

4. **Calculation**. The molar quantity of HCl required to titrate the borate in step 3 is equivalent to number of moles of nitrogen present in the food sample. This can then be converted from molar to mass quantity (g) by multiplying by 14, the atomic mass of nitrogen. Typically, nitrogen content would be expressed in percentage terms (g N per 100 g of food). Then, by assuming that protein is the only substantial source of nitrogen, the nitrogen content can be converted to protein by multiplying the measured nitrogen content by 6.25. This assumes an average nitrogen content of protein of 16 per cent for animal proteins. The equivalent conversion factor for plant proteins is 5.7, based on a protein content of 17.5 per cent.

Other standard assays

There are two other internationally accepted methods for measuring total protein content in foods. These are:

- **Dumas method**. This involves the high temperature combustion of a sample of food, with subsequent quantification of the released nitrogen by gas chromatography (p. 363) using a thermal capture detector (for further details of the practical procedure involved, see Neilsen, 2003).
- **Near infra-red spectroscopy (NIFS)**. This exploits the absorption of infra-red radiation by the peptide bond, e.g. at wavelengths of 3300–3500 nm. It provides a rapid and non-destructive alternative to the Kjeldahl and Dumas methods, although it requires more specialised analytical equipment.

Other protein assays

A number of features are exploited for the quantitative analysis of proteins in science laboratories, and you are likely to carry out these procedures in practical classes in biochemistry and related subjects.

The biochemical features most commonly exploited for quantitative analysis are:

- the peptide bond, e.g. in the biuret reaction (Box 59.1);
- the phenolic group of tyr and the indole group of trp, which react with the oxidising agents phosphotungstic and phosphomolybdic acids in the Folin–Ciocalteau reagent to produce a blue colour. This is combined with the biuret reaction in the Lowry method (Box 59.1);
- dye binding to hydrophobic regions, e.g. Bradford assay (Box 59.1);
- the primary amino groups of lys residues, the guanidino group of arg residues and the N-terminal amino acid residue will react with ninhydrin, allowing colorimetric assay similar to that described above for isolated amino acids. However, a more sensitive method uses the reaction of these groups with fluorescamine to give an intensely fluorescent product, which you can measure by spectrofluorimetry (p. 426).

Box 59.1 How to determine the amount of protein/peptide in an aqueous solution

For all of the following methods, the amounts are appropriate for semi-micro cuvettes (1.5 mL volume, path length 1 cm). Appropriate controls (blanks) must be analysed, to assess possible interference (e.g. due to buffers, etc.).

Biuret method

This is based on the specific reaction between cupric ions (Cu^{2+}) in alkaline solution and two adjacent peptide bonds, as found in proteins and peptides. As such, it is not significantly affected by differences in amino acid composition.

1. **Prepare protein standards over an appropriate range** (typically, between 1 and 10 mg mL^{-1}).
2. **Add 1 mL of each standard solution to separate test tubes. Prepare a reagent blank, using 1 mL of distilled water or an appropriate solution.**
3. **Add 1 mL of each unknown solution to separate test tubes.**
4. **Add 1 mL of biuret reagent (1.5 g $CuSO_4 \cdot 5H_2O$, 6.0 g sodium potassium tartrate in 300 mL of 10 w/v NaOH) to all standard and unknown tubes and to the reagent blank.**
5. **Incubate at 37°C for 15 minutes.**
6. **Read the absorbance of each solution at 520 nm against the reagent blank.** The violet colour is stable for several hours.

The main limitation of the biuret method is its lack of sensitivity – it is unsuitable for solutions with a protein content of less than 1 mg mL^{-1}.

Direct measurement of UV absorbance (Warburg–Christian method)

Proteins and peptides absorb EMR maximally at 280 nm (due to the presence of aromatic amino acids) and this forms the basis of the method. The principal advantages of this approach are its simplicity and the fact that the assay is non-destructive. The most common interfering substances are nucleic acids, which can be assessed by measuring the absorbance at 260 nm: a pure solution of protein will have a ratio of absorption (A_{280}/A_{260}) of approximately 1.8, decreasing with increasing nucleic acid contamination. Note also that any free aromatic amino acids in your solution will absorb at 280 nm, leading to an overestimation of protein content. The simplest procedure, which includes a correction for small amounts of nucleic acid, is as follows (use quartz cuvettes throughout):

1. **Measure the absorbance of your solution at 280 nm (A_{280}):** if A_{280} is greater than 1, dilute by an appropriate amount and remeasure.
2. **Repeat at 260 nm (A_{260}).**
3. **Estimate the approximate protein content using the following relationship:**

$$\text{[protein] mg mL}^{-1} = 1.45\,A_{280} - 0.74\,A_{260} \qquad [59.3]$$

This equation is based on the work of Warburg and Christian (1942) for enolase. For other proteins, it should not be used for quantitative work, since it gives only a rough approximation of the amount present, due to variations in aromatic amino acid composition.

Lowry (Folin–Ciocalteau) method

This is a colorimetric assay, based on a combination of the biuret method, described above, and the oxidation of tyrosine and tryptophan residues with Folin–Ciocalteau reagent to give a blue–purple colour. The method is extremely sensitive (down to a protein/peptide content of 20 μg mL^{-1}), but is subject to interference from a wide range of non-protein substances, including many organic buffers (e.g. TRIS, HEPES), EDTA, urea and certain sugars. The choice of an appropriate standard is important, as the intensity of colour produced for a particular protein/peptide is dependent on the amount of aromatic amino acids present.

1. **Prepare protein standards within an appropriate range for your samples** (the method can be used from 0.02 to 1.00 mg mL^{-1}).
2. **Add 1 mL of each standard solution to separate test tubes. Prepare a reagent blank, using 1 ml of distilled water, or an appropriate solution.**
3. **Add 1 mL of each of your unknown solutions to separate test tubes.**
4. **Then, add 5 mL of 'alkaline solution' (prepared by mixing 2% w/v Na_2CO_3 in 0.1 mol L^{-1} NaOH, 1% w/v aqueous $CuSO_4$ and 2% w/v aqueous NaK tartrate in the ratio 100 : 1 : 1.** Mix thoroughly and allow to stand for at least ten minutes.
5. **Add 0.5 ml of Folin–Ciocalteau reagent (commercial reagent, diluted 1 : 1 with distilled water on the day of use).** Mix rapidly and thoroughly and then allow to stand for 30 minutes.
6. **Read the absorbance of each sample at 600 nm.**

(continued)

Box 59.1 (continued)

Dye-binding (Bradford) method

Coomassie brilliant blue combines with proteins and peptides to give a dye–protein complex with an absorption maximum of 595 nm. This provides a simple and sensitive means of measuring protein content, with few interferences. However, the formation of dye–protein complex is affected by the number of basic amino acids within a protein, so the choice of an appropriate standard is important. The method is sensitive down to a protein content of approximately 5 μg mL^{-1} but the relationship between absorbance and concentration is often non-linear, particularly at high protein content.

1. **Prepare protein standards over an appropriate range (between 5 and 100 g mL^{-1}).**
2. **Add 100 μL of each standard solution to separate test tubes. Prepare a reagent blank, using 100 μL of distilled water, or an appropriate solution** (note that these small volumes must be accurately dispensed, e.g. using a calibrated pipettor, p. 155).
3. **Add 100 μL of your unknown solutions to separate test tubes.**
4. **Add 5.0 mL of Coomassie brilliant blue G250 solution (0.1 g L^{-1}).**
5. **Mix and incubate for at least 5 min: read the absorbance of each solution at 595 nm.**

Other methods are less widely used. They include determination of the total amount of nitrogen in solution (e.g. using the Kjeldahl technique) and calculating the protein content, assuming a nitrogen content of 16%. An alternative approach is to precipitate the protein (e.g. using trichloracetic acid, tannic acid or salicylic acid) and then measure the turbidity of the resulting precipitate (using a nephelometer, or a spectrophotometer, p. 356).

Advice on preparing standard (calibration) curves – see p. 337 for practical guidance.

Most assays for proteins and peptides do not give absolute values, but require standard solutions, containing appropriate amounts of a particular protein, to be analysed at the same time, so that a standard curve can be constructed. Bovine serum albumin (BSA) is commonly used as a protein standard. However, you may need an alternative standard if the protein you are assaying has an amino acid composition which is markedly different from that of BSA, depending on your chosen method.

Exploiting the biological properties of proteins and peptides – relevant analytical methods include immunoassay, enzymatic analysis and affinity chromatography, based on specific biological interactions.

Physical properties of proteins and peptides

The characteristics that can be used for separation and analysis include:

- **absorbance of aromatic amino acid residues** (phe, tyr and trp) at 280 nm (Warburg–Christian method, Box 59.1);
- **prosthetic groups of conjugated proteins** – these often have characteristic absorption maxima, e.g. the haem group of haemoglobin absorbs strongly at 415 nm, allowing quantitative assay;
- **density** – most proteins have a density of 1.33 kg L^{-1}, but lipoproteins have a lower density. This can be used to separate lipoproteins from other classes of protein, or to subdivide the various classes of lipoproteins;
- **net charge** – proteins and peptides differ in the types and number of amino acids with ionisable groups in their side chains. The ionisation of these side chains is pH dependent, resulting in variation of net charge with pH. This property is exploited in the technique of electrophoresis (Chapter 55);
- **water solubility and surface hydrophobicity** – these are exploited in salt fractionation and hydrophobic interaction chromatography (p. 365).

Definitions

Lipoproteins – globular, micelle-like particles consisting of a non-polar core of triacylglyceryl and cholesteryl esters surrounded by a coating of proteins, phospholipids and cholesterol: involved in the transport of lipids in blood.

Glycoproteins – compounds consisting of proteins covalently attached to carbohydrate residues in post-translational modifications of the protein constituents.

Understanding the techniques of protein purification – most of the core methods are used across the life sciences, e.g. to investigate enzyme activity or to characterise structural proteins in biological systems.

Characterising protein composition

While the methods described above provide information on the overall protein content of a food, if a more detailed understanding is required as to the nutritional and physico-chemical properties of different types of proteins, then it is necessary to separate and isolate the individual protein constituents; once separated, they can be further characterised, e.g. by determining their amino acid composition and/or sequence.

KEY POINT The separation and isolation of proteins from food involves a step-by-step set of procedures, often based on differential solubility, chromatography and/or electrophoresis, to selectively separate the protein(s) of interest from other proteins and constituents. Separation methods make use of the physico-chemical properties of different proteins, including size, charge and solubility in aqueous and non-aqueous solvents.

Following homogenisation (p. 380), proteins and other components will be present in a soluble form. The choice of procedures to be used will depend on:

Monitoring protein purification – the assays described in Box 59.1 can be used to measure the amount of protein at each stage in the procedure, e.g. to calculate the overall yield and the relative amounts of each type.

- **the objectives of the analysis** (e.g. to purify a target protein to homogeneity, and then determine its amino acid sequence);
- **the amount of protein required** (e.g. mg quantities may be needed for structural studies);
- **the desired purity** (typically, high-purity procedures involve greater loss of protein during the purification steps);
- **whether it is important to retain the three-dimensional structure of the protein during purification** (if this is not important, then methods such as denaturing gel electrophoresis can be used, as detailed on p. 368).

A typical sequence of steps would be:

Using ammonium sulfate – while the phenomenon of 'salting out' is seen with other salts, including NaCl, $(NH_4)_2SO_4$ is used because of its high solubility, low cost and ready availability in pure form.

1. **Ammonium sulfate precipitation** ('salting out'). This is the most widely used technique to separate proteins into a series of 'fractions'. It involves the selective precipitation of different types of proteins based on their differential solubility in increasing concentrations of $(NH_4)_2SO_4$, which is used because of its high solubility in water, saturating at around 4 mol L^{-1} (>500 g L^{-1}). After each addition of ammonium sulfate, the mixture is centrifuged (p. 383) to remove precipitated proteins. The supernatant can then be used for further additions/fractions. The precipitated protein is often 'desalted' by dialysis against a suitable buffer, by ultrafiltration, or by gel permeation chromatography (p. 364). Alternative approaches include: precipitation by changing pH, differential solvent/polymer extraction (e.g. using ethanol or polyethylene glycol) and heat denaturation (useful for heat-stable proteins, including some bacterial exotoxins).
2. **Chromatographic separation.** For example, gel permeation (size exclusion) chromatography (p. 364) can be used to further fractionate a protein mixture obtained by 'salting out'. It is a low-capacity method where separation is independent of the composition of the mobile phase. Additional chromatographic stages might be used,

including ion-exchange chromatography (p. 364) and hydrophobic interaction chromatography (HIC).

3. **Electrophoresis**. This is often used to characterise the protein composition of fractions obtained at different stages in the process, typically using polyacrylamide gel electrophoresis (PAGE) of subsamples from each stage to indicate the number of proteins present (p. 370). Thus a protein purified to homogeneity will give a single 'band' after suitable staining (p. 368). SDS-PAGE (p. 368) can be used to provide information on the molecular mass of the individual protein constituents.

For detailed information on separation of proteins, a specialist text should be consulted, e.g. Simpson *et al.* (2008) or Wrolstad *et al.* (2004).

Table 59.1 The 20 amino acids incorporated into protein, grouped according to their side chains

Name	Three letter code	Capital letter code
Aliphatic side chains		
Glycine	gly	G
Alanine	ala	A
Valine	val	V
Leucine	leu	L
Isoleucine	ile	I
Aromatic side chains		
Phenylalanine	phe	F
Tyrosine	tyr	Y
Tryptophan	trp	W
S-containing side chains		
Cysteine	cys	C
Methionine	met	M
Side chains with –OH groups*		
Serine	ser	S
Threonine	thr	T
Basic side chains		
Histidine	his	H
Lysine	lys	K
Arginine	arg	R
Acidic side chains		
Aspartate	asp	D
Glutamate	glu	E
Amide side chains		
Asparagine	asn	N
Glutamine	gln	Q
Cyclic structure (imino acid)		
Proline	pro	P

*tyr also has an OH group.

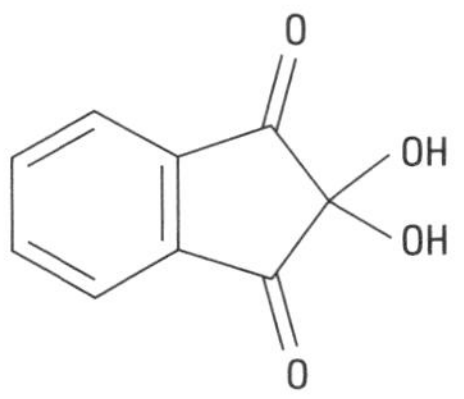

Fig. 59.2 Structure of ninhydrin

SAFETY NOTE Hydrolysis of proteins involves working with hot acids. Wear suitable personal protective equipment (e.g. suitable gloves and safety glasses) and work in a fume hood.

Amino acid assays

The 20 amino acids most commonly found in proteins are typically subdivided into groups, according to the biochemical features of their side chains (see also Table 59.1):

- **Aliphatic** (with linear or branched side chains) – e.g. glycine, alanine, valine, leucine, isoleucine cysteine, methionine.
- **Aromatic** (side chains containing a benzene ring) – e.g. phyenylalanine, tyrosine, tryptophan.
- **Charged** (side chains containing an additional acidic or basic group) – e.g. arginine, lysine histidine, aspartate, glutamate.
- **Polar** (side chains are hydrophilic) – e.g. serine, threonine, asparagine, glutamine.

The remaining example, praline, does not readily fit into the above scheme as it is technically an imino acid, with a five-membered ring structure that includes the amino group. Note also that other amino acids (e.g. hydroxyproline) occur in proteins as a result of post-translational modification.

Detection and quantification of amino acids

The primary amino group of amino acids will react with ninhydrin (2,2 dihydroxyindane-1,3-dione, see Fig. 59.2) to give a purple-coloured product – this reaction can be used for qualitative assay, e.g. to detect individual amino acids during separation, as described below, and for quantitative measurement, e.g. by measuring the absorbance at 570 nm using a spectrophotometer (p. 355). Note that different amino acids give different amounts of coloured product on reaction with ninhydrin, so careful standardisation is required. Proline and hydroxyproline give yellow products that can be assayed at 440 nm.

Quantitative analysis of amino acid composition of proteins

This can be carried out in crude mixtures, or on purified proteins, with the following major steps:

1. **Acid hydrolysis of proteins** – e.g. using 6 mol L^{-1} HCl at 110°C for 24 h. However, a problem with this approach is that it results in the

partial destruction of selected amino acids, for example cysteine and methionine and tryptophan. One means of trying to quantify the extent of destruction is to assay the hydrolysate after 24 h, 48 h and 72 h and extrapolate back to zero time to determine the original concentration of these labile amino acids. Also, as tryptophan is typically destroyed by acid hydrolysis, a separate procedures is required, e.g. alkali hydrolysis. Another approach used to test the destruction of particular amino acids is to measure the quantity of a known amount of added amino acid as a marker; this is broadly equivalent to the 'internal standard' procedure used in chromatography (p. 366).

Dansylated amino acids – these are prepared by reaction with dansyl chloride (5-dimethylamino-1-naphthalene sulfonyl chloride) to produce derivatives that can be readily detected by their fluorescence using a suitable detector.

2. **Separation of amino acids.** Typically, this will involve one or more forms of chromatography, e.g. thin layer chromatography (p. 360) for qualitative analysis, or column chromatography using a suitable solvent and stationary phase for quantitative studies. Various approach can be used including: reverse phase high-performance liquid chromatography (HPLC, p. 362) often in the form of a dedicated amino acid analyser that separates and assays dansylated derivatives of amino acids. An alternative approach is to use ion exchange column chromatography with detection using ninhydrin or fluorescamine reagents. Gas chromatography (GC, p. 353) can also be used, though this requires conversion of amino acids to volatile derivatives prior to their separation; derivatives include trimethlysilyl and other esters, though interference due to cross-reactions with other food constituents can be a problem in crude fractions.

Definition

Auxotroph – an organism that has lost the ability to make a particular metabolite and which therefore requires this substance in order to grow, e.g. a leucine auxotroph would grow only in a medium supplemented with this amino acid.

3. **Calculation of the amount of each amino acid.** Typically, this will involve measurement of the amount of each separate amino acid 'peak' at the end point of the chromatographic separation, using the principles outlined in Chapter 54. The measured response is then converted into a specific amount of the test substance, in mass or molar terms, using either external or internal standardisation (p. 365).

Microbiological bioassay is a completely different approach, sometimes used to quantify the amount of a specific amino acid when other suitable equipment is not available. This type of assay system is often illustrated within practical classes in food microbiology. It is based on the use of an auxotrophic strain of a specific bacterium that requires the test amino acid in order to grow. Consequently, the growth of this organism in a medium in which all other nutrients are present in excess will be directly proportional to the amount of the amino acid.

Amino acid sequence analysis

Having obtained the amino acid composition of a purified protein (polypeptide), the next step is often to determine the order of the amino acid residues along the polypeptide chain, i.e. the *primary sequence* of the protein. This can be achieved by:

1. **Cleavage of the polypeptide into small fragments** (peptides) using a suitable chemical reagent (e.g. cyanogen bromide) or enzyme (e.g. trypsin) to cleave at known points in the sequence (e.g. at methionine and lysine/arginine residues respectively).

2. **Separation of the peptides**, e.g. by column chromatography.
3. **Determination of the amino acid sequence of each peptide.** Typically, this involves a process termed 'Edman degradation', which involves derivatising the amino (N) terminal residue, followed by acid hydrolysis and identification of the specific derivative using chromatography. Repeated cycles of Edman degradation release the N-terminal residue on each occasion and subsequent analysis allows the sequence of peptides of up to 50 amino acids to be determined.
4. **Determination of the amino acid sequence of the entire polypeptide chain.** This is achieved by matching the sequenced, overlapping peptide fragments obtained using different reagents in a process termed 'sequence alignment'.

Establishing the primary structure (amino acid sequence) of a protein – sequence alignment of all but the smallest proteins is best carried out using dedicated software, e.g. Clustal.

Protein purification

Much of the current knowledge about metabolic and physiological events has been gained from *in vitro* studies of purified proteins. Such studies range from investigations into the kinetics and regulation of enzymes to the determination of the structure of a protein and its relationship to function. In addition, certain purified proteins have a role as therapeutic agents, e.g. Factor VIII in the treatment of haemophilia.

Protein purification and proteomics – an important aspect of proteomics is the characterisation of individual proteins, using some of the techniques described in this chapter.

> **KEY POINT** The purification of most proteins involves a series of procedures, based on differential solubility and/or chromatography, that selectively separate the protein of interest from contaminating proteins and other material. It is rare for purification to be a one-step process.

Preliminary considerations for protein purification

Finding out about the protein

Rather than approaching every purification step on a 'trial and error' basis, try to find out as much as you can about the physical and biological properties of the desired protein before you begin your practical work. Even if the protein to be purified is novel, it is likely that similar proteins will be described in the literature. Information on the isoelectric point (pI) and M_r, of both the protein and of the likely major contaminants, is particularly useful for ion-exchange and gel permeation chromatography (p. 364). Knowledge of other factors such as metal ion and co-enzyme requirements, presence of thiol groups and known inhibitors can indicate useful chromatographic steps, and may allow you to take steps to preserve the tertiary structure and biological activity of the desired protein during the purification process.

Choosing the source material

Source material for protein purification – an important advance has been the use of genetically engineered organisms, designed to produce large amounts of a particular protein.

If you have a choice of starting material, use a convenient source in which the protein of interest is abundant. This can vary from animal tissues (e.g. heart, kidney or liver) obtained from an abattoir or from laboratory animals, to plant material, or to microbial cells from a laboratory culture.

Homogenisation and solubilisation

Unless the protein is extracellular, the cells in the source material need to be disrupted and homogenised by one of the methods described in Chapter 57. Following homogenisation, proteins from the cytosol or extracellular fluid will normally be present in a soluble form, but membrane-bound proteins and those within organelles will require further treatment. Isolation of organelles by differential centrifugation (p. 384) will provide a degree of purification, while membrane-bound proteins will need organic solvents or detergents to render them soluble. Once the protein of interest is in a soluble form, any particulate material in the extract should be removed, e.g. by centrifugation or filtration.

Storing protein solutions during a purification procedure – overnight storage at 4°C is acceptable, but it is advisable to include bacteriostatic agents (e.g. azide at 0.5% w/v) and protease inhibitors: freezing may be required for longer-term storage – use liquid nitrogen or dry ice/methanol for rapid freezing. If freezing and thawing might denature the sample, include 20% v/v glycerol in buffers, to allow storage at −20°C in liquid form.

Preserving enzyme activity during purification

Once an enzyme has been released from its intracellular environment, it will encounter potentially adverse conditions that may result in denaturation and permanent inactivation, e.g. a lower pH, an oxidising environment or exposure to lysosomal proteases. During homogenisation, and in some or all of the purification steps, buffers should contain reagents that can counteract these potentially damaging effects. Fewer precautions may be needed with extracellular enzymes. For most proteins, the initial procedures should be carried out at 4°C to minimise the risk of proteolysis.

Devising a strategy for protein purification

Working with recombinant proteins – these are often produced with either N-terminal or C-terminal hexahistidine ‘tags’, enabling them to be purified readily using immobilised metal affinity chromatography (p. 365).

Any successful protein purification scheme will exploit the unique properties of the desired protein in terms of its size, net charge, hydrophobic nature, biological activity, etc. The chromatographic techniques that separate on the basis of these properties are detailed in Chapter 54. However, the *order* in which the various purification steps are carried out needs some thought, and each stage needs to be considered in relation to the following factors:

- **Capacity** – the amount of material (volume or concentration) that the technique can handle. High capacity techniques such as ammonium sulphate precipitation (see later) and ion-exchange chromatography (p. 364) should be used at an early stage, to reduce the sample volume.
- **Resolution** – the efficiency of separating one component from another. At later stages of the purification, the sample volume will be considerably reduced, but any contaminating proteins may well have very similar properties to those of the protein of interest. Therefore, a high resolution (but low capacity) technique will be required, such as covalent chromatography or immobilised metal affinity chromatography (IMAC).
- **Yield** – the amount of protein recovered at each step, expressed as a percentage of the initial amount (p. 403). For example, yields of >80 per cent can be obtained with ammonium sulphate precipitation. With certain types of affinity chromatography, where harsh elution conditions need to be employed, yields may be quite low (e.g. <20 per cent). Some of the other possible reasons for a decreased yield are considered in Table 59.2.

Measuring yield – note that yields of >100 per cent may be obtained if an inhibitor is lost during purification.

Table 59.2 Principal causes of decreased yield in protein purification

Cause	Possible solution
Denaturation	Include EDTA or reducing agents in buffers; avoid extreme temperatures
Inhibition	Check buffer composition for possible inhibitors
Proteolysis	Include protease inhibitors in buffers
Non-elution	Alter salt concentration or pH of the eluting buffer
Co-factor loss	Recombine fractions on a trial-and-error basis

A typical step-wise purification procedure would be:

Step 1 **Ammonium sulphate precipitation** – a high capacity, low resolution technique that yields a protein solution with a greatly reduced volume and a high concentration of ammonium sulphate.

Step 2 **Hydrophobic interaction chromatography** (HIC) – an absorption technique (p. 365) using a starting buffer with a high ammonium sulphate concentration. During development of the column, the ionic strength of the mobile phase is reduced in a gradient elution.

Step 3 **Ion-exchange chromatography** (IEC) – another absorption technique (p. 364). Here the starting buffer has a low ionic strength, and a gradient of increasing ionic strength is used to separate components.

Step 4 **Gel permeation chromatography** (GPC) – a low capacity method (p. 364) where separation is independent of the composition of the mobile phase.

The sample from step 1 may be able to be applied directly to the HIC column, and that from step 2 may be used for IEC without changing the buffer. A concentration step (p. 406) would be required before step 4. The above scheme is very much an idealised approach: in practice, buffers may need to be changed by dialysis or ultrafiltration, as described later. However, the principle of using the minimum number of manipulations to obtain the desired purification remains valid.

Calculating total protein content – multiply the protein concentration by the total volume, making sure that the units are consistent, e.g. mg mL^{-1}×mL. A similar procedure is required to determine the total amount of enzyme.

Using immunoassays – note that inactive or denatured forms of the protein may be detected along with the biologically active form.

Monitoring protein purification

At each step, the separated material is usually collected as a series of 'fractions'. Each fraction must be assayed for the protein of interest, and fractions containing that protein are pooled prior to the next step. The assay performed will depend on the properties of the protein of interest but specific enzyme assays or immunoassays are most commonly used. The protein concentration and the volume of the pooled fraction must be determined, and these values, together with those obtained for the amount of the protein of interest, are used to determine the purity and yield after each step. For enzymes, the biological activity is measured and used to calculate the specific activity (the enzyme activity per unit mass of protein) as described on pages 406–7. By determining the specific activity of the enzyme at each step, the degree of purification, or purification factor (*n*-fold purification), can be obtained from this relationship:

$$\text{Purification factor} = \frac{\text{specific activity after a particular step}}{\text{specific activity of initial sample}} \qquad [59.3]$$

Example For an enzyme extract having a specific activity of 250.2 U mg^{-1} compared with an initial specific activity of 45.5 U mg^{-1}, using eqn [59.1] gives a purification factor of 250.2 ÷ 45.5 = 5.5-fold (to one decimal place).

Increased purification usually represents a decrease in total protein relative to the biological activity of the protein of interest, though in some instances it may reflect the loss of an inhibitor during a purification step.

Calculation of the yield of enzyme at each step is straightforward, since:

$$\text{Yield} = \frac{\text{total enzyme activity after a particular step}}{\text{total enzyme activity in initial sample}} \times 100\ (\%) \qquad [59.4]$$

Example For an enzyme extract having a total activity of 8521 U compared with an initial total activity of 9580 U, using eqn [59.4] gives a yield of [8521 ÷ 9580] × 100 = 88.9% (to one decimal place).

Note that the yield equation uses the total amount of enzyme and is therefore unaffected by the volume of the solutions involved. You should

Table 59.3 Example of a record of purification for an enzyme

Step	Procedure	Total protein (mg)	Total enzyme (U)*	Specific activity (U mg^{-1})	Purification factor (*n*-fold)	Enzyme yield (%)
Initial sample	—	210	1984	9.45	—	100.0
Step 1	'Salting out'	112	1740	15.54	1.6	87.7
Step 2	HIC	30	1701	56.70	6.0	85.7
Step 3	IEC	6.0	1604	267.33	28.3	80.8
Step 4	GPC	3.0	1550	516.66	54.7	78.1

*U = unit (p. 407).

make a record of the progress of your purification procedure at each step, as in Table 59.3.

Assaying enzymes – if possible, use a preliminary screening technique to detect active fractions prior to quantitative assay, e.g. using tetrazolium dyes in microtitre plates to detect oxidoreductases.

Measuring the mass of proteins – this is sometimes expressed in daltons (Da), or kilodaltons (kDa), where 1 dalton = 1 atomic mass unit. M_r is an alternative, numerically equivalent expression.

Using ammonium sulphate – although the phenomenon of salting out is seen with several salts, $(NH_4)_2SO_4$ is the most widely used because it is highly soluble (saturating at ≈4 mol L^{-1}), inexpensive, and can be obtained in very pure form.

Monitoring progress using electrophoresis

Polyacrylamide gel electrophoresis (PAGE) of subsamples from each step will give an indication of the number of proteins still present: ideally, a pure protein will give only one band after silver staining; however, this stain is so sensitive that even trace impurities can be detected in what you expect to be a pure sample, so don't be dismayed by the appearance of the gel after staining.

Carrying out SDS-PAGE (p. 368) and isoelectric focussing on subsamples also gives information on the M_r and pI of any remaining contaminants, and this can be taken into account when planning the next chromatographic step.

Differential solubility separation techniques for protein purification

Often the volume and protein concentration of the initial soluble extract will be quite high. Application of a differential solubility technique at this stage results in the precipitation of selected proteins. These can be recovered by filtration or centrifugation, washed and then resuspended in an appropriate buffer. This will reduce the sample volume and may give a small degree of purification, making the sample more suitable for subsequent chromatographic steps.

Ammonium sulphate precipitation ('salting out')

This is the most widely used differential solubility technique, having the advantage that most precipitated enzymes are not permanently denatured, and can be redissolved with restoration of activity. Precipitation depends on the existence of hydrophobic 'patches' on the surface of proteins, inducing a reorganisation of water molecules in their vicinity. When ammonium sulphate is added to the extract, it dissolves to give ions that become hydrated, leaving fewer water molecules in association with the protein. As a result, the hydrophobic patches become 'exposed', and hydrophobic interactions between different protein molecules lead to their aggregation and precipitation. The basis of fractionation in this method is that, as the salt concentration of the extract is increased, proteins with larger or more abundant hydrophobic patches will precipitate before those with smaller or fewer patches.

Table 59.4 Amount of $(NH_4)_2SO_4$ ($g\,L^{-1}$) required for a particular percentage saturation

Final concentration (%) →	**20**	**30**	**40**	**50**	**60**	**70**	**80**	**90**	**100**
Initial concentration (%) ↓	**Ammonium sulphate added ($g\,L^{-1}$)**								
0	107	166	229	295	366	442	523	611	707
10	54	111	171	236	305	379	458	545	636
20	—	56	115	177	244	316	392	476	565
30		—	57	119	184	253	328	408	495
40			—	59	122	190	262	340	424
50				—	61	127	197	272	353
60					—	63	131	204	283
70						—	66	136	212
80							—	68	141
90								—	71

Unless you can obtain information from the literature about the $(NH_4)_2SO_4$ concentration that will precipitate your target protein, fractionation is done on a trial-and-error basis. The $(NH_4)_2SO_4$ concentration is expressed in terms of the percentage saturation value: Table 59.4 shows the amount of $(NH_4)_2SO_4$ required to give 20–100% saturation. For each percentage saturation chosen, the $(NH_4)_2SO_4$ salt should be added slowly while stirring, and the mixture left at 4°C for 1 h before centrifuging at 3000 g for 40 minutes. For an effective separation, start with the maximum percentage saturation that does not precipitate the protein of interest, then increase the percentage saturation by the minimum amount that will then precipitate it. The proteins precipitated between any two values of percentage saturation (say between 30 and 50 per cent) are referred to as a 'cut'.

Precipitation by changing pH

Purifying bacterial proteins – adjusting the extract to pH 5 can be useful, since many bacterial proteins have pI values in this region.

Proteins are least soluble at their isoelectric points because, at that pH, there is no longer the repulsion that occurs between positively or negatively charged protein molecules at physiological pH values. If the precipitated proteins are required for further purification, it is essential that the protein of interest is not irreversibly denatured. The method is probably best employed by precipitating contaminating proteins, leaving the desired protein in solution. Use citric acid for <pH 3, acetic acid for <pH 4, and sodium carbonate or ethanolamine for >pH 8.

Heat denaturation

Example The HPII catalase of the bacterium *E. coli* remains intact when heated to 55°C for 15 min, aiding its separation from other proteins.

Exposure of most proteins to high temperatures disrupts their conformation through effects on non-covalent interactions such as hydrogen bonds and van der Waals forces. However, different proteins are denatured, and hence precipitated, at different temperatures, and this can provide a basis for the separation of some heat-stable proteins. By incubating small aliquots (≈1 mL) of extract for one minute at a range of temperatures between 45 and 65°C, it is possible to determine the temperature that gives maximum precipitation of contaminating protein with minimal inactivation of the desired protein.

Definition

Dielectric constant ($\mathcal{E}$) – a dimensionless measure of the screening effect on the force (F) between two charges (q_1 and q_2) due to the presence of solvent.

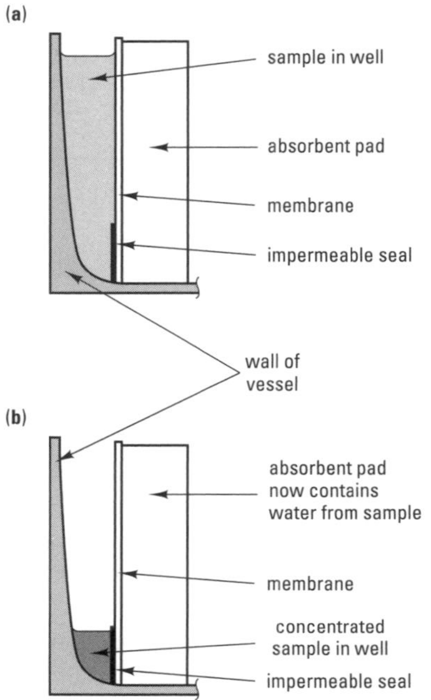

Fig. 59.3 Minicon ultrafiltration system (a) with sample added, (b) after ultrafiltration.

Solvent and polymer precipitation methods

Organic solvents (e.g. acetone, ethanol) cause precipitation of proteins by lowering the dielectric constant of the solution. Performing the precipitation at 0 °C minimises permanent denaturation. Stepwise concentration (% v/v) increments are used, giving 'cuts' of precipitated proteins, as with ammonium sulphate precipitation.

Organic polymers, particularly polyethylene glycol (PEG), also lower the dielectric constant, but at lower concentrations than with acetone or ethanol. The most commonly used PEG preparations have M_r values of 6000 or 20 000, and these can be removed from the sample by ultrafiltration. PEG precipitation does not involve salts, so it may be a useful preliminary step prior to ion-exchange chromatography which starts with low salt buffer. Also, since the techniques of PEG and ammonium sulphate precipitation involve different principles, they can be used sequentially.

Concentration of proteins by ultrafiltration

This involves forcing water and small molecules through a semi-permeable membrane using high pressure or centrifugation. A range of membranes with 'nominal' molecular weight cut-offs between 500 and 300 000 are commercially available (e.g. Amicon, Millipore), with pore sizes of 0.1–10 µm. Concentration of small samples (<5 mL) can be achieved using either a membrane backed by an absorbent pad (e.g. Minicon, Fig. 59.3, available with M_r cut-off from 5000 to 30 000), or by using a centrifugal concentrator (e.g. Vivaspin, Centricon). Larger volumes (up to 400 mL) can be concentrated using a stirred ultrafiltration chamber (an ultrafiltration 'cell') where the liquid is forced through the membrane using nitrogen or an inert gas.

Ultrafiltration not only concentrates the sample, but also may give a degree of purification. It can also be used to change the buffer composition by diafiltration. Note that molecular weight cut-off values are quoted for globular proteins – fibrous proteins of higher M_r may pass through the ultrafiltration membrane.

Assaying enzymes

Enzymes are globular proteins that increase the rate of specific biochemical reactions. Each enzyme operates on a limited number of substrates of similar structure to generate products under well-defined conditions of concentration, pH, temperature, etc. In metabolism, groups of enzymes work together in sequential pathways to carry out complex molecular transformations, e.g. the multi-reaction conversion of glucose to lactate (glycolysis).

Example Enzyme EC 1.1.1.1 is usually known by its trivial name, alcohol dehydrogenase

KEY POINT Enzymes are categorised according to the chemical reactions they catalyse, leading to a four-figure Enzyme Commission (EC) code number and a systematic name for each enzyme. Most enzymes also have a recommended trivial name, often denoted by the suffix 'ase'.

Measuring enzyme activity

This is measured in terms of the rate of enzyme reaction. Activity may be expressed directly as amount of substrate utilised per unit time (e.g. nmol min^{-1}, etc.), or in terms of the non-SI international unit (U or

Example An enzyme preparation that converted 295 μmol of substrate to product in 15 minutes would have an activity, expressed in terms of non-SI units (U) of: 285 ÷ 15 = 19.7 U (to three significant figures). Using eqn [59.5], the same preparation would have an activity, expressed in katals, of: $295 \times 10^{-6} \div 900 = 3.27 \times 10^{-7}$ = 329 nkat (to three significant figures).

Example The hydrolysis of one molecule of maltose to give two glucose molecules by α-glucosidase means that enzyme activity specified in terms of substrate consumption (nmol maltose) would be half the value expressed with respect to product formation (nmol glucose).

Definition

Specific activity – enzyme activity (e.g. kat, U, ng min^{-1}) expressed per unit mass of protein present (e.g. mg, μg).

Example Phosphoenolpyruvate carboxylase (PEP carboxylase) can be assayed by coupling to malate dehydrogenase (MDH): the PEP carboxylase reaction converts phosphoenolpyruvate to oxaloacetate, which is then oxidised to malate by MDH (present in excess), with a stoichiometric (1:1) reduction of NAD^+. The coupled assay is monitored spectophotometrically as an increase in A_{340} against time (p. 409).

sometimes IU), defined as the amount of enzyme which will convert 1 μmol of substrate to product(s) in one minute under specified conditions. However, the recommended (SI) unit of enzyme activity is the katal (kat), which is the amount of enzyme which will convert 1 mol of substrate to product(s) in 1 s under optimal conditions, determined from the following equation:

$$\text{enzyme activity (kat)} = \frac{\text{substrate converted (mol)}}{\text{time (s)}} \qquad [59.5]$$

This unit is relatively large ($1\ \text{kat} = 6 \times 10^7\ \text{U}$) so SI prefixes are often used, e.g. nkat or pkat (p. 199). Note that the units involve amount of substrate (mol), not concentration ($mol\ L^{-1}$ or $mol\ m^{-3}$).

For enzymes with macromolecular substrates of unknown molecular weight (e.g. deoxyribonuclease, amylase), activity can be expressed as the mass of substrate consumed (e.g. ng DNA min^{-1}), or amount of product formed (e.g. nmol glucose min^{-1}). You must ensure that your units clearly specify the substrate or product used, especially when the enzyme transformations involve different numbers of substrate or product molecules. Specific activity, expressed in terms of the amount or mass of substrate or product, is useful for comparing the purity of different enzyme preparations.

The turnover number of an enzyme is the amount of substrate (mol) converted to product in 1 s by 1 mol of enzyme operating under optimum conditions. In practice, this requires information on the molecular weight of the enzyme, the amount of enzyme present and its maximum activity.

The rate of substrate utilisation or product formation must be measured under controlled conditions, using some characteristic which changes in direct proportion to the concentration of the test substance.

Spectrophotometric assays

Many substrates and products absorb visible or UV light and the change in absorbance at a particular wavelength provides a convenient assay method (p. 356). In other cases, a product may be measured by a colorimetric chemical reaction.

Several assays are based on interconversion of the nicotinamide adenine dinucleotide coenzymes NAD^+ or $NADP^+$ which are reversibly reduced in many enzymic reactions. The reduced form (either NADH or NADPH) can be detected at 340 nm, where the oxidised form has negligible absorbance (p. 355). An alternative approach is to use a coupled enzyme assay, where a product of the test enzyme is used as a substrate for a second enzyme reaction which involves oxidation/reduction of nucleotide coenzymes. Such assays are particularly useful for continuous monitoring of enzyme activity and for reactions where the product from the test substance is too low to detect by other methods, since coupled assays are more sensitive. Note that the reaction of interest (test enzyme) must be the rate-limiting process, not the indicator reaction (second enzyme).

Radioisotopic assays

These are useful where the substrate and product can be easily separated, e.g. in decarboxylase assays using a ^{14}C-labelled substrate, where gaseous $^{14}CO_2$ is produced.

Electrochemical assays

Enzyme reactions involving acids and bases can be monitored using a pH electrode (p. 174), though the change in pH will also affect the activity of the test enzyme. An alternative approach is to measure the amount of acid or alkali required to maintain a constant preselected pH in a pH-stat.

An oxygen electrode can be used if O_2 is a substrate or a product. Other ion-specific electrodes can monitor ammonia, nitrate, etc.

Chromogenic and fluorogenic enzyme substrates

Artificial substrates that generate either (i) a coloured product (i.e. they are chromogens) or (ii) a fluorescent product (fluorogens) are used widely for the detection and quantification of enzymes, especially when no suitable spectrophotometric assay is available for the natural product or the natural substrate. In most cases, these artificial substrates are analogues of the natural substrates, composed of a 'core molecule' that is either fluorescent or coloured, coupled to another group by a covalent bond. The addition of this group to the core molecule usually either causes a decrease in fluorescence and a shift in the excitation and emission signal to longer wavelengths, or converts the coloured compound to a non-coloured form. The covalent bond between core molecule and added group is recognised and cleaved by the target enzyme to liberate the core molecule, thereby generating either a fluorescent or a coloured product that can be measured fluorimetrically or spectrophotometrically. The principle is best illustrated using a specific example of each type of substrate (see Fig. 59.4). Such substrates serve as a sensitive and specific means of detecting individual enzymes in the presence of a range of other biomolecules, e.g. in foods. Applications include:

- biochemical identification of bacteria, on the basis of enzyme profiles;
- enzyme-linked immunoassays, e.g. ELISA (p. 350);
- nucleic acid hybridisation and blotting (p. 374);
- quantitative assay for enzyme activity.

Note that care is required for quantitative fluorimetric assay, since impurities in the enzyme preparation can lead to background fluoresence or to quenching (reduction) of the signal: appropriate controls must be run at the same time as test samples. A wide range of chromogenic core molecules (chromophores) are used in artificial substrates (Table 59.5), including: nitrophenols (e.g. ONP, Fig. 59.4); nitroanilines (e.g. Z-arginine-*para*-nitroanilide derivatives for assaying trypsin activity); indoxyl substrates (e.g. 5-bromo-4-chloro-3-indolyl-β-D-galactoside, also known as X-GAL, which is used widely to detect the activity of the *lacZ* gene in molecular biology, p. 378), along with derivatives of alizarin and 3,3′,5,5′-tetramethylbenzidine (Fig. 59.5). Similarly, several different fluorophores are commercially available, including coumarin derivatives (e.g. 4-methylumbelliferone, MU, Fig. 59.4, and 7-amino-4-methylcoumarin, AMC, which fluoresce blue), fluorescein and resorufin (Table 59.5).

Example Detection of enzymes that degrade β-lactam antibiotics in resistant bacteria can be carried out using the chromogenic substrate nitrocefin, which changes from pale yellow to red when hydrolysed by β-lactamases.

Table 59.5 Core molecules used in chromogenic and fluorogenic enzyme substrates

Core molecule	Reaction
Alizarin	Red colour
Aminomethyl coumarin	Blue fluorescence
Fluorescein	Green fluorescence
Indoxyl (and derivatives)	Blue colour (and others)
Methylumbelliferone	Blue fluorescence
Nitroaniline	Yellow colour
Ortho-nitrophenol	Yellow colour
Tetramethylbenzidine	Blue (yellow at low pH)
Rhodamine	Red fluorescence

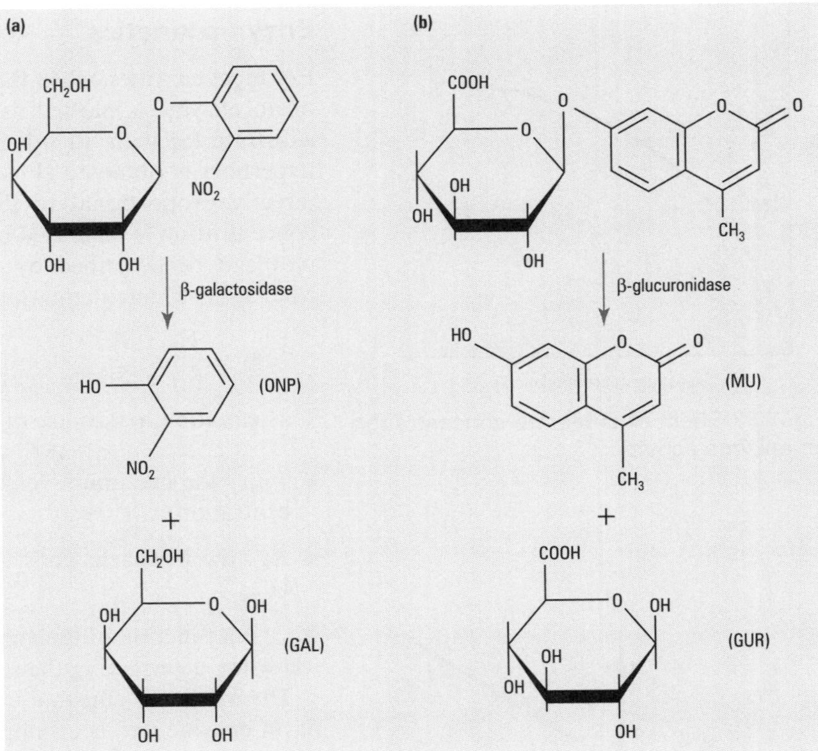

Fig. 59.4 (a) The chromogenic substrate *ortho*-nitrophenyl-β-D-galactoside (ONP-β-GAL); (b) the fluorogenic substrate 4-methylumbelliferyl-β-D-glucuronide (MU-β-GUR). Both substrates show minimal colour and/or fluorescence due to the coupling of the core molecule to a carbohydrate group. ONP-β-GAL is cleaved by the enzyme β-galactosidase to liberate ONP, which is yellow (usually assayed spectrophotometrically at 420 nm). In contrast, the cleavage of MU-β-GUR by the enzyme β-glucuronidase liberates MU, which is strongly fluorescent under UV light (e.g. excitation wavelength 360 nm, emisson wavelength 440 nm). These two substrates are used to detect and differentiate between coliform bacteria (which contain β-galactosidase) and *Escherichia coli* (which contains both β-galactosidase and β-glucuronidase), e.g. in the IDEXX Colilert system for water analysis (see: http://www.idexx.com/water/colilert/).

Fig. 59.5 The chromophore 3,3′,5,5′-tetramethylbenzidine (TMB), used in peroxide assays, e.g. in ELISA (p. 350).

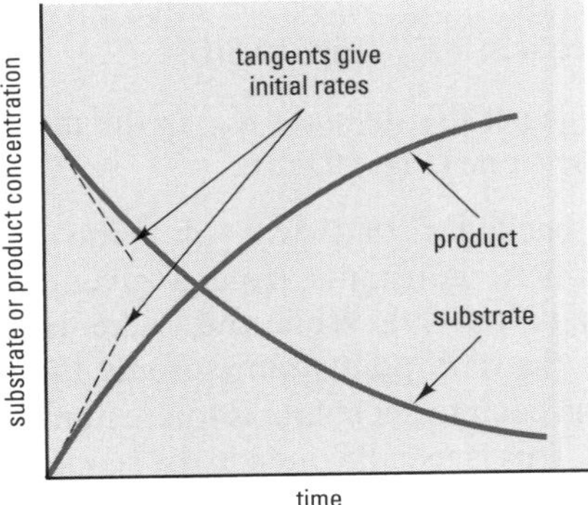

Fig. 59.6 Enzyme reaction progress curve: sub-strate utilisation/product formation as a function of time.

Methods of monitoring substrate utilisation/product formation

Continuous assays (kinetic assays)

The change in substrate or product is monitored as a function of time, to provide a progress curve for the reaction (Fig. 59.6). These curves start off in a near-linear manner, decreasing in slope as the reaction proceeds and substrate is used up. The initial velocity of the reaction (v_0) is obtained by drawing a tangent to the curve at zero time and measuring its slope. Continuous monitoring can be used when the test substance can be assayed rapidly (and non-destructively), e.g. using a chromogenic substrate. Reaction rate analysers allow simultaneous addition of reactant(s) or enzyme, mixing and measurement of absorbance: this enables the initial rate to be determined accurately.

Discontinuous assays (fixed time assays)

It is sometimes necessary to measure the amount of substrate consumed or product formed after a fixed time period, e.g. where the test substance is assayed by a (destructive) colorimetric chemical method. It is vital that the time period is kept as short as possible, with the change in substrate concentration limited to around 10%, so that the assay is within the linear part of the progress curve (Fig. 59.6). A continuous assay may be carried out as a preliminary step, in order to determine whether the reaction is approximately linear over the time period to be used in the fixed time assay.

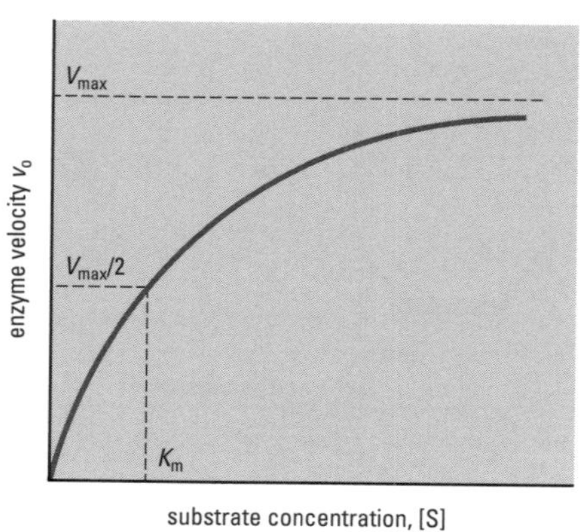

Fig. 59.7 Effect of substrate concentration on enzyme activity.

Enzyme kinetics

For most enzymes, when the initial reaction rate (v_0) using a fixed amount of an enzyme is plotted as a function of the concentration of a single substrate [S] with all other substrates present in excess, a rectangular hyperbola is obtained (Fig. 59.7). At low substrate concentrations v_0 is directly proportional to [S], with a decreasing response as substrate concentration is increased until saturation is achieved. The shape of this plot can be described by a mathematical relationship, known as the Michaelis–Menten equation:

$$v_0 = \frac{V_{max}\,[S]}{K_m + [S]} \qquad [59.6]$$

This equation makes use of two kinetic constants:

- V_{max}, the maximum velocity of the reaction (at infinite substrate concentration).
- K_m, the Michaelis constant, is the substrate concentration where $v_0 = \frac{1}{2}V_{max}$.

V_{max} is a function of the amount of enzyme and is the appropriate rate to use when determining the specific activity of a purified enzyme.

The Michaelis constant can provide a measure of the substrate affinity of an enzyme and is an important characteristic of a particular enzyme. Thus, an enzyme with a large K_m usually has a low affinity for the substrate whereas an enzyme with a small K_m usually has a high affinity.

Your first step in determining the kinetic constants for a particular enzyme is to measure the rate of reaction at several substrate concentrations. There are various ways to obtain K_m and V_{max} from such data, mostly involving drawing a graph representing a linear transformation of eqn [59.6]:

- The Lineweaver–Burk plot: a graph of the reciprocal of the reaction rate ($1/v_0$) against the reciprocal of the substrate concentration (1/[S]) gives $-1/K_m$ as the intercept of the x axis and $1/V_{max}$ as the intercept of the y axis (Fig. 59.8(a)). Note that the slope of the plot is most affected by the least-accurate values, i.e. those measured at low substrate concentration.
- The Eadie–Hofstee plot: v_0 against v_0/[S], where the intercept on the y axis gives V_{max} and the slope equals $-K_m$ (Fig. 59.8(b)).
- The Hanes–Woolf plot: [S]/v_0 against [S], giving $-K_m$ as the intercept of the x axis and $1/V_{max}$ from the slope (Fig. 59.8(c)).

(a)

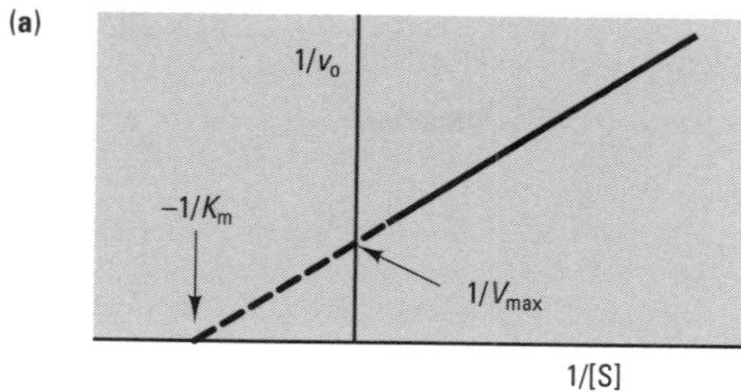

(b)

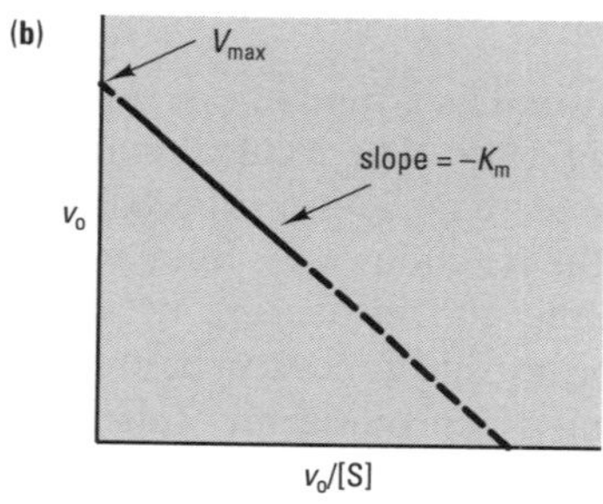

(c)

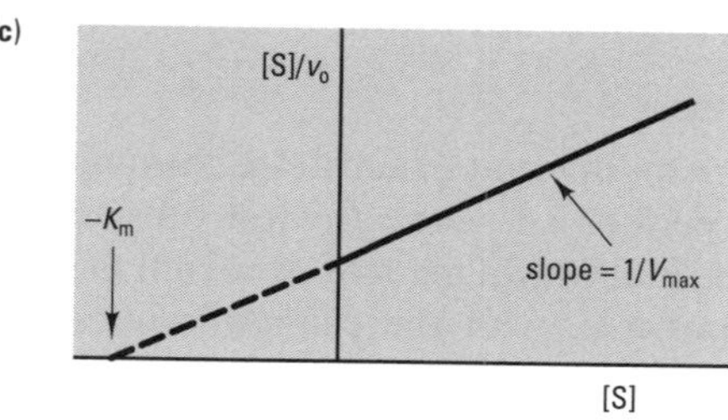

Fig. 59.8 Graphical transformations for deter-mining the kinetic constants of an enzyme. (a) Lineweaver–Burk plot. (b) Eadie–Hofstee plot. (c) Hanes–Woolf plot.

There are several computer packages that will plot the above relationships and calculate the kinetic constants from a given set of data using linear regression analysis (p. 339). While the Eadie–Hofstee and Hanes–Woolf plots distribute the data points more evenly than the Lineweaver–Burk plot, the best approach to such data is to use non-linear regression on untransformed data. This is usually outside the scope of the simpler computer programs, although tailor-made commercial packages can carry out such analyses. Note also that some enzymes do not show Michaelis–Menten kinetics, e.g. those with more than one active site per molecule (allosteric enzymes), which often give sigmoid curves of v_0

against [S]. Such enzymes are usually involved in the control of metabolism.

Factors affecting enzyme activity

If you want to measure the maximum rate of a particular enzyme reaction, you will need to optimise the following:

Substrate concentration

The substrates must be present in excess, to ensure maximum reaction velocity. Equation [59.6] shows that a substrate concentration equivalent to $10 \times K_m$ will give 91% of V_{max}, while a concentration of $100 \times K_m$ will give 99 per cent of the maximum rate.

Cofactors

Many enzymes require appropriate concentrations of specific cofactors for maximum activity. These are subdivided into coenzymes (soluble, low molecular weight organic compounds that are actively involved in catalysis by accepting or donating specific chemical groups, i.e. they are co-substrates of the enzyme; examples include NAD^+ and ADP); and activators (inorganic metal ions, required for maximal activity, e.g. Mg^{2+}).

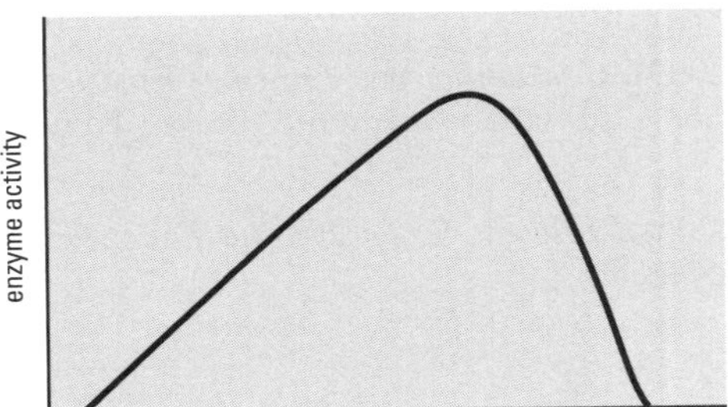

Fig. 59.9 Effect of temperature on enzyme activity.

Temperature

Enzyme activity increases with temperature, until an optimum is reached. Above this point, activity decreases as a result of protein denaturation (Fig. 59.9). Note that the optimum temperature for enzyme *activity* may not be the same as that for maximum *stability* (enzymes are usually stored at temperatures near to or below 0°C, to maximise stability).

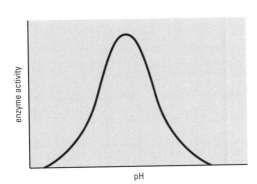

Using stored enzymes – When removing a bottle of freeze-dried enzyme from a freezer or fridge, do not open it until it has been warmed to room temperature or water may condense on the contents – this will make weighing inaccurate and may lead to loss of enzyme activity.

pH

Enzymes work best at a particular pH, due to changes in ionisation of the substrates or of the amino acid residues within the enzyme (Fig. 59.10). Most enzyme assays are performed in buffer solutions (p. 177), to prevent changes in pH during the assay.

KEY POINT Note that temperature and pH optima are dependent upon reaction conditions – you should therefore specify the experimental conditions under which such optima are determined.

enzyme activity

pH

Fig. 59.10 Effect of pH on enzyme activity.

Inhibitors

Many compounds can reduce the rate of an enzyme reaction, e.g. substances that compete for the active site of the enzyme due to similarities in chemical structure with the natural substrate (competitive inhibitors), or substances that bind to the enzyme at other sites, inhibiting normal function (non-competitive inhibitors). Most non-competitive inhibitors are not chemically related to the natural substrate, e.g. heavy-metal ions, such as Hg^{2+} and Cd^+, acting as irreversible, non-competitive inhibitors. Competitive inhibitors reduce K_m but have no effect on V_{max} whereas non-competitive inhibitors reduce V_{max} but have no effect on K_m.

Text references

Nielsen, S.S. (2003) *Food Analysis*, 3rd edn. Springer, New York.

Simpson, R.J., Adams, P.D. and Golemis, E.A. (2008) *Basic Methods in Protein Purification and Analysis: Laboratory Manual*. Cold Spring Harbor Laboratory Press, New York.

Warburg, O. and Christian, W. (1942) 'Isolierung und Kristallisation des Garungferments Enolase', *Biochemische Zeitschrift*, **310**, 384–421.

Wrolstad, R.E., Decker, E.A., Schwartz, S.J. and Sporns, P. (2004) *Handbook of Food Analytical Chemistry, Water, Proteins, Enzymes, Lipids and Carbohydrates*. Wiley, Chichester.

Sources for further study

Bonner, P.R.L (2002) *Protein Purification: The Basics*. Taylor and Francis, New York.

Copeland, R.A. (2000) *Enzymes: A Practical Introduction to Structure, Mechanism, and Data Analysis*. Wiley, New York.
[Covers basic principles, kinetics and experimental systems.]

Li, L.C. (2006) *Protocols Online*. Available: http://www.protocol-online.org/
Last accessed: 21/12/10.
[Links to a range of resources, procedures, background information and commercial examples – search the site for protein purification protocols.]

Nielsen, S.S. (2003) *Food Analysis: Laboratory Manual,* 3rd edn. Springer, New York.

Purich, D.L. (2009) *Contemporary Enzyme Kinetics and Mechanism*. 3rd edn. Academic Press, New York.

Whitford, D. (2005) *Proteins: Structure and Function*. Wiley, Chichester.

60 Assaying lipids

The term 'lipid' is used to describe a broad group of compounds with a wide variety of chemical structures and physical properties. Dietary lipids represent a major source of energy, as well as providing precursors for the synthesis of important biomolecules, including some hormones. Lipids are also important components of foods in terms of their positive effects on organoleptic characteristics, including texture, mouthfeel and appearance (see Chapter 65).

While the biochemical convention is to describe those lipids that are solid at room temperature as 'fats' and those that are liquid are 'oils', in food analysis the term 'fat' is often used interchangeably with 'lipid'; as an example, the lipid content of a food is typically described in terms of the amount of fat per 100 g, as shown on the label (p. 138). In biological systems, lipids are often combined with either proteins (e.g. in lipoproteins) or carbohydrates (e.g. in glycolipids), leading to further complexity.

Table 60.1 Some examples of fatty acids

No. of carbon atoms	Systematic name	Trivial name
Saturated fatty acids (no C═C bonds)		
12	*n*-Dodecanoic	Lauric
14	*n*-Tetradecanoic	Mystiric
16	*n*-Hexadecanoic	Palmitic
18	*n*-Octadecanoic	Stearic
20	*n*-Eicosanoic	Arachidic
22	*n*-Docosanoic	Behenic
Mono-unsaturated fatty acids (one C═C bond)		
12	*cis*-9-Dodecenoic	Lauroleic
14	*cis*-9-Tetradecenoic	Myristoleic
16	*cis*-9-Hexadecenoic	Palmitoleic
18	*cis*-9-Octadecenoic	Oleic
20	*cis*-9-Eicosenoic	Gadoleic
22	*cis*-9-Docosenoic	Erucic

Note: palmitic, stearic, palmitoleic and oleic acids are quantitatively the most common fatty acids in the majority of organisms. Major polyunsaturated fatty acids include linoleic acid (C_{18}, two double bonds), linolenic acid (C_{18}, three double bonds) and arachidonic acid (C_{20}, four double bonds).

> **KEY POINT** The defining feature of lipids is their relative insolubility in water: consequently, they are extracted from biological material using organic solvents, e.g. acetone, ether and chloroform. Because of their diversity and complexity, they are often referred to by more than one name, and the various types of non-systematic names can be confusing.

In terms of food analysis, the key aspects to consider are:

- measurement of total lipid content;
- determining the types of lipids present and their structural organisation;
- assessing the physico-chemical properties of individual constituent lipids, including melting point and rheology (see p. 334).

Lipids are often subdivided into two main types, each of which contains fatty acids as a major structural component:

Simple or neutral lipids

These are esters of fatty acids and an alcohol (e.g. Fig. 60.1). Fatty acids are straight-chain carboxylic acids, typically with an even number of carbon atoms and chain lengths of C_{12}–C_{22}, which may be saturated or unsaturated (Table 60.1). The greater the chain length and the fewer the number of double bonds, the higher the melting point, making most long chain saturated fatty acids solids at room temperature. Glycerol is the most common alcohol found in simple lipids, though higher M_r alcohols occur in waxes, and cyclic alcohols (sterols, e.g. cholesterol) occur in bile acids, steroid hormones and vitamins (e.g. vitamin K). While glycerol is a liquid at room temperature, cholesterol remains solid up to 150°C.

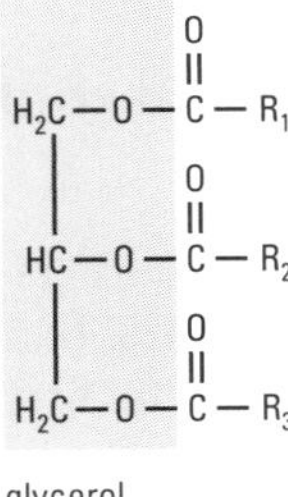

Fig. 60.1 General structure of a neutral lipid: the structure of a triacylglycerol. R = remainder of fatty acid.

Neutral fats (triglycerides or triacylglycerol) are esters of fatty acids and glycerol, as shown in Fig. 60.1. Many animal triglycerides ('fats') contain mainly saturated fatty acids and are solids at room temperature, while plant triglycerides ('oils') often have shorter chain lengths and a greater degree of unsaturation, and are liquids at room temperature. Waxes are esters of fatty acids with alcohols of higher M_r than glycerol. The major biological functions of simple lipids include (i) energy storage, e.g. oils in plant seeds, (ii) insulation, e.g. subcutaneous fat deposits in whales, and (iii) waterproofing, e.g. waxes in the cuticles of plant leaves.

Complex, compound or conjugated lipids

These are acyl esters of glycerol, or the amino alcohol sphingosine, that also include a hydrophilic group (e.g. a phosphoryl or carbohydrate group). They are often described in terms of this hydrophilic group, e.g. phospholipids contain a phosphoryl group while glycolipids contain a carbohydrate group.

KEY POINT An important feature of complex lipids is their amphipathic nature, i.e. each molecule has a polar (hydrophilic) and a non-polar (hydrophobic) region.

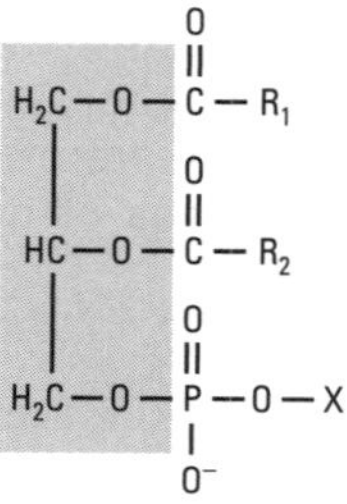

Fig. 60.2 General structure of a phospholipid. R = remainder of fatty acid, X = hydrophilic group, e.g. $-CH_2CH_2NH_3^+$ in phosphatidyl ethanolamine and $-CH_2CH_2N(CH_3)_3^+$ in phosp3hatidyl choline.

The two major types of complex lipid are:

1. **Phospholipids** – the most common types (phosphoglycerides) are based on phosphatidic acid, with two fatty acids esterified to glycerol. Most phospholipids also contain a hydrophilic amino alcohol or a similar group, attached to the phosphoryl group (Fig. 60.2). The principal phospholipid classes are: (i) phosphatidyl cholines (or lecithins), which form stable emulsions with water and dissolve completely on addition of bile salts – these are insoluble in acetone, a feature that enables lecithins to be separated from most other lipids; (ii) phosphatidyl ethanolamines (or cephalins) – unlike lecithins, they are insoluble in ethanol and methanol; (iii) phosphatidyl serine; (iv) phosphatidyl inositol; and (v) plasmalogens.
2. **Sphingolipids** – these incorporate the amino dialcohol, sphingosine, rather than glycerol. Fatty acids are linked to sphingosine via an amide bond to form ceramides, which include: (i) cerebrosides (glyco-sphingolipids); (ii) sulphatides (sulphated cerebrosides); (iii) gangliosides (glycosphingolipids containing sialic acid residues); and (iv) phospho-sphingolipids, including sphingomyelins, which are esters of a ceramide and phosphoryl choline.

Complex lipids have important roles in biological membranes: phospholipids are major structural components while sphingolipids are involved in cell–cell recognition and similar membrane features, e.g. glycosphingolipids are determinants of human ABO blood groups.

SAFETY NOTE Working with solvents – chloroform and benzene must be used with care, due to their high toxicity: they are often replaced by dichloromethane and toluene respectively. All mixing and pouring steps should be carried out in a spark-free fume cupboard. Note that ethers may form explosive peroxides on prolonged storage. Lipid extracts extracted in flammable solvents must be stored in a spark-proof refrigerator, not in routine lab fridges.

Extraction and measurement of total lipid content

Solvent extraction

Following drying and grinding of food, lipid extraction is then carried out using a suitable organic solvent. For foods with a high proportion of lipoproteins or glycolipids, it may be necessary to first carry out acid hydrolysis, e.g. in 3 mol L^{-1} HCl for 1 h, to release the lipid components. For assays of total lipid content, the choice of solvent depends on the expected proportions of non-polar and polar lipids, with non-polar constituents (e.g. triglycerides) being highly soluble in hexane, while polar lipids (e.g. phospholipids) will be extracted more effectively using a polar solvent such as ethanol.

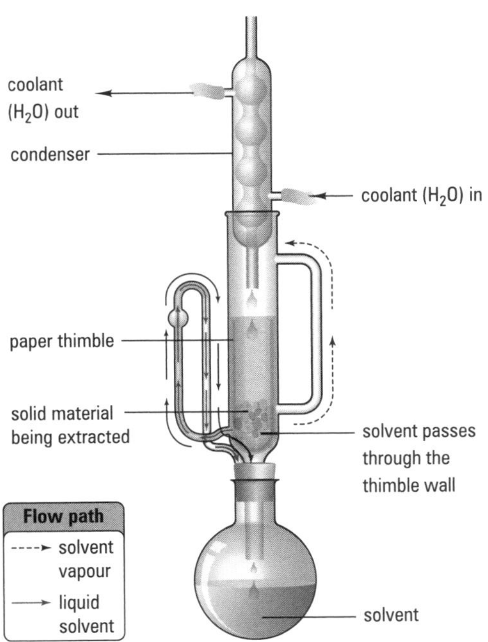

Fig. 60.3 Soxhlet apparatus (heat applied to lower vessel, containing solvent).

Soxhlet method

This is one of the most widely used methods and is based on semi-continuous extraction of lipids using a suitable solvent – this approach greatly increases the efficiency of the extraction and separation of lipids from other components when compared to standard batch solvent extraction and filtration. Following drying and grinding, a known mass of food (Mfood) is added to a porous 'thimble', which is then placed in a Soxhlet apparatus (Fig. 60.3); this comprises a central extraction chamber suspended above a flask containing the solvent with a condenser above. On heating, the solvent evaporates and is distilled into liquid droplets that drip down into the extraction chamber, causing it to fill. Lipids will dissolve from the food sample into the solvent at this stage. Once the extraction chamber is full of solvent, a siphon side-arm automatically empties the solvent back into the flask. This cycle is allowed to repeat for several hours (typically 6 h). Then, the flask is removed, the solvent evaporated (e.g. by rotary evaporation) and the remaining mass of lipid (M_{lipid}) is determined by weighing. The percentage lipid content is then calculated from the relationship:

$$\%\ \text{lipid} = M_{lipid}/M_{food} \times 100 \qquad [60.1]$$

Commercial examples of this apparatus include the Soxtec® and Edutec systems. An alternative approach is the Godlfisch system, which is faster but less efficient at extracting lipids from foodstuffs (see Wrolstad *et al.*, 2004 for procedural details).

Separation and analysis of lipids

Adsorption chromatography

Preventing oxidative rancidity – all lipids will oxidise (become rancid) when exposed to air in daylight, due to hydrolysis and/or photo-oxidation. Keep extracted lipids in the dark, and add an antioxidant such as butylated hydroxytoluene for longer-term storage. Minimise oxidation by flushing vessels with nitrogen gas and bubbling solvents with (oxygen-free) nitrogen during analysis.

Silica gel, octadecylsilane-bonded silica or ion-exchange resins can be used to bind solvent-extracted lipids by a combination of polar, ionic and van der Waals forces. In practice, a glass column is packed with a slurry of adsorbent in an appropriate organic solvent, and the lipid extract (dissolved in the same solvent) is applied to the top of the column. Lipids can then be selectively eluted; a mixture can be broadly separated into neutral lipids, glycolipids and phospholipids using solvents of increasing polarity, e.g. chloroform → acetone → methanol. It is possible to further separate the lipid subfractions on silica gel columns, as follows:

- **Neutral lipids** can be separated and eluted using hexane containing increasing proportions of diethyl ether (0 → 100% v/v) in the order: hydrocarbons, cholesterol esters, triacylglycerols, free fatty acids, cholesterol, diacylglycerols, then monoacylglycerols.
- **Glycolipids** and sulpholipids can be separated by first eluting the glycolipids with chloroform : acetone (1 : 1 v/v), then using acetone to elute the sulpholipids.

Understanding silica gel codes – silica gel is manufactured as particles of 10–15 μm diameter, containing pores of diameter 40, 60, 80, 100 or 150 Å (Ångstrom, p. 201), where Å = 10^{-10} m (thus, silica gel 60 has a pore size of 60 Å). Silica gel G contains calcium sulphate to assist binding to the glass support, while silica gel H has no binder.

- **Phospholipids** can be eluted using chloroform containing increasing proportions of methanol (5% → 50% v/v) in the order: phosphatidic acid, phosphatidyl ethanolamine, phosphatidyl serine, phosphatidyl choline, phosphatidyl inositol, sphingomyelin.

Thin layer chromatography (TLC) of lipids

This can be used to separate lipid mixtures into their constituents, or to quantify particular lipids, as part of an analytical procedure. The principle of the technique is described in Chapter 54.

Silica gel G60 is the most frequently used stationary phase, acting as a polar absorbent. When the separation is carried out with a non-polar mobile phase, non-polar lipids will migrate more rapidly (i.e. they will have high R_F values, p. 360), while polar lipids will migrate more slowly. By increasing the polarity of the solvent, the R_F values of the polar lipids can be increased. Your choice of solvent will depend on the lipids in the extract:

- For a broad range of neutral lipids, a typical solvent system is hexane : diethyl ether : glacial acetic acid at 80 : 20 : 2 (v/v/v), separating in the following order of decreasing R_F: steryl esters, wax esters, fatty acids, methyl esters, triacylglycerols, fatty acids, fatty alcohols, sterols, 1,2-diacylglycerols, monoacylglycerols.
- For polar lipids, silica gel H is preferred, as silica gel G prevents the separation of acidic phospholipids. Most solvent systems for polar lipids are based on chloroform : methanol : water, e.g. at 65 : 25 : 4 (v/v/v). In this system, the relative order of migration is: monogalactosyldiacylglycerol, cerebrosides, phosphatidic acid, cardiolipin, lysophosphatidyl ethanol-amine, phosphatidyl ethanolamine and digalactosyldiacylglycerol, sulphatides, phosphatidyl choline, phosphatidyl inositol, sphingomyelin and phosphatidyl serine.

SAFETY NOTE Working with lipid stains – these often contain corrosive or toxic reagents, so all manipulations should be performed in a fume cupboard.

KEY POINT No single solvent system will completely separate all lipid components in a single TLC procedure. However, 2D TLC may separate up to 200 individual constituents from a sample.

Lipids can also be separated by reversed-phase TLC (RP-TLC), where the silica gel is made non-polar (e.g. by silanisation), and highly polar solvents are used as the mobile phase: here, the polar lipids will have the highest R_F values and the non-polar lipids the lowest R_F values. Lipids separated by TLC can be located by staining. Several stains are non-specific, enabling almost all types of lipid to be visualised, while others will locate particular lipid classes. Staining can be carried out by immersion of the TLC plate in the stain, or by spraying. Non-specific staining methods include:

SAFETY NOTE Spraying TLC plates with sulphuric acid solutions – this must be carried out on a suitable support (e.g. disposable paper or card) within a fume hood or a commercial spray cabinet. Wear safety glasses and gloves throughout the procedure.

- Spraying with iodine solution (e.g. 1–3% w/v in chloroform) – most lipids appear as brown spots on a yellow background, although glycolipids stain weakly by this method.
- Treatment with a strong oxidising agent followed by charring (e.g. spray with 5% v/v sulphuric acid in ethanol, followed by heating in an oven at 180°C for 30–60 minutes) – lipids appear as black deposits; the detection limit is 1–2 μg. Scanning densitometry can provide quantitative information.

- Fluorescent stains (e.g. the widely used 2′,7′-dichlorofluorescein, DCF, at 0.1–0.2% w/v in ethanol). Under UV, lipids appear as bright yellow spots on a yellow–green background; the limit of detection is 5 µg. Other fluorescent stains such as 1-anilo-8-naphthalene sulphonate (ANS, 0.1% w/v in water) can detect ng quantities of lipid.

Quantitative assay of lipids and their components

Definitions

Saponification – the hydrolysis of an ester under alkaline conditions, to form an alcohol and the salt of the acid.

Saponification value – the amount of KOH (in mg) required to completely saponify 1 g of fat. Typically within the range 150–300.

Iodine number – the amount of iodine (g) absorbed by 100 g of fat, due to the reaction of iodine with C=C bonds within the fat. Typically within the range 30–200.

While TLC can be used to quantify particular lipids, it is more common to assay the compounds released on hydrolysis of simple or complex lipids, namely the alcohols, fatty acids or other components, rather than the native lipid. Alkali hydrolysis of lipids containing fatty acids results in the formation of a soap (i.e. saponification), e.g. the incubation of tripalmitin with KOH yields potassium palmitate and glycerol. Acid hydrolysis of triacylglycerols releases 'free' fatty acids and glycerol.

Basic information about the relative size of the fatty acid component of oils and fats is given by the saponification value, determined by titration against 0.8 mol L^{-1} KOH; the lower the saponification value, the higher the M_r of the fatty acids. The degree of unsaturation of the fatty acids is given by the iodine number; the higher the iodine number, the greater the content of unsaturated fatty acids.

To obtain the iodine number for a particular fat, the free iodine remaining after reaction is titrated against 0.1 mol L^{-1} sodium thiosulphate ($Na_2S_2O_3$) using a trace amount of starch as an indicator, giving a titration volume for the test solution, V_t. A blank containing no fat is also titrated, to establish the volume of sodium thiosulphate required to titrate the initial free iodine, V_o. This allows the amount of iodine that reacts with the fat to be calculated according to the formula:

Example A 0.15 g sample of cod liver oil was titrated against 0.1 moL l^{-1} $Na_2S_2O_3$, giving a titration volume of 31.7 ml (V_t), compared with a blank of 49.8 ml (V_o). Substituting into eqn [60.2] gives an iodine number of [1.27 (49.8 – 31.7)] ÷ 0.15 = 153 (to three significant figures).

$$\text{iodine number (in g)} = \frac{1.27(V_0 - V_t)}{m} \qquad [60.2]$$

where m is the mass of test fat (in g), and V_o and V_t are expressed in ml.

Measurement of glycerol content

Glycerolipids can be quantified by measuring the glycerol released on hydrolysis. A widely used method involves a coupled enzyme assay (p. 350), with the following reactions:

Understanding coupled enzyme assays based on NADH/NAD$^+$ interconversion – these are explained in more detail on p. 407, while the procedure required to convert changes in A_{340} to [NADH] is given on p. 356.

$$\text{triacylglycerol} \xrightarrow{\textit{lipase}} \text{glycerol} + \text{fatty acids} \qquad [60.3]$$

$$\text{glycerol} + \text{ATP} \xrightleftharpoons{\textit{glycerol kinase, Mg}^{2+}} \text{glycerol-3-phosphate} + \text{ADP} \qquad [60.4]$$

$$\text{ADP} + \text{phosphoenolpyruvate} \xrightleftharpoons{\textit{pyruvate kinase}} \text{ATP} + \text{pyruvate} \qquad [60.5]$$

$$\text{pyruvate} + \text{NADH} + \text{H}^+ \xrightleftharpoons{\textit{lactate dehydrogenase}} \text{lactate} + \text{NAD}^+ \qquad [60.6]$$

The glycerol concentration is determined by measuring the decrease in absorbance of NADH at 340 nm (A_{340}), compare with that of a blank with no added lipase.

Alternative approaches to measuring H_2O_2 – the hydrogen peroxide produced from the oxidation of cholesterol in [eqn 60.7] can be converted to water and oxygen by the enzyme catalase ($2H_2O_2 \rightarrow 2H_2O + O_2$) and the oxygen produced can be quantified, e.g. using an oxygen electrode.

Measurement of cholesterol content

Cholesterol in cholesterol esters can be estimated after hydrolysis to free cholesterol using cholesterol esterase. The subsequent assay is as follows:

$$\text{cholesterol} + O_2 \xrightarrow{\textit{cholesterol oxidase}} \text{cholest-4-en-3-one} + H_2O_2 \qquad [60.7]$$

The hydrogen peroxide produced as a result of the action of cholesterol oxidase can be measured amperometrically or colorimetrically, via the peroxidase-catalysed reaction:

$$2H_2O_2 + \text{phenol} + \text{4-aminoantipyrene} \xrightarrow{\textit{peroxidase}} \text{quinoneimine} + 4H_2O \qquad [60.8]$$

This is the basis of many commercially available cholesterol testing kits. An alternative approach is to measure the cholest-4-en-3-one directly, *via* its absorbance maximum at 240 nm.

Gas chromatography (GC) of lipids and lipid components

Example Fatty acids are derivatised to their methyl esters by boiling for two–three minutes with boron trifluoride (BF_3) solution (14% w/v in methanol), prior to GC analysis, to increase their volatility and stability in the gas phase. The separated components are normally detected by flame ionisation (p. 358).

This technique is used for the quantitative and qualitative analysis of a broad range of lipids. Volatile lipids may be analysed without modification, while non-volatile lipids must first be converted to a more volatile form, either by degradation (e.g. phospholipids), or derivatisation. The most effective GC columns are support-coated open tubular (SCOT) capillary columns, with a thin film (0.1–10 μm) of the stationary phase coated onto the internal wall. The choice of stationary phase depends on the components to be separated, e.g. some non-polar stationary phases cannot resolve methyl esters of saturated and mono-unsaturated fatty acids. Many stationary phases are based on silicone greases or polysiloxane, ranging from the non-polar dimethyl polysiloxanes (e.g. OV-101) to the polar trifluoropropyl methyl polysiloxanes (e.g. OV-210). Other polar stationary phases are based on polyethylene glycol (e.g. Carbowax 20M).

Text reference

Wrolstad, R.E., Decker, E.A., Schwartz, S.J. and Sporns, P. (2004) *Handbook of Food Analytical Chemistry, Water, Proteins, Enzymes, Lipids and Carbohydrates*. Wiley, Chichester.

Sources for further study

Anon. *Cyberlipid Center Homepage*. Available: http://www.cyberlipid.org
Last accessed: 21/12/10.

Byrdwell, W.C. (2006) *Modern Methods for Lipid Analysis by Liquid Chromatography/Mass Spectrometry and Related Techniques*. American Oil Chemists Society, Boulder, Colorado.

Gunstone, F.D. and Padley, F.B. (1997) *Lipid Technologies and Applications*. Marcel Dekker, New York.

Gunstone, F.D. and Harwood, J.L. (2007) *The Lipid Handbook*, 3rd edn. CRC Press, Boca Raton, Florida.

Gurr, M.I., Harwood, J.L. and Frayn, K.N. (2002) *Lipid Biochemistry*. Blackwell, Oxford.

Hemming, F.W., Hawthorne, J.N. and White, D.A. (1996) *Lipid Analysis*. Bios, Cambridge.

61 Assaying carbohydrates

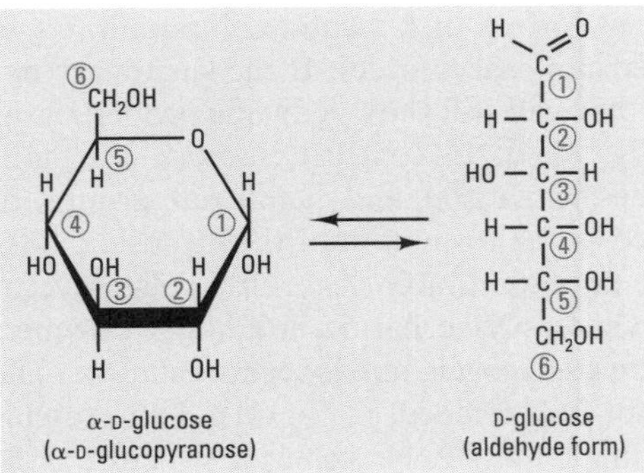

Fig. 61.1 Ring and straight chain forms of glucose. Note that β-D-glucose has H and OH groups reversed at C-1 in the pyranose form.

These are compounds with a formula based on $C_x(H_2O)_x$. They play a key role in energy metabolism, and are essential constituents of cell walls and membranes. They may exist individually, or as heteropolymers, e.g. linked to protein in glycoproteins (where carbohydrates form the minor component), or proteoglycans (where they form the major component).

> **KEY POINT** The identification and quantitative analysis of carbohydrates in food samples can be difficult, due to structural and physicochemical similarities between related compounds. Several routine analytical methods cannot distinguish between isomeric forms of a particular carbohydrate.

Monosaccharides

The simplest carbohydrates are the monosaccharide sugars, which are polyhydroxy aldehydes and polyhydroxy ketones (so-called aldoses and ketoses), typically with three to seven carbon atoms per molecule. The common names end with the suffix '-ose', e.g. glucose (Fig. 61.1), which contains six carbon atoms and is a hexose. The simplest aldose is the three-carbon compound (triose) glyceraldehyde (Fig. 61.2a), and the other aldoses can be considered to be derived from glyceraldehyde by the addition of successive secondary alcohol groups (H–C–OH). In a similar manner, all ketoses can be considered to be structurally related to the triose dihydroxyacetone (Fig. 61.2b).

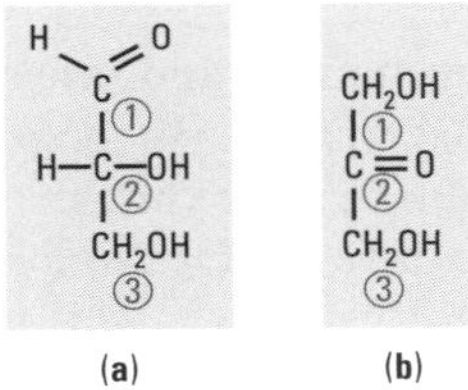

Fig. 61.2 Structure of (a) glyceraldehyde and (b) dihydroxyacetone.

Monosaccharides are assigned as D or L isomers according to a convention based on D-glyceraldehyde as the reference compound, with the carbon atoms numbered from the end of the chain containing the reactive group. Most of the carbohydrates found in biological systems are D isomers. While trioses and tetroses exist in linear form, pentoses and larger monosaccharides can be represented either as a linear structure (Fischer form) or as a ring structure (Haworth form), as in Fig. 61.1. The cyclic structure results from the reaction of the carbonyl group (aldehyde or ketone) at one end of the molecule with a hydroxyl group at the other end of the chain, forming a hemiacetal or hemiketal, as shown for glucose (Fig. 61.1) and fructose (Fig. 61.3).

The formation of a hemiacetal or hemiketal creates another asymmetric carbon atom, so that two ring forms exist – one with the –OH group on this asymmetric carbon positioned below the plane of the ring (α) and another with the –OH group above the plane of the ring (β). These different isomeric forms are called anomers and are particularly difficult to separate by chromatographic methods. The six-membered ring structure shown for glucose has a structure similar to that of pyran and is termed glucopyranose. A few monosaccharides have a five-membered ring structure similar to that of furan, e.g. fructofuranose.

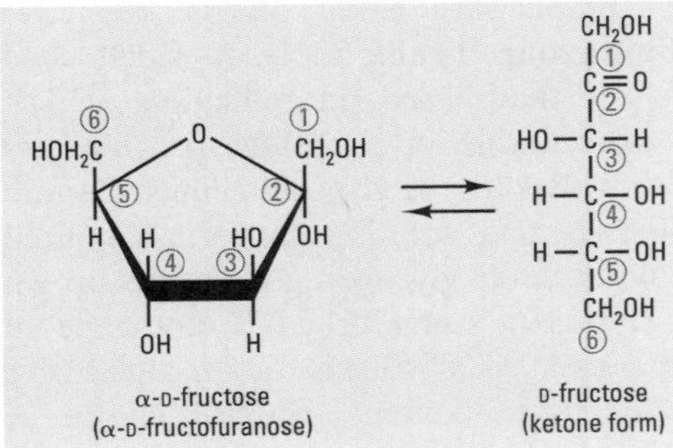

Fig. 61.3 Ring and straight chain forms of fructose. Note that β-D-fructose has OH and CH_2OH groups reversed at C-2 in the furanose form.

While the open chain form is present in very small amounts in aqueous solution it has a very reactive carbonyl group which is responsible for the reducing properties of sugars: several analytical methods are based on this feature (see p. 420).

Studying the biological roles of glycosides – these compounds are often formed in plants during the detoxification of certain compounds, or to control plant hormone activity.

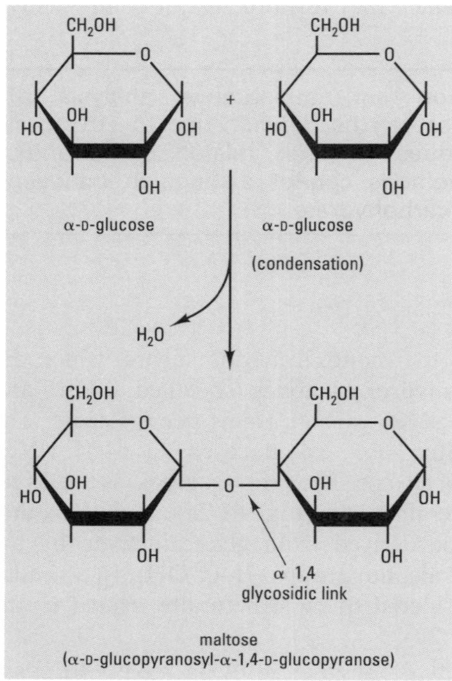

Fig. 61.4 Formation of a glycosidic link ($\alpha 1 \rightarrow 4$) in the disaccharide maltose.

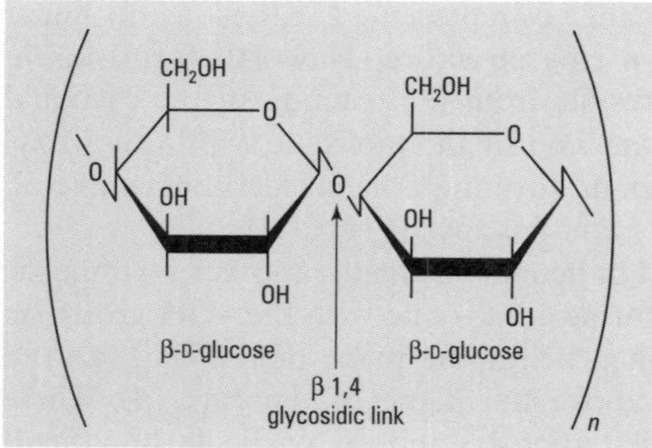

Fig. 61.5 The repeating structure of cellulose ($\beta 1 \rightarrow 4$ glycosidic links) $n \approx 5000$.

SAFETY NOTE Using the anthrone method – note that this involves hot H_2SO_4 – wear gloves, carry out the heating step in a fume hood and rinse all spillages thoroughly with excess water.

Glycosides

These are formed when a covalent bond, or glycosidic link, is created between the hemiacetal or hemiketal group of a carbohydrate (e.g. a monosaccharide) and the hydroxyl group of a second compound (e.g. a polyhydroxy alcohol, or another monosaccharide). If the sugars are not joined via their reactive groups, they will still show reducing properties, as in the disaccharide maltose (Fig. 61.4). In contrast, sucrose is a non-reducing disaccharide, since the hemiacetal and hemiketal groups of glucose and fructose are involved in the formation of the glycosidic link.

Polymeric carbohydrates have important biological roles – oligosaccharides contain up to 10 sugars, while polysaccharides are larger polymers with M_r of up to several million, e.g. glycogen, amylopectin, cellulose. The structure of polysaccharides is not fully defined, in contrast with proteins (Chapter 59) and nucleic acids (Chapter 62). Polysaccharide structure is the result of the separate actions of a number of biosynthetic enzymes, producing a range of molecules that may vary in the number of monosaccharide residues and the types of glycosidic bond. The terminology used to describe the glycosidic links in such compounds denotes the anomer involved (α or β) and the C atoms involved in the link, e.g. $\alpha\ 1 \rightarrow 4$ in maltose (Fig. 61.4), $\beta\ 1 \rightarrow 4$ in cellulose (Fig. 61.5).

Extraction and analysis of carbohydrates

While most low M_r carbohydrates are soluble in water, ethanol : water (80% v/v) is more often used, since polysaccharides and other biological macromolecules are insoluble in aqueous ethanol. In contrast, polysaccharides are more diverse, and the isolation procedures vary greatly, e.g. boiling water, mild acid or mild alkali can be used to solubilise storage polysaccharides, e.g. starch, while more vigorous treatment is required for structural polysaccharides, e.g. 24% w/v KOH for cellulose. Among the techniques used to purify extracted polysaccharides are gel permeation chromatography (p. 364) and ultracentrifugation.

Identification and quantification of carbohydrates

This can be achieved by a variety of procedures, including:

Chemical methods

Several monosaccharide assay methods are based on the reductive capacity of the aldehyde or ketone groups (Table 61.1). A widely used method for quantitative analysis is that based on reduction of 3,5-dinitrosalicylate. This method is also suitable for glycosides, provided the reducing carbonyl groups are not involved in the glycosidic links. Certain polysaccharides react with iodine in acid solution to form coloured complexes: starch gives a blue colour, while glycogen gives a red–brown colour. The anthrone method (typically using $0.1\ g\ L^{-1}$ anthrone in H_2SO_4, at 100°C for 10 min, then assayed at 630 nm) is an alternative assay for estimating total carbohydrate content. Careful choice of calibration standards is required for quantitative work – try to use a standard that matches the likely composition of the samples. While such chemical methods can give a general indication of the relative amount of carbohydrate in a sample, they can provide little useful information on the types of carbohydrate present.

Table 61.1 Methods for carbohydrate analysis

Method	Principle	Comments
Chemical assay		
Benedict's test (and Fehling's test)	Reduction of Cu^{2+} to Cu^{+} in presence of reducing sugar; alkaline solution plus heat results in formation of Cu_2O; solution turns from blue, through yellow, to red	Usually presence/absence test; quantitative assay involves measurement of Cu_2O formed
Dinitrosalicylate (DNS)	Reduction of DNS (yellow) to orange-red derivative; alkaline solution plus heat (100 °C, 10 min)	Quantitative: read at 540 nm
Enzymatic assay		
Glucose oxidase (coupled reaction)	β-D-glucose + O_2 $\xrightarrow{\text{glucose oxidase}}$ gluconic acid + H_2O_2 H_2O_2 + reduced dye $\xrightarrow{\text{peroxidase}}$ H_2O + oxidised dye	Mutarotation allows reaction to reach completion; hydrogen peroxide formed may be measured using peroxidase; ABTS* is a suitable dye Assay at 437 nm

*ABTS = 2,2′-azino-di-[3-ethylbenzthiazoline]-6-sulphonate.

Enzymatic methods

These offer a higher degree of specificity in monosaccharide assay, and may allow differentiation between the various stereoisomeric and anomeric forms. Hydrogen peroxide may be assayed using peroxidase and a suitable chromogenic substrate, or by electrochemical methods, e.g. using an amperometric sensor. Alternatively, the consumption of oxygen in the initial reaction can be measured using an oxygen electrode. Glycosidases can be used to hydrolyse specific disaccharides or polysaccharides into their constituent monosaccharides, which can then be identified and quantified, e.g. α-glucosidase, which hydrolyses the $\alpha(1 \rightarrow 4)$ linkage between the glucose residues in maltose (Fig. 61.4). However, you should note that it is rare for such enzymes to show absolute specificity for a particular substrate, so you should be alert to the possibility of interference due to related compounds in the food sample, or to impurities in the enzyme preparation.

Other applications of glycosidase assays – some microbial identification schemes are based on the detection of specific glycosidase enzymes, e.g. β-glucuronidase (p. 409).

Chromatographic methods

Traditional methods include paper and thin layer chromatographic procedures. Suitable supports for TLC include microcrystalline cellulose (in which the sugars partition between the mobile phase and the cellulose-bound water complex) and silica gel. A wide range of traditional solvent systems can be used (Stahl, 1965, gives details) and your choice of mobile phase will depend upon the expected composition of the mixture, e.g. cellulose with ethyl acetate : pyridine : water (100 : 35 : 25 v/v/v) gives a good separation of pentoses and hexoses, and will resolve glucose and galactose, as well as some disaccharides. Staining methods usually exploit the reducing properties of carbohydrates. Of the high resolution techniques, HPLC (p. 262) is the preferred method for the analysis of simple monosaccharide mixtures, and for oligosaccharide analysis and purification. Ion-exchange columns are often used for these purposes, with refractive index, or electrochemical, detection of separated components. GC is more suitable for complex monosaccharide mixtures, and can analyse subnanomolar amounts of carbohydrates and their derivatives, e.g. polyols,

Measuring carbohydrate migration in TLC systems – the distance migrated by glucose is taken as a reference ($R_F = 100$), and the migration of other carbohydrates is given as the R_G value, where R_G *is:*

$$\frac{\text{distance moved by carbohydrate}}{\text{distance moved by glucose}} \times 100$$

SAFETY NOTE Working with trimethylsilylating reagents – these compounds are extremely reactive and must be handled with care, using gloves and a fume hood. Since these reagents react violently with water, samples are dried and then redissolved in an organic solvent before trimethylsilylation.

SAFETY NOTE Working with acids – HCl gas is produced when concentrated HCl is heated: always wear gloves and work inside a fume hood.

Complex heteropolymers that contain carbohydrates – these include:

- **lignins** – structural polymers in plants: composed of amino acids (phenylalanine and tyrosine) together with carbohydrates, and extremely resistant to hydrolysis;
- **peptidoglycan** – found in the bacterial cell wall: composed of amino acids and carbohydrates, providing strength and rigidity;
- **proteoglycans** – found in animal/human connective tissues: composed of a core protein with many carbohydrate side chains attached, responsible for hydration, lubrication, resistance to compressive forces and mediation of cellular interactions.

including glycerol (p. 417). However, a preliminary step is necessary to produce volatile derivatives of the carbohydrates in the mixture, e.g. methylation or, more often, trimethylsilation (adding hexamethyldisilazane and trimethylchlorosilane at 2 : 1 v/v at room temperature rapidly produces trimethylsilyl ethers – TMS derivatives). Efficient separations can be obtained with a non-polar stationary phase, e.g. methylpolysiloxy gum (OV-1): the use of SCOT columns (p. 418) can enable over 30 components to be resolved. A disadvantage of GC methods is that the carbohydrates are first converted to a derivative form, then quantified by destructive means, preventing further analysis.

Capillary electrophoresis

This is a powerful technique for the separation of carbohydrates, although it is usual to modify uncharged carbohydrates by reductive amination to form primary amines to allow their separation within the capillary.

Characterisation of polysaccharides

Polysaccharides are characterised according to the relative proportions of the constituent sugar residues, the various types of glycosidic links and the M_r. To investigate the composition of a given type of polysaccharide, its glycosidic links must be hydrolysed (e.g. by heating with concentrated HCl at 60°C for 30 minutes), followed by separation, identification and quantification of the individual components. The position of glycosidic linkages can be determined by methylation of all free hydroxyl groups of the polysaccharide followed by complete hydrolysis to give a mixture of partially methylated monosaccharides. These are then reduced to alditols, and acetylated. The partially methylated alditol acetates can be identified by GC-mass spectrometry, allowing the types of glycosidic links to be deduced from the positions of the acetylated groups. The α and β configuration of the glycosidic linkages can be determined by enzymic assay or by NMR spectroscopy.

Text references and sources for further study

Bemiller, J.N. (2007) *Carbohydrate Chemistry for Food Scientists*, 2nd edn. American Association of Cereal Chemists, Saint Paul.

Coultate, T.P. (2009) *Food: The Chemistry of its Components*, 5th edn. Royal Society of Chemistry, Cambridge.

Davis, B.G. and Fairbanks, A.J. (2002) *Carbohydrate Chemistry*. Oxford University Press, Oxford. [A primer on carbohydrate chemistry].

Lindhorst, T.K. (2007) *Essentials of Carbohydrate Chemistry and Biochemistry*, 3rd edn. Wiley-VCH, Berlin.

Rassi, Z.E. (2002) *Carbohydrate Analysis by Modern Chromatography and Electrophoresis*. Elsevier, Amsterdam.

Stahl, E. (1965) *Thin Layer Chromatography – A Laboratory Handbook*. Springer, Berlin.

62 Assaying nucleic acids and nucleotides

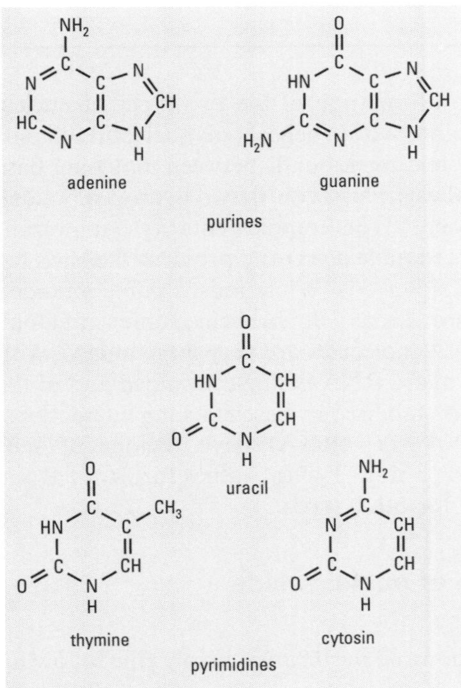

Fig. 62.1 Nitrogenous bases in nucleic acids.

nucleoside

nucleotide (nucleoside monophosphate)

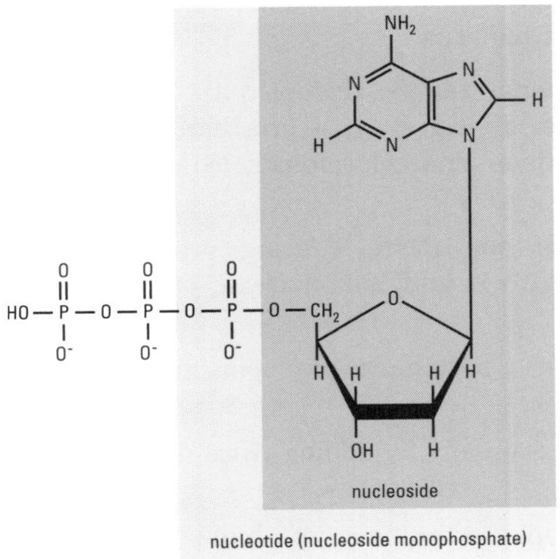

Fig. 62.2 A nucleoside triphosphate – deoxy-adenosine 5′ triphosphate (dATP).

Nucleic acids are nitrogen-containing compounds of high M_r, often found within nucleic acid–protein (nucleoprotein) complexes in biological materials. The two main groups of nucleic acids are:

1. **Deoxyribonucleic acid** (DNA) – found in chromosomes and the principal molecule responsible for the storage and transfer of genetic information.
2. **Ribonucleic acid** (RNA) – involved with the DNA-directed synthesis of proteins in cells. Three principal types of RNA exist: messenger RNA (mRNA), ribosomal RNA (rRNA) and transfer RNA (tRNA). In some viruses, RNA acts as the genetic material. In eukaryotes, mRNA molecules initially synthesised in the nucleus from genomic DNA (nascent mRNA) will contain several sequences – called introns – that are not transcribed into protein. These are successively excised in the nucleus, leaving only coding sequences (exons) in the mRNA that migrates to the cytoplasm, to be translated at the ribosome.

KEY POINT Nucleic acids are important in the transmission of information within cells, and one of the most important aspects of contemporary nucleic acid analysis is to decipher the coded information within these molecules as part of genome analysis in molecular biology (Chapter 56).

You are most likely to encounter the techniques of nucleic acid extraction and analysis within courses dealing with molecular biology (see Chapter 56), and the examples given in this chapter are focused towards this aspect. However, the same general principles also apply to food analysis for overall nucleic acid content.

The structure of nucleic acids

Nucleic acids are polymers of nucleotides (polynucleotides), where each nucleotide consists of:

- a nitrogenous base, of which there are five main types. Two have a purine ring structure, i.e. adenine (A) and guanine (G), and three have a pyrimidine ring, i.e. thymine (T), uracil (U) and cytosine (C), as shown in Figure 62.1. Their carbon atoms are numbered C-1, C-2, etc.;
- a pentose sugar, which is ribose in RNA and deoxyribose in DNA. The carbon atoms are denoted as C-1′, C-2′, etc. and deoxyribose has no hydroxyl group on C-2′ (Fig. 62.2). The C-1′ of the sugar is linked either to the N-9 of a purine or the N-1 of a pyrimidine;
- a phosphate group, which links with the sugars to form the sugar–phosphate backbone of the polynucleotide chain.

A compound with sugar and base only is called a nucleoside (Fig. 62.2) and the specific names given to the various nucleosides and nucleotides are listed in Table 62.1. The individual nucleotides within nucleic acids are linked by phosphodiester bonds between the 3′ and 5′ positions of the sugars (Fig. 62.3).

Table 62.1 Nomenclature of nucleosides and nucleotides

Base	Nucleoside	Nucleotide
Adenine	Adenosine	Adenylic acid
Guanine	Guanosine	Guanylic acid
Uracil*	Uridine	Uridylic acid
Cytosine	Cytidine	Cytidylic acid
Thymine†	Thymidine	Thymidylic acid

*In DNA.
†In RNA.

KEY POINT RNA and DNA differ both in the nature of the pentose sugar residue, and in their base composition: both types contain adenine, guanine and cytosine, but RNA contains uracil while DNA contains thymine.

Differences also exist in the conformation of the two types of nucleic acid. DNA typically exists as two interwoven helical polynucleotide chains, with their structure stabilised by hydrogen bonds between matching base pairs on the adjacent strands: A always pairs with T (two hydrogen bonds), and G with C (three hydrogen bonds). This complementarity is important since it stabilises the DNA duplex (double helix) and provides the basis for replication and transcription (Chapter 56). While most double stranded (ds) DNA molecules are in this form, i.e. as a double helix, some viral DNA is single stranded (ss). Intact DNA molecules are very large indeed, with high M_r values (e.g. 10^9). In the main, RNA is single stranded and in the form of a gentle right-handed helix stabilised by base-stacking interactions, although some sections of RNA (i.e. tRNA) have regions of self-complementarity, leading to base pairing. Typical values for M_r of RNA range from 10^4 for tRNA to 10^6 for other types.

Minimising damage to chromosomes – chromosomes vary in size from 0.3 to 200 megabase pairs (Mb), so some breaks in DNA inevitably occur during manipulation. Shear effects can be minimised by using wide-mouthed pipettes, gentle mixing, by avoiding rotamixing, and by precipitating DNA with ethanol at −20°C.

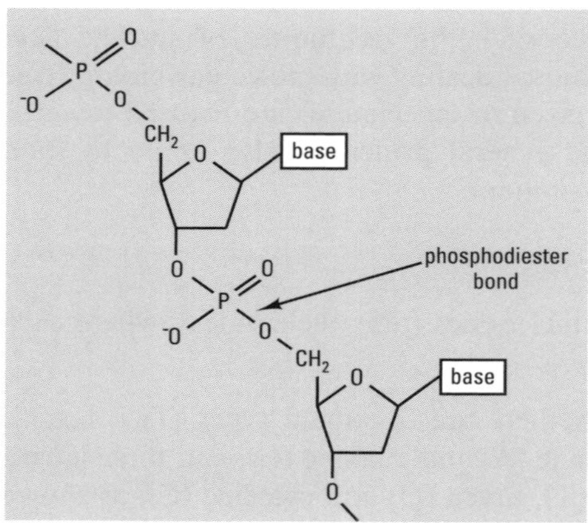

Fig. 62.3 Linkage of nucleotides in nucleic acids.

Working with precipitated DNA – this can be spooled from solution by winding it around a glass rod.

Extraction and purification of nucleic acids from tissues and cells

Irrespective of the source, extraction and purification involve the following broad stages, in sequence:

- **disruption of biological material** to release the contents;
- **removal of non-nucleic acid components** (e.g. proteins), leaving DNA and/or RNA;
- **concentration of the remaining nucleic acids.**

DNA isolation procedures

Specific details for the preparation of DNA from bacterial cells are given in Chapter 56. For other sources of DNA, such as animal tissues or plant material, the following steps are required:

1. **Homogenisation** – food material can be disrupted by the methods described in detail in Chapter 57, e.g. by lysis in a buffered solution containing the detergent sodium dodecyl sulphate (SDS) or Triton X-100.
2. **Enzymic removal of protein and RNA** – using proteinase (e.g. proteinase K at 0.1 mg mL^{-1}) and ribonuclease (typically at 0.1 µg mL^{-1}) for 1–2 h.
3. **Phenol–chloroform extraction** – to remove any remaining traces of contaminating protein.
4. **Precipitation of nucleic acids** – usually by adding twice the volume of ethanol.
5. **Solubilisation** in an appropriate volume of buffer (pH 7.5) – ribo-nuclease is often added to remove any traces of contaminating RNA.

Avoiding contamination of glassware by RNases – autoclave all glassware before use, to denature RNases. These enzymes are present in skin secretions: use gloves at all times and use plasticware wherever possible.

Density gradient centrifugation (p. 385) is an alternative approach to the separation of DNA from contaminating RNA, as described below.

RNA isolation procedures

A typical mammalian cell contains about 10 pg of RNA, made up of rRNA (80–85 per cent), tRNA (10–15 per cent) and mRNA (1–5 per cent). While rRNA and tRNA components are of discrete sizes, mRNA is heterogeneous and varies in length from several hundred to several thousand nucleotides.

Avoiding RNase degradation of RNA – endogenous RNases can be inhibited by including diethyl pyrocarbonate (DEP) (at 0.1% v/v g ml^{-1}) in solutions used for RNA extraction.

> **KEY POINT** RNA is more difficult to purify than DNA, partly because of degradation during the extraction process due to the action of contaminating ribonucleases, and partly because the rigorous treatment required to dissociate the RNA from protein in ribosomes may fragment the polyribonucleotide strands.

RNA can be prepared either from the cytoplasm of cells (to give mainly rRNA, tRNA and fully processed mRNA), or from whole cells, in which case nascent mRNA from the nucleus will also be present (p. 423):

Purification of mRNA – following wholecell extraction of RNA, mRNA can be separated from other types by affinity chromatography (p. 365) using poly (U)-Sepharose which binds to the poly (A) 'tail' sequence at the 3′ end of mRNA molecules.

- Preparation of cytoplasmic RNA involves lysis of cells or protoplasts with a hypotonic buffer, leaving the nuclei intact. Cell debris and nuclei are then removed by centrifugation (Chapter 57), and sodium dodecyl sulphate (SDS) is added to the supernatant (the cytoplasmic fraction) to inhibit ribonuclease. Proteinase K can be added to release rRNA from ribosomes. Phenol–chloroform extraction removes contaminating proteins, as for DNA preparation (p. 424), and the RNA present in the aqueous phase can then be precipitated by addition of twice the volume of ethanol.
- Preparation of whole-cell RNA requires more vigorous cell lysis, e.g. with a solution containing 6 mol L^{-1} guanidinium chloride and 2-mercaptoethanol: this effectively denatures any ribonuclease present. Caesium chloride is added to the extract to give a final concentration of 2.4 mol L^{-1}, and the solution is processed by density gradient centrifugation (Chapter 57) at 100 000 *g* for 18 h, using a cushion of 5.7 mol L^{-1} CsCl. DNA and protein remain in the upper layers of CsCl, while RNA forms a pellet at the base of the tube. The RNA pellet is redissolved in buffer and precipitated in cold ethanol.

Measuring nucleic acids by spectrophotometry – ideally, the nucleic acid extracts should be prepared to give A_{260} values of between 0.10 and 0.50, for maximum accuracy and precision.

Separating nucleic acids

Electrophoresis is the principal method used for separating nucleic acids. At alkaline pH values, linear DNA and RNA molecules have a uniform net negative charge per unit length due to the charge on the phosphoryl group of the backbone. Electrophoresis using a supporting medium that acts as a molecular sieve (e.g. agarose or polyacrylamide, p. 368) enables DNA fragments or RNA molecules to be separated on the basis of their relative sizes.

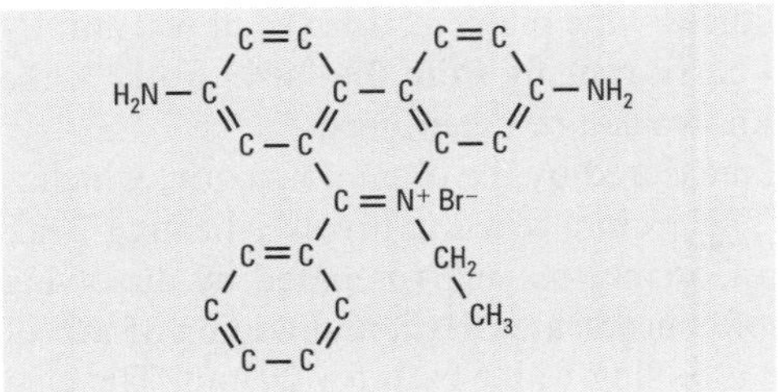

```
        C=C         C=C
       /   \       /   \
 H2N—C      C — C       C—NH2
      \\   //    \\    //
        C—C        C—C
           \      /
            C = N+ Br−
           /       \
        C=C         CH2
       /   \          \
      C     C          CH3
       \\  //
        C—C
```

Fig. 62.4 Ethidium bromide (EtBr), a fluorescent molecule used for the detection and assay of DNA.

Quantitative analysis of nucleic acids

Measuring nucleic acid content

The concentration of reasonably pure samples of DNA or RNA can be measured by spectrophotometry (p. 356). In contrast, measurement of the nucleic acid content of whole cell or tissue homogenates requires chemical

SAFETY NOTE Working with ethidium bromide – this compound is highly toxic and mutagenic. Avoid skin contact (wear gloves) and avoid ingestion. Use a safe method of disposal (e.g. adsorb from solution using an appropriate adsorbant, e.g. activated charcoal).

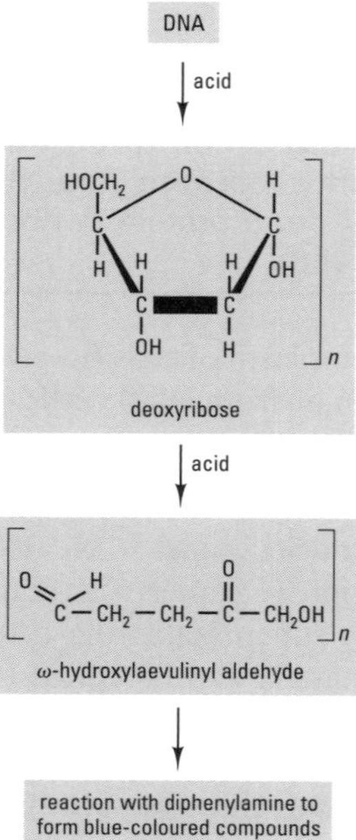

Fig. 62.5 The diphenylamine reaction for assay of DNA.

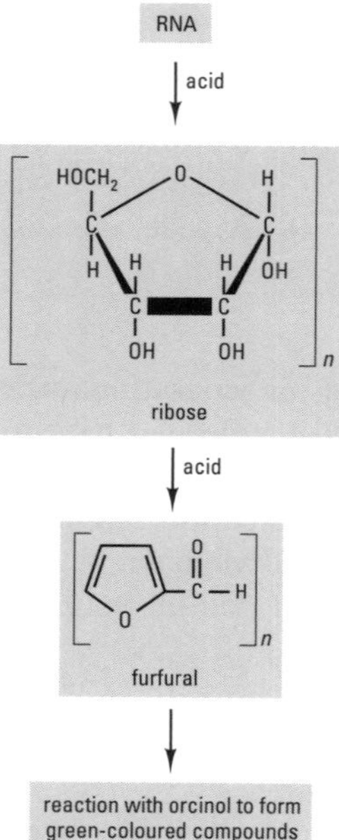

Fig. 62.6 The orcinol reaction for assay of RNA.

methods, since the homogenates will contain many interfering substances. The principles involved in each technique are as follows:

- **Spectrophotometry** – DNA and RNA both show absorption maxima at ≈260 nm, due to the conjugated double bonds present in their constituent bases. At 260 nm, an A_{260} value of 1.0 is given by a 50 µg mL^{-1} solution of dsDNA, or a 40 µg mL^{-1} solution of ssRNA. If the absorbance at 280 nm is also measured, protein contamination can be quantified. Pure nucleic acids give A_{260}/A_{280} ratios of 1.8–2.0, and a value below 1.6 indicates significant protein contamination. Further purification steps are required for contaminated samples, e.g. by repeating the phenol–chloroform extraction step. RNA contamination of a DNA preparation is indicated if A_{260} decreases when the sample is treated with 2.5 µL of RNase at 20 µg µL^{-1}. DNA contamination of an RNA preparation might be suspected if the sample is very viscous, and this can be confirmed by electrophoresis.
- **Spectrofluorimetry** – this is the best approach for samples where the DNA concentration is too low to allow direct assay by the spectrophotometric method described above. The method uses the fluorescent dye ethidium bromide (Fig. 62.4), which binds to dsDNA by insertion between stacked base pairs, a phenomenon termed 'intercalation'. The fluorescence of ethidium bromide is enhanced 25-fold when it interacts with dsDNA. Since ssDNA gives no significant enhancement of fluorescence, dsDNA can be quantified in the presence of denatured DNA. The concentration of dsDNA in solution, $[\text{dsDNA}]_x$, can be calculated by comparing its fluorescence (excitation, 525 nm; emission, 590 nm) with that of a standard of known concentration, $[\text{dsDNA}]_{std}$, using the relationship:

$$[\text{dsDNA}]_x = \frac{[\text{dsDNA}]_{std} \times \text{fluorescence of unknown}}{\text{fluorescence of standard}} \qquad [62.1]$$

- **Chemical methods** – these are mostly based on colorimetric reactions with the pentose groups of nucleic acids. The total DNA concentration can be measured by the diphenylamine reaction (Fig. 62.5), which is specific for 2-deoxypentoses. The diphenylamine reaction involves heating 2 mL of DNA solution with 4 mL of freshly prepared diphenylamine reagent (diphenylamine, at 10 g L^{-1} in glacial acetic acid, plus 25 mL concentrated sulphuric acid) for ten minutes in a boiling water bath. The acids cleave some of the phosphodiester bonds, and hydrolyse the glycosidic links between the deoxyribose and purines. Deoxyribose residues are converted to ω-hydroxylaevulinyl aldehyde (Fig. 62.5), which reacts with the diphenylamine to produce a blue pigment, assayed at 600 nm. By constructing a DNA standard curve ranging from 0 to 400 µg mL^{-1}, the DNA concentration of the unknown can be determined.

  RNA concentration can be measured by the orcinol reaction, which is a general assay for pentoses. The orcinol reaction involves heating 2 mL of RNA solution with 3 mL of orcinol reagent (prepared by dissolving 1 g of $FeCl_3 \cdot 6H_2O$ in 1 Litre of concentrated HCl, and adding 35 mL of 6% w/v orcinol in ethanol) in a boiling water bath for 20 min. The acid cleaves some phosphodiester bonds and hydrolyses the glycosidic links between the ribose and purines. The hot acid also converts the ribose to furfural (Fig. 62.6), which reacts with orcinol in the presence of ferric ions to produce green-coloured compounds, assayed at 660 nm. A standard curve for RNA ranging from 0 to 400 $\mu g\,mL^{-1}$ is used to determine the RNA concentration of the unknown. The orcinol reaction is less specific than the diphenylamine reaction, as deoxyribose reacts to some extent and DNA gives about 10% of the colour given by the same concentration of RNA. If the DNA concentration of the extract is known (e.g. from a diphenylamine assay), the contribution of DNA to A_{660} can be measured for a standard prepared to have the same DNA concentration as the sample: the A_{660} due to DNA can then be subtracted from the result of the orcinol reaction, and the remaining A_{660} value is then used to determine RNA concentration from the RNA standard curve.

Sources for further study

Blackburn, G.M., Gait, M.G., Loakes, D. and Williams, D.M. (2005) *Nucleic Acids in Chemistry and Biology*, 3rd edn. Royal Society of Chemistry London.

Neidle, S. (2002) *Nucleic Acid Structure and Recognition.* Oxford University Press, Oxford.

Walker, J. and Rapley, R. (2000) *The Nucleic Acids Protocols Handbook.* Humana Press, New Jersey.

63 Assaying active phytochemicals in functional foods

Definitions

Phytochemicals – plant secondary metabolites which, when consumed, may be health-promoting in some cases (e.g. polyphenol antioxidants) and toxic in other cases (e.g. some alkaloids).

Functional foods – foods that have demonstrable benefits beyond their basic nutritional functions, e.g. in terms of physiological benefits and/or reduction in disease risks or symptoms.

Antioxidant – substance that reduces or eliminates oxidative damage. In the context of nutrition and cell function, antioxidants will inactivate highly reactive free radicals within cells and tissues. Free radicals are unstable molecules having one or more unpaired electrons.

Polyphenols – organic compounds of plant origin containing more than one phenol group (p. 429; Fig. 63.1); they can act as potent antioxidants and may protect the body from some aspects of oxidative stress.

Phytooestrogens – plant compounds of similar structure to human oestrogen, with potentially similar effects in the human body.

Understanding the role of advertising – functional foods are often termed 'superfoods' in promotional advertising, even though human studies on their benefits may not have been carried out, or may be conflicting. For example, goji berries have been heavily promoted as a 'superfood', yet most of the current published data, although positive, are from animal studies and cell cultures based on the use of isolated components from the berries. A search on Google for 'superfoods' will give you some indication of their over-promotion by the companies selling them. It is important that you maintain a healthy scepticism of claims that are not fully supported by appropriate scientific evidence, e.g. human trials.

Phytochemicals and the functional foods containing them aim to go beyond the simple treatment of nutritional deficiencies to provide significant health benefits and protection against the onset of disease. As an example, antioxidants in foods may act to reduce vascular inflammation and thereby decrease the risk of cardiovascular disease.

This is a highly controversial area, with claims and counter-claims for the health benefits of various foods. It is also often difficult to extrapolate from laboratory results using specific nutritional components to estimate their potential health benefits when consumed as part of a balanced diet.

Functional foods

A balanced and varied diet incorporating the five major food groups (p. 273) provides the necessary nutrients required for metabolism. However, research also suggests that including some specific foods within the diet can have positive effects in relation to certain diseases, e.g. red wine and olive oil, associated with the Mediterranean diet, appear to reduce the risk of cardiovascular disease.

KEY POINT The notion of 'functional foods' is a relatively recent one, and more research needs to be carried out on many of the claims associated with functional foods to establish their true effects on human health.

Some important questions to consider in relation to so-called functional foods include:

- **What is the strength of evidence for a particular functional food?** Are there sufficient published human trials to conclude that there is a likely benefit, or is the data based on laboratory experimentation (e.g. cell culture studies) and conjecture? How large is the likely benefit, and has it been statistically tested (p. 259)?
- **What is the role of such foods in the overall diet?** Their effectiveness or otherwise might also be affected by other components of the diet.
- **What is the likely concentration of phytochemicals in the food?** Nutrient levels of plants will vary depending on the type of soil and the climate in which the plant was grown or raised, the variety used, etc.
- **Are these foods safe to eat, and in what amounts?** Safety is not always known or easily established, particularly when a new food is introduced from another country. Can they interact with medical drugs? (For example, some people on warfarin have shown an increased tendency to bleed when consuming goji berry products.) Foods are not subject regulated by the same procedures as pharmaceutical compounds, so their potential negative health effects are often unquantified.
- **Could functional foods lead to dietary complacency?** For example, people may think that by consuming green tea with antioxidants this means that they can continue to eat unhealthful foods with impunity.

Definitions

Nutraceutical – food or component of a food that purportedly provides medicinal or health benefits, including the prevention and treatment of disease. A nutraceutical may be a food or herb believed to be medicinally active, e.g. the herb St John's Wort (*Hypericum perforatum*) has been shown in human trials to have an anti-depressant effect; or it may be a specific component of a food, e.g. omega-3 oil from fish.

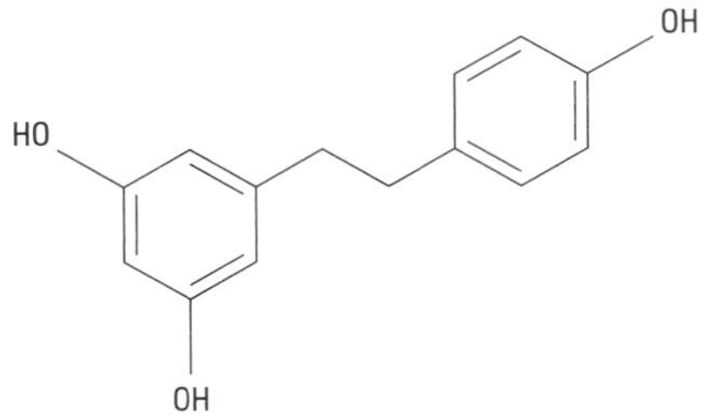

Fig. 63.1 Structure of resveratrol – a plant component with possible anti-aging and anti-cancer properties.

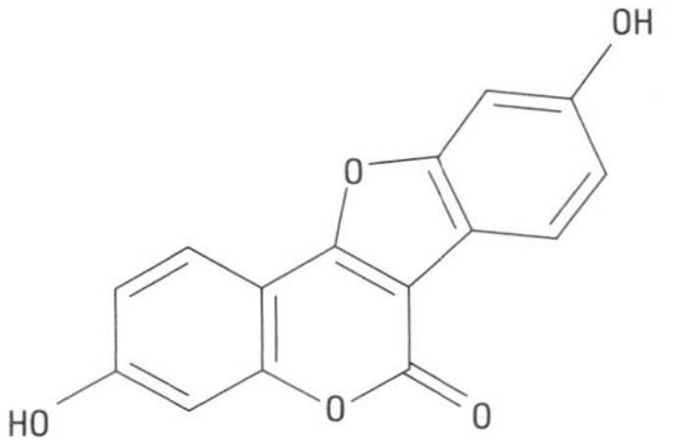

coumestrol

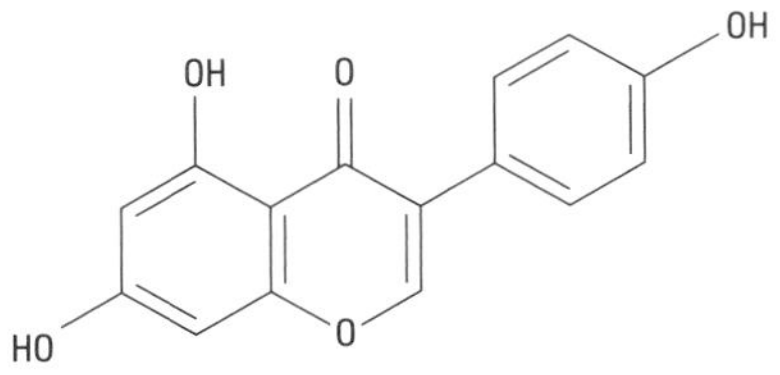

genistein

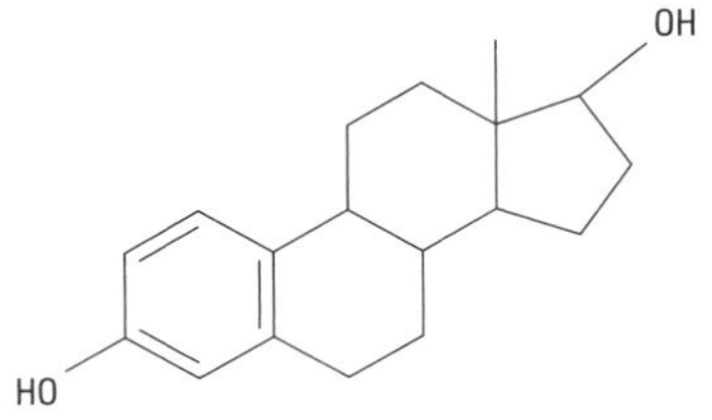

17β-oestradiol

Fig. 63.2 The human hormone 17β-oestradiol compared structurally with two phytoestrogens (coumestrol and genistein).

- **Are there vested interests in claiming functional food status?** For example, suppliers or nutritionists involved in marketing a particular product. Do they provide value for money? Has the price been inflated simply because of the 'superfood' or 'nutraceutical' categorisation?

Assaying phytochemicals

KEY POINT There are literally millions of chemical constituents within plants, and many of these have several different potential physiological actions in addition to their role as nutrients, making them candidates for phytochemical status.

Table 63.1 gives selected examples of different types of plant constituents, their potential physiological effects and typical analysis methods. It should be noted that the list includes compounds with negative as well as positive physiological effects.

Well over 10 000 individual phytochemicals are currently being researched. This is a rapidly evolving field of research with new evidence of the health-related effects of phytochemicals being demonstrated weekly. Research includes studies on bioavailability (% absorbed and concentration of compound or metabolite at target organ), bioefficacy, and metabolomic studies to identify new phytochemical metabolites. Appropriate extraction, purification and assay methods need to be applied for different types of chemical compounds (see Chapters 49–55 for details of methodologies used). As an example, high-performance liquid chromatography (HPLC, p. 362), sometimes in combination with mass spectrometry, has been used to separate and determine the concentrations of the specific phytochemical resveratrol in food (Sakkiadi *et al.*, 2007). Resveratrol is a biphenol (Fig. 63.1) found in red grapes (approx. 0.31 mg/100 g; Zamora-Ros *et al.*, 2008), red wine (approx. 0.85 mg/100 g), some fruits and nuts (e.g. peanuts, approx. 0.01 mg/100 g); it has been shown to extend the life of yeast and animals such as nematodes and fruit flies, but not yet humans. It has also been shown to prevent skin cancer in mice.

Soy products are usually rich in phytoestrogens and preliminary research indicates they may be beneficial to the cardiovascular system. They resemble human oestrogen (17β-oestradiol) in structure (Fig. 63.2). Phytoestrogen supplements should currently be viewed with caution as they are being studied with regard to the possibility that they may stimulate oestrogen-dependent cancers. Likewise, until more research is done, recommendations suggest that foods containing high levels of phytoestrogens may be best consumed in moderation as part of an otherwise balanced and varied diet.

Phytosterols have a structure similar to cholesterol and may help in reducing cholesterol absorption. They are marketed in some margarines and other products on this basis. Phytoestrogens and phytosterols typically are separated and analysed by HPLC (p. 362).

Antioxidant capacity assays

Some human studies have indicated an association between the consumption of fruits and vegetables and a reduction in morbidity and

Table 63.1 Selected plant constituents, potential physiological actions and methods of analysis

Name (example)	Plant sources	Possible actions	Analysis method
Carotenoids (β-carotene)	Yellow and green fruits, vegetables	Antioxidants, may reduce cellular damage and systemic inflammation	HPLC (p. 362)
Indoles (indole-3-carbonyl)	Broccoli, Brussel sprouts, cabbage, cauliflower	May activate production of enzymes that block DNA damage and prevent resultant cancers	GC (p. 353)
Phenols (chlorogenic acid)	Coffee (approx. 150 mg chlorogenic acid per 200 mL cup of coffee)	May decrease the risk of type II diabetes mellitis	HPLC
Biphenols and polyphenols (resveratrol)	Grapes, black and green tea, olives, walnuts	Antioxidants, may reduce risk of cardiovascular diseases	HPLC
Flavonoids (specific class of polyphenols; quercetin)	Berries, citrus fruit, apples, olives, green tea, vegetables, whole wheat	Antioxidants, may reduce cellular damage and systemic inflammation	HPLC Spectrophotometry (p. 363) Capillary electrophoresis (p. 371)
Lignans (specific class of polyphenols; pinoresinol)	Flax seed, sesame seed, rye	Antioxidants; may have oestrogenic or anti-oestrogenic activity in the body depending on structure; require conversion by gut microflora to become oestrogenic (e.g. enterodial)	HPLC
Phytosterols (cholestatin)	Rice bran, corn, wheat germ, and as food additive	May reduce cholesterol absorption and therefore benefit the cardiovascular system	HPLC
Volatile oils (eucalyptus)	*Eucalyptus globulus* (Eucalyptus)	May have antiseptic, anti-inflammatory and decongestant properties	GC GC/MS (p. 363)
Isoprenoids (artemisinin)	*Artemisia annua* [Sweet Wormwood]	May have anti-cancer properties; medicinal – anti-malarial drug	GC GC-MS
Tannins (gallotannin)	Pomegranate, grapes, strawberries, hazelnuts, walnuts, tea, wine	May have anti-bacterial and anti-viral properties	Spectrophotometry HPLC
Polysaccharides (LBP, from *L. barbarum*)	*Lycium barbarum* [goji berries]	May have anti-aging, anti-cancer properties; possible increase in actions of blood thinning agents	HPLC
Alkaloids (scopolamine)	*Atropa belladonna* [Deadly Nightshade]	Deadly Nightshade – *highly toxic*; medicinal – atropene is a sedative extracted from *A. belladonna*	GC/HPLC/TLC (p. 360)
Glycosides (digoxin)	*Digitalis purpurea* [Foxglove]	Foxglove – *highly toxic*; medicinal – digoxin is a cardiac glycoside used for treating heart failure	HPLC Spectrophotometry

Caveat emptor – some food companies use total antioxidant capacity to promote their food products as 'healthy', but there may be no scientifically validated reason to believe that such measures of antioxidant capacity are directly correlated to the healthfulness of any particular product, when all factors are considered.

mortality due to certain degenerative illnesses. The relatively high antioxidant content of fruits and vegetables has led to the hypothesis that antioxidants may be one of the major contributors to the protection offered by fruits and vegetables.

Fruits and vegetables contain many different classes and types of antioxidants. Determining the total antioxidant capacity of a food is important in ongoing research into possible associations with health benefits. However, the total antioxidant values often provided in advertising of commercial foods need to be viewed with considerable caution as to their nutritional significance.

Example of two-stage extraction – pea plant leaves were ground with a mortar and pestle in aqueous buffer. The homogenate was then centrifuged (p. 383) and the resultant supernatant analysed by FRAP for water-soluble antioxidant capacity. The pellet was then extracted with acetone for 30 minutes at room temperature to determine the antioxidant capacity of the water-insoluble lipophilic fraction. The supernatant contained 56.4 per cent of the antioxidant activity and the pellet 43.6 per cent (Kerchev and Ivanov, 2008)

Sample extraction for antioxidant capacity assays

Preparation of material for a total antioxidant capacity assay varies according to the type of food product or biological sample to be tested. For example, milk or plasma can be assayed without any processing, while berries are typically assayed following solvent extraction, e.g. using 70:30 v/v ethanol/water or 70:30 v/v acetone/water, with the latter giving the highest extraction efficiency when assayed by ferric reducing antioxidant power assay.

Particular groups of antioxidants can be targeted by choice of extraction solvent, and often multiple extraction procedures are required, depending on which types of antioxidants are to be assayed. Solvent selection depends upon both the chemical nature of the component(s) to be extracted, and the physio-chemical nature of the matrix. For the general extraction of overall antioxidants from wheat, a 70:30 v/v ethanol/water mixture gave a higher extraction efficiency than water alone, particularly in regard lipophilic antioxidants.

Ferric reducing antioxidant power (FRAP) assay

Total antioxidant capacity can be determined by the FRAP assay. It is inexpensive compared with other assays and analysis can be carried out using a spectrophotometer (p. 356). It involves a redox reaction between a ferric reagent and any antioxidants present in the food, converting the reagent from Fe^{3+} (oxidised form) to Fe^{2+} (reduced form), resulting in the formation of an intense blue colour with maximum absorption at 593 nm. Details of the FRAP assay are given in Box 63.1.

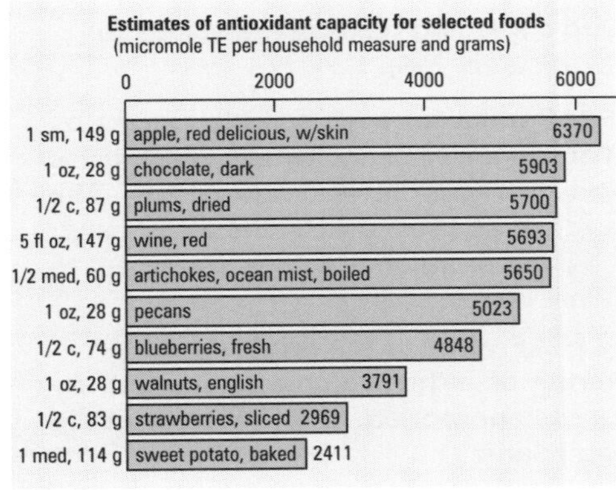

Fig. 63.3 Estimates of antioxidant capacity from USDA database of ORAC values for selected foods.

Oxygen radical absorbance capacity (ORAC) assay

The ORAC test is another approach to measuring the total antioxidant capacity of an extract of food or food supplement. ORAC test results can be used to compare the antioxidant capacities of different foods (Fig. 63.3), provided a consistent methodology is used. The ORAC assay can also be used to measure antioxidant activity in biological fluids, tissues and cells. It is based on fluorescence detection (p. 426). Test kits are available, e.g. the Zen-Bio® ORAC Antioxidant Kit, which uses 96-well assay plates.

The principle of the assay is the formation of a peroxyl radical in the reaction which oxidises fluorescein to give a non-fluorescent product. Antioxidants suppress this reaction. Fluorescence is then measured over time and plotted as a graph. The concentration of antioxidants in the samples can then be determined by the area under the curve, relative to a known antioxidant standard.

Locating reliable ORAC information – the USDA have a database of ORAC values for foods at:

http://www.ars.usda.gov/Services/docs.htm?docid=15866.

Overview of ORAC and FRAP

A comparison of the ORAC versus FRAP methods is shown in Table 63.2.

Other antioxidant capacity assays

DPPH free radical inhibition assay

This assay determines the antioxidant capacity by utilising the unpaired electron in the diphenyl-1-picrylhydrazyl (DPPH) free radical. The reaction gives a strong absorption maximum at 517 nm. The DPPH in methanol/water solution is initially a purple colour. In reaction with the sample the colour turns from purple to yellow as the molar absorptivity of

Box 63.1 How to measure total antioxidant capacity using the ferric reducing antioxidant power (FRAP) assay

Preparation of reagents and samples

1. **Prepare sample extracts in sufficient volume to run each in triplicate** (p. 431).
2. **Make solutions for a standard curve (p. 336)** – first make a stock solution of ascorbic acid (M.W. 176.13) at a concentration of 1000 μmol L^{-1}; use this to prepare a series of standards in distilled water, to include 100, 200, 400, 600, 800 and 1000 μmol L^{-1} ascorbic acid. Prepare on day of use.
3. **Prepare a dilute hydrochloric acid solution at 40 mmol L^{-1}** – add 1.46 mL of concentrated HCl to a 1 Litre volumetric flask half filled with distilled water (while mixing well and following appropriate safety procedures, pp. 173–4). Then make up to 1L with distilled water. Store at room temperature.
4. **Prepare acetate buffer at 300 mmol L^{-1}** – to a 1-L volumetric flask, first add 3.1 g sodium acetate trihydrate and then carefully add 16 mL of glacial acetic acid. Top up to 1 L with distilled water, then adjust pH to 3.6. Store at 4°C until required.
5. **Prepare a solution of TPTZ (2, 4, 6-tri[pyridyl]-s-triazine) at 10 mmol L^{-1}** – weigh 0.031 g TPTZ and make up to 10 mL using 40 mmol L^{-1} HCl. Dissolve using a water bath at 50°C. Prepare on day of assay.
6. **Prepare a solution of ferric chloride at 20 mmol L^{-1}** – weigh 0.054 g $FeCl_3.6H_2O$ and make up to 10 mL of solution using distilled water. Prepare on day of assay.
7. **Prepare the working FRAP reagent** – mix solutions from steps 4, 5 and 6 above in a clean reagent bottle in the ratio 10:1:1 immediately before use. This solution should be straw-coloured (if blue-coloured, discard and remake in a clean reagent bottle thoroughly rinsed with distilled water). Mix, and place bottle in 37°C water bath before use.

FRAP assay

1. **Prepare the spectrophotometer** – switch on, set the wavelength to 593 nm and leave to warm up for at least ten minutes. Zero using the working FRAP reagent as a blank.
2. **Bring the FRAP reagent to the correct temperature** – set a water bath to 37°C, add the FRAP reagent (from step 7, above) and leave for 30 minutes, to come to assay temperature.
3. **Start the assay** - mix 100 μL of sample extract or standard solution with 3.0 mL of FRAP reagent. Vortex then *immediately* measure absorbance (T_0).
4. **Incubate for exactly four minutes in a 37°C water bath** – then re-measure the absorbance at 593 nm (T_4).
5. **Repeat for subsequent samples and standards**.

Calculation of results

1. **Calculate the change in absorbance over the four-minute assay period** – for each sample and standard, subtract the absorbance at 0 minutes from the absorbance at four minutes ($T_4 - T_0$).
2. **Plot a standard curve** and read off the values for the sample extracts (p. 219).

Determine the FRAP units for each sample extract by multiplying each absorbance change by 2.0, which gives FRAP units in μmol/g of ascorbic acid equivalents. (The stoichiometric factor of ascorbic acid in the FRAP assay is 2.0; therefore, 1 μmol of ascorbic acid is equivalent to 2 μmol of antioxidant power.)

3. **Calculate the FRAP value of the original food sample, in terms of either volume or weight**, by correcting for the appropriate dilution factor in the extraction procedure.

(Adapted from Benzie and Strain, 1999)

Table 63.2 Some advantages and disadvantages of ORAC compared with FRAP.

ORAC	FRAP
Measures biologically relevant free radicals	Measuring reducing capacity using Fe is not biologically relevant
Slower to perform	Faster to perform
More expensive	Less expensive
Requires fluorescence reader	Requires spectrophotometer
Measures all lipophilic and hydrophilic antioxidant components	Cannot detect species that act by radical quenching (H transfer), particularly SH-containing antioxidants, giving an underestimation of activity
Usually standardised for comparisons between laboratories	Not always standardised for comparisons between laboratories

The measured total antioxidant capacity of foods can also vary, depending on a number of factors, including –

- the colloidal properties of the substrates;
- the current status of oxidation within the particular food item, e.g. degree of oxidation may change in a piece of fruit as it ages;
- the localisation of antioxidants in different phases, e.g. aqueous and lipid phases;
- the method and equipment used for analysis, including which oxidant or free radical agent/generator is used in the assay and its biological relevance.

the DPPH radical reduces when the unpaired electron of the DPPH becomes paired with a hydrogen from a free radical scavenging antioxidant to form the reduced DPPH-H. Following 30 min incubation of sample and standard curve (e.g. ascorbic acid) with DPPH at room temperature the absorbance is measured at 517 nm on a spectrophotometer (p. 356). The advantages of this method are its simplicity, and accuracy comparable with other methods.

Trolox equivalent antioxidant capacity (TEAC)

This assay is based on the suppression of absorbance of radical cations of the substrate 2,2′-azino-bis(3-ethylbenzthiazoline-6-sulphonic acid) (ABTS) by antioxidants in the test sample. ABTS reacts with peroxidase and hydrogen peroxide to produce a blue–green colour with maximal absorbance at 730 nm. The production of the colour is suppressed by antioxidants proportionally to the amount present. TEAC requires 30 minutes of inhibition time to give results that correlate well with ORAC. Protocols giving shorter inhibition times tend to give a lack of correlation with other methods. Advantages of TEAC include: easy to do and relatively inexpensive; pH stable so can be used to research pH effects on activity; and relatively quick reaction. A commercial kit is available from Randox Laboratories Ltd, UK.

Text references

Benzie, I.F.F. and Strain, J.J. (1999) Ferric reducing/antioxidant power assay: direct measure of total antioxidant activity of biological fluids and modified version for simultaneous measurement of total antioxidant power and ascorbic acid concentration. *Methods in Enzymology*, **299**, 15–23.

Kerchev, P. and Ivanov, S. (2008) Influence of extraction techniques and solvents on the antioxidant capacity of plant material. *Biotechnology and Biotechnological Equipment*, **22**, 556–559. Available: http://www.diagnosisp.com/dp/journals/view_pdf.php?journal_id=1&archive=0&issue_id=17&article_id=462
Last accessed: 21/12/10.

Sakkiadi, A-V., Georgiou, C.A. and Haroutounian, S.A. (2007) A standard addition method to assay the concentration of biologically interesting polyphenols in grape berries by reversed-phase HPLC. *Molecules*, **12**, 2259–2269.

Zamora-Ros, R., Andres-Lacueva, C., Lamuela-Raventós, R.M., *et al.* (2008) Concentrations of resveratrol and derivatives in foods and estimation of dietary intake in a Spanish population: European prospective investigation into cancer and nutrition (EPIC)-Spain cohort. *British Journal of Nutrition*, **100**, 188–196.

Sources for further study

Goldacre, B. (2009) *Bad Science*. Fourth Estate, HarperCollins, London.
[Provides counterbalance to some of the more outlandish claims of some nutritionists and the food industry.]

Higdon, J. (2007) *An Evidence-Based Approach to Dietary Phytochemicals*. Thieme Medical Publishers, New York.

Hurst, W.J. (ed.) (2008) *Methods of Analysis for Functional Foods and Nutraceuticals*. CRC Press, Boca Raton, Florida.

ARS. *Phytochemical Search Database*. Agricultural Research Service, USDA. Available: http://www.pl.barc.usda.gov/usda_chem/achem_home.cfm
Last accessed: 21/12/10.

Waksmundzka-Hajnos, M. and Sherma, J. (2010) *High Performance Liquid Chromatography in Phytochemical Analysis*. Chromatographic Science Series. CRC Press, Boca Raton, Florida.

64 Sensory analysis techniques

Understanding sensory analysis of a food – this may involve:

- **sight** – appearance, size, shape, colour, texture;
- **taste** – sweet, sour, bitter, salty, spicy;
- **smell** – aroma, flavour;
- **hearing** – sound, e.g crunch of raw vegetables, sizzle of frying food;
- **touch** – texture, tactile properties, mouth feel.

Fig. 64.1 Example of a sensory testing booth for six participants.

SAFETY NOTE Tasters must always be asked if they have any food allergies or insensitivities (p. 312) before commencing tasting sessions.

Sensory analysis uses people as 'instruments' to evaluate food and food products in the food industry and in food research. The senses are used to determine the impacts or otherwise of a food product for consumers by measuring, analysing and interpreting the sensory characteristics of food (Stone and Sidel, 1993). Sensory analysis is useful in different stages of product development and can be used to compare a product with a competitor's similar product. This chapter describes some of the basic procedures that you may use during your degree programme.

Requirements for sensory testing

The following are some of the equipment and requirements for industry-standard sensory testing:

- a quiet, comfortable, temperature-regulated area, good shadowless illumination, around 50 per cent relative humidity;
- plenty of space and an absence of crowding and confusion;
- booths for six to eight testers (Fig. 64.1);
- no odours – air exhausts may be required for each booth if there are strong odours associated with the food being tested, positive pressure in the room, recirculated air through activated carbon filters;
- wood and materials used for making the furniture and booths should be odour-free, e.g. some types of particle board can give off a formaldehyde odour; carpets may contain mould and dirt giving an interfering odour; odourless cleaners should be used;
- surfaces should be easy to clean, e.g. Laminex® or stainless steel;
- off-white walls and furniture colours; no bright colours in the room;
- red, blue and green lighting options using coloured bulbs or filters if required to mask food colours;
- fluorescent lights with an option of types – warm white, cool white, simulated daylight;
- a separate preparation room; a hatch may be provided in the wall for entrance and removal of trays;
- round tables for discussion of descriptive analyses, if required.

Guidelines for testing in the sensory laboratory or other suitable room are provided in Box 64.1. Students should also familiarise themselves with basic food hygiene principles and methods, detailed in Box 64.2, and apply these when preparing foods and beverages in the sensory laboratory and also doing sensory testing.

Training participants

For practical exercises, you are likely to carry out initial training in taste identification and taste intensity, as described below.

Table 64.1. Preparation of solutions per student for taste identification training

Cup	Preparation	Taste
∇	250 mL water + 5.0 g sugar	Sweet
✱	250 mL water + 10 mL lemon juice	Sour
□	250 mL water + 100 mL decarbonated tonic water	Bitter
◆	250 mL water + 2.5 g salt	Salty

Rinse water should also be made available

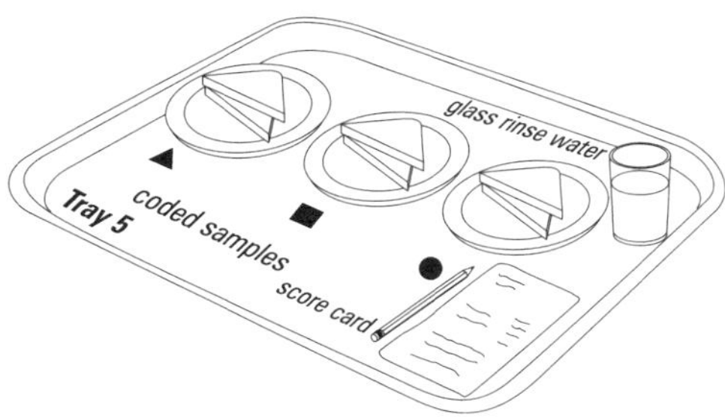

Fig. 64.2 Example tray layout for a taster testing three food products.

Taste identification training

Cups are coded with symbols for each taster and prepared as shown in Table 64.1. Tasters should not be involved in preparing or coding the solutions.

Coded samples are then provided on trays for each tester together with a score card and pencil (Fig. 64.2). The taster then records the taste of each sample, which is then scored. An example of a completed score card is shown in Fig. 64.3.

Score card – taste identification

Tray number... *3*...... Name:... *John Smith* ....

Instructions: on your tray are four coded samples of solutions representing four basic tastes – **sweet**, **bitter**, **sour** and **salty**.
In any order choose a cup, take a sip of liquid and hold in the mouth for 10 seconds or more and note the taste, then record it next to the corresponding code below. Proceed similarly with the other samples, rinsing your mouth well with water between the samples.

Solution	Taste	Correct ✓/ Incorrect ×
□	*sour*	×
✱	*bitter*	×
∇	*sweet*	✓
◆	*salty*	✓

Fig. 64.3 Example completed score card for taste identification test.

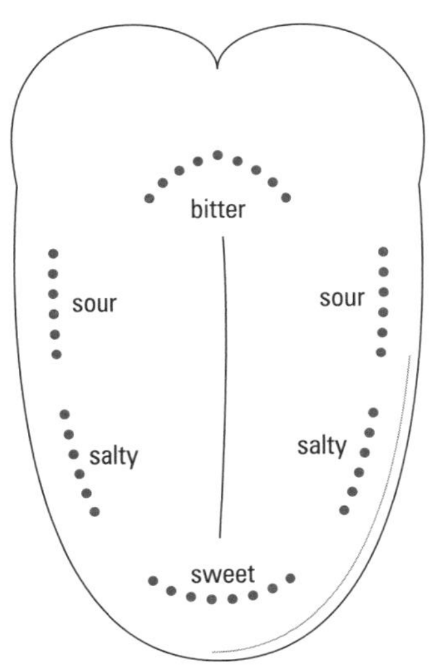

Fig. 64.4 Diagram showing approximate locations on the upper tongue surface of taste buds for sweet, salty, sour and bitter tastes.

Tasters should review the tastes where incorrect until they become familiar with the tastes. Fig. 64.4 shows the approximate positions on the tongue where the different tastes are sensed. When tasting, the solutions should be swilled around in the whole mouth to expose all taste buds to the solution.

> **KEY POINT** Training is required for all food tasters in order to provide accurate and repeatable assessments.

Taste intensity training

Typically, tasters are trained in discriminating between three different concentrations of specific tastes (tasters should not be involved in preparing or coding the solutions). Tasters are provided with three coded samples of different concentrations of a particular taste and then indicate the order of intensity of the samples on a score card.

For example, the following coded solutions might be prepared for each taster:

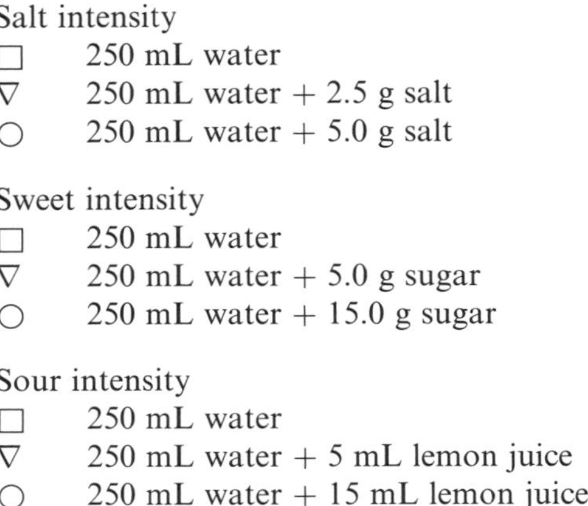

Salt intensity
□ 250 mL water
▽ 250 mL water + 2.5 g salt
○ 250 mL water + 5.0 g salt

Sweet intensity
□ 250 mL water
▽ 250 mL water + 5.0 g sugar
○ 250 mL water + 15.0 g sugar

Sour intensity
□ 250 mL water
▽ 250 mL water + 5 mL lemon juice
○ 250 mL water + 15 mL lemon juice

Fig. 64.5 gives an example of a completed score card.

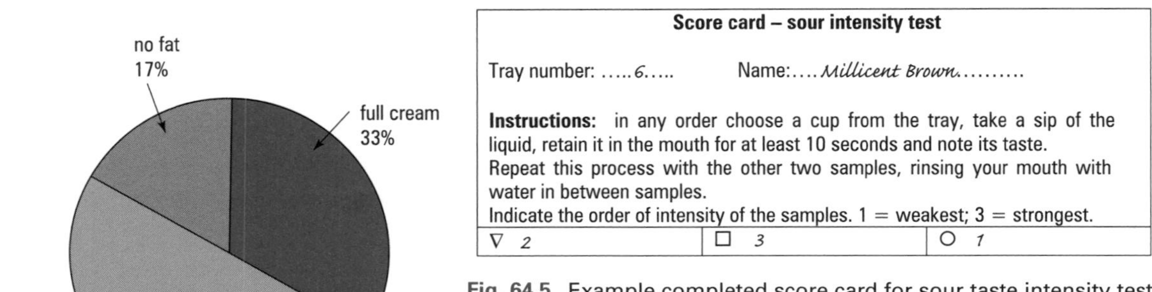

Score card – sour intensity test

Tray number: .....*6*.....   Name:....*Millicent Brown*.........

Instructions: in any order choose a cup from the tray, take a sip of the liquid, retain it in the mouth for at least 10 seconds and note its taste.
Repeat this process with the other two samples, rinsing your mouth with water in between samples.
Indicate the order of intensity of the samples. 1 = weakest; 3 = strongest.

▽ *2*	□ *3*	○ *1*

Fig. 64.5 Example completed score card for sour taste intensity test.

Fig. 64.6 Example of a pie chart showing preference for three types of milk.

Preference tests

This is used to determine people's preference between different, but related, products, e.g. two different samples of potato chips, or three types of milk as shown in the example in Fig. 64.6. Note that you should plot such results (Chapter 35) before you carry out any basic statistical analyses (Chapter 38) to gain an overview of the data and trends.

Preparing for testing – participants on a testing panel should have everything involved in the forthcoming process clearly explained:

- test procedures and how to taste;
- how to rinse between samples;
- use of the score card, and how samples are coded;
- type of evaluation required, e.g. preference, difference;
- need for quiet.

Ranking test

This is undertaken by tasters who each list the foods in order of overall preference, or on specific, defined characteristics, e.g. sweetness. From four to seven samples are usually used. Each taster is provided a tray with symbol-coded foods, utensils for each food as required, a score card and pencil. Rinse water should be provided for use between samples. Each level of rank is then allocated a number by the tester, e.g. in the example score card for four cooked fish products shown in Fig. 64.7 there are four foods being tested, so the ranking is 1 to 4 with 1 being the most liked and 4 being the least liked.

The data can be summarised by tallying up the numbers of times each rank was scored for each food, by placing data into a table, and determine the overall rank (e.g. – Fig. 64.8). Two products obviously ranked higher than two others. Testers' comments may help to clarify why there is such a large difference, e.g. perhaps the latter two containing bones or being too salty.

Understanding factors that may influence your sensory analysis results –

- **Coding error** – samples labelled numerically or alphabetically may subconsciously be associated with order in the taster's mind, hence use of coded symbols for labels to avoid such bias.
- **Expectation error** – testers should not receive any accidental information about a sample.
- **Stimulus error** – avoid any possible stimulus cues, e.g. mixed colours for containers, unmasked colour differences in foods.
- **Logic error** – two or more attributes of the sample become associated, e.g. a darker ginger beer may be deemed to have stronger flavour than a lighter one.
- **Halo effect** – scoring several attributes at the same time can result in cross-influence which can be avoided if attributes are studied separately.
- **Physical condition** – stress and various illnesses, e.g. colds and flu, may affect sensory awareness.

(from Meilgaard *et al.*, 2007)

Score card – ranking test

Tray number... *5*... Name...... *S. Holmes* ....

Instructions: taste the samples and rank them in the order from 1 to 4 that you like most. 1 = first choice; 4 = fourth choice.

Sample code	Order	Comments
□	*2*	*oily, may be good for omega 3*
✱	*1*	*nice texture and flavour*
∇	*3*	*contains bones*
◆	*4*	*Far too salty*

Fig. 64.7 Example ranking test score card for tasting cooked fish samples. Note the association of oily for one sample was linked in the tester's mind to omega 3 lipids being healthy, which could result in a stimulus error.

Sample	1st choice (4 points)	2nd choice (3 points)	3rd choice (2 points)	4th choice (1 point)	Score	Rank
□	4,4	3,3,3	2		19	2
✱	4,4,4	3,3,3			21	1
∇			2,2,2	1,1,1	9	3
◆			2,2	1,1,1,1	8	4

Fig. 64.8 Example chart for summarising results for six testers for four different fish products.

Rating test

This test is used to determine how much a person likes or dislikes a food product, based on a semantic differential scale. An example completed rating test score card is shown in Fig. 64.9. A tally chart (p. 196) can then be produced by addition of participants' results for each food product to determine the one most liked.

Definition

Semantic differential scale – scale to rate a product based upon a five-point rating scale that has bi-polar adjectives at each end.

Score card – rating test

Tray number... *2* ...... Name... *Jenny Williams*.......

Instructions: in any order, taste each of the lemonade samples and rate the samples on a scale of 5 to 1, 5 being excellent, 1 being very poor. Use the rinse water between samples.

Sample code	5☺	4	3😐	2	1☹
□	✓				
✱		✓			
∇		✓			
◆			✓		

Fig. 64.9 Example rating test score card for taster of four different brands of lemonade.

Difference tests

Difference tests are used to compare two types of the same food, e.g. regular baked beans compared with salt-reduced baked beans. The outcome for each tester will be that they are perceived as either (i) the same or (ii) different.

The triangle test

This test is used to determine if there is a sensory difference between two comparable products:

1. The tester is provided with three coded samples.
2. Two of the samples are the same, the third is different.
3. The tester tries to determine which sample is different from the other two.

Preparation for the triangle test is carried out as follows:

1. Trays are labelled for the testers (e.g. 1 to 6 for six testers).
2. Food containers are coded and set up three on each tray.
3. Foods are added to containers on trays.
4. Separate cutlery is provided (e.g. spoon) for each container.
5. A glass of rinse water is provided on each tray.
6. A pencil and score card are added to each tray.

This is followed by tasting of foods and completing score cards as per instructions (Fig. 64.10) and then analysis and presentation of results (Figs. 64.11 and 64.12)

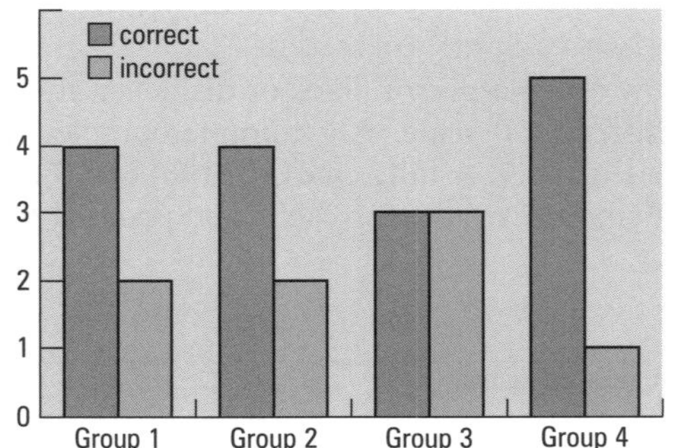

Fig. 64.11 Example class results plotted as a bar chart for four groups of six students undertaking the triangle test.

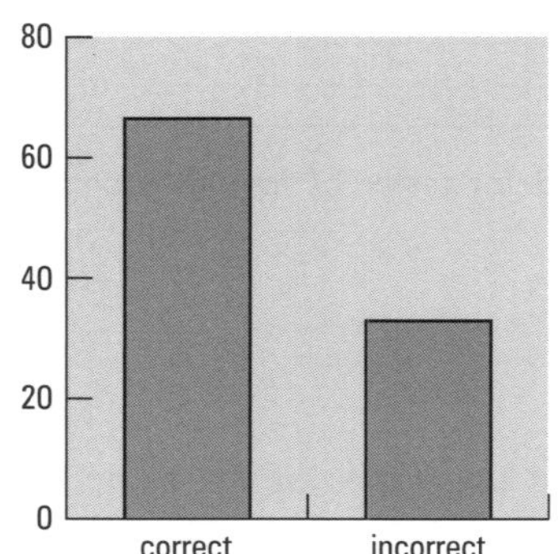

Fig. 64.12 Combined class results from Fig. 64.11 as percentages.

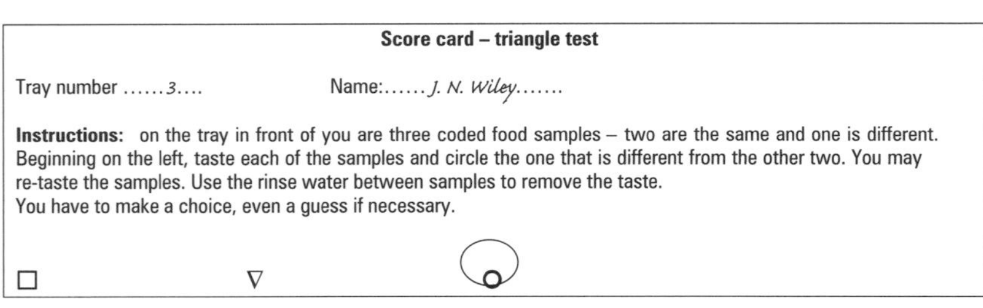
Score card – triangle test

Tray number ......3....   Name:......J. N. Wiley.......

Instructions: on the tray in front of you are three coded food samples – two are the same and one is different. Beginning on the left, taste each of the samples and circle the one that is different from the other two. You may re-taste the samples. Use the rinse water between samples to remove the taste.
You have to make a choice, even a guess if necessary.

□   ▽   ○

Fig. 64.10 Example completed score card for a triangle test.

Example arrangement of samples for six testers (e.g. X = standard cola, Y = diet cola; Fig. 64.13). There should be equal numbers of each sample overall in all possible combinations, with random order allocation. Several groups of six may be used for enhanced statistical accuracy. Score for analysis is either correct or incorrect.

Tray	XYX
1	□▽○

Tray	YXY
2	□▽○

Tray	XYY
3	□▽○

Tray	YXX
4	□▽○

Tray	YYX
5	□▽○

Tray	XXY
6	□▽○

Fig. 64.13 Arrangement of samples for six tasters doing a triangle test.

Descriptive tests

These are used to produce a written sensory profile on a food product.

> **Definitions**
>
> **Likert item** – respondents specify their level of agreement to a statement.
>
> **Likert scale** – sum of responses on several Likert items.

Line scales

These may be used for a descriptive rating test for sensory profiling of a single product:

1. Each taster is provided with same food product.
2. Using the five-point Likert items provided, the tasters rate the intensity of the indicated attributes.

An example of a completed score card is shown in Fig. 64.14 and a plot of example results is shown in Fig. 64.15.

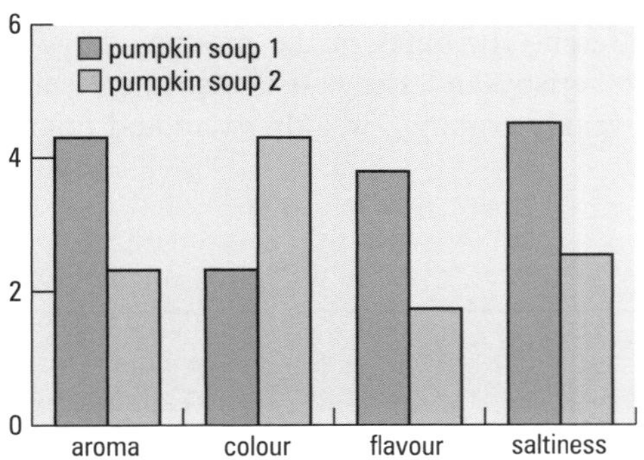

Fig. 64.15 Example average results of a descriptive rating test – in this case, for comparing two types of pumpkin soup. Resulting description: pumpkin soup 1 has more aroma, flavour and saltiness than pumpkin soup 2; however, pumpkin soup 2 has more colour.

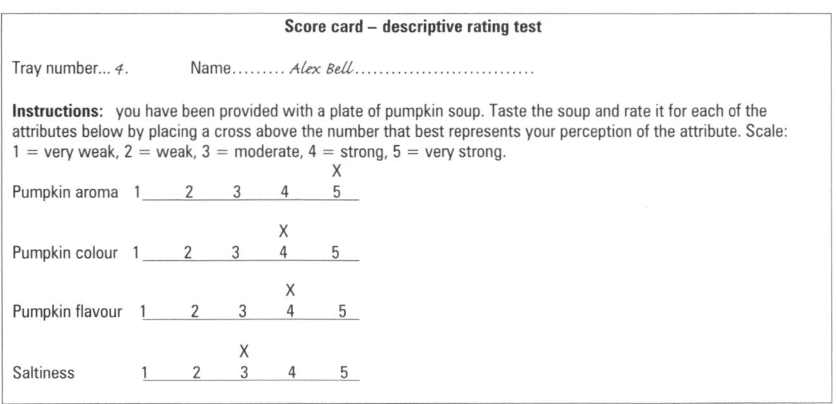

Score card – descriptive rating test

Tray number... 4. Name......... Alex Bell..............................

Instructions: you have been provided with a plate of pumpkin soup. Taste the soup and rate it for each of the attributes below by placing a cross above the number that best represents your perception of the attribute. Scale: 1 = very weak, 2 = weak, 3 = moderate, 4 = strong, 5 = very strong.

Attribute	1	2	3	4	5
Pumpkin aroma					X
Pumpkin colour				X	
Pumpkin flavour				X	
Saltiness			X		

Fig. 64.14 Example of a completed score card for a descriptive rating test for four sensory attributes on a sample of pumpkin soup.

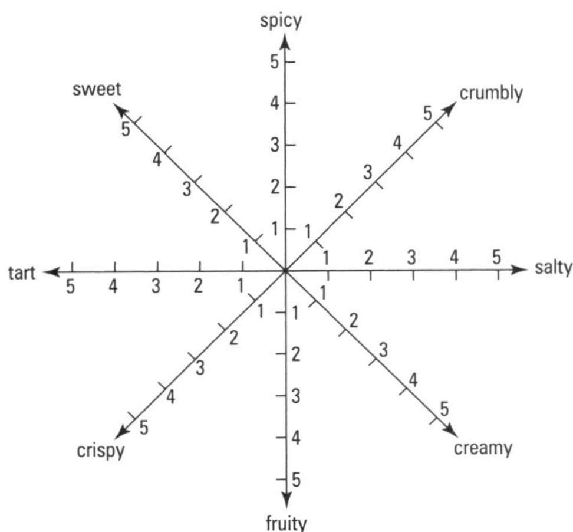

Fig. 64.16 Example four-line star template with attributes added.

If the data shown in Fig. 64.14 represented the average scores of a number of testers (rounded to the nearest scores) the food product could be described as: 'this pumpkin soup has strong pumpkin flavor and colour, a very strong pumpkin aroma and is moderately salty'.

Star profile

These are used to rate specific aspects of one or more foods. Completed star profiles can then be used to compare two or more foods. An example of a star diagram ready for a tester to use is provided in Fig. 64.16 and completed in Fig. 64.17. To prepare and use a star profile:

1. Choose a number of attributes that describe the important characteristics of the product.
2. Construct a star profile with the appropriate number of intersecting lines. You may find some templates through an Internet search.
3. Label the end of each line with the particular attribute to be rated. For each attribute the relative intensity increases as it goes further from the centre point (0–5 scale).

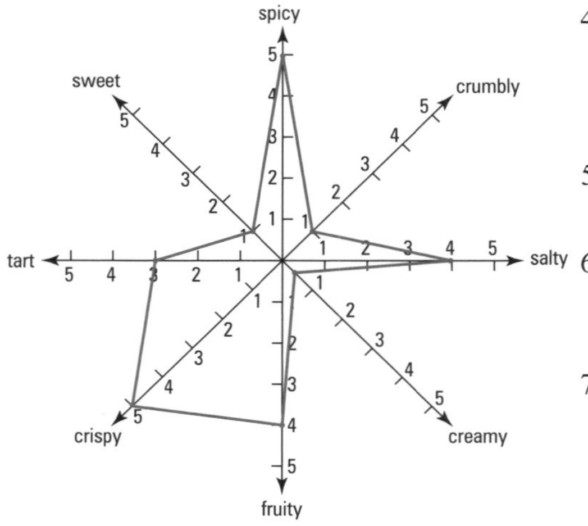

Fig. 64.17 Example of a completed star profile for a biscuit.

4. Each tester decides for themselves the intensity of each attribute, using the scale of 0 to 5, e.g. 0 = not at all, 1 = weak, 2 = fairly, 3 = moderate, 4 = quite, 5 = very. The scale should be designed according to the information being researched.
5. The tester places a cross on each line at the perceived point of intensity.
6. The averages of all testers' scores for each attribute can then be placed on another similar template and straight lines drawn between the crosses.
7. A product profile can then be written (using the exact same attribute words) that describes the different attributes of the product, e.g. for Fig. 64.17 'this biscuit is very crispy and spicy. It is also quite salty and fruity, moderately tart, weakly crumbly, weakly sweet and not at all creamy'.

Box 64.1 Guidelines for sensory testing

Your university or college may already have facilities for food testing. Designated booths may be available and coloured lighting available to mask food colours. However, if these facilities are not in place, a quiet room with sufficient ventilation and lighting might be used. There should be sufficient room for reasonable space between testers. Ensure there are no regulations or policies excluding foods in the room, and ensure any ethical requirements for testing are met (p. 143).

- **Choose a time when the room is certain to be free of cooking odours.** Testers should also not eat any strongly flavoured foods for at least a half an hour before the testing.
- **Requirements for food hygiene must be understood and followed** (Box 64.2).
- **Check for possible allergies.** Students who are to participate should be asked first about whether they have allergies or insensitivities to the foods being tested, including possible trace amounts of allergen in a product.
- **Allocate numbers according to the class size and the type of test.** In the food industry a large number of testers are usually used, and tests are repeated a number of times for statistical power and accuracy of the results.
- **Testers should not be involved in setting up samples.** Sample codes, arranging the food samples and collating results should be done by others to avoid unconscious bias by testers. During testing the testers must not discuss or verbally express their findings.
- **Use accurate measurements.** Gram scales and/or volumetric cylinders should be used for accuracy in preparing foods and beverages, and for measuring portion sizes to ensure the same amount is provided to each tester; stopwatch for timing preparation, thermometers for measuring food temperature and serving temperature; logbook for record keeping.
- **Use standard shaped and sized containers for presentation.** When possible use glass and/or glazed china containers and stainless steel equipment. Disposable plates and cups and plastic cutlery may release volatile compounds that can alter the flavor of a food product.
- **Ensure there are sufficient food samples prepared for the number of testers.**
- **Ensure consistency in size, shape, colour and temperature of the samples provided to the testers.**
- **Samples should be coded with different geometric symbols,** e.g. circle, square, triangle, and these can be produced using a Microsoft Word. Avoid use of 1, 2, 3, etc. or A, B, C, etc., as testers may subconsciously think A is better than B is better than C, etc. Records have to also be kept of the arrangement of coded samples provided to each tester.
- **Make sure a glass of water is available on each tray.** Testers rinse their mouths with water to remove the taste of the previous food between tastings.
- **After testing is complete clean up as indicated by your instructor.**

Box 64.2 How to apply basic food hygiene methods

Cleanliness

- **Cover any skin lesions with an appropriate dressing**, e.g. a plaster.
- **Always wash your hands thoroughly**, including scrubbing under nails, with soapy water before preparing and serving food. Also wash hands after touching animals, handling soil, using pesticides and other household chemicals, and after visiting the toilet, to avoid contaminating the household.
- **Make sure the tap for washing hands is kept clean** – ideally, use a tap that can be turned off with your arm or elbow, otherwise you might re-contaminate your hands after washing them.
- **Keep the kitchen floor, surfaces and kitchen sinks clean** with appropriate detergents. Be sure to carefully clean any crevices, or arrange to have them permanently sealed, if appropriate.
- **Clean countertops before preparing food** and between tasks to avoid cross-contamination.
- **Wash all utensils, dishes and cutting boards well,** after use and between tasks associated with different types of foods to prevent cross-contamination. Use hot soapy water and when finished with utensils store in a clean, sealed cupboard.
- **Change dish cloths, sponges and towels** used for washing and cleaning up regularly.
- **Wherever possible mix foods with clean utensils instead of your hands**, and avoid touching the 'working end' of any utensil to parts of the body, particularly mouth, nose or hair.
- **Avoid coughing or sneezing over food** and avoid preparing food for others while you have an infectious illness.
- **Clean up food spills as soon as possible.**
- **Cover foods on the bench or table** if there are flies or other insects around. Do not spray insecticides in the presence of foods.

Food storage

- **Store raw foods low in the refrigerator.** Do not store raw meat products where juices may drip onto items below. Thaw frozen meats on a plate, low in the refrigerator, rather than at room temperature.
- **When out on tasks do the food shopping last.** When food shopping collect cold items last and refrigerate as soon as possible. If there may be delays take an icebox with you with a couple of ice bricks inside for cold items, e.g. it doesn't take long for bacteria to multiply in milk at room temperature.
- **Check that frozen food is solidly frozen before purchase**, and that refrigerated food feels suitably cold.
- **Do not refreeze meats or fish that have been thawed.** Discard if not to be consumed soon.
- **Avoid preparing or eating foods that smell 'wrong'.** Remember the saying: 'if in doubt, throw it out'.
- **Freeze raw meat, chicken or fish immediately** if not planning to use it by the use-by date.
- **Refrigerate leftovers as soon as possible** and use within the next couple of days.
- **Keep hot foods hot and cold foods cold until ready to consume.** Bacteria multiply faster at lukewarm or room temperature. When preparing foods that need refrigeration return unused portions to the refrigerator immediately.

Packaged foods

- **Always look for the 'best before' or 'use-by' dates** on the labels of packaged foods. If the date is not clear or smudged, bypass this item. Do not use items after the date.
- **Follow advice on food labels**. Some packaged products as well as having expiry dates also have notes on the label saying, e.g., 'Use within 3 days of opening' or 'Refrigerate after opening and use within 7 days'. Note that this advice takes precedence over any 'use-by' date.
- **Check packages at purchase to ensure they are not broken or unsealed.**
- **Do not purchase cans of food that are bulging or misshapen.** Canned foods are usually sterile and can be stored for long periods. Once opened contents not immediately used should be transferred to an airtight container and stored in the refrigerator for up to a few days. Throw away any such cans that you find.

Text reference

Meilgaard, M.C., Civille, G.V. and Carr, B.T. (2007) *Sensory Evaluation Technique's* 4th edn. CRC Press, Taylor & Francis Group, Boca Raton, Florida.

Stone, H. and Sidel, J.L. (1993) *Sensory Evaluation Practice's* 2nd edn. Academic Press, San Diego, California.

Sources for further study

Home Economics Downloads. *Sensory Analysis Teacher's Manual.* Available: http://www.homeeconomics.ie/homeeconomics/Main/downloads.htm
Last accessed: 21/12/10.

Worksheetworks.com. *Star Diagram Creator.* Available: http://www.worksheetworks.com/miscellanea/graphic-organizers/star.html
Last accessed: 21/12/10.

65 Product development and evaluation

Reasons for innovative product development –

- increase sales;
- lower market prices by competitors;
- new technology becomes available;
- consumer demand;
- change in economic and social conditions, e.g. multicultural market develops;
- streamlining production;
- healthier food;
- improving packaging;
- improving product appearance;
- improving sensory attributes.

Example (new product ideas) Conduct a competition, perhaps printed on a current product label, inviting consumers to submit new product ideas with the incentive of a prize for the winner, e.g. Smith's® Potato Chips had a 2009 competition titled 'Do us a flavour'. The incentive was 1 per cent of future sales and $30 000 for the winner.

Example (name a new product) – Kraft Foods Limited®, Australia in 2009 released a new Vegemite™ product. Instead of having a name on the container the label said 'Name Me'. There was no monetary prize incentive, the incentive was being becoming part of history. The company indicates that the original Vegemite was named by the public in 1923, and would, of course, continue to be available.

Food product development is a broad topic that involves everything from genetic engineering of new foods, through to on-the-shelf food products targeting a significant risk group such as people with obesity, to just providing a product more delicious and appealing to a market. For a food product to be successful it has to be developed with careful market research throughout. Product innovation in food companies needs to be a more or less ongoing process as food fashions and people's diets change over time and other competitive products appear. At key points of development there have to be evaluations and quality-control checks, with the target market in mind.

Ten steps in development

KEY POINT In producing a new food product it is imperative to follow a logical process. It is estimated that about 90 per cent of new food products fail within 12 months of launching.

1. **Develop a new idea**, e.g. other food companies claim they have put out a healthier potato chip product. With a current focus on community health in the market area we want to investigate how our product can become more competitive to those interested in healthier foods.
2. **Carry out market research** to check and confirm that the market is actually there for the planned, updated product. We should first check what is actually in our competitor's product compared with our own. Perhaps ours is already healthier than theirs and we only need to re-focus our advertising? Market research could involve surveys and questionnaires, telephone interviews and studying market trend information. Is there a niche not being met, e.g. what if we included currently popular healthy omega 3 oils in our potato chips, if other companies are not? Is there a market for a smaller/larger serving size package?
3. **Set design specifications** of the proposed product, e.g. size, weight, proposed ingredients and amounts, preservatives or natural, required sensory attributes, shelf life, costs, packaging, equipment availability and any further requirements.
4. **Determining the cost of producing the product** and see if this will allow sufficient profit.
5. **Develop the product** on a small scale, followed by sensory analysis by consumer panels (Chapter 64) and instrumentation.
6. **Modify the product to improve appeal**, followed by further testing. Repeat steps 5 and 6 as required, e.g. the panel found our new healthy potato chips were too strongly flavoured so we need to revisit this aspect.
7. **Prepare the final manufacturing product specifications**, including control checks and consistent standards.

8. **Test the target market** with the trial product along with a simple questionnaire to complete. This may be done with shoppers in supermarkets.
9. **Launch the product** by advertising and marketing.
10. **Monitor sales and market trends** as future changes in the market or by competitors may lead back to step 1.

Quality control

The internationally recognised Hazard Analysis and Critical Control Points (HACCP) system is used to determine and identify where potential food safety problems might occur, and introduce appropriate checks (p. 137). The aims of HACCP are to eliminate or reduce the risks of chemical, biological and physical hazards in food products. This risk analysis applies to the product design stage, manufacturing stages and to the final product.

A quality management system (QMS) is used in the UK food industry to document and ensure product quality. Companies correctly using a QMS to achieve high standards of quality control receive a recognised certificate called ISO 9000 (p. 138).

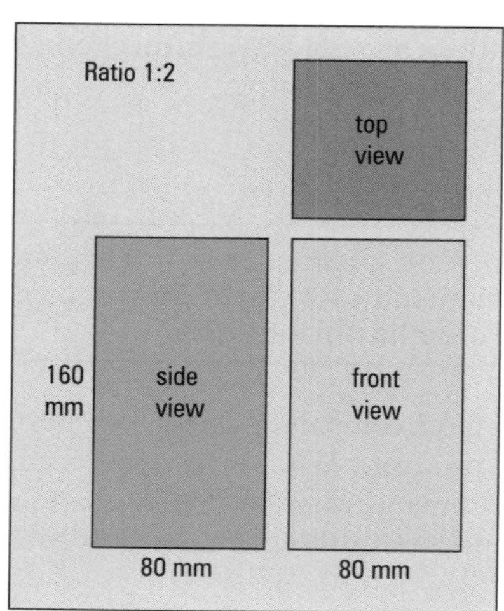

Fig. 65.1 Example template for designing an orthographic projection for a custard carton.

Package graphic design

Package design can have a significant bearing as to whether the product is successful or not. Food package and advertising designers use computer software packages for graphic design. Examples of software that you may use in class exercises are Adobe® Illustrator™, Adobe® Photoshop™, Corel® Paint Shop Pro™, or Microsoft® Paint™ (under Accessories). Orthographic projections are often used to illustrate a potential product package in three dimensions for evaluation. This involves an accurate scale drawing showing front, side and top of the product each as a ratio of actual size, e.g. 1:2. Figure 65.1 shows an example layout for developing an orthographic projection of a custard carton. Once illustrated, the software can then be used to join the sections and convert this to a three-dimensional view.

Packaging design – the main considerations are:

- **design and graphics** – can be market tested during development;
- **physical protection** – to prevent crushing, shock, tearing or other damage;
- **barrier protection** – against water vapour, atmospheric gases, microbes, dust;
- **convenience** – ease of handling, stacking, opening, reclosing, multiple compartments for serves;
- **security** – tamper-resistant and tamper-evident features, e.g. authentication seals, 'intelligent packaging' that changes colour if punctured or tampered with;
- **environmentally friendly** – reduced amount of packaging, particularly non-biodegradable materials;
- **labelling** – nutritional information, use, storage, recycling, etc. (p. 138).

Shelf life analysis

Locate and familiarise yourself with the food standards regulations for shelf life in your region, e.g. the Australian Food Standards Code reads 'the Standard requires packaged food, with some exceptions, to be date marked, and prohibits the sale of packaged food after the expiration of the use-by date, where such a mark is required'.

Responsibility for determining the shelf life of a food product usually falls upon the manufacturer and packager.

Factors that affect food shelf life

- **Microbial changes.** In food products that are not sterile the rate of reproduction of spoilage microorganisms is usually the main factor influencing shelf life. However, some foods are sterilised and therefore do not have microbial growth, e.g. canned foods. Some food products have a low water activity (p. 489) that prevents growth of microbes,

Example (shelf life analysis) –
1. 50 consumers comprise the testing panel.
2. Raw minced meat samples are stored at −20°C, 4°C, 10°C and 25°C.
3. Ten different storage time lengths are set.
4. Panel members observe the samples and indicate whether or not they would accept the samples.
5. Microbial cultures are done at each time point.
6. Data analysed with repeated measures/accelerated failure time models.

e.g. sucrose. Lowering water activity in a product is one method for lengthening shelf life. Shelf life related to microbial growth may be detected with appropriate culture procedures (Chapters 68 and 69) and also, where safe, by sensory testing, e.g. the aroma of sour milk.

- **Moisture transfer.** Some foods may either gain or lose water from or to the environment, respectively, e.g. some types of biscuits may absorb atmospheric moisture and then be more prone to spoilage. Some types of baked goods can lose moisture as water vapour, affecting their sensory attributes. Therefore moisture-proof packaging is required to stop water vapour transfer and lengthen shelf life. Sensory panels can be used to evaluate shelf life under different conditions and packaging (Chapter 64).
- **Biochemical and chemical changes.** Many reactions can occur that limit the shelf life of a food product, e.g. rancidity (oxidation) of foods containing fats and oils. This reaction is slowed by freezing but still continues, requiring shelf life considerations for frozen fish and meats. Sensory testing of odour can be done to determine shelf life related to rancidity. Vitamins may also become oxidised and degraded over time and this has to be taken into consideration for package fortified, vitamin content labelling and shelf life. Analysis methods for vitamins are covered in Chapter 45.

Instrumental analysis methods

Gas chromatography olfactory analysis

This method of analysing volatile aromas in food products involves a GC or GC/MS (Chapter 54) outlet connected to an olfactory detector. An example is the Antech Olfactory Detector Outlet shown in Fig. 65.2. This method provides a link between instrumental methods and human sensing to profile a product. As components of a food product are sequentially eluted by chromatography they are transferred in a humidified air stream to a nose cone. A trained operator can then record the aromas and their quality as they elute and a product profile can be developed.

Electronic sensor analysers

As well as sensory analysis panels (Chapter 64) food product developers can now also utilise instrumentation calibrated to evaluate flavour and taste. An example is the Alpha MOS FOX E-Nose Analyzer, which incorporates gas and liquid array sensors. An advantage over human sensory analysis is its greater objectivity and quicker results. Applications include flavour assessment, product matching, quality control, benchmarking, process monitoring, identifying spoilage, aroma quality over time and developing new formulations. An electronic tongue is used for analysing the taste of liquids and emulsions, while an electronic nose detects and analyses aerosols and volatile organic compounds. Reproducibility of the sample injection volume is important for accuracy.

The sample is analysed by chemical-sensitive array layers which are specifically modified on interaction with the food molecules resulting in measurable electrical conductance changes in conjunction with reference electrodes. Pattern recognition software is then used in the analysis. The

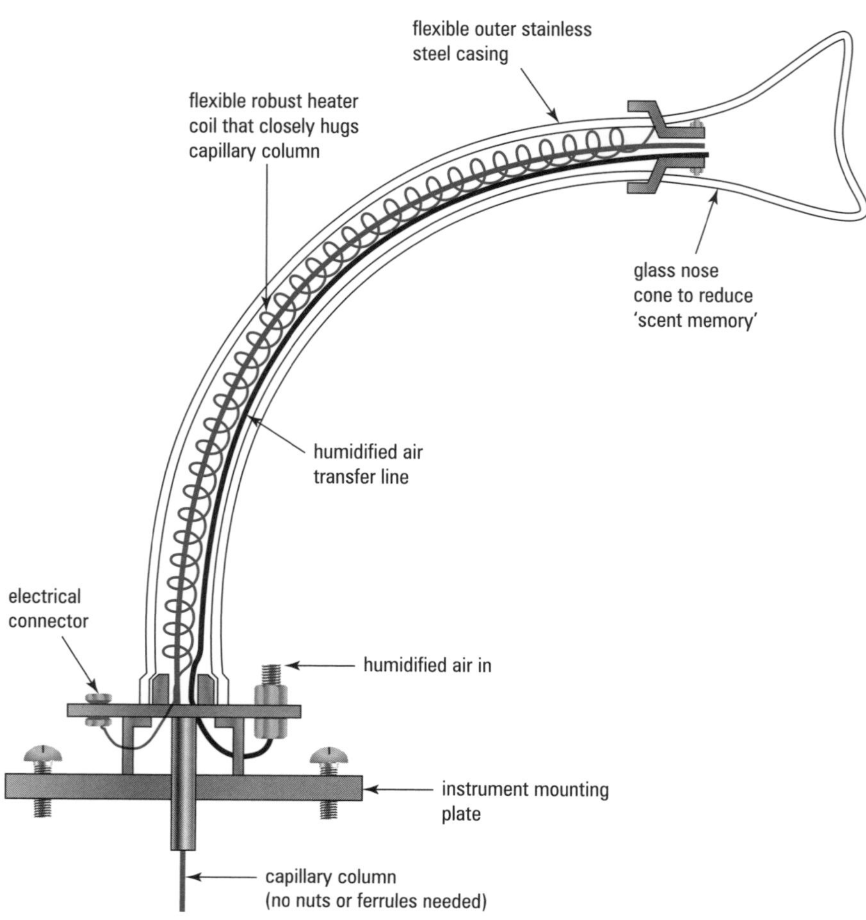

Fig. 65.2 Antech Olfactory Detector Outlet.

> **Example** – Ten batches of apples are sampled for aroma by a sensory testing panel with instructions to discriminate good versus bad ones. After analysis of panel results, six batches are declared good and four bad. Profiles of both the good batches and bad batches can then be produced with the electronic sensor and qualitative models determined. Unknown batches of apples can subsequently be tested with the apparatus for quality of aroma.

instruments can detect a range of substances accountable for taste and odor, and produce a unique fingerprint for a product.

Basic analysis procedure:

1. Product samples are first qualified by other methods such as panel sensory analysis, or chromatography to define a mathematical model to incorporate in analyses.
2. The model is then incorporated and validated (or otherwise) on the electronic sensor analyser.
3. Once validated the analyser can subsequently be used for routine tests on this product to provide quick and objective results.

Food texture measurements

Commercial equipment is available for testing different aspects of food texture, e.g. the Food Technology Corporation's Computer Controlled Food Texture Measurement System, TMS-Pro. A detailed range of test cells, probes and fixtures is available for studying different texture attributes, with real time graphing of results. If suitable equipment is not available at your university perhaps an industry visit or placement might be arranged by your instructor. For some examples of types of food texture measurements refer to p. 334.

Sources for further study

CSIRO Food and Nutritional Sciences Fact Sheet (2005) *Shelf Life of Foods*. http://www.foodscience.csiro.au/shelf-life.htm
Last accessed: 21/12/10.

Earle, M.D. and Earle, R.L. (eds) (2007) *Case Studies in Food Product Development*. Woodhead Publishing, Abington, Cambridge.

Fuller, G.W. (2004) *New Food Product Development – From Concept to Marketplace*. CRC Press, Boca Raton, Florida.

Steele, R. (ed.) (2004) *Understanding and Measuring the Shelf-life of Food*. Woodhead Publishing, Abington, Cambridge.

Food microbiology and processing

66 Sterile technique and microbial culture

Achieving a sterile state – you should assume that all items of laboratory equipment have contaminating microbes on their surfaces, unless they have been destroyed by some form of sterilisation. Such items will only remain sterile if they do not come into contact with the non-sterile environment.

Fig. 66.1 International symbol for a biohazard. Usually red on a yellow background, or black on a red background.

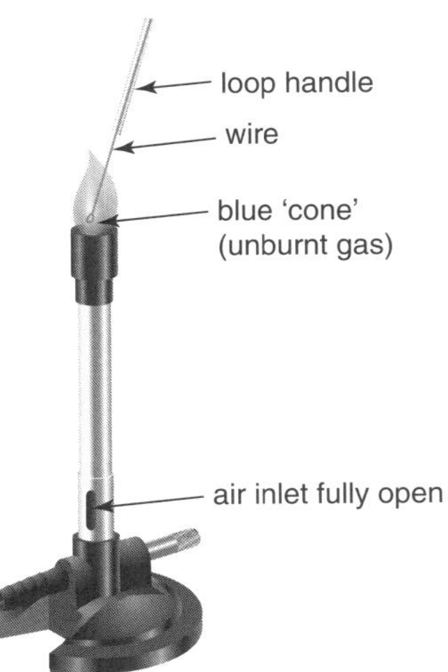

Fig. 66.2 'Flaming' a wire loop. Keep the loop in the hottest part of the Bunsen flame (just outside the blue 'cone') until the wire is red-hot.

Sterile technique (aseptic technique) is the name given to the procedures used in cell culture. Although the same general principles apply to all cell types, you are most likely to learn the basic procedures, within a food microbiology course, using bacteria, and therefore most of the examples given in this section refer to bacterial culture.

Sterile technique serves two main purposes:

1. **To prevent accidental contamination of laboratory cultures** due to microbes from external sources, e.g. skin, clothing or the surrounding environment.

2. **To prevent microbial contamination of laboratory workers**, in this instance you and your fellow students.

KEY POINT *All* microbial cell cultures should be treated as if they contained potentially harmful organisms. Sterile technique forms an important part of safety procedures, and must be followed whenever cell cultures are handled in the laboratory.

Care is required because:

- You may accidentally isolate a harmful microbe as a contaminant when culturing a relatively harmless strain.
- Some individuals are more susceptible to infection and disease than others – not everyone exposed to a particular microbe will become ill.
- Laboratory culture involves purifying and growing large numbers of microbial cells – this represents a greater risk than small numbers of the original microbe.
- A microbe may change its characteristics, perhaps as a result of gene exchange or mutation.

The international biohazard symbol, shown in Fig. 66.1, is used to indicate a significant risk due to a pathogenic microbe (pp. 487–8).

Sterilisation procedures

Given the ubiquity of microbes, the only way to achieve a sterile state is by their destruction or removal. Several methods can be used to achieve this objective:

Heat treatment

This is the most widespread form of sterilisation and is used in several basic laboratory procedures including the following:

- **Red-heat sterilisation.** Achieved by heating metal inoculating loops, forceps, needles, etc. in a Bunsen flame (Fig. 66.2). This is a simple and effective form of sterilisation as no microbe will survive even a brief exposure to a naked flame. Flame sterilisation using alcohol is used for glass rods and spreaders (see below).
- **Dry-heat sterilisation.** Here, a hot-air oven is used at a temperature of at least 160°C for at least two hours. This method is used for the

Fig. 66.3 Autoclave tape – the bottom sample is untreated while the upper sample (with dark diagonal lines) has been autoclaved.

routine sterilisation of laboratory glassware. Dry-heat procedures are of little value for items requiring repeated sterilisation during use.

- **Moist-heat sterilisation.** This is the method of choice for many laboratory items, including most fluids, apart from heat-sensitive media. It is also used to decontaminate liquid media and glassware after use. The laboratory autoclave is used for these purposes. Typically, most items will be sterile after 15 minutes at 121°C, although large items may require a longer period. The rapid killing action results from the latent heat of condensation of the pressurised steam, released on contact with cool materials in the autoclave. Although special heat-sensitive tape (Fig. 66.3) is sometimes used to check that the autoclave is operating correctly, a better approach is to use spores of *Bacillus stearothermophilus*.

Radiation

Many disposable plastic items used in microbiology and cell biology are sterilised by exposure to UV or ionising radiation. They are supplied commercially in sterile packages, ready for use. Ultraviolet radiation has limited use in the laboratory, while ionising radiation (e.g. γ-rays) requires industrial facilities and cannot be operated on a laboratory scale.

Filtration

Using a sterile filter – most filters are supplied as pre-sterilised items. Make sure you follow a procedure that does not contaminate the filter on removal from its protective wrapping.

Heat-labile solutions (e.g. complex macromolecules, including proteins, antibiotics, serum) are particularly suited to this form of sterilisation. The filters come in a variety of shapes, sizes and materials, usually with a pore size of either 0.2 µm or 0.45 µm. The filtration apparatus and associated equipment is usually sterilised by autoclaving or by dry heat. Passage of liquid through a sterile filter of pore size 0.2 µm into a sterile vessel is usually sufficient to remove bacteria but not viruses, so filtered liquids are not necessarily virus-free.

Chemical agents

SAFETY NOTE When working with biocides, take care to avoid skin contact or ingestion, as most are toxic and irritant. If contact does occur, rinse with plenty of water.

These are known as disinfectants, or biocides, and are most often used for the disposal of contaminated items following laboratory use, e.g. glass slides and pipettes. They are also used to treat spillages. The term ‘disinfection’ implies destruction of disease-causing bacterial cells, although spores and viruses may not always be destroyed. Remember that disinfectants require time to exert their killing effect – any spillage should be covered with an appropriate disinfectant and left for at least ten minutes before mopping up.

Use of laboratory equipment

Working area

One of the most important aspects of good sterile technique is to keep your working area as clean and tidy as possible. Start by clearing all items from your working surface, wipe the bench down with disinfectant and then arrange the items you need for a particular procedure so that they are close at hand, leaving a clear working space in the centre of your bench.

Using molten agar – a water bath (at 45–50°C) can be used to keep an agar-based medium in its molten state after autoclaving. Always dry the outside of the container on removal from the water bath, to reduce the risk of contamination from microbes in the water, e.g. during pour plating (p. 471).

Media

Cells may be cultured in either a liquid medium (broth), or a solidified medium (p. 476). The gelling agent used in most solidified media is agar, a complex polysaccharide from red algae that produces a stiff transparent gel when used at 1–2% (w/v). Agar is used because it is relatively resistant to degradation by most bacteria and because of its rheological properties – an agar medium melts at 98°C, remaining solid at all temperatures used for routine laboratory culture. Once melted however, it does not solidify again until the temperature falls to about 44°C. This means that heat-sensitive constituents (e.g. vitamins, blood, cells, etc.) can be added aseptically to the medium after autoclaving.

Working with plastic disposable loops – these are used in many research laboratories: pre-sterilised and suitable for single use, they avoid the hazards of naked flames and the risk of aerosol formation during heating. Discard into a disinfectant solution after use.

Inoculating loops

The initial isolation and subsequent transfer of microbes between containers can be achieved by using a sterile inoculating loop. Most teaching laboratories use nichrome wire loops in a metal handle. A wire loop can be repeatedly sterilised by heating the wire, loop downwards and almost vertical, in the hottest part of a Bunsen flame until the whole wire becomes red-hot. Then the loop is removed from the flame to minimise heat transfer to the handle. After cooling for eight–ten seconds (without touching any other object), it is ready for use.

When re-sterilising a contaminated wire loop in a Bunsen flame after use, do not heat the loop too rapidly, as the sample may spatter, creating an aerosol: it is better to soak the loop for a few minutes in disinfectant than to risk heating a fully charged (contaminated) inoculating loop.

Using a Bunsen burner to reduce airborne contamination – working close to the updraught created by a Bunsen flame reduces the likelihood of particles falling from the air into an open vessel.

Containers

There is a risk of contamination whenever a sterile bottle, flask or test tube is opened. One method that reduces the chance of airborne contamination is quickly to pass the open mouth of the glass vessel through a flame. This destroys any microbes on the outer surfaces nearest to the mouth of the vessel. In addition, by heating the air within the neck of the vessel, an outwardly directed air flow is established, reducing the likelihood of microbial contamination.

It is general practice to flame the mouth of each vessel immediately after opening and then repeat the procedure just before replacing the top. Caps, lids and cotton wool plugs must not be placed on the bench during flaming and sampling: they should be removed and held using the smallest finger of one hand, to minimise the risk of contamination. This also leaves the remaining fingers free to carry out other manipulations. With practice, it is possible to remove the tops from two tubes, flame each tube and transfer material from one to the other while holding one top in each hand.

Using glass pipettes – these are plugged with cotton wool at the top before being autoclaved inside a metal can. Flame the open end of the can on removal of a pipette, to prevent contamination of the remaining pipettes. Autopipettors and sterile disposable tips (p. 155) offer an alternative approach.

Table 66.1 Classification of microbes on the basis of hazard. The following categories are recommended by the UK Advisory Committee on Dangerous Pathogens (ACDP).

Hazard group	Comments
1	Unlikely to cause human disease
2	May cause disease: possible hazard to laboratory workers, minimal hazard to community
3	May cause severe disease: may be a serious hazard to laboratory workers, may spread to community
4	Causes severe disease: is a serious hazard to laboratory workers, high risk to community

Laminar-flow cabinets

These are designed to prevent airborne contamination, e.g. when preparing media or subculturing microbes or tissue cultures. Sterile air is produced by passage through a high-efficiency particulate air (HEPA) filter: this is then directed over the working area, either horizontally (towards the operator) or downwards. The operator handles specimens, media, etc., through an opening at the front of the cabinet. Note that standard laminar-flow cabinets do *not* protect the worker from contamination and must not be used with pathogenic microbes: special safety cabinets are used for work with ACDP hazard groups 3 and 4 microbes (Table 66.1) and for samples that might contain such pathogens.

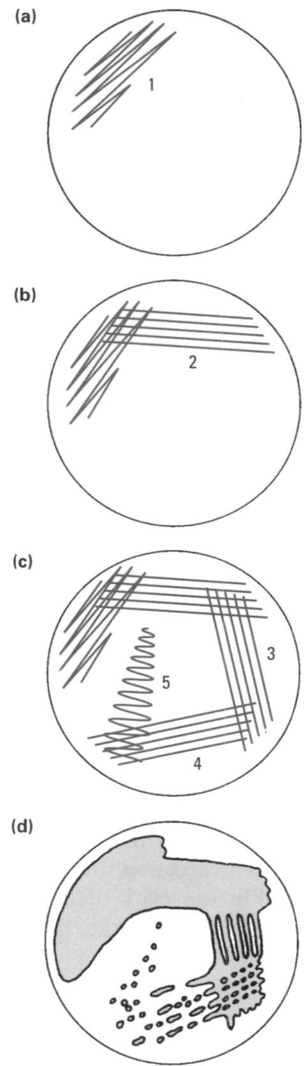

Fig. 66.4 Preparation of a streak plate for single colonies. (a) Using a sterile metal loop, take a small sample of the material to be streaked. Distribute the sample over a small sector of the plate (area 1), then flame the loop and allow to cool (approximately 8–10 s). (b) Make several small streaks from the initial sector into the adjacent sector (area 2), taking care not to allow the streaks to overlap. Flame the loop and allow to cool. (c) Repeat the procedure for areas 3 and 4, re-sterilising the loop between each step. Finally, make a single, long streak, as shown for area 5. (d) The expected result after incubation at the appropriate temperature (e.g. 37°C for 24 h): each step should have diluted the inoculum, giving individual colonies within one or more sectors on the plate. Further subculture of an individual colony should give a pure (clonal) culture.

Microbiological hazards

KEY POINT The most obvious risks when handling microbial cultures are those due to ingestion or entry via a cut in the skin – all cuts should be covered with a plaster or disposable plastic gloves. A less obvious source of hazard is the formation of aerosols of liquid droplets from microbial suspensions, with the risk of inhalation, or surface contamination of other objects.

The following steps will minimise the risk of aerosol formation:

- **Use stoppered tubes when shaking**, centrifuging or mixing microbial suspensions.
- **Pour solutions gently**, keeping the difference in height to a minimum.
- **Discharge pipettes on to the side of the container.**

Other general rules that apply in all laboratories include:

- **Avoid all 'hand-to-mouth' operations**, e.g. chewing the end of your pen while thinking.
- **Take care with sharp instruments**, including needles and glass Pasteur pipettes.
- **Do not pour waste cultures down the sink** – they must be autoclaved.
- **Put other contaminated items** (e.g. slides, pipettes) **into disinfectant after use.**
- **Wipe down your bench with disinfectant when practical work is complete.**
- **Always wash your hands before leaving the laboratory.**

Plating methods

Many culture methods make use of a solidified medium within a Petri plate. A variety of techniques can be used to transfer and distribute the organisms prior to incubation. The three most important procedures are described below.

Streak-dilution plate

Streaking a plate for single colonies is one of the most important basic skills in microbiology, since it is used in the initial isolation of a cell culture and in maintaining stock cultures, where a streak-dilution plate with single colonies all of the same type confirms the purity of the strain. A sterile inoculating loop is used to streak the organisms over the surface of the medium, thereby diluting the sample (Fig. 66.4). The aim is to achieve single colonies at some point on the plate: ideally, such colonies are derived from single cells (e.g. in the case of unicellular bacteria, animal and plant cell lines) or from groups of cells of the same species (in filamentous or colonial forms). Single colonies, containing cells of a single species and derived from a single parental cell, form the basis of all pure culture methods (p. 459).

Note the following:

- Keep the lid of the Petri plate as close to the base as possible to reduce the risk of aerial contamination.

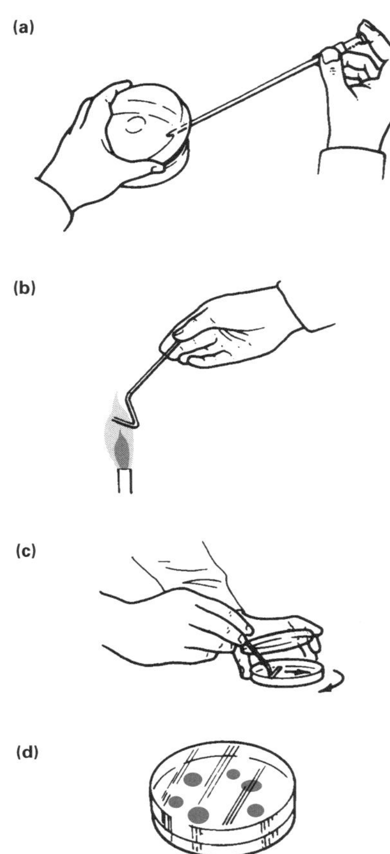

Fig. 66.5 Preparation of a spread plate. (a) Transfer a small volume of cell suspension (0.05–0.5 ml) to the surface of a solidified medium in a Petri plate. (b) Flame sterilise a glass spreader and allow to cool (8–10 s). (c) Distribute the liquid over the surface of the plate using the sterile spreader. Make sure of an even coverage by rotating the plate as you spread: allow the liquid to be absorbed into the agar medium. Incubate under suitable conditions. (d) After incubation, the microbial colonies should be distributed over the surface of the plate.

- Allow the loop to glide over the surface of the medium. Hold the handle at the balance point (near the centre) and use light, sweeping movements, as the agar surface is easily damaged and torn.
- Work quickly, but carefully. Do not breathe directly on to the exposed agar surface and replace the lid as soon as possible.

Spread plate

This method is used with cells in suspension, either in a liquid growth medium or in an appropriate sterile diluent. It is one method of quantifying the number of viable cells (or colony forming units) in a sample, after appropriate dilution.

An L-shaped glass spreader is sterilised by dipping the end of the spreader in a beaker containing a small amount of 70% v/v alcohol, allowing the excess to drain from the spreader and then igniting the remainder in a Bunsen flame. After cooling, the spreader is used to distribute a known volume of cell suspension across the plate (Fig. 66.5). *There is a significant fire risk associated with this technique*, so take care not to ignite the alcohol in the beaker, e.g. by returning an overheated glass rod to the beaker. The alcohol will burn with a pale blue flame that may be difficult to see, but will readily ignite other materials (e.g. a laboratory coat). Another source of risk comes from small droplets of flaming alcohol shed by an overloaded spreader on to the bench and this is why you *must* drain excess alcohol from the spreader *before* flaming. Some laboratories now provide plastic disposable spreaders for student use, to avoid the risk of fire.

Pour plate

This procedure also uses cells in suspension, but requires molten agar medium, usually in screw-capped bottles containing sufficient medium to prepare a single Petri plate (i.e. 15–20 mL), maintained in a water bath at 45–50°C. A known volume of cell suspension is mixed with this molten agar, distributing the cells throughout the medium. This is then poured without delay into an empty sterile Petri plate and incubated, giving widely spaced colonies (Fig. 66.6). Furthermore, as most of the colonies are formed within the medium, they are far smaller than those of the surface streak method, allowing higher cell numbers to be counted (e.g. up to 1000 colonies per plate): some workers pour a thin layer of molten agar onto the surface of a pour plate after it has set, to ensure that no surface colonies are produced. Most bacteria and fungi are not killed by brief exposure to temperatures of 45–50°C, though the procedure may be more damaging to microbes from low temperature conditions, e.g. psychrophilic bacteria.

One disadvantage of the pour-plate method is that the typical colony morphology seen in surface-grown cultures will not be observed for those colonies that develop within the agar medium. A further disadvantage is that some of the suspension will be left behind in the screw-capped bottle. This can be avoided by transferring the suspension to the Petri plate, adding the molten agar, then swirling the plate to mix the two liquids. However, even when the plate is swirled repeatedly and in several directions, the liquids are not mixed as evenly as in the former procedure.

Working with phages

Bacterial viruses ('bacteriophages', or simply 'phages') are often used to illustrate the general principles involved in the detection and enumeration

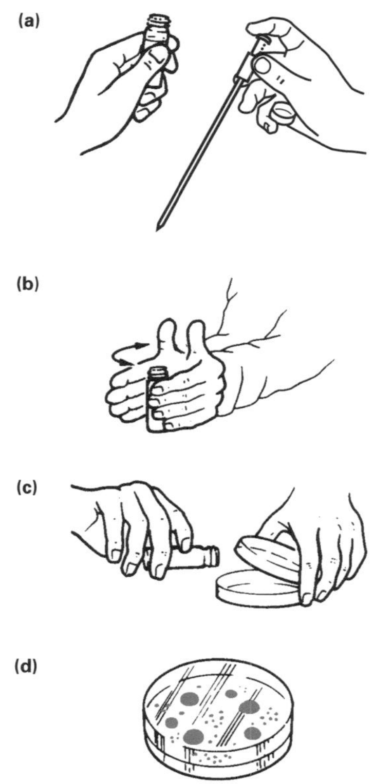

Fig. 66.6 Preparation of a pour plate. (a) Add a known volume of cell suspension (0.05–1.0 mL) to a small bottle of molten agar medium from a 45°C water bath. (b) Mix thoroughly, by rotating between the palms of the hands: do not shake or this will cause frothing of the medium. (c) Pour the mixture into an empty, sterile Petri plate and allow to set. Incubate under suitable conditions. (d) After incubation, the microbial colonies will be distributed throughout the medium: any cells deposited at the surface will give larger, spreading colonies.

of viruses. They also have a role in genome mapping of bacteria. Individual phage particles (virions) are too small to be seen by light microscopy, but are detected by their effects on susceptible host cells:

- Virulent phages will infect and replicate within actively growing host cells, causing cell lysis and releasing new infective phages – this 'lytic cycle' takes ≈30 min for T-even phages of *E. coli*, e.g. T4.
- Temperate phages are a specialised group, capable either of lytic growth or an alternative cycle, termed lysogeny – the phage becomes latent within a host cell (lysogen), typically by insertion of its genetic information into the host cell genome, becoming a 'prophage'. At a later stage, termed induction, the prophage may enter the lytic cycle. A widely used example is λ phage of *E. coli*.

The lytic cycle can be used to detect and quantify the number of phages in a sample. A known volume of sample is mixed with susceptible bacterial cells in molten soft agar medium (45–50°C), then poured on top of a plate of the same medium, creating a thin layer of 'top agar'. The upper layer contains only half the normal amount of agar, to allow phages to diffuse through the medium and attach to susceptible cells. On incubation, the bacteria will grow throughout the agar to produce a homogeneous 'lawn' of cells, except in those parts of the plate where a phage particle has infected and lysed the cells to create a clear area, termed a plaque (Fig. 66.7). Each plaque is due to a single functional phage (i.e. a plaque-forming unit, or PFU). A count of the number of plaques can be used to give the number of phages in a particular sample (e.g. as PFU mL^{-1}), with appropriate correction for dilution and the volume of sample counted in an analogous manner to a bacterial plate count (p. 476). When counting plaques in phage assays you should view them against a black background to make them easier to see: mark each plaque with a spirit-based marker to ensure an accurate count. Temperate phages often produce cloudy plaques, because many of the infected cells will be lysogenised rather than lysed, creating turbidity within the plaque. Samples of material from within the plaque can be used to subculture the phage for further study, perhaps in a broth culture where the phages will cause widespread cell lysis and a decrease in turbidity. Alternatively, phages can be stored by adding chloroform to aqueous suspensions – this will prevent contamination by cellular micro-organisms. A similar approach can be used to detect and count animal or human viruses, using a monolayer of susceptible host cells.

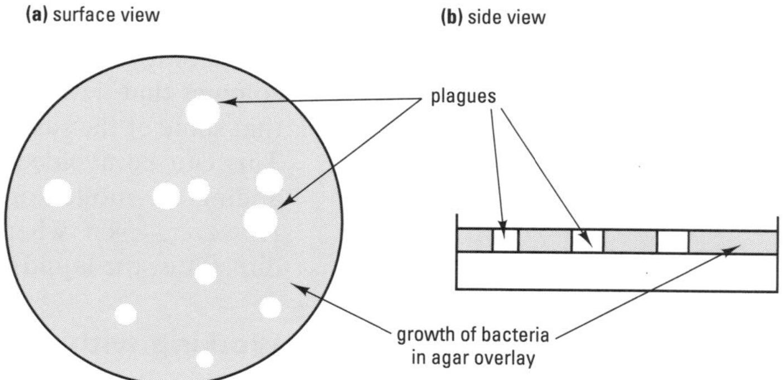

Fig. 66.7 Phage plaques in a 'lawn' of susceptible host bacterium.

Labelling Petri plates – the following information should be recorded on the base:

- date;
- the growth medium;
- your name or initials;
- brief details of the experimental treatment.

Electron microscopy (EM) provides an alternative approach to the detection of viruses, avoiding the requirement for culture of infected host cells, and giving a faster result. However, it requires specialised equipment and expertise. EM counts are often higher than culture-based methods, for similar reasons to those described for bacteria (p. 477). Another approach is to use immunoassays for virus detection (Chapter 52).

Labelling your plates and cultures

Petri plates should always be labelled on the *base*, rather than the lid. Restrict your labelling to the outermost region of the plate, to avoid problems when counting colonies, assessing growth, etc. After labelling, Petri plates usually are incubated upside down in a temperature-controlled incubator (often at 37°C) for an appropriate period (usually 18–72 hours). Plates are also usually kept upside down on the lab bench – following incubation, the base (containing medium and microbes) can then be lifted from the lid and examined.

Sources for further study

Anon. (1995) *Advisory Committee on Dangerous Pathogens: Categorisation of Biological Agents According to Hazard and Categories of Containment*, 4th edn. HSE Books, London.

Anon. (2004) *Advisory Committee on Dangerous Pathogens: Approved List of Biological Agents.* Available: http://www.hse.gov.uk/pubns/misc208.pdf Last accessed 21/12/10.

Collins, C.H., Lyne, P.M., Grange, J.M. and Falkinham, J. (2004) *Collins and Lyne's Microbiological Methods*, 8th edn. Hodder-Arnold, London.

Mantville. T.J. and Mathews, K.R. (2009) *Food Microbiology: An Introduction,* 2nd edn. ASM Press, Herndon, VA.

Jay, J.M., Loessner, M.J. and Golden, D.A. (2005) *Modern Food Microbiology*, 7th edn. Springer, Berlin.

67 Isolating, identifying and naming microbes

Microorganisms have a broad range of applications in the food and nutritional sciences:

- Microbes are widely used as model systems, as they are often easier to study under controlled laboratory conditions than 'higher' organisms.
- Bacteria and fungi are used as sources of particular biomolecules, e.g. for the characterisation of a specific enzyme *in vitro*.
- Pathogenic microbes are isolated and studied at the molecular level, to find new methods of identification using biochemical 'markers' and to investigate the molecular basis of their pathogenicity.
- Microbes are used in biotechnology and food microbiology as sources of particular biomolecules.

Alternatives to traditional culture-based methods – microbes can now be investigated by molecular methods, including the amplification of specific nucleic acid sequences by PCR (p. 375), and by immunoassay, e.g. ELISA (p. 350).

In your food microbiology practical classes, you are likely to gain experience of a wide range of laboratory exercises involving microbes, particularly bacteria such as *Escherichia coli* (*E. coli*), using the procedures of sterile technique described in Chapter 66.

Sampling microbes

Microbes can be studied in one of the following ways:

- by direct examination of samples for individual cells of a particular microbe, e.g. using fluorescence microscopy;
- by isolating/purifying a particular species or related individuals of a taxonomic group, e.g. the faecal indicator bacterium *E. coli* in water;
- by studying microbial processes, rather than individual microbes, either *in situ* or in the laboratory.

Avoiding contamination during sampling – always remember that you are the most important source of contamination of field samples: components of the oral or skin microflora are the most likely contaminants.

Sampling techniques for food include the use of swabs, Sellotape strips and agar contact methods for sampling surfaces, and tubes or bottles for beverages and water.

KEY POINT An important feature of all microbiological sampling protocols is that the sampling apparatus must be sterile; strict aseptic technique must be used throughout the sampling process (see Chapter 66).

SAFETY NOTE When working with newly isolated microbes, you should always treat them as potentially harmful until they have been identified.

The sampling method must minimise the chance of contamination with microbes from other sources, especially the exterior of the sampling apparatus and the operator. A portable Bunsen burner or spirit lamp can be used to assist sterile technique during field sampling, e.g. while flaming a loop (p. 451). Alternatively, use disposable single-use plastic loops.

Process the sample as quickly as possible to minimise any changes in microbiological status. As a general guideline, many procedures require food and beverage samples to be analysed within 6 h of collection. Changes in aeration, pH and water content may occur after collection. Some microbes are more susceptible to such effects, e.g. anaerobic bacteria may not survive if the sample is exposed to air. Sunlight can also inactivate bacteria; samples should be shielded from direct sunlight during collection and transport to the laboratory.

Food and water samples are often kept cool (at 0–5°C) during transport to the laboratory. In contrast, some microbes adapted to grow in association with warm-blooded animals may be damaged by low temperatures. An alternative approach is to keep the sample near the ambient sampling temperature using an insulated vessel (e.g. Thermos flask).

Sub-sampling – to minimise the effects of changes in temperature, aeration and water status during transportation, a primary sample may be returned to the laboratory, where the working sample (sub-sample) is then taken (e.g. from the centre of a food item).

Isolating a particular microbe

Several different approaches may be used to obtain microbes in pure culture. The choice of method will depend upon the microbe to be isolated: some organisms are relatively easy to isolate, while others require more involved procedures.

Separation methods

Most microbial isolation procedures involve some form of separation to obtain individual microbial cells. The most common approach is to use an agar-based medium for primary isolation, with streak dilution, spread plating or pour plating to produce single colonies, each derived from a single type of microbe (pp. 454–6). It is often necessary to dilute samples before isolation, so that a small number of individual microbial cells are transferred to the growth medium. Strict serial dilution (p. 149) of a known amount of sample is needed for quantitative work.

Obtaining a pure culture – if a single colony from a primary isolation medium is used to prepare a streak dilution plate and all the colonies on the second plate appear identical, then a pure culture has been established. Otherwise, you cannot assume that your culture is pure and you should repeat the subculture until you have a pure culture.

KEY POINT If your aim is to isolate a particular microbe, perhaps for further investigation, you will need to subculture individual colonies from the primary isolation plate to establish a pure culture, also known as an axenic culture.

Pure cultures of most microbes can be maintained indefinitely, using sterile technique and microbial culture methods (Chapter 66).

Other separation techniques include:

- **Dilution to extinction.** This involves diluting the sample to such an extent that only one or two microbes are present per millilitre: small volumes of this dilution are then transferred to a liquid growth medium (broth). After incubation, most of the tubes will show no growth, but some tubes will show growth, having been inoculated with a single viable microbe at the outset. This should give a pure culture, although it is quite wasteful of resources.
- **Sonication/homogenisation.** This is useful for separating individual microbial cells from each other and from particles of food, prior to isolation. However, some decrease in viability is likely.
- **Filtration.** This can be useful for liquid samples where the number of microbes is low. Samples can be passed through a sterile cellulose ester filter (pore size 0.2 μm), which is then incubated on the surface of an appropriate solidified medium.
- **Micromanipulation.** It may be possible to separate a microbe from contaminants using a micropipette and dissecting microscope. The microbe can then be transferred to an appropriate growth medium, to give a pure culture. However, this is rarely an easy task for the novice.

Using a sonicator – minimise heat damage with short treatment 'bursts' (typically up to one minute), cooling the sample between bursts, e.g. using ice.

Definitions

Psychrophile – a microbe with an optimum temperature for growth of <20°C (literally 'cold-loving').

Psychrotroph – a microbe with an optimum temperature for growth of ≥ 20°C, but capable of growing at lower temperature, typically 0–5°C (literally 'cold-feeding').

Thermophile – a microbe with an optimum growth temperature of >45°C (literally 'heat-loving').

Mesophile – a microbe with an optimum growth temperature of 20–45°C (literally 'middle-loving').

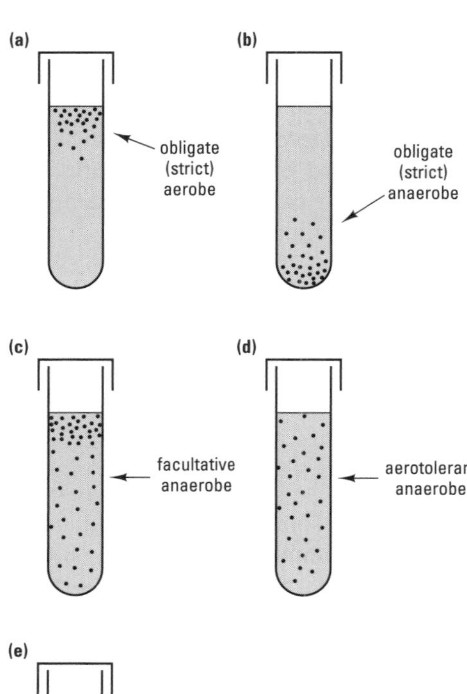

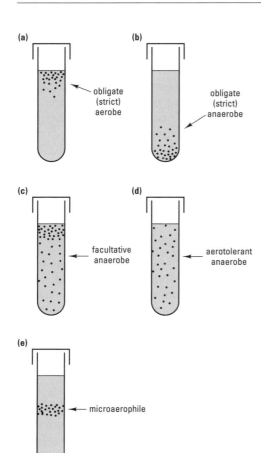

Fig. 67.1 Agar shake tubes. Bacteria are suspended in molten agar at 45–50°C and allowed to cool. The growth pattern after incubation reflects the atmospheric (oxygen) requirements of the bacterium.

Definitions

Facultative anaerobe – a microbe that grows by aerobic respiration when oxygen is present, switching to fermentation under anaerobic conditions.

Aerotolerant anaerobe – a microbe that grows by fermentation, but which is insensitive to air/oxygen (in contrast with strict anaerobes, which are typically killed by exposure to air/oxygen).

Capnophile – a microbe that thrives in the presence of high levels of atmospheric carbon dioxide.

- **Motility.** Heterotrophic flagellate bacteria will move through a filter of appropriate pore size into a nutrient solution, or away from unfavourable conditions (chemotaxis).

Selective and enrichment methods

Laboratory incubation under selective/enrichment conditions will allow particular microbes to be isolated in pure culture. Selective and enrichment techniques can be considered together, since they both enhance the growth of a particular microbe when compared with its competitors and they are often combined in specific media (Box 67.1).

The fundamental difference between selective and enrichment techniques is that the former use growth conditions unfavourable for competitors while the latter provide improved growth conditions for the chosen microbes.

KEY POINT Selective methods are based on the use of physico-chemical conditions that will permit the growth of a particular group of microbes while inhibiting others. Enrichment techniques encourage the growth of certain bacteria, usually by providing additional nutrients in the growth medium.

Methods based on specific physical conditions include:

- **Temperature.** Psychrophilic and psychrotrophic microbes can be isolated by incubating the growth medium at 4°C, while thermophilic microbes require temperatures above 45°C for isolation. Short-term heat treatment of samples can be used to select for endospore-forming bacteria, e.g. 70–80°C for 5–15 minutes, prior to isolation.
- **Atmosphere.** Many eukaryotic microbes are obligate aerobes, requiring an adequate supply of oxygen to grow. Bacteria vary in their responses to oxygen: obligate anaerobes are the most demanding, growing only under anaerobic conditions (e.g. in an anaerobic cabinet or jar). Oxygen requirements can be determined using the agar shake tube method as part of the isolation procedure (Fig. 67.1). Some pathogenic bacteria grow best in an atmosphere with a reduced oxygen status and increased CO_2 concentration: such carboxyphilic bacteria (capnophiles) are grown in an incubator where the gas composition can be adjusted.
- **Centrifugation.** This can be used to separate buoyant microbes from their non-buoyant counterparts – on centrifugation, such organisms will collect at the surface while the remaining microbes will sediment. Alternatively, density gradient methods may be used (p. 385). Centrifugation can be combined with repeated washing, to separate microbes from contaminants.

Chemical methods form the mainstay of bacteriological isolation techniques and various media have been developed for the isolation of specific groups of bacteria. The chemicals involved can be subdivided into the following groups:

- **Nutrients that encourage the growth of certain microbes:** including the addition of a particular carbon source, or specific inorganic nutrients.
- **Selectively toxic substances:** for example, salt-tolerant, Gram-positive cocci can be grown in a medium containing 7.5% w/v NaCl, which

Box 67.1 Differential media for bacterial isolation – an example

MacConkey agar is both a selective and a differential medium, useful for the isolation and identification of intestinal Gram-negative bacteria. Each component in the medium has a particular role:

- **Peptone:** (a meat digest) provides a rich source of complex organic nutrients, to support the growth of non-exacting bacteria.
- **Bile salts:** toxic to most microbes apart from those growing in the intestinal tract (selective agent).
- **Lactose:** present as an additional, specific carbon source (enrichment agent).
- **Neutral red:** a pH indicator dye, to show the decrease in pH which accompanies the breakdown of lactose (differential agent).
- **Crystal violet:** selectively inhibits the growth of Gram-positive bacteria. (This component is only present in certain formulations of MacConkey agar.)

Any intestinal Gram-negative bacterium capable of fermenting lactose will grow on MacConkey agar to produce large red–purple colonies, the red colouration being due to the neutral red indicator under acidic conditions while the purple colouration, often accompanied by a metallic sheen, is due to the precipitation of bile salts and crystal violet at low pH.

In contrast, enteric Gram-negative bacteria unable to metabolise lactose will give colonies with no obvious pigmentation. This differential medium has been particularly useful in medical microbiology, since many enteric bacteria are unable to ferment lactose (e.g. *Salmonella*, *Shigella*) while others metabolise this carbohydrate (e.g. *Escherichia coli*, *Klebsiella* spp.). Colonial morphology (Fig. 67.2) on such a medium can give an experienced bacteriologist important clues to the identity of an organism, e.g. capsulate *Klebsiella* spp. characteristically produce large, convex, mucoid colonies with a weak pink colouration, due to the fermentation of lactose, while *E. coli* produces smaller, flattened colonies with a stronger red colouration and a metallic sheen.

Example Cellulolytic bacteria can be isolated from soil or water using a growth medium that contains cellulose as the major source of carbon.

Example Slow-growing *Legionella pneumophila* can be isolated from water samples using media containing the antibiotics vancomycin, polymyxin and cycloheximide to suppress other microbes.

prevents the growth of most common heterotrophic bacteria. Several media include dyes as selective agents, particularly against Gram-positive bacteria.

- **Antibiotics:** for example, the use of antibacterial agents (e.g. penicillin, streptomycin, chloramphenicol) in media designed to isolate fungi, or the use of antifungal agents (e.g. cycloheximide, nystatin) in bacterial media. Some antibacterial agents show a narrow spectrum of toxicity and these can be incorporated into selective isolation media for resistant bacteria, e.g. metronidazole for anaerobic bacteria.
- **Substances that affect the pH of the medium:** for example, the use of alkaline peptone water at pH 8.6 for the isolation of *Vibrio* spp.

KEY POINT Note that subcultures from a primary isolation medium must be grown in a non-selective medium, to confirm the purity of the isolate.

Many of the selective and enrichment media used in bacteriology are able to distinguish between different types of bacteria: such media are termed differential media or diagnostic media and they are often used in the preliminary stages of an identification procedure. Box 67.1 gives details of the constituents of MacConkey medium, a selective, differential medium used in clinical microbiology (e.g. for the isolation of certain faecal bacteria), while Table 67.1 gives details of other selective agents.

Further details on methods can be found in Collins *et al.* (2004) or Roberts and Greenwood (2002). Note that isolation procedures for a particular microbe often combine several of the techniques described

Table 67.1 Selective agents in bacteriological media

Substance	Selective for
Azide salts	*Enterococcus* spp.
Bile salts	Intestinal bacteria
Brilliant green	Gram-negative bacteria
Gentian violet	Gram-negative bacteria
Lauryl sulphate	Gram-negative bacteria
Methyl violet	*Vibrio* spp.
Malachite green	*Mycobacterium*
Polymyxin	*Bacillus* spp.
Sodium selenite	*Salmonella* spp.
Sodium chloride	Halotolerant bacteria *Staphylococcus aureus*
Sodium tetrathionate	*Salmonella* spp.
Tergitol/surfactant	Intestinal bacteria
Trypan blue	*Streptococcus* spp.

above. For instance, a protocol for isolating a food-poisoning bacterium from a foodstuff might involve:

1. **Homogenisation of a known amount of sample** in a suitable diluent (p. 380).
2. **Serial decimal dilution.**
3. **Separation procedures** using spread or pour plates to quantify the number of bacteria of a particular type present in the foodstuff and provide a plate count (p. 476).
4. **Selective/enrichment procedures,** e.g. specific media/temperatures/atmospheric conditions, depending on the bacteria to be isolated.
5. **Confirmation of identity:** any organism growing on a primary isolation medium would require subculture and further tests, to confirm the preliminary identification as discussed below.

Identifying a particular microbe

Most of the methods described in this chapter were developed for the identification of bacteria, and bacterial examples are used to illustrate the principles involved. While the basic techniques are equally applicable to other types of microbe, the identification systems for some protozoa, fungi and algae rely predominantly on microscopic appearance. Identification of viruses requires immunological techniques or electron microscopy.

KEY POINT Identification of bacteria is often based on a combination of a number of different features, including growth characteristics, microscopic examination, physiological or biochemical characterisation, and, where necessary, immunological tests.

Direct observation

Once a microbe has been isolated and cultured in the laboratory (Chapter 66), the visual appearance of individual colonies on the surface of a solidified medium may provide useful information. Bacteria typically produce smooth, glistening colonies, varying in diameter from <1 mm to >1 cm. Actinomycete colonies are often <1 cm, with a shrivelled, powdery surface. Filamentous fungi usually grow as large, spreading mycelia with a matt appearance and are identified by microscopy, using the morphological characteristics of their reproductive structures. Yeasts produce small, glistening colonies: identification usually involves microscopy, combined with physiological and biochemical tests similar to those used for bacteria.

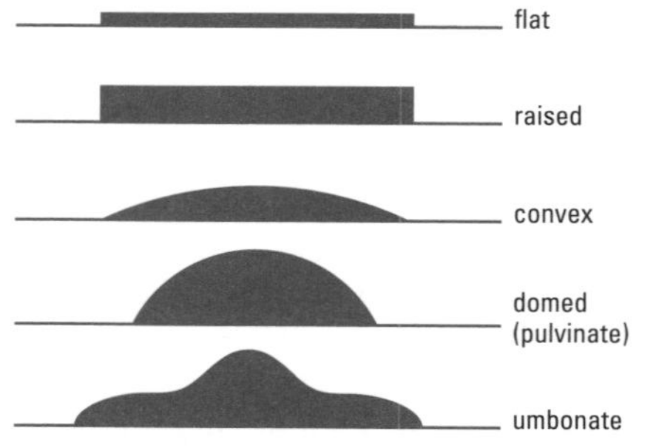

Fig. 67.2 Colony elevations (cross-sectional profile).

Colony characteristics

When measuring colony size, choose a typical colony, well spaced from any others as colony size is affected by competition for nutrients. The characteristics of a microbial colony on a particular medium include:

- **Size:** some bacteria produce punctiform colonies, with a diameter of less than 1 mm, while motile bacteria may spread over the entire plate.
- **Form:** colonies may be circular, irregular, lenticular (spindle-shaped) or filamentous.

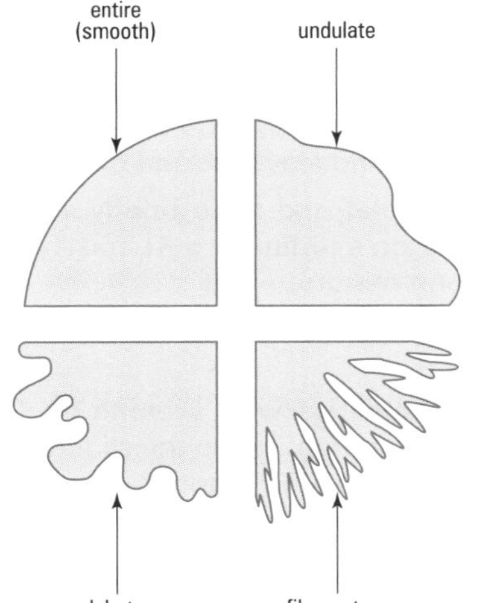

Fig. 67.3 Colony margins (surface view).

SAFETY NOTE Working with mould cultures – never attempt to smell mould cultures, because of the risk of inhaling large numbers of spores.

Using cell shape in microbial identification – many bacteria are pleomorphic, varying in size and shape according to the growth conditions and the age of the culture: thus, other characteristics are required for identification.

SAFETY NOTE The Gram staining procedure involves toxic dyes and flammable solvents: avoid skin contact and extinguish any naked flames (e.g. nearby Bunsens).

Assessing the Gram status of an unknown bacterium – if a pure culture gives both Gram-positive and Gram-negative cells, identical in size and shape, it can be regarded as a Gram-positive organism that is demonstrating Gram-variability.

- **Elevation:** colonies may be flat, raised, convex, etc. (see Fig. 67.2).
- **Margin:** the edge of a colony may be entire (smooth) or more distinctive, e.g. undulate, lobate or filamentous (Fig. 67.3).
- **Consistency:** colonies may be viscous (or mucoid), butyrous (of similar consistency to butter) or friable (dry and granular), etc.
- **Colour:** some bacteria produce characteristic pigments. A few pigments are fluorescent under UV light.
- **Optical properties:** colonies may be translucent or opaque.
- **Haemolytic reactions on blood agar:** many pathogenic bacteria produce characteristic zones of haemolysis. Alpha haemolysis is a partial breakdown of the haemoglobin from the erythrocytes, producing a green zone around the colony, while beta haemolysis is the complete destruction of haemoglobin, producing a clear zone.
- **Odour:** some actinomycetes and cyanobacteria produce earthy odours, while certain bacteria and yeasts produce fruity or 'off' odours. However, odour is not a reliable characteristic in bacterial identification and smelling plates creates a risk of inhaling potentially harmful microbes.

Microscopic examination – cell shape

Bacteria are usually observed using an oil immersion objective at a total magnification of ×1000 (p. 187). Bacteria are subdivided into the following groups, based on their cell shape:

- **Cocci** (singular, coccus): spherical, or almost spherical, cells, sometimes growing in pairs (diplococci), chains or clumps.
- **Rods:** straight, cylindrical cells of variable length with flattened, tapered or rounded ends – sometimes termed bacilli. Short rods are sometimes called cocco-bacilli.
- **Curved rods:** the curvature varies according to the organism, from short curved rods, sometimes tapered at one end, to spiral shapes.
- **Branched filaments:** characteristic of actinomycete bacteria.

Gram staining

This is the most important differential staining technique in bacteriology (Box 67.2 gives details). It enables us to divide bacteria into two distinct groups, Gram-positive and Gram-negative, according to a particular staining procedure (the technique is given a capital letter, since it is named after its originator, H.C. Gram). The basis of the staining reaction is the different structure of the cell walls of Gram-positive and Gram-negative bacteria. Heat fixation of air-dried bacteria causes some shrinkage, but cells retain their shape: to measure cell dimensions use a chemical fixative.

Gram staining should be carried out using light smears of young, active cultures, since older cultures may give variable results. In particular, certain Gram-positive bacteria may stain Gram-negative if older cultures are used. This Gram-variability is due to autolytic changes in the cell wall of Gram-positive bacteria. Developing spores are often visible as unstained areas

Box 67.2 Preparation of a heat-fixed, Gram-stained smear

Preparation of a heat-fixed smear

The following procedure will provide you with a thin film of bacteria on a microscope slide, for staining.

1. **Take a clean microscope slide and pass it through a Bunsen flame twice** to ensure it is free of grease. Allow to cool.
2. **Using a sterile inoculating loop, place a single drop of water in the centre of the slide and then mix in a small amount of sample** from a single bacterial colony with the drop, until the suspension is slightly turbid. Smear the suspension over the central area of the slide, to form a thin film. For liquid cultures, use a single drop of culture fluid, spread in a similar manner.
3. **Allow to air-dry at room temperature**, or high above a Bunsen flame: air-drying must proceed gently or the cells will shrink and become distorted.
4. **Fix the air-dried film by passage through a Bunsen flame.** Using a slide holder or forceps, pass the slide, film side up, rapidly through the hottest part of the flame (just above the blue cone). The temperature of the slide should be just too hot for comfort on the back of your hand: note that you must not overheat the slide or you may burn yourself (you will also ruin the preparation).
5. **Allow to cool**: the smear is now ready for staining.

Gram-staining procedure

The version given here is a modification of the Hucker method, since acetone is used to decolourise the smear. Note that some of the staining solutions used are flammable, especially the acetone decolourising solvent: you must make sure that all Bunsens are turned off during staining. The procedure should be carried out with the slides suspended over a sink, using a staining rack.

1. **Flood a heat-fixed smear with 2% w/v crystal violet in 20% v/v ethanol : water** and leave for one minute.
2. **Pour off the crystal violet and rinse briefly with tap water. Flood with Gram's iodine** (2 g KI and 1 g L_2 in 300 mL water) for one minute.
3. **Rinse briefly with tap water** and leave the tap running gently.
4. **Tilt the slide and decolourise with acetone** for 2–3 s: acetone should be added dropwise to the slide until no colour appears in the effluent. This step is critical, since acetone is a powerful decolourising solvent and must not be left in contact with the slide for too long.
5. **Immediately immerse the smear in a gentle stream of tap water** to remove the acetone.
6. **Pour off the water and counterstain for 10–15 s using 2.5% w/v safranin** in 95% v/v ethanol : water.
7. **Pour off the counterstain, rinse briefly with tap water, then dry the smear** by blotting gently with absorbent paper: all traces of water must be removed before the stained smear is examined microscopically.
8. **Place a small drop of immersion oil on the stained smear: examine directly** (without a coverslip) using an oil-immersion objective (p. 187).

Gram-positive bacteria retain the crystal violet (primary stain) and appear purple, while Gram-negative bacteria are decolourised by acetone and counterstained by the safranin, appearing pink or red when viewed microscopically.

Other decolourising solvents are sometimes used, including ethanol : water, ethyl ether : acetone and acetone : alcohol mixtures. The time of decolourisation must be adjusted, depending upon the strength of the solvents used, e.g. 95% v/v ethanol : water is less powerful than acetone, requiring around 30 s to decolourise a smear.

within older vegetative cells of *Bacillus* and *Clostridium*. Other stains are required to demonstrate spores, capsules or flagella (p. 487).

Using the hanging drop technique – place a drop of bacterial suspension on a coverslip and invert over a cavity slide so that the drop does not make contact with the slide: motile aerobes are best observed at the edge of the droplet, where oxygen is most abundant.

Motility

Wet mounts can be prepared by placing a small drop of bacterial suspension on a clean, degreased slide, adding a coverslip and examining the film by light microscopy without delay. For aerobes, areas near air bubbles or by the edge of the coverslip give best results, while anaerobes show greatest motility in the centre of the preparation, with rapid loss of motility due to oxygen toxicity.

Prepare wet mounts using young cultures in exponential growth in a liquid medium (p. 472). It is best to work with cultures grown at 20 or

Assessing motility – if you have not seen bacterial motility before, it is worth comparing your unknown bacterium with a positive and a negative control.

Performing the oxidase test – *never* use a nichrome wire loop, as this will react with the oxidase reagent, giving a false positive result.

SAFETY NOTE The catalase and oxidase reagents are irritants and could be harmful if swallowed. Avoid skin contact and ingestion.

Avoiding false negatives – ensure you use sufficient material during oxidase and catalase testing, otherwise you may obtain a false negative result: a clearly visible 'clump' of bacteria should be used.

25°C, since those grown at 37°C may not be actively motile on cooling to room temperature. It is essential to distinguish between the following:

- **True motility**, due to the presence of flagella: bacteria dart around the field of view, changing direction in zigzag, tumbling movements.
- **Brownian motion:** non-motile bacteria show a localised, vibratory, random motion, due to bombardment of bacterial cells by molecules in the solution.
- **Passive motion**, due to currents within the suspension: all cells will be swept in the same direction at a similar rate of movement.
- **Gliding motility:** a slower, intermittent movement, parallel to the longitudinal axis of the cell, requiring contact with a solid surface.

Basic laboratory tests

At least two simple biochemical tests are usually performed:

1. Oxidase test

This identifies cytochrome *c* oxidase, an enzyme found in obligate aerobic bacteria. Soak a small piece of filter paper in a fresh solution of 1% (w/v) *N*-*N*-*N*′-*N*′-tetramethyl-*p*-phenylenediamine dihydrochloride on a clean microscope slide. Rub a small amount from the surface of a young, active colony onto the filter paper using a glass rod, a *plastic* loop or a wooden applicator stick: a purple-blue colour within 10 s is a positive result.

2. Catalase test

This identifies catalase, an enzyme found in obligate aerobes and in most facultative anaerobes, which catalyses the breakdown of hydrogen peroxide into water and oxygen ($2H_2O_2 \rightarrow 2H_2O + O_2$). Transfer a small sample of your unknown bacterium onto a coverslip using a disposable plastic loop or glass rod. Invert onto a drop of hydrogen peroxide: the appearance of bubbles within 30 s is a positive reaction. This method minimises the dangers from aerosols formed when gas bubbles burst.

The oxidase and catalase tests effectively allow us to subdivide bacteria on the basis of their oxygen requirements, without using agar shake cultures (p. 460) and overnight incubation, since, for the most part:

- **obligate aerobes** will be oxidase and catalase positive;
- **facultative anaerobes** are generally oxidase negative and catalase positive;
- **microaerophilic bacteria,** aerotolerant anaerobes and strict (obligate) anaerobes will be oxidase and catalase negative – the latter group will grow only under anaerobic conditions (p. 460).

Once you have reached this stage (colony characteristics, motility, shape, Gram reaction, oxidase and catalase status) it may be possible to make a tentative identification, at least for certain Gram-positive bacteria, at the generic level. To identify Gram-negative bacteria, particularly the oxidase-negative, catalase-positive rods, further tests are required.

Table 67.2 Identification table for selected Gram-negative rods

Bacterium	Biochemical test 1	2	3	4	5	6	7	8	9
Escherichia coli	v	+	−	+	−	v	v	+	−
Proteus mirabilis	−	−	v	−	+	−	+	−	+
Morganella morganii	−	−	−	−	+	−	+	+	−
Vibrio parahaemolyticus	−	+	v	−	−	+	+	+	−
Salmonella spp.	−	+	v	−	−	+	+	−	+

Key to biochemical tests and symbols:
1. sucrose utilisation
2. mannitol utilisation
3. citrate utilisation
4. β-galactosidase activity
5. urease activity
6. lysine decarboxylase activity
7. ornithine decarboxylase activity
8. indole production
9. H_2S production

+, >90% of strains tested positive
−, <10% of strains tested positive
v, 10–90% of strains tested positive

Carbohydrate utilisation tests and isolation media – many diagnostic agar-based media incorporate one or more specific carbohydrates and pH indicator dyes, thereby providing additional information as part of the isolation procedure (Box 67.1).

Identification tables: further laboratory tests

Bacteria are asexual organisms and strains of the same species may give different results for individual biochemical/physiological tests. This variation is allowed for in identification tables, based on the results of a large number of tests. Identification tables are often used for particular subgroups of bacteria, after Gram staining and basic laboratory tests have been performed: an example is shown in Table 67.2.

A large number of specific biochemical and physiological tests are used in bacterial identification, including:

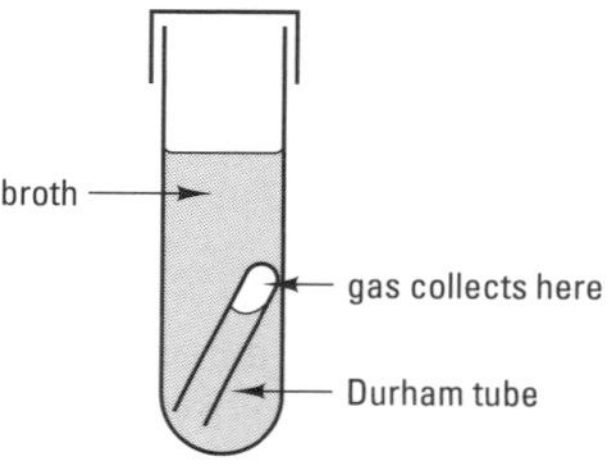

Fig. 67.4 Durham tube in carbohydrate utilisation broth. Air within the Durham tube is replaced by broth during the autoclaving procedure.

- **Carbohydrate utilisation tests**. Some bacteria can use a particular carbohydrate as a carbon and energy source. Acidic end-products can be identified using a pH indicator dye (p. 174) while CO_2 is detected in liquid culture using an inverted small test tube (Durham tube, Fig. 67.4). Aerobic breakdown (via respiration) is termed oxidation while anaerobic breakdown is known as fermentation. Identification tables usually incorporate tests for several different carbohydrates, e.g. Table 67.2.
- **Enzyme tests.** Most of these incorporate a substance which changes colour if the enzyme is present, e.g. a pH indicator, or a chromogenic or fluorogenic substrate (p. 408).
- **Tests for specific end-products of metabolism,** e.g. the production of indole due to the metabolic breakdown of the amino acid tryptophan, or H_2S from sulphur-containing amino acids.

Molecular approaches to microbial identification – several novel methods of detection and identification are based on nucleic acid techniques, including use of the PCR and Southern blotting to detect particular microbes (pp. 374–6).

Identification kits

Some biochemical tests are now supplied in kit form, e.g. the API 20E system incorporates 20 tests within a sterile plastic strip (Fig. 67.5). After inoculation and overnight incubation, the results of the tests are

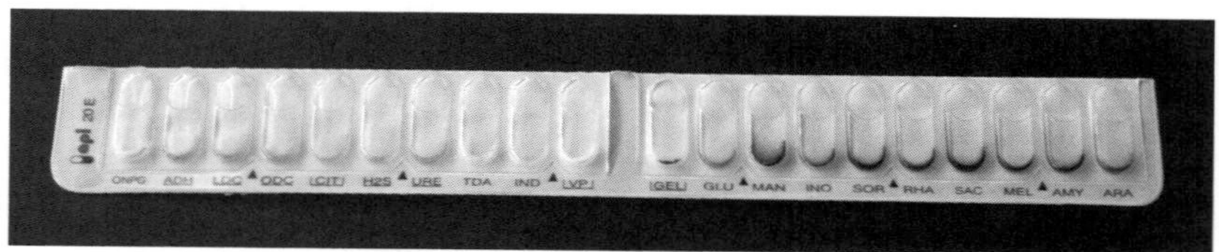

Fig. 67.5 Example of a bacterial identification kit (API 20E). Property of bioMérieux S.A./photographer: Andrea Bannuscher.

converted into a seven-digit code, for comparison with known bacteria using either a reference book (the Analytical Profile Index), or a computer program. While kit identification systems save time and labour, they are more expensive and less flexible than conventional biochemical tests.

Immunological tests

Tests used in diagnostic microbiology include:

- **Agglutination tests:** based on the reaction between specific antibodies and a particular bacterium (p. 348). These tests are particularly useful for subdividing biochemically similar bacteria.
- **Fluorescent antibody tests:** the reaction between a labelled antibody and a particular bacterium can be visualised using UV microscopy. The direct fluorescent antibody test uses fluorescein isothiocyanate as the label.
- **Enzyme-linked immunoassay tests** using antibodies labelled with a particular enzyme, e.g. the double antibody sandwich ELISA or competitive ELISA tests (p. 350).

While such tests can give specific and accurate confirmation of the identity of a bacterium under controlled laboratory conditions, they are often too expensive and time-consuming for routine identification purposes, especially when large numbers of tests are required.

Typing methods

Practical applications of bacterial typing – *E. coli* O157:H7 is a serotype of this bacterium that is capable of causing severe human disease: it can be identified on the basis of an agglutination reaction with an appropriate antiserum. Other applications include tracking the development and spread of particular types of antibiotic-resistant bacteria in hospitals and the community.

The identification of bacteria at subspecies level is known as typing: this is usually done in a specialist laboratory, e.g. as part of an epidemiological study to establish the source of an outbreak of foodborne disease. Various methods are used:

- **Antigen typing or serotyping** is based on immunological testing (p. 347).
- **Phage typing** is based on the susceptibility of different strains to certain bacterial viruses (phages).
- **Biotyping** is based on biochemical differences between different strains, e.g. enzyme profiles or antibiotic resistance screening ('antibiograms').
- **Bacteriocin typing:** bacteriocins are proteins released by bacteria which inhibit the growth of other members of the same species.

Naming microbes

The use of scientific names is fundamental to all aspects of biological science since it aims to provide a system of identification which is precise, fixed and of universal application. Without such a system, comparative studies would be impossible.

There are two possible bases for such classification:

- **Phenetic taxonomy**, which involves grouping on the basis of phenotypic similarity, frequently using complex statistical techniques to obtain objective measures of similarity. The characters used have been largely morphological and anatomical, but biochemical, cytological and other characters are increasingly used, especially for microbes, where structural characteristics are few.

- **Phylogenetic (= phyletic) taxonomy**, which involves grouping on the basis of presumed evolutionary, and therefore genetic, relationships.

These two systems are sometimes similar in outcome, since closely related organisms are usually fairly similar to each other and because judgements of evolutionary relationships are usually themselves based upon similarities.

The basis of classification

No single, simple definition of a species is possible, but there are two generally used definitions:

- A group of organisms capable of interbreeding and producing fertile offspring – this, however, excludes all asexual organisms, such as bacteria and some other microbes.
- A group of organisms showing a close similarity in phenotypic characteristics – this would include morphological, anatomical, biochemical, ecological and life history characters or, increasingly, similarities in nucleic acid sequence.

KEY POINT The basic unit of classification is the species, which represents a group of recognisably similar individuals, clearly distinct from other such groups.

When species are compared, groups of species may show a number of features in common; they are then arranged into larger groupings known as genera (singular, genus). Cellular microbes (e.g. bacteria, archaea, fungi, protozoa and algae) are given two Latin terms to identify their genus and species (a Latin 'binomial'), for example *Escherichia coli*. All scientific names of organisms are either Latin or are treated as Latin, written in the Latin alphabet and subject to the rules of Latin grammar. Consequently, you must be very precise in your use of such names. When used in a formal scientific context, the specific name can be followed by the authority on which the name is based, i.e. the name of the person describing that species, and the year in which it was first described. For example, the full, formal name of baker's yeast is *Saccharomyces cerevisiae* Meyen ex Hansen 1883. Where specified, additional terms may follow the species name to indicate the type or subspecies, e.g. *Shigella dysenteriae* type 1. Many microorganisms are now referred to by their generic and specific names followed by a culture collection reference number, e.g. *Bacillus subtilis* NCTC 10400, where NCTC stands for the National Collection of Type Cultures and 10400 is the reference number of that strain in the collection.

After first use in a text, the genus name may be abbreviated to a single letter, e.g. *E. coli*, as long as this does not cause confusion with other genera; for example, with *Escherichia coli* and *Enterococcus faecalis*, these can be abbreviated to *Esch. coli* and *Ent. faecalis*, to avoid confusion. Where the species name is unknown, the (non-italicised) abbreviation 'sp.'

Definitions

Systematics – the study of the diversity of living organisms and of the evolutionary relationships between them.

Classification (taxonomy) – the study of the theory and methods of organisation of taxa and, therefore, a part of systematics.

Taxon (plural taxa) – an assemblage of organisms sharing some basic features.

Nomenclature – the allocation of names to taxa.

Identification – the placing of organisms into taxa.

DNA-based definition of a species – in bacteriology, members of a single species can be characterised by sharing at least 70 per cent DNA, based on nucleic acid hybridisation studies.

Writing taxonomic names – always underline (handwritten text) or italicise (word-processed text) genus and species names to avoid confusion: thus bacillus is a descriptive term for a rod-shaped bacterium, while Bacillus is a generic name.

Understanding microbiological terms – the term *strain* is widely used, particularly in the context of the practice of lodging microbiological strains with culture collections, while the term *isolate* is often used for a pure culture derived from a natural (wild) population. Cell lines are often given code names and/or reference numbers.

Understanding the hierarchical taxonomic system – the taxonomic groups, in decreasing level, are: kingdom, phylum or division, class, order, family, genus, species. An example of a full hierarchical classification would be: Bacteria (monera); Gracilicutes; Scotobacteria; Pseudomonadales; Pseudomonadaceae; Pseudomonas; Pseudomonas aeruginosa. Note that the names of orders normally end in '-ales' while family names usually end in '-aceae' and that all names of taxa apart from species begin with a capital letter.

Example The virus that causes tobacco mosaic disease belongs to the *Tobamovirus* group and can be referred to as tobacco mosaic tobamovirus, tobacco mosaic virus or TMV.

(singular) or 'spp.' (plural) should be used, e.g. *Enterococcus* sp. denotes a single unknown species of the genus *Enterococcus*.

KEY POINT While the use of taxa above those of the genus are somewhat subjective and may vary between different authorities, the use of genus and species names are governed by strict internationally agreed conventions called Codes of Nomenclature (e.g. Lapage, 1992, for bacteria).

The classification and nomenclature of viruses are less advanced than for cellular organisms and the current nomenclature has been arrived at on a piecemeal, *ad hoc* basis. The International Committee for Virus Taxonomy proposed a unified classification system, dividing viruses into 50 families on the basis of: host preference, nucleic acid type (i.e. DNA or RNA), whether the nucleic acid is single or double stranded and the presence or absence of a surrounding envelope. Many viruses are still referred to by trivial names or by code-names (sigla), e.g. the bacterial viruses ϕX174, T4, etc. Many of the names used reflect the diseases caused by the virus. Often, a three-letter abbreviation is used, e.g. HIV (for human immunodeficiency virus), TMV (for tobacco mosaic virus).

Text references

Collins, C.H., Lyne, P.M. and Grange, J.M. (2004) *Microbiological Methods*, 8th edn. Hodder-Arnold, London.

Lapage, S.P. (1992) *International Code of Nomenclature for Bacteria: 1990 Revision*. American Society for Microbiology, Washington, DC.

Roberts, D. and Greenwood, M. (2002) *Practical Food Microbiology*, 3rd edn. Wiley, New York.

Sources for further study

Atlas, R.M. (2010) *Handbook of Microbiological Media*, 4th edn. CRC Press, Boca Raton, Florida. [Gives details of culture media for a broad range of microbes.]

Barrow, G.I. and Feltham, R.K.A. (2004) *Cowan and Steel's Manual for the Identification of Medical Bacteria*, 3rd edn. Cambridge University Press, Cambridge. [A standard reference work for the identification of clinically important bacteria.]

Buchen-Osmond, C. (2002) Universal Virus Database of the International Committee on Taxonomy of Viruses. Available: http://www.ncbi.nlm.nih.gov/ICTVdb/. Last accessed: 21/12/10.

Eaton, A.D., Clesceri, L.S., Rice, E.W. and Greenberg, A.E. (2005) *Standard Methods for Examination of Water and Wastewater*, 21st edn. American Public Health Association, Washington, DC. [Gives standard US protocols for a range of indicator bacteria.]

Fauquet, C.M., Mayo, M.A., Maniloff, J., Desselberger, U. and Ball, L.A. (2005) *Virus Taxonomy. Eighth Report of the International Committee on the Taxonomy of Viruses*. Elsevier, London.

Fisher, F., Cook, N.B., Fisher, F.W. and Kaszczuk, S. (1998) *Fundamentals of Diagnostic Mycology*. Saunders, Philadelphia, Pennsylvania. [Covers identification techniques for medically important fungi.]

Fox, A. (ed.) *Journal of Microbiological Methods.* Elsevier, London (available through Science Direct at: http://www.sciencedirect.com/) Last accessed: 21/12/10.
[Provides information on novel developments in all aspects of microbiological methods, including microbial culture/isolation and molecular approaches.]

Institute of Biology (2000) *Biological Nomenclature: Recommendations on Terms, Units and Symbols.* Institute of Biology, London.
[Includes sections on taxonomy and classification of organisms.]

Levett, P.N. (ed.) (1991) *Anaerobic Microbiology: A Practical Approach.* IRL Press, Oxford.
[Provides details of the methods used to isolate and culture anaerobic microorganisms.]

MacFaddin, J.F. (2007) *Biochemical Tests for the Identification of Medical Bacteria*, 3rd edn. Williams and Wilkins, Philadelphia, Pennsylvania.
[Explains the operating principles underlying most of the biochemical tests in routine use in diagnostic bacteriology.]

68 Culture systems and growth measurements

Definitions

Heterotroph – an organism that uses complex organic carbon compounds as a source of carbon and energy.

Photoautotroph – an organism that uses light as a source of energy and CO_2 as a carbon source (photosynthetic metabolism).

Chemoautotroph – an organism that acquires energy from the oxidation of simple inorganic compounds, fixing CO_2 as a source of carbon (chemosynthetic metabolism).

Microbial, animal and plant cell culture methods are based on the same general principles, requiring:

- **a pure culture** (also known as an axenic culture), perhaps isolated as part of an earlier procedure, or from a culture collection;
- **a suitable nutrient medium** to provide the necessary components for growth. This medium must be sterilised before use;
- **satisfactory growth conditions** including temperature, pH, atmospheric requirements, ionic and osmotic conditions;
- **sterile technique** (p. 451) to maintain the culture in pure form.

Heterotrophic, fungi and many bacteria require appropriate organic compounds as sources of carbon and energy. Non-exacting bacteria can utilise a wide range of compounds and they are often grown in media containing complex natural substances (including meat extract, yeast extract, soil, blood). For chemoautotrophic bacteria, the light source is replaced by a suitable inorganic energy source, e.g. H_2S for sulphur-oxidising bacteria, NH_3/NH_4^+ for nitrifying bacteria, etc. Animal cells have more stringent growth requirements.

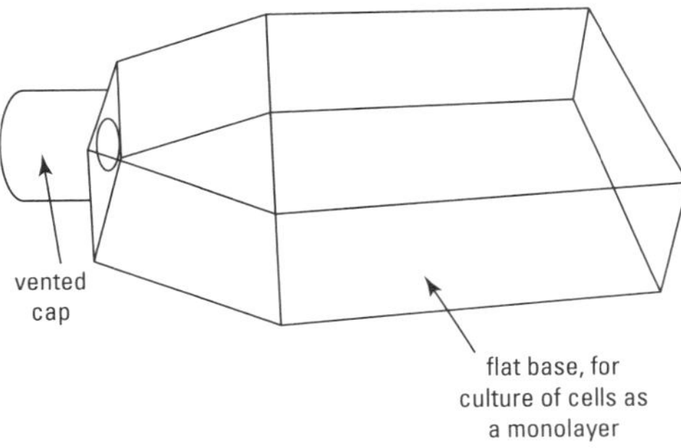

Fig. 68.1 Plastic flask for animal cell culture – this design provides a large surface area for growth of an adherent monolayer of cells.

Growth on solidified media

Many organisms can be cultured on an agar-based medium (p. 452).

KEY POINT An important benefit of agar-based culture systems is that an individual cell inoculated onto the surface can develop to form a visible colony: this is the basis of most microbial isolation and purification methods, including the streak dilution, spread plate and pour plate procedures (p. 476).

Harvesting bacteria from an agar plate – colonies can be harvested using a sterile loop, providing large numbers of cells without the need for centrifugation. The cells are relatively free from components of the growth medium; this is useful if the medium contains substances that interfere with subsequent procedures.

Animal cells are often grown as an adherent monolayer on the surface of a plastic or glass culture vessel (Fig. 68.1), rather than on an agar-based medium (p. 454).

Several types of culture vessel are used:

- **Petri plates** (Petri dishes): usually the presterilised, disposable plastic type, providing a large surface area for growth.
- **Glass bottles or test tubes:** these provide sufficient depth of agar medium for prolonged growth of bacterial and fungal cultures, avoiding problems of dehydration and salt crystallisation. Inoculate aerobes on the surface and anaerobes by stabbing down the centre, into the base (stab culture).
- **Flat-sided bottles:** these are used for animal cell culture, to provide an increased surface area for attachment and allow growth of cells as a surface monolayer. Usually plastic and disposable (Fig. 68.1).

Subculturing – when subculturing microbes from a colony on an agar medium, take your sample from the growing edge, so that viable cells are transferred.

The dynamics of growth are usually studied in liquid culture, apart from certain rapidly growing filamentous fungi, where increases in colony diameter can be measured accurately, e.g. using Vernier callipers.

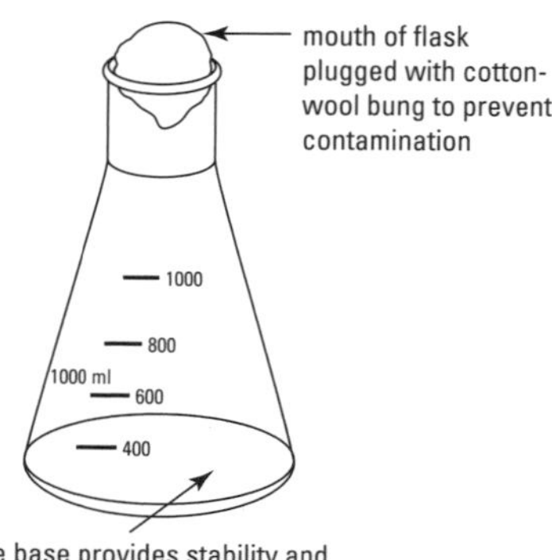

Fig. 68.2 Conical (Erlenmeyer) flask.

Growth in liquid media

Many bacteria and yeasts, can be grown as a homogeneous unicellular suspension in a suitable liquid medium, where growth is usually considered in terms of cell number (population growth) rather than cell size. Most liquid culture systems need agitation, to ensure adequate mixing and to keep the cells in suspension. A conical (Erlenmeyer) flask of 100–2000 mL capacity (Fig. 68.2) can be used to grow a batch culture on an orbital shaker, operating at 20–250 cycles per minute. For aerobic organisms, the surface area of such a culture should be as large as possible: restrict the volume of medium to not more than 20 per cent of the flask volume. Larger cultures may need to be gassed with sterile air and mixed using a magnetic stirrer rather than an orbital shaker. The simplest method of air sterilisation is filtration, using glass wool, non-absorbent cotton wool or a commercial filter unit of appropriate pore size (usually 0.2 μm). Air is introduced via a sparger (a glass tube with many small holes, so that small bubbles are produced) near the bottom of the culture vessel to increase the surface area and enhance gas exchange. More complex systems have baffles and paddles to further improve mixing and gas exchange.

Liquid culture systems may be subdivided under two broad headings:

Batch culture

This is the most common approach for routine liquid culture. Cells are inoculated into a sterile vessel containing a fixed amount of growth medium. Your choice of vessel will depend upon the volume of culture required: larger-scale vessels (e.g. 1 Litre and above) are often called 'fermenters' or 'bioreactors', particularly in biotechnology. Growth within the vessel usually follows a predictable S-shaped (sinusoidal) curve when plotted in log–linear format (Fig. 68.3), divided into four components:

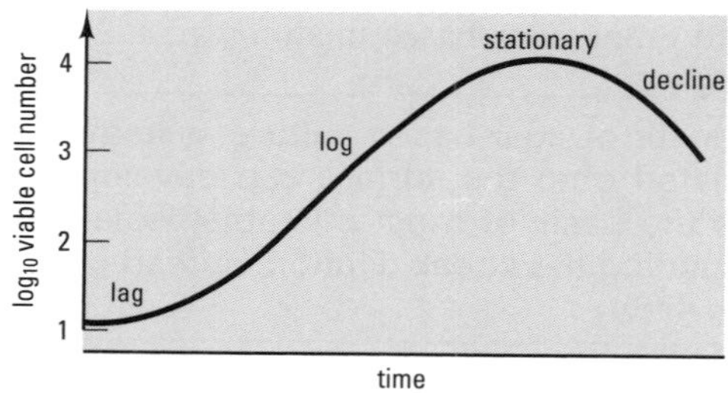

Fig. 68.3 Population growth curve for cells in batch culture (liquid medium).

1. **Lag phase:** the initial period when no increase in cell number is seen. The larger the inoculum of active cells the shorter the lag phase will be, provided the cells are transferred from similar growth conditions.
2. **Log phase**, or exponential phase: where cells are growing at their maximum rate. This may be quantified by the specific growth rate (μ or k), where:

$$\mu \text{ (or) } k = \frac{2.303\,(\log N_x - \log N_0)}{(t_x - t_0)} \qquad [68.1]$$

where N_0 is the initial number of cells at time t_0 and N_x is the number of cells at time t_x. For times specified in hours, μ is expressed as h^{-1}.

Prokaryotes grow by binary fission while eukaryotes grow by mitotic cell division; in both cases each cell divides to give two identical offspring. Consequently, the doubling time or generation time (g, or T_2) is:

$$g = \frac{0.301\,(t_x - t_0)}{\log N_x - \log N_0} \qquad [68.2]$$

Cells grow at different rates, with doubling times ranging from under 20 min for some bacteria to 24 h or more for animal and plant

Example Suppose you counted 2×10^3 cells ($\log_{10} = 3.30$) per unit volume at t_0 and 6.3×10^4 cells ($\log_{10} = 4.80$) after 2 h (t_x).

Substitution into eqn [68.1] gives $[2.303(4.8 - 3.3)] \div 2 = 1.727\ h^{-1}$ (or $0.0288\ min^{-1}$).

Substituting the same values into eqn [68.2] gives: $[0.301 \times 2] \div [4.8 - 3.3] = 0.40$ h (or 24 min).

Working with logarithms – note that there is no log value for zero, so you cannot plot zero on a log–linear growth curve or on a death curve.

cells in culture. Exponential phase cells are often used in laboratory experiments, since growth and metabolism are nearly uniform.

3. **Stationary phase:** growth decreases as nutrients are depleted and waste products accumulate. Any increase in cell number is offset by death. This phase is usually termed the 'plateau' in animal cell culture.
4. **Decline phase**, or death phase: this is the result of prolonged starvation and toxicity, unless the cells are subcultured. Like growth, death often shows an exponential relationship with time, which can be characterised by a rate (specific death rate), equivalent to that used to express growth or, more often, as the decimal reduction time (d, or T_{90}), the time required to reduce the population by 90 per cent:

Example Suppose you counted 5.2×10^5 cells (log10 = 5.716) per unit volume at t_0 and 3.7×10^3 cells (log10 = 3.568) after 60 min (t_x). Substitution into eqn [68.3] gives 60 ÷ [2.148] = 27.9 min. To the nearest minute, this gives a value for d of 28 min.

$$d = \frac{t_x - t_0}{\log N_x - \log N_0} \qquad [68.3]$$

Some cells undergo rapid autolysis at the end of the stationary period while others show a slower decline.

Batch culture methods can be used to maintain stocks of particular organisms; cells are subcultured onto fresh medium before they enter the decline phase. However, primary cultures of animal cells have a finite life unless transformed to give a continuous cell line, capable of indefinite growth.

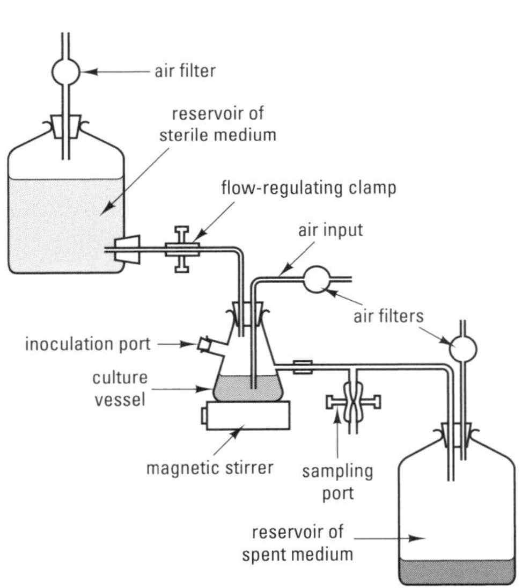

Fig. 68.4 Example of a two-dimensional lab equipment diagram of components of a chemostat.

Continuous culture

This is a method of maintaining cells in exponential growth for an extended period by continuously adding fresh growth medium to a culture vessel of fixed capacity. The new medium replaces nutrients and displaces some of the culture, diluting the remaining cells and allowing further growth.

After inoculating the vessel, the culture is allowed to grow for a short time as a batch culture, until a suitable population size is reached. Then medium is pumped into the vessel: the system is usually set up so that any increase in cell number due to growth will be offset by an equivalent loss due to dilution, i.e. the cell number within the vessel is maintained at a steady state. The cells will be growing at a particular rate (μ or κ), counterbalanced by dilution at an equivalent rate (D):

$$D = \frac{\text{flow rate}}{\text{vessel volume}} \qquad [68.4]$$

where D is expressed per unit time (e.g. h^{-1}). In a chemostat, the growth rate is limited by the availability of some nutrient in the inflowing medium, usually either carbon or nitrogen (see Fig. 68.4). In a turbidostat, the input of medium is controlled by the turbidity of the culture, measured using a photocell. A turbidostat is more complex than a chemostat, with additional equipment and controls.

Example Suppose a continuous culture system of 2000 mL volume had a flow of 600 mL over a period of 40 min (flow rate 600 ÷ 40 = 15 mL min^{-1}. Substitution into eqn [68.4] gives a dilution rate D of 15 ÷ 2000 = 0.00075 min^{-1} or 0.00075 × 60 = 0.45 h^{-1}.

Example Suppose you wanted to convert a doubling time of 20 minutes to a specific growth rate. Rearrangement of eqn [68.5] gives $\mu = 0.693 \div 20 = 0.03465$ min^{-1} ($= 2.08\,h^{-1}$). For the example given on page 234, with a growth rate of $0.45\,h^{-1}$, substitution into eqn [68.5] gives $0.693/0.45 = 1.54$ h, (approx. 92 minutes).

To determine the specific growth rate (μ or k) of a continuous culture:

1. Measure the flow of medium through the vessel over a known time interval (e.g. connect a sterile measuring cylinder or similar volumetric device to the outlet), to calculate the flow rate.
2. Divide the flow rate by the vessel volume (eqn [68.4]) to give the dilution rate (D).
3. This equals the specific growth rate, since $D = \mu$ at steady state.
4. If you want to know the doubling time (g), calculate using the relationship:

$$g = \frac{0.693}{\mu} \qquad [68.5]$$

(Note that eqn [68.5] also applies to exponential phase cells in batch culture and is useful for interconverting g and μ.)

Continuous culture systems are more complex to set up than batch cultures. They are prone to contamination, having additional vessels for fresh medium and waste culture: strict aseptic technique is necessary when the medium reservoir is replaced, and during sampling and harvesting. However, they offer several advantages over batch cultures, including the following:

- The physiological state of the cells is more clearly defined, since actively growing cells at the same stage of growth are provided over an extended time period. This is useful for biochemical and physiological studies.
- Monitoring and control can be automated and computerised.
- Modelling can be carried out for biotechnology/fermentation technology.

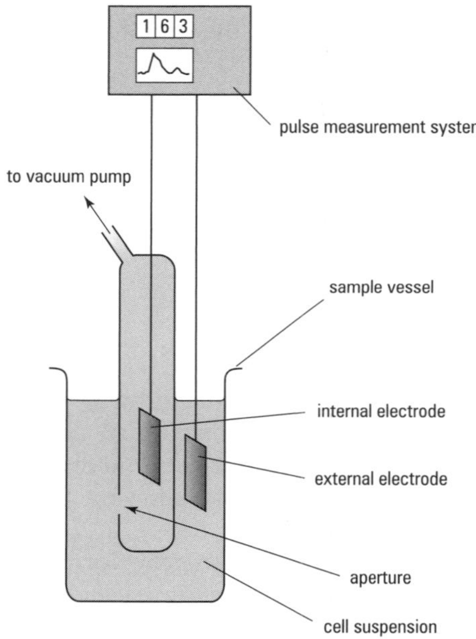

Fig. 68.5 Components of an electronic particle counter. During operation, the cell suspension is drawn through the aperture by the vacuum, creating a 'pulse' of resistance between the two electrodes as each cell passes through the aperture.

Measuring growth in cell cultures

The most widely used methods of measuring growth are based on cell number:

Direct microscopic counts

One of the simplest methods is to count the cells in a known volume of medium using a microscope and a counting chamber or haemocytometer (Box 68.1). While this gives a rapid assessment of the total cell number, it does not discriminate between living and dead cells. It is also time-consuming as a large number of cells must be counted for accurate measurement. It may be difficult to distinguish individual cells, e.g. for cells growing as clumps.

Alternative approaches to measuring growth – these include biomass, dry weight, turbidity, absorbance or any major cellular component, e.g. protein, nucleic acid, ATP, etc.

Electronic particle counters

These instruments can be used to give a direct (total) count of a suspension of microbial cells. The Coulter counter detects particles due to change in electrical resistance when they pass through a small aperture in a glass tube (Fig. 68.5). It gives a rapid count based on a larger number of cells than direct microscopy. It is well-suited for repeat measurements or large sample numbers and can be linked to a microcomputer for data processing. If

Box 68.1 How to use a counting chamber or haemocytometer

A counting chamber is a specially designed slide containing a chamber of known depth with a grid etched onto its lower surface. When a flat coverslip is placed over the chamber, the depth is uniform. Use as follows:

1. **Place the special coverslip over the chamber.** Press the edges firmly, to ensure that the coverslip makes contact with the surface of the slide, but take care that you do not break the slide or coverslip by using too much force. When correctly positioned, you should be able to see interference rings (Newton's rings) at the edge of the coverslip.
2. **Add a small amount of your cell suspension to fill the central space above the grid.** Place on the microscope stage and allow the cells to settle (two–three minutes).
3. **Examine the grid microscopically,** using the $\times 10$ objective lens first, since the counting chamber is far thicker than a standard microscope slide. Then switch to the $\times 40$ objective: take care not to scratch the surface of the objective lens, as the special coverslip is thicker than a normal coverslip. For a dense culture, the small squares are used, while the larger squares are used for dilute suspensions. You may need to dilute your suspension if it contains more than 30 cells per small square.
4. **Count the number of cells in several squares**: at least 600 cells should be counted for accurate measurements. Include those cells that cross the upper and left-hand boundaries, but not those that cross the lower and right-hand rulings. A hand tally may be used to aid counting. Motile cells must be immobilised prior to counting (e.g. by killing with a suitable biocide).
5. **Divide the total number of cells (*C*) by the number of squares counted (*S*)**, to give the mean cell count per square.
6. **Determine the volume (in mL) of liquid corresponding to a single square (*V*)**, e.g. a Petroff–Hausser chamber has small squares of linear dimension 0.2 mm, giving an area of $0.04\,mm^2$; since the depth of the chamber is 0.02 mm, the volume is $0.04 \times 0.02 = 0.0008\,mm^3$; as there are $1000\,mm^3$ in 1 mL, the volume of a small square is 8×10^{-7} mL; similarly, the volume of a large square (equal to 25 small squares) is 2×10^{-5} ml. Note that other types of counting chamber will have different volumes: check the manufacturer's instructions. For example, the improved Neubauer chamber (Fig. 68.6) has small squares of volume $0.000\,25\,mm^3 = 2.5 \times 10^{-7}$ mL.

Fig. 68.6 Haemocytometer grid (Improved Neubauer rulings) viewed microscopically. The large square (delimited by triple etched lines) has a volume of 1/250 mm^3 ($0.004\,mm^3$ = 4 nL) while each small square (16 contained within the large square) has a volume of 1/4000 mm^3 ($0.000\,25\,mm^3$ = 0.25 nL). Note that the boundary line for squares delimited by triple-etched lines is the *middle* line, so this line must be used when counting (count cells straddling the top and left-hand gridlines and ignore those straddling the bottom and right-hand gridlines).

7. **Calculate the cell number per ml by dividing the mean cell count per square by the volume of a single square (in mL).**
8. **Remember to take account of any dilution of your original suspension** in your final calculation by multiplying by the reciprocal of the dilution (*M*), e.g. if you counted a 1 in 20 dilution of your sample, multiply by 20, or if you diluted to 10^{-5}, multiply by 10^5.

The complete equation for calculating the total microscopic count is:

$$\text{Total cell count (per ml)} = (C \div S \div V) \times M \quad [68.6]$$

e.g. if the mean cell count for a 100-fold dilution of a cell suspension, counted using a Petroff–Hausser chamber, was 12.4 cells in 10 small squares, the total count would be

$$(12.4 \div 10 \div 8 \times 10^{-7}) \times 10^2 = 1.55 \times 10^8\,mL^{-1}$$

A simpler, less accurate approach is to use a known volume of sample under a coverslip of known area on a standard glass slide, counting the number of cells per field of view using a calibrated microscope of known field diameter, then multiplying up to give the cell number per mL.

correctly calibrated, the counter can also measure cell sizes. A major limitation of electronic counters is the lack of discrimination between living cells, dead cells, cell clumps and other particles (e.g. dust). In addition, the instrument must be set up and calibrated by trained personnel. Flow cytometry is a more specialised alternative, since particles can be sorted as well as counted.

Definition

CFU – colony-forming unit: a cell or group of cells giving rise to a single colony on a solidified medium.

Culture-based counting methods

A variety of culture-based techniques can be used to determine the number of microbes in a sample. A major assumption of such methods is that, under suitable conditions, an individual viable microbial cell will be able to multiply and grow to give a visible change in the growth medium, i.e. a colony on an agar-based medium, or turbidity ('cloudiness') in a liquid medium. You are most likely to gain practical experience using bacterial cultures, counted by one or more of the following methods:

- **Spread or pour plate methods** ('plate counts', p. 455). The most widespread approach is to transfer a suitable amount of the sample to an agar medium, incubate under appropriate conditions and then count the resulting colonies (Box 68.2).

Box 68.2 How to make a plate count of bacteria using an agar-based medium

1. **Prepare serial decimal dilutions of the sample in a sterile diluent (p. 149).** The most widely used diluents are 0.1% w/v peptone water or 0.9% w/v NaCl, buffered at pH 7.3. Take care that you mix each dilution before making the next one. For soil, food or other solid samples, make the initial decimal dilution by taking 1 g of sample and making this up to 10 mL using a suitable diluent. Gentle shaking or homogenisation may be required for organisms growing in clumps. The number of decimal dilutions required for a particular sample will be governed by your expected count: dilute until the expected number of viable cells is around 100–1000 mL^{-1}.
2. **Transfer an appropriate volume (e.g. 0.05–0.5 mL) of the lowest dilution to an agar plate** using either the spread plate method or the pour plate procedure (p. 455). At least two, and preferably more, replicate plates should be prepared for each sample. You may also wish to prepare plates for more than one dilution, if you are unsure of the expected number of viable cells.
3. **Incubate under suitable conditions for 18–72 h, then count the number of colonies on each replicate plate at the most appropriate dilution.** The most accurate results will be obtained for plates containing 30–300 colonies. Mark the base of the plate with a spirit-based pen each time you count a colony. Determine the mean colony count per plate at this dilution (C).
4. **Calculate the colony count per mL of that particular dilution** by dividing by the volume (in mL) of liquid transferred to each plate (V).
5. **Now calculate the count per mL of the original sample** by multiplying by the reciprocal of the dilution: this is the multiplication factor (M); e.g. for a dilution of 10^{-3}, the multiplication factor would be 10^3. For food or other solid samples, the count should be expressed per g of sample.

The complete equation for calculating the viable count is:

$$\text{Count per mL (or per g)} = (C \div V) \times M \qquad [68.7]$$

e.g. for a sample with a mean colony count of 5.5 colonies per plate for a volume of 0.05 mL at a dilution of 10^{-7}, the count would be:

$$(5.5 \div 0.05) \times 10^7 = 1.1 \times 10^9 \text{ CFU mL}^{-1}$$

The count should be reported as colony-forming units (CFU) per mL, rather than as cells per mL, since a colony may be the product of more than one cell, particularly in filamentous microbes or in organisms with a tendency to aggregate. You should also be aware of the problems associated with counts of zero – these are best recorded as '<1', and you should then apply the appropriate correction factors for dilution and volume to obtain the detection limit. For example, a zero count (<1) of 100 μL of a five-fold dilution gives a detection limit of $(<1 \div 0.1) \times 5 = <50$ CFU mL^{-1}.

Alternative approaches in plate counting – when large numbers of samples have to be counted, a single agar plate can be divided into segments and a single droplet of each dilution placed into the appropriate segment ('Miles and Misra' droplet counting).

Counting injured or stressed microbes – a resuscitation stage may be required, to allow cells to grow under selective conditions, p. 460.

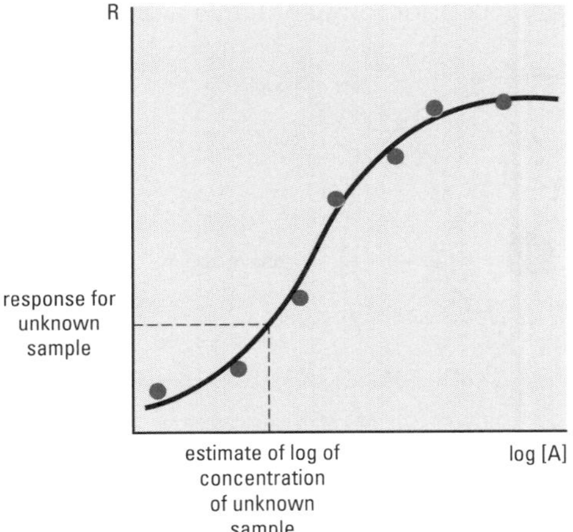

Fig. 68.7 Typical bioassay response curve, showing estimation of an unknown sample. Closed circles represent responses with standard samples. R = response; A = analyte.

Examples The following are typical bioassays:

- measuring the amount of a vitamin or antibiotic in a food sample using a bacterial growth assay,
- measuring the mutagenic properties of a chemical compound in the Ames test (see Box 68.3).

- **Membrane filtration.** For bacterial samples where the expected cell number is lower than 10 CFU mL^{-1}, pass the sample through a sterile filter (pore size 0.2 or 0.45 μm). The filter is then incubated on a suitable medium until colonies are produced, giving a count by dividing the mean colony count per filter by the volume of sample filtered.
- **Multiple tube count,** or most probable number (MPN). A bacteriological technique where the sample is diluted and known volumes are transferred to several tubes of liquid medium (typically, five tubes at three volumes), chosen so that there is a low probability of the smallest volumes containing a viable cell. After incubation, the number of tubes showing growth (turbidity) is compared to tabulated values to give the most probable number (MPN per mL).

The principal advantage of culture-based counting procedures is that dead cells will not be counted. However, for such techniques the incubation conditions and media used may not allow growth of all cells, underestimating the true viable count. Further problems are caused by cell clumping and dilution errors. In addition, such methods require sterile apparatus and media and the incubation period is lengthy before results are obtained. An alternative approach is to use direct microscopy, combined with 'vital' or 'mortal' staining. For example, the direct epifluorescence technique (DEFT) uses acridine orange and UV epifluorescence microscopy to separate living and dead bacteria. A further approach is to use DNA-based methods, such as the polymerase chain reaction (PCR, p. 376).

Bioassays and their applications

A bioassay is a method of quantifying a chemical substance (analyte) by measuring its effect on a biological system under controlled conditions. The hypothetical underlying phenomena are summarised by the relationship:

$$A + Rec \rightleftharpoons ARec \rightarrow R \qquad [68.8]$$

where A is the analyte, Rec the receptor, ARec the analyte–receptor complex and R the response. This relation is analogous to the formation of product from an enzyme–substrate complex and, using similar mathematical arguments with those of enzyme kinetics (p. 410), it can be shown that the expected relationship between [A] and rate of response is hyperbolic (sigmoidal in a log–linear plot like Fig. 68.7). This pattern of response is usually observed in practice if a wide enough range of [A] is tested.

To carry out the assay, the response elicited by the unknown sample is compared with the response obtained for differing concentrations of the substance, as shown in Fig. 68.7. When fitting a curve to standard points and estimating unknowns, the available methods, in order of increasing accuracy, are:

1. fitting by eye;
2. using linear regression on a restricted 'quasi-linear' portion of the assay curve;
3. linearisation followed by regression (e.g. by probit transformation);
4. non-linear regression (e.g. to the Morgan–Mercer–Flodin equation).

In general, bioassay techniques have more potential faults than physicochemical assay techniques. These may include the following:

- **A greater level of variability:** error in the estimate of the unknown compound will result because no two organisms will respond in exactly the same way. Assay curves vary through time, and because they are non-linear, a full standard curve is required each time the assay is carried out.

Box 68.3 Mutagenicity testing using the Ames test – an example of a widely used bioassay

Chemical carcinogens can be identified by the formation of tumours in laboratory animals exposed to the compound under controlled conditions. However, such animal bioassays are time-consuming and expensive. Dr B.N. Ames and co-workers have shown that most carcinogens are also mutagens, i.e. they will induce mutational changes in DNA. The Ames test makes use of this correlation to provide a simple, rapid and inexpensive bioassay for the initial screening of potential carcinogens. The test makes use of particular strains of *Salmonella typhimurium* with the following characteristics:

- histidine auxotrophy – the tester strains are unable to grow on a minimal medium without added histidine: this characteristic is the result of specific mutational changes to the DNA of these strains, including base substitutions and frame shifts;
- increased cell envelope permeability, to permit access of the test compound to the cell interior;
- defects in excision repair systems and enhanced error-prone repair systems, to reduce the likelihood of DNA repair after treatment with a potential mutagen.

When grown in the presence of a chemical mutagen, the bacteria may revert to prototrophy as a result of back mutations that restore the wild-type phenotype: such revertants grow independently of external histidine and are able to form colonies on minimal medium, unlike the original test strains. The extent of reversion can be used to assess the mutagenic potential of a particular chemical compound. Since many chemicals must be activated *in vivo*, the test incorporates a rat liver homogenate (so-called 'S-9 activator', containing microsomal enzymes) to simulate the metabolic events within the liver. The tester strains are mixed with S-9 activator and a small amount of molten soft agar, then poured as a thin agar overlay (top agar) on a minimal medium plate. The top agar layer contains a very small amount of histidine, to allow the tester strains to divide a few times and express any mutational changes (i.e. prototrophy).

The test can be performed in one of two ways:

1. **Spot test:** a concentrated drop of the test compound is placed at the centre of the plate, either directly on the agar surface, or on a small filter paper disc. The test compound will diffuse into the agar and revertants appear as a ring of colonies around the site of inoculation, as shown in Fig. 68.8. The distance of the ring of colonies from this site provides a measure of the toxicity of the compound, while the number of colonies within the ring gives an indication of the relative mutagenicity of the test substance. The spot test is often carried out in a simplified form, without added S-9 activator, as a rapid preliminary test prior to quantitative analysis.

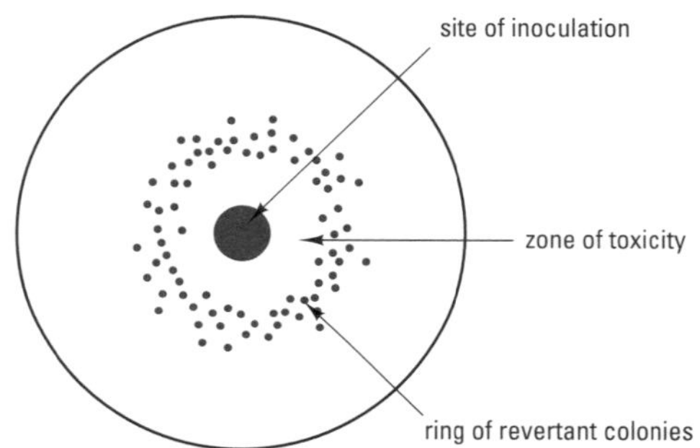

Fig. 68.8 Typical outcome of spot test (Ames test).

2. **Agar incorporation test:** known amounts of the test compound are mixed separately with molten top agar and the other constituents and poured onto separate minimal agar plates. After incubation for 48 h, revertants will appear as evenly dispersed colonies throughout the agar overlay. The number of colonies reflects the relative mutagenicity of the test compounds, with a direct relationship between colony count and the amount of mutagen. Agar incorporation tests can be used to generate dose–response curves similar to that shown in Fig. 68.7. The simplicity, sensitivity and reproducibility of the Ames test has resulted in its widespread use for screening potential carcinogens in many countries, though it is not an infallible test for carcinogenicity.

SAFETY NOTE Correct handling procedures must be followed at all times, as the test substances may be carcinogenic – testers should wear gloves and avoid skin contact or ingestion.

- **Lack of chemical information:** bioassays provide information about *biological* activity; they say little about the chemical structure of an unknown compound. The presence of a specific compound may need to be confirmed by a physicochemical method (e.g. mass spectrometry).
- **Possibility of interference:** while many bioassays are very specific, it is possible that different chemicals in the extract may influence the results.

Despite these problems, bioassays are still used, especially in food microbiology classes. They are 'low-tech' and generally cheap to set up. They often allow detection at very low concentrations. Bioassays also provide the means to assess the biological activity of chemicals and to study changes in sensitivity to a chemical, which physicochemical techniques cannot do. Changes in sensitivity may be evident in the shape of the dose–response curve and its position on the concentration axis.

Bioassays are the basis for characterising the efficacy of drugs and the toxicity of chemicals. Here, response is often treated as a quantal (all-or-nothing) event. The E_{50} is defined as that concentration of a compound causing 50 per cent of the organisms to respond. Where death is the observed response, the LD_{50} describes the concentration of a chemical that would cause 50 per cent of the test organisms to die within a specified period under a specified set of conditions. Box 68.3 presents details of the Ames test, a widely used bioassay used to assess the mutagenicity of chemicals.

Legal use of bioassays – in the UK, where bioassays involving 'higher' (vertebrate) animals are controlled by the Animals Scientific Procedures Act (1986), they can only be carried out under the direct supervision of a scientist licensed by the Home Office.

Considerations when setting up a bioassay

- The response should be easily measured.
- The experimental conditions should mimic the *in vivo* environment.
- The calibration standards (p. 338) should be chemically identical to the compound being measured and spread over the expected concentration range being tested.
- The samples should be purified if interfering compounds are present and diluted so the response will be on the 'linear' portion of the assay curve.

To check for interference, the bioassay may be standardised against another method (preferably physicochemical). Related compounds known to be present in the analyte solution should be shown to have minimal activity in the bioassay. If an interfering compound is present, this may show up if a known amount of standard is added to sample vials – the result will not be the sum of independently determined results for the standard and sample.

Sources for further study

Ball, A.S. (1997) *Bacterial Cell Culture: Essential Data.* Wiley, New York.

Cann, A.J. (1999) *Virus Culture: A Practical Approach.* Oxford University Press, Oxford.

Cartledge, T.G. (ed.) (1992) *In Vitro Cultivation of Micro-organisms.* Butterworth-Heinemann, Oxford. [Includes in-depth coverage of the mathematical principles underlying growth in batch and continuous culture.]

Hewitt, W. and Vincent, S. (1989) *Theory and Application of Microbiological Assay.* Academic Press, New York.

Rhodes, P.M. and Stanbury, P.F. (1997) *Applied Microbial Physiology: A Practical Approach.* Oxford University Press, Oxford. [Provides details of growth requirements and culture methods applicable to a range of different microbes.]

69 Microbiological analysis of food

Food microbiology is a broad subject, covering a range of topics including aspects of food production, food spoilage, food-borne infection and food poisoning. While Chapters 66–68 cover fundamental microbiological principles and basic techniques, this chapter outlines some of the more specific aspects of microbiological analysis that you are likely to encounter in your degree programme. Chapter 70 also provides further details of the microbiology of food preservation.

Microbiological analysis is required to assess the safety and quality of foods and drinks. Culture-based methods (p. 471) are often used to detect, identify and enumerate particular microbes, since it has the following advantages:

- A positive result gives a visible change, e.g. as a colony on an agar medium, or as a turbid broth, often also accompanied by other visible changes when differential growth media are used (p. 461).
- A positive result demonstrates the viability of the microbes present (for this reason, culture-based counts are sometimes termed 'viable counts', though this assumes that the microbes will grow under the test conditions, which may not always be true for injured microbes).
- The cultured microbes are alive and available for further analysis, e.g. to determine their serotype (p. 467), growth and survival characteristics, or antibiotic resistance profile.
- Culture methods are cost-effective, especially for large-scale screening, e.g. drinking water analysis. They also have a long track record of use.

However, a number of alternative approaches have been developed in recent years and these are often grouped together under the heading of 'rapid methods' (p. 485).

> **KEY POINT** Conventional culture-based methods are often regarded as a 'gold standard' against which newer methods are evaluated.

Definitions

Food spoilage – any negative change, e.g. a decrease in organoleptic or nutritional quality of food.

Food-borne illness – any type of disease caused by food.

Food-borne infection – an illness caused by the consumption of living pathogenic microbes in food.

Food poisoning – an illness resulting from the ingestion of microbial toxins or other poisonous substances, e.g. naturally harmful substances or toxic chemicals (sometimes colloquially used to include food-borne infection).

SAFETY NOTE Understanding the hazards of food microbiology - remember that some microbes present in food can cause food poisoning or food-borne infections. Appropriate attention to safe working (p. 133) is *essential* during laboratory classes where these microbes may be cultured.

Food-borne microbes

Organisms from all of the major microbial groups are found in foods, including:

- **Bacteria:** single-celled prokaryotes, some are involved in food fermentation (e.g. *Lactobacillus acidophilus* is used to make yogurt), as well as food spoilage (e.g. species of *Pseudomonas* in refrigerated foods), food poisoning (e.g. *Staphylococcus aureus* and *Bacillus cereus*) and food-borne infection (e.g. *Salmonella enterica* and *Listeria monocytogenes*). Many of the practical techniques that you will encounter in your degree programme are likely to be illustrated using bacteria, although the techniques also apply to other cellular microbes.

Definitions

Prokaryote – a cellular organism without a membrane-bound nucleus (a bacterium).

Hypha (plural: hyphae) – a long multinucleate filament that forms the basic structural unit of a mould.

Mycelium – the entire mass of hyphae of a single mould. The 'body' of a filamentous fungus.

Mycotoxin – a fungal toxin.

Definitions

Cyst – a resistant, dormant stage, formed by some protozoa and microalgae to enable them to survive adverse environmental conditions.

Biomagnification – an increase in concentration of a substance through successively levels of the food chain.

Ciguatera – food poisoning due to ingestion of heat-resistant ciguatoxin, typically in 'top carnivore' tropical reef fish including barracuda and snapper.

- **Fungi**: broadly divided into (i) single-celled yeasts (e.g. *Saccharomyces cerevisiae*, used in baking and brewing) and (ii) filamentous moulds (e.g. food spoilage moulds, including species of *Aspergillus*, *Penicillium* and *Mucor*). Yeasts form colonies on agar media while the hyphal filaments of moulds spread and branch, forming a visible mycelium. Mycotoxins (including aflatoxins from *Aspergillus flavus*) produced in food are of concern for their long-term health effects – they are usually detected using immunoassay, e.g. ELISA (p. 350), or chromatography, e.g. HPLC (p. 362).
- **Protozoa**: single-celled heterotrophs, they include a number of important food-borne and water-borne pathogens such as *Cryptosporidium parvum* and *Giardia lamblia*, both of which form resistant cysts that are able to survive outside the host. They are often detected using specialised immunofluoresence microscopy or antibody tests (p. 347).
- **Algae**: these single-celled photosynthetic organisms include a number of toxic species of dinoflagellates (e.g. *Gonyaulax tamarensis* and *Gambierdiscus toxicus*) and other microalgae, whose toxins are accumulated and biomagnified in filter-feeding shellfish and fish, causing paralytic shellfish poisoning and ciguatera when consumed. While the algae have distinctive microscopic appearances (Fig. 69.1), the toxins are usually detected by either immunoassay, e.g. ELISA (p. 350), or bioassay (p. 477).
- **Viruses**: unable to grow in food, though they can be transmitted between people via food or water, causing gastroenteritis (e.g. rotavirus and norovirus) and other illnesses (e.g. hepatitis A virus). Since they are obligate intracellular parasites, they are either studied using a suitable host cell culture system (pp. 455–6), by electron microscopy or molecular methods (p. 372).

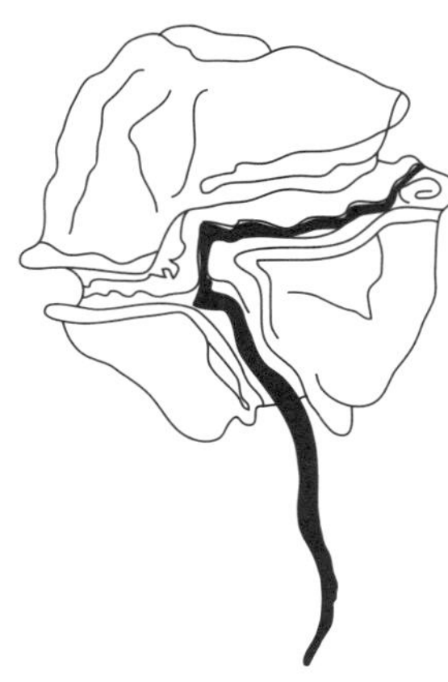

Fig. 69.1 *Gonyaulax.*

Sampling and enumeration procedures

Food sampling must be carried out prior to laboratory analysis. Most sampling protocols within the food industry are designed to meet formal codes of practice and statutory regulations, in accordance with specific sampling plans. However, in addition to the general aspects of microbiological sampling covered in Chapter 67 (p. 458), the following features apply across all procedures:

- **A minimum sample size is usually specified** – typically 100 g for foods and 100 mL for drinks/water.
- **The sample should reach the laboratory in the same condition as when it was taken**. This means (i) it must be taken with due regard for sterile technique (Chapter 66), to prevent contamination; (ii) it must be transported/stored appropriately (e.g. samples of fresh food or drinking water would be refrigerated); and (iii) it must be processed as soon as possible, preferably within 1–2 h for food samples and within 6 h for water samples.
- **A process for tracking samples and recording results is required**. While this is subject to more formal control in the food industry, it is also

Legislation for microbiological sampling – in the UK this includes the *Food Safety Act 1990, General Food Regulations 1994* and the *Food Safety (Sampling and Qualifications) Regulations,* available at: http://www.opsi.gov.uk.

important to pay very careful attention when writing details of sample identity and results within your lab book, to develop your skills in data recording.

Sampling equipment

Depending on the type of food involved, one or more of the following may be used:

Example Diluents for sampling/swabbing – these are designed to minimise osmotic stress to the target microbes. They include:
- 0.9% w/v NaCl solution;
- quarter-strength Ringer's solution (a mix of several salts);
- maximum recovery diluent (MRD), a peptone-saline mix.

- **Swabs** – used for surface sampling of some foods, e.g. cuts of meat, and for food preparation surfaces, e.g. kitchen workbenches. Typically, a sterile template with a central cut-out of a known area is used. A sterile swab is dipped in a container with a known volume of sterile diluent, then swabbed over the test area and the tip of the swab is then returned to the diluent container. A second (dry) swab is then used to remove any remaining visible moisture from the test area and this tip is then added to the container, which is then shaken vigorously to release any microbes into the liquid phase (adding glass beads within the container can help to disrupt the fibres of the swab and dislodge microbes). Alternative approaches for direct sampling of surfaces include the use of 'contact' plates/slides.
- **Membrane filtration** – for water samples. A known volume of liquid sample is filtered through a sterile membrane of suitable pore size, which is then either (i) incubated and counted as described on p. 477, or (ii) subjected to direct microscopy or fluorescent antibody staining. The advantage of membrane filtration is that it can be used to enumerate microbes at extremely low densities, e.g. the detection of protozoan cysts in 1–10 L of water (Wohlson *et al.*, 2004).
- **Paddle blenders** – for solid foods. Often referred to as Stomachers^{TM} (Stomacher^{TM} is a registered trademark of Seward Ltd.). A sterile plastic bag is filled with a known amount of food (excluding any sharp component, such as bone fragments) and a known volume of diluent, then placed inside the stomacher chamber (Fig 69.2). The reciprocating paddles then homogenise this mixture, typically in a 30–60 s process cycle. Alternatively, a sterile blender may be used, if a stomacher is not available. The liquid homogenate is then treated as a liquid suspension for counting purposes.

Fig. 69.2 Stomacher.

- **Rinsing apparatus** – used for surface sampling, e.g. vegetables/fruits. A known mass (weight) of food is added to a container with a known volume of diluent, shaken, left to stand for a fixed time, and re-shaken, to dislodge microbes into the rinse fluid, which is then cultured.
- **Dip slides** – for liquid samples. These are made with a layer of agar medium on a 'paddle' that is dipped into the test liquid, then removed and excess liquid is drained from the slide before returning it to its container. After incubation, the number of colonies per mL of liquid can be estimated either (i) by comparison with manufacturer's charts (Fig 69.3) or (ii) by counting the number of individual colonies on the slide and correcting for volume sampled.

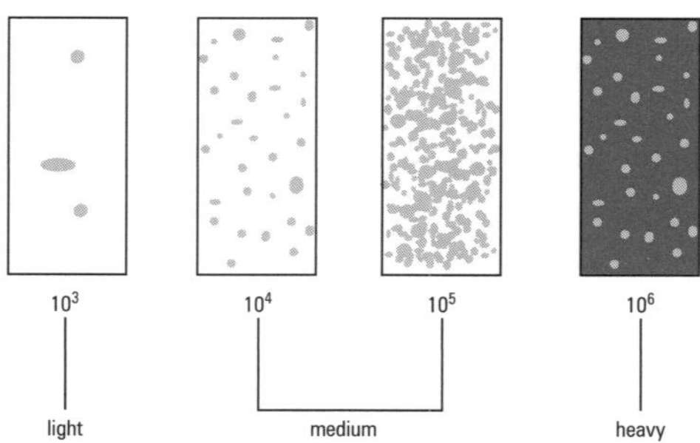

Fig. 69.3 Example of dipslide comparison chart. Values shown are approximate counts per mL.

Example A count of 42 colonies for 0.2 mL of a 10^{-2} dilution of fruit juice would give a corrected count for the original juice of $42 \div 0.2 \times 10^2 = 2.1 \times 10^3$ CFU/mL (see also Box 68.2).

Culture-based enumeration

While 'contact' plates, dip slides and membrane filters enable direct counting of colonies after suitable incubation, the end result of the other methods described is a homogeneous suspension of microbes in a liquid diluent. This fluid is then typically processed by:

- **serial decimal dilution** (p. 149), if the number of microbes is expected to be high, then either
- **spread or pour plate methods** (p. 455), to provide a colony count, or
- **most probable number** (MPN), a method involving multiple tubes of broth for each sample (p. 477), or
- **roll tubes** – here, a known amount of the suspension is mixed with molten agar medium in a tube which is the rolled to create a thin layer of solidified agar on the inside of the tube. Following incubation, colonies within the agar are counted by rotating the tube. This approach is particularly useful for anaerobic bacteria (p. 458).

Suitable correction for (i) the amount of material sampled and (ii) the volume of diluent, plus any subsequent serial dilution factor, will give a count expressed in terms of the original mass of food (typically per g or per 100 g) or volume of liquid (e.g. per mL or per 100 mL).

Understanding the direct epifluorescence technique – this is based on the differential interaction of acridine orange with DNA, which stains green, and RNA, which stains red. Thus living cells (containing RNA and DNA) will appear orange-red while dead cells (containing DNA alone) will be green.

The main alternative to culture-based counting methods uses direct microscopy (p. 474) to provide a cell count. The Breed Smear technique was an early example of this approach, with 0.01 mL of milk spread over 1 cm^2 of a glass slide, then a standard number of fields of view counted and converted to cells per mL using a suitable formula. This approach has been further developed in the more recent direct epifluorescence filter technique (DEFT). This method provides a direct microscopic count of membrane filters stained with acridine orange and then examined by UV epifluorescence microscopy. Originally developed for rapid enumeration of spoilage bacteria in milk samples, it is now used across a broader range of food types, including vegetables and fermented foods. Microcolony-DEFT is a derivative method that aims to conclusive demonstrate viable cells by the formation of microcolonies following short-term (3–6 h) incubation of membranes on a suitable growth medium, with subsequent epifluorescent microscopy. Another format uses fluorescent antibodies to enumerate target microbes.

Food industry standards for microbiological analysis – given the significance of microbiological criteria, laboratories that carry out such procedures must meet certain minimum requirements, in terms of technical competence and quality systems, typically validated through a process of accreditation (e.g. in the UK, through a body such as the UK Accreditation Service, or in Australia, the National Association of Testing Laboratories, NATA).

Microbiological criteria and standards

It is important to appreciate that interpretation of microbiological food analysis in dependent on the establishment of suitable microbiological criteria in advance of the test procedure. Microbiological criteria specify how the outcomes of food analysis will be interpreted, in terms of the safety and quality of the foodstuff. Typically, microbiological criteria specify:

- **the type of food;**
- **the target microbe, or their toxins;**
- **the sampling and analytical procedures** used;

Definitions

Microbiota – microbes in a particular environmental location.

CFU – colony-forming unit(s), obtained using spread or pour plates (see Chapter 66 for details of methodology).

Faecal (thermotolerant) coliforms – facultatively anaerobic Gram-negative non-sporing rods capable of growth and β-galactosidase activity at 44°C in the presence of bile or other surfactants. The **total coliform** group is broader, and includes environmental (non-thermotolerant) organisms.

E. coli – the major thermotolerant coliform in human faeces. It is differentiated from other coliforms typically by additional biochemical features, such as β-glucuronidase activity, or indole production from tryptophan.

Detecting and enumerating faecal indicators – detailed protocols vary by country, e.g. for the UK consult Roberts and Greenwood (2002) for foods or Environment Agency UK (2002) for water samples; for the USA, refer to Eaton *et al.* (2005).

Example (quantitative criterion) The European Union regulation 2073/2005 specifies that butter made from raw milk is unsatisfactory if it contains *E. coli* at >100 CFU/g^{-1}.

- **the acceptable microbiological limit**, typically in quantitative terms, or in terms of presence/absence of specific microbes;
- **guidance on interpreting results.**

Microbiological criteria are usually specified in terms of the following:

- **Standards** – legislative mandatory criteria. Failure to comply constitutes a violation of the law and may result in prosecution.
- **Guidelines** – advisory criteria applied by industry and/or regulatory agencies. Often used as a component of good manufacturing practice (GMP) and as part of HACCP analysis (p. 137).
- **Specifications** – criteria stipulated by a purchaser, to monitor food quality or hygiene.

Indicator microbes

Since tests for specific microbial pathogens may be time-consuming, slow and/or expensive, it has proved useful to take an alternative approach based on the concept of an indicator organism, whose presence or absence can provide information on the past history of a sample. Perhaps the most widely used indicator organism is *Escherichia coli* (*E. coli*), whose presence in food or water is generally interpreted as indicating faecal contamination and the potential presence of pathogenic faecal bacteria. Sometimes the term 'index organism' is used to more closely link a test microbe with an ecologically similar pathogen, as in the use of *E. coli* as an index organism for *Salmonella enterica*.

To illustrate the principle from the perspective of faecal contamination, an ideal faecal indicator should have the following characteristics:

- it should be a member of the normal intestinal microbiota of mammals (including humans), typically present in high numbers;
- it should be present whenever faecal contamination is present (indicating the potential presence of faecal pathogens);
- it should be present in equal or greater number than the pathogens;
- it should have equal or greater survival in the environment than the pathogens;
- it should not reproduce in the environment, outside the host organism;
- detection and identification of the indicator should be simpler, faster, and less expensive than for the pathogens.

E. coli satisfies the above criteria, since it is typically found at 10^8–10^9 CFU per g of freshly voided faeces, enabling it to be detected in very high dilution (e.g. in excess of one-billion-fold in water, using membrane filtration). It is also relatively easy to detect and quantify, e.g. using conventional differential media (p. 461) or through the enzymic hydrolysis of chromogenic/fluorogenic substrates (p. 408). Alternative faecal indicator bacteria include the faecal (thermotolerant) coliforms (a broader group that (also includes *E. coli*), enterococci (formerly termed faecal streptococci) and the spore-forming sulphite-reducing clostridia (including *Clostridium perfringens*). Coliphages (p. 456) are sometimes used as viral indicators of faecal contamination.

Presence-absence tests – these are often performed as simple 'one bottle' broth tests, with a presumptive positive result being demonstrated by growth (turbidity) along with a colour change, due to specific metabolic activity of the target microbe.

Improved terminology – while heterotrophic plate counts were once termed 'total viable counts' (TVCs) it is now appreciated that no single set of conditions will enable all bacteria to grow, and consequently, this term is no longer recommended.

Understanding colony counts – these follow the Poisson distribution (p. 261), where the variance is equal to the mean, which results in an increase in the confidence limits associated with samples means (= decreased precision) as counts are reduced.

Understanding the action of pyruvate – this peroxide scavenger protects injured cells from any peroxides present in the growth medium as well as from hydrogen peroxide formed as a by-product of respiration in the cells themselves, thereby enabling them to grow.

Working with methylene blue – this dye is known to be phototoxic, so you should incubate all experimental tubes in darkness to avoid any inhibitory effects caused by exposure to sunlight. Also avoid skin contact for the same reason.

Quantitative criteria

In some instances, the microbiological criterion will specify the complete absence ('zero tolerance') of the test microbe in a specified amount of sample, as is the case for faecal indicator bacteria in municipal drinking water, where *E. coli* should be absent from test samples of 100 mL, or for *Salmonella* spp. and *Listeria monocytogenes* in 25 g samples of ready-to-eat meals. Consequently, so-called 'presence–absence' tests can be used, since any positive growth can be interpreted as failing to meet the criterion. In other cases, the acceptable limit is specified in terms of the maximum number of viable organisms per sampling unit, assessed by quantitative methods such as plate counting. This approach recognises that most fresh foods are not sterile. However, processed foods, such as heat-treated and canned products, would be required to have few, or no, test microbes (Chapter 70).

Heterotrophic plate counts (sometimes termed standard plate counts, or aerobic colony counts) provide information on the overall number of bacteria in a sample. Typically, the methods specific a particular non-selective medium (e.g. plate count agar) and growth temperature (often either 22–25°C for environmental bacteria or 35–37°C for bacteria that may be of human or other mammalian origin). Heterotrophic plate counts are used to investigate bottled and natural waters and some foods, e.g. raw minced beef and also ready-to-eat foods.

It is important to appreciate the limitations of conventional enumeration by spread or pour plating (p. 455), especially at low counts. While the *limit of detection* is a single viable colony-forming unit on an agar plate, the *limit of quantitative measurement*, which is based on statistical principles, is typically 20 colonies. Colony counts below this limit should be regarded (and reported) as estimates. It is also important to understand that these considerations apply to the plate count itself, irrespective of the amount of sample used, or of any prior dilution. Thus a colony count of 5 CFU in 0.1 mL of a 10^{-3} dilution gives only an *estimated count* of 5.0×10^4 CFU/mL^{-1} when corrected for amount and dilution (p. 476). A different problem would occur for a 'zero' count of the same sample, which would be better recorded as '< 1', thereby allowing this to be expressed as $< 1.0 \times 10^4$ CFU/mL when corrected for amount and dilution.

Another important consideration in quantitative analysis is that the test microbe must be able to grow under the enumeration conditions used. While uninjured bacteria readily cope with standard conditions, this is not necessarily so for sublethally injured cells. Since food-processing protocols such as heat treatment or freezing can cause injury to microbes, it is often necessary to incorporate a resuscitation step into the enumeration protocol, to enable injured cells to repair themselves. As an example, incubation for several hours in buffered peptone water can be used to resuscitate injured *Salmonella* spp. An alternative approach is to include resuscitative agents such as sodium pyruvate within the culture medium, as in Baird–Parker agar (p. 486). If such steps are not taken, then culture-based enumeration can substantially underestimate the number of bacteria, since injured cells will not grow to form colonies and will therefore not be counted.

Alternatives to culture-based assays

These include:

- **Dye reduction assays.** While not strictly a formal counting procedure, this provides information on the metabolic activity of microbes in milk and other nutrient drinks. A known amount of either (i) methylene blue or (ii) resazurin is added to a liquid sample (e.g. milk) and the time required to decolourise (reduce) the dye provides an indication of microbial activity, since the decolourisation time is inversely proportional to microbial number. The advantage of such tests is that they are simple, inexpensive and rapid, typically taking 1–2 h. However, they cannot be used with foods that naturally contain reducing agents (e.g. homogenised raw meats), as these will decolourise the dyes. A more specific approach is to use chromogenic or fluorogenic enzyme substrates (p. 408).

Using ATP-based bioluminescent swabs (e.g. Biotrace®) – these provide a simple, rapid hygiene test for food preparation surfaces, since cleaned surfaces should not contain ATP from (i) food residues or (ii) microbes.

- **ATP assays**. These are often based on luminescence, using the luciferin–luciferase assay system. Any ATP in a sample will react with luciferin in the presence of oxygen to emit light, which can be measured rapidly using a luminometer, giving a value in relative light units (RLUs). However, because ATP is also present in fresh foods, its application is limited, e.g. to heat-processed foods such as pasteurised milk. An alternative approach is to preferentially extract and assay microbial ATP from the sample, though this reduces the simplicity and rapidity of the method.

Understanding immunoassays – ELISA provides a sensitive means of detecting microbial toxins in food. The specificity of the antibody–antigen reaction enables the target antigen (toxin) to be detected within a complex food matrix, avoiding the need to purify the toxin beforehand.

- **Immunological methods**. This approach is based on the use of specific antibodies to recognise and bind to target antigens, as described in detail in Chapter 52. For microbiological assays, an antibody must first be raised to a target organism (e.g. *Listeria monocytogenes*) or to its products using either (i) a laboratory animal (p. 347) or (ii) a hybridoma culture for monoclonal antibodies (p. 348). Once extracted and purified, such antibodies are then incorporated in various assay formats, including agglutination and precipitin reactions (p. 348) and the more sensitive enzyme-linked immunosorbent assays (ELISA, p. 350). This approach is widely used for the detection of microbial toxins.

Using immunomagnetic separation – one of the most widely used assays is for *E. coli* O157, which can be difficult to detect by conventional culture-based techniques alone.

- **Immunomagnetic separation**. This is an enrichment technique, designed to purify microbes or toxins from food. Small magnetic beads (e.g. Dynabeads®) are pre-coated with antibodies to the target organism/toxin and then mixed with a homogenised sample of food. The target microbe or toxin is bound by the antibodies on the surface of the beads, and then removed from the homogenate using a powerful magnet. Conventional detection then follows.

Understanding the advantages of gene probes – The GENE TRAK system for *E. coli* detection can detect a wide range of strains, which can show different biochemical reactions in conventional culture.

- **Nucleic acid detection**. DNA probes (p. 374) enable the identification of specific nucleic acid sequences found in particular microbes within a food sample. Typically sold in kit form, their advantages include their speed and sensitivity, while their drawbacks include the inability to easily distinguish between living and dead cells.
- **Nucleic acid amplification**. The polymerase chain reaction (PCR, p. 375) provides a rapid way to increase the amount of target nucleic acid in a sample, thereby enabling its detection. The approach is described in detail in Chapter 56. It is particularly useful for the assay of viruses in foods, as conventional culture-based detection methods are too specialised and laborious to be used for routine assay.

Example Baird-Parker agar is used as a primary isolation medium for *Staphylococcus aureus*. It contains:

- pyruvate to encourage the growth of sub-lethally injured cells;
- lithium and potassium salts to inhibit the growth of other bacteria;
- egg yolk, to visualise lethicinase as a zone of clearing around each colony;
- tellurite, which colours the colonies black when reduced to metallic tellurium.

(See also Table 68.1.)

Detection and identification of pathogens in food and water

The following provides a brief outline of general principles, using selected bacterial examples; for detailed information, consult Roberts and Greenwood (2002) or Correy *et al*. (2003).

The conventional approach is a two-step process:

1. **Primary isolation**, using an appropriate differential medium (p. 461) that has been designed to preferentially recover the target microbe, with diagnostic features being regarded as a presumptive positive (tentative identification).
2. **Confirmatory (secondary) testing** is then carried out on presumptive positive isolates from step 1, to minimise the likelihood of a false-positive result. The number of tests required to confirm identity will depend upon the nature of the target microbe, e.g. more tests may be required to confirm pathogens such as *Salmonella* spp. or *Listeria* spp. than for coliforms/*E. coli*.

In some circumstances, it may be necessary to include a resuscitation step for injured organisms (p. 485), or a pre-enrichment step, if the number of target microbes is likely to be low, or if they are weakly competitive in comparison to the rest of the natural microbiota.

Key features of food-borne bacteria are used in primary isolation media and confirmatory tests. Specific details for some of the organisms that you are most likely to encounter in food microbiology classes are:

- ***Staphylococcus aureus*:** salt-tolerance is used as a selective feature in primary isolation media, sometimes combined with mannitol utilization (e.g. in mannitol salt agar). Confirmation is typically by detection either of the enzyme coagulase, which clots blood plasma, or of heat-stable nuclease.
- ***Pseudomonas* spp.:** food spoilage organisms, usually isolated by incorporating the quaternary ammonium compound cetrimide into the primary medium. Different species produce characteristic pigments, e.g. *P. aeruginosa* produces pyocyanin, giving colonies a blue-green appearance while *P. fluorescens* produces pyoverdine, which is fluorescent under UV light. Confirmation of *Pseudomonas* is by Gram staining (Gram-negative rods), plus positive catalase and oxidase tests (p. 465), with identification to species level by biochemical profiling.
- ***Salmonella* spp.:** whereas traditional isolation media included bile salts, bismuth or other inhibitors, modern methods are based on the detection of specific enzymes using chromogenic media. Conventional confirmation is by biochemical profiling, e.g. using the API20E test system (p. 466), or by rapid methods, including immunoassay or nucleic acid methods.
- ***Vibrio* spp.:** typically associated with seafood, thiosulfate citrate bile sucrose (TCBS) agar is used as a primary isolation medium for *V. parahaemolyticus* and others, disintinguished by their differential use of sucrose. Confirmation of *Vibrio* is by Gram staining (Gram-negative curved rods and comma-shaped cells) plus typically positive oxidase and catalase reactions (p. 464).

SAFETY NOTE Working with food-borne bacteria. It is important to remember that many of these organisms are capable of causing gastroenteritis, so care must be taken to avoid ingestion during lab classes (e.g. no 'hand-to-mouth' operations).

Example Harlequin® ABC medium for *Salmonella* uses chromogenic substrates to demonstrate the production of a-galactosidase in the absence of β-galactosidase, which distinguishes *Salmonella* spp. from closely related bacteria.

Assaying foods for bacterial spores – since spores of *Bacillus* and *Clostridium* are heat-resistant, their presence in food can be evaluated by heating samples (e.g. at 70°C for 15 minutes) before primary isolation. Each colony formed can be interpreted as a spore present in the original foodstuff.

> **Definition**
>
> **Probiotic** – a dietary supplement containing living microbes (typically *Lactobacillus* spp.), which are thought to be beneficial to intestinal health.

- ***Bacillus cereus*:** a spore-forming aerobe, primary isolation media often contain the enzyme poylmyxin, to which *Bacillus* spp. are resistant, plus pyruvate as a food source. Confirmation is by Gram staining (Gram-positive rods with unstained spores, p. 464), with additional biochemical or serological tests if required.
- ***Clostridium perfringens*:** a spore-forming anaerobe, it can be isolated on media containing polymyxin, oleandomycin and/or sulfadiazine to inhibit competing bacteria, plus sulfite which is converted to metal sulfide, giving distinctive black colonies. Confirmation is by biochemical testing, e.g. for acid phosphatase. As with *B. cereus*, first-stage confirmation is also by Gram staining (Gram-positive rods, with unstained spores) or by immunoassay for perfringens α-toxin.
- ***Lactobacillus* spp.:** important in fermented foods and as probiotics. Primary isolation typically uses the selectivity of a low pH medium (e.g. pH 5-6), plus the antibiotic vancomycin. Confirmation of species identity typically involves a number of conventional biochemical tests, or by contemporary molecular assays (e.g. gene probes or PCR).

Text references

Correy, J.E.L., Curtis, G.D.W. and Baird, R.M. (2003) *Handbook of Culture Media for Food Microbiology*, 2nd edn. Elsevier, Amsterdam.
[A reference text on methodology and media.]

Eaton, A.D., Clesceri, L.S., Rice, E.W., Greenberg, A.E. and Franson, M.A.H. (2005) *Standard Methods for Examination of Waters and Wastewaters*, 21st edn. APHA, Washington, DC.

Environment Agency UK (2002) *The Microbiology of Drinking Water (2002): Methods for the Examination of Waters and Associated Materials.* Available: http://www.dwi.gov.uk/regs/pdf/micro.htm
Last accessed: 21/12/10.

Roberts, D. and Greenwood, M. (2002) *Practical Food Microbiology*, 3rd edn. Blackwell, Reading, Massachusetts.
[A detailed guide to UK methods and protocols.]

Wohlson, T., Bates, J., Gray, B. and Katouli, M. (2004) Evaluation of five membrane filtration methods for recovery of *Cryptosporidium* and *Giardia* isolates from water samples. *Applied and Environmental Microbiology*, **70**, 2318–2322.

Sources for further study

Jay, J.M., Loessner, M.J. and Golden, D.A. (2005) *Modern Food Microbiology*, 7th edn. Springer, New York.
[Comprehensive coverage of a range of food types and pathogens.]

Pitt, J.I. and Hocking, A.D. (2009) *Fungi and Food Spoilage*. Springer, Dordrecht.

Ray, B. and Bhunia, A. (2007) *Fundamental Food Microbiology*, 4th edn. CRC Press, Boca Raton, Florida.

United States Environmental Protection Agency (2009) *Microbiology Homepage*. Available: http://www.epa.gov/nerlcwww/index.html
Last accessed: 21/12/10.
[Includes standard methods for a range of food-borne and water-borne microbes.]

United States Food and Drug Administration (1998) *Bacteriological Analytical Manual Online*. Available: http://www.fda.gov/Food/ScienceResearch/LaboratoryMethods/BacteriologicalAnalyticalManualBAM/default.htm
Last accessed: 21/12/10
[Contains protocols for a range of foods and bacteria.]

United States Food and Drug Administration (2009) *The Bad Bug Book*. Available: http://vm.cfsan.fda.gov/mow/intro.html
Last accessed: 21/12/10.
[Provides general background information on food-borne pathogens.]

70 Food processing

Definitions

Food processing – the various stages by which raw ingredients are converted to a final product.

Food preservation – processing that slows down or prevents spoilage due to microbial activity and endogenous metabolic activity, thereby extending shelf life.

While a few foods such as fresh fruit and salad vegetables are consumed in their natural (raw) state, many other foods are subjected to various procedures designed to make them more attractive, extend their 'shelf life' and/or reduce the risks of food poisoning/spoilage. While the term strictly covers all practical techniques following harvest of the original material such as washing and chopping, this chapter focusses on the technological and microbiological aspects of processed foods, since these are most likely to be covered in practical classes during your degree programme.

KEY POINT While food processing is a broad topic, the most important aspects from the perspective of food science and technology are those concerned with its effects on (i) microbial growth and (ii) nutritional quality.

Food preservation techniques aim to limit microbial activity and growth in food, while maintaining its nutritional value. This is typically achieved either by:

1. **Making conditions unfavourable for microbial growth,** e.g. by refrigeration, or by adding agents such as salt or chemical preservatives. Such actions often also reduce endogenous biological (enzymic) activity that would degrade the foodstuff.
2. **Applying a short-term treatment to destroy microbes,** e.g. canning or irradiation, where, the objective is to reduce or eliminate the target microbes.

KEY POINT It is important to understand the fundamental difference between preservation methods that destroy microbes in food with those that prevent microbial growth/multiplication. The former are time-limited treatments and subsequent storage must not allow recontamination and spoilage, whereas the latter are only effective while the treatment remains in place.

Physical methods of food preservation

Drying and dehydration

Understanding water activity – essentially, the a_w scale runs from 1.0 (pure water) to 0.0 (completely dry).

Measuring water activity – methods typically determine the equilibrium water content in an enclosed space within a measuring chamber containing a known amount of food, e.g. by psychrometry or hygrometry (see Kress-Rogers and Brimelow, 2002, for further information).

Since water is the essential solvent in all biological systems, its reduction or removal is an important means of food preservation, used in traditional methods (e.g. sun-drying) and contemporary procedures (e.g. freeze-drying). The moisture content of fresh food is typically measured as % water content, which may be determined as part of overall physico-chemical analysis (p. 329). However, for dried/preserved foods, this measure provides little useful information on the availability of water, which is best quantified in terms of water activity, a_w. This term is an expression of the mole fraction of unbound water (eqn [26.7], p. 165). Fresh foods typically have a_w values of ≥ 0.98 and readily support microbial growth, whereas dried foods have lower a_w levels. Traditional drying requires increased temperature (typically $\geq 60°C$) and an effective

Table 70.1 a_w values for microbial growth and in foods

a_w	Microbes growing above this value	Typical foods
≥0.98	All	Fresh raw fruit/vegetables/meats
0.95	*Escherichia coli, Pseudomonas* spp.	Bread, cooked meat
0.9	*Salmonella, Clostridium botulinum*	Cheeses, sausages
0.8	*Staphylococcus aureus*	Jams, jellies
0.7	*Saccharomyces bisporus*, yeasts	Peanut butter
0.6	*Zygosaccharomyces rouxii*, xerotolerant yeasts	Dried fruits
0.3	None	Biscuits, crackers
0.2	None	Milk powder, spices, freeze-dried foods

Note that minimum a_w values are affected by the solutes (salts/sugars) involved, as well as other factors.
Based on information from http://foodsafety.psu.edu/Foodpreservation/Water_activity_of_foods.htm and elsewhere.

Definitions

Intermediate moisture foods – typical in the a_w range 0.60–0.85.

Low-moisture foods – typically with a_w values of ⩽ 0.30

Halotolerant – literally 'salt-tolerant'.

Xerotolerant – literally 'dry-tolerant'.

Definitions

Q_{10} (temperature coefficient) – a measure of the change in the rate of a reaction due to a change in temperature of 10°C.

Autolysis – literally self-destruction, typically due to endogenous enzyme activity, resulting in food spoilage.

Psychrophilic – cold-loving; growth optimum is <20°C, e.g. *Pseudomonas syringae*.

Psychrotrophic – cold-tolerant; growth optimum is >20°C but capable of growth at refrigeration temperatures of <5°C, e.g. *Listeria monocytogenes*.

air flow, either using a drying over or solar radiation. Freeze-drying (lyophilisation) is a modern alternative to heat-drying for the rapid removal of water by sublimation from food in its frozen state, causing fewer changes to the composition of the food than conventional drying procedures. Once dried, the food items must be stored in a sealed container, to prevent water uptake and maintain a low a_w. The addition of salts/sugars also effectively reduces water availability and this approach is used in the preservation of salted meats, jams, etc. Table 70.1 gives examples of the minimum a_w required for growth of selected microbes alongside typical foods with these values. While intermediate moisture foods prohibit the growth of most bacteria, halotolerant and xerotolerant microbes grow slowly. This, together with undesirable biochemical changes such as lipid oxidation and Maillard browning reactions gives these products a finite shelf-life. In contrast, low-moisture foods such as biscuits and dried cereals can be stored for considerably longer without spoilage, typically for months–years under suitable conditions (see Chen and Mujumdar, 2008, for more detailed information on drying techniques).

Cold storage

These are based on the principle that biological activity is slowed with decreasing temperature. With Q_{10} values around 2 for most biological systems, a decrease in temperature of 10°C results in a halving of the rate of activity, thereby slowing the rate of microbial growth and of autolytic spoilage. This principle applies while the food remains unfrozen: freezing causes further changes, described later. Refrigeration is used for short-to-medium term storage of food on industrial and domestic scales. An important aspect of the industrial process is the maintenance of an unbroken 'cold chain' during storage and distribution, which requires food manufacturers to establish quality monitoring systems, e.g. as part of a HACCP protocol (p. 137). Typically, chilled foods, including ready-to-eat meals, should be kept at ⩽ 7°C, and industrial refrigeration equipment is often set at 1–5°C (in contrast, many domestic users do not monitor refrigerator operating temperature). From a microbiological perspective, the most important organisms are psychrophilic and psychrotrophic spoilage bacteria and fungi, with both groups being involved in spoilage of refrigerated foods. The growth of psychrotrophic

Working with psychrophiles and psychrotrophs – these organisms can be selectively cultured in a refrigerated incubator, although their growth rate will be slow when compared with conventional microbes grown at higher temperature.

Effects of freezing on microbial spores – while bacterial spores survive repeated cycles of freezing – thawing, fungal spores are somewhat less resistant, in line with their reduced tolerance of elevated temperatures.

Definitions

Pasteurisation – a form of disinfection, using only enough heat to reduce the number of pathogenic microbes in foods/drinks to the point where they cause no harm.

Sterilisation – complete destruction of all microbes.

Commercial sterilisation – high-level heat treatment specifically designed to eliminate the risks associated with spore-forming bacteria in food, typically defined in terms of the destruction of spores of *Clostridium botulinum* (p. 501).

Listeria monocytogenes is a particular concern, with the elderly and pregnant women at particular risk of listeriosis, which is why US authorities favour a 'zero tolerance' approach for this bacterium in chilled food. In contrast, European authorities specify an upper limit for *L. monocytogenes* of 100 CFU g^{-1} (regulation EC 2073/2005), advising at-risk groups to avoid meat-based patés, unpasteurised dairy products and soft cheeses, coleslaw, raw fish products and ready-to-eat foods (for more information on cold storage technologies, see Dellino, 1997, or Forsythe and Hayes, 1998).

Freezing

The process of freezing food to at least −18°C causes damage to microbes due to a combination of the negative effects of (i) cold shock damage, especially to cell membranes, which is maximised in rapid freezing processes; (ii) osmotic effects, with increasing solute concentrations in any unfrozen water; and (iii) ice crystal formation, which comprises initial nucleation and subsequent crystal growth, disrupting structures within microbial cells as well as those of the food itself. The other important practical consideration is that food should reach a low temperature as quickly as possible, to minimise the risk of microbial growth during the cooling process. For this reason, blast freezing is the method of choice of the food industry. While rapid freezing has the advantage of reducing the overall size of ice crystals and therefore causing less damage to food structure, it also leads to better survival of contaminant microbes. The other aspect that is important is the rate of thawing; microbial survival is reduced in slowly thawed food, which has been linked to oxidative damage (Stead and Park, 2000). Frozen and thawed food is more vulnerable to subsequent microbial spoilage, presumably due to the damage caused to the tissue structure of the food itself, which often causes nutrient-rich fluids to be released onto the surface of the food.

Heating

This is one of the most important and widely used methods of food preservation, due to the effectiveness of heat treatment in eliminating microbes and inactivating enzymes. Treatments such as boiling, baking and roasting are used in domestic and commercial cooking to enhance the organoleptic properties of food. In terms of food preservation, there are two broad approaches: (i) 'pasteurisation' – used to describe mild heat treatments designed to destroy non-spore-forming pathogenic/spoilage bacteria; and (ii) 'commercial sterilisation' – applied to harsher heating procedures designed to reduce the number of spore-forming bacteria to such an extent that they are of no consequence. Pasteurisation, originally developed by Louis Pasteur to prevent the spoilage of wine, traditionally involved heating at 63°C for 30 minutes (so-called 'low temperature long time', LTLT), whereas more recent developments include shorter times at higher temperatures, e.g. 73°C for 15 s ('high temperature short time', HTST) and 135°C for 1 s ('ultra-high temperature', UHT). The latter treatment is also sufficiently harsh to inactivate most spores in milk. Canning is the most widely used form of commercial sterilisation. It is sometimes referred to as 'appertisation' after its originator, Nicholas Appert.

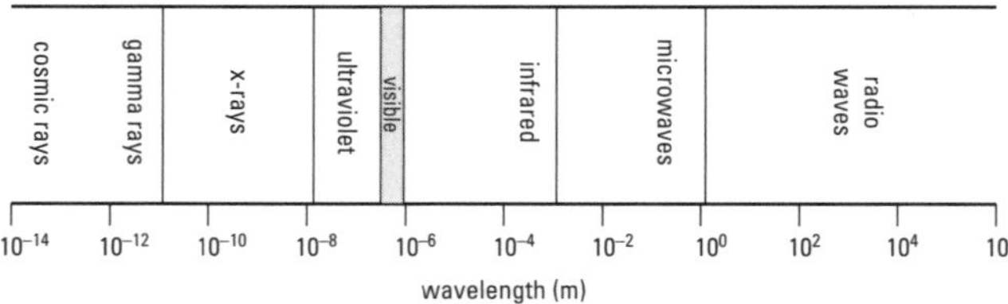

Fig. 70.1 The electromagnetic spectrum.

KEY POINT An important practical aspect of heat treatment is the relationship between heating time and the rate/extent of inactivation of a particular target group of microbes, since this underpins the scientific approach to thermal processing of food. This is considered in detail in Chapter 71 (p. 498).

Various approaches are used to apply heat, including: (i) conventional cooking/heating methods, using a variety of fuels; (ii) microwave treatment, which uses electromagnetic radiation at a frequency of 2.45 GHz equivalent to a wavelength of 0.122 m. (Fig. 70.1) to heat water molecules within the food; and (iii) ohmic heating (or electrical resistance heating), which works by passing a current through the food. A major advantage of microwave and ohmic heating is that heat is generated within the foodstuff, which results in rapid, even heating with less deterioration in food quality compared to conventional heating processes.

Effects of modified atmosphere on bacteria – these conditions extend the initial lag phase (p. 472), as well as reducing the rate of logarithmic growth. Aerobic spoilage bacteria, e.g. *Pseudomonas* spp., are particularly sensitive to lower O_2 and higher CO_2.

Modified/controlled atmosphere

Most of the methods involve a combination of (i) increased CO_2 and (ii) reduced O_2 concentration, as in the controlled storage of fruit and vegetables in rooms containing 2–4% O_2 and 8–10% CO_2, which reduces over-ripening and oxidative reactions as well as inhibiting microbial respiratory activity. This is likely to be due in part to the formation of carbonic acid when CO_2 dissolves in water:

$$CO_2 + H_2O \rightleftharpoons H_2CO_3 \rightleftharpoons H^+ + HCO_3^- \rightleftharpoons 2H^+ + CO_3^{2-} \qquad [70.1]$$

The process works on a smaller scale in modified atmosphere packaging (MAP), though there are several variants to conventional low O_2 MAP; including (i) high-O_2 MAP for red meats, where the elevated O_2 helps to maintain red colouration, and (ii) 'vacuum packaging' (VP), which creates an atmosphere within the sealed package with decreased O_2 and increased CO_2 as a result of the respiratory activity of food tissues as well as any contaminant microbes. One potential disadvantage of modified atmosphere methods is that the growth of obligately aerobic spoilage bacteria is preferentially inhibited, which can then favour the multiplication of anaerobes, e.g. *Clostridium botulinum* type E in fish products.

Vacuum-packed and modified atmosphere foods – while these items have an extended shelf life, this method does not eliminate microbial activity/ growth, especially for acid-tolerant microbes such as the lactic acid bacteria.

Hydrostatic pressure

Sometimes termed 'pascalisation', this involves the rapid application and subsequent removal of high pressures (up to 1000 MPa) at rates of 2–3 MPa/s, causing the denaturation of proteins, dissociation of intracellular organelles such as ribosomes, membrane damage and leakage of intracellular solutes. Effectiveness increases with (i) pressurisation time and (ii) the number of cycles of pressurisation–depressurisation. High-pressure treatment seems to cause minimal damage to most foods, and is used for the semi-continuous flow treatment of milk and fruit juices.

High-pressure treatment of bacterial spores – this is only effective in combination with a second treatment method, e.g. elevated temperature (80–95°C).

Irradiation

This is conventionally considered in terms of (i) non-ionising (UV) radiation and (ii) ionising radiation (comprising γ radiation, X-rays and high-energy electrons). UV radiation at a wavelength of 260 nm (Fig 70.1) is preferentially absorbed by DNA, leading eventually to irreversible damage and inactivation. UV light is widely applied to the disinfection of air, e.g. to reduce airborne spoilage microbes in food preparation areas. It can be used to disinfect drinking water and some fruit juices, leaving no chemical residue in contrast to chlorine and other oxidising agents. However, its poor penetration into solid foods restricts its practical application to surfaces. Similar limitations apply to the use of high-intensity pulsed light technology (PLT) in food processing. In contrast, ionising radiation is able to inactivate microbes within food and has broader applications, particularly in circumstances where other food preservation methods cannot readily be applied, e.g. to herbs and spices. Gamma (γ) radiation, X-rays (Fig. 70.1) or high-energy electrons are used, resulting in direct and indirect (oxygen-dependent) damage to microbial DNA that causes irreversible inactivation when used at a high level for sufficient time (radioappertisation). Ionising radiation can also be used at low dose to inhibit microbial spoilage (radurisation), to limit metabolic activity in foods, e.g. sprouting of vegetables, and for the destruction of insect pests. It is important to appreciate that while ionising radiation is produced by the radioactive decay of isotopes such as Co^{60}, the irradiated food items do not become radioactive. At a practical level, you are more likely to gain experience in the use of UV radiation than ionising radiation, due to the latter's requirement for a suitable source of radiation. As with other time-limited treatments, the intensity and duration of irradiation are critical, with doses being measure in grays (where 1 gray (Gy) = energy absorption of 1 J kg^{-1}).

Definitions

Radappertisation – commercial sterilisation of food using high dose ionising radiation.

Radurisation – radiological equivalent of pasteurisation, reducing the number of microbes in food.

Ultrasound

Ultrasound waves ($\geq$100 kHz) have potential applications in food preservation, since ultrasonic waves create micro-bubbles that can destroy microbial cells through the shock waves produced when they collapse (cavitation). However, it is impractical for use as a single processing method, since the times/energies involved also cause substantial damage to food structure. Its future value is likely to be in combination with other methods, such as heat treatment or high pressure processing. For more details see Povey and Mason (1998) or Sun (2005).

High-voltage electrical fields

Repeated cycles (10–100) comprising short pulses (1–10 μs) of high-voltage electric fields, applied between two electrodes at 30–50 kV/cm, can cause irreversible disruption to microbial membranes and has the potential to replace some thermal processing methods such as pasteurisation, since it is causes less damage to food quality and organoleptic properties. Pulsed electric fields (PEF) can also be used to aid the extraction of bioactive compounds from plant tissues. Continuous flow systems have been developed for the processing of fruit juices and drinks.

Pulsed electrical fields and molecular biology – electroporation of cell membranes is sometimes used in genetic engineering (p. 372), where a single sub-lethal pulse can facilitate the uptake of recombinant DNA.

Definition

Synergism – where two factors work together to produce an effect that is greater than the sum of their individual effects.

High magnetic fields

Oscillating high-intensity magnetic fields can inactivate microbes in food, though the technology is still at the development stage. Other emerging technological for food processing are considered in detail by Tapia and Cano (2004) and Sun (2005).

It is also worth noting that physical methods are often combined to give a synergistic effect, as in the combination of high pressure and temperature, leading to reduced thermal degradation of the treated food.

Definitions

Preservative – an additive that prevents decay/decomposition.

Antibiotic – natural compounds produced by one microbe to inhibit or inactivate other microbes.

Disinfectant – any agent (including physical processes and chemical compounds) used to remove the risk of transmission of pathogenic microbes.

Sanitiser – an agent used to create hygienic/healthy conditions, which includes the killing of microbes.

Chemical methods of food preservation

Antimicrobial agents fall into three categories:

1. **those naturally present,** e.g. lactoferrin and lysozyme in milk or allicin in garlic;
2. **those produced by microbial action,** e.g. organic acids and alcohols produced by fermentation;
3. **those deliberately added,** including preservatives and antibiotics.

Irrespective of their origin, these agents are usually categorised into two broad groups, based on their mode of action:

1. **microbiocidal agents** – kill their target microbes;
2. **microbiostatic agents** – prevent the growth of target microbes.

However, the above classification is overly simplistic, since microbiocidal agents may cause sub-lethal damage at low concentration, while microbiostatic agents can have lethal effects at very high concentration. Also, since these treatments are not transient, in contrast to physical techniques such as heat treatment, it is less important whether the microbes are killed or simply prevented from reaching their minimum infectious dose in food.

A wide range of natural and synthetic substances are used as antimicrobial agents in food. Those generally regarded as safe (GRAS) at their effective concentration are described below.

Salts and sugars as food preservatives – it is important to be aware of the broader aspects of the use of these food additives, including positive organoleptic effects (e.g. improved taste and texture), as well as potentially negative health impacts (e.g. the effect of high salt intake in raising blood pressure, or of high sugar consumption on energy intake and weight control).

Salt (NaCl)

This is probably the oldest known preservative, acting by lowering the water activity (a_w) of foods (p. 490), causing plasmolysis of microbial cells and thereby preventing their growth. Additionally, high concentrations of Na^+ are directly toxic to most biological systems, inhibiting enzyme activity. However, a limited number of halotolerant bacteria, e.g. *S. aureus*, are able to grow in some salted foods (Table 70.1).

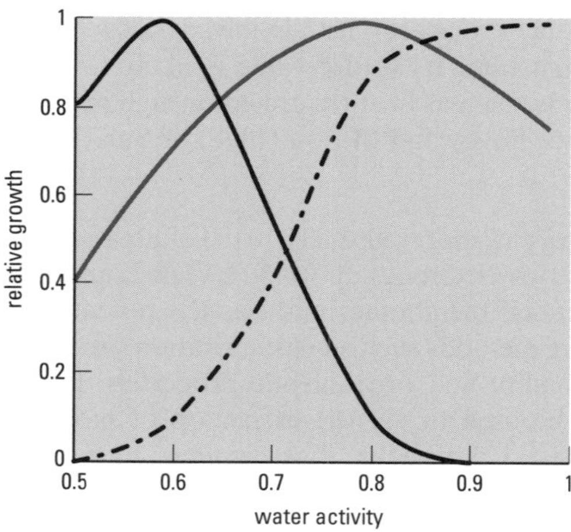

Fig. 70.2 Representative growth curves of xerosensitive (· – · –), xerotolerant (—) and xerophilic (**—**) microbes.

Sugars

High concentrations of sugars in food act in a broadly similar manner to salt, preventing the growth of most microbes through osmotic effects and the lowering of a_w, though without the effects of direct toxicity. Microbes capable of growth at high sugar concentrations are mostly yeasts, such as *Z. rouxii* (Table 70.1). Typically these organisms are xerotolerant, (capable of growth at low a_w) rather than xerophilic (requiring low a_w to grow). This distinction is shown in general terms in Fig. 70.2.

Oxidising agents

These include various forms of chlorine (hypochlorite, chlorite chlorine dioxide and hypochlorous acid), ozone and peroxides, including H_2O_2. They are used for the treatment of drinking water, enabling it to reach the required microbiological specification (p. 485), and as food sanitisers, e.g. for surface treatment of fruits, vegetables and meats and for cleaning of food preparation areas. They are readily studied in the laboratory, and you are likely to gain practical experience in assessing their effectiveness in practical classes, using the techniques described in Chapter 71. The mode of action of these agents is to oxidise biopolymers through a variety of chemical reactions, often involving reactive oxygen species (ROS) such as singlet oxygen, superoxide, hydroxyl radicals and/or peroxides. Oxidised biopolymers include microbial proteins, lipids and nucleic acids, leading eventually to irreversible inactivation of the target microbe, once sufficient oxidative damage has been accumulated (the 'multi-hit' hypothesis, p. 499). However, such agents also cause oxidative damage to foodstuffs and may generate mutagenic residues, limiting their use to the examples previously listed.

Understanding the preservative action of organic acids – because the permeable form of these compounds is the undissociated acid, HA (p. 173), rather than the corresponding salt, A^- (e.g. benzoic acid rather than benzoate), they are most effective in low pH foods near or below their pK_a values (e.g. 4.9 for propionic acid, 4.8 for sorbic acid, 4.7 for acetic acid, 4.2 for benzoic acid and 3.9 for lactic acid).

Weak organic acids

These include the traditional products of food fermentations involving lactic acid bacteria (e.g. yogurt) and acetic acid bacteria (e.g. vinegar). Others in this category are propionic, sorbic and benzoic acids, which also occur naturally in some tissues/foods. Irrespective of their source, they all share a common mode of action, being permeable to cell membranes in their undissociated (uncharged) form, where they cause intracellular acidification and disrupt ATP synthesis and transmembrane proton gradients, leading eventually to membrane disruption and cell death. Their addition to foods at high concentrations also has an acidulant action, which limits the growth of pathogenic and spoilage microbes, as well as giving the food a distinctive acid/sour taste. An alternative approach is used with probiotic foods, where living lactic acid bacteria are consumed with the intention of maintaining intestinal balance.

Nitrites in food – first applied in cured meat to inhibit growth and toxin production of *Clostridium botulinum*, the formation of carcinogenic nitrosamine by-products has led to the use of alternative preservatives, and the reduction or removal of nitrites from some foods.

Ethanol

This is a traditional product of yeast fermentation, and has important antimicrobial effects in wine and beer. However, on its own it is unable to guarantee long-term preservation, so additional agents are used, including SO_2/sulfites.

Nitrites

Sodium and potassium nitrite are used to cure meat, due to their beneficial effects in stabilising its colouration as well their antimicrobial action, due to their inhibitory effects on enzymes, including transport systems and components of metabolism, e.g. cytochrome oxidase.

Antibiotics in food – because of the risk of cross-resistance, most of the antibiotics used in food are not used for clinical treatment of human/animal diseases, though the tetracyclines are notable exceptions, potentially contributing to increased tetracycline resistance in clinically important pathogens.

Phosphates

Sometimes used to reduce nitrite levels in bacon and other foods, phosphates have general antimicrobial and antioxidant activity. Their mode of action remains poorly understood, though it appears to be linked to disruption of ion metabolism.

Sulfites

While sulphur dioxide (SO_2) is a traditional food preservative, the group also includes the sodium and potassium salts of sulfite (SO_3^{2-}), bisulfite (HSO_3^-) and metabisulfite ($S_2O_5^{2-}$). All of the sulfites share common

(a)

(b)

(c)

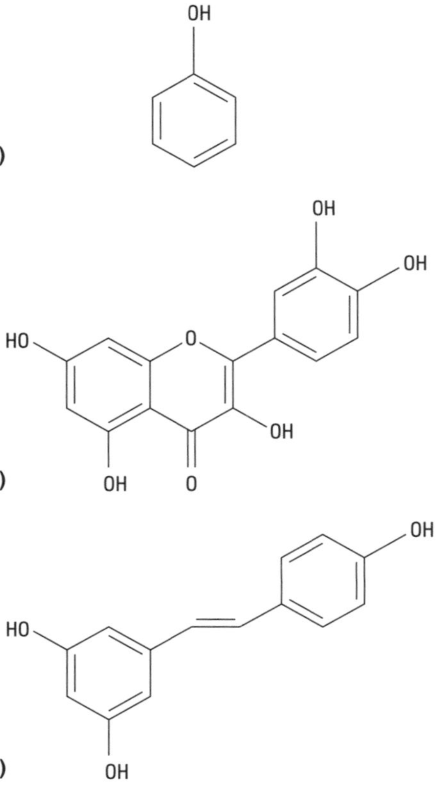

Fig. 70.3 Structure of:
(a) phenol;
(b) quercetin (a polyphenol/flavonoid);
(c) resveratrol (a diphenol).

applications, principally to control spoilage fungi and bacteria on fruit products, including wine and vegetables. They inhibit a wide range of cellular activities, including membrane functions, protein synthesis and metabolic pathways.

Antibiotics

These are naturally occurring secondary metabolites of microbes that inhibit or kill other microbes. Their use in food is tightly regulated. They include nisin and the related compound subtilin, which are produced by *Lactococcus lactis* and *Bacillus subtilis* respectively. They are active against other Gram-positive bacteria. Nisin is used to prevent microbial spoilage of some dairy products and cheeses. Tylosin is another antibiotic added to food to prevent the growth of Gram-positive bacteria, including *C. botulinum* while natamycin is licensed for food use to prevent the growth of yeasts and moulds.

Activated lactoferrin

Lactoferrin is an iron-binding protein that forms part of the innate immune system of mammals that has potential applications in the food industry, including in infant milk formula and as a surface spray for meat. Future developments may include the potential use of genetic engineering (p. 372) to produce recombinant lactoferrin in foodstuffs.

Phenolics and polyphenolics

This broad grouping includes a range of different compounds, all of which possess one or more six-membered benzene rings with one or more hydroxyl groups (Fig. 70.3). They include the benzoates (p. 495) and their ester derivatives, the parabens, as well as some essential oils (see below) and flavonoids. Simple phenolics (Fig. 70.3a) are the basis of food preservation by smoking of foods such as salmon, herring (kippers) and lapsang souchong tea. Phenolics disrupt membrane structure and cause leakage of intracellular contents. Many of the naturally occurring polyphenolic compounds act as phytoalexins, protecting plants against pathogenic microbes. Flavonoids such as quercetin and rutin are three-ringed polyphenolics (Fig. 70.3b) with antiviral and antifungal properties (Cushnie and Lamb, 2005) while the diphenolic resveratrol has antimicrobial effects as well as beneficial antioxidant features in foods. For further information on flavonoids in food, consult Anderson and Markham (2005).

Spices and essential oils

These are primarily used to flavour foods, but they also have some antimicrobial activity, due to a broad range of compounds including: allicin in garlic and onions, formed from the precursor aliin by the enzyme aliinase which is released when the plant tissues are crushed during food preparation; eugenol in cinnamon and cloves; curcumin in turmeric; and terpenes in oregano, sage, rosemary and thyme. Many of these compounds are phenolic, and are likely to have similar mode of action to terpenoid disinfectants, which act on the lipid components of membranes, disrupting function and increasing permeability. Additionally, the toxicity of some essential oils and spice constituents is light-dependent (phototoxic), limiting their major effect to the surface of foods.

Antimicrobial effects of spices in food – this is difficult to assess, since it depends on the quantity used. However, since many spices are used sparingly, it would be easy to overestimate their practical effectiveness in controlling microbial growth in foods, though they are likely to be effective in preventing spoilage of the spices themselves.

Isothiocyanates

These compounds have the general formula R-N = C = S and are found in highest amounts in plants of the family Cruciferae, which includes Brussel sprouts, broccoli, cabbage, cauliflower and mustard. They inhibit Gram-negative bacteria and some fungi, though Gram-positive bacteria are less sensitive.

Other dietary benefits of isothiocyanates in food – there is evidence that these components may reduce the risk of cancer when included in a balanced diet (e.g. Stan *et al.*, 2008).

It is often the case that physical and chemical methods may be combined and used at lower levels than for single treatments, to reduce any negative side effects while gaining maximum benefit overall. This approach has been developed into the so-called 'hurdle concept' (Leistner, 2000), which states that several sequential factors can act as a series of hurdles to collectively limit microbial growth in food. As an example, low-temperature storage of food of low pH and low water activity with natural flavonoids and added benzoate combines three physical methods and two chemical agents to limit microbial growth/spoilage.

Text references and sources for further study

Anderson, Ø.M. and Markham, K.R. (2005) *Flavonoids: Chemistry, Biochemistry and Applications*. CRC Press, Boca Raton, Florida.

Chen, Z.D. and Mujumdar, A.S. (2008) *Drying Technologies in Food Processing*. Blackwell, Chichester.

Cushnie, T.P.T. and Lamb, A.J. (2005) Antimicrobial activity of flavonoids. *Journal of Antimicrobial Agents*, **26**, 343–356.

Dellino, C.V.J. (1997) *Cold and Chilled Storage Technology*. Springer, New York.

Forsythe, S.J. and Hayes, J.P. (1998) *Food Hygiene Microbiology and HACCP*. Aspen, Galthersburg.

Kress-Rogers, E. and Brimelow, C.J.B. (2002) *Instrumentation and Sensors in the Food Industry*, 2nd edn. CRC Press, Boca Raton, Florida.

Leistner, L. (2000) Basic aspects of food preservation by hurdle technology. *International Journal of Food Microbiology*, **55**, 181–186.

Povey, M.J. and Mason, T.J. (1998) *Ultrasound in Food Processing*. Blackie, London.

Russell, N.J and Gould, G.W. (2003) *Food Preservatives*, 2nd edn. Springer, Berlin.

Stan, S.D., Kar, S., Stoner, G.D. and Singh, S.V. (2008) Bioactive food components and cancer risk reduction. *Journal of Cellular Biochemistry*, **104**, 339–356.

Stead, D. and Park S.F. (2000) Roles of Fe, superoxide dismutase and catalase in resistance of *Campylobacter coli* to freeze–thaw stress. *Applied and Environmental Microbiology*, **66**, 3110–3112.

Tapia, G.V.B. and Cano, M.S. (2004) *Novel Food Processing Technologies*. CRC Press, Boca Raton, Florida.

Sun, D.W. (2005) *Emerging Technologies for Food Processing*. Elsevier Academic Press, San Diego, California.

71 Measuring the microbiological effectiveness of food processing

Understanding the broader effects of food processing – in addition to inhibiting/inactivating microbes, processing may also reduce organoleptic quality and/or micronutrient levels. These are studied using the techniques described in Chapter 64 and Chapters 57–63, respectively.

The most important aspects of processing are those concerned with the preservation of food (Chapter 70). Consequently, performance is typically assessed in microbiological terms, using the culture-based techniques described in Chapters 66–69 to study the responses of selected pathogens.

KEY POINT For preservation methods that work by making conditions unsuitable for microbial growth, it is usually sufficient to monitor counts over time, to establish 'shelf life'. For methods that inactivate microbes in food, it is important to establish the dose and time of treatment, based on laboratory experiments with appropriate test microbes, typically using food-borne bacterial pathogens.

Dynamics of inactivation

The effects of processing on microbes are best understood in terms of the four-phase population growth curve described in Chapter 68.

KEY POINT Techniques that work by preventing growth, e.g. refrigeration, aim to lengthen the lag phase (p. 472), thereby extending storage time. In contrast, time-limited treatments e.g. heating, induce a decline/death phase (p. 473).

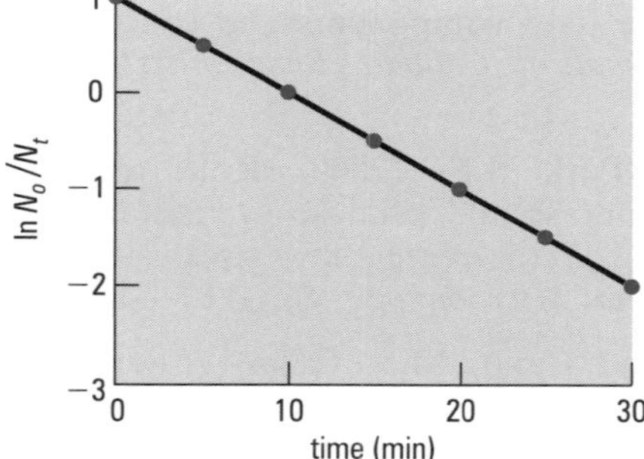

Fig. 71.1 Plot of ln N_o/N_t against time.

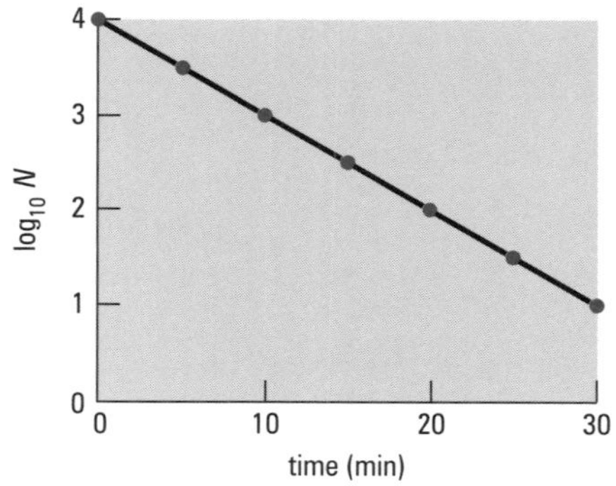

Fig. 71.2 Plot of $\log_{10}N$ against time.

Exponential inactivation kinetics

In its simplest form, the irreversible inactivation of microbes resulting from a particular treatment, e.g. high temperature, can be described by the exponential relationship often termed 'Chick's law', first used to describe the kinetics of inactivation of bacteria in the presence of chlorine-based disinfectants. Chick's law is typically expressed in terms of the initial number of cells (N_o) and the number of cells after time t (N_t):

$$N_t/N_0 = e^{-kt} \qquad [71.1]$$

where k is the rate constant for inactivation, which can be determined by plotting ln N_0/N_t against time, giving $-k$ as the slope (Fig. 71.1). However, since k is expressed as 1/time (e.g. min^{-1}) its practical application is sometimes difficult to grasp and alternative expressions are widely used in the food industry. The two most important are:

1. ***D*-value** (decimal reduction time). This is the time required to reduce the number of a particular microbes by 90 per cent under a specified set of conditions, e.g. at a particular temperature:

$$D = (t_x - t_0)/(\log_{10}N_0) - \log_{10}N_x \qquad [71.2]$$

where N_0 is the number of cells of the target microbe at time t_0 and N_x the number of cells at time t_x. In the water treatment industry, this is usually termed T_{90} (the time for 90 per cent reduction), rather

Determining counts of microbes – since physical and chemical treatments are likely to cause injury, a non-selective resuscitation medium is likely to better quantify the survivors than a conventional primary isolation medium (p. 487).

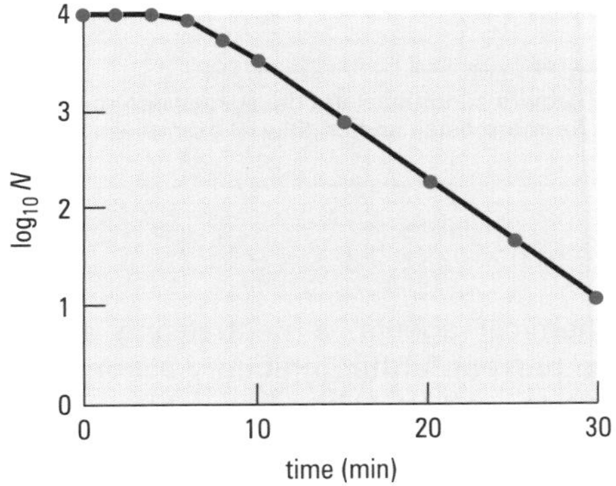

Fig. 71.3(a) Convex inactivation curve.

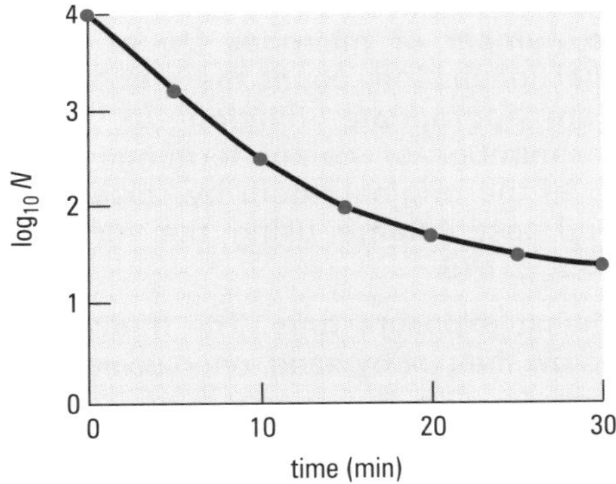

Fig. 71.3(b) Concave inactivation curve.

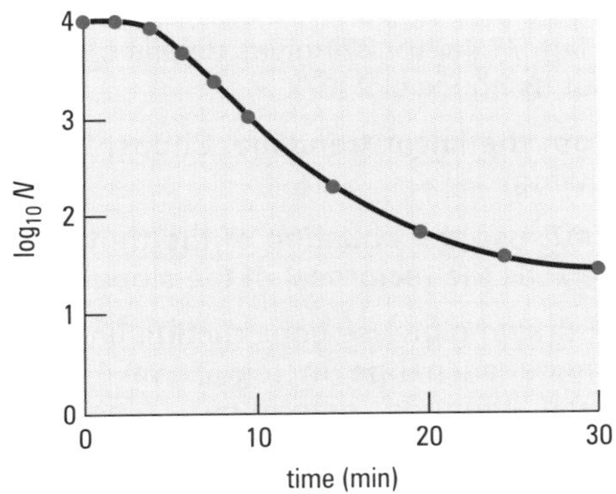

Fig. 71.3(c) S-shaped inactivation curve.

than D, although the principle is identical. One advantage of D (T_{90}) is that it is more readily interpreted than k, since it is expressed directly in terms of time. At a practical level, rather than calculate D using eqn [71.2], based on only two time points during the processing treatment, it is more usual to taking culture-based counts (e.g. by spread or pour plating, p. 455) at a number of time points and determine D from a 'survivor plot' (inactivation curve) of $\log_{10} N$ against time (a 'log-linear' plot; Fig. 71.2). Box 71.1 explains how to determine a D-value for a particular microbe while Table 71.1 gives representative D-values for selected examples of food-borne bacteria.

2. **Thermal death time** (TDT). This is the time required to kill a specified number of a particular microbe at a set temperature. TDT is somewhat less useful that D, since it is dependent upon inoculum size (a larger inoculum will give a longer TDT). Perhaps more importantly, because it is an 'end-point' method, it relies on accurate determination of the point at which no organisms can be cultured from a sample, e.g. using a simple presence/absence method rather than a formal quantitative count and this end point is not readily measured with accuracy, due to the increased variability associated with counts close to the detection limit (p. 485). Consequently, TDT measurements should be carried out using multiple replicates, to increase precision.

At a practical level, for any experiment involving the exposure of microbial cells to an inactivating treatment it is important to decide in advance how you intend to start and stop the process – Box 71.1 gives examples for heat and chlorine treatments.

Complex inactivation kinetics

While the standard approach described above assumes exponential inactivation, with a simple linear relationship between $\log_{10} N$ and time [71.1], this is not always so in laboratory tests. A number of different relationships have been observed:

- **Convex curves**. These display a 'shoulder' or lag period prior to the linear phase (Fig. 71.3a). One explanation for the lag period is that inactivation is unlikely to result from one event (a single 'hit'), but is more likely to results from accumulated damage to several biomolecules (multiple 'hits') leading eventually to death, and that this accumulated damage will take a finite amount of time to occur. This is also consistent with the sub-lethal injury seen in bacteria exposed to adverse conditions.
- **Concave curves**. Here a 'tail' of survivors extends the curve (Fig. 71.3b). This is typically interpreted in terms of the increased resistance of a fraction of the population to the treatment.
- **S-shaped curves**. A combination of a shoulder and a tail can give rise to sigmoid-type curves, as shown in Fig. 71.3c.

Table 71.1 Typical D-values for selected microbes at specific temperatures

Organism	Temperature(°C)	*D*(min)
Escherichia coli (saline solution)	55	4
Salmonella enterica (buffer)	65	0.3
Listeria monocytogenes (meat emulsion)	66	2
Spores of *Clostridium botulinum* type A (buffer)	121	0.2
Spores of *Bacillus stearothermophilus* (water)	121	3-4

Note: *D*-values are not absolute, and vary with strain, initial growth conditions and the suspension medium used (see Hauschild and Dodds, 1992, or Doyle *et al.*, 2007, for more detail and additional examples).

Box 71.1 How to determine the *D*-value of a microbe

These are typically measured for either (i) vegetative cells or (ii) spores of bacteria. While the principles are the same, the level of treatment required to inactivate spores will be higher than that for cells.

1. **Prepare an inoculum.** Typically this will be a suspension of microbes in water, saline solution or broth. Depending on your experimental design, you will use either multiple tubes, introduced to the treatment at the same time (t_0) and then removed at different times, or a single larger tube, from which samples will be removed at timed intervals.
2. **Prepare the experimental treatment.** For high temperature, this will typically be a water bath rather than an oven, as direct contact between the water and the test vessel ensures rapid heat transfer. For chemical agents, e.g. chlorine, a stock solution will be prepared at a higher concentration than is required in the experimental treatment. Ideally, this stock should be prepared at $\geq 10\times$ the level required, to minimise the dilution effect due to its addition. An alternative approach is to add a small volume of concentrated cell suspension to the test solution.
3. **Start the experiment.** Either by transferring the sample to the appropriate physical conditions or, for chemical agents, by adding an appropriate volume of stock solution to the microbial suspension, with appropriate rapid mixing (e.g. using a vortex mixer) to ensure homogeneity. Heat treatments are best carried out using narrow, thin-walled glass tubes containing small volumes of microbial suspension, to enable rapid equilibration at the outset.
4. **Take samples at timed intervals.** For high temperature treatments, samples should be rapidly cooled (e.g. using a water bath at room temperature), whereas for chemical agents, a neutralising agent must be used to prevent further inhibitory action, e.g. sodium thiosulfate reacts with free chlorine in solution, neutralising its effect.
5. **Prepare culture-based counts.** For example, using serial decimal dilution (p. 149), followed by spread or pour plating (p. 455), or, less frequently, by the multiple tube (MPN) method (p. 477).
6. **Determine the number of microbes (*N*) at each time point.** After incubation, count the number of colonies at an appropriate dilution (p. 476) calculate the number of colony-forming units (CFUs) per mL of the original liquid by correcting for dilution and test volume using the formula given in Box 68.2, p. 476.
7. **Plot $\log_{10}N$ against exposure time.** The traditional approach is to draw the plot by hand, often using log-linear graph paper (p. 220). However, it is now easily carried out using a spreadsheet, e.g. in Excel 2007 by first selecting the *Formulas* tab, then the *Math and trig* group, and finally the *LOG10* function to log-transform all counts ($\log_{10}N$). Then prepare a *scatter plot* (p. 221) of $\log_{10}N$ against time – the plot should show the linear *trendline*, plus the *equation* of the trendline, as in Fig. 71.4a (p. 224 gives details of how this is carried out using Excel 2007).
8. **Determine *D* from the linear trendline.** Choose from one of the following:
   (i) In Excel 2007, use the *equation* of the trendline to calculate D as the reciprocal of the slope.
   (ii) For a hand-drawn plot, use construction lines to first determine a change in $\log_{10}N$ of 1 unit (*y* axis) and then to determine *D* as the corresponding change in time (x axis), as shown in Fig 71.4b.

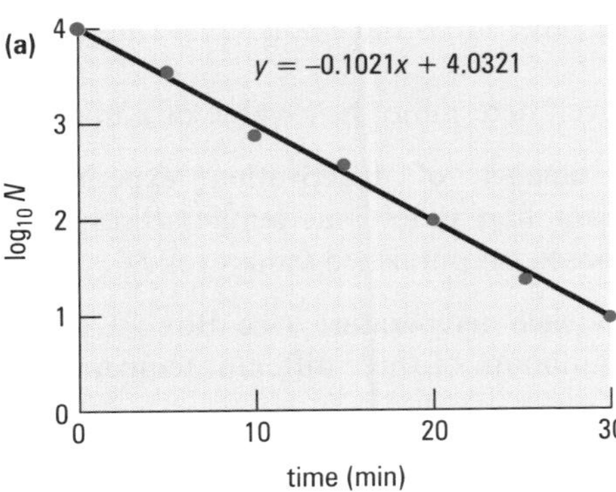

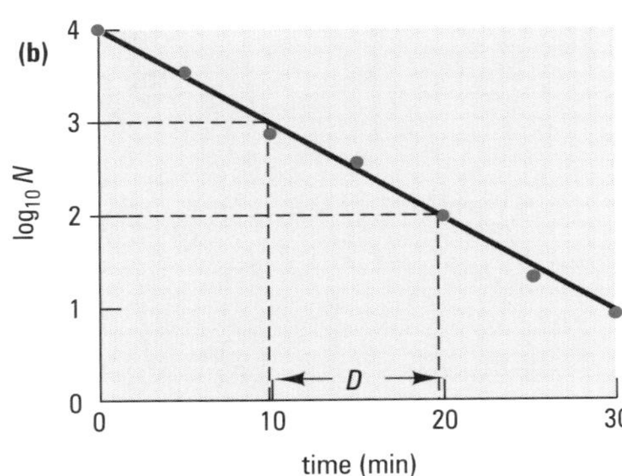

Fig. 71.4 Two methods of determining *D*: (a) Excel plot displaying trendline equation on chart ($D = 1/0.1021 = 9.8$ min); (b) hand-drawn version, with construction lines ($D \approx 9.5$ minutes).

Understanding the use of *D*-values in the food industry – since the relationship between survival and time is exponential, a plot of *N* against *t* never truly reaches zero. This is why it is usual to express treatment in terms of the time required to reduce N by a specified amount, e.g. $2D = 99\%$ reduction, $3D = 99.9\%$ reduction, etc. A similar approach is used in the water treatment industry, using T_{90} values (e.g. $T_{99.9} = 3 \times T_{90}$).

Applications in food processing

D-*values*

The food industry typically determines the time required for thermal processing of foods by applying a multiple of *D*, where the temperature is often shown as a subscript; thus D_{70} is the decimal reduction time at 70°C while D_{121} is the equivalent time at 121°C.

Probably the best example is for the commercial sterilisation of low-acid canned foods, where a 'botulinum cook' involves a minimum heating period equivalent to $12D_{121}$, based on data for spores of *Clostridium botulinum*. The logic is that, should a single *C. botulinum* spore be present in a can at the start of the process, the chance of its survival after this time period would be 1×10^{-12} (one in a thousand billion) and this is sufficiently rigorous to be effective. The greater heat resistance of spores of other bacteria, e.g. *Bacillus stearothermophilus* (Table 71.1), means that commercially sterilised foodstuff are more likely to spoil due to the survival and subsequent growth of spores of such bacteria than to cause botulism due to the growth of *C. botulinum*. However, even such thermoduric spores are unlikely to survive in most commercially sterilised items.

In terms of the practical application of *D*-values in the food industry, any substantial deviation from simple linear kinetics (Fig. 71.3) makes *D* less useful as a predictor of microbial inactivation, since it will only describe either (i) the overall average response, if calculated from a linear trendline (Fig. 71.4a) or (ii) the response during the linear part of the inactivation curve, if construction lines are used to determine *D* during this phase (Fig. 71.4b).

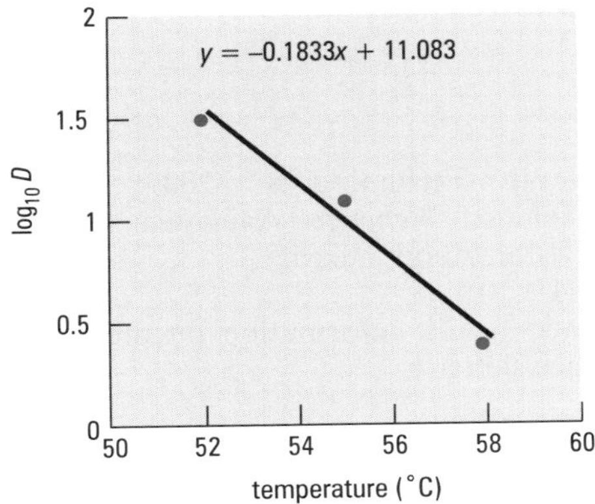

Fig. 71.5 Plot of $\log_{10}D$ against temperature. $Z = 1/0.1833 = 5.5$°C.

z-*values*

Simple comparisons between two treatment processes can be made by comparing their *D*-values; for example, a bactericidal UV light that has a *D*-value of 3 min is more effective than one with a *D*-value of 5 min, provided they have been tested with the same microbe under equivalent conditions. In contrast, the relationship between temperature and *D* can be mathematically defined by a further exponential function, since $\log_{10}D$ is inversely related to temperature, as shown in Fig. 71.5. The temperature difference required to give a ten-fold change in D ($\log_{10}D = 1$) is known as the z-value.

At a practical level, determine the z-value of a microbe as follows:

1. **Carry out a number of inactivation experiments at different temperatures**. Sampling times and temperatures should be selected appropriately, based on published values for D.
2. **Determine D at each temperature** (see Box 71.1 for details of the procedure). A minimum of three different temperatures is required.
3. **Calculate $\log_{10}D$** using a calculator or spreadsheet.
4. **Plot $\log_{10}D$ against temperature**. Either using software, e.g. Excel or by hand.
5. **Determine z** either (i) as the reciprocal of the slope of the trendline (e.g. in Excel, Fig. 71.5) or (ii) by drawing construction lines at two points that differ by 1 in $\log_{10}D$.

Understanding the principles of time-dependent food processing – while this chapter mostly illustrates the general principles with reference to temperature, they apply broadly to other treatments including UV and ionising radiation, and to chemical agents, e.g. disinfectants and sanitisers.

Note that z has units of temperature (°C). As an example, a z-value of 7°C has been reported for both *Listeria monocytogenes* and *Salmonella enterica* serovar Typhi (Sallami *et al.*, 2006), which means that both bacteria would be equally affected by an increase in processing temperature.

z-values can be used to reduce the overall thermal processing time of a food item: thus a rise in temperature from a particular value of x°C, to a value equivalent to $x + z$°C will enable the same thermal inactivation to be achieved in one-tenth of the time required at x°C, while $x + 2z$°C will achieve the same effect in one-hundredth of the time. This is the basis of so-called high-temperature short-time (HTST) pasteurisation processing of milk and other liquids (p. 491).

Text references

Doyle, M.P., Beuchat, L.R. and Montville, T.J. (2007) *Food Microbiology: Fundamentals and Frontiers*. American Society for Microbiology Press, Washington, DC.

Hauschild, A.H.W. and Dodds, K.L. (1992) *Clostridium botulinum: Ecology and Control in Foods*. CRC Press, Boca Raton, Florida.

Sallami, L., Marcotte, M., Naim, F., Ouattara, B., Leblanc, C. and Saucier, L. (2006) Heat inactivation of *Listeria monocytogenes* and *Salmonella enterica* serovar Typhi in a typical bologna matrix during an industrial cooking–cooling cycle. *Journal of Food Protection*, **69**, 3025–3030.

Sources for further study

Health Protection Agency, UK (2009) *National Standard Methods*. Available: http://www.hpa-standardmethods.org.uk/
Last accessed: 21/12/10.
[A comprehensive database of standard procedures for working with pathogenic microbes.]

Institute of Food Technologists (2009) *Food Microbiology Division Webpage*. Available: http://www.ift.org/divisions/food_micro/
Last accessed 21/12/10.
[Includes contact details, information on meetings and a regular newsletter.]

Jørgensen, K., Hilberg, C., Vintov, J. and Aalbaek, B. (2000) *Online Photo Atlas: Food Microbiology*. Available: http://www.microbiologyatlas.kvl.dk/biologi/english/default.asp
Last accessed: 21/12/10.
[Includes a wide range of high-quality images of food-borne bacteria, media and biochemical reactions.]

McKellar, R. and Lu, X. (2003) *Modelling Microbial Responses in Food*. CRC Press, Boca Raton, Florida.

Montville T.J. and Matthews, K.R. (2008) *Food Microbiology: An Introduction*. American Society for Microbiology Press, Washington, DC.

Pflug, I.J. (2003) *Microbiology and Engineering of Sterilization Processes*. Environmental Sterilization Laboratory, University of Minnesota, Minneapolis.

Ryser, E.T. and Marth, E.H. (2007) *Listeria, Listeriosis and Food Safety*, 3rd edn. CRC Press, Boca Raton, Florida.

Sperber, W.H. and Doyle, M.P. (2009) *Compendium of the Microbiological Spoilage of Foods and Beverages*. Springer, New York.

Wehr, M. and Frank, J.F. (2004) *Standard Methods for the Examination of Dairy Products,* 17th edn. American Public Health Association, Washington, DC.

Index

Index